FOOD SECURITY AND PLANT DISEASE MANAGEMENT

FOOD SECURITY AND PLANT DISEASE MANAGEMENT

Edited by

Ajay Kumar

Samir Droby

ELSEVIER

WP
WOODHEAD PUBLISHING
An imprint of Elsevier

Woodhead Publishing is an imprint of Elsevier
The Officers' Mess Business Centre, Royston Road, Duxford, CB22 4QH, United Kingdom
50 Hampshire Street, 5th Floor, Cambridge, MA 02139, United States
The Boulevard, Langford Lane, Kidlington, OX5 1GB, United Kingdom

British Library Cataloguing-in-Publication Data
A catalogue record for this book is available from the British Library

Library of Congress Cataloging-in-Publication Data
A catalog record for this book is available from the Library of Congress

ISBN: 978-0-12-821843-3 (print)

ISBN: 978-0-12-821855-6 (online)

For information on all Woodhead Publishing publications
visit our website at https://www.elsevier.com/books-and-journals

Publisher: Charlotte Cockle
Acquisitions Editor: Nancy J. Maragioglio
Editorial Project Manager: Cole Newman
Production Project Manager: Anitha Sivaraj
Cover Designer: Limbert Matthew

Typeset by MPS Limited, Chennai, India

Contents

5. Principle, diversity, mechanism, and potential of practical application of plant probiotic bacteria for the biocontrol of phytopathogens by induced systemic resistance

Musrat Zahan Surovy and Tofazzal Islam

6. RNA interference as a promising strategy for plant disease management

Ganeshamoorthy Hariharan, Nishanthi Sivasubramaniam and Kandeeparoopan Prasannath

7. Safflower disease—a sustainable protection against *Alternaria carthami* L.

Sudheeran Pradeep Kumar

8. Mycotoxin-associated food safety concerns of agriculture crops

Nitesh S. Litoriya and Arpan Modi

9. CRISPR/Cas in food security and plant disease management

Sopan Ganpatrao Wagh, Manoj Baliram Pohare and Ravindra Ramrao Kale

20. Molecular techniques used in plant disease diagnosis

Swapnil Sapre, Iti Gontia-Mishra, Vishwa Vijay Thakur, Sumana Sikdar and Sharad Tiwari

21. Biomanagement of *Fusarium* spp. associated with pulse crops

Sampat Nehra, Raj Kumar Gothwal, Alok Kumar Varshney, Pooran Singh Solanki, Shivani Chandra, Poonam Meena, P.C. Trivedi and P. Ghosh

List of contributors

Mohd Aamir Laboratory of Mycopathology and Microbial Technology, Centre of Advanced Study in Botany, Institute of Science, Banaras Hindu University, Varanasi, India; Division of Crop Improvement and Biotechnology, Indian Institute of Vegetable Research, Varanasi, India

Leandro Afonso Microbial Ecology Laboratory, Department of Microbiology, State University of Londrina, Londrina, Brazil

Galdino Andrade Microbial Ecology Laboratory, Department of Microbiology, State University of Londrina, Londrina, Brazil

Matheus Felipe de Lima Andreata Microbial Ecology Laboratory, Department of Microbiology, State University of Londrina, Londrina, Brazil

Waqar Akhtar Ansari Division of Crop Improvement and Biotechnology, Indian Institute of Vegetable Research, Varanasi, India

Rahul Bhadouria Department of Botany, University of Delhi, New Delhi, India

Antonio Biasi Department of Postharvest Science, Agricultural Research Organization (ARO), Volcani Center, Rishon LeZion, Israel

Shivani Chandra Amity Institute of Biotechnology, Amity University, Noida, India

Andreas Lazaros Chryssafidis Agroveterinary Sciences Center, University of the State of Santa Catarina, Lages, Brazil

Shivika Datta Department of Zoology, Doaba College Jalandhar, Jalandhar, India

Mickely Liuti Dealis Microbial Ecology Laboratory, Department of Microbiology, State University of Londrina, Londrina, Brazil

Vaishali Dhaka Department of Biotechnology, Lovely Professional University, Phagwara, India; Climate Mitigation and Sustainable Agriculture Research lab (CMaSAR), Division of Research and Development, Lovely Professional University, Phagwara, India

Daljeet Singh Dhanjal Department of Biotechnology, Lovely Professional University, Phagwara, India

Ritu Dixit National Botanical Research Institute, Lucknow, India; Department of Botany, Lucknow University, Lucknow, India

Samir Droby Department of Postharvest Science, Agricultural Research Organization (ARO), Volcani Center, Rishon LeZion, Israel

Ajinath Dukare ICAR-Central Institute of Post-Harvest Engineering & Technology, Ludhiana, India

Oleg Feygenberg Department of Postharvest Science, Agricultural Research Organization (ARO), Volcani Center, Rishon LeZion, Israel

P. Ghosh Birla Institute of Scientific Research, Statue Circle, Jaipur, India

Sougata Ghosh Department of Microbiology, School of Science, RK University, Rajkot, India

Bárbara Gionco-Cano Microbial Ecology Laboratory, Department of Microbiology, State University of Londrina, Londrina, Brazil

Deen Dayal Giri Department of Botany, Maharaj Singh College, Saharanpur, India

Ved Prakash Giri National Botanical Research Institute, Lucknow, India; Department of Botany, Lucknow University, Lucknow, India

Iti Gontia-Mishra Biotechnology Centre, Jawaharlal Nehru Agriculture University, Jabalpur, India

Raj Kumar Gothwal Birla Institute of Scientific Research, Statue Circle, Jaipur, India; Department of BioEngineering, Birla Institute of Technology, Mesra, Ranchi, Jaipur Campus, Jaipur, India

Vishal Gupta Centre of Advanced Study in Botany, Institute of Science, Banaras Hindu University, Varanasi, India

P.N. Guru ICAR-Central Institute of Post-Harvest Engineering & Technology, Ludhiana, India

Ganeshamoorthy Hariharan Department of Agricultural Biology, Faculty of Agriculture, Eastern University, Chenkalady, Sri Lanka

Tofazzal Islam Institute of Biotechnology and Genetic Engineering (IBGE), Bangabandhu Sheikh Mujibur Rahman Agricultural University, Gazipur, Bangladesh

Chelliah Jayabhaskaran Department of Biochemistry, Indian Institute of Science, Bengaluru, Karnataka, India

Ravindra Ramrao Kale Crop improvement division, ICAR-IIRR, Hyderabad, India

Subban Kamalraj Department of Biochemistry, Indian Institute of Science, Bengaluru, Karnataka, India

Siya Kamat Indian Institute of Science, Bengaluru, India

Yogesh Khade ICAR-Directorate of Onion and Garlic Research, Pune, India

Ajay Kumar Department of Postharvest Science, Agricultural Research Organization (ARO), Volcani Center, Rishon LeZion, Israel

Akshay Kumar Centre of Advanced Study in Botany, Institute of Science, Banaras Hindu University, Varanasi, India

Shambhu Kumar Forest Pathology Department, KSCSTE - Kerala Forest Research Institute, Peechi, Thrissur, Kerala, India

Vijay Kumar Regional Ayurveda Research Institute for Drug Development, Gwalior, India

Madhuree Kumari Indian Institute of Science, Bengaluru, India; National Botanical Research Institute, Lucknow, India; Academy of Scientific and Innovative Research (AcSIR), Ghaziabad, India

Nitesh S. Litoriya AINP on Pesticide Residues, Anand Agricultural University, Anand, India

Hemant Maheshwari ICAR-Indian Institute of Soybean Research, Indore, India

Poonam Meena Department of Botany, University of Rajasthan, Jaipur, India

Aradhana Mishra National Botanical Research Institute, Lucknow, India; Academy of Scientific and Innovative Research (AcSIR), Ghaziabad, India

Manisha Mishra University Department of Botany, T.M. Bhagalpur University, Bhagalpur, India

Arpan Modi Department of Fruit Tree Sciences, Institute of Plant Sciences, Agricultural Research Organization, Volcani Center, Rishon LeZion, Israel

Soma Mondal Forest Protection Division, ICFRE - Institute of Wood Science and Technology, Bengaluru, Karnataka, India

Sampat Nehra Birla Institute of Scientific Research, Statue Circle, Jaipur, India

Erika Tyemi Goya Niekawa Microbial Ecology Laboratory, Department of Microbiology, State University of Londrina, Londrina, Brazil

Shipra Pandey National Botanical Research Institute, Lucknow, India; Academy of Scientific and Innovative Research (AcSIR), Ghaziabad, India

Ramalingam Parthasarathy Department of Biochemistry, Indian Institute of Science, Bengaluru, Karnataka, India

Guilherme Edison Almeida Pega Microbial Ecology Laboratory, Department of Microbiology, State University of Londrina, Londrina, Brazil

Manoj Baliram Pohare Department of Infection, Immunity and Cardiovascular Diseases, University of Sheffield, Sheffield, United Kingdom

Sudheeran Pradeep Kumar Department of Postharvest Science, Institute of Postharvest and Food Sciences, Agriculture Research Organisation, The Volcani Center, Rishon LeZion, Israel

Uma Prajapati ICAR-Indian Agricultural Research Institute, New Delhi, India

Bhanu Prakash Centre of Advanced Study in Botany, Institute of Science, Banaras Hindu University, Varanasi, India

Kandeeparoopan Prasannath Department of Agricultural Biology, Faculty of Agriculture, Eastern University, Chenkalady, Sri Lanka

H.R. Raghavendra ICAR-Indian Agricultural Research Institute, New Delhi, India

Ashutosh Rai Division of Crop Improvement and Biotechnology, Indian Institute of Vegetable Research, Varanasi, India

Shoshana Salim Department of Postharvest Science, Agricultural Research Organization (ARO), Volcani Center, Rishon LeZion, Israel

Seema Sangwan ICAR-Indian Agricultural research Institute, New Delhi, India

Swapnil Sapre Biotechnology Centre, Jawaharlal Nehru Agriculture University, Jabalpur, India

Kankan Sharma Department of Biotechnology, Lovely Professional University, Phagwara, India

Ram Roshan Sharma ICAR-Indian Agricultural Research Institute, New Delhi, India

Sumana Sikdar Biotechnology Centre, Jawaharlal Nehru Agriculture University, Jabalpur, India

Ane Stéfano Simionato Microbial Ecology Laboratory, Department of Microbiology, State University of Londrina, Londrina, Brazil

Divya Singh Central Sericulture Research and Training Institute, Mysore, India

Joginder Singh Department of Biotechnology, Lovely Professional University, Phagwara, India; Climate Mitigation and Sustainable Agriculture Research lab (CMaSAR), Division of Research and Development, Lovely Professional University, Phagwara, India

Prem Pratap Singh Centre of Advanced Study in Botany, Institute of Science, Banaras Hindu University, Varanasi, India

Rishikesh Singh Institute of Environment and Sustainable Development, Banaras Hindu University, Varanasi, India

Sandeep Kumar Singh Department of Botany, Centre of Advanced Study, Institute of Science, Banaras Hindu University, Varanasi, India

Satyender Singh Regional Advance Water Testing Lab (RAWTL), Department of Water Supply and Sanitation, S.A.S. Nagar, India

Simranjeet Singh Department of Biotechnology, Lovely Professional University, Phagwara, India; Climate Mitigation and Sustainable Agriculture Research lab (CMaSAR), Division of Research and Development, Lovely Professional University, Phagwara, India; Punjab Biotechnology Incubator (PBTI), S.A.S. Nagar, India; Regional Advance Water Testing Lab (RAWTL), Department of Water Supply and Sanitation, S.A.S. Nagar, India

Sushil Kumar Singh Department of Agricultural Biotechnology, Assam Agricultural University, Jorhat, India

Vipin Kumar Singh Department of Botany, Centre of Advanced Study, Institute of Science, Banaras Hindu University, Varanasi, India

Nishanthi Sivasubramaniam Department of Agricultural Biology, Faculty of Agriculture, Eastern University, Chenkalady, Sri Lanka

Pooran Singh Solanki Birla Institute of Scientific Research, Statue Circle, Jaipur, India; Department of BioEngineering, Birla Institute of Technology, Mesra, Ranchi, Jaipur Campus, Jaipur, India

R. Sundararaj Forest Protection Division, ICFRE - Institute of Wood Science and Technology, Bengaluru, Karnataka, India

Musrat Zahan Surovy Institute of Biotechnology and Genetic Engineering (IBGE), Bangabandhu Sheikh Mujibur Rahman Agricultural University, Gazipur, Bangladesh

Larissa Ayumi Taira Microbial Ecology Laboratory, Department of Microbiology, State University of Londrina, Londrina, Brazil

Vishwa Vijay Thakur ICAR-Indian Institute of Natural Resins and Gums, Namkum-Ranchi, India

Sharad Tiwari Department of Plant Breeding and Genetics, Jawaharlal Nehru Agriculture University, Jabalpur, India

P.C. Trivedi Department of Botany, University of Rajasthan, Jaipur, India

Ram S. Upadhyay Laboratory of Mycopathology and Microbial Technology, Centre of Advanced Study in Botany, Institute of Science, Banaras Hindu University, Varanasi, India

Alok Kumar Varshney Birla Institute of Scientific Research, Statue Circle, Jaipur, India; Department of BioEngineering, Birla Institute of Technology, Mesra, Ranchi, Jaipur Campus, Jaipur, India

Hariom Verma Department of Botany, Centre of Advanced Study, Institute of Science, Banaras Hindu University, Varanasi, India

S. Vijay Rakesh Reddy ICAR-Indian Institute of Horticultural Research, Bengaluru, Karnataka, India

Rajesh Kumar Vishwakarma ICAR-Central Institute of Post-Harvest Engineering & Technology, Ludhiana, India

Sopan Ganpatrao Wagh Department of Bio-Science and Technology, Agri-biotech College, VNMK University, Aurangabad, India

James Francis White Department of Plant Biology, Rutgers University, New Brunswick, NJ, United States

Michael Wisniewski Appalachian Fruit Research Station, Agricultural Research Service, United States Department of Agriculture, Kearneysville, WV, United States

H.C. Yashavantha Rao Department of Biochemistry, Indian Institute of Science, Bengaluru, Karnataka, India

Andleeb Zehra Laboratory of Mycopathology and Microbial Technology, Centre of Advanced Study in Botany, Institute of Science, Banaras Hindu University, Varanasi, India

V. Yeka Zhimo Department of Postharvest Science, Agricultural Research Organization (ARO), Volcani Center, Rishon LeZion, Israel

CHAPTER

1

Recent advancement in plant disease management

Prem Pratap Singh, Akshay Kumar, Vishal Gupta and Bhanu Prakash

Centre of Advanced Study in Botany, Institute of Science, Banaras Hindu University, Varanasi, India

1.1 Introduction

Plant pathology is an applied science concern about the nature, causes, and management of plant diseases in order to ensure the food safety and food security for the world. Old historical documents confer the plant disease reports since the first light of the agriculture. However, its only 19th century, after which infectious agents of plant diseases became identifiable and made researchers combat these plant diseases with the development of fungicides, pesticides, and disease-resistant plant varieties. All the management practices and development of control methods against plant diseases contribute to the vital processes of improvement of crop productivity (Lucas, 2017). Parallel to the advancements, there is another problem, of which researchers have to take care of, the significant economic loss. It was preassumed that the crops were the severe victims of the phytopathogens in comparison to wild plant varieties. Diverse genotype, mixed-age structures, dispersed distribution, and host–pathogen equilibrium in the coevolutionary span with pathogens of wild plant varieties made crops more vulnerable to infection (Lucas, 2009). There is a loss of 8%–21% to six major crops (wheat, rice, maize, potatoes, soybeans, and cotton) caused by only microbial and pest infestation annually (Oerke, 2006). Wastage caused by microbial and pest spoilage along with loss at transport and storage makes it a significant for crops at their postharvest phase. These losses count for a significant economic loss as they are closer to the produce sale time, and also these are not only the wastage of agro-crop products but also the wastage of resources put into its production.

Plant disease (impairment in the normal physiological function of the plant) can be caused by both the factors, inanimate (abiotic) and animate (biotic). However, the

Food Security and Plant Disease Management
DOI: https://doi.org/10.1016/B978-0-12-821843-3.00012-X

potential of animate factors to grow, reproduce, and disperse, enables them to diffuse rapidly over a wide range of landscape, entire continent sometimes (Singh, 2018). Therefore their management practices must comprise at the preharvest, harvest, and postharvest phases to impact the complicate relationship of host–pathogen–environment. The triangle of host–pathogen–environment is the most elementary notion of plant pathology to deal with plant diseases (Scholthof, 2007). With recent technological advancements, this notion enlarged up to a whole new level of the host–pathogen–environment complex where the simple sum of these three components, that is, host, pathogen, and environment, gets upgraded up to their interactive ventures.

A suitable inducive combination of the susceptible host, virulent pathogen, and environmental conditions is necessary for successful establishment of disease and completion of its cycle (all the event represent the biology of pathogen through the interconnected stages of development like dormancy, reproduction, dispersal, and pathogenesis) (De Wolf and Isard, 2007). Therefore the experimental aspect of the host–pathogen–environment makes this domain a significant territory for the researchers of plant pathology. Several experimental analyses have been going on in the field of plant recognition, signal pathways, and stress physiology related to the fundamental aspects of the plant diseases (Dodds and Rathjen, 2010). However, in recent trends, the use of molecular techniques along with the molecular-genetics approach allocated some new prospects of pathogenicity and plant defense. The molecular techniques not only advance the understanding of molecular plant–microbe interactions but also remarkably influence their taxonomic relationship and methods of disease control. In turn, these new understandings propose novel methods for the management of plant diseases (Lucas, 2011).

In the conventional disease-management practices, visual examination is the only way to conclude plant pathogenesis, and after that identification of phytopathogen can be made. At that phase, the major loss has already been made to crop fields making management practices limited or of no use. Hence, the diagnostic process must be *accurate*, that is, specific for the pathogen, *sensitive*, that is, able to detect pathogen at the initial stage, *simple*, that is, can be performed at small laboratory conditions, *rapid*, that is, takes minimal processing time, *cheap* and *safe* for the operating person (Chandrashekara et al., 2012). The advancement in molecular biology of plant diseases accelerates the process of disease detection and pathogen identification. Molecular biology techniques enable the researchers to develop such "kits" that detect the plant diseases at the early phase of infection by detecting the presence of pathogenic DNA or protein molecules. These "kits" are developed in such a handy way that they can be used in the fields without the need for laboratory facilities and give more accurate findings in minimal processing time (Boonham et al., 2008).

The chapter provides an overview of the basic and recent developments in the field of plant pathology, focusing on the advancements and their shortcomings. Along with it, the improved strategies for disease management and their limitations have been discussed. This chapter highlights the key points of pathogen's infection and transmission for fungi, bacteria, viruses, and pests. Finally, the chapter summarizes the sustainable strategies and techniques for disease management. Some common plant diseases, along with their symptoms and causal organisms, were presented in Table 1.1.

TABLE 1.1 Some mon plant diseases along with their symptoms and causal organisms.

Disease	Plant	Pathogen	Symptoms	References
Citrus canker	Citrus	*Xanthomonas axonopodis pv. Citri*	Distinctively raised, necrotic lesions on fruits, stems, and leaves	Graham et al. (2004)
Crown gall	Euonymus	*Agrobacterium tumefaciens*	Overgrowths appearing as galls on roots	Kado (2002)
Fire blight	Pear	*Erwinia amylovora*	Leaves on diseased shoots often show blackening along the midrib and veins before becoming fully necrotic	Stockwell et al. (1998)
Stewart's wilt	Maize	*Erwinia stewartii*	New leaves emerging from the plant whorl and cavities may form in the stalks near the soil line	Pataky et al. (2004)
Root rot	Chickpea	*Ectophoma multirostrata*	Foliar chlorosis, dry and brittle root system and prominent black necrotic lesions on the taproot	Chobe et al. (2020)
Leaf spot	Horseradish	*Alternaria brassicicola*	Spots are dark brown; a chlorotic halo is often visible if surrounding leaf area is still green	Blagojević et al. (2019)
Wilt disease	Queen palm	*Fusarium oxysporum*	Trees displayed rapid death of the crown with dead leaves remaining attached and upright. Symptomatic leaves presented necrotic leaflets unilaterally and petiole cross-sections with unilateral vascular discoloration. Trunk cross-sections revealed inner tissue discoloration	Gomez and Pryor (2019)
Fruit rot	Cuiguan pear	*Fusarium proliferatum*	Initial symptoms were shallow black spots on the fruit surface that later enlarged into circular brown lesions with water-soaked margins and a lesion diameter of 20–40 mm. White hyphae grew on the pear lesions and slowly, infected fruit became soft and necrotic	Zhao et al. (2020)
Pokkah boeng	Sugarcane	*Fusarium sacchari*	Chlorosis at the base of young leaves, wrinkling, twisting and/or shortening leaves, or rotting of apical part of stalk	Meng et al. (2020)
Powdery mildew	Japanese clover	*Erysiphe lespedezae*	White powdery colonies on the upper leaf surfaces. As the disease progressed, infected leaves were covered with a dense white to greyish white mycelia	Lee and Nguyen (2019)
Gray mold	Golden berry	*Botryotinia fuckeliana*	Earlier small, round, pale-brown lesions on the calyx of the fruit. Later, the calyx and fruit became completely rotted and covered with gray mycelium	Gümrükcü et al. (2016)
Leaf curl	Cucumber	*Tomato yellow leaf curl virus*	Upward leaf curling, chlorosis, plant dwarfing, mosaic, yellowing, and fruit and leaf deformation	Al-Ali et al. (2016)

(Continued)

TABLE 1.1 (Continued)

Disease	Plant	Pathogen	Symptoms	References
Yellow spot	Garlic	*Iris yellow spot virus*	Straw-colored, spindle-shaped spots with poorly defined ends on leeks, and diamond-shaped lesions with green chlorotic islands on garlic. These coalesced to form large, brown, necrotic patches on the leaves	Karavina et al. (2016)
Potato cyst nematode	Potato	*Globodera rostochiensis*	Inefficient roots system, that is, poor growth, wilting during periods of water stress, early senescence and reduced tuber yield up to levels in excess of 80%	Inacio et al. (2020)
Aphanomyces root rot	Lentil	*Aphanomyces euteiches*	Yellowing and wilting of lower leaves, discoloration of roots	Zitnick-Anderson et al. (2020)
Leaf spot	Strawberry	*Botrytis cinerea*	Circular lesions measuring 2–3 cm in diameter, concentric rings with a darker brown outer ring and fungal sporulation in the center of the lesion	Mansouripour and Holmes (2020)
Leaf spot	Hemp	*Cercospora cf. flagellaris*	Older leaves and eventually spread throughout the canopy. Infections began with small yellow, individual flecks; the center had fascicles of conidiophores visible to naked eye, were light tan, or even white, with yellow halos	Marin et al. (2020)
Rice bacterial blight	Rice	*Xanthomonas oryzae pv. oryzae*	Leaves of young plants as pale-green to grey-green, water-soaked streaks near the leaf tip and margins, wilting, desiccation of leaves and death, particularly of young transplanted plants	Xian et al. (2020)
Sharp eyespot	Wheat	*Rhizoctonia cerealis*	Oval-shaped lesions that develop on basal leaf sheaths, margins are dark brown with pale, straw-colored centers and whiteheads, and shriveled kernels	Ren et al. (2020)
Powdery mildew	*Capparis spinosa subsp. himalayensis*	*Leveillula taurica*	Circular to irregular white colonies, which subsequently develop into an abundant growth on the upper and lower surface of the leaves	Song et al. (2020)
Bacterial leaf scorch	American elm	*Xylella fastidiosa*	Leaf discoloration and browning, marginal scorching, and dieback	Ali et al. (2020)
Leaf blight	Taro	*Alternaria alternata*	Yellow to brown, stripped to irregular-shaped, followed by necrosis spreading into interveinal areas. Ultimately, the affected leaves turned dark brown and became cracked and withered	Liu et al. (2020)

(*Continued*)

TABLE 1.1 (Continued)

Disease	Plant	Pathogen	Symptoms	References
Powdery mildew	*Ipomoea obscura L.*	*Erysiphe alphitoides*	White, superficial mycelia and conidia covered the leaf surfaces of affected plants, resulting in leaf curling, discoloration, and defoliation	Pan et al. (2020)
Bot rot	*Ficus carica*	*Botryosphaeria dothidea*	First observed on green fruit, and infected fruit initially showed soaked symptoms on the surface, which subsequently became covered with fluffy mycelia that ultimately caused fruit decay	Wang et al. (2020)
Leaf spot	*Areca catechu*	*Diaporthe limonicola*	The leaves exhibited small yellow spots, as the symptom progressed, the middle of the lesions appeared black with distinct yellow halos. Lesions were long oval, sometimes irregular with black to brown, small black spots, and distinct yellow halos. As the lesions continued to expand, the necrotic spots were enlarged, gradually gray, and then combined to form larger necrotic areas	Xu et al. (2020)
Leaf blight wilt	*Dracaena sanderiana*	*Pantoea stewartii subsp. indologenes*	Both young and older leaves developed necrotic lesions with small water-soaked spots at the tip or at the margin of the leaves, which then enlarged and were bordered by chlorosis	Zhang et al. (2020)
Blight disease	Cotton	*Pantoea anthophila*	Angular leaf spots, necrotic lesions, and defoliation	Tufail et al. (2020)
Leafroll disease	Cotton	Cotton leafroll dwarf virus	Leaf yellowing, discoloration, reddening, and short internodes with reduced or small boll sets	Ali and Mokhtari (2020)
Ringspot disease	Red mulberry	Prunus necrotic ringspot virus	Chlorotic line pattern on the fully expanded leaves	Dey et al. (2020)
Calico disease	potato	Alfalfa mosaic virus	Bright chlorotic spots in the foliage and is often associated with tuber necrosis in susceptible cultivars	Vologin et al. (2019)

1.2 Disease: A dynamic phenomena

The literature review revealed that plant disease is a dynamic phenomenon, as even after 100 years of research and innovation for the study of cause and control of plant disease, the losses due to pathogen infestation have been witnessed in the modern era. The actual picture regarding the solutions acquainted with the plant diseases isn't achieved as per the expectation from the crop improvement and chemical treatment. Several control measures viz. breeding programs and chemical treatments had a remarkable impact

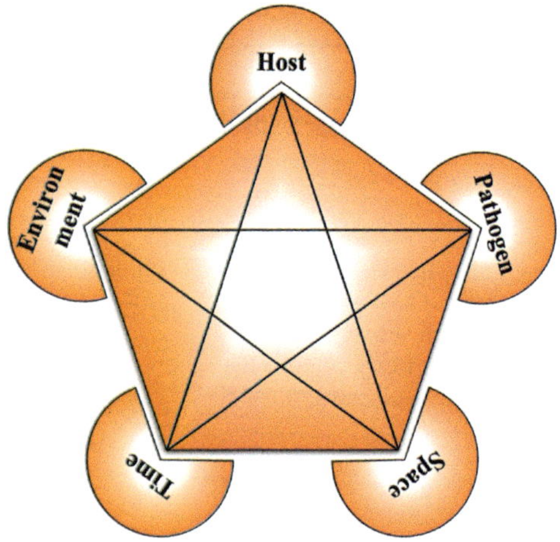

FIGURE 1.1 Representation of factors associated with the plant pathogenesis.

against the plant diseases but, the pathogenic machinery with the premise of "booming and busting" have successfully override those impacts. The "boom and bust" cycle confers the coevolutionary process of host and pathogen. Improved host-defense machinery, against prevalent pathogen pathotype, serves as the selection approach for the "mutated/novel" pathotype that can overrule the plant disease resistance (Brown and Tellier, 2011; Lucas, 2017). Coevolution is a natural phenomenon that balances the host–pathogen equilibrium over a long period, but the plant breeders led the equilibrium disordered (Burdon et al., 2013). Also, the factors viz. inception of crops in new geographical regions, reduced genetic diversity (negligence of crop rotation), and the uniform age structure of crops aid the pathogens every time, to set-aside the control measures and cause the infection.

Furthermore, excessive use of fertilizers (in ample space with shorter rotation), chemicals (growth factors, herbicides, fungicides, and pesticides), and improved irrigation facilities are some of the plant-beneficiary elements of which; excessive use cherishes the pathogens (Lucas et al., 2015). All these factors merge at a point and serve as the selection pressures for the pathogen's adaptation and evolution. Field understandings in the management of plant diseases aided with the modern genetic analysis of pathogens enlightened the researchers. They understand that the threat, they are encountering, is not a single one; instead, it was an assemblage of heterogeneous pathotypes showing dynamic evolution in both space and time (Fig. 1.1).

1.3 Pathogenicity: Facet of a successful infection

Pathogenicity (the potential of pathogens to cause the disease) requires the transmission or spreading of pathogenic pustulates from the reservoir (infected host) to a fresh host (uninfected). The pathogen must have the potential of virulence (to break all the host-defense mechanisms and harm it) to infect a new host, as only the possession of virulence differentiates the pathogens from nonpathogenic organisms. Moreover, based on the potency of virulence, pathogens can be categorized as *opportunistic* or *primary*. Primary

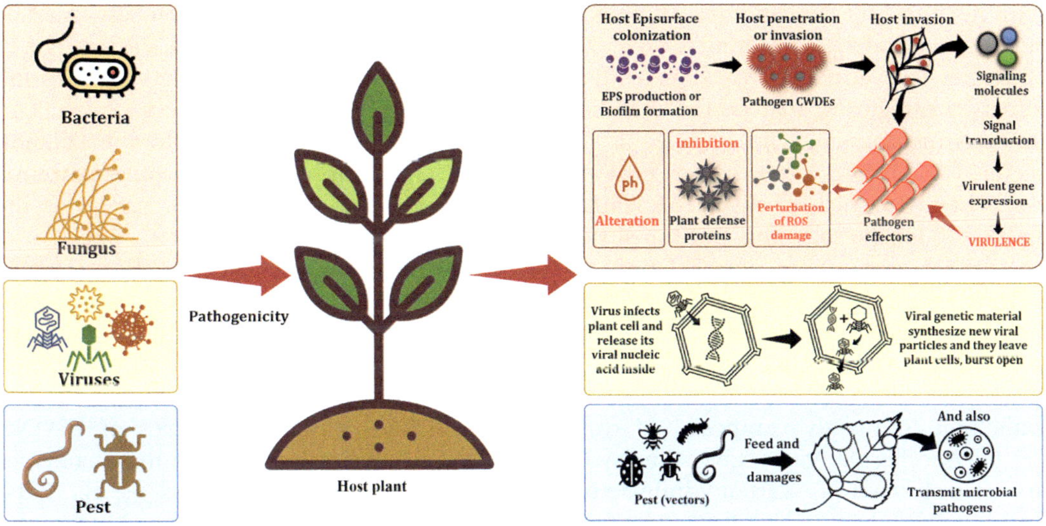

FIGURE 1.2 Schematic representation of elucidated mode of actions of different plant-pathogens.

pathogens have enough virulence to breach a healthy host and cause the disease, while opportunistic pathogens require a distortion in host defense to cause the disease. Pathogens can have a broad spectrum of virulence capacity, where they can infect a wide range of hosts, or they must be highly specialized and adaptive to infect a particular group of hosts. Schematic representation of the elucidated mode of actions of different plant-pathogens is described in Fig. 1.2.

Pest, which causes damage to plants, also serve as important vectors for plant-pathogens. They feed on plants, and all the time during their feeding, they move from plant to plant, dispersing the pathogens from one plant to another. Usually, pest includes insects viz. aphids, leaf miners, gall-forming sawflies, leafhoppers, and some nematodes, but larger animals can also serve the functions of a pest. Pest causes severe damages to plant tissues in order to get nutrition (aphids extract sap from the phloem, caterpillars consume large areas of the leaf lamina). There are two types of pathogen–vector interaction can be seen in dispersing the pathogens. In the first case, the pathogen does not cross the membrane barrier of the vectors and also does not multiply inside them. In contrast, in another case, pathogen crosses the membrane barrier to enter the body of the vectors and multiply within them. Former case of interaction is nonspecific, and the pathogen can be spread sporadically by a variety of vectors while later one is quite specific and requires a close interaction between the partners for effective translocation.

1.3.1 Fungal pathogenicity

Fungal pathogens affect the plant in three ways, that is, *biotrophy* where fungi feed on the living host, *necrotrophy*, wherein fungi triggers cell death in host to ensure its nutrition supply, and *Hemibiotrophy*, where host–pathogen relation starts with biotrophy but ends with necrotrophy. Nevertheless, before all these procedures, the infection process must be successful (Rodriguez-Moreno et al., 2018). Phytopathogenic fungi use a vast range of strategies to evade and hijack plant defense machinery. All the pathogenic strategies fall into two categories, that is, the establishment of infection structure and breaching host-defense mechanism. The pathogenic fungus brings into play the effectors in planta-secreted molecules that cause host colonization and suppression of their immune responses (Rovenich et al., 2014). Haustoria, the fungal feeding structures, serves as the active site for the secretion and translocation of effector molecules (Rafiqi et al., 2012).

In the process of the establishment of fungal infection, the first barrier that pathogens have to breach is the host plant penetration. The fungal phytopathogens have a specialized structure, called appressoria, which is used to penetrate inside the plant by applying the turgor pressure (Ryder and Talbot, 2015). Furthermore, fungi have an arsenal of specialized chemicals, called cell-wall-degrading enzymes (CWDEs) that are used to dissolve the cell wall, that is, mainly composed of carbohydrates. Some common examples of CWDEs are polysaccharide lyases, glycoside hydrolases, and carbohydrate esterases; these enzymes are also termed as carbohydrate-active enzymes (CAZymes) (Lombard et al., 2014; Rodriguez-Moreno et al., 2018). Along with modulating the plant cell walls, CWDEs also modulate the fungal cell walls for attaining the beneficial morphological changes and release nutrients for the pathogens (King et al., 2017). After that, at the second level of breaching, phytopathogens have to prevent themselves from getting recognized by the plant defense system. In due course of evolution, plant evolved plenty of membrane-localized immune receptors that recognize pathogen-derived ligands and these ligands, in the downstream signaling, leads to the activation of plant defense mechanisms (Bolton, 2009; Couto and Zipfel, 2016; Cook et al., 2015). The pathogen-derived ligands comprise chiefly the fungal cell wall structural molecules, glucans, and chitin. In turn, fungi also evolved strategies to override the host immune responses and cause pathogenicity. Fungi use to do alterations in their cell wall compositions to confuse plant immune receptors and secrete effectors to protect the cell wall (Sanchez-Vallet et al., 2015). Examples of the cell wall alterations can be seen in *Magnaporthe oryzae*, a hemibiotrophic fungal pathogen causing rice blast disease, where α-1,3-glucans get accumulated at the wall surface and inhibit chitinase action (Fujikawa et al., 2012). Furthermore, fungi do not get over only with the cell wall modification, but they also secrete effector proteins to suppress host immunity as seen in the case of *Cladosporium fulvum*, the tomato leaf mold fungus, which secretes a carbohydrate-binding effector protein, Ecp6, that outcompete host receptors for chitin-binding and inhibits chitin-triggered host-defense responses (Sanchez-Vallet et al., 2013).

The most important thing that pathogens have to take care of is the host-defense responses that started due to their recognition by plant receptors. There are various tactics that fungi use to subvert the host-defense responses, that is, perturbing the reactive oxygen species damage (oxidative damage) (*Ustilago maydis* secrets effector Pep1, that inhibits the maize peroxidase activity; Hemetsberger et al., 2012), host tissue pH manipulation (*Sclerotinia sclerotiorum* causes rapid death of host tissues by acidifying the infection area; Bolton et al., 2006), host proteases suppression (*C. fulvum* effector Avr2 causes conformational distortion of protease Rcr3 in tomato; Rooney et al., 2005), damaging host hormone signaling (*U. maydis* secretes Cmu1 that damages the salicylic acid biosynthesis hence inhibiting the salicylic acid-induced resistance; Djamei et al., 2011).

1.3.2 Bacterial pathogenicity

For a successful infection, survival adaptation over plant external surfaces and responsiveness to the environment, are the critical phenomena for the bacterial pathogenicity. Only after the integration of the environment factor, pathogens are able to trigger the appropriate responses for invading the host plant tissue. With the help of community-level signal transduction pathways, pathogens breach the physical and chemical barriers lined in the plant defense system and enter the plant apoplast to release their effector proteins or toxins to execute pathogenic functions (Pfeilmeier et al., 2016a,b).

If we talk about the survival adaptions over the plant's external surfaces, pathogens adapted for the cold, osmotic shock, and desiccation stresses so for (Freeman et al., 2013; Wu et al., 2012). They have metabolic responses against these stresses for their survival and population maintenance. Exopolysaccharide (EPS) production is one of the major metabolic responses that aid phytopathogenic bacteria in their epiphytic survival (Dunger et al., 2007). To date, a plethora of EPS molecules has been reported that contribute to plant surface colonization and infection. Some examples are as follows—trehalose in *Pseudomonas syringae* (Freeman et al., 2010) and *Pseudomonas aeruginosa* (Djonovic et al., 2013); acetylated cellulose in *Pseudomonas fluorescens* (Gal et al., 2003) and *P. syringae* pv. *tomato* (Pto) DC3000 (Prada-Ramirez et al., 2016); Xanthan in *Xanthomonas* spp. (Dunger et al., 2007); alginate and levan in *P. syringae* (Laue et al., 2006).

Signaling pathways capacitate the pathogens residing on the epiphytic surfaces to switch between the mechanisms of surviving on the surfaces to breaching in the host apoplast. Signaling molecules regulates pathogen's movement, biofilm formation, virulence expression, and many other critical pathogenic functions. The main signaling pathways include quorum sensing (QS) and second messenger such as cyclic guanosine 3′,5′-monophosphate (cdG) signaling (Pfeilmeier et al., 2016a; Perez-Mendoza et al., 2014). QS, also called autoinducers, helps bacteria to communicate with each other and assess their cell density. Once they attain a critical level of cell density, these autoinducers affect the gene expression of bacteria to make them pathogenic. Some important QS molecules reported in the pathogenic bacteria are *acyl-homoserine lactones (AHLs)*, the major class of QS molecules, and vast distribution in gram-negative bacteria; *diffusible signal factor (DSF)*, fatty acid QS signal which has linkage with the cdG signaling and *Ax21*, small sulfated protein. The QS molecules help bacteria to cause virulence by controlling their motility (by controlling flagellum assembly), expression of virulence genes, toxin secretion, and EPS production. cdG signaling has been reported to control various aspects of bacterial virulence (biofilm formation, regulation of flagellum and T3SS activity, siderophore, and toxin production). It works in a negative control manner, that is, high levels of cdG inhibit virulence and promote a sessile mode of living in bacteria, while its degradation causes virulence. The proteins involved in the cdG signaling, are controlled by environmental factors; hence, modulate bacteria to respond accordingly (Romling et al., 2013; Pfeilmeier et al., 2016b). The regulatory gene systems and function of some QS molecules of bacterial pathogens were given in Table 1.2.

Now, for initiating the infection process, bacteria have to breach the plant defense mechanisms to gain entry inside the plant. The most obvious place for the entry inside the plant is stomatal opening. They secrete several phytotoxins (viz. coronatine and syringolin A) and

TABLE 1.2 The QS molecules of some bacterial pathogens along with their regulatory gene systems and function they renders.

QS class	Bacteria	Gene system	QS molecule	Rendering function	References
AHLs	*Pseudomonas syringae*	ahlI/ahlR	3-oxo-C6-HSL (HSL, homoserine lactone)	Motility, EPS production, colonization, and virulence	Quinones et al. (2004, 2005),
	Agrobacterium tumefaciens	Ti plasmid	3-oxo-C8-HSL	To control virulence	Pappas and Winans (2003)
	Burkholderia glumae	TofI (LuxI homolog)	C8-HSL	Phytotoxin secretion and motility (control of flagellum assembly)	Kim et al. (2004)
	Pseudomonas aeruginosa	LasI/LasR, RhlI-RhlR, and QscR (orphan regulator)	3-oxoC12-HSL and C4-HSL	Toxin secretion, motility, and biofilm formation	Schuster et al. (2003)
		LasR	2-Heptyl-3-hydroxyl-4-quinolone (PQS, pseudomonas quinolone signal)	Production of virulence factors and the control of biofilm formation	Allesen-Holm et al. (2006), Pesci et al. (1999)
Ax21	*Xanthomonas oryzae* pv. *oryzae*	RaxR/H and PhoP/Q	Ax21 protein	Virulence gene expression	Han et al. (2011)
	X. oryzae pv. *oryzicola*	*rax* genes (RaxST)	Ax21 like protein	Virulence and cell motility	Qian et al. (2013)
	Stenotrophomonas maltophilia strain K279a	*rax* genes	Ax21 protein	Virulence, biofilm formation, and motility	McCarthy et al. (2011)
DSF	*Xanthomonas campestris* pv. *campestris*	Rpf gene cluster (RpfF)	Cis-11-methyl-dodecenoic acid	Virulence, EPS (xanthan), and protease production	Wang et al. (2004), Barber et al. (1997)
	Xylella fastidiosa	Rpf gene cluster (RpfF)	Cis-2-tetradecenoic acid	Virulence regulation, adhesion, and colonization	Beaulieu et al. (2013), Chatterjee et al. (2008b)
	X. oryzae pv. *oryzae*	Rpf gene cluster (RpfC)	Cis-2-dodecenoic acid and cis, cis-11-methyldodeca-2, 5-dienoic acid	Virulence factor production	He et al. (2010), Tang et al. (1996)

antistomate defense factors to overcome stomatal immunity and prevents closure of the stomatal pores (Melotto and Kunkel, 2013; Zeng and He, 2010). Alongside to the stomatal pores, phytopathogen uses wounds on the plant episurfaces to enter the apoplast. For that, pathogens release enzymes to break cell wall structural molecules and dissolve the connective lamellae between the cells, making a passage inside the host. Some bacterial pathogens (such as *Xylella fastidiosa*) (Chatterjee et al., 2008a) also use the sap-feeding insects as vectors for the penetrating of host tissue.

Phytopathogenic bacteria ooze out diverse kind of virulence factors called as effectors which include enzymes and toxins that subvert host-defense responses and causes a successful infection. These effectors not only manipulate the host responses, but they also modulate the host metabolic and signaling pathways causing chlorosis and tissue necrosis. They hamper the host hormone system by producing hormone analogs (viz. coronatine; Zeng and He, 2010) to impede the internal signaling pathways in order to suppress the plant immune responses and maximize the virulence.

1.3.3 Viral pathogenicity

The viral disease is one of the causes that impart a significant challenge to food security. There are thousands of plant-infecting viruses that are known till date. As well said by the Pallas and García (2011) that diseases are the anomaly in-spite of being a common effect as plants potentially resist the pathogenic agents. Viruses use the biotrophic nature of pathogenesis, where they require the living host for their propagation. Their competitive interference for the host resources causes the symptomatic morphology of hosts that finally leads to the disease occurrence. This competitive interference damages the host physiology and exploits the specific host genetic mechanisms belonging to the hormonal system, cell cycle control, and transportation system of macromolecules. Like the other pathogen, viruses also have to cross many defense fortresses of the plant. Firstly, they have to break the plant's preformed barriers for entering the host. If the viruses successfully breach in the first layer of host-defense, after that, they have to encounter the two-tiered immunity of plants, that is, PAMP-triggered immunity [PTI; pathogen-associated molecular patterns (PAMP)] and Effector-Triggered Immunity (ETI) (Jones & Dangl, 2006). Viruses produce several virulence factors in order to cause disease, and in response, the plant creates a new, more specific resistance layer. Finally, the virus will only be able to cause successful infection only when they scourge and break this defense layer.

Besides this, when the virus penetrates the host cells, another layer of defense mechanism starts. Viral RNA induces specific plant defense responses, in which complex set of antiviral plant proteins are involved to fend off the viruses; this mechanism is commonly termed as RNA interference (RNAi). In order to protect themselves, viruses not only require escape from this antiviral plant response but also have to suppress them. Viruses adapt to different machinery, which is commonly known as the production of a silencing suppressor strategy. These suppressors are various viral proteins that not only suppress the RNAi mechanism but also interfere in plants' various cellular and physiological processes. Thus it is a kind of tug of war between viral pathogens and host plant, whose outcome of disease outbreak solely depends on the defense responses of both the partners.

The symptoms of major pathogens and their common diseases have been summarized in Table 1.3.

1.4 Technological upliftment: An integrated approach

With the emergence of molecular techniques such as polymerase chain reaction, genetic tools, the picture of plant diseases has been transformed significantly (Hadidi et al., 1995).

TABLE 1.3 Symptoms of major pathogens and common diseases they cause.

Pathogen	Common symptoms	Common diseases	Common causal pathogen genera
Fungi	Damping off of seedlings; leaf spot, may develop giving the spot a "bull's eye" effect; chlorosis; round, oval, or irregular necrotic lesions; grey to white mycelia and conidia on leaves; development of black specks (cleistothecia) in mycelia; presence of rusty brown (rust) to black spores (smut); needle dropping (in case of conifers); fruit rotting; necrotic and sunken lesion (ulcer-like)	Mildew diseases (powdery or downy); rust disease; smut diseases; damping off disease, fungal blight disease; scab disease; anthracnose disease; sclerotium rots	*Cryptosporiopsis, Fusarium, Rhizoctonia, Pythium, Venturia, Phytophthora, Armaillaria, Puccinia, Colletotrichum, Ustilago, Alternaria, Pyricularia*, and *Pestalotiopsis*
Bacteria	Leaf spots, often angular due to limitation by leaf veins; presence of bacterial ooze; chlorotic halo around the lesion; cankers and gum exudation; presence of small; whitish; soft round galls on the plant crown or root; wilting of aerial plant parts; slimy foul smelling ooze from infected parts; fruit rotting; presence of localized scabby lesions on outer surface; formation of corky tissue below and around the lesion	Bacterial blight disease; spot disease; canker disease; bacterial galls; bacterial wilts; soft rot disease; scab disease	*Erwinia, Pectobacterium, Pantoea, Agrobacterium, Pseudomonas, Ralstonia, Burkholderia, Acidovorax, Xanthomonas, Clavibacter, Streptomyces, Xylella, Spiroplasma*, and *Phytoplasma*
Viruses	Vein banding; mosaic; ringspot; distorted leaves (from normal shape and size); fruit discoloration; shortened internodes (stunting); bark scaling	Curly top disease; mosaic disease; psorosis disease; spotted wilt disease	Potyvirus, Alfamovirus, Cucumovirus, Luteovirus, Tospovirus, Begomovirus, Sobemovirus, Comovirus, Tobamovirus, Necrovirus, Mastrevirus, Badnavirus, Nepovirus, and Tobravirus
Nematodes	Excessive root branching, injured root tips; stunted root; shortened internodes; bud distortions; galls	Root knot disease; cyst disease; root lesion disease; pine wilt disease; reniform disease	*Meloidogyne, Heterodera, Globodera, Pratylenchus, Helicotylenchus, Xiphinema, Aphelenchoides*, and *Rotylenchulus*

However, the solutions adopted for disease management proven to be temporary. The reappearance of problems (pathogens) in the plant protection is because of the "quick-fix" approach (management programs based on single major R gene resistance or single-site inhibiting chemical compounds) adopted for it. The "quick-fix" approach gets easily busted and ends up to the "boom and bust" cycle of the pathogen because of its simple process. Therefore the required solution must be an "integrated" approach, that is, a combination of different other control measures (Lucas, 2009). The integrated approach aims toward the potential as well as the sustainable management of the diseases. Potential management counts for the higher efficacy of host resistance and agrochemicals, whereas sustainable management checks out its cost and impact on environmental systems.

With the development of molecular biology and techniques associated with it, it is now easier to track down the signaling cascades of host–pathogen interaction. The information related to the host receptors, host-defense genes, and their functions became apparent in a very detailed manner (Dodds and Rathjen, 2010) along with the interactive venture of pathogens with host and their mechanism of pathogenicity (Guttman et al., 2014). The integration of molecular biology with the plant breeding led to breeding more efficient cultivars. The advent of molecular marker techniques led to the identification of such genetic region, which shows significant phenotypic variance on diseases. Plant breeders by introgression of such quantitative traits' loci (which is also durable than qualitative) will produce crop varieties with improved resistance (Arraiano and Brown, 2017).

The same quantitative approach can be employed in chemical disease-management procedures. Despite using a single chemical compound in large amount, integration of low doses of different chemical compounds with different modes of action, have better potential. Ideally, they will impact low-risk to the environment and serves higher-risk single-site inhibitor for pathogens (Van den Bosch et al., 2014). Also, chemical disease control aided with computer-based disease forecasting systems (viz. SimCast/BLITECAST—Disease forecasting system generated for the late blight of the potato in United Kingdom) requires less amount of chemicals and gives better results.

1.5 Sustainable intensification: A sustainable way of plant disease management

The challenges arising after the advent of the technological advancements is the durability of the approaches available for the management of plant diseases in the context of its sustainability and the evolutionary pressure excreting on the pathogens. Therefore in this notion, sustainable intensification (SI) has been put forwarded to fulfill the requirement of the disease-management systems without causing any collateral impact on finite resources, that is, land, water, soil, and biodiversity (Royal Society, 2009). SI embraces the utilization of natural constrains over the external inputs of chemicals for increasing the plant's resistance against the pest and pathogens. SI pleads for the concept of diversified agriculture, that is, growing a range of crops simultaneously, crop rotation, intercropping, organic manure, and management of plant microbiome. For disease management, because systems with more biodiversity have shown high resistance and stability against the pest and pathogens (Seddon et al., 2016; Pretty and Bharucha, 2014). Diversified agriculture practices are well-suited concepts under the integrated pest management program for disease management as they have their own automated management systems viz. legumes for boosting biologically fixed nitrogen, ecological constrains for eliminating the pathogens, an array of the beneficial microbial population as biofertilizers.

In general, to boost the immunity of plant against the range of phytopathogen the following condition must be provided such as nutrients availability, soil management, the beneficial microbiome in the environment; resistance breeding, genetic diversification, endophytes associated with the plant and screening, exclusion, quarantine, soil sterilization, and biological control of pathogen (Lucas, 2017). The recent convergence of molecular studies on revealing the integrated picture of the plant–pathogen interaction changed the perception of crop breeding. Nowadays, plant breeding is the most environmentally acceptable and most effective way out for crop protection. The phenomenon of plant–pathogen interaction provides a picture of how these two coexists together in nature. With these understandings, we can identify the factors affecting the evolution of pathogens at the individual level as well as the population level.

Plant immunity is packed with the two strategies to detect pathogens and eliminate them. The primary one operates in the external environment where plant receptors recognize the conserved microbial elicitors called PAMPs and trigger PTI in the plant. While the secondary strategy operates when the plant's internal receptors recognize the pathogen virulence molecules called effectors and activate ETI. PTI and ETI cause same similar responses, and the only difference is the potential and speed of the responses. ETI poses relatively more reliable and rapid responses in comparison to PTI because ETI brings immunity against the adapted pathogens while PTI concerns for the nonadapted pathogens. The mode of recognition of effectors in ETI is the dynamic one, as effectors are characteristically variable in comparison to PAMPs. The diversification of these two, ETI receptors and pathogen effectors, confers the coevolutionary disease dynamics (Dodds and Rathjen, 2010).

Hence the extensive deployment of technological advancements viz. next-generation sequencing in the widespread genome sequencing unravels the effector's identification and its expression patterns (Derevnina and Michelmore, 2015). This knowledge can be exploited in enabling the different plant species to access different pathogens recognizing receptors (also from sexually incompatible plants), which multiply the number of resistance genes. This advancement is environmentally acceptable and could cause potential control over pathogens.

1.6 Conclusion

The developmental context of plant pathology concerns about the practical solution to the disease problems in the prospects of crop protection. Still, in the unprecedented technological advancement, many crops are not doing well and losing their productivity. This happens only because of the lack of knowledge on the disease epidemiology and their poorly defined diagnostic procedures. As comparing all the pathological events, the scenario suggests that there is no accurate formula for disease management. Hence, the adopted management approach must be according to the circumstances prevailing in the agriculture field and markets. Till date, there is a plethora of literature and reports present for the different disease-management strategies. However, very few of them render the combinatorial effect of chemical, biological, and technological factors with less number of shortcomings. The newly developed resistant crop varieties should be commercially available to reduce the chemical application without sustaining economic losses. Hence, there is an urgent need to search for a well-coordinated and more integrated disease-management system to derive the notion of reduction of plant diseases without the collateral damage to the environment and resources.

References

Al-Ali, E., Al-Hashash, H., Ben Heji, A., Al-Aqeel, H., 2016. First report of tomato yellow leaf curl virus infecting cucumber in Kuwait. Plant. Dis. 100, 656-656.

Ali, M.E., Hudson, O., Waliullah, S., Ji, P., Williams-Woodward, J., Oliver, J.E., 2020. First report of bacterial leaf scorch disease of American elm caused by *Xylella fastidiosa* in Georgia, USA. Plant. Dis. Available from: https://doi.org/10.1094/PDIS-11-19-2367-PDN.

Ali, A., Mokhtari, S., 2020. First report of cotton leafroll dwarf virus infecting cotton (*Gossypium hirsutum* L.) in Kansas. Plant. Dis. Available from: https://doi.org/10.1094/PDIS-12-19-2589-PDN.

Allesen-Holm, M., Barken, K.B., Yang, L., Klausen, M., Webb, J.S., Kjelleberg, S., et al., 2006. A characterization of DNA release in *Pseudomonas aeruginosa* cultures and biofilms. Mol. Microbiol. 59, 1114–1128.

Arraiano, L.S., Brown, J.K.M., 2017. Sources of resistance and susceptibility to *Septoria tritici* blotch of wheat. Mol. Plant. Pathol. 18, 276–292.

Barber, C.E., Tang, J.L., Feng, J.X., Pan, M.Q., Wilson, T.J., Slater, H., et al., 1997. A novel regulatory system required for pathogenicity of *Xanthomonas campestris* is mediated by a small diffusible signal molecule. Mol. Microbiol. 24, 555–566.

Beaulieu, E.D., Ionescu, M., Chatterjee, S., Yokota, K., Trauner, D., Lindow, S., 2013. Characterization of a diffusible signalling factor from *Xylella fastidiosa*. MBio. 4, e00539-12.

Blagojević, J., Vukojević, J., Ivanović, B., Ivanović, Ž., 2019. Characterization of *Alternaria* species associated with leaf spot disease of *Armoracia rusticana* in Serbia. Plant. Dis. Available from: https://doi.org/10.1094/PDIS-02-19-0289-RE.

Bolton, M.D., 2009. Primary metabolism and plant defence-fuel for the fire. Mol. Plant. Microbe Interact. 22, 487–497.

Bolton, M.D., Thomma, B.P., Nelson, B.D., 2006. *Sclerotinia sclerotiorum* (Lib.) de Bary: biology and molecular traits of a cosmopolitan pathogen. Mol. Plant. Pathol. 7, 1–16.

Boonham, N., Glover, R., Tomlinson, J., Mumford, R., 2008. Exploiting generic platform technologies for the detection and identification of plant pathogens. Eur. J. Plant. Pathol. 121, 355–363.

Brown, J.K., Tellier, A., 2011. Plant-parasite coevolution: bridging the gap between genetics and ecology. Annu. Rev. Phytopathol. 49, 345–367.

Burdon, J.J., Thrall, P.H., Ericson, L., 2013. Genes, communities & invasive species: understanding the ecological and evolutionary dynamics of host–pathogen interactions. Curr. Opin. Plant. Biol. 16, 400–405.

Chandrashekara, K.N., Raju, S.R., Chandrashekara, C., 2012. Molecular and biotechnological approaches in plant disease management. In: Singh, Vaibhav K., Singh, Yogendra, Singh, Akhilesh (Eds.), Ecofriendly Innovative Approaches in Plant Disease Management. International Book Distributors, India, pp. 271–290.

Chatterjee, S., Almeida, R.P., Lindow, S., 2008a. Living in two worlds: the plant and insect lifestyles of *Xylella fastidiosa*. Annu. Rev. Phytopathol. 46, 243–271.

Chatterjee, S., Newman, K.L., Lindow, S.E., 2008b. Cell-to-cell signaling in *Xylella fastidiosa* suppresses movement and xylem vessel colonization in grape. Mol. Plant Microbe Interact. 21, 1309–1315.

Chobe, D.R., Tarafdar, A., Chandran, U.S., Rani, S., Sharma, M., 2020. First report of *Ectophoma multirostrata* causing root rot of chickpea. Plant. Dis. Available from: https://doi.org/10.1094/PDIS-01-19-0121-PDN.

Cook, D.E., Mesarich, C.H., Thomma, B.P., 2015. Understanding plant immunity as a surveillance system to detect invasion. Annu. Rev. Phytopathol. 53, 541–563.

Couto, D., Zipfel, C., 2016. Regulation of pattern recognition receptor signalling in plants. Nat. Rev. Immunol. 16, 537–552.

De Wolf, E.D., Isard, S.A., 2007. Disease cycle approach to plant disease prediction. Annu. Rev. Phytopathol. 45, 203–220.

Derevnina, L., Michelmore, R.W., 2015. Wheat rusts never sleep but neither do sequencers: will pathogenomics transform the way plant diseases are managed? Genome Biol. 16, 44.

Dey, K., Velez-Climent, M., Soria, P., Hart, S., 2020. First report of prunus necrotic ringspot virus infecting red mulberry (*Morus rubra*) in the USA. Plant. Dis. Available from: https://doi.org/10.1094/PDIS-09-19-2048-PDN.

Djamei, A., Schipper, K., Rabe, F., Ghosh, A., Vincon, V., Kahnt, J., et al., 2011. Metabolic priming by a secreted fungal effector. Nature 478, 395–398.

Djonovic, S., Urbach, J.M., Drenkard, E., Bush, J., Feinbaum, R., Ausubel, J.L., et al., 2013. Trehalose biosynthesis promotes *Pseudomonas aeruginosa* pathogenicity in plants. PLoS Pathog. 9, e1003217.

Dodds, P.N., Rathjen, J.P., 2010. Plant immunity: towards an integrated view of plant–pathogen interactions. Nat. Rev. Genet. 11, 539–548.

Dunger, G., Relling, V.M., Tondo, M.L., Barreras, M., Ielpi, L., Orellano, E.G., et al., 2007. Xanthan is not essential for pathogenicity in citrus canker but contributes to *Xanthomonas epiphytic* survival. Arch. Microbiol. 188, 127–135.

Freeman, B.C., Chen, C., Beattie, G.A., 2010. Identification of the trehalose biosynthetic loci of *Pseudomonas syringae* and their contribution to fitness in the phyllosphere. Env. Microbiol. 12, 1486–1497.

Freeman, B.C., Chen, C., Yu, X., Nielsen, L., Peterson, K., Beattie, G.A., 2013. Physiological and transcriptional responses to osmotic stress of two *Pseudomonas syringae* strains that differ in epiphytic fitness and osmotolerance. J. Bacteriol. 195, 4742–4752.

Fujikawa, T., Sakaguchi, A., Nishizawa, Y., Kouzai, Y., Minami, E., Yano, S., et al., 2012. Surface a-1,3-glucan facilitates fungal stealth infection by interfering with innate immunity in plants. PLoS Pathog. 8, e1002882.

Gal, M., Preston, G.M., Massey, R.C., Spiers, A.J., Rainey, P.B., 2003. Genes encoding a cellulosic polymer contribute toward the ecological success of *Pseudomonas fluorescens* SBW25 on plant surfaces. Mol. Ecol. 12, 3109–3121.

Gomez, B.A., Pryor, B., 2019. First report of *Fusarium* wilt of queen palm in Arizona caused by *Fusarium oxysporum* f. sp. *palmarum*. Plant. Dis. Available from: https://doi.org/10.1094/PDIS-03-19-0551-PDN.

Graham, J.H., Gottwald, T.R., Cubero, J., Achor, D.S., 2004. *Xanthomonas axonopodis* pv. citri: factors affecting successful eradication of citrus canker. Mol. Plant. Pathol. 5, 1–15.

Gümrükcü, E., Polat, İ., Sülü, G., Kitapcı, A., Baysal, Ö., 2016. First report of grey mold caused by *Botryotinia fuckeliana* on golden berry. Plant. Dis. 100, 656-656.

Guttman, D.S., McHardy, A.C., Schulze-Lefert, P., 2014. Microbial genome-enabled insights into plant–microorganism interactions. Nat. Rev. Genet. 15, 797–813.

Hadidi, A., Levy, L., Podleckis, E.V., 1995. Polymerase chain reaction technology in plant pathology. Mol. Method. Plant. Pathol. 167–187.

Han, S.W., Lee, S.W., Ronald, P.C., 2011. Secretion, modification, and regulation of Ax21. Curr. Opin. Microbiol. 14, 62–67.

He, Y.W., Wu, J., Cha, J.S., Zhang, L.H., 2010. Rice bacterial blight pathogen *Xanthomonas oryzae* pv. oryzae produces multiple DSF-family signals in regulation of virulence factor production. BMC Microbiol. 10, 187.

Hemetsberger, C., Herrberger, C., Zechmann, B., Hillmer, M., Doehlemann, G., 2012. The *Ustilago maydis* effector Pep1 suppresses plant immunity by inhibition of host peroxidase activity. PLoS Pathog. 8, e1002684.

Inacio, M.L.S., Camacho, M.J., Serra, C., Cordovil, C., Cordeiro, L., Andrade, E., 2020. First report of the potato cyst nematode *Globodera rostochiensis* on potato in Azores, Portugal. Plant. Dis. Available from: https://doi.org/10.1094/PDIS-07-19-1429-PDN.

Jones, J.D.G., Dangl, J.L., 2006. The plant immune system. Nat 444, 323–329.

Kado, C.I., 2002. Crown gall. Plant Health Instr. Available from: https://doi.org/10.1094/PHI-I-2002-1118-01.

Karavina, C., Ibaba, J.D., Gubba, A., Pappu, H.R., 2016. First report of Iris yellow spot virus infecting garlic and leek in Zimbabwe. Plant. Dis. 100, 657-657.

Kim, J., Kim, J.G., Kang, Y., Jang, J.Y., Jog, G.J., Lim, J.Y., et al., 2004. Quorum sensing and the LysR-type transcriptional activator ToxR regulate toxoflavin biosynthesis and transport in *Burkholderia glumae*. Mol. Microbiol. 54, 921–934.

King, R., Urban, M., Lauder, R.P., Hawkins, N., Evans, M., Plummer, A., et al., 2017. A conserved fungal glycosyltransferase facilitates pathogenesis of plants by enabling hyphal growth on solid surfaces. PLoS Pathog. 13, e1006672.

Laue, H., Schenk, A., Li, H., Lambertsen, L., Neu, T.R., Molin, S., et al., 2006. Contribution of alginate and levan production to biofilm formation by *Pseudomonas syringae*. Microbiol 152, 2909–2918.

Lee, H.B., Nguyen, T.T., 2019. First report of powdery mildew caused by *Erysiphe lespedezae* on *Kummerowia striata* in Korea. Plant. Dis. Available from: https://doi.org/10.1094/PDIS-10-19-2236-PDN.

Liu, Y.L., Tang, J.R., Wang, C.L., Zhou, Y.H., 2020. First report of *Alternaria alternata* causing leaf blight of taro in Zhanjiang, China. Plant. Dis. Available from: https://doi.org/10.1094/PDIS-12-19-2653-PDN.

Lombard, V., Golaconda Ramulu, H., Drula, E., Coutinho, P.M., Henrissat, B., 2014. The carbohydrate-active enzymes database (CAZy) in 2013. Nucl. Acids Res. 42, 490–4955.

Lucas, J.A., 2009. Plant Pathology and Plant Pathogens. John Wiley & Sons.

Lucas, J.A., 2011. Advances in plant disease and pest management. J. Agric. Sci. 149, 91–114.

Lucas, J.A., 2017. Fungi, food crops, and biosecurity: advances and challenges, Advances in Food Security and Sustainability, 2. Elsevier, pp. 1–40.

Lucas, J.A., Hawkins, N.J., Fraaije, B.A., 2015. The evolution of fungicide resistance. Adv. Appl. Microbiol. 90, 29–92.

Mansouripour, S., Holmes, G., 2020. First report of *Botrytis cinerea* causing leaf spot on strawberry in California. Plant. Dis. Available from: https://doi.org/10.1094/PDIS-06-19-1287-PDN.

Marin, M.V., Coburn, J., Desaeger, J., Peres, N.A., 2020. First report of *Cercospora* leaf spot caused by *Cercospora* cf. *flagellaris* on industrial hemp in Florida. Plant. Dis. Available from: https://doi.org/10.1094/PDIS-11-19-2287-PDN.

McCarthy, Y., Dow, J.M., Ryan, R.P., 2011. The Ax21 protein is a cell–cell signal that regulates virulence in the nosocomial pathogen *Stenotrophomonas maltophilia*. J. Bacteriol. 193, 6375–6378.

Melotto, M., Kunkel, B.N., 2013. Virulence strategies of plant pathogenic bacteria. In: Rosenberg, E. (Ed.), The Prokaryotes – Prokaryotic Physiology and Biochemistry. Springer-Verlag, Heidelberg, pp. 61–75.

Meng, J.R., Haung, H.J., Li, Y.X., Li, Y.J., Li, J.Q., Chen, B.S., 2020. First report of *Fusarium sacchari* causing sugarcane Pokkah Boeng in China. Plant. Dis. Available from: https://doi.org/10.1094/PDIS-05-19-0906-PDN.

Oerke, E.C., 2006. Crop losses to pests. J. Agric. Sci. 144, 31–43.

Pallas, V., García, J.A., 2011. How do plant viruses induce disease? Interactions and interference with host components. J. Gen. Virol. 92, 2691–2705.

Pan, Y., Wu, H., Di, R., 2020. First report of powdery mildew caused by *Erysiphe alphitoides* on *Ipomoea obscura* L. in China. Plant. Dis. Available from: https://doi.org/10.1094/PDIS-07-19-1513-PDN.

Pappas, K.M., Winans, S.C., 2003. A LuxR-type regulator from *Agrobacterium tumefaciens* elevates Ti plasmid copy number by activating transcription of plasmid replication genes. Mol. Microbiol. 48, 1059–1073.

Pataky, J.K., Block, C.C., Michener, P.M., Shepherd, L.M., McGee, D.C., White, D.G., 2004. Ability of an ELISA-based seed health test to detect *Erwinia stewartii* in maize seed treated with fungicides and insecticides. Plant. Dis. 88, 633–640.

Perez-Mendoza, D., Aragon, I.M., Prada-Ramirez, H.A., Romero-Jimenez, L., Ramos, C., Gallegos, M.T., et al., 2014. Responses to elevated c-di-GMP levels in mutualistic and pathogenic plant-interacting bacteria. PLoS One. 9, e91645.

Pesci, E.C., Milbank, J.B., Pearson, J.P., McKnight, S., Kende, A.S., Greenberg, E.P., et al., 1999. Quinolone signaling in the cell-to-cell communication system of *Pseudomonas aeruginosa*. Proc. Natl. Acad. Sci. U.S.A. 96 (20), 11229–11234.

Pfeilmeier, S., Caly, D.L., Malone, J.G., 2016a. Bacterial pathogenesis of plants: future challenges from a microbial perspective: challenges in bacterial molecular plant pathology. Mol. Plant. Pathol. 17, 1298–1313.

Pfeilmeier, S., Saur, I.M., Rathjen, J.P., Zipfel, C., Malone, J.G., 2016b. High levels of cyclic-di-GMP in plant-associated *Pseudomonas correlate* with evasion of plant immunity. Mol. Plant. Pathol. 17, 521–531.

Prada-Ramirez, H.A., Perez-Mendoza, D., Felipe, A., Martinez-Granero, F., Rivilla, R., Sanjuan, J., et al., 2016. AmrZ regulates cellulose production in *Pseudomonas syringae* pv. tomato DC3000. Mol. Microbiol. 99, 960–977.

Pretty, J., Bharucha, Z.P., 2014. Sustainable intensification in agricultural systems. Ann. Bot. 114, 1571–1596.

Qian, G., Zhou, Y., Zhao, Y., Song, Z., Wang, S., Fan, J., et al., 2013. Proteomic analysis reveals novel extracellular virulence-associated proteins and functions regulated by the diffusible signal factor (DSF) in *Xanthomonas oryzae* pv. oryzicola. J. Proteome Res. 12, 3327–3341.

Quinones, B., Dulla, G., Lindow, S.E., 2005. Quorum sensing regulates exopolysaccharide production, motility, and virulence in *Pseudomonas syringae*. Mol. Plant Microbe Interact. 18, 682–693.

Quinones, B., Pujol, C.J., Lindow, S.E., 2004. Regulation of AHL production and its contribution to epiphytic fitness in *Pseudomonas syringae*. Mol. Plant Microbe Interact. 17, 521–531.

Rafiqi, M., Ellis, J.G., Ludowici, V.A., Hardham, A.R., Dodds, P.N., 2012. Challenges and progress towards understanding the role of effectors in plant-fungal interactions. Curr. Opin. Plant. Biol. 15, 477–482.

Ren, Y., Yu, P., Hou, W., Wang, Y., Yang, X., Fan, J., et al., 2020. Development of a rapid approach for detecting sharp eyespot resistance in seedling-stage wheat and its application in Chinese wheat cultivars. Plant. Dis. Available from: https://doi.org/10.1094/PDIS-12-19-2718-RE.

Rodriguez-Moreno, L., Ebert, M.K., Bolton, M.D., Thomma, B.P., 2018. Tools of the crook-infection strategies of fungal plant pathogens. Plant. J. 93, 664–674.

Romling, U., Galperin, M.Y., Gomelsky, M., 2013. Cyclic di-GMP: the first 25 years of a universal bacterial second messenger. Microbiol. Mol. Biol. Rev. 77, 1–52.

Rooney, H.C., Van't Klooster, J.W., van der Hoorn, R.A., Joosten, M.H., Jones, J.D., de Wit, P.J., 2005. *Cladosporium* Avr2 inhibits tomato Rcr3 protease required for Cf-2-dependent disease resistance. Science 308, 1783–1786.

Rovenich, H., Boshoven, J.C., Thomma, B.P., 2014. Filamentous pathogen effector functions: of pathogens, hosts and microbiomes. Curr. Opin. Plant. Biol. 20, 96–103.

Royal Society, 2009. Reaping the Benefits: Science and the Sustainable Intensification of Global Agriculture. Royal Society, London.

Ryder, L.S., Talbot, N.J., 2015. Regulation of appressorium development in pathogenic fungi. Curr. Opin. Plant. Biol. 26, 8–13.

Sanchez-Vallet, A., Mesters, J.R., Thomma, B.P., 2015. The battle for chitin recognition in plant-microbe interactions. FEMS Microbiol. Rev. 39, 171–183.

Sanchez-Vallet, A., Saleem-Batcha, R., Kombrink, A., Hansen, G., Valkenburg, D.J., Thomma, B.P., et al., 2013. Fungal effector Ecp6 outcompetes host immune receptor for chitin binding through intrachain LysM dimerization. Elife. 2, e00790.

Scholthof, K.B.G., 2007. The disease triangle: pathogens, the environment and society. Nat. Rev. Microbiol. 5, 152–156.

Schuster, M., Lostroh, C.P., Ogi, T., Greenberg, E.P., 2003. Identification, timing, and signal specificity of *Pseudomonas aeruginosa* quorum-controlled genes: a transcriptome analysis. J. Bacteriol. 185, 2066–2079.

Seddon, N., Mace, G.M., Naeem, S., Tobias, J.A., Pigot, A.L., Cavanagh, R., et al., 2016. Biodiversity in the anthropocene: prospects and policy. Proc. Biol. Sci. 283, 20162094.

Singh, R.S., 2018. Plant diseases. Oxford and IBH Publishing.

Song, J.G., Xu, B., Yang, H., Zhang, S.H., 2020. First report of Powdery mildew caused by *Leveillula taurica* on *Capparis spinosa* subsp. *himalayensis* from China. Plant. Dis. Available from: https://doi.org/10.1094/PDIS-03-19-0581-PDN.

Stockwell, V.O., Johnson, K.B., Loper, J.E., 1998. Establishment of bacterial antagonists of *Erwinia amylovora* on pear and apple blossoms as influenced by inoculum preparation. Phytopathol. 88, 506–513.

Tang, J.L., Feng, J.X., Li, Q.Q., Wen, H.X., Zhou, D.L., Wilson, T.J., et al., 1996. Cloning and characterization of the rpfC gene of *Xanthomonas oryzae* pv. oryzae: involvement in exopolysaccharide production and virulence to rice. Mol. Plant Microbe Interact. 9, 664–666.

Tufail, M.R., Fayyaz, A., Sarfraz, S., Waqas, M., Ahmed, M.Z., Hameed, A., et al., 2020. First report of pantoea blight of cotton caused by *Pantoea* spp. in Punjab, Pakistan. Plant. Dis. Available from: https://doi.org/10.1094/PDIS-10-19-2267-PDN.

Van den Bosch, F., Paveley, N., van den Berg, F., Hobbelen, P., Oliver, R., 2014. Mixtures as a fungicide resistance management tactic. Phytopathol 104, 1264–1273.

Vologin, S.G., Zamalieva, F.F., Stasevski, Z., Karasev, A.V., 2019. Occurrence of alfalfa mosaic virus in potato (*Solanum tuberosum*) in Middle Volga Region of Russia. Plant. Dis. 103, 3289.

Wang, L.H., He, Y., Gao, Y., Wu, J.E., Dong, Y.H., He, C., et al., 2004. A bacterial cell–cell communication signal with cross-kingdom structural analogues. Mol. Microbiol. 51, 903–912.

Wang, X., Zhang, X., Li, M., Ji, X., Feng, C., Wang, F., 2020. First report of *Ficus carica* bot rot caused by *Botryosphaeria dothidea* in China. Plant. Dis. Available from: https://doi.org/10.1094/PDIS-09-19-2039-PDN.

Wu, Z., Kan, F.W., She, Y.M., Walker, V.K., 2012. Biofilm, ice recrystallization inhibition and freeze–thaw protection in an epiphyte community. Prikl. Biokhim Mikrobiol. 48, 403–410.

Xian, C., Laborda, P., Liu, F., 2020. Exogenous melatonin enhances rice plant resistance against *Xanthomonas oryzae* pv. oryzae. Plant. Dis. Available from: https://doi.org/10.1094/PDIS-11-19-2361-RE.

Xu, G., Qiu, F., Li, X., Zheng, F.Q., Zheng, L., Miao, W., et al., 2020. *Diaporthe limonicola* causing leaf spot disease on *Areca catechu* in China. Plant. Dis. Available from: https://doi.org/10.1094/PDIS-11-19-2324-PDN.

Zhang, S., Xu, Z., Ren, L., Hu, H., 2020. First report of leaf blight wilt on *Dracaena sanderiana* by *Pantoea stewartii* subsp. *indologenes* in China. Plant. Dis. Available from: https://doi.org/10.1094/PDIS-06-19-1143-PDN.

Zeng, W., He, S.Y., 2010. A prominent role of the flagellin receptor FLAGELLINSENSING2 in mediating stomatal response to *Pseudomonas syringae* pv tomato DC3000 in *Arabidopsis*. Plant. Physiol. 153, 1188–1198.

Zhao, X.Y., Chen, M., Pan, L.Y., Fu, Y.Q., Chen, J.Y., Xiang, M.L., 2020. First report of fruit rot on 'Cuiguan' pear caused by *Fusarium proliferatum* in China. Plant. Dis. Available from: https://doi.org/10.1094/PDIS-04-19-0875-PDN.

Zitnick-Anderson, K., Pasche, J.S., Gargouri-Jbir, T., Kalil, A., 2020. First report of *Aphanomyces euteiches* causing Aphanomyces root rot on lentil in North Dakota. Plant. Dis. Available from: https://doi.org/10.1094/PDIS-11-19-2422-PDN.

CHAPTER

2

Current status of plant diseases and food security

Vipin Kumar Singh[1], *Rishikesh Singh*[2], *Ajay Kumar*[3] *and Rahul Bhadouria*[4]

[1]Department of Botany, Centre of Advanced Study, Institute of Science, Banaras Hindu University, Varanasi, India [2]Institute of Environment and Sustainable Development, Banaras Hindu University, Varanasi, India [3]Department of Postharvest Science, Agricultural Research Organization (ARO), Volcani Center, Rishon LeZion, Israel [4]Department of Botany, University of Delhi, New Delhi, India

2.1 Introduction

Worldwide, plants diseases are among challenging threats to food security as well as ecosystem integrity and have considerable potential in modulating the economic condition of both developed and developing countries (Kamoun, 2003). Plant diseases are considered as one of the major factors responsible for decline in more than 50% of crop yield (Oerke, 2006) giving rise to food scarcity and starvation (Pimentel, 2009). As per an estimate, 10%−16% of total productivity is devastated annually because of plant diseases (Strange and Scott, 2005). Further, postharvest losses equivalent to 6%−12%, particularly in tropical regions of developing nations, is also contributed by storage molds and mycotoxins secreted by them. The postharvest biodeterioration leading to qualitative and quantitative losses of food items, further makes the conditions more severe in regions with limited resources. Interaction of pathogens with host plants is the primary event for disease development and early identification of these pathogens could be an important strategy to delimit the disease progression and pathogen virulence. The primary events during disease development caused by *Ralstonia solanacearum* infection involve attachment with root cells, multiplication, transportation, and finally penetration of xylem (Vasse et al., 1995). The interacting microbial pathogens may secrete plant hormones like cytokinins and indole acetic acids to modulate the vital physiological status of plants ultimately responsible for plant disease development (Yamada, 1993). The genes associated with

Food Security and Plant Disease Management
DOI: https://doi.org/10.1016/B978-0-12-821843-3.00019-2

hormone synthesis are located in extra-chromosomal genetic element like plasmids. Although, cellular entry via roots appears as an important limiting step during diseases development by most of the bacterial pathogens, effectors Gala7 and AvrA facilitating entry of bacterium *Ralstonia solanacearum* into root cortical cells depending upon the genetic ability of host cells is presented in *Medicago truncatula* (Turner et al., 2009). Detection of already known pathogens causing plant diseases as early as possible has immense opportunities in limiting the magnitude of overall crop productivity losses and advancement of sustainable management approaches. Moreover, precise recognition of newly emerging pathogens and restricting their access to large numbers of susceptible host is also necessary for effective disease management. Application of methods relying on nucleic acids and protein features are recognized as important techniques for precise identification of plant diseases, however, these techniques suffer from challenges in case of pathogen spreading systemically and diseased host not expressing visible symptoms (Martinelli et al., 2015). The application of efficient newly designed sensors possessing the ability to identify molecules of host response and remote sensing integrated with spectroscopy techniques could provide a strategy for rapid and early stages of disease progression. Soil is the natural adobe for many disease causing agents (pathogens) like fungi, bacteria, viruses, and nematodes responsible for various soil borne plant diseases (Huang et al., 2020; Hong et al., 2020; Kroese et al., 2020; Fanning et al., 2020). For instance, soil borne fungal pathogens *Rhizoctonia solani*, *Macrophomina phaseolina*, and *Fusarium solani* infect sunflower, declining seedling viability and subsequently reduction in yield (Harveson et al., 2016). The further progression of disease is facilitated under altered soil characteristics creating stress. In comparison to other plant pathogens, fungal pathogens belonging to class oomycetes are of greater concern causing huge losses in productivity of crops having much economic importance (Panabieres et al., 2016). Heavy reductions in yield and tuber quality of potato as a result of soil borne diseases induced by bacterial pathogens have also been reported by Larkin (2020). Most notably, changes in climatic conditions may influence plant disease like leaf rust of wheat (Launay et al., 2020) with the resultant modifications in global food productivity. Although plant diseases are known to diminish crop yield, influence on overall quality characteristics of crop produce is also evident (Siuda et al., 2010; Shi et al., 2020), thereby affecting its use in respective food products. Moreover, fungal infection introduces unacceptable contents of hazardous mycotoxins in grains (Castañares et al., 2020), making unsuitable for feed and food purposes.

Plant diseases can be controlled by numerous physical, chemical, and biological methods. The application of chemicals for controlling plant pathogens is most widely used disease management method across the world. The expensiveness, toxicity to human health and natural ecosystem, and induction of resistance in target pathogens (Miyamoto et al., 2020), however, limits their wide acceptability among farmers for disease management. Further, the evolution of resistance in target pathogens against chemicals may provoke the severity of diseases as noted for heavy loss in productivity of potato and tomato (Kato et al., 1997; Shattock, 2002), making the disease control even more challenging. The emergence of resistance, for example, in bacterial pathogens, has been ascribed the resultant of selection pressure as well as existence of plasmid residing resistance genes (Sundin and Bender, 1993). Deployment of biological control agents synthesizing a large array of bioactive substances, for managing diseases, on the other hand is more attractive, economically feasible, environmentally safe, and devoid of possible resistance emergence in plant pathogens. Some of the well-recognized biocontrol agents are *Trichoderma* (Coninck et al., 2020), and *Aspergillus flavipes* (El-Sayed and Ali, 2020). The plant products like essential oils have also been used to control diseases both in preharvest and postharvest conditions (Basaid et al., 2020; Rguez et al., 2020). Nevertheless, the integration of conventional control practices together with the use of chemicals, biocontrol agents, and gene transfer technology is more meaningful disease control strategy. Thus plant diseases are among major constraints in achieving the goals of food security. The combined action of ever rising population and changing climatic conditions will further impose restriction in accomplishing the target of global food security. Food security can be simply defined as the accessibility of safe and nutrient enriched food to every people in adequate quantity to comply with dietary requirement, essential for healthier biological processes (FAO, 2003). In other words, food security is the challenge associated with diverse food productivity, availability, consumption, etc. and is influenced by varied factors like environmental contamination, changes in economy, business terms, and climatic conditions. The present chapter has tried to address huge losses in crop productivity occurring from plant diseases caused by various fungi, bacteria, viruses, and nematodes, modern techniques of pathogens detection and important strategies for controlling plant diseases in order to attain the targets of food security.

2.2 Worldwide plant diseases and loss of crop productivity

Plant diseases caused by various pathogens are the outcome of profound qualitative and quantitative productivity losses in many economically important crops, worldwide, leading to considerable loss in economy as well as declined food security both nationally and internationally. The difficulty in management of plant pathogens much frequently arises due to substantial differences in spatiotemporal existence and genetic constitution (Strange and Scott, 2005). Huge losses in some of the staple crops have given rise to broad scale starvation in many parts of the globe. For instance, notoriously known Irish famine during 1840s, the resultant of widespread infection by *Phytophthora infestans* has been recorded for overwhelming impacts of plant diseases on large human population in Ireland (Dehnen-Schmutz et al., 2010). Worldwide, the average decline in productivity by plant diseases and multiple pests have been documented to reach 11%–30% (Oerke and Dehne, 2004). Most remarkably, the yield declines are much prominent in locations across the globe previously challenged with problems of food security (Savary et al., 2019) and generally linked with newly recognized and reemerging pests and disease causing agents. Further, providing accounts on disease induced losses at comparative basis for various crops, geographical regions, and agricultural systems is quite challenging. The productivity loss potential of pathogens has been estimated to reach 15% in major crops including rice, maize, barley, wheat, potato, cotton, etc., while for virus induced reduction in productivity of crops like potato and sugar beet is registered in the range of 6%–7% and less than 1%–3% for additional crops (Oerke and Dehne, 2004). The differences in loss of different crops are affected largely by pathogen's characteristics. Therefore strengthening crop yield for supporting the ever rising demand of food for humans and livestock must be integrated with sustainable disease management practices.

2.3 Plant disease detection methods

So far, various approaches have been introduced in the field of agriculture to diagnose numerous devastating plant diseases. Early stage detection of plant diseases using technologically advanced real-time approaches like computer vision technology (Cheng, 2020), machine learning algorithms (Panigrahi et al., 2020), nanosensors (Yusof and Isha, 2020), biosensors (Chalupowicz et al., 2020), and image processing (Khairnar and Goje, 2020) in field as well as postharvest storage condition may be considered as an important tool for effective management of crop productivity losses so that goals of food security could be achieved in a sustainable manner. Nevertheless, the early detection of plant diseases remains a challenge rendered by lack of technological access and limitations of field experts. Although, appropriate disease detection model with maximum precision has potential application to be effectively used by farmers for encountering problems in crop productivity losses; pathogen identification at asymptomatic stage poses another substantial challenge and can be overcome through employment of biophotonics methodologies (Wijesinghe et al., 2017, 2018). For instance, nondestructive approach possessing sharp resolution named as optical coherence tomography relying on biophonics is described to decipher the initial stage visible changes in fruit anthracnose induced by pathogen infection (Wijesinghe et al., 2017) and diseases of other agricultural produce including apple and cucumber. Characteristic modulations in plant's basic structural components and functional integrity upon challenged with pathogens is well documented. Therefore plant tissue analysis to reveal diseases based on imaging techniques like magnetic resonance and positron emission tomography, microscopy, and electromagnetic radiation have been regularly used (Barreiro et al., 2008; Shahin et al., 2002; Jahnke et al., 2009; Urban et al., 2018), but shortcomings like time consuming nature, lower depth invasion, and restricted instrumental resolution make field scale application inappropriate. In addition, plants encountered with disease causing agents have also been demonstrated to emit volatile organic compounds (VOCs) with considerably altered composition and quantity (Jansen et al., 2011), hence detailed chemical analysis of such biomarker components having high vapor pressure, increased volatility, and low molecular weight offers immense potential to reveal the interaction of pathogens with host at an early stage. But, the detection of diseases solely depending on VOCs is challenging. Furthermore, the cost associated with field scale application of the technology for diseases diagnosis at an initial stage is not much practically feasible but could have limited opportunities in identifying pathogens associated with postharvest food produce losses under storage.

2.4 Plant diseases caused by fungi, bacteria, viruses, and nematodes

Different plant pathogens such as fungi, bacteria, viruses, protozoa, and nematodes have long been recognized for large scale destruction of crop produce, thereby imposing severe threats to food security. Fungal pathogens inducing lesion, rot, blight, wilting, etc. are responsible for varied plant diseases and may result in huge amount damage to many economically important crops worldwide due to features like (1) rapid formation of spores possessing the ability to attack other plants, (2) low duration latent periods, ability to transport as a source of copious inoculum via water, precipitation, and air currents, (3) synthesis of extracellular enzymes or chemical components with inhibitory action over plants, and (4) acquisition of nutrients from economically important parts of host plants (Strange and Scott, 2005), and (5) long duration survival even under adverse environment as observed for sclerotia. As an example, one of the most studied fungus, the blast fungus *Pyricularia oryzae*, (teleomorph *Magnaporthe grisea*) alone has been marked with potential to destruct 10%−35% of rice crops annually from 85 countries (Fisher et al., 2012), and may attain the yield loss percentage to 50% in case of epidemic diseases (Asibi et al., 2019), implying the catastrophic nature. Astonishingly, during optimum environmental condition, the rice blast fungus can completely destroy the crop between 15 and 20 days (Araújo et al., 2000). The fungus *Rhizoctonia solani* [teleomorph: *Thanatephorus cucumeris* (Frank) Donk], responsible for rice sheath blast has destructive potential to reduce the productivity by 50% in susceptible crops (Tiwari et al., 2017). Moreover, the broad range host infection by a particular fungus may aggravate the severity of plant diseases and the geographical distribution of a fungus even may be much higher than expected as supposed to be facilitated by changing environmental conditions and human interventions. Some of the important fungal pathogens causing plant diseases are *Penicillium* (Coutinho et al., 2020), *Alternaria* (Attia et al., 2020), *Colletotrichum* (Gorman et al., 2020), *Puccinia graminis tritici* (Mengesha, 2020), *Phytophthora* (Peterson et al., 2020), *Fusarium verticillioides* (Veenstra et al., 2019), *Rhizoctonia solani* (Wu et al., 2019), *Albugo* (Bhayana et al., 2020), and *Pythium* (Martin and Loper, 1999). Conclusively, broad range host specificity and wider distribution put a great risk in controlling crop productivity losses caused by phytopathogenic fungi.

Bacterial pathogens have been found associated with plethora of plant diseases substantially affecting the productivity of many economically important crops worldwide (Martins et al., 2018). The annual losses of crop produce by phytopathogenic bacteria valued more than one billion dollar have been documented (Mansfield et al., 2012; Kannan et al., 2015). Bacterial pathogens can infect multiple host plants from seedling germination to heading stage both in field and postharvest storage conditions. Some of the common strategies deployed by bacterial phytopathogens for survival in a given environment involve the formation of spores, release of effector proteins, biofilm generation, and resistance to toxic chemicals by using efflux mechanisms (Martins et al., 2018). A single genus *Xanthomonas* is evidenced to attack host plant giving rise to minimum 350 plant diseases (Leyns et al., 1984). Rice blight disease, majorly identified in tropical regions of Asia due to *Xanthomonas oryzae* pv. *Oryzae* is observed to infect basmati rice varieties. The pathogen is recorded to exist in different parts of India cultivating basmati rice posing great limitation in productivity. The pathogenic action of bacterial pathogens is associated with the plethora of virulence factors and secretion of effector proteins known to be regulated by different genes. Some fungal genera causing plant diseases include *Xanthomonas* (Zhou et al., 2020), *Pseudomonas* (Collmer, 2020), *Xylella fastidiosa* (Chen et al., 2019), *Ralstonia* (Dadrasnia et al., 2020), *Dickeya solani*, *Pectobacterium atrosepticum* and *Pectobacterium carotovorum* (Rodríguez et al., 2020), *Agrobacterium tumefaciens* (Caparrotta et al., 2019), and *Erwinia amylovora* (Kron et al., 2020).

Viruses causing plant diseases have been identified as an important biotic constraint and noted to decline the crop productivity several orders of magnitude both qualitatively and quantitatively (Zhao et al., 2020). Surprisingly, many of the emerging plant diseases are the resultant of host interaction with viral pathogens (McLeish et al., 2020). Although the crop productivity losses by viral pathogens are between 3% and 7% (Oerke and Dehne, 2004), in developing regions of the world, the host plant infection by a single virus may lead to potential yield losses reaching maximally 90% as observed for potato (Wang et al., 2011; Palukaitis, 2012) and a single host may be challenged with more than one type of viruses probably contributed by insect vectors. More importantly, the synergistic association of a particular virus with different virus in a host system is described to further increase the disease severity (Mascia and Gallitelli, 2016). The annual economic losses rendered by viral pathogens are expected to reach USD 60–80 billion, mainly due to lack of any agrochemical directly acting on target viral pathogens. The challenges of agricultural practices for controlling plant diseases caused by viruses are further enhanced because of fast genetic modification and unexpected rise in target hosts they can attack. Viral phytopathogens may be transmitted with and without aid of vectors. Important vectors transmitting viral pathogens include insects like whiteflies, aphids, grasshoppers, and soil inhabiting fungi and nematodes. However, in absence of any vector, viruses may also be transmitted by affected plant remains, infected seeds, and vegetative propagules such as bulbs. Some of the important viral pathogens of plants are cucumber green mottle mosaic virus and tomato leaf curl New Delhi virus (Crespo et al., 2020), tobacco mosaic virus (Ellis et al., 2020), yellow vein mosaic virus (Jamir et al., 2020), tomato leaf curl virus (Pérez-Padilla et al., 2020), tomato spotted wilt virus (Xu et al., 2020a), lettuce necrotic yellow virus (Zhang et al., 2020a), potato virus Y (Torrance et al., 2020), and papaya ring spot virus (Branham et al., 2020).

Soil inhabiting nematodes present a major threat to economically important crops in many parts of the world because of parasitic nature. Such plant parasitic nematodes (PPNs) with more than 4100 species known so far (Decraemer and Hunt, 2006), are cosmopolitan in distribution and are capable to parasitize large numbers of host plants. The parasitizing nematodes may be ectoparasite or endoparasite and endoparasite may be of sedentary or migratory nature (Hussey, 1989). Nematodes, more specifically, the root knot nematodes have strong negative influences on cultivated vegetables under a protected agroecosystem. The annual crop loss reaching 173 billion USD is accounted for root knot nematodes having the potential to affect more than 2000 hosts (Elling, 2013). Generally, infection by PPNs reduces plant fitness rendered by decline in nutritional contents of cell and substantial tissue damage in vasculature perturbing water and nutrient acquisition potential of root system (Costa et al., 2019). Some of the nematodes parasitizing plants are *Heterodera glycines* (Masonbrink et al., 2019), *Meloidogyne* (Singh, 2020), *Globodera* (Inacio et al., 2020), *Pratylenchus* (Fanning et al., 2020), *Radopholus* (Yang et al., 2020), *Ditylenchus* (Haraguchi and Yoshiga, 2020), *Bursaphelenchus* (Xie et al., 2020), and *Aphelenchoides* (Xu et al., 2020b). The control of plant diseases by different attacking pests is quite complex and is considerably determined by nature of pathogens as well as prevailing environmental conditions in field. The early detection of pathogens using modern agricultural science tools and application of integrated pest management practices could be helpful in overcoming destructive effects of plant pests in order to eliminate the factors negatively concerned with food security goals.

2.5 Effect of plant diseases on yield and nutritional attributes of crop produce

The devastating effect of various plant diseases may result into considerable reduction in yield and alterations in nutritional attributes of important crop produce. Generally, the qualitative and quantitative yield losses occurring through low seed weight and increased scabby kernels, and health hazardous toxin accumulation (Mukanga et al., 2010) lead to reduced market value. Study on influence of powdery mildew and rust disease on wheat attributes useful in baking industry, thereby reducing food value for humans and livestock is registered by Everts et al. (2001). In addition, changes in nutrient and mineral composition by infestation with fungi are elaborated by Bandara et al. (2017). In comparison to control sets, fungal pathogen inoculation led to significant reductions in protein (15%), fat (12%), carbohydrate (13%), and ash content (13%) based on unit grain weight. The observed differences in grain quality characteristics were inferred to be influenced by host genetic setup and environmental conditions. Most notably, considerable changes in quality attributes like minerals and nutrients content, and hardness of crop products, compromise its application for intended purposes in food and related industries, implying the application of appropriate control measure. Infestation of crops and crop produce with pests, especially fungi is illustrated to modulate the quality in field condition (Conner et al., 2003; Salgado et al., 2011; Bandara et al., 2018) as well as in postharvest storage facilities (OuYang et al., 2020) chiefly due to secretion of enzymes and mycotoxins contamination (Martínez et al., 2020) causing heavy degradation and deterioration. The decreased crop productivity and compromised quality attributes rendered by diverse pest infestation is regulated by different environmental parameters like growing season, cultivars used, grain moisture content, water activity, pH, temperature, soil attributes, and nutritional status. Therefore management strategies considering important factors should involve plausible sustainable approaches to protect crop productivity losses both at qualitative and quantitative level without influencing natural characteristics in view of eliminating additional cost necessary for food fortification as well as for attaining the goals of food security and human health.

2.6 Plant disease management approaches

Plant disease management is very much important for preventing productivity losses of various crops. Most of the management approaches can be classified into physical, chemical, and biological approaches. Although physicochemical approaches are quite useful in managing crop productivity losses caused by plant pathogens, nevertheless, their expensive nature, hazardous impacts on human health and environment, agrochemical accumulation in food and feed, increased consumer awareness about chemical hazards through food, infestation with secondary pest, and most strikingly the induction of resistance development in pests (Silva et al., 2019) do not advocate application in agroecosytem. Further, agrochemicals exert deleterious impact on naturally abundant beneficial soil microorganisms, conferring host susceptibility to plant pathogens (Heydari and Pessarakli, 2010). In contrast, strategies based on target specific action of microbial agents (biocontrol agents, BCA) and their metabolic products as well as plant products against potential pathogens (Kashyap et al., 2020) are much attractive and promising from environment point of view. Some of the widely used biocontrol agents include *Trichoderma* (Coninck et al., 2020), *Aspergillus* (El-Sayed and Ali, 2020), *Pichia* (Zhang et al., 2020b), *Albifimbria* (Li et al., 2020), *Penicillium* (Rojas et al., 2020), *Bacillus* (Xu et al., 2020c), *Cupriavidus* (Ye et al., 2020), *Streptomyces* (Ling et al., 2020), *Burkholderia* (Kim et al., 2020), and *Paenibacillus* (Chávez-Ramírez et al., 2020). Microbial volatiles, the low molecular weight secondary metabolic products of microorganisms have also been documented as important substances against numerous plant diseases (Wonglom et al., 2020; Sridharan et al., 2020). The application of plant products like essential oils (Basaid et al., 2020; Rguez et al., 2020) and crude extracts (Mulugeta et al., 2020; Ramírez et al., 2020) are also remarkably documented against numbers of phytopathogens. Effectiveness of each of the biocontrol agents and their metabolic products as well as phytochemicals to control plant diseases possesses certain advantages and disadvantages and are substantially affected by type of BCA used, soil characteristics, plant extracts, and prevailing environmental conditions. However, the field scale application of BCA and phytochemicals is still a major challenge for managing crop productivity losses caused by various plant pests.

2.7 Genetic modification and genome editing tactics for plant disease management

Application of conventional practices relying on employment of various synthetic chemicals and plant breeding approaches has largely been applied in agroecosystem for diseases management. However, for instance, the strict regulatory laws regarding the use of agrochemicals and induction of resistance development in pests have necessitated the introduction of suitable alternative strategies for sustainable management of plant diseases (Scott et al., 2016). In this perspective, genetic engineering principles directed toward improving disease resistance in host plant (van Esse et al., 2020), are ecologically sound, safer, and possess enormous potential to control worldwide huge crop losses occurring through destructive plant pathogens. The resistance in host system can be improved via interferences (van Esse et al., 2020) at the level of (1) pathogen identification and effector molecules, (2) signaling pathways, (3) susceptibility genes, (4) dominant genes conferring plant resistance, (5) molecules with antimicrobial potential and, (6) RNA interference. In spite of characteristic potential in offering diseases resistance to hosts, the employment of genetically modified crops are the subject of global controversy (Scott et al., 2016) and are suspected to exert negative influences on environment and human health. Many of such assumptions, however, have no foundation based on experimental facts. On the other hand, recently, the technique of clustered regularly interspaced short palindromic repeat (CRISPR)/CRISPR-associated protein 9 (CRISPR/Cas9) as a means of genome editing with considerable specificity and promising potential in delivering resistance to host systems in order to manage crop yield losses with an aim to attain the objectives of food security (Ahmad et al., 2020) has also been proposed. Although, the CRISPR/Cas9 gene editing technology has great precision and specificity, there still has some challenge that needs to be eliminated. For instance, (1) the technique is quite expensive and time demanding,

(2) targeting particular S gene of a host may influence the associated genes of host characteristically controlling growth and development. Additionally, the product synthesis of a particular pathway may also be influenced by gene editing technology causing considerable changes in necessary nutrient and minerals. Since, disease resistant crops developed through application of conventional genetic modification and genome editing approaches are largely expected to suffer from challenges of widespread consumer acceptance (Mandal et al., 2020), environmental integrity and national/international policies, multi-level field based trials for any possible negative effect on natural ecosystem, and human health is an important and necessary step before commercialization.

2.8 Food security and climate change

In general, food security refers to availability of desired food in sufficient quantity to all persons. The world population is expected to reach beyond 9 billion by 2050 (Hickey et al., 2019) and the continuously rising human population, reinforcing the enhancement in agricultural productivity, has aggravated the challenges of global food security. Currently, out of total seven billion global human population, 800 millions are devoid of adequate food quantity, mostly associated with developing nations (FAO, 2015). The loss in crop yield is also noticed because of small scale expenditure for future research and diverse infrastructure development, in addition to rising water shortage (Rosegrant and Cline, 2003). As per an estimate, 15% of total crop produce are destructed annually by devastating plant diseases (Oerke and Dehne, 2004) culminating into increased food crisis, again posing considerable threat to food security. Furthermore, changing climatic conditions as a consequence of greenhouse gas emission has also influenced the major components of food security (Fig. 2.1) (Lobell et al., 2011). However, definite impact of climate change is unpredictable, general consensus pointing out negative outcomes is much apprehended. The climate change will exert much pressure on developing nations. Climate change would possibly lead to increased atmospheric carbon dioxide concentrations, temperature (Zhao et al., 2017), drought (Hay et al., 2016; Jha et al., 2020), and compromised quality characteristics of crop produce (Sreenivasulu et al., 2015). Therefore trio challenges

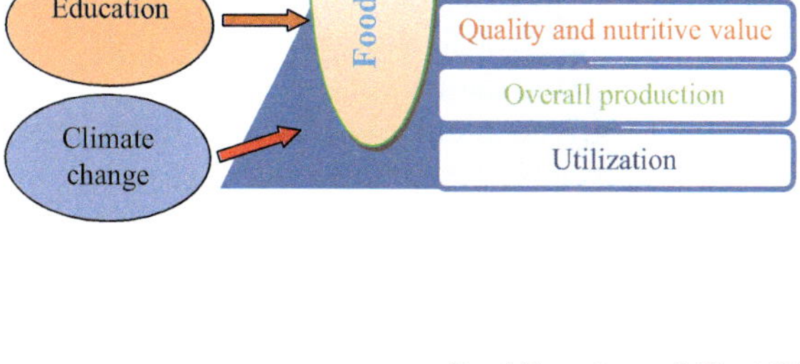

FIGURE 2.1 Major factors influencing food security.

of food security comprising of demand, supply, and crop productivity considering the facts of sustainability would be experienced multiple orders of magnitude in coming generation (Smith, 2013), suggesting alterations in food system (Reardon and Timmer, 2014). Modulations in climatic conditions may lead to heavy precipitation in tropical regions during dry durations or declined water supply, causing upsurge in pest population as well as increased crop and animal diseases (Lake et al., 2012), in addition to significant perturbations in soil characteristics. As per estimates, one degree increase in average global temperature would suppress the productivity of economically important crops like rice, wheat, maize, and soybean by 3.2%, 6.0%, 7.4%, and 3.1%, respectively (Zhao et al., 2017). Rising value of agricultural commodities will possibly decline the nutritional characteristics of food and support greater differences in human health. Altered uses of agrochemicals application in soil system and pharmaceuticals will considerably modify the transport features, thereby influencing food safety.

Current investigations have demonstrated rising population of undernourished people after 2014. For some regions around globe, numbers of people with food crisis are even very high as observed for previous 10 years. Although, adaptation is a meaningful, sustainable, and promising strategy to deal with the food crisis resulting from global climate change, there is a lack of uniformity in implementation. The adaptation could be achieved by three ways including (1) rising crop productivity, (2) improvement in distribution of agricultural commodities, and (3) economic access to crop products (Gregory et al., 2005). The improvement in crop productivity is determined critically by employment of innovative agricultural technologies like modified crop varieties resistant to environmental perturbations, pest infestation, farm management, etc. as well as widespread distribution. Accelerated breeding has also been advocated to overcome the challenges of food productivity, although in partial manner (Lenaerts et al., 2019). Genetic improvement in agricultural crops based on 5Gs is highlighted recently (Varshney et al., 2020). Nevertheless, most of the approaches are not implemented successfully so far in many developing regions across the globe, thereby restricting likely amendments in future crop productivity potential. The rise in economy has capability to improve food accessibility, nevertheless, it could not be appropriately helpful in absolutely declining hunger and undernourishment. Continuously rising population has accelerated the risks of food security multifold because of declined agricultural areas fueled by urbanization. Land degradation facilitated by changing climatic conditions and human interventions has additionally raised the future risks of food insecurity (Lenaerts et al., 2019). Extensification (modulation of natural ecosystem for cropping) and intensification (increased crop productivity in available land resources) are the plausible ways for enhancing the crop productivity (Gregory and Ingram, 2000). Current investigations have shown inability of modern agricultural practices in achieving the goals of food security, hence modulation in food system must be thought. Acceptable changes in diet (Davis et al., 2016), diminution of food loss (Godfray et al., 2010) by deterioration, and minimization of yield differences via agronomic development (Mueller et al., 2012; Harahagazwe et al., 2018) are plausible measures to enhance food accessibility. Most strikingly, the diverse adaptation strategies directed toward minimizing the impacts of global climate change on food security must be in sustainable manner.

Attaining food security requires improvement in government policies and investment sectors oriented toward human resource development, laboratory and field research, conservation of soil and water resources, and infrastructure development. Policies formulated must be capable to improve the crop productivity together with the enhancement in income and alleviation of poverty, especially in villages harboring most of the poor people. This strategy would facilitate in increasing the productivity of cereals and meat by 71% and 131%, respectively during 1997 to 2050 (Rosegrant and Cline, 2003). The incorporation of policies concerned with regulating the emission of greenhouse gases could have important contribution toward approaching the goals of food security. In addition, the early identification of threats to food safety is also important within the perspectives of global climate change. The integrated corporation of people working in the field of environment and human health would be highly appreciable for achieving the goals of food security.

2.9 Conclusion

Plant diseases are the major limiting factor influencing the crop productivity worldwide. Globally, very large quantities of agricultural products are lost annually due to destructive action of fungi, bacteria, viruses, and nematodes. Also, the infested crops lose important nutritional characteristics necessary for human health. The application of conventional processes through employment of agrochemicals to control plant diseases are not considered safe due to toxicity, chemical accumulation in crop produce, expensiveness, and resistance development in targeted pests. Biological methods relying on the utilization of microbes and their metabolic products, natural products, and plant extract possess promising potential in protecting crop losses caused by plant diseases. Microbiological agents often designated as biocontrol agents are ecofriendly and have immense opportunities for future application in disease management. However, field scale application of biocontrol agents and plant products is not commendable. The application of genetic engineering principles and recently evolved CRISPR/Cas9 technology is quite attractive in developing disease resistance crops, but they also have certain limitations like conventional breeding strategies. The rising human population and changing climatic conditions have imposed severe restrictions in attaining the goals of food security, in spite of increased food products at the cost of available soil and water resources. The changing climatic conditions, therefore demands the formulation of sustainable strategies helpful in securing the objectives of food availability without influencing natural ecosystem.

References

Ahmad, S., Wei, X., Sheng, Z., Hu, P., Tang, S., 2020. CRISPR/Cas9 for development of disease resistance in plants: recent progress, limitations and future prospects. Brief. Funct. Genom. Available from: https://doi.org/10.1093/bfgp/elz041.

Araújo, L.G.D., PRabhu, A.S., Freire, A.D.B., 2000. Development of blast resistant somaclones of the upland rice cultivar Araguaia. Pesqui. Agropec. Bras. 35 (2), 357–367.

Asibi, A.E., Chai, Q., Coulter, J.A., 2019. Rice blast: a disease with implications for global food security. Agronomy 9 (8), 451.

Attia, M.S., El-Sayyad, G.S., Abd Elkodous, M., El-Batal, A.I., 2020. The effective antagonistic potential of plant growth-promoting rhizobacteria against *Alternaria solani*-causing early blight disease in tomato plant. Sci. Hortic. 266, 109289.

Bandara, Y.M.A.Y., Tesso, T.T., Bean, S.R., Dowell, F.E., Little, C.R., 2017. Impacts of fungal stalk rot pathogens on physicochemical properties of sorghum grain. Plant. Dis. 101 (12), 2059–2065.

Bandara, Y.M.A.Y., Tesso, T.T., Zhang, K., Wang, D., Little, C.R., 2018. Charcoal rot and Fusarium stalk rot diseases influence sweet sorghum sugar attributes. Ind. Crops. Prod. 112, 188–195.

Barreiro, P., Zheng, C., Sun, D.W., Hernandez-Sanchez, N., Perez-Sanchez, J.M., Ruiz-Cabello, J., 2008. Non-destructive seed detection in mandarins: comparison of automatic threshold methods in FLASH and COMSPIRA MRIs. Postharvest Biol. Technol. 47 (2), 189–198.

Basaid, K., Chebli, B., Mayad, E.H., Furze, J.N., Bouharroud, R., Krier, F., et al., 2020. Biological activities of essential oils and lipopeptides applied to control plant pests and diseases: A review. Int. J. Pest. Manag. 1–23.

Bhayana, L., Paritosh, K., Arora, H., Yadava, S.K., Singh, P., Nandan, D., et al., 2020. A mapped locus on LG A6 of *Brassica juncea* line Tumida conferring resistance to white rust contains a CNL type R gene. Front. Plant. Sci. 10, 1690.

Branham, S.E., Wechter, W.P., Ling, K.S., Chanda, B., Massey, L., Zhao, G., et al., 2020. QTL mapping of resistance to *Fusarium oxysporum* f. sp. niveum race 2 and papaya ringspot virus in *Citrullus amarus*. Theor. Appl. Genet. 133 (2), 677–687.

Caparrotta, S., Comparini, D., Marone, E., Kimmenfield, R., Luzzietti, L., Taiti, C., et al., 2019. Correlation between VOC fingerprinting and antimicrobial activity of several essential oils extracted by plant resins against *A. tumefaciens* and *P. savastanoi*. Flavour. Fragr. J. 34 (5), 377–387.

Castañares, E., Pavicich, M.A., Dinolfo, M.I., Moreyra, F., Stenglein, S.A., Patriarca, A., 2020. Natural occurrence of *Alternaria* mycotoxins in malting barley grains in the main producing region of Argentina. J. Sci. Food Agric. 100 (3), 1004–1011.

Chalupowicz, D., Veltman, B., Droby, S., Eltzov, E., 2020. Evaluating the use of biosensors for monitoring of *Penicillium digitatum* infection in citrus fruit. Sens. Actuators B Chem. 311, 127896.

Chávez-Ramírez, B., Kerber-Díaz, J.C., Acoltzi-Conde, M.C., Antonio Ibarra, J., Vásquez-Murrieta, M.-S., Estrada-de los Santos, P., 2020. Inhibition of *Rhizoctonia solani* RhCh-14 and *Pythium ultimum* PyFr-14 by *Paenibacillus polymyxa* NMA1017 and *Burkholderia cenocepacia* CACua-24: A proposal for biocontrol of phytopathogenic fungi. Microbiol. Res. 230, 126347.

Chen, C., Bock, C.H., Brannen, P.M., 2019. Novel primers and sampling for PCR detection of Xylella fastidiosa in peach. Phytopathology 109 (2), 307–317.

Cheng, Y., 2020. Plant disease detection method based on computer vision technology. Rev. Fac. Agron. LUZ 37, 2.

Collmer, A., 2020. James Robert Alfano, A Giant in Phytopathogenic Bacteria Effector Biology. Mol. Plant Microbe Interact. 33 (3), 377–381.

Coninck, E., Scauflaire, J., Gollier, M., Liénard, C., Foucart, G., Manssens, G., et al., 2020. Trichoderma atroviride as a promising biocontrol agent in seed coating for reducing Fusarium damping-off on maize. J. Appl. Microbiol. Available from: https://doi.org/10.1111/jam.14641.

Conner, R.L., Kuzyk, A.D., Su, H., 2003. Impact of powdery mildew on the yield of soft white spring wheat cultivars. Can. J. Plant. Sci. 83 (4), 725–728.

Costa, C.D.S.R., da Silva Costa, A.E., dos Santos, A.M.M., de Lima Pereira, J.W., e Carvalho, R.R.D.C., de Carvalho Filho, J.L.S., 2019. Current status of the occurrence and reaction root-knot nematodes in the main botanical families of medicinal plants. J. Exp. Agric. Int. 1–21.

Coutinho, T.C., Ferreira, M.C., Rosa, L.H., de Oliveira, A.M., de Oliveira Júnior, E.N., 2020. *Penicillium citrinum* and *Penicillium mallochii*: new phytopathogens of orange fruit and their control using chitosan. Carbohydr. Polym. 234, 115918.

Crespo, O., Robles, C., Ruiz, L., Janssen, D., 2020. Antagonism of cucumber green mottle mosaic virus against tomato leaf curl New Delhi virus in zucchini and cucumber. Ann. Appl. Biol. 176 (2), 147–157.

Dadrasnia, A., Usman, M.M., Omar, R., Ismail, S., Abdullah, R., 2020. Potential use of Bacillus genus to control of bananas diseases: approaches toward high yield production and sustainable management. J. King Saud. Univ. Sci. Available from: https://doi.org/10.1016/j.jksus.2020.03.011 (in Press).

Davis, K.F., Gephart, J.A., Emery, K.A., Leach, A.M., Galloway, J.N., D'Odorico, P., 2016. Meeting future food demand with current agricultural resources. Glob. Environ. Change 39, 125–132.

Decraemer, W., Hunt, D.J., 2006. Structure and classification. In: Perry, R.N., Moens, M. (Eds.), Plant Nematology. CAB International, Wallingford, Oxfordshire, pp. 3–32.

Dehnen-Schmutz, K., MacLeod, A., Reed, P., Mills, P.R., 2010. The role of regulatory mechanisms for control of plant diseases and food security—case studies from potato production in Britain. Food Secur. 2 (3), 233–245.

Elling, A.A., 2013. Major emerging problems with minor *Meloidogyne* species. Phytopathology 103 (11), 1092–1102.

Ellis, M.D., Hoak, J.M., Ellis, B.W., Brown, J.A., Sit, T.L., Wilkinson, C.A., et al., 2020. Quantitative real-time PCR analysis of individual flue-cured tobacco seeds and seedlings reveals seed transmission of tobacco mosaic virus. Phytopathology 110 (1), 194–205.

El-Sayed, A.S., Ali, G.S., 2020. *Aspergillus flavipes* is a novel efficient biocontrol agent of Phytophthora parasitica. Biol. Control. 140, 104072.

Everts, K.L., Leath, S., Finney, P.L., 2001. Impact of powdery mildew and leaf rust on milling and baking quality of soft red winter wheat. Plant. Dis. 85 (4), 423–429.

Fanning, J.P., Reeves, K.L., Forknall, C.R., McKay, A.C., Hollaway, G.J., 2020. *Pratylenchus thornei*: The relationship between presowing nematode density and yield loss in wheat and barley. Phytopathology 110 (3), 674–683.

FAO (2003),< http://www.fao.org/3/a-y4671e.pdf.>

FAO (2015). The Millennium Development Goals Report 2015. <http://www.un.org/millenniumgoals/2015_MDG_Report/pdf/MDG%202015%20rev%20(July%201).pdf>.

Fisher, M.C., Henk, D.A., Briggs, C.J., Brownstein, J.S., Madoff, L.C., McCraw, S.L., et al., 2012. Emerging fungal threats to animal, plant and ecosystem health. Nature 484 (7393), 186–194.

Godfray, H.C.J., Beddington, J.R., Crute, I.R., Haddad, L., Lawrence, D., Muir, J.F., et al., 2010. Food security: the challenge of feeding 9 billion people. Science 327 (5967), 812–818.

Gorman, Z., Christensen, S.A., Yan, Y., He, Y., Borrego, E., Kolomiets, M.V., 2020. Green leaf volatiles and jasmonic acid enhance susceptibility to anthracnose diseases caused by *Colletotrichum graminicola* in maize. Mol. Plant. Pathol. Available from: https://doi.org/10.1111/mpp.12924.

Gregory, P.J., Ingram, J.S.I., 2000. Global change and food and forest production: future scientific challenges. Agric. Ecosyst. Environ. 82 (1-3), 3–14.

Gregory, P.J., Ingram, J.S., Brklacich, M., 2005. Climate change and food security. Philos. Trans. R. Soc. B: Biol. Sci. 360 (1463), 2139–2148.

Haraguchi, S., Yoshiga, T., 2020. Potential of the fungal feeding nematode *Aphelenchus avenae* to control fungi and the plant parasitic nematode *Ditylenchus* destructor associated with garlic. Biol. Control. 143, 104203.

Harahagazwe, D., Condori, B., Barreda, C., Bararyenya, A., Byarugaba, A.A., Kude, D.A., et al., 2018. How big is the potato (*Solanum tuberosum* L.) yield gap in Sub-Saharan Africa and why? A participatory approach. Open. Agric. 3 (1), 180–189.

Harveson, R.M., Markell, S.G., Block, C.C., Gulya, T.J., 2016. Compendium of sunflower diseases and pests. APS press, St Paul, MN. Available from: https://doi.org/10.1094/9780890545096.

Hay, J.E., Easterling, D., Ebi, K.L., Kitoh, A., Parry, M., 2016. Introduction to the special issue: Observed and projected changes in weather and climate extremes. Weather. Clim. Extremes 11, 1–3.

Heydari, A., Pessarakli, M., 2010. A review on biological control of fungal plant pathogens using microbial antagonists. J. Biol. Sci. 10 (4), 273–290.

Hickey, L.T., Hafeez, A.N., Robinson, H., Jackson, S.A., Leal-Bertioli, S.C., Tester, M., et al., 2019. Breeding crops to feed 10 billion. Nat. Biotechnol. 37 (7), 744–754.

Hong, S., Jv, H., Lu, M., Wang, B., Zhao, Y., Ruan, Y., 2020. Significant decline in banana Fusarium wilt disease is associated with soil microbiome reconstruction under chilli pepper-banana rotation. Eur. J. Soil. Biol. 97, 103154.

Huang, L.Q., Niu, Y.C., Su, L., Deng, H., Lyu, H., 2020. The potential of endophytic fungi isolated from cucurbit plants for biocontrol of soilborne fungal diseases of cucumber. Microbiol. Res. 231, 126369.

Hussey, R.S., 1989. Disease-inducing secretions of plant-parasitic nematodes. Annu. Rev. Phytopathol. 27 (1), 123–141.

Inacio, M.L.S., Camacho, M.J., Serra, C., Cordovil, C., Cordeiro, L., Andrade, E., 2020. First report of the potato cyst nematode *Globodera rostochiensis* on potato in azores, Portugal. Plant. Dis. Available from: https://doi.org/10.1094/PDIS-07-19-1429-PDN.

Jahnke, S., Menzel, M.I., Van Dusschoten, D., Roeb, G.W., Bühler, J., Minwuyelet, S., et al., 2009. Combined MRI–PET dissects dynamic changes in plant structures and functions. Plant. J. 59 (4), 634–644.

Jamir, I., Mandal, A.K., Devi, A.P., Bhattacharjee, T., Maurya, P.K., Dutta, S., et al., 2020. Screening of genotypes against viral diseases and assessment of yield loss due to yellow vein mosaic virus in okra grown in the eastern part of India. Indian. Phytopathol. 1–9.

Jansen, R.M.C., Wildt, J., Kappers, I.F., Bouwmeester, H.J., Hofstee, J.W., Van Henten, E.J., 2011. Detection of diseased plants by analysis of volatile organic compound emission. Annu. Rev. Phytopathol. 49, 157–174.

Jha, U.C., Bohra, A., Nayyar, H., 2020. Advances in "omics" approaches to tackle drought stress in grain legumes. Plant. Breed. 139 (1), 1–27.

Kamoun, S., 2003. Molecular genetics of pathogenic oomycetes. Eukaryot. Cell 2, 191–199.

Kannan, V., Bastas, K., Devi, R., 2015. Scientific and economic impact of plant pathogenic bacteria. In: Kannan, R. V., Bastas, K.K. (Eds.), Sustainable Approaches to Controlling Plant Pathogenic Bacteria. CRC Press, Boca Raton, FL, pp. 369–392. Available from: https://doi.org/10.1201/b18892-21.

Kashyap, P.L., Solanki, M.K., Kushwaha, P., Kumar, S., Srivastava, A.K., 2020. Biocontrol potential of salt-tolerant *Trichoderma* and *Hypocrea* isolates for the management of tomato root rot under saline environment. J. Soil. Sci. Plant. Nutr. 20 (1), 160–176.

Kato, M., Mizubuti, E.S., Goodwin, S.B., Fry, W.E., 1997. Sensitivity to protectant fungicides and pathogenic fitness of clonal lineages of *Phytophthora infestans* in the United States. Phytopathology 87 (9), 973–978.

Khairnar, K., Goje, N., 2020. Image processing based approach for diseases detection and diagnosis on cotton plant leaf. Techno-Societal 2018. Springer, Cham, pp. 55–65.

Kim, H., Mohanta, T.K., Park, Y.H., Park, S.C., Shanmugam, G., Park, J.S., et al., 2020. Complete genome sequence of the mountain-cultivated ginseng endophyte *Burkholderia stabilis* and its antimicrobial compounds against ginseng root rot disease. Biol. Control. 140, 104126.

Kroese, D.R., Schonneker, L., Bag, S., Frost, K., Cating, R., Hagerty, C.H., 2020. Wheat soil-borne mosaic: Yield loss and distribution in the US Pacific Northwest. Crop. Prot. 132, 105102.

Kron, A.S., Zengerer, V., Bieri, M., Dreyfuss, V., Sostizzo, T., Schmid, M., et al., 2020. *Pseudomonas orientalis* F9 pyoverdine, safracin, and phenazine mutants remain effective antagonists against *Erwinia amylovora* in Apple Flowers. Appl. Environ. Microbiol. 86, 8.

Lake, I.R., Hooper, L., Abdelhamid, A., Bentham, G., Boxall, A.B., Draper, A., et al., 2012. Climate change and food security: health impacts in developed countries. Environ. Health Perspect. 120 (11), 1520–1526.

Larkin, R.P., 2020. Biological control of soilborne diseases in organic potato production using hypovirulent strains of *Rhizoctonia solani*. Biol. Agric. Hortic. 1–11.

Launay, M., Zurfluh, O., Huard, F., Buis, S., Bourgeois, G., Caubel, J., et al., 2020. Robustness of crop disease response to climate change signal under modeling uncertainties. Agric. Syst. 178, 102733.

Lenaerts, B., Collard, B.C., Demont, M., 2019. Improving global food security through accelerated plant breeding. Plant. Sci. 110207.

Leyns, F., De Cleene, M., Swings, J.G., De Ley, J., 1984. The host range of the genus *Xanthomonas*. Bot. Rev. 50 (3), 308–356.

Li, Z., Chang, P., Gao, L., Wang, X., 2020. The endophytic fungus *Albifimbria verrucaria* from wild grape as an antagonist of *Botrytis cinerea* and other grape pathogens. Phytopathology 110 (4), 843–850.

Ling, L., Han, X., Li, X., Zhang, X., Wang, H., Zhang, L., et al., 2020. A *Streptomyces* sp. NEAU-HV9: isolation, identification, and potential as a biocontrol agent against *Ralstonia solanacearum* of tomato plants. Microorganisms 8 (3), 351.

Lobell, D.B., Schlenker, W., Costa-Roberts, J., 2011. Climate trends and global crop production since 1980. Science 333 (6042), 616–620.

Mandal, A., Sarkar, B., Owens, G., Thakur, J.K., Manna, M.C., Niazi, N.K., et al., 2020. Impact of genetically modified crops on rhizosphere microorganisms and processes: A review focusing on Bt cotton. Appl. Soil. Ecol. 148, 103492.

Mansfield, J., Genin, S., Magori, S., Citovsky, V., Sriariyanum, M., Ronald, P., et al., 2012. Top 10 plant pathogenic bacteria in molecular plant pathology. Mol. Plant. Pathol. 13 (6), 614–629.

Martin, F.N., Loper, J.E., 1999. Soil borne plant diseases caused by Pythium spp.: ecology, epidemiology, and prospects for biological control. Crit. Rev. Plant. Sci. 18 (2), 111–181.

Martinelli, F., Scalenghe, R., Davino, S., Panno, S., Scuderi, G., Ruisi, P., et al., 2015. Advanced methods of plant disease detection. A review. Agron. Sustain. Dev. 35 (1), 1–25.

Martínez, M., Ramírez Albuquerque, L., Arata, A.F., Biganzoli, F., Fernández Pinto, V., Stenglein, S.A., 2020. Effects of *Fusarium graminearum* and *Fusarium poae* on disease parameters, grain quality and mycotoxins contamination in bread wheat (Part I). J. Sci. Food Agric. 100 (2), 863–873.

Martins, P.M., Merfa, M.V., Takita, M.A., De Souza, A.A., 2018. Persistence in phytopathogenic bacteria: do we know enough? Front. Microbiology 9, 1099.

Mascia, T., Gallitelli, D., 2016. Synergies and antagonisms in virus interactions. Plant. Sci. 252, 176–192.

Masonbrink, R., Maier, T.R., Muppirala, U., Seetharam, A.S., Lord, E., Juvale, P.S., et al., 2019. The genome of the soybean cyst nematode (*Heterodera glycines*) reveals complex patterns of duplications involved in the evolution of parasitism genes. BMC Genomics 20 (1), 119.

McLeish, M.J., Fraile, A., García-Arenal, F., 2020. Trends and gaps in forecasting plant virus disease risk. Ann. Appl. Biol. 176 (2), 102 108.

Mengesha, G.G., 2020. Management of yellow rust (*Puccinia striiformis* f. sp. tritici) and stem rust (*Puccinia graminis* f. sp tritici) of bread wheat through host resistance and fungicide application in Southern Ethiopia. Cogent Food Agric. 6 (1), 1739493.

Miyamoto, T., Hayashi, K., Ogawara, T., 2020. First report of the occurrence of multiple resistance to flutianil and pyriofenone in field isolates of *Podosphaera xanthii*, the causal fungus of cucumber powdery mildew. Eur. J. Plant. Pathol. 156 (3), 953–963.

Mueller, N.D., Gerber, J.S., Johnston, M., Ray, D.K., Ramankutty, N., Foley, J.A., 2012. Closing yield gaps through nutrient and water management. Nature 490 (7419), 254 257.

Mukanga, M., Derera, J., Tongoona, P., Laing, M.D., 2010. A survey of pre-harvest ear rot diseases of maize and associated mycotoxins in south and central Zambia. Int. J. Food Microbiol. 141 (3), 213–221.

Mulugeta, T., Muhinyuza, J.B., Gouws-Meyer, R., Matsaunyane, L., Andreasson, E., Alexandersson, E., 2020. Botanicals and plant strengtheners for potato and tomato cultivation in Africa. J. Integr. Agric. 19 (2), 406–427.

Oerke, E.C., 2006. Crop losses to pests. J. Agric. Sci. 144 (1), 31–43.

Oerke, E.C., Dehne, H.W., 2004. Safeguarding production—losses in major crops and the role of crop protection. Crop. Prot. 23 (4), 275–285.

OuYang, Q., Okwong, R.O., Chen, Y., Tao, N., 2020. Synergistic activity of cinnamaldehyde and citronellal against green mold in citrus fruit. Postharvest Biol. Technol. 162, 111095.

Palukaitis, P., 2012. Resistance to viruses of potato and their vectors. Plant. Pathol. J. 28 (3), 248–258.

Panabieres, F., Ali, G.S., Al, E., 2016. Phytophthora nicotianae diseases worldwide: new knowledge of a long-recognised pathogen. Phytopathol. Mediterr. 55, 20–40.

Panigrahi, K.P., Das, H., Sahoo, A.K., Moharana, S.C., 2020. Maize leaf disease detection and classification using machine learning algorithms. Progress in Computing, Analytics and Networking. Springer, Singapore, pp. 659–669.

Pérez-Padilla, V., Fortes, I.M., Romero-Rodríguez, B., Arroyo-Mateos, M., Castillo, A.G., Moyano, C., et al., 2020. Revisiting seed transmission of the type strain of tomato yellow leaf curl virus in tomato plants. Phytopathology 110 (1), 121–129.

Peterson, E.K., Rupp, F., Eberhart, J., Parke, J.L., 2020. Root rot of juniperus and microbiota by *Phytophthora lateralis* in Oregon horticultural nurseries. Plant. Dis. pp. PDIS-04.

Pimentel, D., 2009. Pesticides and Pest Control. In: Peshin, R., Dhawan, A.K. (Eds.), Integrated Pest Management: Innovation-Development Process. Springer Netherlands, Dordrecht, pp. 83–87.

Ramírez, P.G., Ramírez, D.G., Mejía, E.Z., Ocampo, S.A., Díaz, C.N., Martínez, R.I.R., 2020. Extracts of *Stevia rebaudiana* against *Fusarium oxysporum* associated with tomato cultivation. Sci. Hortic. 259, 108683.

Reardon, T., Timmer, C.P., 2014. Five inter-linked transformations in the Asian agrifood economy: food security implications. Glob. Food Secur. 3 (2), 108–117.

Rguez, S., Slimene, I.B., Abid, G., Hammemi, M., Kefi, A., Elkahoui, S., et al., 2020. *Tetraclinis articulata* essential oil reduces *Botrytis cinerea* infections on tomato. Sci. Hortic. 266, 109291.

Rodríguez, M., Torres, M., Blanco, L., Béjar, V., Sampedro, I., Llamas, I., 2020. Plant growth-promoting activity and quorum quenching-mediated biocontrol of bacterial phytopathogens by *Pseudomonas segetis* strain P6. Sci. Rep. 10 (1), 1–12.

Rojas, E.C., Jensen, B., Jørgensen, H.J., Latz, M.A., Esteban, P., Ding, Y., et al., 2020. Selection of fungal endophytes with biocontrol potential against *Fusarium* head blight in wheat. Biol. Control. 144, 104222.

Rosegrant, M.W., Cline, S.A., 2003. Global food security: challenges and policies. Science 302 (5652), 1917–1919.

Salgado, J.D., Wallhead, M., Madden, L.V., Paul, P.A., 2011. Grain harvesting strategies to minimize grain quality losses due to *Fusarium* head blight in wheat. Plant. Dis. 95 (11), 1448–1457.

Savary, S., Willocquet, L., Pethybridge, S.J., Esker, P., McRoberts, N., Nelson, A., 2019. The global burden of pathogens and pests on major food crops. Nat. Ecol. Evol. 3 (3), 430–439.

Scott, P., Thomson, J., Grzywacz, D., Savary, S., Strange, R., Ristaino, J.B., et al., 2016. Genetic modification for disease resistance: a position paper. Food Secur. 8 (4), 865–870.

Shahin, M.A., Tollner, E.W., McClendon, R.W., Arabnia, H.R., 2002. Apple classification based on surface bruises using image processing and neural networks. Trans. ASAE 45 (5), 1619.

Shattock, R.C., 2002. *Phytophthora infestans*: populations, pathogenicity and phenylamides. Pest. Manag. Sci. 58 (9), 944–950.

Shi, W.Q., Xiang, L.B., Yu, D.Z., Gong, S.J., Yang, L.J., 2020. Impact of the biofungicide tetramycin on the development of *Fusarium* head blight, grain yield and deoxynivalenol accumulation in wheat. World Mycotoxin J. 13 (2), 235–246.

Silva, R.N., Monteiro, V.N., Steindorff, A.S., Gomes, E.V., Noronha, E.F., Ulhoa, C.J., 2019. *Trichoderma*/pathogen/plant interaction in pre-harvest food security. Fungal Biol. 123 (8), 565–583.

Singh, N., 2020. Emerging problem of guava decline caused by *Meloidogyne enterolobii* and *Fusarium oxysporum* f. sp. psidii. Indian Phytopathol. 1–2.

Siuda, R., Grabowski, A., Lenc, L., Ralcewicz, M., Spychaj-Fabisiak, E., 2010. Influence of the degree of fusariosis on technological traits of wheat grain. Int. J. Food Sci. Technol. 45 (12), 2596–2604.

Smith, P., 2013. Delivering food security without increasing pressure on land. Glob. Food Secur. 2 (1), 18–23.

Sreenivasulu, N., Butardo Jr, V.M., Misra, G., Cuevas, R.P., Anacleto, R., Kavi Kishor, P.B., 2015. Designing climate-resilient rice with ideal grain quality suited for high-temperature stress. J. Exp. Bot. 66 (7), 1737–1748.

Sridharan, A.P., Thankappan, S., Karthikeyan, G., Uthandi, S., 2020. Comprehensive profiling of the VOCs of *Trichoderma longibrachiatum* EF5 while interacting with *Sclerotium rolfsii* and *Macrophomina phaseolina*. Microbiol. Res. 126436.

Strange, R.N., Scott, P.R., 2005. Plant disease: a threat to global food security. Annu. Rev. Phytopathol. 43, 83–116.

Sundin, G.W., Bender, C.L., 1993. Ecological and genetic analysis of copper and streptomycin resistance in *Pseudomonas syringae* pv. syringae. Appl. Environ. Microbiol. 59 (4), 1018–1024.

Tiwari, I.M., Jesuraj, A., Kamboj, R., Devanna, B.N., Botella, J.R., Sharma, T.R., 2017. Host delivered RNAi, an efficient approach to increase rice resistance to sheath blight pathogen (*Rhizoctonia solani*). Sci. Rep. 7 (1), 1–14.

Torrance, L., Cowan, G.H., McLean, K., MacFarlane, S., Al-Abedy, A.N., Armstrong, M., et al., 2020. Natural resistance to Potato virus Y in *Solanum tuberosum* Group Phureja. Theor. Appl. Genet. 133 (3), 967–980.

Turner, M., Jauneau, A., Genin, S., Tavella, M.J., Vailleau, F., Gentzbittel, L., et al., 2009. Dissection of bacterial wilt on *Medicago truncatula* revealed two type III secretion system effectors acting on root infection process and disease development. Plant. Physiol. 150 (4), 1713–1722.

Urban, M.A., Barclay, R.S., Sivaguru, M., Punyasena, S.W., 2018. Cuticle and subsurface ornamentation of intact plant leaf epidermis under confocal and superresolution microscopy. Microsc. Res. Tech. 81 (2), 129–140.

van Esse, H.P., Reuber, T.L., van der Does, D., 2020. Genetic modification to improve disease resistance in crops. N. Phytol. 225 (1), 70–86.

Varshney, R.K., Sinha, P., Singh, V.K., Kumar, A., Zhang, Q., Bennetzen, J.L., 2020. 5Gs for crop genetic improvement. Curr. Opin. Plant. Biol. Available from: https://doi.org/10.1016/j.pbi.2019.12.004.

Vasse, J., Frey, P., Trigalet, A., 1995. Microscopic studies of intercellular infection and protoxylem invasion of tomato roots by *Pseudomonas solanacearum*. Mol. Plant. Microbe Interact. 8 (2), 241–251.

Veenstra, A., Rafudeen, M.S., Murray, S.L., 2019. *Trichoderma asperellum* isolated from African maize seed directly inhibits *Fusarium verticillioides* growth in vitro. Eur. J. Plant. Pathol. 153 (1), 279–283.

Wang, B., Ma, Y., Zhang, Z., Wu, Z., Wu, Y., Wang, Q., et al., 2011. Potato viruses in China. Crop. Prot. 30 (9), 1117–1123.

Wijesinghe, R.E., Lee, S.Y., Ravichandran, N.K., Shirazi, M.F., Moon, B., Jung, H.Y., et al., 2017. Bio-photonic detection method for morphological analysis of anthracnose disease and physiological disorders of Diospyros kaki. Opt. Rev. 24 (2), 199–205.

Wijesinghe, R.E., Lee, S.Y., Ravichandran, N.K., Shirazi, M.F., Kim, P., Jung, H.Y., et al., 2018. Biophotonic approach for the characterization of initial bitter-rot progression on apple specimens using optical coherence tomography assessments. Sci. Rep. 8 (1), 1–10.

Wonglom, P., Ito, S.I., Sunpapao, A., 2020. Volatile organic compounds emitted from endophytic fungus *Trichoderma asperellum* T1 mediate antifungal activity, defense response and promote plant growth in lettuce (*Lactuca sativa*). Fungal Ecol. 43, 100867.

Wu, Z.M., Yang, Y., Li, K.T., 2019. Antagonistic activity of a novel antifungalmycin N2 from *Streptomyces* sp. N2 and its biocontrol efficacy against *Rhizoctonia solani*. FEMS Microbiol. Lett. 366 (3), fnz018.

Xie, W., Liang, G., Huang, A., Zhang, F., Guo, W., 2020. Comparative study on the mRNA expression of *Pinus massoniana* infected by *Bursaphelenchus xylophilus*. J. For. Res. 31 (1), 75–86.

Xu, M., Chen, J., Huang, Y., Shen, D., Sun, P., Xu, Y., et al., 2020a. Dynamic transcriptional profiles of *Arabidopsis thaliana* infected by tomato spotted wilt virus. Phytopathology 110 (1), 153–163.

Xu, S., Xie, X., Zhao, Y., Shi, Y., Chai, A., Li, L., et al., 2020c. Whole-genome analysis of *Bacillus velezensis* ZF2, a biocontrol agent that protects *Cucumis sativus* against *Corynespora* leaf spot diseases. 3 Biotech. 10 (4), 186.

Xu, X., Qing, X., Xie, J.L., Yang, F., Peng, Y.L., Ji, H.L., 2020b. Population structure and species delimitation of rice white tip nematode, *Aphelenchoides besseyi* (Nematoda: Aphelenchoididae), in China. Plant. Pathol. 69 (1), 159–167.

Yamada, T., 1993. The role of auxin in plant-disease development. Annu. Rev. Phytopathol. 31 (1), 253–273.

Yang, S.H., Zhou, W.Q., Wang, D.W., Xu, C.L., Xie, H., 2020. Evaluation of *Neoseiulus barkeri* (Acari: Phytoseiidae) for the control of plant parasitic nematodes, *Radopholus similis* (Tylenchida: Pratylenchidae) and *Meloidogyne incognita* (Tylenchida: Heteroderidae). Biocontrol Sci. Technol. 30 (3), 201–211.

Ye, T., Zhou, T., Li, Q., Xu, X., Fan, X., Zhang, L., et al., 2020. *Cupriavidus* sp. HN-2, a novel quorum quenching bacterial isolate, is a potent biocontrol agent against *Xanthomonas campestris* pv. campestris. Microorganisms 8 (1), 45.

Yusof, N.A., Isha, A., 2020. Nanosensors for early detection of plant diseases. Nanomaterials for Agriculture and Forestry Applications. Elsevier, pp. 407–419.

Zhang, H., Deng, L., Yao, S., Ming, J., Zeng, K., 2020b. Optimization of a vacuum-drying protectants for the biocontrol agent *Pichia membranifaciens* and its influence on viability and efficacy. Biol. Control. 142, 104155.

Zhang, Y., Xie, Z., Fletcher, J.D., Wang, Y., Wang, R., Guo, Z., et al., 2020a. Rapid and sensitive detection of lettuce necrotic yellows virus and cucumber mosaic virus infecting lettuce (*Lactuca sativa* L.) by reverse transcription loop-mediated isothermal amplification. Plant. Pathol. J. 36 (1), 76.

Zhao, C., Liu, B., Piao, S., Wang, X., Lobell, D.B., Huang, Y., et al., 2017. Temperature increase reduces global yields of major crops in four independent estimates. Proc. Natl Acad. Sci. 114 (35), 9326–9331.

Zhao, Y., Yang, X., Zhou, G., Zhang, T., 2020. Engineering plant virus resistance: from RNA silencing to genome editing strategies. Plant. Biotechnol. J. 18 (2), 328.

Zhou, L., Wang, C., Wang, G.H., Wei, Z.W., Fu, Q.X., Hang, X.H., et al., 2020. Chemical targeting and manipulation of type III secretion in the phytopathogen *Xanthomonas campestris* for control of disease. Appl. Environ. Microbiol. 86, 3.

CHAPTER

3

Microbial bioactive compounds in plant disease management

Leandro Afonso[1], *Bárbara Gionco-Cano*[1], *Ane Stéfano Simionato*[1], *Erika Tyemi Goya Niekawa*[1], *Guilherme Edison Almeida Pega*[1], *Larissa Ayumi Taira*[1], *Mickely Liuti Dealis*[1], *Matheus Felipe de Lima Andreata*[1], *Andreas Lazaros Chryssafidis*[2] *and Galdino Andrade*[1]

[1]**Microbial Ecology Laboratory, Department of Microbiology, State University of Londrina, Londrina, Brazil** [2]**Agroveterinary Sciences Center, University of the State of Santa Catarina, Lages, Brazil**

3.1 Introduction

Agriculture is one of the main sources, directly or indirectly, of the world income. In constant evolution, agriculture looks for new technologies every day, for achieving the food demand that comes from the increasing global population, with the least environmental damage.

A major problem faced by producers is the control of phytopathogens that can lead to impact on the quality and quantity of the agricultural products, resulting in losses over a billion dollars per year (Martins et al., 2018). The continued preventive use of antimicrobial products in high concentrations selects resistant strains of different microorganisms, increasing the difficulty in controlling pathogens. In addition, the misuse of antimicrobials can jeopardize food security, causing serious consequences to human and environmental health, affecting the maintenance of ecological balance and the conservation of soil biodiversity (Mandal et al., 2020). Due to these factors, investigations targeting the development of new effective products with low toxicity and natural origin have been a priority for this sector.

By definition, natural products are chemical compounds or substances with biological activity produced by living beings found in nature (Bohlin et al., 2010). Until the

Food Security and Plant Disease Management
DOI: https://doi.org/10.1016/B978-0-12-821843-3.00013-1

19th century, 80% of such obtained products were derived from plants. But when Louis Pasteur (1822–95) demonstrated the antagonism between two microorganisms in his studies, the history began to change (Oliveira et al., 2015; Pham et al., 2019).

In the stationary phase of growth, microorganisms can produce several secondary metabolites that comprise a high number of bioactive compounds. These substances can inhibit or kill other microbial species even at low concentrations, conferring a competitive advantage over other organisms (Madigan et al., 2017). Such compounds aroused the interest of several researchers, like Alexander Fleming (1881–1955), who dedicated his life to the research of antimicrobials and discovered penicillin in 1928, a molecule produced by the fungus *Penicillium notatum*, and marked the beginning of the golden age of antibiotics (Gould, 2016; Saga and Yamaguchi, 2009).

These microbial bioactive compounds do not have only antibiotic activity. They can promote growth, induce resistance in plants and present insecticidal and nematicidal activities. Therefore due to these beneficial effects for the plant health, many studies have been carried out for the identification of these substances and their possible use in agriculture.

In this chapter, a revision about some of these microbial bioactive compounds and how they help to promote plant health will be presented.

3.2 Bioactive compounds produced by *Pseudomonas* spp

Pseudomonas is a genus of Gram-negative bacteria phenotypically characterized by bacillar cells generally with polar flagellum. Also, most of these bacteria are obligate aerobes (some species are facultative aerobes) and this genus is widely distributed in the most diverse environments (Sun et al., 2017).

Pseudomonas are widely studied as bioremediators in regions contaminated by pesticides (Hassen et al., 2018), heavy metals (O'Brien et al., 2014), and oil derivatives (Ramadass et al., 2018). They are excellent producers of antifungals (Bajpai et al., 2018), antibacterial (Oliveira et al., 2016), insecticides (Flury et al., 2017), nematicides (Lee et al., 2011), and can also promote induced systemic resistance (ISR) (Preston, 2004).

Due to its biopesticidal and biofertilizer potential, *Pseudomonas* is the most frequently used bacterial genus in studies for the discovery and development of alternatives to agrochemicals, in the pursuit of sustainable agriculture (Shahid et al., 2018). In this section, some of the main compounds produced by *Pseudomonas* will be described.

3.2.1 Phenazines

Phenazines are an extensive group of intensely colored aromatic nitrogen-containing heterocyclic molecules, structurally characterized by presence of three rings, a pyrazine core ring flanked by two benzenes (Guttenberger et al., 2017). Besides a basic molecular skeleton, phenazines feature different types of chemical functional groups, which nature and position in the molecule defines its chemical, physical, and biological characteristics (Price-Whelan et al., 2006). Excepting archaebacterium *Methanosarcina mazei*, phenazines are secondary metabolites exclusive from bacteria occurring in several genera as *Nocardia*, *Sorangium*, *Brevibacterium*, *Burkholderia*, *Erwinia*, *Pantoea*, *Vibrio*, *Pelagiobacter*, *Xanthomonas*, and *Streptomyces* (Pierson and Pierson, 2010; Price-Whelan et al., 2006).

Numerous applications of phenazine are reported, mainly based on their antimicrobial (in bacteria, fungi, and protists), insecticide, antiparasitic, and antitumoral activities (Guttenberger et al., 2017). In nature, their characteristics support that phenazines plays an important role as a competitive overcoming mechanism in microbial relationships also being an important virulence factor (Laursen and Nielsen, 2004).

The main mode of action of phenazines is related to the molecular nature of these compounds, which allows them to accept and donate electrons, a mechanism that provides them with the capacity to generate reactive oxygen species (ROS) (Blankenfeldt and Parsons, 2014; Guttenberger et al., 2017). Additionally, investigations suggest that phenazines can perform intracellular signaling and activate transcription factors (Dietrich et al., 2008), mediate NADH reoxidation under oxygen-limiting conditions (Maddula et al., 2008) and promote mineral reduction and electron shuttle in soil, increasing the availability of minerals for plants and bacteria (Hernandez et al., 2004).

Phenazine-1-carboxylic acid (PCA) is a bright lemon yellow phenazine that is an active metabolite and also a precursor of other phenazine derivatives (Pierson et al., 2013). PCA activity against phytopathogens is extensively studied. *In planta* studies have shown PCA effectiveness against important plant diseases, such as take-all (*Gaeumannomyces graminis* var. *tritici*) in wheat (Tomashow et al., 1990), Phytophthora blight (*Phytophthora capsici*) in pepper, late blight (*Phytophthora infestans*) in potato plants (Morrison et al., 2016), and foliar anthracnose (*Colletotrichum orbiculare*) in cucumber (Lee et al., 2003).

Simionato et al. (2017) evaluated in vitro the PCA antifungal properties against *Botrytis cinerea*, reporting that the compound damaged hyphae morphology and organization, completely inhibiting production of exopolysaccharides with 12.5 μg/mL. Puopolo et al. (2013) studied in vitro the susceptibility of 25 strains of phytopathogenic fungi and oomycetes to PCA. With 25 μg/mL of the phenazine in the culture medium, the authors demonstrated the total growth inhibition of *G. graminis* var. *tritici* and *Phytophthora nicotianae*. In a concentration of 50 μg/mL, PCA completely ceased the development of *Pythium ultimum* and decreased the growth of 16 of the 25 microorganisms tested by over 50%.

Phenazine-1-carboxamide (PCN) is a yellowish-green phenazine and, like PCA, its applications are mostly related to biocontrol of phytopathogenic fungi (Peng et al., 2018). This compound demonstrated antifungal activity against *Fusarium oxysporum* f. sp. *radicis lycopersici* (Chin-A-Woeng et al., 1998) and *B. cinerea* (Zhang et al., 2015), also against the oomycete *Pythium myriotylum* (Tambong and Hofte, 2001) and *Xanthomonas oryzae* pv. *oryzae* bacteria (Shanmugaiah et al., 2010). PCN in low concentration (18.18 μg/mL) generated critical damages to *Rhizoctonia solani* cell wall and marked alterations in its cytoplasm, such as the disappearance of mitochondrias and ribosomes, contraction of vacuoles and cell plasmolysis, indicating that these are the main sites of action of this phenazine (Xiang et al., 2018). PCN acts as electron shuttle as well, promoting mineral reduction of iron and manganese in soil, increasing its bioavailability (Hernandez et al., 2004). Ma et al. (2016) reported that PCN triggered ISR in rice infected by *Magnaporthe oryzae*.

A bright blue phenazine, 5-methyl-1-hydroxyphenazine, also called pyocyanin, had its antifungal activity tested against *Cyrtomium falcatum*, *F. oxysporum*, *Sclerotium rolfsii* (Rane et al., 2008), and *Macrophomina phaseolina* (Khare and Arora, 2011). ISR promoted by this compound was described against *B. cinerea* in tomato (Audenaert et al., 2002) and *Magnaporthe grisea* in rice (Vleesschauwer et al., 2006).

Another phenazine, 1-hydroxyphenazine, had its antifungal activity tested in vitro by Dharni et al. (2012) against important plant fungal pathogens, such as *Alternaria alternata, Alternaria solani, Balanoglossus australiensis, Colletotrichum acutatum, Curvularia andropogonis, F. oxysporum, Fusarium moniliforme,* and *R. solani*. The compound reached the maximum inhibition zone of 24.8 mm against *A. alternata* with 40 μg concentration in disk diffusion assay.

3.2.2 Siderophores

Siderophores are low molecular weight organic molecules (<10 KD) produced by microorganisms under iron-limiting conditions. Approximately 270 types of siderophores had their chemical structure determined, and more than 500 molecules have already been reported as siderophores (Shanmugaiah et al., 2015). Siderophores classes vary according to the iron-coordinating functional groups, structural features, and types of ligands (Saha et al., 2016).

These molecules are metal ion chelating agents with a special affinity for ferric iron (Fe^{3+}), an essential element for various metabolic processes. However, its bioavailability is very low, being rarely found in free form under aerobic conditions (Drehe et al., 2018). Iron sequestration by siderophores decreases the bioavailability of this element, suppressing the growth of competitors (Santoyo et al., 2012).

Siderophores produced by *Pseudomonas* promote several benefits to plants, such as growth stimulation, ISR, biocontrol of phytopathogens, and protection against heavy metals toxicity (Ali and Vidhale, 2013). Among these, the most studied are pyoverdine (PVD), produced by all fluorescent *Pseudomonas*, pyochelin (PCH), enantio-pyochelin (E-PCH), quinolobactin (QB), thioquinolobactin (TQB) (Gasser et al., 2019).

PVDs, also called pseudobactins, are hydroxamate siderophores (Ambrosi et al., 2000). These fluorescent yellow-green molecules are soluble in water, have a high affinity for Fe^{3+} and lower affinity for Ni^{2+}, Cd^{2+}, and Cu^{2+} (Ringel and Brüser, 2018). They are important ISR elicitors, but it is believed that its antimicrobial activity only occurs in synergism with other antimicrobials (Duijff et al., 1999). A study from 2005 demonstrated that inoculation of *Pseudomonas putida* WCS358r and *Pseudomonas fluorescens* WCS374r PVD-positive, elicited ISR in *Eucalyptus urophylla* against bacterial wilt caused by *Ralstonia solanacearum* (Ran et al., 2005). Also, the application of purified PVD from the WCS374r strain increased the resistance of rice (*Oryza sativa*) against *M. oryzae* by 88% when compared to control (Vleesschauwer et al., 2008).

PCH and E-PCH are derived from a salicylate molecule, a hydroxy acid, and two molecules of cysteine. They differ only in the stereochemical configuration of the incorporated cysteine molecule (Youard et al., 2007). They are poorly soluble in water, have affinity for different metal ions, have antimicrobial activity and promote growth in plants (Raaijmakers et al., 2010). Their activity is related to the generation of ROS, although the mechanisms of action are still not well understood (Ong et al., 2017). *Pseudomonas protegens* CS1, isolated from the surface of a lemon leaf (*Citrus limon* cv. *limoneira*), demonstrated in vitro activity against *Xanthomonas citri* subsp. *citri*, etiological agent of citrus canker disease. E-PCH was isolated from the culture of CS1 strain and the results confirmed that its activity involves oxidative stress and it is independent from iron deprivation (Michavila et al., 2017).

The siderophores QB and TQB were identified in *P. fluorescens* pyoverdin-negative mutants, as an alternative to PVD. They present low affinity for metal ions (Dhardemare et al., 2004). TQB is a poorly stable compound in aqueous solution, being rapidly hydrolyzed into QB. However, TQB was reported as the promoter of antagonist activity (Matthijs et al., 2007; Sazinas et al., 2019).

Matthijs et al. (2007) used TQB and PVD produced by *P. fluorescens* ATCC 17400 for in vitro testing their inhibiting potential against *Pythium debaryanum* hyphae growth, an oomycete that causes diseases in several plants of agronomic interest. The activity of these siderophores was analyzed using three mutant strains, deficient for TQB, PVD, and both. TQB producing strain and PVD deficient strain inhibited the oomycete, while the PVD producing strain and deficient for TQB decreased mycelial density but did not inhibit its growth. In contrast, PVD and TQB positive strain (wild) showed better activity compared to the TQB-only producer. This study highlighted the antifungal activity of TQB and the synergistic action of PVD in the control of fungal growth.

QB and TQB are quinolone derivatives, molecules with known antimicrobial activity. Unlike synthetic quinolones, natural quinolones do not have the 3-carboxy group, responsible for binding and blocking DNA type IIA topoisomerase complexes. Thus the exact mechanism of action of TQB remains unknown (Heeb et al., 2011).

Plants can also acquire iron from Fe–siderophore complexes produced by plant growth-promoting rhizobacteria (David et al., 2018). It is considered that bacterial siderophore is not absorbed by the plant, but reduced in rhizosphere, where the released iron is rapidly assimilated by roots (Sah et al., 2017). Nagata et al. (2013) demonstrated that tomato seedlings supplemented with the Fe–PVD complex presented similar weight when compared to the control treatment, with free ferric iron supplementation. In addition, plants supplemented with Fe–PVD had higher iron concentration in leaves and roots (Nagata et al., 2013).

3.3 Bioactive compounds produced by *Burkholderia* spp

Burkholderia are Gram-negative bacteria found in a wide range of environments, due to its capacity of interactions with other living forms that are known to be either beneficial or harmful, or both. For instance, *Burkholderia cepacia* complex (Bcc) is a large group of bacteria that include *B. cepacia*, *Burkholderia metallica*, *Burkholderia contaminans*, *Burkholderia pyrrocinia*, and more.

Many compounds highly relevant to agriculture, like antimicrobials and molecules with the ability to break down environmental contaminants, can be found in the secondary metabolism of Bcc (Castrejón-Godínez et al., 2019; Depoorter et al., 2016; Lassie et al., 2018; Moon et al., 1996; Morya et al., 2020). In the following topics, some compounds produced by *Burkholderia* spp. are presented.

3.3.1 Pyrrolnitrin

Pyrrolnitrin (PRN) is a halogenated light-sensitive bacterial secondary metabolite that presents antifungal and antibacterial properties and serves as a primordial structure of synthetic fungicides (Gordee and Westhead, 1972). One of the original bacterial strains that produce PRN is *B. pyrrocinia* 2327 strain, using tryptophan as a precursor molecule (Pée and Ligon, 2000).

El-Banna and Winkelmann (1998) observed that PRN action mechanism in *Neurospora crassa* fungus is based on interferences in electron transport chain. However, Okada et al. (2005) concluded that the primary mechanism of action against the fungus is to promote interferences in osmotic signal transduction.

In 2018 Jung and collaborators (2018) presented *B. cepacia* JBK9, a PRN producer capable of controlling *P. capsici*, *F. oxysporum*, and *R. solani*. Kim et al. (2020) showed PRN produced by *Burkholderia stabilis* EB159 as effective against ginseng root rot disease caused by *Cylindrocarpon destructans*.

PRN can lose stability within 30 days after application, due to its light sensitivity. Based on it, more photostable analogs were developed, like fenpiclonil and fludioxonil (Corran et al., 2011; Howell and Stipanovic, 1979).

3.3.2 Phenazines

The secondary metabolism of *Burkholderia* spp. also produces phenazine molecules. In *B. cepacia* 5.5B a purple pigment exhibited activity against *R. solani*. Later, it was isolated and identified as 4,9-dihydroxyphenazine-1,6-dicarboxylic acid dimethyl ester (Cartwright et al., 1995). Another molecule, produced by *Burkholderia glumae*, was identified as phencomycin, a phenazine with proven activity against many phytopathogenic fungi as *Aspergillus oryzae*, *B. cinerea*, *Cladosporium cucumerinum*, *Sclerotinia sclerotiorum*, and some bacteria like *Pseudomonas syringae*, *R. solanacearum*, and *Xanthomonas campestris* (Han et al., 2014).

3.3.3 Siderophores

Burkholderia species are able to survive in iron-deficient environments by producing different siderophores, such as PCH (Meyer et al., 1995), ornibactin (Deng et al., 2017), malleobactin (Franke et al., 2015), cepaciachelin (Barelmanna et al., 1996), azurechelin (Sokol et al., 1992), and cepabactin (Meyer et al., 1989).

Ornibactin is the only compound that had its microbiological effects investigated. Deng et al. (2017) isolated it from *B. contaminans* MS14 and described it as a bacteriostatic agent. Herrera (2017), reported the ornibactin activity against *X. citri* pv. *malvacearum*, *Pectobacterium carotovorum* WSCH1, *R. solanacearum*, *P. syringae* B301, *Erwinia amylovora*, *B. glumae* 291, *Escherichia coli*, and *Clavibacter michiganensis* subsp. *michiganensis*.

3.3.4 Xylocandins

Xylocandins, also known as cepacidins, are a set of peptide molecules with proven strict antifungal activity and were first isolated from *B. cepacia* ATCC 39277 (Meyers et al., 1987). Lim et al. (1994) described the protective effects of cepacidine A from *B. cepacia* AF2001 in field conditions against *Aspergillus niger*, *F. oxysporum*, and *Rhizopus stolonifer*. Lee et al. (2000) evidenced the suppressive effect of this compound against *P. ultimum* in cotton and cucumber, but had minor activity against *R. solani* in these plants.

3.3.5 Burkholdines

Burkholdines are cyclic lipopeptides with antimicrobial activity. The burkholdines were analyzed through the isolation of two compounds from *Burkholderia ambifaria* 2.2N, Bk-1229, and Bk-1097 (Tawfik et al., 2010). Both compounds presented strong effects against pathogens, including *P. infestans* (oomycete), *A. solani*, *B. cinerea*, and *Mycosphaerella fijiensis*.

3.3.6 Quinolone and quinoline derivatives

Two new 2-alkylquinolones isolated from *Burkholderia* sp. MBAF1239 demonstrated a nonstrict antifungal activity. Both inhibited the growth of *Tenacibaculum maritimum* and *Rhizopus oryzae* fungi (Li et al., 2018).

B. cepacia RB425 produces two quinoline derivatives, 2(2-heptenyl)-3-methyl-4-quinolinol and 2(2-nonenyl)-3-methyl-4-quinolinol, which present strong antibiotic activity against *P. capsici*, *R. solani*, *F. oxysporum*, and *C. michiganensis* (Yoshihisa et al., 1989). Moon et al. (1996) also reported a growth promotion effect of 2(2-heptenyl)-3-methyl-4-quinolinol in red pepper seeds treatments.

3.3.7 Altericidins

Altericidins are a complex of antibiotic oligopeptides with strong activity against *Alternaria kikuchiana*, the etiological agent of black spot in pears (Kirinuki and Ichiba, 1986). The altericidins selectively inhibit the growth of a wide range of filamentous fungi by change and damage cytoplasmic membrane permeability (Kirinuki and Ichiba, 1986).

3.3.8 Phenylacetic acid, hydrocinnamic acid, 4-hydroxyphenylacetic acid, and 4-hydroxyphenylacetate methyl ester

Four antifungals of low molecular weight were identified through the isolation of *Burkholderia* sp. strain MP-1 (Mao et al., 2006). These molecules displayed strong activity against a wide range of fungi, such as *Alternaria brassicicola*, *F. oxysporum*, *R. solani*, and others. They were detected, characterized and identified as phenylacetic acid, hydrocinnamic acid, 4-hydroxyphenylacetic acid, and 4-hydroxyphenylacetate methyl ester (Mao et al., 2006).

3.3.9 Hydrogen cyanide

Pseudomonas, *Burkholderia*, and recently *Chromobacterium* are known as hydrogen cyanide producers. This volatile compound is poisonous against most eukaryotic cells due to the inhibition of cytochrome *c* oxidase and severe other metalloenzymes (Blumer and Haas, 2000; Gilchrist et al., 2013; Short et al., 2018).

3.3.10 New compounds

Numerous *Burkholderia* spp. produce antimicrobial compounds that are unknown or are still under investigation. Silva (2019) recently isolated a new metabolite FRV7 from *B. metallica* RV7S3 strain. This compound presented strong antifungal activity against *R. solani*, *A. alternata*, and *S. sclerotiorum*. Lassie et al. (2018) demonstrated that some fractions of RV7S3 extracts also have antiyeast effects. *B. pyrrocinia* RV1R2 strain compounds exhibited antifungal activity against *S. sclerotiorum* and *R. solani* and are in isolation process (Silva, 2018). The author found that the purer is the compound, the fewer concentrations are needed for fungal inhibition, constituting a promising bioproduct against phytopathogens.

3.4 Bioactive compounds produced by *Bacillus* spp

The *Bacillus* genus, belonging to the Bacillaceae family, comprises Gram-positive bacilli, endospore-forming with aerobic or facultative anaerobic metabolism. To date, according to List of Prokaryotic names with Standing in Nomenclature (www.bacterio.net), the genus has 440 described species that can be found in the most diversified environments, possibly being the oldest and most diverse bacterial genus (Gopal et al., 2015).

From a biotechnological point of view, the most important aspects of the genus are the production of several secondary metabolites and antagonistic substances. Secondary metabolites produced by *Bacillus* include antibiotics, antifungals, and siderophores (Fira et al., 2018). The production of antimicrobials with activity against fungi and pathogenic bacteria, easily forming spores, longer survival than vegetative cells, and wide distribution in the soil are characteristics that reinforce it as excellent agents of biological control (Estibaliz and Ortiz, 2011).

Among *Bacillus* species we can mention efficient bioactive producers such as *Bacillus subtilis, Bacillus amyloliquefaciens, Bacillus cereus, Bacillus licheniformis, Bacillus megaterium, Bacillus mycoides, Bacillus sphaericus, Bacillus mojavensis, Bacillus pasteurii*, and *Bacillus pumilus* (Bacon et al., 2015). The lipopeptides group stands out in the secondary metabolites produced by *Bacillus* for the biological control of agricultural pathogens (Fira et al., 2018). In this section, the lipopeptides (surfactins, iturins, and fengycins) produced in the different species of *Bacillus* and their effect against phytopathogens will be described.

3.4.1 Lipopeptides

It is possible to classify the numerous bioactive compounds produced by *Bacillus* in molecules with ribosomal or nonribosomal origin. The group of ribosomal origin comprises bacteriocins, characterized by being small, heat-stable, and amphiphilic molecules (Fira et al., 2018). They also present antimicrobial activity (Chauhan et al., 2017; Salazar et al., 2017; Sharma et al., 2018; Zidour et al., 2019), act as probiotics (Abdhul et al., 2015; Khochamit et al., 2015; Shobharani et al., 2015) and as plant growth promoters (Lee et al., 2009; Gautam et al., 2016). Bacteriocin-like inhibitory, peptides that do not correlate to bacteriocins group, can be included in the group of ribosomal compounds as well (Minnaard and Alippi, 2016). These peptides are synthesized by nonribosomal peptide synthetases, large enzyme complexes of several domains with a modular structure (Schneider and Marahiel, 1988).

Lipopeptides, the main class with effective action against plant pathogens, are classified into three families, depending on their amino acid sequence: surfactins, iturins, and fengycins (Fira et al., 2018). These compounds have a cyclic structure with 7 to 10 amino acid residues connected with derivatives of fatty acids. They can present many isometric forms, due to different lengths of the fatty acids hydrocarbon chain or to variations in the amino acid compositions (Ongena and Jacques, 2008).

The surfactins class includes cyclic lipoheptapeptides that have seven residues of D- and L-amino acids in their structure, around a beta-hydroxyl fatty acid. *Bacillus* species that produce surfactins have greater mobility and easily form biofilms (Kinsinger et al., 2003). These compounds interact with phospholipids, interfering with the membrane structure. According to an established model, there is a hydrophobic integration between the fatty substances present in the molecules and the hydrocarbon chains of the membrane phospholipids, causing a destabilization of its structure. After this first interaction, there is a conformational change in the peptide present in the surfactin, facilitating its integration with the membrane (Maget-Dana et al., 1995).

However, another study describes a different mechanism of action, involving the dehydration of polar heads of the phospholipid membrane, caused by the integration of compound dimers in the membrane. This mechanism is altered by a structural change in the molecule and results in the rupture of the lipidic bilayer and the overflow of cytoplasmic content (Carrillo et al., 2003).

Various studies correlated surfactin as biocontrol agent against plants, like the one carried out by Sarwar et al. (2018). The authors showed the effectiveness of surfactin A, produced by *B. subtilis* and *B. amyloliquefaciens*, against bakanae disease in rice plants, caused by species of the *Fusarium* complex. According to that study, surfactin was effective in *Fusarium solani*, *F. oxysporum*, and *F. moniliforme* biocontrol, reducing the disease up to 80%.

Another study developed an infection model demonstrating that the pathogen *P. syringae* pv. *tomato* is capable of infecting *Arabidopsis* roots (Bais et al., 2004). For this infection model, a wild *B. subtilis* and a mutated strain to the gene that controls surfactin biosynthesis were applied against *P. syringae* infection. The mutant strain proved to be ineffective in controlling the pathogen, and it did not form robust biofilms on the roots or inert surfaces. The wild strain, after root colonization, produced enough surfactin and generated a stable biofilm. The biofilm and the surfactin acted simultaneously to protect the plant from the phytopathogen.

The iturins are molecules formed by two main parts: a peptide ring, which has 7 amino acid residues, attached to a hydrophobic tail of 11 to 12 carbons. Thus the structure has an amphiphilic character, and hence the cell membrane is the probable site of action (Aranda et al., 2005). In addition, these lipopeptides exhibit antifungal (Lee et al., 2019) and hemolytic activities (Luo et al., 2015). The mechanism of action relates to the formation of ion-conducting pores, resulting from aggregates in the lipid membrane (Maget-Dana et al., 1995; Sheppard et al., 1991). Still, iturins are able to change the membrane structure, modifying the curvature of phospholipid vesicles (Grau et al., 2000).

About the activity of these compounds against plant pathogens, a study by Lee et al. (2017a,b) demonstrated the effectiveness of iturin A, produced by *B. amyloliquefaciens*, inhibiting mycelium formation of *Fusarium asiaticum*, *Fusarium graminearum*, *Fusarium proliferatum* and *Fusarium verticillioides*, both in vitro and *in planta* (maize). Chen et al. (2014) isolated a strain of *B. amyloliquefaciens* that is able to inhibit the growth of *F. oxysporum* and *R. solanacearum* by the production of iturin. The experiments reported the control of eggplant wilt with an efficiency of 25.3%. Surprisingly, when the bacteria were inoculated with organic fertilizer, the effectiveness of control against eggplant wilt achieved 70.7%.

Fengycins comprise another promising group of lipopeptides. They are characterized by 10 amino acid residues linked to a chain of beta-hydroxyl fatty acids of 16 and 19 carbon atoms (Yaseen et al., 2016). These compounds have antifungal, antibacterial, and antiviral activity (Fracchia et al., 2012; Huang et al., 2006).

Their mechanism of action is based on two transition states controlled by concentration. The first state is a monomeric lipopeptide, which does not anchor deeply in the membrane. The second state interacts firmly with the cell membrane, altering its structure and permeability. Thus the mechanism relates to the amphiphilic character of the lipopeptide and its affinity for the lipid bilayer (Deleu et al., 2008).

The proven antimicrobial activity of fengycins supports their application in the biocontrol of phytopathogens. An example is the study developed by Hanif et al. (2019), which identified a fengycin-producing strain of *B. amyloliquefaciens* with activity against *F. graminearum*, an important cereal pathogen. The compound deformed the fungal hyphae structures, suppressing their growth and the plant colonization. Furthermore, fengycin did inhibit the mycotoxins produced by the pathogenic fungus. Fan et al. (2017) demonstrated the effectiveness of fengycin, produced by a *B. subtilis* strain, in controlling apple rot, caused by *Botryosphaeria dothidea*. The researchers evaluated *B. subtilis*, wild and mutant strains for depleted fengycin production, demonstrating that the biocontrol was only promoted by the wild strain, which produced the compound.

3.5 Bioactive compounds produced by actinobacteria

Actinobacteria are filamentous Gram-positive bacteria, and most of them are saprophytic, decomposing especially lignocellulose, starch, and chitin (Solanki et al., 2016). These spore-forming bacteria (Palaniyandi et al., 2013) grow mainly in aerobic environments (Olanrewaju and Babalola, 2019) and can be found in soil, freshwater, and marine ecosystems, in organic matter, in the rhizosphere, they can be endophytic, live on animals, among other habitats (Dinesh et al., 2017; Golinska et al., 2015; Yadav et al., 2018). Actinobacteria constitute 10%–50% of the soil microbial community (Olanrewaju and Babalola, 2019) and their interaction with plants in the rhizosphere is essential for the microenvironment (Gomes et al., 2018).

Since the discovery of streptomycin, this group has been extensively studied as they can synthesize several bioactive compounds with marketable value, both in medical and agricultural sectors (Solanki et al., 2016). Currently, *Streptomyces* genus is responsible for 75% of commercial antibiotics, being the most explored source of bioactive compounds (Olanrewaju and Babalola, 2019).

Actinobacteria present biotechnological potential as antimicrobial producers, plant growth promoters, herbicides, anthelmintics, insecticides, and stress resistance inducer, being used in biocontrol, competing with conventional agrochemicals (Gomes et al., 2018; Newitt et al., 2019; Olanrewaju and Babalola, 2019).

Avermectins are pesticide compounds produced by *Streptomyces avermitilis* that are frequently used in agriculture and animal health. They cause high mortality of nematodes in roots (Huang et al., 2014) and in cereal cysts (Zhang et al., 2016). A great number of compounds with nematicide and arthropocide action can also be isolated from actinobacteria, such as abamectin, evermectin, emamectin, milbemycin, polynactins, and spinosyns (Yadav et al., 2018).

As bioinoculants, actinobacteria have reduced effects on native microbial balance and enrich beneficial microbial populations, protecting plants from diseases, and stimulating their growth (Palaniyandi et al., 2013). In this section, direct suppression of diseases by antimicrobials, volatiles compounds, *quorum sensing* inhibitors, and hydrolytic enzymes are approached.

3.5.1 Hydrolytic enzymes

Microbial hydrolytic enzymes production, such as chitinase, cellulase, glucanase, protease, amylase, xylanase, pectinase, and phospholipase, is a common antagonistic mechanism (Palaniyandi et al., 2013; Solanki et al., 2016). These enzymes act in cell wall degradation, promoting great antifungal effect (Solanki et al., 2016). Chitinase is also used for chitin degradation, which allows insect control (Gomes et al., 2018).

Combinations of antimicrobial and enzyme-producing strains are often sought for commercial purposes, enhancing disease control, since they can improve the susceptibility of pathogens to antibiotics after the enzymatic action (Palaniyandi et al., 2013). *Streptomyces* bacteria are extensively studied regarding their potential for producing extracellular enzymes, with the purpose of applying them in the development of inoculants that can be applied in biocontrol (Palaniyandi et al., 2013). In Table 3.1, *Streptomyces* spp. and other actinobacteria that can be applied on the control of plants diseases, mediated by hydrolytic enzymes production, are summarized.

TABLE 3.1 Plant disease suppression by hydrolytic enzymes produced by actinobacteria against phytopathogens.

Bioactive compound	Producing microorganisms	Inhibits	Reference
Chitinase	*Streptomyces cavourensis* SY224 *Streptomyces* sp. 80 *Streptomyces halstedii* AJ-7 *Streptomyces violaceusniger* YCED-9 *Micromonospora carbonacea* *Actinoplanes missouriensis* *Streptomyces lydicus* WYEC108	*Colletotrichum gloeosporioides* *Sclerotinia sclerotiorum* *Fusarium oxysporum* *Phytophthora infestans* *Sclerotinia minor* *Plectosporium tabacium* *Pythium* sp. *Pythium aphanidermatum*	Lee et al. (2012), Fróes et al. (2012), Joo (2005), Trejo-Estrada et al. (1998), Solanki et al. (2016), Palaniyandi et al. (2013)
Glucanase	*S. cavourensis* SY224 *Streptomyces* sp. 80	*C. gloeosporioides* *S. sclerotiorum*	Lee et al. (2012), Fróes et al. (2012)
β-glucanase	*Amorphophallus campanulatus* *S. violaceusniger* YCED-9	*P. aphanidermatum* *P. infestans*	El-Tarabily et al. (2009), Trejo-Estrada et al. (1998)
Cellulase	*M. carbonacea*	*Phytophthora cinnamomi*	Palaniyandi et al. (2013)
Protease	*Streptomyces* sp. 80	*S. sclerotiorum*	Fróes et al. (2012)
Endochitinase	*S. violaceusniger* XL-2	*Phanerochaete chrysosporium* *Postia placenta* *Coriolus versicolor* *Gloeophyllum trabeum*	Palaniyandi et al. (2013)

3.5.2 Antimicrobial compounds

Actinobacteria produce a wide spectrum of antimicrobials, with over 10,000 described compounds, which about 7600 are produced by *Streptomyces* genus (Golinska et al., 2015). Several of these molecules are effective in phytopathogenic fungal and bacterial control (Palaniyandi et al., 2013), as presented in Table 3.2. Most of these antimicrobials are inhibitors of protein synthesis, compromising the integrity of cell membrane and the cell wall biosynthesis (Gwinn, 2018).

Several new compounds with antimicrobial potential have been identified by the bioprospection of actinobacteria in different sites, such as marine environments, endophytic, and unhealthy plants, but they have been little explored (Betancur et al., 2017; Chaurasia et al., 2018; Dinesh et al., 2017; Liu et al., 2019). Endophytic actinobacteria can present multiple mechanisms for pathogen suppression. One of them is *quorum sensing* inhibition (discussed later in this section) (Dinesh et al., 2017), a mechanism with lower toxicity to higher organisms (Aamir et al., 2020).

TABLE 3.2 Plant disease suppression by antimicrobial compounds produced by actinobacteria against phytopathogens.

Bioactive compound	Producing microorganisms	Inhibits	Reference
Actinolactomycin	*Streptomyces pratensis*	*Sclerotinia sclerotiorum*	Liu et al. (2019)
Actinomycin D	*Streptomyces panaciradicis* 1MR-8T	*Ralstonia solanacearum*	Ling et al. (2020)
Actinomycin X2	*Streptomyces galbus* R-5	*Rhizoctonia solani*	Hasegawa et al. (2006)
Aloesaponarin 2	*Streptomyces termitum* ATC-2	*Xanthomonas oryzae*	Solanki et al. (2016)
Azalomycin F4a and F5a	*Streptomyces sporoclivatus*	*S. sclerotiorum* *Fusarium moliniforme* *Fusarium oxysporum* *Botrytis* sp.	Liu et al. (2019)
Bafilomycin B1	*Streptomyces cavourensis*	*S. sclerotiorum* *R. solani* *Aspergillus fumigatus* *Botrytis cinerea* *Penicillium roqueforti*	Liu et al. (2019)
Blasticidin S	*Streptomyces griseochromogenes* 2A-327	*Alternaria kikuchiana* *Botrytis* spp *Corticium sasaki* *Gloeosporium kaki* *Glomerella* spp. *Macrosporium bataticola* *Ophiobolus miyabeanus* *Penicillium chrysogenum* *Piricularia oryzae*	Kono et al. (1968)
Dapiramicin	*Micromonospora* sp. SF-1917	*R. solani*	Shomura et al. (1983)
Dinactin	*S. pratensis*	*S. sclerotiorum*	Liu et al. (2019)
Faeriefungin	*Streptomyces griseus*	*Fusarium* spp.	Palaniyandi et al. (2013)

(Continued)

TABLE 3.2 (Continued)

Bioactive compound	Producing microorganisms	Inhibits	Reference
Fungichromin	*Streptomyces padanus* *Streptomyces galbus* R-5	*R. solani*	Solanki et al. (2016), Hasegawa et al. (2006)
Geldanamicin	*Streptomyces melanosporofaciens* *Streptomyces violaceusniger* YCED-9 *Streptomyces hygroscopicus var. geldanus Streptomyces griseus*	*Streptomyces scabies* *R. solani*	Agbessi et al. (2003), Trejo-Estrada et al. (1998), Rothrock and Gottlieb (1984)
Globopeptin	*Streptomyces* sp. MA-23	*P. oryzae* *B. cinerea* *A. kikuchiana*	Tanara et al. (1987)
Gopalamycin	*Streptomyces hygroscopicus*	*Erysiphe cichoracearum* *Puccinia recondita* *Plasmopara viticola* *Erysiphe. graminis tritici* *Fusarium culmorum* *B. cinerea* *P. oryzae* *Helmenthosporium teres*	Nair et al. (1994)
Guanidylfungin A	*Streptomyces violaceusniger* YCED-9	*R. solani*	Trejo-Estrada et al. (1998)
Malayamycin	*Streptomyces malaysiensis*	*Stagonospora nodorum*	Li et al. (2008)
Maremycin G	*Streptomyces capitiformicae*	*S. sclerotiorum*	Liu et al. (2019)
Neopeptin A and B	*Streptomyces* sp. KNF2047	*Sphaerotheca fusca*	Kim et al. (2007)
Nigericin	*S. violaceusniger* YCED-9	*R. solani*	Trejo-Estrada et al. (1998)
Oligomycin A and C	*Streptomyces diastaticus*	*Aspergillus niger* *Alternaria alternata* *B. cinerea* *Phytophthora capsici*	Yang et al. (2010)
Phenylacetic Acid	*Streptomyces humidus*	*P. capsici* *Pseudomonas syringae*	Hwang et al. (2001)
Polyoxin B and D	*Streptomyces cacaoi* var. *asoensis*	*R. solani*	Isono et al. (1965)
Prodiginin	*Streptomyces lividans*	*Verticillium* spp.	Meschke et al. (2012)
Tetracontano	*Streptomyces capoamus*	*R. solanacearum*	Kawuri and Darmayasa (2019)
Tetranactin	*S. pratensis*	*S. sclerotiorum* *B. cinerea*	Liu et al. (2019)
Tubercidin	*Streptomyces violaceoniger*	*Phytophthora infestans* *P. capsici*	Hwang and Kim (1995)
Undecylprodigiosin	*Streptomyces lividans*	*Verticillium dahlia*	Meschke et al. (2012)
Validamycin	*Streptomyces hygroscopicus* var. *limoneus n.* T-7545	*Pellicularia sasaki* *R. solani*	Palaniyandi et al. (2013)

3.5.3 Volatile antimicrobials

Actinobacteria, mainly from *Streptomyces* genus, can produce volatile substances capable of promoting morphological changes in phytopathogenic fungi, such as several species of *Aspergillus, F. oxysporum, Penicillium viridicatum, Trichoderma viride,* and *Zygorhynchus vuillemin* (Palaniyandi et al., 2013).

Boukaew et al. (2013) analyzed the inhibitory activity of volatile compounds produced by *Streptomyces philanthi* RM-1−138 over time, and reported its effects against *R. solani, Pyricularia grisea, Bipolaris oryzae,* and *Gibberella fujikuroi*. Growth inhibition was more intense at 14 days, because at 7 days the *S. philanthi* RM-1−138 strain produced only 17 compounds, while in 14 days it produced 36 compounds. These volatile compounds produced by *Streptomyces* have great potential as biofumigants in agriculture, as an alternative to chemical products, although further studies are necessary (Newitt et al., 2019).

3.5.4 *Quorum sensing* inhibitory compounds

In several strains of *Streptomyces*, the production of compounds capable of inhibiting pathogen *quorum sensing* systems was observed. Even below the minimum inhibitory concentration, these compounds inhibit virulence factors of phytopathogens. They can compromise the pathogen ability to cause disease by generating enzymes that decrease the production of signaling molecules, directly degrade these signaling molecules or inhibit the *quorum sensing* activation receptors (Betancur et al., 2017).

Kang et al. (2016) observed that piericidin A and glucopiericidin A, compounds produced by *Streptomyces xanthocidicus* KPP01532 strain, were efficient in reducing the expression of *quorum sensing* regulated virulence genes in *Erwinia carotovora* subsp. *atroseptica*, a phytopathogen that causes rot in potato stems and tubers.

In vitro studies with endophytic *Streptomyces* revealed the degradation of *N*-acyl-L-homoserine signaling molecules by quenching enzymes produced by *Streptomyces* sp. LPC029 strain, resulting in the loss of pathogenicity in tests against potato soft rot caused by *P. carotovorum* (Chankhamhaengdecha et al., 2013).

3.6 Bioactive compounds produced by Fungi

Plant growth-promoting fungi (PGPF) are a group of nonpathogenic saprophagous fungi that live in the rhizosphere and are associated with plant roots. PGPF improve plant growth and health, biocontrolling several phytopathogens by different mechanisms, such as direct competition for nutrients, mycoparasitism, predation, and production of compounds with antimicrobial effect or inducing plant resistance. This group comprises highly heterogeneous and phylogenetically distant fungi. Some of the most frequently reported genera are *Aspergillus, Trichoderma, Fusarium, Penicillium, Rhizoctonia, Piriformospora, Phoma,* and also *Pythium* and *Phytophthora*, oomycetes previously included in the kingdom Fungi (Hossain et al., 2017).

Secondary metabolites produced by PGPF are efficient against different phytopathogens, being frequently related to the biocontrol potential of each strain. *Trichoderma* and *Fusarium* genus has shown great applicability in agricultural management (Al-Ani, 2019a,b).

3.6.1 *Trichoderma* spp

Trichoderma is a fungal genus found in diverse ecosystems. The availability of nutrients and prey microorganisms near roots are probably the main attractions for the establishment of positive interactions of this fungus with plants in the rhizosphere. The metabolites produced by different *Trichoderma* strains have antifungal, antibacterial, antiviral, insecticidal, nematicidal, and herbicidal properties (Al-Ani, 2019b; Vinale et al., 2014).

The pyrone 6-pentyl-2H-pyran-2-one is a metabolite commonly purified from culture filtrates of different *Trichoderma* species (Jeleń et al., 2014). This compound has in vivo and in vitro activities against phytopathogenic fungi such as *R. solani*, *F. oxysporum* (Scarselletti and Faull, 1994), *F. moniliforme* (El-Hasan et al., 2008), and *F. graminearum* (El-Hasan et al., 2018). Another pyrone, 6-*n*-pentyl-α-pyrone secreted by *Trichoderma harzianum* and *Trichoderma longibrachiatum* showed high inhibitory effect against *Eremothecium coryli* (Tarus et al., 2004). The same compound, secreted by *Trichoderma gamsii* KUC1747, inhibited spore germination of *Ophiostoma piceae, Ophiostoma piliferum, Ophiostoma ips, and Gongronella koreana* in vitro (Hong et al., 2014).

Viridiofungin A, isolated from the T23 strain of *T. harzianum* affected the germination of *Verticillium dahliae* conidia, *P. infestans* sporangia, and *S. sclerotiorum* sclerotia (El-Hasan et al., 2009). Other *T. harzianum* strain has produced antimicrobial metabolites. Harzianopyridone, proved to be a potent growth inhibitor of *R. solani*, *S. rolfsii*, *G. graminis* var. *tritici*, *P. ultimum*, and *F. oxysporum* (Ahluwalia et al., 2015; Vinale et al., 2006). Harzianic acid inhibited hyphae growth of *Pythium irregulare*, *S. sclerotiorum*, and *R. solani* in vitro (Vinale et al., 2014, 2009). T22azaphilone also inhibited in vitro cultures of *R. solani*, *P. ultimum*, and *G. graminis* var. *tritici* (Vinale et al., 2006).

Peptaibols produced by two *T. harzianum* strains demonstrated insecticidal activity in direct treatments of insect cuticles, along with some enzymes, or by their addition to insect larval foods (Shakeri and Foster, 2007). Antiviral compounds have also been described. Trichodermin and trichoderminol, extracted from *Trichoderma albolutescens*, were effective in inactivating phytoviruses, like tobacco mosaic virus, pepper spot virus, cucumber spot virus, and light pepper spot virus (Lee et al, 2017a,b).

3.6.2 Nonpathogenic *Fusarium* spp

Fusarium is a genus of endophytic and saprophytic fungi that live in soil. They may cause different diseases in plants, such as root rot, vascular wilt, yellowing, and leaf necrosis. However, nonpathogenic strains have already shown to be effective in activating hormone production, controlling phytopathogens through competition, mycoparasitism, and induction of host resistance (Al-Ani, 2019a).

Nonpathogenic *Fusarium* secrete a variety of bioactive compounds. Many of these compounds are highly effective in controlling pathogenic *Fusarium*, such as α-humulene, secreted by the MSA35 strain, which is capable of inhibiting mycelial growth and virulence genes of *F. oxysporum* (Minerdi et al., 2009).

Nonpathogenic *F. oxysporum* metabolites proved to be efficient against filamentous fungi. Compounds produced by CanR-46 strain of *F. oxysporum*, including limonene, octanoic acid, 3,4-2H-dihydropyran, and 5-hexenoic acid, were able to suppress *V. dahliae* in cotton (Zhang et al., 2015). Veloso et al. (2016) also reported *V. dahliae* control by nonpathogenic *F. oxysporum* through the secretion of caffeic acid, ferulic acid, and chlorogenic acid. A filtrate of the nonpathogenic *F. oxysporum* strain F221-BO, tested in vitro against 10 phytopathogens, slightly inhibited mycelial growth of all phytopathogens, but completely inhibited the spore germination of *Fusarium semitectum* (F113) and *F. oxysporum* f. sp. *lactucae* (F221-R, F422-G strains). Under hydroponic conditions, this compound reduced the severity of *Fusarium* root rot disease in lettuce by 90% (Thongkamngam and Jaenaksorn, 2016). Cyclosporine produced by *F. oxysporum* S6 strain inhibited the formation of sclerotia from *S. sclerotiorum* (Rodríguez et al., 2006).

Fusarium isolates can also present nematicidal activity. Isolated compounds from nonpathogenic *F. oxysporum* caused high mortality and immobility in *Meloidogyne exigua* second stage juveniles (Costa et al., 2015) and filtered from *F. solani* cultures, also presented activity against same stage on *M. javanica* (Qureshi et al., 2012).

Endophytic *Fusarium* strains have been reported as well efficient mycoherbicides, inhibiting the growth of *Orobanche ramosa* and *Orobanche aegyptiaca* holoparasites (Cohen et al., 2002; Zonno and Vurro, 2002).

3.7 Conclusion

Microbial metabolites that can protect plants against phytopathogens are one of the main focus of microbiology research. The variety of compounds produced by microorganisms and their multiple mechanisms of action seem to be an endless study subject, in which new understandings and applications are being constantly brought to light. These bioactive natural compounds can replace agrochemicals in the near future, promoting efficient biocontrol, reducing damages to the environment, and enhancing food safety. They can also help tackling the microbial resistance problem and all these characteristics reinforce the need for these compounds in modern times.

References

Aamir, M., Rai, K.K., Zehra, A., Dubey, M.K., Samal, S., Yadav, M., et al., 2020. Endophytic actinomycetes in bioactive compounds production and plant defense system. In: Kumar, A., Singh, V.K. (Eds.), Microbial Endophytes: Prospects for Sustainable Agriculture. Elsevier, pp. 189–229.

Abdhul, K., Ganesh, M., Shanmughapriya, S., Vanithamani, S., Kanagavel, M., Anbarasu, K., et al., 2015. Bacteriocinogenic potential of a probiotic strain *Bacillus coagulans* [BDU3] from Ngari. Int. J. Biol. Macromol. 79 (8), 800–806.

Agbessi, S., Beauséjour, J., Déry, C., Beaulieu, C., 2003. Antagonistic properties of two recombinant strains of *Streptomyces melanosporofaciens* obtained by intraspecific protoplast fusion. Appl. Microbiol. Biotechnol. 62 (2–3), 233–238.

Ahluwalia, V., Kumar, J., Rana, V.S., Sati, O.P., Walia, S., 2015. Comparative evaluation of two *Trichoderma harzianum* strains for major secondary metabolite production and antifungal activity. Nat. Prod. Res. 29 (10), 914–920.

Al-Ani, L.K.T., 2019a. Secondary metabolites of non-pathogenic *Fusarium*: scope in agriculture. In: Singh, H.B., Keswani, C., Reddy, M.S., Sansinenea, E., García-Estrada, C. (Eds.), Secondary Metabolites of Plant Growth Promoting Rhizomicroorganisms. Springer Nature Singapore, Singapore, pp. 59–76.

Al-Ani, L.K.T., 2019b. Bioactive secondary metabolites of *Trichoderma* spp. for efficient management of phytopathogens. In: Singh, H.B., Keswani, C., Reddy, M.S., Sansinenea, E., García-Estrada, C. (Eds.), Secondary

Metabolites of Plant Growth Promoting Rhizomicroorganisms. Springer Nature Singapore, Singapore, pp. 125–143.

Ali, S.S., Vidhale, N.N., 2013. Bacterial siderophore and their application: a review. Int. J. Curr. Microbiol. App. Sci. 2 (12), 303–312.

Ambrosi, C., Leoni, L., Putignani, L., Orsi, N., Visca, P., 2000. Pseudobactin biogenesis in the plant growth-promoting rhizobacterium *Pseudomonas* strain B10: identification and functional analysis of the L-ornithine N5-oxygenase (psbA) gene. J. Bacteriol. 182 (21), 6233–6238.

Aranda, F.J., Teruel, J.A., Ortiz, A., 2005. Further aspects on the hemolytic activity of the antibiotic lipopeptide iturin A. Biochim. Biophy. Acta. 1713 (1), 51–56.

Audenaert, K., Pattery, T., Cornelis, P., Höfte, M., 2002. Induction of systemic resistance to *Botrytis cinerea* in tomato by *Pseudomonas aeruginosa* 7NSK2: role of salicylic acid, pyochelin, and pyocyanin. Mol. Plant. Microbe. Interact. 15 (11), 1147–1156.

Bacon, C.W., Palencia, E.R., Hinton, D.M., 2015. Antibiotic and biotic plant stress-tolerant and beneficial secondary metabolites produced by endophytic *Bacillus* species. In: Arora, N. (Ed.), Plant Microbes Symbiosis: Applied Facets. Springer, New Delhi, pp. 163–177.

Bais, H.P., Fall, R., Vivanco, J.M., 2004. Biocontrol of *Bacillus subtilis* against infection of *Arabidopsis* roots by *Pseudomonas syringae* is facilitated by biofilm formation and surfactin production. Plant. Physiol. 134 (1), 307–319.

Bajpai, A., Singh, B., Joshi, S., Johri, B.N., 2018. Production and characterization of an antifungal compound from *Pseudomonas protegens* strain W45. Proc. Natl. Acad. Sci. 88 (3), 1081–1089.

Barelmanna, I., Taraza, K., Budzikiewicza, H., Meyer, J.M., 1996. Cepaciachelin, a new catecholate siderophore from *Burkholderia* (*Pseudomonas*) *cepacia*. Z. Naturforsch. C 51 (9), 627–630.

Betancur, L.A., Naranjo-Gaybor, S.J., Vinchira-Villarraga, D.M., Moreno-Sarmiento, N.C., Maldonado, L.A., Suarez-Moreno, Z.R., et al., 2017. Marine actinobacteria as a source of compounds for phytopathogen control: an integrative metabolic-profiling/bioactivity and taxonomical approach. PLoS One 12 (2), 1–25.

Blankenfeldt, W., Parsons, J.F., 2014. The structural biology of phenazine biosynthesis. Curr. Opin. Struc. Biol. 29, 26–33.

Blumer, C., Haas, D., 2000. Mechanism, regulation, and ecological role of bacterial cyanide biosynthesis. Arch. Microbiol. 173 (3), 170–177.

Bohlin, L., Göransson, U., Alsmark, C., Wedén, C., Backlund, A., 2010. Natural products in modern life science. Phytochem. Rev. 9 (2), 279–301.

Boukaew, S., Plubrukam, A., Prasertsan, P., 2013. Effect of volatile substances from *Streptomyces philanthi* RM-1-138 on growth of *Rhizoctonia solani* on rice leaf. BioControl 58, 471–482.

Carrillo, C., Teruel, J.A., Aranda, F.J., Ortiz, A., 2003. Molecular mechanism of membrane permeabilization by the peptide antibiotic surfactin. Biochim. Biophys. Acta. 1611 (1–2), 91–97.

Cartwright, D.K., Chilton, W.S., Benson, D.M., 1995. Pyrrolnitrin and phenazine production by *Pseudomonas cepacia*, strain 5.5B, a biocontrol agent of *Rhizoctonia solani*. Appl. Microbiol. Biotechnol. 43, 211–216.

Castrejón-Godínez, M.L., Ortiz-Hernández, M.L., Salazar, E., Encarnación, S., Mussali-Galante, P., Tovar-Sánchez, E., et al., 2019. Transcriptional analysis reveals the metabolic state of *Burkholderia zhejiangensis* CEIB S4-3 during methyl parathion degradation. PeerJ. 7, 1–30.

Chankhamhaengdecha, S., Hongvijit, S., Srichaisupakit, A., Charnchai, P., Panbangred, W., 2013. Endophytic actinomycetes: a novel source of potential acyl homoserine lactone degrading enzymes. Biomed. Res. Int. 2013, 1–8.

Chauhan, A.K., Maheshwari, D.K., Bajpai, V.K., 2017. Isolation and preliminary characterization of a bacteriocin-producer *Bacillus* strain inhibiting methicillin resistant *Staphylococcus aureus*. Acta. Biol. Hung. 68 (2), 208–219.

Chaurasia, A., Meena, B.R., Tripathi, A.N., Pandey, K.K., Rai, A.B., Singh, B., 2018. Actinomycetes: an unexplored microorganisms for plant growth promotion and biocontrol in vegetable crops. World J. Microbiol. Biotechnol. 34 (9), 1–16.

Chen, D., Liu, X., Li, C., Tian, W., Shen, Q., Shen, B., 2014. Isolation of *Bacillus amyloliquefaciens* S20 and its application in control of eggplant bacterial wilt. J. Environ. Manage. 137, 120–127.

Chin-A-Woeng, T.F., Bloemberg, G.V., Bij, A.J., Drift, K.M.G.M., Schripsema, J., Kroon, B., et al., 1998. Biocontrol by phenazine-1-carboxamide-producing *Pseudomonas chlororaphis* PCL1391 of tomato root rot caused by *Fusarium oxysporum* f. sp. *radicis-lycopersici*. Mol. Plant Microbe Interact. 11 (11), 1069–1077.

Cohen, B.A., Amsellem, Z., Lev-Yadun, S., Gressel, J., 2002. Infection of tubercles of the parasitic weed *Orobanche aegyptiaca* by mycoherbicidal *Fusarium* species. Ann. Bot. 90 (5), 567–578.

Corran, A., Knauf-Beiter, G., Zeun, R., 2011. Fungicides acting on signal transduction. In: Krämer, W., Schirmer, U., Jeschke, P., Witschel, M. (Eds.), Modern Crop Protection Compounds. Wiley-VCH, pp. 561–580.

Costa, L.S.A.S., Campos, V.P., Terra, W.C., Pfenning, L.H., 2015. Microbiota from *Meloidogyne exigua* egg masses and evidence for the effect of volatiles on infective juvenile survival. Nematology 17 (6), 715–724.

David, B.V., Chandrasehar, G., Selvam, P.N., 2018. *Pseudomonas fluorescens*: a plant growth-promoting rhizobacterium (PGPR) with potential role in biocontrol of pests of crops. In: Prasad, R., Gill, R.S., Tuteja, N. (Eds.), Crop Improvement Through Microbial Biotechnology. Elsevier, pp. 221–243.

Deleu, M., Paquot, M., Nylander, T., 2008. Effect of fengycin, a lipopeptide produced by *Bacillus subtilis*, on model biomembranes. Biophys. J. 94 (7), 2667–2679.

Deng, P., Foxfire, A., Xu, J., Baird, S.M., Jia, J., Delgado, K.H., et al., 2017. Siderophore product ornibactin is required for the bactericidal activity of *Burkholderia contaminans* MS14. Appl. Environ. Microbiol. 83 (8), 1–13.

Depoorter, E., Bull, M.J., Peeters, C., Coenye, T., Vandamme, P., Mahenthiralingam, E., 2016. *Burkholderia*: an update on taxonomy and biotechnological potential as antibiotic producers. Appl. Microbiol. Biotechnol. 100 (12), 5215–5229.

Dhardemare, A.M., Serratrice, G., Pierre, J.L., 2004. Synthesis and iron-binding properties of quinolobactin, a siderophore from a pyoverdine-deficient *Pseudomonas fluorescens*. Biometals 17 (16), 691–697.

Dharni, S., Alam, M., Kalani, K., Abdul-Khaliq, Samad, A., Srivastava, S.K., et al., 2012. Production, purification, and characterization of antifungal metabolite from *Pseudomonas aeruginosa* SD12, a new strain obtained from tannery waste polluted soil. J. Microbiol. Biotechnol. 22 (5), 674–683.

Dietrich, L.E., Teal, T.K., Price-Whelan, A., Newman, D.K., 2008. Redox-active antibiotics control gene expression and community behavior in divergent bacteria. Science 321 (5893), 1203–1206.

Dinesh, R., Srinivasan, V., Sheeja, T.E., Anandaraj, M., Srambikkal, H., 2017. Endophytic actinobacteria: diversity, secondary metabolism and mechanisms to unsilence biosynthetic gene clusters. Crit. Rev. Microbiol. 43 (5), 546–566.

Drehe, I., Simonetti, E., Ruiz, J.A., 2018. Contribution of the siderophores pyoverdine and enantio-pyochelin to fitness in soil of *Pseudomonas protegens* Pf-5. Curr. Microbiol. 75 (12), 1560–1565.

Duijff, B.J., Recorbet, G., Bakker, P.A.H.M., Loper, J.E., Lemanceau, P., 1999. Microbial antagonism at the root level is involved in the suppression of *Fusarium* wilt by the combination of nonpathogenic *Fusarium oxysporum* Fo47 and *Pseudomonas putida* WCS358. Phytopathology 89 (11), 1073–1079.

El-Banna, N., Winkelmann, G., 1998. Pyrrolnitrin from *Burkholderia cepacia*: antibiotic activity against fungi and novel activities against streptomycetes. J. Appl. Microbiol. 85 (1), 69–78.

El-Hasan, A., Schöne, J., Höglinger, B., Walker, F., Voegele, R.T., 2018. Assessment of the antifungal activity of selected biocontrol agents and their secondary metabolites against *Fusarium graminearum*. Eur. J. Plant. Pathol. 150 (1), 91–103.

El-Hasan, A., Walker, F., Buchenauer, H., 2008. *Trichoderma harzianum* and its metabolite 6-pentyl-alpha-pyrone suppress fusaric acid produced by *Fusarium moniliforme*. J. Phytopathol. 156 (2), 79–87.

El-Hasan, A., Walker, F., Schöne, J., Buchenauer, H., 2009. Detection of viridiofungin A and other antifungal metabolites excreted by *Trichoderma harzianum* active against different plant pathogens. Eur. J. Plant. Pathol. 124 (3), 457–470.

El-Tarabily, K.A., Nassar, A.H., Hardy, G.E.S.J., Sivasithamparam, K., 2009. Plant growth promotion and biological control of *Pythium aphanidermatum*, a pathogen of cucumber, by endophytic actinomycetes. J. Appl. Microbiol. 106, 13–26.

Estibaliz, S., Ortiz, A., 2011. Secondary metabolites of soil *Bacillus*. Biotecnol. Lett. 33 (1), 1523–1538.

Fan, H., Ru, J., Zhang, Y., Wang, Q., Li, Y., 2017. Gengycin produced by *Bacillus subtilis* 9407 plays a major role in the biocontrol of apple ring rot disease. Microbial. Res. 199, 89–97.

Fira, D., Dimkić, I., Berić, T., Lozo, J., Stanković, S., 2018. Biological control of plant pathogens by *Bacillus* species. J. Biotechnol. 285, 44–55.

Flury, P., Vesga, P., Péchy-Tarr, M., Aellen, N., Dennert, F., Hofer, N., et al., 2017. Antimicrobial and insecticidal: cyclic lipopeptides and hydrogen cyanide produced by plant-beneficial *Pseudomonas* strains CHA0, CMR12a, and PCL1391 contribute to insect killing. Front. Microbiol. 8 (100), 1–15.

Fracchia, L., Cavallo, M., Giovanna, M., Banat, I.M., 2012. Biosurfactants and bioemulsifiers biomedical and related applications, present status and future potent. In: Ghista, D.N. (Ed.), Biomedical Science, Engineering and Technology. InTech, Croatia, pp. 325–370.

Franke, J., Ishida, K., Hertweck, C., 2015. Plasticity of the malleobactin pathway and its impact on siderophore action in human pathogenic bacteria. Chem. Eur. J. 21 (22), 8010–8014.

Fróes, A., Macrae, A., Rosa, J., Franco, M., Souza, R., Soares, R., et al., 2012. Selection of a *Streptomyces* strain able to produce cell wall degrading enzymes and active against *Sclerotinia sclerotiorum*. J. Microbiol. 50 (5), 798–806.

Gasser, V., Malrieu, M., Forster, A., Mély, Y., Schalk, I.J., Godet, J., 2019. *In cellulo* FRET-FLIM and single molecule tracking reveal the supra-molecular organization of the pyoverdine bio-synthetic enzymes in *Pseudomonas aeruginosa*. Q. Rev. Biophys. 53, 1–11.

Gautam, K., Schwinghamer, T.D., Smith, D.L., 2016. The response of soybean to nod factors and a bacteriocin. Plant. Signal. Behav. 11 (10), 1–9.

Gilchrist, F.J., Sims, H., Alcock, A., Jones, A.M., Bright-Thomas, R.J., Smith, D., et al., 2013. Is hydrogen cyanide a marker of *Burkholderia cepacia* complex? J. Clin. Microbiol. 51 (11), 3849–3851.

Golinska, P., Wypij, M., Agarkar, G., Rathod, D., Dahm, H., Rai, M., 2015. Endophytic actinobacteria of medicinal plants: diversity and bioactivity. Antonie Van Leeuwenhoek 108 (2), 267–289

Gomes, E.B., Dias, L.R.L., Miranda, R.C.M., 2018. Actinomycetes bioactive compounds: biological control of fungi and phytopathogenic insect. Afr. J. Biotechnol. 17 (17), 552–559.

Gopal, N., Hill, C., Ross, P.R., Beresford, T.P., Fenelon, M.A., Cotter, P.D., 2015. The prevalence and control of *Bacillus* and related spore-forming bacteria in the dairy industry. Front. Microbiol. 6 (1418), 1418–1436.

Gordee, R.S., Westhead, J., 1972. Pyrrolnitrin. In: Kavanagh, F. (Ed.), Analytical Microbiology. Academic Press, pp. 329–338.

Gould, K., 2016. Antibiotics: from prehistory to the present day. J. Antimicrob. Chemother. 71 (3), 572–575.

Grau, A., Ortiz, A., Godos, A., Gómez-Fernández, J.C., 2000. A biophysical study of the interaction of the lipopeptide antibiotic iturin A with aqueous phospholipid bilayers. Arch. Biochem. Biophys. 377 (2), 315–323.

Guttenberger, N., Blankenfeldt, W., Breinbauer, R., 2017. Recent developments in the isolation, biological function, biosynthesis, and synthesis of phenazine natural products. Bioorgan. Med. Chem. 25 (22), 6149–6166.

Gwinn, K.D., 2018. Bioactive natural products in plant disease control. In: Rahman, A. (Ed.), Studies In Natural Products Chemistry. Elsevier, pp. 229–246.

Han, J.W., Kim, J.D., Lee, J.M., Ham, J.H., Lee, D., Kim, B.S., 2014. Structural elucidation and antimicrobial activity of new phencomycin derivatives isolated from *Burkholderia glumae* strain 411gr-6. J. Antibiot. 67 (10), 721–723.

Hanif, A., Zhang, F., Li, P., Li, C., Xu, Y., Zubair, M., et al., 2019. Fengycin produced by *Bacillus amyloliquefaciens* FZB42 inhibits *Fusarium graminearum* growth and mycotoxins biosynthesis. Toxins 11 (5), 295–306.

Hasegawa, S., Meguro, A., Shimizu, M., Nishimura, T., Kunoh, H., 2006. Endophytic actinomycetes and their interactions with host plants. Actinomycetologica 20 (2), 72–81.

Hassen, W., Neifar, M., Cherif, H., Najjari, A., Chouchane, H., Driouich, R.C., et al., 2018. *Pseudomonas rhizophila* S211, a new plant growth-promoting rhizobacterium with potential in pesticide-bioremediation. Front. Microbiol. 9 (34), 1–17.

Heeb, S., Fletcher, M.P., Chhabra, S.R., Diggle, S.P., Williams, P., Cámara, M., 2011. Quinolones: from antibiotics to autoinducers. FEMS Microbiol. Rev. 35 (2), 247–274.

Hernandez, M.E., Kappler, A., Newman, D.K., 2004. Phenazines and other redox-active antibiotics promote microbial mineral reduction. Appl. Environ. Microbiol. 70 (2), 921–928.

Herrera, K., 2017. Purifying an unknown *Burkholderia contaminans* MS14 Bactericidal. Undergraduate Research Scholars Thesis. Texas A&M University [viewed 11 April 2019]. Available from: <http://hdl.handle.net/1969.1/164551>.

Hong, J.H., Lee, J., Min, M., Ryu, S.M., Lee, D., Kim, G.H., et al., 2014. 6-Pentyl-α-pyrone as an anti-sapstain compound produced by *Trichoderma gamsii* KUC1747 inhibits the germination of ophiostomatoid fungi. Holzforschung 68 (7), 769–774.

Hossain, M.M., Sultana, F., Islam, S., 2017. Plant growth-promoting fungi (PGPF): phytostimulation and induced systemic resistance. In: Singh, D.P., Singh, H.B., Prabha, R. (Eds.), Plant-microbe Interactions in Agro-ecological Perspectives. 2. Springer Nature Singapore, Singapore, pp. 135–191.

Howell, C.R., Stipanovic, R.D., 1979. Control of *Rhizoctonia solani* on cotton seedlings with *Pseudomonas fluorescens* and with an antibiotic produced by the bacterium. Phytopathology 69 (5), 480–482.

Huang, Q.X., Lu, Z.X., Zhao, H.Z., Bie, X.M., Lu, F.X., Yang, S.J., 2006. Antiviral activity of antimicrobial lipopeptide from *Bacillus subtilis* fmbj against pseudorabies virus, porcine parvovirus, Newcastle disease virus and infectious bursal disease virus *in vitro*. Int. J. Pept. Res. Therap. 12 (13), 373–377.

Huang, W.K., Sun, J.H., Cui, J.K., Wang, G.F., Kong, L.A., Peng, H., et al., 2014. Efficacy evaluation of fungus *Syncephalastrum racemosum* and nematicide avermectin against the root-knot nematode *Meloidogyne incognita* on cucumber. PLoS One 9 (2), 1–6.
Hwang, B.K., Kim, B.S., 1995. *In-vivo* efficacy and *in-vitro* activity of tubercidin, an antibiotic nucleoside, for control of *Phytophthora capsici* blight in *Capsicum annuum*. Pestic. Sci. 44 (3), 255–260.
Hwang, B.K., Lim, S.W., Kim, B.S., Lee, J.Y., Moon, S.S., 2001. Isolation and *in vivo* and *in vitro* antifungal activity of phenylacetic acid and sodium phenylacetate from *Streptomyces humidus*. Appl. Environ. Microbiol. 67 (8), 3739–3745.
Isono, K., Nagatsu, J., Kawashima, Y., Suzuki, S., 1965. Studies on polyoxins, antifungal antibiotics - Part I. Isolation and characterization of Polyoxins A and B. Agric. Biol. Chem. 29 (9), 848–854.
Jeleń, H., Błaszczyk, L., Chełkowski, J., Rogowicz, K., Strakowska, J., 2014. Formation of 6-n-pentyl-2H-pyran-2-one (6-PAP) and other volatiles by different *Trichoderma* species. Mycol. Prog. 13 (3), 589–600.
Joo, G.J., 2005. Purification and characterization of an extracellular chitinase from the antifungal biocontrol agent *Streptomyces halstedii*. Biotechnol. Lett. 27 (19), 1483–1486.
Jung, B.K., Hong, S.J., Park, G.S., Kim, M.C., Shin, J.H., 2018. Isolation of *Burkholderia cepacia* JBK9 with plant growth-promoting activity while producing pyrrolnitrin antagonistic to plant fungal diseases. Appl. Biol. Chem. 61 (2), 173–180.
Kang, J.E., Han, J.W., Jeon, B.J., Kim, B.S., 2016. Efficacies of quorum sensing inhibitors, piericidin A and glucopiericidin A, produced by *Streptomyces xanthocidicus* KPP01532 for the control of potato soft rot caused by *Erwinia carotovora* subsp. *atroseptica*. Microbiol. Res. 184, 32–41.
Kawuri, R., Darmayasa, I.B.G., 2019. Bioactive compound of *Streptomyces capoamus* as biocontrol of bacterial wilt disease on banana plant. IOP Conf. Ser. Earth Environ. Sci. 347 (012054), 1–8.
Khare, E., Arora, N.K., 2011. Dual activity of pyocyanin from *Pseudomonas aeruginosa* - antibiotic against phytopathogen and signal molecule for biofilm development by rhizobia. Can. J. Microbiol. 57 (9), 708–713.
Khochamit, N., Siripornadulsil, S., Sukon, P., Sirornadulsil, W., 2015. Antibacterial activity and genotypic-phenotypic characteristics of bacteriocin-producing *Bacillus subtilis* KKU213: potential as a probiotic strain. Microbiol. Res. 170 (1), 36–50.
Kim, H., Mohanta, T.K., Park, Y., Park, S., Shanmugam, G., Park, J., et al., 2020. Complete genome sequence of the mountain-cultivated ginseng endophyte *Burkholderia stabilis* and its antimicrobial compounds against ginseng root rot disease. Biol. Control. 140, 1–7.
Kim, Y.S., Kim, H.M., Chang, C., Hwang, I.C., Oh, H., Ahn, J.S., et al., 2007. Biological evaluation of neopeptins isolated from a *Streptomyces* strain. Pest. Manag. Sci. 63 (12), 1208–1214.
Kinsinger, R., Shirk, M., Fall, R., 2003. Rapid surface motility in *Bacillus subtilis* is dependent on extracellular surfactin and potassium ion. J. Bacteriol. 185 (9), 5627–5631.
Kirinuki, T., Ichiba, T., 1986. Effects of altericidins on the leakage of cellular constituents from *Alternaria kikuchiana* and *Ustilago maydis*. J. Pest. Sci. 11 (3), 453–458.
Kono, Y., Takeuchi, S., Yonehara, H., 1968. Studies on blasticidin A. J. Antibiot. 21 (7), 433–438.
Lassie, F., Emiliano, J., Simionato, A., Navarro, M., Gionco, B., Silva, C., et al., 2018. Bioactive substances produced by *Burkholderia* sp. with antifungal action in *Candida* spp. Agron. Sci. Biotechnol. 4 (2), 70–78.
Laursen, J.B., Nielsen, J., 2004. Phenazine natural products: biosynthesis, synthetic analogues, and biological activity. Chem. Rev. 104 (3), 1663–1685.
Lee, C.H., Kempf, H.J., Lim, Y.H., Cho, Y.H., 2000. Biocontrol activity of *Pseudomonas cepacia* AF2001 and anthelmintic activity of its novel metabolite, cepacidine A. J. Microbiol. Biotechnol. 10 (4), 568–571.
Lee, D.H., Kim, J.J., Ryu, K.H., Kim, B.S., Ryu, S.M., 2017a. Antiviral composition and method for controlling plant viruses using the same. US Patent No. US20170265473A1.
Lee, J.H., Ma, K.C., Ko, S.J., Kang, B.R., Kim, I.S., Kim, Y.C., 2011. Nematicidal activity of a nonpathogenic biocontrol bacterium, *Pseudomonas chlororaphis* O6. Curr. Microbiol. 62, 746–751.
Lee, J.Y., Moon, S.S., Hwang, B.K., 2003. Isolation and *in vitro* and *in vivo* activity against *Phytophthora capsici* and *Colletotrichum orbiculare* of phenazine-1-carboxylic acid from *Pseudomonas aeruginosa* strain GC-B26. Pest. Manag. Sci. 59 (8), 872–882.
Lee, K.D., Gray, E.J., Mabood, F., Jung, W.J., Charles, T., Clark, S.R., et al., 2009. The class IId bacteriocin thuricin-17 increases plant growth. Planta 229 (4), 747–755.
Lee, S., Zhao, H., Pang, B., Qu, R., Lian, Z., Jiang, C., et al., 2019. Capability of iturin from *Bacillus subtilis* to inhibit *Candida albicans in vitro* and *in vivo*. Appl. Microbiol. Biotechnol. 103 (11), 4377–4392.

Lee, S.Y., Tindwa, H., Lee, Y.S., Naing, K.W., Hong, S.H., Nam, Y., et al., 2012. Biocontrol of anthracnose in pepper using chitinase, β-1,3 glucanase, and 2-furancarboxaldehyde produced by *Streptomyces cavourensis* SY224. J. Microbiol. Biotechnol. 22 (10), 1359–1366.

Lee, T., Park, D., Kim, K., Lim, S.M., Yu, N.H., Kim, S., et al., 2017b. Characterization of *Bacillus amyloliquefaciens* DA12 showing potent antifungal activity against mycotoxigenic *Fusarium* species. Plant. Pathol. J. 33 (5), 499–507.

Li, D., Oku, N., Hasada, A., Shimizu, M., Igarashi, Y., 2018. Two new 2-alkylquinolones, inhibitory to the fish skin ulcer pathogen *Tenacibaculum maritimum*, produced by a rhizobacterium of the genus *Burkholderia* sp. J. Org. Chem. 14 (1), 1446–1451.

Li, W., Csukai, M., Corran, A., Crowley, P., Solomon, P.S., Oliver, R.P., 2008. Malayamycin, a new streptomycete antifungal compound, specifically inhibits sporulation of *Stagonospora nodorum* (Berk) Castell and Germano, the cause of wheat glume blotch disease. Pest. Manag. Sci. 64 (12), 1294–1302.

Lim, Y., Suh, J.W., Kim, S., Hyun, B., Kim, C., Lee, C.H., 1994. Cepacidine A, a novel antifungal antibiotic produced by *Pseudomonas cepacia*. II. Physico-chemical properties and structure elucidation. J. Antibiot. 47 (12), 1406–1416.

Ling, L., Han, X., Li, X., Zhang, X., Wang, H., Zhang, L., et al., 2020. A *Streptomyces* sp. NEAU-HV9: isolation, identification, and potential as a biocontrol agent against *Ralstonia solanacearum* of tomato plants. Microorganisms 8 (3), 1–15.

Liu, C., Zhuang, X., Yu, Z., Wang, Z., Wang, Y., Guo, X., et al., 2019. Community structures and antifungal activity of root-associated endophytic actinobacteria of healthy and diseased soybean. Microorganisms 7 (8), 1–15.

Luo, C., Lu, X., Zhou, H., Wang, X., Chen, Z., 2015. Nonribosomal peptide synthase gene clusters for lipopeptide biosynthesis in *Bacillus subtilis* 916 and their phenotypic functions. Appl. Environ. Microbiol. 81 (1), 422–431.

Ma, Z., Hua, G.K.H., Ongena, M., Höfte, M., 2016. Role of phenazines and cyclic lipopeptides produced by *Pseudomonas* sp. CMR12a in induced systemic resistance on rice and bean. Environ. Microbiol. Rep. 8 (5), 896–904.

Maddula, V.S., Pierson, E.A., Pierson, L.S., 2008. Altering the ratio of phenazines in *Pseudomonas chlororaphis* (*aureofaciens*) strain 30–84: effects on biofilm formation and pathogen inhibition. J. Bacteriol. 190 (8), 2759–2766.

Madigan, M.T., Martinko, J.M., Bender, K.S., Buckley, D.H., Stahl, D.A., 2017. Brock Biology of Microorganisms. Pearson Education.

Maget-Dana, R., Ptak, M., Peypoux, F., Michel, G., 1995. Pore-forming properties of iturin A, a lipopeptide antibiotic. Biochim. Biophys. Acta. 815 (3), 405–409.

Mandal, A., Sarkar, B., Mandal, S., Vithanage, M., Patra, A.K., Manna, M.C., 2020. Impact of agrochemicals on soil health. In: Prasad, M.N.V. (Ed.), Agrochemicals Detection, Treatment and Remediation. Elsevier, Hyderabad, pp. 161–187.

Mao, S., Lee, S.J., Hwangbo, H., Kim, Y.W., Park, K.H., Cha, G.S., et al., 2006. Isolation and characterization of antifungal substances from *Burkholderia* sp. culture broth. Curr. Microbiol. 53 (5), 358–364.

Martins, P.M.M., Merfa, M.V., Takita, M.A., Souza, A.A., 2018. Persistence in phytopathogenic bacteria: do we know enough? Front. Microbiol. 9 (1099), 1–14.

Matthijs, S., Tehrani, K.A., Laus, G., Jackson, R.W., Cooper, R.M., Cornelis, P., 2007. Thioquinolobactin, a *Pseudomonas* siderophore with antifungal and anti-*Pythium* activity. Environ. Microbiol. 9 (2), 425–434.

Meschke, H., Walter, S., Schrempf, H., 2012. Characterization and localization of prodiginines from *Streptomyces lividans* suppressing *Verticillium dahliae* in the absence or presence of *Arabidopsis thaliana*. Environ. Microbiol. 14 (4), 940–952.

Meyer, J.M., Hohnadel, D., Hallé, F., 1989. Cepabactin from *Pseudomonas cepacia*, a new type of siderophore. J. Gen. Microbiol. 135 (6), 1479–1487.

Meyer, J.M., Van, T.V., Stintzi, A., Berge, O., Winkelmann, G., 1995. Ornibactin production and transport properties in strains of *Burkholderia vietnamiensis* and *Burkholderia cepacia* (formerly *Pseudomonas cepacia*). Biometals 8, 309–317.

Meyers, E., Bisacchi, G.S., Dean, L., Liu, W.C., Minassian, B., Slusarchyk, D.S., et al., 1987. Xylocandin: a new complex of antifungal peptides I. Taxonomy, isolation and biological activity. J. Antibiot. 40 (11), 1515–1519.

Michavila, G., Adler, C., Gregorio, P.R., Lami, M.J., Santo, M.C.C., Zenoff, A.M., et al., 2017. *Pseudomonas protegens* CS1 from the lemon phyllosphere as a candidate for citrus canker biocontrol agent. Plant. Biol. 19 (4), 608–617.

Minerdi, D., Bossi, S., Gullino, M.L., Garibaldi, A., 2009. Volatile organic compounds: a potential direct long-distance mechanism for antagonistic action of *Fusarium oxysporum* strain MSA 35. Environ. Microbiol. 11 (4), 844–854.

Minnaard, J., Alippi, A.M., 2016. Partial characterization of bacteriocin-like compounds from two strains of *Bacillus cereus* with biological activity against *Paenibacillus larvae*, the causal agent of American foulbrood disease. Lett. Appl. Microbiol. 63 (6), 442–449.

Moon, S.S., Kang, P.M., Park, K.S., Kim, C.H., 1996. Plant growth promoting and fungicidal 4-quinolinones from *Pseudomonas cepacia*. Phytochemistry 42 (2), 365–368.

Morrison, C.K., Arseneault, T., Novinscak, A., Filion, M., 2016. Phenazine-1-carboxylic acid production by *Pseudomonas fluorescens* LBUM636 alters *Phytophthora infestans* growth and late blight development. Phytopathology 107 (3), 273–279.

Morya, R., Salvachúa, D., Thakur, I.S., 2020. *Burkholderia*: an untapped but promising bacterial genus for the conversion of aromatic compounds. Trends Biotechnol. 38 (9), 963–975.

Nagata, T., Oobo, T., Aozasa, O., 2013. Efficacy of a bacterial siderophore, pyoverdine, to supply iron to *Solanum lycopersicum* plants. J. Biosci. Bioeng. 115 (6), 686–690.

Nair, M.G., Thorogood, D.L., Chandra, A., Ammermann, E., Walker, N., Kiehs, K., 1994. Gopalamicin, an antifungal macrodiolide produced by soil actinomycetes. J. Agric. Food Chem. 42 (10), 2308–2310.

Newitt, J.T., Prudence, S.M.M., Hutchings, M.I., Worsley, S.F., 2019. Biocontrol of cereal crop diseases using streptomycetes. Pathogens 8 (78), 1–25.

O'Brien, S., Hodgson, D.J., Buckling, A., 2014. Social evolution of toxic metal bioremediation in *Pseudomonas aeruginosa*. Proc. Biol. Sci. 281 (1787), 1–6.

Okada, A., Banno, S., Ichiishi, K.M., Yamaguchi, I., Fujimura, M., 2005. Pyrrolnitrin interferes with osmotic signal transduction in *Neurospora crassa*. J. Pestic. Sci. 30 (4), 378–383.

Olanrewaju, O.S., Babalola, O.O., 2019. *Streptomyces*: implications and interactions in plant growth promotion. Appl. Microbiol. Biotechnol. 103, 1179–1188.

Oliveira, A.G., Simionato, A.S., Navarro, M.O.P., Andrade, G., 2015. Antimicrobianos naturais produzidos por microrganismos: da busca à identificação. In: Yamada-Ogatta, S.F., Nakazato, G., Furlaneto, M.C., Nogueira, M.A. (Eds.), Tópicos Especiais em Microbiologia. State University of Londrina, Londrina, pp. 180–199.

Oliveira, A.G., Spago, F.R., Simionato, A.S., Navarro, M.O.P., Silva, C.S., Barazetti, A.R., et al., 2016. Bioactive organocopper compound from *Pseudomonas aeruginosa* inhibits the growth of *Xanthomonas citri* subsp. *citri*. Front. Microbiol. 7 (113), 1–12.

Ong, K.S., Cheow, Y.L., Lee, S.M., 2017. The role of reactive oxygen species in the antimicrobial activity of pyochelin. J. Adv. Res. 8 (4), 393–398.

Ongena, M., Jacques, P., 2008. *Bacillus* lipopeptides: versatile weapons for plant disease biocontrol. Trends Microbiol. 16 (3), 115–125.

Palaniyandi, S.A., Yang, S.H., Zhang, L., Suh, J.W., 2013. Effects of actinobacteria on plant disease suppression and growth promotion. Appl. Microbiol. Biotechnol. 97 (22).

Pée, K.H., Ligon, J.M., 2000. Biosynthesis of pyrrolnitrin and other phenylpyrrole derivatives by bacteria. Nat. Prod. Rep. 17 (2), 157–164.

Peng, H., Zhang, P., Bilal, M., Wang, W., Hu, H., Zhang, X., 2018. Enhanced biosynthesis of phenazine-1-carboxamide by engineered *Pseudomonas chlororaphis* HT66. Microb. Cell. Fact. 17 (117), 1–12.

Pham, J.V., Yilma, M.A., Feliz, A., Majid, M.T., Maffetone, N., Walker, J.R., et al., 2019. A review of the microbial production of bioactive natural products and biologics. Front. Microbiol. 10 (1404), 1–27.

Pierson, E.A., Wang, D., Pierson, L.S., 2013. Roles and regulation of phenazines in the biological control strain *Pseudomonas chlorophis* 30–84. In: Chincholkar, S., Thomashow, L. (Eds.), Microbial Phenazines: Biosynthesis, Agriculture and Health. Springer, London, pp. 141–162.

Pierson, L.S., Pierson, E.A., 2010. Metabolism and function of phenazines in bacteria: impacts on the behavior of bacteria in the environment and biotechnological processes. Appl. Microbiol. Biotechnol. 86 (6), 1659–1670.

Preston, G.M., 2004. Plant perceptions of plant growth-promoting *Pseudomonas*. Philos. Trans. R. Soc. Lond. B Biol. Sci. 359 (1146), 907–918.

Price-Whelan, A., Dietrich, L.E.P., Newman, D.K., 2006. Rethinking 'secondary' metabolism: physiological roles for phenazine antibiotics. Nat. Chem. Biol. 2 (2), 71–78.

Puopolo, G., Masi, M., Raio, A., Andolfi, A., Zoina, A., Cimmino, A., et al., 2013. Insights on the susceptibility of plant pathogenic fungi to phenazine-1-carboxylic acid and its chemical derivatives. Nat. Prod. Res. 27 (11), 956–966.

Qureshi, S.A., Ruqqia, Sultana, V., Ara, J., Ehteshamul-Haque, S., 2012. Nematicidal potential of culture filtrates of soil fungi associated with rhizosphere and rhizoplane of cultivated and wild plants. Pak. J. Bot. 44 (3), 1041–1046.

Raaijmakers, J.M., Bruijn, I., Nybroe, O., Ongena, M., 2010. Natural functions of lipopeptides from *Bacillus* and *Pseudomonas*: more than surfactants and antibiotics. FEMS Microbiol. Rev. 34 (6), 1037–1062.

Ramadass, K., Megharaj, M., Venkateswarlu, K., Naidu, R., 2018. Bioavailability of weathered hydrocarbons in engine oil-contaminated soil: impact of bioaugmentation mediated by *Pseudomonas* spp. on bioremediation. Sci. Total. Environ. 636, 968–974.

Ran, L.X., Li, Z.N., Wu, G.J., Loon, L.C., Bakker, P.A.H.M., 2005. Induction of systemic resistance against bacterial wilt in *Eucalyptus urophylla* by fluorescent *Pseudomonas* spp. Eur. J. Plant Pathol 113, 59–70.

Rane, M.R., Sarode, P.D., Chaudhari, B.L., Chincholkar, S.B., 2008. Exploring antagonistic metabolites of established biocontrol agent of marine origin. Appl. Biochem. Biotechnol. 151, 665–675.

Ringel, M.T., Brüser, T., 2018. The biosynthesis of pyoverdines. Microb. Cell 5 (10), 424–437.

Rodríguez, M.A., Cabrera, G., Godeas, A., 2006. Cyclosporine A from a nonpathogenic *Fusarium oxysporum* suppressing *Sclerotinia sclerotiorum*. J. Appl. Microbiol. 100 (3), 575–586.

Rothrock, C.S., Gottlieb, D., 1984. Role of antibiosis in antagonism of *Streptomyces hygroscopicus* var. *geldanus* to *Rhizoctonia solani* in soil. Can. J. Microbiol. 30 (12), 1440–1447.

Saga, T., Yamaguchi, K., 2009. History of antimicrobial agents and resistant bacteria. Japan Med. Assoc. J. 52 (2), 103–108.

Sah, S., Singh, N., Singh, R., 2017. Iron acquisition in maize (*Zea mays* L.) using *Pseudomonas* siderophore. 3 Biotech. 7 (2), 1–7.

Saha, M., Sarkar, S., Sarkar, B., Sharma, B.K., Bhattacharjee, S., Tribedi, P., 2016. Microbial siderophores and their potential applications: a review. Environ. Sci. Pollut. Res. 23 (5), 3984–3999.

Salazar, F., Ortiz, A., Sansinenea, E., 2017. Characterisation of two novel bacteriocin-like substances produced by *Bacillus amyloliquefaciens* ELI149 with broad-spectrum antimicrobial activity. J. Glob. Antimicrob. Resist. 11 (12), 177–182.

Santoyo, G., Orozco-Mosqueda, M.C., Govindappa, M., 2012. Mechanisms of biocontrol and plant growth-promoting activity in soil bacterial species of *Bacillus* and *Pseudomonas*: a review. Biocontrol Sci. Technol. 22 (8), 855–872.

Sarwar, A., Hassan, M.N., Imran, M., Iqbal, M., Majeed, S., Brader, G., et al., 2018. Biocontrol activity of surfactin A purified from *Bacillus* NH-100 and NH-217 against rice bakanae disease. Microbiol. Res. 209 (4), 1–13.

Sazinas, P., Hansen, M.L., Aune, M.I., Fischer, M.H., Jelsbak, L., 2019. A rare thioquinolobactin siderophore present in a bioactive *Pseudomonas* sp. DTU12.1. Genome Biol. Evol. 11 (12), 3529–3533.

Scarselletti, R., Faull, J.L., 1994. *In vitro* activity of 6-pentyl-α-pyrone, a metabolite of *Trichoderma harzianum*, in the inhibition of *Rhizoctonia solani* and *Fusarium oxysporum* f. sp. *lycopersici*. Mycol. Res. 98 (10), 1207–1209.

Schneider, A., Marahiel, M.A., 1988. Genetic evidence for a role of thioesterase domains, integrated in or associated with peptide synthetases, in non-ribosomal peptide biosynthesis in *Bacillus subtilis*. Arch. Microbiol. 169 (5), 404–414.

Shahid, I., Malik, K.A., Mehnaz, S., 2018. A decade of understanding secondary metabolism in *Pseudomonas* spp. for sustainable agriculture and pharmaceutical applications. Environ. Sustain. 1 (1), 3–17.

Shakeri, J., Foster, H.A., 2007. Proteolytic activity and antibiotic production by *Trichoderma harzianum* in relation to pathogenicity to insects. Enzyme Microb. Technol. 40 (4), 961–968.

Shanmugaiah, V., Mathivanan, N., Varghese, B., 2010. Purification, crystal structure and antimicrobial activity of phenazine-1-carboxamide produced by a growth-promoting biocontrol bacterium, *Pseudomonas aeruginosa* MML2212. J. Appl. Microbiol. 108 (2), 703–711.

Shanmugaiah, V., Nithya, K., Harikrishnan, H., Jayaprakashvel, M., Balasubramanian, N., 2015. Biocontrol mechanisms of siderophores against bacterial plant pathogens. In: Kannan, V.R., Bastas, K.K. (Eds.), Sustainable Approaches to Controlling Plant Pathogenic Bacteria. CRC Press, Boca Raton, FL, pp. 167–181.

Sharma, G., Dang, S., Gupta, S., Gabrani, R., 2018. Antibacterial activity, cytotoxicity, and the mechanism of action of bacteriocin from *Bacillus subtilis* GAS101. Med. Princ. Pract. 27 (2), 186–192.

Sheppard, J.D., Jumarie, C., Cooper, D.G., Laprade, R., 1991. Ionic channels induced by surfactin in planar lipid bilayer membranes. Biochim. Biophys. Acta 1064 (1), 13–23.

Shobharani, P., Padmaja, R.J., Halami, P.M., 2015. Diversity in the antibacterial potential of probiotic cultures *Bacillus licheniformis* MCC2514 and *Bacillus licheniformis* MCC2512. Res. Microbiol. 166 (6), 546–554.

Shomura, T., Nishizawa, N., Iwata, M., Yoshida, J., Ito, M., Amano, S., et al., 1983. Studies on a new nucleoside antibiotic, dapiramicin - I. Producing organism, assay method and fermentation. J. Antibiot. 36 (10), 1300–1304.

Short, S.M., Van Tol, S., MacLeod, H.J., Dimopoulos, G., 2018. Hydrogen cyanide produced by the soil bacterium *Chromobacterium* sp. Panama contributes to mortality in *Anopheles gambiae* mosquito larvae. Sci. Rep. 8 (8358), 1–13.

Silva, C.S., 2018. Metabolites Produced by Burkholderia Pyrrocinia With Activity Against Phytopathogenic Fungi (Master thesis). State University of Londrina [viewed 11 april 2019]. Available from: <http://www.biblioteca-digital.uel.br/document/?code = vtls000230109>.

Silva, F.M., 2019. Analysis of Secondary Metabolism of *Burkholderia metallica* RV7S3: Purification and Characterization of Metabolite With Antimicrobial Activity (Master thesis). State University of Londrina [viewed 11 april 2019].

Simionato, A.S., Navarro, M.O.P., Jesus, M.L.A., Barazetti, A.R., Silva, C.S., Simões, G.C., et al., 2017. The effect of phenazine-1-carboxylic acid on mycelial growth of *Botrytis cinerea* produced by *Pseudomonas aeruginosa* LV strain. Front. Microbiol. 8 (1102), 1–9.

Sokol, P.A., Lewis, C.J., Dennis, J.J., 1992. Isolation of a novel siderophore from *Pseudomonas cepacia*. J. Med. Microbiol. 36 (3), 184–189.

Solanki, M.K., Malviya, M.K., Wang, Z., 2016. Actinomycetes bio-inoculants: a modern prospectus for plant disease management. In: Subramaniam, G., Arumugam, S., Rajendran, V. (Eds.), Plant Growth Promoting Actinobacteria. Springer Nature Singapore, Singapore, pp. 63–81.

Sun, Y., Feng, L., Li, A., Zhang, X., Yang, J., Ma, F., 2017. Ammonium assimilation: an important accessory during aerobic denitrification of *Pseudomonas stutzeri* T13. Bioresour. Technol. 234, 264–272.

Tambong, J.T., Hofte, M., 2001. Phenazines are involved in biocontrol of *Pythium myriotylum* on cocoyam by *Pseudomonas aeruginosa* PNA1. Eur. J. Plant. Pathol. 107, 511–521.

Tanara, Y., Hirata, K., Takahashi, Y., Iwai, Y., Omura, S., 1987. Globopeptin, a new antifungal peptide antibiotic. J. Antibiot. 40 (20), 242–244.

Tarus, P.K., Lang'at-Thoruwa, C.C., Wanyonyi, A.W., Chhabra, S.C., 2004. Bioactive metabolites from *Trichoderma harzianum* and *Trichoderma longibrachiatum*. Bull. Chem. Soc. Ethiop. 17 (2), 185–190.

Tawfik, K.A., Jeffs, P., Bray, B., Dubay, G., Falkinham, J.O., Mesbah, M., et al., 2010. Burkholdines 1097 and 1229, potent antifungal peptides from *Burkholderia ambifaria* 2.2N. Org. Lett. 12 (4), 664–666.

Thongkamngam, T., Jaenaksorn, T., 2016. Efficacy of culture filtrate from *Fusarium oxysporum* F221-B against plant pathogenic fungi *in vitro* and *Fusarium* root rot and wilt disease in hydroponics. Int. J. Agric. Technol. 12 (3), 513–526.

Tomashow, L.S., Weller, D.M., Bonsall, R.F., Pierson, L.S., 1990. Production of the antibiotic phenazine-1-carboxylic acid by fluorescent *Pseudomonas* species in the rhizosphere of wheat. Appl. Environ. Microbiol. 56 (4), 908–912.

Trejo-Estrada, S.R., Paszczynski, A., Crawford, D.L., 1998. Antibiotics and enzymes produced by the biocontrol agent *Streptomyces violaceusniger* YCED-9. J. Ind. Microbiol. Biotechnol. 21, 81–90.

Veloso, J., Alabouvette, C., Olivain, C., Flors, V., Pastor, V., García, T., et al., 2016. Modes of action of the protective strain Fo47 in controlling *Verticillium* wilt of pepper. Plant. Pathol. 65 (6), 997–1007.

Vinale, F., Flematti, G., Sivasithamparam, K., Lorito, M., Marra, R., Skelton, B.W., et al., 2009. Harzianic acid, an antifungal and plant growth promoting metabolite from *Trichoderma harzianum*. J. Nat. Prod. 72 (11), 2032–2035.

Vinale, F., Marra, R., Scala, F., Ghisalberti, E.L., Lorito, M., Sivasithamparam, K., 2006. Major secondary metabolites produced by two commercial *Trichoderma* strains active against different phytopathogens. Lett. Appl. Microbiol. 43 (2), 143–148.

Vinale, F., Sivasithamparam, K., Ghisalberti, E.L., Woo, S.L., Nigro, M., Marra, R., et al., 2014. *Trichoderma* secondary metabolites active on plants and fungal pathogens. Open. Mycol. J. 8 (1), 127–139.

Vleesschauwer, D., Cornelis, P., Höfte, M., 2006. Redox-active pyocyanin secreted by *Pseudomonas aeruginosa* 7NSK2 triggers systemic resistance to *Magnaporthe grisea* but enhances *Rhizoctonia solani* susceptibility in rice. Mol. Plant. Microbe Interact. 19 (12), 1406–1419.

Vleesschauwer, D., Djavaheri, M., Bakker, P.A.H.M., Höfte, M., 2008. *Pseudomonas fluorescens* WCS374r-induced systemic resistance in rice against *Magnaporthe oryzae* is based on pseudobactin-mediated priming for a salicylic acid-repressible multifaceted defense response. Plant. Physiol. 148 (4), 1996–2012.

Xiang, Y., Zhang, Y., Wang, C., Liu, S., Liao, X., 2018. Effects and inhibition mechanism of phenazine-1-carboxamide on the mycelial morphology and ultrastructure of *Rhizoctonia solani*. Pestic. Biochem. Phys. 147, 32–39.

Yadav, A.N., Verma, P., Kumar, S., Kumar, V., Kumar, M., Sugitha, T.C.K., et al., 2018. Actinobacteria from rhizosphere: molecular diversity, distributions, and potential biotechnological applications. In: Singh, B.P., Gupta, V.L., Passari, A.K. (Eds.), New and Future Developments in Microbial Biotechnology and Bioengineering. Elsevier, pp. 13–41.

Yang, P.W., Li, M.G., Zhao, J.Y., Zhu, M.Z., Shang, H., Li, J.R., et al., 2010. Oligomycins A and C, major secondary metabolites isolated from the newly isolated strain *Streptomyces diastaticus*. Folia Microbiol. 55 (1), 10–16.

Yaseen, Y., Gancel, F., Drider, D., Béchet, M., Jacques, P., 2016. Influence of promoters on the production of fengycin in *Bacillus* spp. Res. Microbiol. 167 (4), 272–281.

Yoshihisa, H., Zenji, S., Fukushi, H., Katsuhiro, K., Haruhisa, S., Takahito, S., 1989. Production of antibiotics by *Pseudomonas cepacia* as an agent for biological control of soilborne plant pathogens. Soil. Biol. Biochem. 21 (5), 723–728.

Youard, Z.A., Mislin, G.L.A., Majcherczyk, P.A., Schalk, I.J., Reimmann, C., 2007. *Pseudomonas fluorescens* CHA0 produces enantio-pyochelin, the optical antipode of the *Pseudomonas aeruginosa* siderophore pyochelin. J. Biol. Chem. 282 (49), 35546–35553.

Zhang, J., Wang, L.M., Li, Y.H., Ding, S.L., Yuan, H.X., Riley, I.T., et al., 2016. Biocontrol of cereal cyst nematode by *Streptomyces anulatus* isolate S07. Australas. Plant. Pathol. 45, 57–64.

Zhang, Q., Yang, L., Zhang, J., Wu, M., Chen, W., Jiang, D., et al., 2015. Production of anti-fungal volatiles by non-pathogenic *Fusarium oxysporum* and its efficacy in suppression of *Verticillium* wilt of cotton. Plant. Soil. 392 (1–2), 101–114.

Zidour, M., Belguesmia, Y., Cudennec, B., Grard, T., Flahaut, C., Souissi, S., et al., 2019. Genome sequencing and analysis of *Bacillus pumilus* ICVB403 isolated from *Acartia tonsa* copepod eggs revealed surfactin and bacteriocin production: insights on anti-*Staphylococcus* activity. Probiotic. Antimicrob. Proteins. 11 (3), 990–998.

Zonno, M.C., Vurro, M., 2002. Inhibition of germination of *Orobanche ramosa* seeds by *Fusarium* toxins. Phytoparasitica 30 (5), 519–524.

CHAPTER

4

Exploring the molecular signatures of host–pathogen interactions in plant diseases: conflict and cooperation

H.C. Yashavantha Rao[1], *Chelliah Jayabhaskaran*[1], *Subban Kamalraj*[1], *Ramalingam Parthasarathy*[1], *Soma Mondal*[2], *R. Sundararaj*[2] *and Shambhu Kumar*[3]

[1]Department of Biochemistry, Indian Institute of Science, Bengaluru, Karnataka, India [2]Forest Protection Division, ICFRE - Institute of Wood Science and Technology, Bengaluru, Karnataka, India [3]Forest Pathology Department, KSCSTE - Kerala Forest Research Institute, Peechi, Thrissur, Kerala, India

4.1 Introduction

Plants are the rich green source on the earth. Unlike mammals, plants lack defender cells and adaptive immune systems. Indeed, they depend on innate immune system of each cell and systemic signals exhaling from infection areas (Amil-Ruiz et al., 2011). Plant–pathogen microbes adopt many strategies like killing hosts and feeds on the contents (necrotrophs) and biotrophs. Microbial necrotrophy is followed by the production of toxins (Abu-Qamar et al., 2006). Pathogenic microbes proliferate in the intercellular spaces after entering through stomata, hydathodes, or gains access through wounds (Melotto et al., 2008). Pathogenic fungi and symbiotic fungi have haustoria. These various pathogens deliver effects or virulence factors into the plant cell to increase the microbes fitness (Craig et al., 2009; Amil-Ruiz et al., 2011). Plants cell wall has high dynamic structure that provides mechanical support needs to respond various environmental cues and fulfills significant functions in signaling events, the defense against growth, biotic, and abiotic stresses (Bosch et al., 2011). It even represents as a reservoir of bioactive compounds and as an important source for signaling molecules (Almagro et al., 2009).

Food Security and Plant Disease Management
DOI: https://doi.org/10.1016/B978-0-12-821843-3.00003-9

4.2 Pathogens virulence

Pathogen virulence depends on the host factor as represented by the avirulent microorganisms' pathogenicity in immune-compromised hosts and lack of virulent pathogens pathogenicity in immune hosts. Pathogen virulence affirms that pathogens are distinguished by their virulence factors from nonpathogens. Even though this concept applies to certain microorganisms that induce diseases in the normal host, it need not to apply for certain microorganisms which cause diseases in immune-compromised host. To attack a host cell, pathogens must pass over the cell's outer barrier. The cell wall degradation by a battery of polysaccharides produced by invading pathogens is one of the major pathogenic mechanisms involved in plants. The phyllosphere is a harsh environment for the pathogens. The leaves surface is regularly exposed to certain conditions like heat, UV radiation, lack of moisture, and strong winds. The most abundant microbe on leaves surface, that is, bacteria can reach higher population density (Andrews and Harris, 2000; Lindow and Brandl, 2003). Pathogens reach to the leaves surface might have quickly penetrated the leaf or expresses traits which ensure survival till all the environmental conditions are favorable for the penetration.

4.3 How pathogen sense plants?

Plant pathogens have evolved many unique virulence factors to successfully infect a host and the virulence is the possible result of host–microbes interaction (Rao et al., 2017; Jones et al., 2019). The better understanding of host–pathogens interactions has expanded our cognition on the abilities of the plant to increase their microbiota and vice versa (Rao et al., 2015; Rao et al., 2016). Like humans, plants do not have sensory organs, interestingly pathogens have the capability to sense plant by several volatile secretions and root exudates. Zoospores propagates in soil water are appealed to the differentiation zones and elongation of plants. *Phytophthora parasitica* spores clearly do not have specific plant species root predilections, indeed contrasting to *Phytophthora sojae* spores, which are appealed to the root exuding genistein and isoflavones daidzein. *P. parasitica* spores can be passively transferred by splash water to the leaves surface, where they attract to wound site earlier to encystment and germination. Oomycetes spores secrete PcVsv1, a protein that contains multiple thrombospondin type 1 repeats. PcVsv1 or mucin-like glycoproteins and certain surface-binding proteins has been identified during early infection stages of *Phytophthora infestans* and *P. parasitica* are assumed to ensure host adhesion of zoospores. The encysted spores on leaf, roots, and then attracts zoospores to form a clump of cysts. This development of self-attraction seems is a good strategy to enhance the likelihood of infections and known as "homing response" which might be due to the chemical signal including calcium. Once encysted on plants surface, $Ca2^{+}$ dependent signal triggers the germination of *P. parasitica* spores to initiate an infection. The germ tube later enters directly to the intercellular space via the wound openings on leaf or swellings, which allows the epidermal cells penetration on the root surfaces (Allard et al., 2008).

4.4 How pathogens recognize host plant?

Sixty years ago, J.B.S. Haldane anticipated that polymorphism in protein mediated host–pathogen interaction was carried out by the increased rate of microbe's evolution (Haldane, 1949). He predicted that slow evolving host plants can manage to dodge microbial infections by producing high level of polymorphism in immune system genes. These strategies were prompted by observation of variability in wheat resistance to the rust pathogen *Puccinia graminis*. To detect pathogens, plants have evolved 02 strategies (Jones and Dangl, 2006; Chisholm et al., 2006). Host cells on the external surface, plants are equipped with sense evolutionary conserved microbes molecular signatures which collectively known as pathogens-associated molecular patterns (PAMPs) as well as microbes-associated molecular patterns (MAMPs), that are identified by receptor protein known as pattern recognition (Boller and Felix, 2009). PAMPs are the essential agents of pathogens, like fungal chitin and bacterial flagellin. Plant responds to endogenous agents produced by pathogen invasions, like cuticular fragments and cell wall known as danger-associated molecular patterns. The stimulation of pattern recognition leads to PAMPs-triggered immunity which is a form of innate immunity (Jones and Dangl, 2006).

The second perception includes recognition through intracellular receptor of pathogens virulence molecule known as effectors which increase effector-triggered immunity (ETI). This recognition mode leads to the coevolutionary dynamics between pathogens and plants which differ from PAMP-triggered immunity (PTI) and PAMPs, whereas effectors are characteristically dispensable and variable. ETI and PTI exhibits similar responses, even though ETI is faster, stronger, and involves hypersensitive response (HR) (Dodds and Rathjen, 2010). PAMPs were discovered by pattern recognition receptors (PRRs), usually by leucine rich repeat (LRR)-ribulose bisphosphate carboxylase proteins or cell surface-localized receptor kinases (Zhang and Zhou, 2010).

4.5 Elicitors as weapons of plant pathogens

Elicitors are used for the molecules that able to induce the production of phytoalexins, and now it is usually used to stimulate plant defense compounds. Indeed, defense response induction may lead to increased resistance. This definition for elicitors involves both substances (exogenous and microbial) of pathogens origin and compounds produced by plants by pathogens action (D'Ovidio et al., 2004). During evolution, certain invading pathogens have acquired to get over from PAMP-triggered resistant by producing the capability to deliver proteins into the plant cell (Chisholm et al., 2006). This effector protein interferes manipulate, or suppresses disease signaling and thereby increasing growth of pathogens and disease developments. During evolution of plants, they adapt to detect these particular pathogen-derived biomolecules. This gene-for-gene disease resistance system (cultivar-specific) is defined by pathogens-encoded effector protein and plant–derived R-proteins (Amil-Ruiz et al., 2011). Several Gram-negative pathogens have hypersensitive reaction and pathogenicity gene clusters which encode type III secretion system. It is employed by the bacteria for the effector protein injection into the plant cell. Zhang and Zhou (2010) demonstrated that *Pseudomonas syringae* effector AvrRpt2 and AvrRpm1 proteins suppressed PAMP-activated defense response in *Arabidopsis* by suppressing the accumulation of callose. However, *P. syringae,* AvtPto effector, inhibited *Arabidopsis* genes that encode cell wall secretion and defense protein. Certain avirulence (*Avr*) factor acts by inhibiting HR responses by activating some defense response in plants. Even though several effector proteins have been cloned, biochemical functions of many are unknown. AvrPtoB have shown ubiquitin in-vivo ligase activity. The deletion of key residue from AvrPtoB protein has annihilated ubiquitin ligase activity and ability to inhibit cell death. Hence, this protein was advised to target proteins that are responsible for the regulation of programmed cell death (PCD) to degrade host ubiquitin ligase (Chisholm et al., 2006). Indeed, if the protein matches *R*-gene, then becomes a specific elicitor and defense system will be activated by the R protein in the plant.

The cell-surface receptors of plants are the key components which perceive extracellular stimuli from the environment. The capability to separate between nonself, and self is the quality of animals and plants from evolutionary innate immunity. It relies on PAMPs characters of potentially harmful microbes (Nurnberger and Kemmerling, 2006). The pathogen-derived molecules are of various natures like chitins, lipopolysaccharides, glucans, flagellins, and serve as elicitor to trigger defense response independently of the individual pathogen. The flagellin acts as PAMP in plants and animals. The general elicitors are normally molecules which are indispensable in microbes; lifestyle, and provides fitness for pathogens if discerned by the plant surveillance system. Indeed, endogenous plants cell wall–derived structure produced by hydrolytic enzymes activity of microorganisms might act as general elicitors (Nurnberger and Kemmerling, 2006).

4.6 Molecular basis of host–pathogen recognition

Generally, plant response to invading pathogens based on their capability to identify signature molecules produced by the pathogens. Plant lacks mobile defender cell and adaptive immune system. Indeed, it relies on the innate immunity of cells and systematic signals emanating from the infection site. Several plant R-proteins can activate indirectly by pathogens-encoded effectors, and not by direct recognition. This "guard hypothesis" suggests that R-protein recognizes pathogen effector by observing host cellular target of effector actions. The concept of R-protein recognizes "pathogens-induced modified self" is similar to the recognition of "modified self" in "danger signal" model of mammalian immune systems. Therefore there are 02 branches of the plants immune system in essence.

Transmembrane PRRs respond to slow evolving PAMPs or MAMPS. Another one acts inside cells with polymorphic nucleotide binding (NB)-LRR proteins encoded by the *R* gene. It was named after their LRR and NB domains. Pathogens effector from different kingdoms is recognized by NB-LRR proteins, and activates similar defense response. LRR-NB–mediated disease resistance is more efficient against invading pathogens which can grows in living host tissues or hemibiotrophic pathogens and not against pathogen that kills host tissues during colonization (Munch et al., 2008). This interaction mechanism of plant immune system with pathogens was represented as four phased "zigzag" model (Dangl and Jones, 2001).

4.7 Plant fungus gene-for-gene relationship

The gene-for-gene relationship was revealed by Flor (1942). They explored that, the inheritance of resistance genes in hosts and pathogens *Avr* gene capability to cause disease is controlled by a pair of matching genes in *Melampsora lini* of flax (*Linum usitatissimum*). The resistance (*R*) gene is produced by plants and *Avr* gene is produced by the pathogens. Plants produce specific *R* gene product that are resistant toward a pathogen which produce corresponding *Avr* gene products (Robinson and Raoul, 1987). The gene-for-gene relationship is widespread and most significant aspect of plant disease resistance. An example can be seen with *Lactuca serriola*. The race-specific hypersensitive resistance is activated by the product of *Avr* gene, race-specific elicitors after interaction with a receptor and putative product of a resistance gene (Fig. 4.1). This HR is super-imposed on the system that governs basic compatibility (De-Wit, 1987; Keen, 1982).

4.8 Hypersensitive response

The resistance gene and complementary *Avr* gene in genome and plant has the genetic ingredients for HR when both genes are expressed. Indeed, it is produced in transgenic plants and disturbing HR is induced by attacking pathogens. Therefore either resistance gene or both must be regulated. Ideally, one or both genes should be specifically activated upon attack by pathogens. A model for the induction of HR by various plant pathogens using the *avr9* and *Cf9* gene-for-gene system is represented (Fig. 4.2). The two-component sensor system includes in one plant (circle), a sensor (*avr9* gene-pathogen inducible promoter), and an effector (*Cf9* gene). The promoter fused to *avr9* gene will be activated upon induction by nonspecific elicitors originating from various plant pathogens. The *avr9* gene

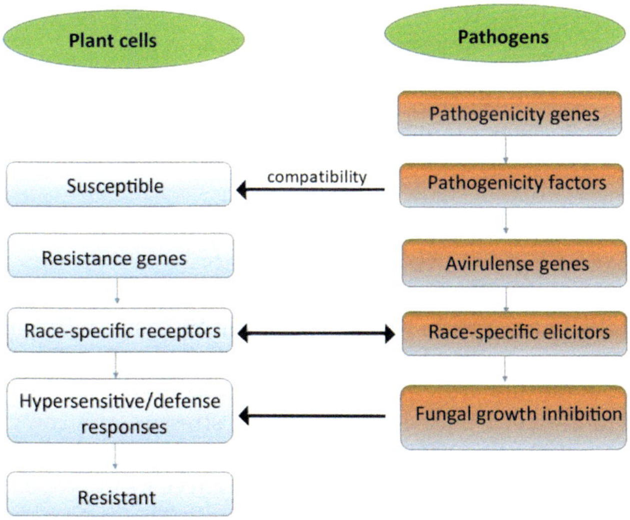

FIGURE 4.1 The interaction between fungal pathogen and their host.

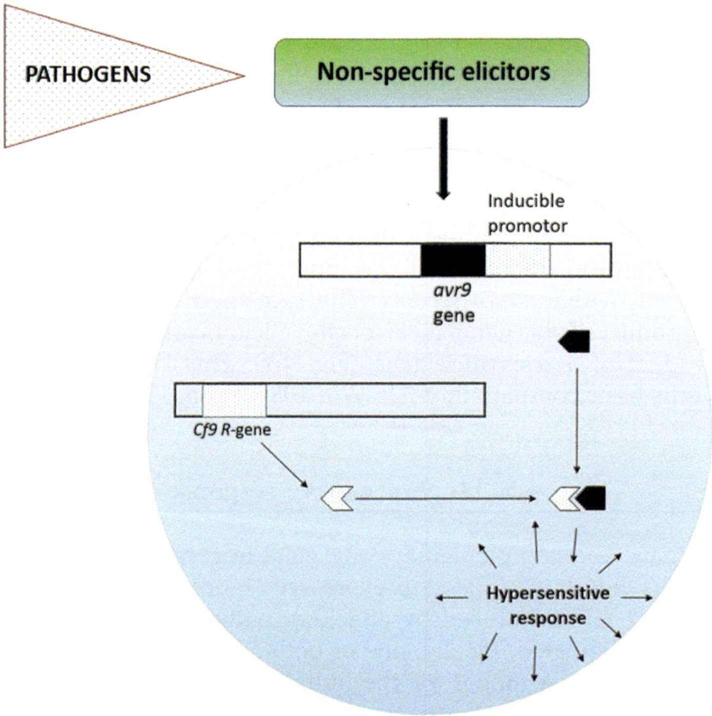

FIGURE 4.2 A representation for the induction of the hypersensitive response by several plant pathogens using gene-for-gene system.

product, the peptide elicitors will interact with *Cf9* resistance gene product leads to HR activation. This requirement suggests the need to isolate both *Avr*, resistance genes, and promoters which specifically regulate these genes. The *Avr* gene, or both resistance and *Avr* genes under the control of pathogen-inducible promoters, which responds to attack by pathogens (Caeiles et al., 1989; Matton and Brisson, 1989; Somssich et al., 1988).

4.9 Induced systemic resistance enhances plant growth through fungal biocontrol agents

Biocontrol fungi (BCF) are agents that control plant diseases, which include well-known *Trichoderma* spp. and the recently described *Sebacinales* spp. (51). They have the ability to control many foliar, root, and fruit pathogens and nematodes (Shoresh et al., 2010). BCF have the capability to perfect a wide range of abiotic stresses, and can improve physiological stresses (Adams and De-Lij, 2007). Indeed, BCF enhance the nutrient uptake in plants and can substantially enhance nitrogen use efficiency (NUE) in crops. These abilities may be more important to agriculture than in disease control. The certain strains have the abilities to improve photosynthetic efficiency and respiratory activities of plants. All of these

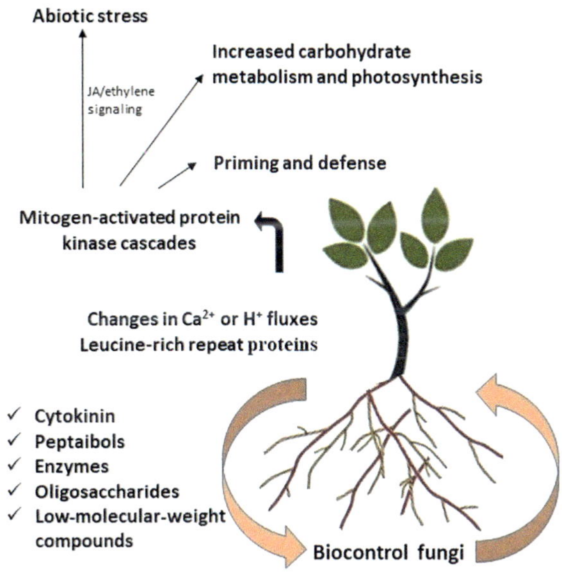

FIGURE 4.3 A representation of biocontrol fungi grows to interact with the roots. Details are given in the text.

capabilities are a consequence of their ability to reprogram plant-gene expression and probably via the activation of limited numbers of general plant pathways.

BCF occupy to interact with the rhizosphere which interactions between plant and BCF by exchange signals. BCF release elicitor of chemical compounds outside and inside the root tissues which activates mitogen-activated protein kinase cascade in the plant system (Shahollari et al., 2007). Indeed, jasmonic acid/ethylene signaling is being activated resulting in increase of plant defense genes that ultimately increase plant resistance to pathogens (Fig. 4.3). Recent findings indicate that BCF control pathogens primarily through the ability of the fungi to reprogram plant-gene expressions. As a consequence, induced systemic resistance (ISR) occurs. The genetic reprogramming induces mechanism in the plant that alleviates physiological and abiotic stresses and improves plant NUE. The significant progress in understanding how BCF interact directly with the plants has been achieved. BCF's abilities and the molecular mechanisms for induction of plant responses.

4.10 Programmed cell death during plant–pathogen interactions

The incident of cell death during host–pathogen interactions in which pathogens can replicates in their hosts. A recent report of infected oat with variety of pathogens (fungi, viruses, and bacteria) established that many virulent pathogens induce cell death (Yao et al., 2002a). The apoptotic-like events were obtained with different time and infectious agents in the directly infected cells or the neighboring cells. Even though morphological view contends that cells may die by the same mechanism. This approach mentioned for the HR studies, in that the specificity of the reagents used must be established. PCD and other disease symptoms results from the host-encoded functions.

The cell death resulted in fungal pathogens attack, in which two genotypes of plant host and the fungal invader or any one cell death pathways can be activated. These pathways might intersect cell death. During the recognition of fungal contest by the plant, host-controlled by HR-PCD leads to a restricted cell death phenotype and which leads the ultimately immunity. Conversely, pathogen may mediate PCD suppresses this host recognition, leading to unrestricted spread of the pathogen accompanied by PCD, susceptibility, and disease (Fig. 4.4).

For example, *Arabidopsis* mutants form disease-like lesions spontaneously have been documented by Greenberg et al. (2000), Pilloff et al. (2002). One of mutants, *ACD5* shows disease symptoms increased and modestly increased growth of *P. syringae* when infected before the formation of lesion. Interestingly, *ACD5* codes a ceramide kinase (CERK) that is induced during virulent *P. syringae* infections (Liang et al., 2003). Ceramide known to be bioactive lipids, which activates apoptosis signals in animals (Hannun and Obeid, 2002). The ceramide is an appropriate to induce apoptotic-like cell death. Finally, *acd5* mutant plants accumulate increased amounts of the ACD5 CERK substrate. Together, these findings suggest that cell death activated by ceramide is important for *P. syringae* virulence. Ceramides are sphingolipid family of bioactive lipids (Hannun and Obeid, 2002). This sphingolipid pathway may be targeted by plant pathogens. The certain fungal pathogens secrete cerebrosides (sphingolipids) that induce HR-like cell death in rice (Koga et al., 1998). Additionally, a number of fungal pathogens secrete related mycotoxins (e.g., fumonisin) that cause PCD and disrupt sphingolipid metabolism (Abbas et al., 1994).

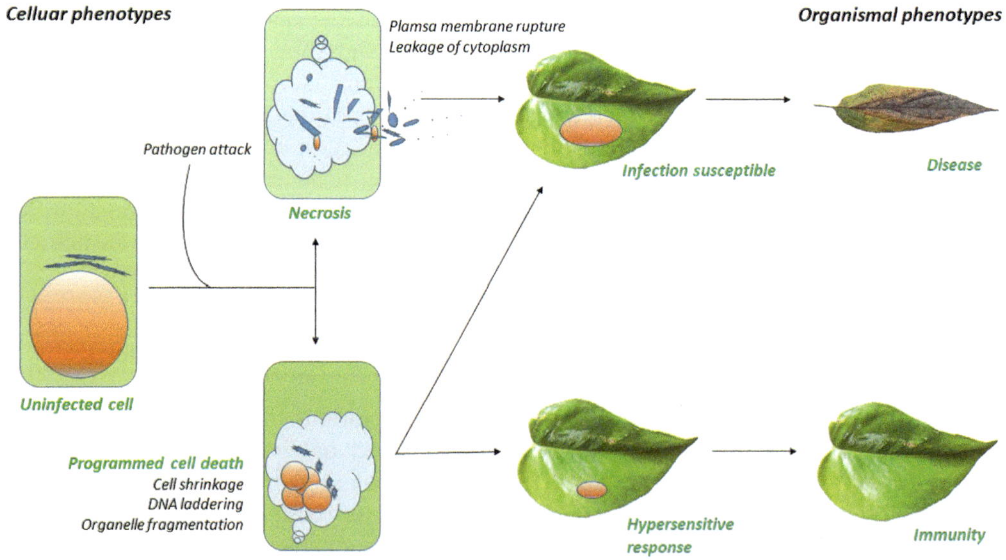

FIGURE 4.4 Recognition of pathogen challenge by plants: host-controlled PCD-HR directs to a cell death phenotype and immunity. Pathogen-mediated PCD suppress host recognition, directs to spread of pathogens, susceptibility, and disease.

4.11 Improvement of biocontrol agents by molecular approaches

Genes involved in biocontrol properties: the genetic basis of mechanism of action can be understood by techniques like gene inactivation and over-expression studies through insertion of mutagenesis, PCR tilling, and RNA silencing studies. This requires a prior selection of genes classical gene mining technique, which involved in biocontrol potential used to gain knowledge on the whole pattern of genes involved in biocontrol. The genes involved in the biocontrol could be identified by assessing the differential gene expressions under varied exposure conditions by the molecular techniques, cDNA amplified fragment. For example, differential expression of thioredoxin h-rice, p23co-chaperone, nucleotide diphosphate kinase, ribulose-bisphosphate carboxylase large chain precursor, proteosome subunit protein, and putative glutathione S-transferase proteins in rice leaves primed with *Pseudomonas fluorescens* (Saveetha et al., 2009; Saveetha et al., 2010). These methods with regards to their putative involvement in biocontrol properties depend heavily on the selected comparison model. The whole genomic sequencing of microbes like *Pseudomonas*, *Bacillus*, and *Trichoderma* species opens up a new avenue to advance the knowledge on beneficial biocontrol agents through genomics approach. The complete genome sequencing of the biocontrol bacterium, *P. fluorescens* Pf-5 revealed that biosynthesis of secondary metabolites including antibiotics and siderophores. Of three orphan gene clusters, new group of bioactive cyclic lipopeptides with a putative role in biological control of plant disease was identified (Paulsen et al., 2005).

4.11.1 Molecular approaches to improvement of biocontrol agents

The molecular techniques allow modification of wild type strains to improve their ability to control diseases. The selected biocontrol agents could be improved by genetic modifications and recombination technique like protoplast fusion and genetic recombination.

4.11.2 Tapping potential biocontrol genes

The biocontrol properties contribute to more than one characters, it is imperative to study the genes involved in different mechanisms of action. An extensive work on biocontrol properties elucidated including siderophore, competition, colonization, antibiosis, lysis, growth promotion, ISR either individually or in combination determines the success of biological control of plant diseases. Thus studies on mechanism of action may offer a pedestal for tapping potential genes for further improvement of biocontrol strains.

4.12 Host–pathogen interactions: cooperation and conflict

The main insight from evolutionary and mechanistic research has both cooperation and conflicts in plants immune system and their interactions. The genetic research on immunity of plant has been highly trigged by Flor's gene-for-gene hypothesis, where one host resistant gene is matched by single effector gene by a specific pathogen. It was unexpected that in *Arabidopsis thaliana* genome two nucleotide-binding leucine-rich repeat (NLR) genes which are located adjacent to each other were resistance to the *Hyaloperonospora arabidopsidis* ex parasitica isolate Cala2.

Plant–microbial conflict frequently begins in apoplastic region at the cellular level which is a significant extracellular habitat for the interaction among proteins produced by plants cell and invading pathogens (Doehlemann and Hemetsberger, 2013). These proteins interactions play a vital role during pathogenesis in the plant system (Bednarek et al., 2010). The immune receptors and host–client interaction is another type of protein–protein interactions in the plant immune system which are direct targets of pathogens effectors. In host view, pathogens-targeting NLR-client might be undesirable when leads to increased pathogen virulence. However, targeting might be suitable, since NLR-clients are simply decoy version of real targets (Williams et al., 2014). The interactions between NLR-guards and their guardee mediate an indirect pathogen recognition of effectors (Eitas and Dangl, 2010).

4.13 Conclusion

This chapter has presented an overview of molecular signatures during the host–pathogen interactions in plant diseases. Understanding of signal cascades involved in disease resistant and the host–defense linked genes can be attained by the new molecular strategies. Plants have innate immune responses to fight against several biotic stresses. Whereas, pathogens equipped with strong effector molecules that can defeat host's plant immunity. Many significant outcomes related to disease resistance are being discovered, like HR by major *R* gene, ROS generation, BCF, PR gene activation, elicitors, hormone biosynthesis, and other signaling pathways. Indeed, the overall connections and cofactors involved in plants immune system utilizing disease-resistant genes has to be studied in depth. The selective information explored in this chapter will pay a way to understand molecular mechanism of host–pathogen interactions which may help in the development of disease-resistant cultivars.

Acknowledgments

H.C. Yashavantha Rao thanks University Grants Commission (UGC), New Delhi for the award of Dr. D.S. Kothari Post-Doctoral Fellowship (No. F.4–2/2006 (BSR)/BL/17–18/0234). Subban Kamalraj thanks Indian Institute of Science, Bengaluru, for his IISc Research Associate Fellowship. R. Parthasarathy thanks DST-SERB, Government of India for NPDF (No.PDF/2017/001184).

Conflicts of interest

The authors declare no conflict of interest.

References

Abbas, H.K., Tanaka, T., Duke, S.O., Porter, J.K., Wray, E.M., Hodges, L., et al., 1994. Fumonisin and AAL-toxin-induced disruption of sphingolipid metabolism with accumulation of free sphingoid bases. Plant Physiol. 106, 1085–1093.

Abu-Qamar, S., Chen, X., Dhawan, R., Bluhm, B., Salmeron, J., Lam, S., et al., 2006. Expression profiling and mutant analysis reveals complex regulatory networks involved in *Arabidopsis* response to *Botrytis* infection. Plant J. 48 (1), 28–44.

Adams, P., De-Lij, A.A.M., 2007. *Trichoderma harzianum* Rifai 1295-22 mediates growth promotion of crack willow (*Salix fragilis*) saplings in both clean and metal-contaminated soil. Microb. Ecol. 54 (2), 306–313.

Allard, A., Gourgues, M., Galiana, E., Panabieres, F., Ponchet, M., Keller, H., 2008. Strategies of attack and defensein plant–oomycete interactions, accentuated for *Phytophthora parasitica* Dastur (syn. P. Nicotianae Breda de Haan). J. Plant Physiol. 165, 83–94.

Almagro, L., Gomez Ros, L.V., Belchi-Navarro, S., Bru, R., Barceló, A.R., et al., 2009. Class III peroxidases in plant defence reactions. J. Exp. Bot. 60 (2), 377–390.

Amil-Ruiz, F., Blanco-Portales, R., Munoz-Blanco, J., Caballero, J.L., 2011. The strawberry plant defense mechanism: a molecular review. Plant Cell Physiol. 52 (11), 1873–1903.

Andrews, J.H., Harris, R.F., 2000. The ecology and biogeography of microorganisms on plant surfaces. Annu. Rev. Phytopathol. 38, 145–180.

Bednarek, P., Kwon, C., Schulze-Lefert, P., 2010. Not a peripheral issue: secretion in plant–microbe interactions. Curr. Opin. Plant Biol 13, 378 387.

Boller, T., Felix, G.A., 2009. Renaissance of elicitors: perception of microbe-associated molecular patterns and danger signals by pattern-recognition receptors. Annu. Rev. Plant Biol. 60, 379–406.

Bosch, M., Mayer, C., Cookson, A., Donnison, I.S., 2011. Identification of genes in cell wall biogenesis in grasses by differential gene expression profiling of elongation and non-elongating maize internodes. J. Exp. Bot. 1–17.

Caeiles, C., Ferrer, A., Balceils, L., Hegardt, F.G., Boronat, A., 1989. Isolation and structural characterization of a cDNA encoding *Arabidopsis* tha Liana 3-hydroxy-3-methylglutaryl coenzyme A reductase. Plant Mol. Biol. 13, 627–638.

Chisholm, S.T., Coaker, G., Day, B., Staskawicz, B.J., 2006. Host–microbe interactions: shaping the evolution of the plant immune response. Cell 124, 803–814.

Craig, A., Ewan, R., Mesmar, J., Gudipati, V., Sadanandom, A., 2009. E3 ubiquitin ligases and plant innate immunity. J. Exp. Bot. 60 (4), 1123–1132.

D'Ovidio, R., Mattei, B., Roberti, S., Bellincampi, D., 2004. Polygalacturonases polygalacturonases-inhibiting proteins and pectic oligomer in plant-pathogen interactions. Biochim. Biophys. Acta 1696, 237–244.

Dangl, J.L., Jones, J.D., 2001. Plant pathogens and integrated defence responses to infection. Nature 411, 826–833.

De-Wit, P.J.G.M., 1987. Specificity of active resistance mechanisms in plant fungus interactions. In: Pegg, G.F., Ayres, P. (Eds.), Fungal Infection of Plants. Cambridge University Press, Cambridge/New York/Sydney, pp. 1–24.

Dodds, P.N., Rathjen, J.P., 2010. Plant immunity: towards an integrated view of plant–pathogen interactions. Nat. Rev. Genet. 11, 539–548.

Doehlemann, G., Hemetsberger, C., 2013. Apoplastic immunity and its suppression by filamentous plant pathogens. N. Phytol. 198, 1001–1016.

Eitas, T.K., Dangl, J.L., 2010. NB-LRR proteins: pairs, pieces, perception, partners, and pathways. Curr. Opin. Plant Biol. 13, 472–477.

Flor, H.H., 1942. Inheritance of pathogenicity in *Melampsora lini*. Phytopath 32, 653–669.

Greenberg, J.T., Silverman, F.P., Liang, H., 2000. Uncoupling salicylic acid-dependent cell death and defense-related responses from disease resistance in the *Arabidopsis* mutant ACD5. Genetics 156, 341–350.

Haldane, J.B.S., 1949. On infectious disease and evolution. Ric. Sci. A19 (Suppl), 68–76.

Hannun, Y.A., Obeid, L.M., 2002. The ceramide-centric universe of lipid-mediating cell regulation: stress encounters of the lipid kind. J. Biol. Chem. 277, 25847–25850.

Jones, J.D., Dangl, J.L., 2006. The plant immune system. Nature 444, 323–329.

Jones, P., Garcia, B.J., Furches, A., Tuskan, G.A., Jacobson, D., 2019. Plant host-associated mechanisms for microbial selection. Front. Plant Sci. 10, 862.

Keen, N.T., 1982. Specific recognition in gene-for-gene host parasite systems. Adv. Plant Pathol. 1, 35–82.

Koga, J., Yamauchi, T., Shimura, M., Ogawa, N., Oshima, K., Umemura, K., et al., 1998. Cerebrosides A and C, sphingolipid elicitors of hypersensitive cell death and phytoalexin accumulation in rice plants. J. Biol. Chem. 273, 31985–31991.

Liang, H., Yao, N., Song, J.T., Luo, S., Lu, H., Greenberg, J.T., 2003. Ceramides modulate programmed cell death in plants. Genes Dev. 17, 2636–2641.

Lindow, S.E., Brandl, M.T., 2003. Microbiology of the phyllosphere. Appl. Env. Microbiol. 69, 1875–1883.

Matton, D.P., Brisson, N., 1989. Cloning, expression and sequence conservation of pathogenesis-related gene transcripts of potato. Mol. Plant Microbe Interact. 2 (6), 325–331.

Melotto, M., Underwood, W., He, S.Y., 2008. Role of stomata in plant innate immunity and foliar bacterial diseases. Annu. Rev. Phytopathol. 46, 101–122.

Munch, S., Lingner, U., Floss, D.S., Nancy, L., Sauer, N., Deising, H.B., 2008. The hemibiotrophic lifestyle of *Colletotrichum* species. J. Plant Physiol. 165, 41–51.

Nurnberger, T., Kemmerling, B., 2006. Receptor protein kinases–pattern recognition hreceptors in plant immunity. Trends Plant Sci. 11, 519–522.

Paulsen, I.T., Press, C.M., Ravel, I., Kobayashi, D., Myers, G.S.A., Mavrodi, D.V., et al., 2005. Complete sequence of the plant commensal *Pseudomonas fluorescens* Pf-5. Nat. Biotechnol. 23, 873–880.

Pilloff, R.K., Devadas, S.K., Enyedi, A., Raina, R., 2002. The Arabidopsis gain-of-function mutant dll1 spontaneously develops lesions mimicking cell death associated with disease. Plant J. 30, 61–70.

Rao, H.C.Y., Rakshith, D., Gurudatt, D.M., Satish, S., 2016. Implication of PKS type I gene and chromatographic strategy for the biodiscovery of antimicrobial polyketide metabolites from endosymbiotic *Nocardiopsis prasina* CLA68. Sci. Nat. 103 (5–6), 45.

Rao, H.C.Y., Rakshith, D., Harini, B.P., Gurudatt, D.M., Satish, S., 2017. Chemogenomics driven discovery of endogenous polyketide anti-infective compounds from endosymbiotic *Emericella variecolor* CLB38 and their RNA secondary structure analysis. PLoS One 12 (2), e0172848.

Rao, H.C.Y., Rakshith, D., Satish, S., 2015. Antimicrobial properties of endophytic actinomycetes isolated from *Combretum latifolium* Blume, a medicinal shrub from Western Ghats of India. Front. Biol. 10 (6), 528–536.

Robinson, R.A., 1987. Host Management in Crop Pathosystems. Macmillan Publishing Company.

Saveetha, K., Karthiba, L., Raveendran, M., Saravanakumar, D., Raguchander, T., Samiyappan, R., 2010. Transcriptional analysis of molecular interactions between *Pseudomonas fluorescens* strains TDK 1, *Oryza sativa* and *Cnaphalocrocis medinalis*. J. Appl. Entomol. 134, 762–773.

Saveetha, K., Karthiba, L., Raveendran, M., Saravanakumar, D., Suresh, S., Raguchander, T., et al., 2009. Understanding the molecular basis of plant growth promotional effect of *Pseudomonas fluorescens* on rice through protein profiling. Proteome Sci. 7, 47.

Shahollari, B., Vadassery, J., Varma, A., Oelmueller, R., 2007. A leucine-rich repeat protein is required for growth promotion and enhanced seed production mediated by the endophytic fungus *Piriformospora indica* in *Arabidopsis thaliana*. Plant J. 50, 1–13.

Shoresh, M., Harman, G.E., Mastouri, F., 2010. Induced systemic resistance and plant responses to fungal biocontrol agents. Ann. Rev. Phytopathol. 48, 21–43.

Somssich, I.E., Schmelzer, E., Kawalleck, P., Hahlbrock, K., 1988. Gene structure and in situ transcript localization of pathogenesis-related protein 1 in parsley. Mol. Genet. Genomics 213, 93–98.

Williams, S.J., Sohn, K.H., Wan, L., Bernoux, M., Sarris, P.F., et al., 2014. Structural basis for assembly and function of a heterodimeric plant immune receptor. Science 344, 299–303. Available from: https://doi.org/10.1126/science.1247357.

Yao, N., Tada, Y., Sakamoto, M., Nakayashiki, H., Park, P., Tosa, Y., et al., 2002a. Mitochondrial oxidative burst involved in apoptotic response in oats. Plant J. 30, 567–579.

Zhang, J., Zhou, J.M., 2010. Plant immunity triggered by microbial molecular signatures. Mol. Plant 3 (5), 783–793.

Further reading

Cesari, S., Kanzaki, H., Fujiwara, T., Bernoux, M., Chalvon, V., et al., 2014. The NB-LRR proteins RGA4 and RGA5 interact functionally and physically to confer disease resistance. EMBO J. 33, 1941–1959.

Mondal, S., Rai, V.R., 2019. Molecular profiling of endophytic *Streptomyces cavourensis* MH16 inhabiting *Millingtonia hortensis* Linn. and influence of different culture media on biosynthesis of antimicrobial metabolites. Sci. Nat. 106 (9-10), 51.

Sinapidou, E., Williams, K., Nott, L., Bahkt, S., Tor, M., et al., 2004. Two TIR: NB: LRR genes are required to specify resistance to *Peronospora parasitica* isolate Cala2 in Arabidopsis. Plant J. 38, 898–909.

CHAPTER

5

Principle, diversity, mechanism, and potential of practical application of plant probiotic bacteria for the biocontrol of phytopathogens by induced systemic resistance

Musrat Zahan Surovy and Tofazzal Islam

Institute of Biotechnology and Genetic Engineering (IBGE), Bangabandhu Sheikh Mujibur Rahman Agricultural University, Gazipur, Bangladesh

5.1 Introduction

Plant-associated bacteria live in different organs of plant (leaf, shoot, and root) and trigger beneficial effects on plant growth and development and boost resistance to disease is termed as plant probiotic bacteria (PPB) (Islam and Hossain, 2012; Surovy et al., 2020). These bacteria play significant roles in the growth and development of plants useful for the sustainable crop production system. *Bacillus*, *Paraburkholderia*, *Pseudomonas*, *Acinetobacter*, *Alcaligenes*, *Arthrobacter*, and *Serratia* are important genera of probiotic bacteria, that promote plant growth and suppress phytopathogens effectively (Saharan, 2011; Khan et al., 2016; Rahman et al., 2018, Surovy et al., 2020). The PPB promote plant growth directly by nitrogen fixation, production of phytohormones, solubilization of nutrient elements in soils and/or by suppressing phytopathogens through secretion of antimicrobial compounds (Babu et al., 2015). The PPB also have immense potential to trigger induce systemic resistance (ISR) in host plants which is an indirect mechanism of action by the PPB to promote and protect plant growth (Abdelrahman et al., 2016). They naturally help to promote highly adaptive resistance in plants under adverse conditions. The association between PPB and host plants form symbiotic relations in the natural environments (Whipps, 2001; Rahman et al., 2018) (Figs. 5.1 and 5.2).

Food Security and Plant Disease Management
DOI: https://doi.org/10.1016/B978-0-12-821843-3.00004-0

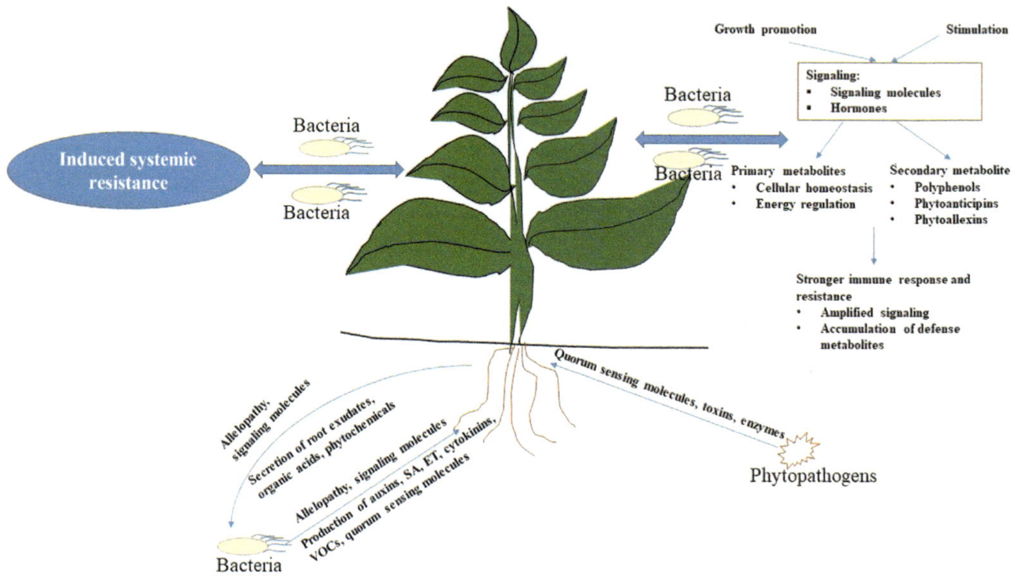

FIGURE 5.1 Interactions between plant probiotic bacteria (PPB) and phytopathogens for the development of induced systemic resistance (ISR).

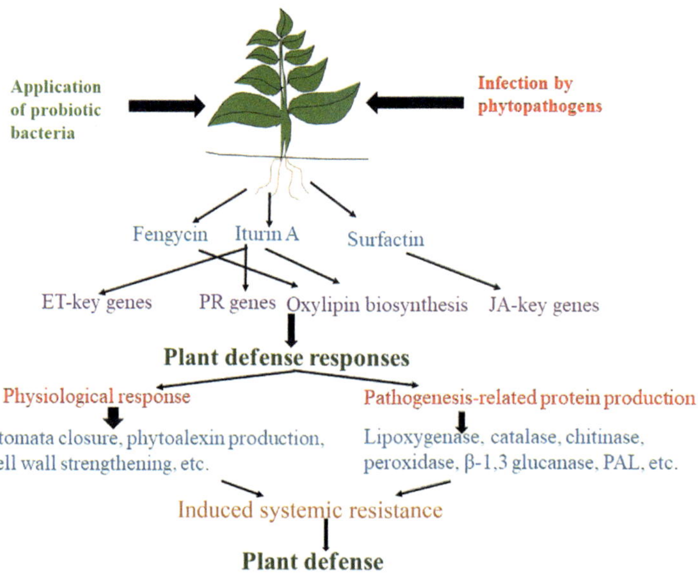

FIGURE 5.2 Mechanisms of induced systemic resistance in plants.

Outbreaks of fungal and other microbial diseases in plants are increasingly recognized as significant constrains in agricultural crop production and economic development (Fisher et al., 2012). There is an urgent need to develop control strategies against phytopathogens since the disease outbreaks are treated as important constrains in agricultural crop production that have become a matter of great concern of food security. The use of probiotic bacteria as microbial control agents is one of the best alternatives to control notorious phytopathogens (Kumari and Khanna, 2020). Biological control of phytopathogens is an important alternative to reduce the use of synthetic chemicals in agriculture (Gerhardson, 2002; Postma et al., 2003; Welbaum et al., 2004; Zohara et al., 2016). Priming seeds or seedlings with PPB provides systemic resistance against a broad spectrum of phytopathogens. By ISR, plants successfully develop defense mechanisms in the response to primary infection by the pathogen (Walters et al., 2013). Sometimes visible symptoms are not appeared in plants by the infection of phytopathogens. In this case, plant develops resistance against different phytopathogens, which is called ISR (Choudhary et al., 2007; Nguvo and Gao, 2019; Sharma et al., 2019). The PPB can develop ISR in host plants. Development of ISR was first observed in carnation plants against *Fusarium* spp. (Nguvo and Gao, 2019). The expression of ISR depends on the combination of host plant and the PPB. Activation of flagellins, production of siderophores, lipopolysaccharides, volatile organic compounds, and accumulation of salicylic acid (SA) by the PPB trigger ISR in plants (Muslim et al., 2019; Jiang et al., 2019). There is no clear evidence about how PPB can actively trigger ISR in the plant. The majority of PPB activates jasmonic acid (JA) and ethylene (ET) signaling pathways to trigger ISR (Fu and Dong, 2013). Nevertheless, some PPB also trigger SA-dependent pathway by inducing a very minute amount of SA in plants. Sensitivity or production of SA, JA, and ET in plants leads to the activation of different defense-related genes. As a result, ISR triggered by PPB fortifies the strength of plant cell wall, alters the physiology and metabolic responses of plants, resulting in increased synthesis of plant defense regulatory compounds under abiotic stress or pathogenic attack. Understanding detailed mechanisms of PPB for induction of ISR in the plant opens a new door to design strategies for improving the efficacy of biocontrol agents (Morrissey et al., 2002; Morrissey et al., 2004; Compant et al., 2005). Identification of key antimicrobial compounds, determination of the role of possible parameters for disease suppression, antibiotic production, and identification of different mechanisms of actions to suppress pathogens with a broader spectrum of microbial weapons produced by the PPB at the molecular level are highly needed. Continued work with PPB also holds the potential for developing biocontrol strategies against phytopathogens and induces resistance to plants.

A large body of literature is available on the roles of PPB on induction of ISR in the host plants, however, no comprehensive review has so far been published on principle, mechanism, and practical application of PPB for biocontrol of plant diseases through triggering ISR in the host plants. This chapter updates our understanding of basic principle, mechanism of action, and practical application of PPB to control phytopathogens by triggering ISR in the host plants.

5.2 Induced systemic resistance in plants

The ISR is a very complex mechanism occurred in the plant (Choudhary et al., 2007). The production of pathogenesis-related (PR) proteins is generally produced in plants by the attack of a phytopathogen, and also produced because of wounding, and/or necrosis which induced by the plant pathogens or other causes. These PR proteins alter the functional mechanism for the induction of systemic resistance (van Loon and Glick, 2004). The pathogenic mechanism relies on the production of SA (Hartmann and Zeier, 2019). The plant produces SA as a signaling molecule. On the other hand, the wounding or necrosis relies on the production of jasmonic acid (JA) and ET as the signaling molecules (Geilfus, 2019). These plant signaling compounds and their derivatives induce the same responses when they applied externally. The JA- and ET-induced pathways have been considered as ISR (Liu et al., 2019a,b). A cascade of PR proteins including antifungal enzymes (chitinases, glucanases, and thaumatins), and oxidative enzymes [viz., peroxidases, polyphenol oxidases (PPOs), and lipoxygenases] produce by salicylate- and jasmonic acid-induced pathways (Choudhary et al., 2007). Phytoalexins, low-molecular-weight compounds also accumulate in plants (Altinok and Yildiz, 2019). Nonpathogenic plant-associated bacteria also produce ISR, and it referred to bacteria-induced systemic resistance (BISR) (Kannojia et al., 2019). The BISR helps to develop systemic resistance against plant diseases. However, this mechanism is very different. The PR proteins and phytoalexins are not induced by probiotic bacteria in the absence of an attack of phytopathogen (Choudhary et al., 2007). Pattern-recognition receptors (PRRs) of plants are developed as special patterns to recognize common microbial compounds, such as bacterial flagellin or fungal chitin. Then, this pattern translated into the first line of defense against pathogens by pathogen-triggered immunity (PTI) (Mishina and Zeier, 2007). Effector triggered immunity (ETI) also a type of gene-for-gene-resistance, which is supplement by a programmed cell death at the site of infection that prevents further progression of pathogen in the host. The PTIs and ETIs both trigger induced resistance in healthy tissues of the host from the site of infection and provide signals that produce high defense capacity in undamaged plant parts (Mishina and Zeier, 2007). If a pathogen attacks a plant, then the degree of the plant responses to attack is also increased and the disease infestation is naturally reduced.

5.3 Plant probiotic bacteria–mediated induced systemic resistance

Haas and Keel first introduced the term "plant probiotic bacteria (PPB)" (Haas and Keel, 2003; Menendez and Garcia-Fraile, 2017). The PPB are a group of microorganisms that benefiting plants by effective and competitive colonization, induce systemic resistance in plants and play a role in suppressing the notorious phytopathogens. The PPB, which found in plant rhizosphere and soil, is termed as plant probiotic rhizospheric bacteria and the PPB, which found inside the plant cell are endophytic bacteria (Bhattacharyya and Jha, 2012; Afzal et al., 2019). Well-known PPB genera include *Pseudomonas, Bacillus, Serratia, Streptomyces, Azotobacter, Paraburkholderia, Rhizobium,* etc. (Compaqnt et al., 2010; Bhattacharyya and Jha, 2012; Rahman et al., 2018; Siyar et al., 2019;).

Biological or chemical inducers also trigger induced systemic resistance (ISR) in plants. The ISR protects plants from future attack and trigger a defense mechanism known as ISR, which is phenotypically similar to systemic acquired resistance (SAR). The SAR develops when plants successfully activate their defense mechanism in response to primary infection by a pathogen (Ryan et al., 2008). It induces a hypersensitive reaction through which it becomes limited to a local necrotic lesion of brown desiccated tissue (Van Loon et al., 1998), while ISR differs from SAR in that the inducing bacterium does not cause visible symptoms on the host plant (Van Loon et al., 1998). The ISR induces the defense mechanisms and guards unattacked portion of plants from future attack by the plant pathogens. Besides this, numerous endophytic bacteria induce SA along with jasmonic acid (JA) and ET-mediated ISR regulation (Kloepper and Ryu, 2006). The cytosol-specific transcription coregulator non-expression of pathogenesis-related gene 1 (*NPR1*) and transcription factors (TFs) MYB72 and MYC2, well studied for the induction of defense genes in the JA- and ET-mediated ISR regulation (Mihute et al., 2015). The study of cabbage colonized by endophytes in the greenhouse had season-long reduced black rot due to the induction of defense mechanisms (Jetiyanon, 1994). Further, Banana bunchy top disease caused by banana bunchy top virus (BBTV) is the most severe virus infection of the banana. The endophytic bacterial strains, *Pseudomonas* and *Bacillus* species isolated from the corm and roots of banana showed the ability to induce systemic resistance against BBTV in tissue-cultured banana plantlets with the help of enzyme-linked immunosorbent assay, dot immunobinding assay, and polymerase chain reaction.

The expression of defense-related enzymes and PR proteins was more on the plants treated with endophytic bacterial formulations than the control plants (Harish et al., 2008). Similarly, a wheat endophytic bacterium, *Pseudomonas aeruginosa* PW09, triggers an ISR in cucumber plants challenged with biotic stress of fungal pathogen, *Sclerotium rolfsii*. The endophytes alleviate the stress by protecting the plant mortality due to pathogen infection as well as the physiological state through phenylpropanoid metabolism, antioxidant activities, and proline accumulation. They also reduce 60% seedling mortality, accumulate higher proline, total phenolics, PPO, phenylalanine ammonia-lyase, and superoxide dismutase in *S. rolfsii* stressed plant (Pandey et al., 2012 and 2019). The SA transported to the roots, leaves from the inoculation point of PPB, and induces ISR (Khan et al., 2020). Root colonization with *Pseudomonas fluorescens* strain CHAO, and leaf inoculation with tobacco mozaic virus (TMV) in tobacco plants enhanced SA content in leaves (Maurhofer et al., 1994). Scientists found that PPB increase accumulation of phenol compounds in plants and activate carbohydrate bound phenolic compounds to defense against phytopathogens (Peer et al., 1991; Widnyana et al., 2019). Mutant strains of *Bacillus velezensis* FZB42 synthesize surfactin. When *Rhizoctonia solani* invades in lettuce plant, the FZB42 activates *CH1* and *CH5*, surfactin regulatory genes impaired in triggering of JA/ET-dependent ISR (Chowdhury et al., 2015). The lower expression of PDF1.2, inducible plant defense factor regulates the expression of the JA/ET and involved in triggering plant response (Borriss et al., 2019). This bacterium also produces bacillomycin D. Bacillomycin D triggers ISR against plant pathogens by expressing defense-related genes, mitogen-activated protein kinases phosphorylation, and alters pathogenicity against phytopathogens (Wu et al., 2018). Tomato seeds treated with PPB, performed better against *Alternaria solani*, which causes early blight infection in tomato (Babu et al., 2015). The PPB help in the accumulation of proline oxidase (POX) and PPO enzymes. These types of responses were also associated with the siderophores, HCN, chitinase, and glucanase production by the PPB (Table 5.1).

5.4 Role of induced systemic resistance in controlling phytopathogen

Plants harbor diverse group of beneficial bacteria in their leaf, root, and stem tissues. These bacteria play beneficial roles in the improvement of the plant health (Choudhary et al., 2007). When beneficial PPB colonize in plants, ISR elicits in distal parts of plant

TABLE 5.1 Summary of representative PPB, their mode of action in controlling phytopathogens.

Bacterial isolate	Host plant	Controlled phytopathogen	Mechanism of action	References
Bacillus subtilis SQR9	Cucumber	*Fusarium oxysporum* f. sp. *cucumerinum*	Biofilm formation, competition for root niches and nutrients, antibiosis, and induced systemic resistance in host plants	Cao et al. (2011); Wu et al. (2018)
B. subtilis ABS-S14	Mandarin	*Penicillium digitatum*	Activation of plant defensive system at transcriptional and proteomic levels	Tunsagool et al. (2019)
B. subtilis PTA-271	Grapevine	*Neofusicoccum parvum*	Expression of PR2 (encoding a β-1,3-glucanase) and NCED2 (9-cis-epoxycarotenoid dioxygenase involved in abscisic acid biosynthesis) genes upon pathogen challenge. And triggers the expression of SA and JA responsive genes, including *GST1* (encoding a glutathione-*S*-transferase)	Trotel-Aziz et al. (2019)
*Pseudomonas*WCS417	Carnation plants	*F. oxysporum* f. sp. *dianthi*	Secretion of elicitors like-lipopolysaccharides, peptidoglycans, flagellin, quorum-sensing molecules, cyclic lipopeptides, and iron-chelating siderophores	Bakker et al. (2007); Verhagen et al. (2011)
Bacillus cereus	*Arabidopsis thaliana*	*Pseudomonas syringae* pv. *tomato* DC3000	Expression of ET/JA responsive marker gene PDF1.2 and the salicylic acid-responsive genes (i.e., PR1, PR2, PR5).	Rabbee et al. (2019)
B. amyloliquefaciens FZB42	Lettuce	*Rhizoctonia solani*	Upregulate PDF1.2 gene	Chowdhury et al. (2015)
B. amyloliquefaciens SQRT3	Tomato	*Ralstonia solanacearum*	Upregulate the expression of several defense-related marker genes (i.e., Pin2, PR-1a, and Osmotin-like)	Li et al. (2017)
Pseudomonas aeruginosa PW09	Cucumber	*Sclerotium rolfsii*	Phenylpropanoid metabolism, antioxidant activities, and proline accumulation	Pandey et al. (2012)
Klebsiella spp. MBE02	Peanut plant	*Aspergillus* spp.	Modulate plants immunity by activation of jasmonic acid, ethylene, and pathogen defense-related signaling genes	Sharma et al. (2019)
B. subtilis UMAF6614	Cucurbit	*Podosphaera fusca*	Induce the secretion of SA and JA defense-related responses	Romero et al. (2007)
Enterobacter HA02	*Gossypium hirsutum* L.	*Verticillium dahliae*	Production of siderophore and protease enzyme	Li et al. (2012)
Streptomyces virginiae Y30 and E36	Tomato	*R. solanacearum*	Siderophores production and ACC deaminase activity	Tan et al. (2011)
		V. dahliae	Quorum sensing (signal interference)	

(*Continued*)

TABLE 5.1 (Continued)

Bacterial isolate	Host plant	Controlled phytopathogen	Mechanism of action	References
Serratia plymuthica HRO-C48	Oilseed rape			Müller et al. (2009)
S. marcescens UPM39B3	Banana	*F. oxysporum*	Induced systemic resistance in host plants	Ting et al. (2008)
Pseudomonas putida	Pigeon pea and chickpea	*F. oxysporum* f. sp. *melonis*	Produces phenazine-1-carboxylic acid and oxychlororaphin antibiotics and also induces defense-related genes	Planchamp et al. (2015)

body (Pieterse et al., 2014). To combat phytopathogens and herbivores, ISR has been widely reported. The JA- and ET-dependent signaling pathways express JA (e.g., PDF1.2) or ET responsive genes in response to pathogenic attack (Van der Ent et al., 2009). The ISR boost protective mechanism of resistance by improving the defense capacity of plants (Walters et al., 2013; Pieterse et al., 2014). It reduces disease incidence (Choudhary et al., 2007) and develops defense responses by forming structural barriers (thorns and trichomes), and toxin production (nicotine, phenolics, and terpenoids) (Goellner and Conrath, 2008). Probiotic bacteria can induce resistance in plant through hormonal regulation, control of JA and ET level, triggerg SAR pathway, augment structural barriers, and regulation of TFs for priming, etc. (Pieterse et al., 2014). *P. fluorescens* WCS417r develops ISR in radish against *Fusarium oxysporum* without accumulation of PR proteins (Hoffland et al., 1995). Without the accumulation of PR proteins, *P. fluorescens* WCS417r also develops ISR in *Arabidopsis* (Pieterse et al., 1996; Pieterse et al., 2014). The JA and ET are also important plant growth regulators of the plant immune system (Thomma *et al.*, 2001). In the *Arabidopsis* model system, it is clearly understood that JA and ET are key players in the regulation of PPB-mediated ISR (Nishimura and Dang, 2010). When *Serratia marcescens* 90–166, *Pseudomonas protegens* CHA0, and *P. fluorescens* Q2–87 applied to *Arabidopsis* plant, they regulate the JA and ET pathways and induce resistance in *Arabidopsis* against pathogenic attack (Iavicoli et al., 2003; Hossain et al., 2008; Stein et al., 2008; Weller et al., 2012). The JA and ET are also major players in the regulation of ISR in tomato and rice plants (De Vleesschauwer et al., 2008; Hase et al., 2008, Van der Ent et al., 2009). These JA and ET boost structural barriers in plants and help in the expression of defense-related genes.

P. fluorescens enhances cell wall appositions at the infection area by *Pythium ultimum* (Pieterse et al., 2014), also increases callose deposition in *Arabidopsis* when infected by downy mildew pathogen (Van der Ent et al., 2009). *Bacillus subtilis* FB17 induces ISR by closing stomata in the response to pathogen attack (Kumar et al., 2012).

The preparation of plant to better fighting against pathogen attack is termed as priming. It is considered as a strong stimulation of cellular defenses upon invasion, which results in a better level of resistance (Conrath et al., 2006). Van Peer et al. (1991) found that ISR in carnation is associated with the accumulation of phytoalexins at the site of pathogen infection (Kleopper et al., 2004; Wang et al., 2005; Ahn et al., 2007; Van Wees et al., 2008). A large number of probiotic bacteria supported the ISR based on priming. Enhanced transcriptional changes occurred in primed plants after pathogen attack and induced resistance in plants by improving the plant immune system (Conrath et al., 2006; Frost et al., 2008). Jain et al. (2015) found that the combination of *P. aeruginosa* PJHU15 and *B. subtilis* BHHU100 induced SA contents in pea leaves and inhibited the growth of phytopathogens. Through transcriptome analysis, *Pseudomonas* spp. mediating ISR in *Arabidopsis thaliana* have been investigated and transcriptomic data have shown that the systemic responses that are typically seen after attack by necrotizing pathogens. Verhagen et al. (2004) found that inoculation of plants with *Pseudomonas syringae* pv. *tomato*, showed that ISR-positive plants were countered faster and more intensely to pathogen attack by inducing defense mechanisms. The ISR elicited by *Pseudomonas putida* strain in bean associated with elevated levels of hexenal, a volatile antifungal compound (Ongena et al., 2004). *Pseudomonas* sp. GRP3 produced mono- and dirhamnolipids, which showed antimycelial activity against damping-off pathogen *Pythium* and *Phytophthora* (Sharma et al., 2007). *Pseudomonas* spp. produce rhamnolipids, that cause lysis of the plasma membrane of zoospores of *Phytophthora* sp. to facilitate control of damping-off plant pathogens (Arora et al., 2019). The SA-overproducing recombinant *P. fluorescens* strain P3 showed enhanced protection of tobacco against TMV by overproducing SA, which induced the defense capacity of tobacco plants (Maurhofer et al., 1998). In tomato plants, *Pseudomonas* sp. produces siderophores, pyocyanin, and pyochelin. *P. fluorescens* GRP3 showed ISR against rice sheath blight disease in rice plant (Singh and Sinha, 2005).

5.5 Molecular mechanism of induced systemic resistance by plant probiotic bacteria

Plant develops PAMP-triggered immunity (PTI) and ETI to prevent pathogenic invasion in plant body (Macho and Zipfel, 2014). Pathogen associated molecular patterns (PAMPs) and/or damage-associated molecular patterns (DAMPs) help in the early detection of phytopathogen in plants. When flagellation, oligogalacturonides, and chitin from fungal cell wall are detected by plant body then, PAMPs or DAMPs signaling pathways are activated in plants (Boller and Felix, 2009). NPR1, the cytosol-specific transcription coregulator and its TFs, MYB72 and MYC2 are responsible for the induction of JA–ET-dependent ISR regulatory defense genes (Miliute et al., 2015). This complicated mechanism is essential to ward-off notorious phytopathogens (Gómez-Gómez, 2004). The SAR and ISR are developed to fight distal invaders in plants (Nguvo and Gao, 2019). The SA, jasmonic acid (JA), and ET pathways are involved in triggering SAR and ISR (Heil and Bostock, 2002; Fu and Dong, 2013). The *NPR1*-mediated signaling pathway controlling the immune system to activate systemic resistance against distal phytopathogenic invasion (Walters and Heil, 2007; Robert-Seilaniantz et al., 2011; Nguvo and Gao, 2019). From extensive study on *Pseudomonas* spp., it is clear about their ability to trigger broad-spectrum ISR (Pieterse et al., 2014). The beneficial microbe triggers ISR on a specific host (Leeman et al., 1996; Berendsen et al., 2015). The type III secretion system (T3SS) that delivers immune-suppressive effector molecules into the host cell, emerged as a conserved mechanism in Gram-negative bacteria employed for effective colonization of their host (Alfano and Collmer, 2004; Deslandes and Rivas, 2012). The T3SS is a protein secretion nanomachine composed of approximately 30 proteins (Stringlis et al., 2019). Gray leaf spot disease caused by *Magnaporthe oryzae* is a serious disease in perennial ryegrass (*Lolium perenne*). A mutant strain of *B. velezensis* FZB42 (AK3) able to produce only surfactin but no other lipopeptides (bacillomycin D, fengycin), to induce ISR in ryegrass. A similar effect as in live cells obtained in the root-drench application of solid-phase extraction (SPE)-enriched surfactin. Treatment with *B. velezensis* FZB42 reduced disease incidence and severity on perennial ryegrass (Borriss et al., 2019). The ISR defense response also characterized by enhanced hydrogen peroxide (H_2O_2), elevated cell wall/apoplastic peroxidase activity, and deposition of callose and phenolic/polyphenolic compounds underneath the fungal appressoria in naïve leaves. A hypersensitive response (HR)-type reaction and enhanced expression of LpPrx (Prx, peroxidase), LpOXO4 (OXO, oxalate oxidase), LpPAL (PAL, phenylalanine ammonia-lyase), LpLOXa (LOX, lipoxygenase), LpTHb (putative defensin), and LpDEFa (DEFa, putative defensin) in perennial ryegrass were associated with SPE-enriched surfactin. Live AK3 cell treatments, acting as a second layer of defense when preinvasive defense responses failed (Rahman et al., 2015). Despite the importance of surfactin function, *B. velezensis* strains with deletions within the surfactin gene cluster are unable to synthesize this lipopeptide (He et al., 2012; Borriss et al., 2019).

Application of *P. fluorescens* and *Azospirillum brasilense* in oregano roots induced the accumulation of monoterpenes (thymol, carvacrol, sabinene hydrate, and γ-terpinene) in shoots. It modulates the gene expression of the host-plant terpene synthase and helps in chemical communication and defense against a broad spectrum of microbes (Banchio et al., 2010). *Klebsiella* spp. MBE02 strain alters JA homeostasis to trigger the ISR in peanut by induction of JA biogenesis genes (Sharma et al., 2019). Levels of total phenols, PAL, POX, LOX, and lipoxygenase enhanced by *Bacillus* spp., that indicate the involvement of ISR in PPB-mediated disease control (Xie et al., 2015). These bacteria also increase auxin levels in host plants and promote growth (Vacheron et al., 2013). The volatile organic compounds secreted by *B. subtilis* and *B. amyloliquefaciens* induce ISR in *Arabidopsis* by producing volatile organic compounds against a bacterial phytopathogen, *Erwinia carotovora* (Ryan et al., 2001; Ryu et al., 2004). 2, 3-butanediol (BDO), a volatile compound from *B. subtilis* GB03 bacterium elicits defense gene *CsLOX* in cucumber plants. This *CsLOX* gene is responsible for JA signaling and induced resistance against *P. syringae*
pv. *Lachrymans* in cucumber plants (Song et al., 2019). Similarly, volatiles from the *B. velezensis* FZB42 activate signal pathways positively to induce plant systemic resistance. Genes responsible for JA response (*VSP1* and *PDF1.2*) and SA response (*PR1* and *FMO1*) triggered in the leaves or roots of *Arabidopsis* plantlets after incubation with the volatiles. Volatiles from *B. velezensis* FZB42 in *Arabidopsis* roots also suppresses the infection by root nematodes (Hao et al., 2016).

Proteome analysis revealed that iturin A produced by *Bacillus* sp. attached to 12-oxophytodienoate reductase 2. This compound is solely involved in the oxylipin biosynthetic process produced JA in plants, and implicates ISR (Tunsagool et al., 2019). *Bacillus* sp. also produces fengycin and iturin in rice plants and elicits *ACS1* transcription levels. *ACS* and *ACO* gene expressions are prerequisite for ET biosynthesis and cause ISR induction in plants (Chandler et al., 2015). Chamam et al. (2013) found that Cigalon rice cultivar inoculated with *Azospirillum* strains causes' reduction of accumulation of feruloyl quinic acid in roots and enhanced lignification of root tissues triggered by ISR in rice. *Azospirillum* strains also cause the alternation of benzoxazinoid contents in maize and manipulate the induction of DIMBOA (2,4-dihydroxy-7-methoxy-2H-1, 4-benzoxazin-3

(4H)-one) as a biocontrol tool against notorious phytopathogens (Couillerot et al., 2013). *P. aeruginosa* isolates control chili anthracnose by producing phenylalanine ammonia-lyase (PAL) and PPO enzymes. The PAL regulates the biosynthesis of phenol in plants. The activation of PAL associated with disease resistance in plants. The PPO oxidizes phenolics to highly toxic quinones and terminates oxidation of diseased plant tissues. As a result, it also plays an important role in disease resistance (Jisha et al., 2018). The transcriptome sequencing of *B. velezensis* F21 revealed that it differentially expressed near about one thousand ripening-related genes and triggering ISR against *F. oxysporum* f. sp. *niveum* (Fon) in watermelon. It also activates and validates expression patterns of numerous TFs and disease resistance genes in the roots of watermelon. The mitogen-activated protein kinase signaling pathway and phytohormone signaling pathway were also modulated phytohormone signal factors and induce systemic resistance to Fon in watermelon (Jiang et al., 2019). *Brevibacillus laterosporus* A60 secreted protein elicitor, PeBL2, which induced a HR in tobacco (*Nicotiana benthamiana*). The PeBL2 induces HR-related transcription of defense genes along with the production of reactive oxygen species (ROS) and induce systemic resistance against *Botrytis cinerea* (Jatoi et al., 2019).

5.6 Expression pattern of induced systemic resistance in plant against phytopathogen

Tunsagool et al. (2019) demonstrated that *B. subtilis* ABS-S14 elicits activation of wounded mandarin fruit defensive system against *Penicillium digitatum* at transcriptional and proteomic levels. *B. subtilis* strain ABS-S14 also produces fengycin which has strong elicitation of defense-related gene transcription and produces PR protein and also increases the activities of glucanase (GLU), chitinase (CHI), peroxidase (POD), and LOX in Valencia fruit (Waewthongrak et al., 2014). Transcriptomic analysis showed that the PAL gene of *B. subtilis* ABS-S14 plays a crucial role in SA biosynthesis. After infection by *P. digitatum* in mandarin fruit, *B. subtilis* ABS-S14 triggers *ACS1* and *ACO* transcriptions and plays a vital role in the ET-signaling pathway (Tunsagool et al., 2019). ET-signaling pathway in rice also elicited by fengycin and the iturin family through *ACS1* transcription and helps in expression of the PR genes, which helps host plants to protect itself from plant pathogens (Chandler et al., 2015). For ISR induction, *ACS* and *ACO* genes were responsible for ET biosynthesis (Wang et al., 2002). Together with bacilysin, difficidin produced by *B. velezensis* FZB42 caused downregulated expression of genes involved in *Xanthomonas* virulence, cell division, and protein and cell wall synthesis (Wu et al., 2015). The PPB causes drought stress tolerance in wheat plants and induced systemic tolerance (Vurukonda et al., 2016). *B. subtilis* PTA-271 significantly protects grapevine from *Neofusicoccum parvum* and expresses *PR2* (encoding a b-1, 3-glucanase) and *NCED2* (9-cis-epoxycarotenoid dioxygenase involved in abscisic acid biosynthesis) genes in plants after attack by a pathogen. These two genes trigger the expression of SA and JA responsible genes, including GST1 (encoding a glutathione-*S*-transferase) helps in detoxification of (−)-terremutin and (*R*)-mellein phytotoxin produced by the pathogen. As a result, the *Bacillus* spp. helps in enhancing host immune responses and detoxifying phytotoxins produced by *N. parvum* (Trotel-Aziz et al., 2019).

P. fluorescens N21.4 induces systemic resistance in *A. thaliana* against *P. syringae* DC3000. *P. fluorescens* N21.4 involved in the expression of SA signaling pathway (*NPR1, PR1, ICS,* and *PR2*) and the JA/ET-signaling pathways (*PDF1, MYC2, LOX2,* and *PR3*). It also helps in overexpression of *ICS, PR1, PR2, PDF1,* and *LOX2* genes and reduces ROS as well as decreases malondialdehyde (Martin-Rivilla et al., 2019). Through the qPCR method, it was demonstrated that *B. subtilis* RSS-1 had significantly increased the expression levels of *PR1, PR10,* and *LOX7* genes in soybean plants. The *PR2* and *PR3* genes also slightly expressed in soybean due to RSS-1. *B. amyloliquefaciens* JDF3 probiotic bacteria showed a significant upregulated expression of all five genes. However, the *PR2* gene expressed after 3 h and *LOX7* gene after 12 h of inoculation. As a result, these two-probiotic bacteria effectively control *Phytophthora* rot in soybean and stimulate resistance in soybean (Upchurch and Ramirez, 2011; Liu et al., 2019a,b). *Bacillus* Fcl1 has the ability for PAMP or fungal elicitors to elicit miRNAs, which target auxin receptors such as *TIR, AFB2,* and *ABF3* with the upregulation of plant defense genes such as PR and HR genes (Jayakumar et al., 2019).

5.7 Relationship of induced systemic resistance with plant growth and biocontrol

A large body of evidence suggests that PPB is capable of managing phytopathogens in crops via inducing ISR and promotes plant growth. Type IV pili encoded by *pilAB* (Dorr et al., 1998), lipopolysaccharide (Balsanelli et al., 2010), and exopolysaccharide (Meneses et al., 2011) along with other bacterial surface components are essential for bacterial colonization in plants (Pandey et al., 2019). After successful colonization into the plant tissue, PPB indirectly induced ISR to control phytopathogens (Niu et al., 2011). Combined application of *Corynebacterium flavescens* and *Bacillus pumilus* in rice enhance growth and control rice pathogen *A. brasilense* (Bacilio-JimeÂnez et al., 2001). The PPB isolated from root and stem of tomato is capable to control *Ralstonia solanacearum*, in tomato plants and increase the tomato yield (Purnawati et al., 2014). Pea seeds priming with *P. fluorescens* strain 63-28 protect pea plants from infection of *P. ultimum* and *F. oxysporum* f. sp. *pisi*. This probiotic bacterium causes the apposition of the cell wall, deposition of newly formed callose and accumulates phenolic compounds at the penetration sites of invading hyphae. As a result, the formation of structural barriers produced control phytopathogen biologically (Ramamoorthy et al., 2001). Association of *Klebsiella* spp. MBE02 with peanut plants significantly reprograms the expression patterns of hormone signaling genes, which involve in growth promotion of peanut. This MBE02 strain also helps in differentially express genes related to plant defense as well as provides immunity to the plants (Sharma et al., 2019). *P. fluorescens* induced resistance in rice plants against the sheath blight disease-causing fungus and played a double role in reducing disease severity and plant growth promotion (Nanda Kumar et al., 2001).

Pseudomonas spp. WCS14r protects carnation systemically from Fusarium wilt. *Serratia* spp. protect cucumber plants systemically from Fusarium wilt and *P. fluorescens* strain FP7 protects rice from sheath blight pathogen upon systemic seed treatment, and protects sugarcane by the same strain against the red rot fungus systemically. Microcapsules of *B. velezensis* NH-1 suppress *F. oxysporum*, causing Fusarium wilt in cucumber. In the field experiment, alginate-coated monolayer microcapsules of *B. velezensis* NH-1 are capable to control 100% *Fusarium* wilt. Monolayer alginate helps in the slow release of *B. velezensis* NH-1 in soils and significantly improves its biocontrol efficiency on cucumber *Fusarium* wilt (Luo et al., 2019). *B. amyloliquefaciens* JDF3 effectively controls Phytophthora rot disease caused by *Phytophthora sojae. B. amyloliquefaciens* JDF3 induces active oxygen burst, callose deposition, and lignification, which results in inhibition of mycelial growth, effective or no germination of the cysts, and stop swimming behavior of the motile zoospores. *B. subtilis* RSS-1 also capable to induce systemic resistance against *P. sojae* pathogen by lignification and production of phytoalexin (Liu et al., 2019a,b). *Pseudomonas* spp., and *Bacillus* spp. are well reported for their ability to produce the phenolic compounds (Korenblum and Aharoni, 2019) and these phenolic compounds can inhibit the growth of phytopathogens (Chomel et al., 2016). The *Bacillus* Fcl1 produces lipopeptide, which controls the phytopathogens and synthesizes indole acetic acid (IAA) and 1-aminocyclopropane-1-carboxylate (ACC) deaminase, that promotes plant growth and development (Jayakumar et al., 2019). Tomato seed coating with PPB can reduce 8%–9% of pathogen attack in the greenhouse condition (Pandey et al., 2019). Coinoculation of *Pseudomonas* sp. R41805 with mycorrhizal fungi, *Rhizophagus irregularis* MUCL 41833 in potato plants activates the systemic defense genes against *R. solani* infection through the ET resistance network (Enebe and Babalola, 2019). *Pseudomonas* sp. is effective in increasing shoot length with increased activity of peroxidase and PAL. Peroxidase and PAL are two well-known indicators for the lignification process (van Loon and Glick, 2004). *Bacillus* spp. produces plant growth stimulating volatile 2,3-butanediol and initiates ISR (Petrova et al., 2019). *Pseudomonas* strain GRP3 produces siderophores in mung bean and reduces chlorotic symptoms and enhanced chlorophyll level in mung bean plants. It also enhances peroxidase activity and decreases catalase activity in roots. It improves iron availability to plants and reduces the use of synthetic fertilizers (Sharma et al., 2003) (Table 5.2).

5.8 Conclusion and future perspectives

Plants maintain a complex physiological process for acclimatization in nature. Different environmental stresses influence plant growth and develop the disease and retard the growth of plants. Naturally, plants harbor a complex microbiota to fulfill important plant functions (growth, development, vigor, and defense). The PPB can stimulate plant immunity and ward-off phytopathogens as well as the development of resistance system in plants. Plant health–supporting probiotic bacteria may hold the key to the next generation green revolution in agriculture. Interaction of PPB with plant stimulates induce systemic resistance in plants and modulates plant immune system. The ISR is an important subject area for further research. There is a major gap of knowledge on the recognition of beneficial microbes by plant toward plant growth and develop resistance against stresses. However, the molecular understanding between PPB and ISR have already been expanded, but still, some major questions need to be resolved. It is high time to understand the knowledge of recognition of beneficial microbes by the plant cells toward enhanced growth, development, and stress resistance in plants. Metagenomic, genomics, and postgenomics and quantitative plant genetics studies may play a vital role in molecular understanding of PPB and ISR relationship. These studies will facilitate better

TABLE 5.2 Role PPB in biocontrol and plant growth promotion.

PPB	Functions	Mechanisms of action	References
Pseudomonas fluorescens SS101	Increase the biomass of *Arabidopsis thaliana*	Activation of VOCs	Park et al. (2015)
Bacillus subtilis SQR9	Control of *Fusarium* wilt of cucumber	Biofilm formation	Cao et al. (2011)
B. subtilis HJ5	Control of *Verticillium* wilt of cotton	Root colonization and biofilm formation	Li et al. (2013)
Bacillus amyloliquefaciens L3	Control *Fusarium* wilt of watermelon	Production of 2-nonanone, 2-heptanone, and induce systemic resistance	Wu et al. (2019)
P. fluorescens UM270	Promotes plant growth	Production of dimethylhexadecylamine	Hernandez-Leon et al. (2015)
Bacillus velezensis ZSY-1,	Inhibits growth of *Botrytis cinerea*	Production of pyrazine and benzothiazole	Yuan et al. (2012)
B. subtilis GB03	Promote *Arabidopsis* growth	Production of 2,3-butanediol and acetoin and induce systemic resistance	Wu et al. (2019)
B. subtilis ABS-S14	Stimulate plants immunity by ISR	Production of cyclic lipopeptides (CLPs)	Tunsagool et al. (2019)
B. amyloliquefaciens	Induce immunity and control phytopathogens in vegetables	Produce lipopeptide	Saravanakumar et al. (2019)
Bacillus Fcl1	Plant growth and suppression of phytopathogens	Production of lipopeptides, IAA, and ACC deaminase	Jayakumar et al. (2019)

understanding the relationship between PPB and plant health. However, knowledge of the molecular and genetic basis of plant-beneficial bacterial communication in the context of its evolutionary and environmental relevance will be highly influential for the development of sustainable future crop production.

Acknowledgments

The authors would like to thank DAAD (German Academic Exchange Service) for providing financial assistance as a Doctoral fellowship to Musrat Zahan Surovy. The authors also thankful to the Bangladesh Academy of Sciences for funding through a BAS-USDA project to Tofazzal Islam.

References

Abdelrahman, M., Abdel-Motaal, F., El-Sayed, M., Jogaiah, S., Shigyo, M., et al., 2016. Dissection of *Trichodermafv longibrachiatum*-induced defense in onion (Allium cepa L.) against *Fusarium oxysporum* f. sp. cepa by target metabolite profiling. Plant Sci. 246, 128–138. Available from: https://doi.org/10.1016/j.plantsci.2016.02.008. PMID: 26993243.

Afzal, I., Shinwari, Z.K., Sikandar, S., Shahzad, S., 2019. Plant beneficial endophytic bacteria: mechanisms, diversity, host range and genetic determinants. Microbiol. Res 221, 36–49.

Ahn, I.-P., Lee, S.-W., Suh, S.-C., 2007. Rhizobacteria-induced priming in *Arabidopsis* is dependent on ethylene, jasmonic acid, and NPR1. Mol. Plant Microbe Interact. 20, 759–768.

Alfano, J.R., Collmer, A., 2004. Type III secretion system effector proteins: double agents in bacterial disease and plant defense. Annu. Rev. Phytopathology 42, 385–414.

Altinok, H.H., Yildiz, H.N., 2019. Induced systemic resistance by plant growth promoting rhizobacteria in control of plant disease. Curr. Trends Nat. Sci. 8 (16), 125–133.

Arora, N.K., Fatima, T., Mishra, I., Verma, S., 2019. Microbe-based inoculants: role in next green revolution. Environmental Concerns and Sustainable Development. pp. 191–246.

Babu, N., Jogaiah, A., Ito, S., Kestur Nagaraj, A., Tran, L.-S.P., 2015. Improvement of growth, fruit weight and early blight disease protection of tomato plants by rhizosphere bacteria is correlated with their beneficial traits and induced biosynthesis of antioxidant peroxidase and polyphenol oxidase. Plant Sci. 231, 62–73.

Bacilio-JimeÂnez, M., Aguilar-Flores, S., del Valle, M.V., PeÂrez, A., Zepeda, A., Zenteno, E., 2001. Endophytic bacteria in rice seeds inhibit early colonization of roots by *Azospirillum brasilense*. Soil Biol. Biochem. 33, 167–172. Available from: https://doi.org/10.1016/S0038-0717(00)00126-7.

Bakker, P.A.H.M., Pieterse, C.M.J., van Loon, L.C., 2007. Induced systemic resistance by fluorescent *Pseudomonas* spp. Phytopathology 97, 239–243.

Balsanelli, E., Serrato, R.V., de Baura, V., Sassaki, G., Yates, M.G., Rigo, L.U., et al., 2010. *Herbaspirillum seropedicae* rfbB and rfbC genes are required for maize colonization. Environ. Microbiol. 12, 2233–2244. Available from: https://doi.org/10.1111/j.1462-2920.2010.02187.x.

Banchio, E., Bogino, P.C., Santoro, M., Torres, L., Zygadlo, J., Giordano, W., 2010. Systemic induction of monoterpene biosynthesis in *origanum* × *majoricum* by soil bacteria. J. Agric. Food Chem. 58, 650–654.

Berendsen, R.L., Verk, M.C.V., Stringlis, I.A., Zamioudis, C., Tommassen, J., Pieterse, C.M.J., et al., 2015. Unearthing the genomes of plant-beneficial *Pseudomonas* model strains WCS358, WCS374 and WCS417. BMC Genomics 16, 539. Available from: https://doi.org/10.1186/s12864-015-1632-z.

Bhattacharyya, P.N., Jha, D.K., 2012. Plant growth-promoting rhizobacteria (PGPR): emergence in agriculture. World J. Microbiol. Biotechnol. 28 (4), 1327–1350.

Boller, T., Felix, G., 2009. A renaissance of elicitors: perception of microbe-associated molecular patterns and danger signals by pattern-recognition receptors. Annu. Rev. Plant Biol. 60, 379–406. Available from: https://doi.org/10.1146/annurev.arplant.57.032905.105346.

Borriss, R., Wu, H., Gao, X., 2019. Secondary metabolites of the plant growth promoting model rhizobacterium *Bacillus velezensis* FZB42 are involved in direct suppression of plant pathogens and in stimulation of plant induced systemic resistance. In: Singh, H.B., Keswani, C., Reddy, M.S., Royano, E.S., Garcia-Estrada, C. (Eds.), Secondary Metabolites of Plant Growth Promoting Rhizomicroorganisms. Springer Nature, Singapore, pp. 147–168.

Cao, Y., Zhang, Z., Ling, N., Yuan, Y., Zheng, X., Shen, B., et al., 2011. *Bacillus subtilis* SQR 9 can control Fusarium wilt in cucumber by colonizing plant root. Biol. Fertil. Soils 47, 495–506.

Chamam, A., Sanguin, H., Bellvert, F., Meiffren, G., Comte, G., Wisniewski-Dyé, F., et al., 2013. Plant secondary metabolite profiling evidences strain-dependent effect in the *Azospirillum-Oryza sativa* association. Phytochemistry. 87, 65–77.

Chandler, S., Van Hese, N., Coutte, F., Jacques, P., Höfte, M., De Vleesschauwer, D., 2015. Role of cyclic lipopeptides produced by *Bacillus subtilis* in mounting induced immunity in rice (*Oryza sativa* L.). Physiol. Mol. Plant Pathol. 91, 20–30. Available from: https://doi.org/10.1016/j.pmpp.2015.05.010.

Chomel, M., Guittonny-Larchevêque, M., Fernandez, C., Gallet, C., DesRochers, A., Paré, D., et al., 2016. Plant secondary metabolites: a key driver of litter decomposition and soil nutrient cycling. J. Ecol. 104, 1527–1541.

Choudhary, D.K., Prakash, A., Johri, B.N., 2007. Induced systemic resistance (ISR) in plants: mechanism of action. Indian J. Microbiol. 47, 289–297.

Chowdhury, S.P., Uhl, J., Grosch, R., Alquéres, S., Pittroff, S., Dietel, K., et al., 2015. Cyclic lipopeptides of *Bacillus amyloliquefaciens* subsp. plantarum colonizing the lettuce rhizosphere enhance plant defense responses toward the bottom rot pathogen *Rhizoctonia solani*. Mol. Plant Microbe Interact. 28, 984–995.

Compant, S., Clément, C., Sessitsch, A., 2010. Plant growth-promoting bacteria in the rhizo-and endosphere of plants: their role, colonization, mechanisms involved and prospects for utilization. Soil Biol. Biochem. 42 (5), 669–678.

Compant, S., Duffy, B., Nowak, J., Clement, C., Barka, E.A., 2005. Use of plant growth-promoting bacteria for biocontrol of plant diseases: principles, mechanisms of action, and future prospects. App. Environ. Microbiol. 4951–4959.

Conrath, U., Beckers, G.J.M., Flors, V., García-Agustín, P., Jakab, G., et al., 2006. Priming: getting ready for battle. Mol. Plant Microbe Interact. 19, 1062–1071.

Couillerot, O., Ramírez-Trujillo, A., Walker, V., Von Felten, A., Jansa, J., Maurhofer, M., et al., 2013. Comparison of prominent *Azospirillum* strains in *Azospirillum-Pseudomonas-Glomus* consortia for promotion of maize growth. Appl. Microbiol. Biotechnol. 97, 4639–4649.

Deslandes, L., Rivas, S., 2012. Catch me if you can: bacterial effectors and plant targets. Trends Plant. Sci. 17, 644–655.

De Vleesschauwer, D., Djavaheri, M., Bakker, P.A.H.M., Höfte, M., 2008. Pseudomonas fluorescens WCS374r–induced systemic resistance in rice against *Magnaporthe oryzae* is based on pseudobactin-mediated priming for a salicylic acid–repressible multifaceted defense response. Plant Physiol. 148, 1996–2012.

Dorr, J., Hurek, T., Reinhold-Hurek, B., 1998. Type IV pili are involved in plant-microbe and fungus-microbe interactions. Mol. Microbiol. 30, 7–17. Available from: https://doi.org/10.1046/j.1365-2958.1998.01010.x.

Enebe, M.C., Babalola, O.O., 2019. The impact of microbes in the orchestration of plants resistance to biotic stress: a disease management approach. Appl. Microbiol. Biotechnol. 103, 9–25.

Fisher, M.C., Henk, D.A., Briggs, C.J., Brownstein, J.S., Madoff, L.C., Mccraw, S.L., et al., 2012. Emerging fungal threats to animal, plant and ecosystem health. Nature 484, 186–194.

Frost, C.J., Mescher, M.C., Carlson, J.E., De Moraes, C.M., 2008. Plant defense priming against herbivores: getting ready for a different battle. Plant Physiol. 146, 818–824.

Fu, Z.Q., Dong, X., 2013. Systemic acquired resistance: turning local infection into global defense. Annu. Rev. Plant Biol. 64, 839–863.

Geilfus, C.-M., 2019. Wounding. In: Controlled Environment Horticulture, pp. 113-120. Available from: https://doi.org/10.1007/978-3-030-23197-2_10.

Gerhardson, B., 2002. Biological substitutes for pesticides. Trends Biotechnol. 20, 338–343.

Goellner, K., Conrath, U., 2008. Priming: it's all the world to induced disease resistance. In: Sustainable Disease Management in a European Context, pp. 233-242. Available from: https://doi.org/10.1007/978-1-4020-8780-6_3.

Gómez-Gómez, L., 2004. Plant perception systems for pathogen recognition and defense. Mol. Immunol. 41, 1055–1062. Available from: https://doi.org/10.1016/j.molimm.2004.06.008.

Haas, D., Keel, C., 2003. Regulation of antibiotic production in root-colonizing *Pseudomonas* spp. and relevance for biological control of plant disease. Ann. Rev. Phytopathol. 41, 117–153.

Harish, S., Kavino, M., Kumar, N., Saravanakumar, D., Soorianathasundaram, K., Samiyappan, R., 2008. Biohardening with plant growth promoting rhizosphere and endophytic bacteria induces systemic resistance against banana bunchy top virus. Appl. Soil. Ecol. 39, 187–200.

Hartmann, M., Zeier, J., 2019. N-hydroxypipecolic acid and salicylic acid: a metabolic duo for systemic acquired resistance. Curr. Opin. Plant Biol. 50, 44–57.

Hao, H.-T., Zhao, X., Shang, Q.-H., Wang, Y., Guo, Z.-H., Zhang, Y.-B., et al., 2016. Comparative digital gene expression analysis of the Arabidopsis response to volatiles emitted by *Bacillus amyloliquefaciens*. PLoS ONE 11 (8), e0158621. Available from: https://doi.org/10.1371/journal.pone.0158621.

Hase, S., Takahashi, S., Takenaka, S., Nakaho, K., Arie, T., et al., 2008. Involvement of jasmonic acid signaling in bacterial wilt disease resistance induced by biocontrol agent *Pythium oligandrum* in tomato. Plant Pathol. 57, 870–876.

He, P., Hao, K., Blom, J., Rückert, C., Vater, J., Mao, Z., et al., 2012. Genome sequence of the plant growth promoting strain *Bacillus amyloliquefaciens* subsp. *plantarum* B9601-Y2 and expression of mersacidin and other secondary metabolites. J. Biotechnol. 164 (2), 281–291. Available from: https://doi.org/10.1016/j.jbiotec.2012.12.014.

Heil, M., Bostock, R.M., 2002. Induced systemic resistance (ISR) against pathogens in the context of induced plant defences. Ann. Bot. 89, 503–512. Available from: https://doi.org/10.1093/aob/mcf076.

Hernandez-Leon, R., Rojas-Solis, D., Contreras-Perez, M., et al., 2015. Characterization of the antifungal and plant growth-promoting effects of diffusible and volatile organic compouns produced by *Pseudomonas fluorescens* strains. Biol. Control 81, 83–92.

Hoffland, E., Pieterse, C.M.J., Bik, L., Van Pelt, J.A., 1995. Induced systemic resistance in radish is not associated with accumulation of pathogenesis-related proteins. Physiol. Mol. Plant Pathol. 46, 309–320.

Hossain, M.M., Sultana, F., Kubota, M., Hyakumachi, M., 2008. Differential inducible defense mechanisms against bacterial speck pathogen in *Arabidopsis thaliana* by plant-growth-promoting-fungus *Penicillium* sp. GP16-2 and its cell free filtrate. Plant Soil 304, 227–239.

Iavicoli, A., Boutet, E., Buchala, A., Métraux, J.-P., 2003. Induced systemic resistance in *Arabidopsis thaliana* in response to root inoculation with *Pseudomonas fluorescens* CHA0. Mol. Plant Microbe Interact. 16, 851–858.

Islam, M.T., Hossain, M.M., 2012. Plant probiotics in phosphorus nutrition in crops, with special reference to rice. In: Maheshwari, D.K. (Ed.), Bacteria in Agrobiology: Plant Probiotics. Springer, Berlin, pp. 325–363.

Jain, A., Singh, A., Singh, S., Singh, H.B., 2015. Phenols enhancement effect of microbial consortium in pea plants restrains *Sclerotinia sclerotiorum*. Biol. Control 89, 23–32.

Jatoi, G.H., Lihua, G., Xiufen, Y., Gadhi, M.A., Keerio, A.U., Abdulle, Y.A., et al., 2019. A novel protein elicitor PeBL2, from *Brevibacillus laterosporus* A60, induces systemic resistance against *Botrytis cinera* in tobacco plant. Plant Pathol. J. 35 (3), 208–218.

Jayakumar, A., Krisna, A., Mohan, M., Nair, I.C., Radhakrishnan, E.K., 2019. Plant growth enhancement, disease resistance, and elemental modulatory effects of plant probiotic endophytic *Bacillus* sp. Fcl 1. Probiotics Antimicrob. Proteins 11, 526–534.

Jetiyanon K., Immunization of cabbage for long-term resistance to black rot. M.S. Thesis, 1994, Plant Pathology, Auburn University, Auburn, Alabama.

Jiang, C.-H., Yao, X.-F., Mi, D.-D., Li, Z.-J., Yang, B.-Y., Zheng, Y., et al., 2019. Comparative transcriptome analysis reveals the biocontrol mechanism of *Bacillus velezensis* F21 against Fusarium wilt on watermelon. Front. Microbiol. 10, 652. Available from: https://doi.org/10.3389/fmicb.2019.00652.

Jisha, M.S., Linu, M.S., Sreekumar, J., 2018. Induction of systemic resistance in chilli (*Capsicum annuum* L.) by *Pseudomonas aeruginosa* against anthracnose pathogen *Colletotrichum capsici*. J. Trop. Agric. 56 (2), 153–166.

Kannojia, P., Choudhary, K.K., Srivastava, A.K., Singh, A.K., 2019. PGPR bioelicitors: induced systemic resistance (ISR) and proteomic perspective on biocontrol. Food Sec. Environ. Manag. 67–84. Available from: https://doi.org/10.1016/B978-0-12-815879-1.00004-5.

Khan, N., Bano, A., Ali, S., Babar, M.A., 2020. Crosstalk amongst phytohormones from planta and PGPR under biotic and abiotic stresses. Plant Growth Regul. Available from: https://doi.org/10.1007/s10725-020-00571-x.

Khan, M.M.A., Khatun, A., Islam, M.T., 2016. Promotion of plant growth by phytohormone producing bacteria. In: Garg, N., Aeron, A. (Eds.), Microbes in Action. Nova Science Publishers, New York, pp. 45–76.

Kloepper, J.W., Ryu, C.-M., Zhang, S.A., 2004. Induced systemic resistance and promotion of plant growth by *Bacillus* spp. Phytopathology 94, 1259–1266.

Kloepper, J.W., Ryu, C.M., 2006. Bacterial Endophytes as Elicitors of Induced Systemic Resistance Microbial Root EndophytesIn: Schulz, B.J.E., Boyle, C.J.C., Sieber, T.N. (Eds.), Soil Biology, 9. Springer, Berlin, Heidelberg. Available from: https://doi.org/10.1007/3-540-33526-9_3.

Korenblum, E., Aharoni, A., 2019. Phytobiome metabolism: beneficial soil microbes steer crop plants' secondary metabolism. Pest. Manag. Sci. 75, 2378–2384.

Kumar, A.S., Lakshmanan, V., Caplan, J.L., Powell, D., Czymmek, K.J., et al., 2012. Rhizobacteria *Bacillus subtilis* restricts foliar pathogen entry through stomata. Plant J. 72, 694–706.

Kumari, S., Khanna, V., 2020. Induction of systemic resistance in chickpea (*Cicer arietinum* L.) against *Fusarium oxysporum* f. sp. *ciceris* by antagonistic rhizobacteria in assistance with native mesorhizobium. Curr. Microbiol. 77, 85–98.

Leeman, M., 1996. Iron availability affects induction of systemic resistance to fusarium wilt of radish by *Pseudomonas fluorescens*. Phytopathology 86 (1996), 149–155.

Li, C.Y., Hu, W.C., Pan, B., Liu, Y., Yuan, S.F., Ding, Y.Y., et al., 2017. Rhizobacterium *Bacillus amyloliquefaciens* strainSQRT3-mediated induced systemic resistance controls bacterial wilt of tomato. Pedosphere 27, 1135–1146.

Li, C.-H., Shi, L., Han, Q., Hu, H.-L., Zhao, M.-W., Tang, C.-M., et al., 2012. Biocontrol of verticillium wilt and colonization of cotton plants by an endophytic bacterial isolate. J. Appl. Microbiol. 113, 641–651.

Li, S., Zhang, N., Zhang, Z., 2013. Antagonist *Bacillus subtilis* HJ5 controls Verticillium wilt of cotton by root colonization and biofilm formation. Biol. Fert. Soils 49 (3), 295–303. Available from: https://doi.org/10.1007/s00374-012-0718-x.

Liu, D., Li, K., Hu, J., Wang, W., Liu, X., Gao, Z., 2019a. Biocontrol and action mechanism of *Bacillus amyloliquefaciens* and *Bacillus subtilis* in soybean phytophthora blight. Int. J. Mol. Sci. 20, 2908.

Liu, N., Luo, X., Tian, Y., Lai, D., Zhang, L., Lin, F., et al., 2019b. The stereoisomeric *Bacillus subtilis* HN09 metabolite 3,4-dihydroxy-3-methyl-2-pentanone induces disease resistance in Arabidopsis via different signaling pathways. BMC Plant Biol. 19, 384.

Luo, W., Liu, L., Qi, G., Yang, F., Shi, X., Zhao, X., 2019. Embedding *Bacillus velezensis* NH-1 in microcapsules for biocontrol of cucumber *Fusarium* wilt. Appl. Environ. Microbiol. 85, e03128-18. Available from: https://doi.org/10.1128/AEM.03128-18.

Macho, A.P., Zipfel, C., 2014. Plant PRRs and the activation of innate immune signaling. Mol. Cell 54, 263-272. Available from: https://doi.org/10.1016/j.molcel.2014.03.028.

Martin-Rivilla, H., Garcia-Villaraco, A., Ramos-Solano, B., Gutierez-manero, F.J., Lucas, J.A., 2019. Extracts from cultures of *Pseudomonas fluorescens* induce defensive patterns of gene expression and enzyme activity while depressing visible injury and reactive oxygen species in *Arabidopsis thaliana* challenged with pathogenic *Pseudomonas syringae*. AoB Plants 11, 1–9.

Maurhofer, M., Hase, C., Meuwly, P., Metraux, J.P., Defago, G., 1994. Induction of systemic resistance of tobacco to tobacco necrosis virus by the root-colonizing *Pseudomonas fluorescens* strain CHA0: influence of the gacA gene and of pyoverdine production. Phytopathology 84, 139–146.

Maurhofer, M., Reimmann, C., Schmidli-Sacherer, P., Heeb, S., Haas, D., Defago, G., 1998. Salicylic acid biosynthetic genes expressed in *Pseudomonas fluorescens* strain P3 improve the induction of systemic resistance in tobacco against tobacco necrosis virus. Phytopathology 88, 678–684.

Menendez, E., Garcia-Fraile, P., 2017. Plant probiotic bacteria: solutions to feed the world. AIMS Microbiol. 3 (3), 502–524.

Meneses, C.H.S.G., Rouws, L.F.M., Simoes-Araujo, J.L., Vidal, M.S., Baldani, J.I., 2011. Exopolysaccharide production is required for biofilm formation and plant colonization by the nitrogen-fixing endophyte *Gluconacetobacter diazotrophicus*. Mol. Plant Microbe Interact. 24, 1448–1458. Available from: https://doi.org/10.1094/MPMI-05-11-0127.

Miliute, I., Buzaite, O., Baniulis, D., Stanys, V., 2015. Bacterial endophytes in agricultural crops and their role in stress tolerance: a review. Zemdirbyste 102 (4), 465–478. Available from: https://doi.org/10.13080/z-a.2015.102.060.

Mishina, T.E., Zeier, J., 2007. Pathogen-associated molecular pattern recognition rather than development of tissue necrosis contributes to bacterial induction of systemic acquired resistance in *Arabidopsis*. Plant J. 50, 500–513.

Morrissey, J.P., Dow, J.M., Mark, G.L., O'Gara, F., 2004. Are microbes at the root of a solution to world food production? EMBO Rep. 5, 922–926.

Morrissey, J.P., Walsh, U.F., O'Donnell, A., Moënne-Loccoz, Y., O'Gara, F., 2002. Exploitation of genetically modified inoculants for industrial ecology applications. Antonie van Leeuwenhoek 81, 599–606.

Müller, H., Westendorf, C., Leitner, E., Chernin, L., Riedel, K., Schmidt, S., et al., 2009. Quorum-sensing effects in the antagonistic rhizosphere bacterium *Serratia plymuthica* HRO-C48. FEMS Microbiol. Ecol. 67, 468–478.

Muslim, A., Hyakumachi, M., Kageyama, K., Suwandi, S., 2019. Induction of systemic resistance in cucumber by hypovirulent binucleate Rhizoctonia against anthracnose caused by *Colletotrichum orbiculare*. Trop. Life Sci. Res. 30 (1), 109–122.

Nanda Kumar, R., Babu, S., Vishwanathan, R., Raghuchander, T., Samiyappan, R., 2001. Induction of systemic resistance in rice against sheath blight disease by *Pseudomonas fluorescens*. Soil Biol. Biochem. 33, 603–612.

Nguvo, K.J., Gao, X., 2019. Weapons hidden underneath: bio-control agents and their potentials to activate plant induced systemic resistance in controlling crop. J. Plant Dis. Prot. 126, 177–190.

Nishimura, M.T., Dang, J.L., 2010. Arabidopsis and the plant immune system. Plant J. 61 (6), 1053–1066.

Niu, D.-D., Liu, H.-X., Jiang, C.-H., Wang, Y.-P., Wang, Q.-Y., Jin, H.-L., et al., 2011. The plant growth-promoting rhizobacterium *Bacillus cereus* AR156 induces systemic resistance in *Arabidopsis thaliana* by simultaneously activating salicylate and jasmonate/ethylene-dependent signaling pathways. Mol. Plant Microbe Interact 24, 533–542. Available from: https://doi.org/10.1094/MPMI-09-10-0213.

Ongena, M., Duby, F., Rossignol, F., Fauconnier, M.-L., Dommes, J., Thonart, P., 2004. Stimulation of the lipoxygenase pathway is associated with systemic resistance induced in bean by a nonpathogenic *Pseudomonas* strain. Mol. Plant Microbe Interact. 17, 1009–1018.

Pandey, P.K., Samanta, R., Yadav, R.N.S., 2019. Inside the plants: addressing bacterial endophytes in biotic stress alleviation. Arch. Microbiol. 201, 415–429.

Pandey, P.K., Yadav, S.K., Singh, A., Sarma, B.K., Mishra, A., Singh, H.B., 2012. Cross-species alleviation of biotic and abiotic stresses by the endophyte *Pseudomonas aeruginosa* PW09. J. Phytopathol. 160, 532–539. Available from: https://doi.org/10.1111/j.1439-0434.2012.01941.x.

Park, Y.-S., Dutta, S., Ann, M., Raaijmakers, J.M., Park, K., 2015. Promotion of plant growth by *Pseudomonas fluorescens* strain SS101 via novel volatile organic compounds. Biochem. Biophysic. Res. Commun. 461 (2), 361–365.

Peer, V., Niemann, R.G.J., Schippers, B., 1991. Induced resistance and phytoalexin accumulation in biological control of Fusarium wilt of carnation by *Pseudomonas* sp. strain WCS 417 r. Phytopathology 81 (7), 728–734.

Petrova, P., Vellkova, P., Petrov, K., 2019. Genome sequence of *Bacillus velezensis* 5RB, an overproducer of 2, 3-Butanol. Microbiol. Resour. Announc. 8, e01475-18. Available from: https://doi.org/10.1128/MRA.01475-18.

Pieterse, C.M.J., Van Wees, S.C.M., Hoffland, E., Van Pelt, J.A., Van Loon, L.C., 1996. Systemic resistance in *Arabidopsis* induced by biocontrol bacteria is independent of salicylic acid accumulation and pathogenesis related gene expression. Plant Cell 8, 1225–1237.

Pieterse, C.M.J., Zamioudis, C., Berendsen, R.L., Weller, D.M., van Wees, S.C.M., Bakker, P.A.H.M., 2014. Induced systemic resistance by beneficial microbes. Annu. Rev. Phytopathol. 52, 347–375. Available from: https://doi.org/10.1146/annurev-phyto-082712-102340.

Planchamp, C., Glauser, G., Mauch-Mani, B., 2015. Root inoculation with *Pseudomonas putida* KT2440 induces transcriptional and metabolic changes and systemic resistance in maize plants. Front. Plant. Sci. 5, 719. Available from: https://doi.org/10.3389/fpls.2014.00719.

Postma, J., Montanari, M., van den Boogert, P.H.J.F., 2003. Microbial enrichment to enhance the disease suppressive activity of compost. Eur. J. Soil Biol. 39, 157–163.

Purnawati, A., Sastrahidayat, I.R., Abadi, A.L., Hadiastono, T., 2014. Endophytic bacteria as biocontrol agents of tomato bacterial wilt disease. J. Trop. Life Sci. 4 (1), 33–36.

Rabbee, M., Ali, M., Choi, J., Hwang, B., Jeong, S., Baek, K.-H., 2019. *Bacillus velezensis*: A valuable member of bioactive molecules within plant microbiomes. Molecules 24, 1046.

Rahman, M., Sabir, A.A., Mukta, J.A., Khan, M.M.A., Mohi-ud-Din, M., Miah, M.G., et al., 2018. Plant probiotic bacteria *Bacillus* and *Paraburkholderia* improve growth, yield and content of antioxidants in strawberry fruit. Sci. Rep. 8, 2504.

Rahman, A., Uddin, W., Wenner, N.G., 2015. Induced systemic resistance responses in perennial ryegrass against *Magnaporthe oryzae* elicited by semi-purified surfactin lipopeptides and live cells of *Bacillus amyloliquefaciens*. Mol. Plant Pathol. 16 (6), 546–558. Available from: https://doi.org/10.1111/mpp.12209.

Ramamoorthy, V., Viswanathan, R., Raguchander, T., Prakasam, V., Samiyappan, R., 2001. Induction of systemic resistance by plant growth promoting rhizobacteria in crop plants against pests and diseases. Crop Protec 20, 1–11.

Robert-Seilaniantz, A., Grant, M., Jones, J.D.G., 2011. Hormone crosstalk in plant disease and defense: more than just jasmonate-salicylate antagonism. Annu. Rev. Phytopathol. 49, 317–343. Available from: https://doi.org/10.1146/annurev-phyto-073009-114447.

Romero, D., Vicente, A.D., Zeriouh, H., Cazorla, F.M., Fernandez-Ortuno, D., Tores, J.A., et al., 2007. Evaluation of biological control agents for managing cucurbit powdery mildew on greenhouse-grown melon. Plant Pathol. 56, 976–986.

Ryan, R.P., Germaine, K., Franks, A., Ryan, D.J., Dowling, D.N., 2008. Bacterial endophytes: recent developments and applications. FEMS Microbiology Lett. 278, 1–9.

Ryan, P.R., Delhaize, E., Jones, D.L., 2001. Function and mechanism of organic anion exudation from plant roots. Annu. Rev. Plant Physiol. Plant Mol. Biol. 52, 527–560.

Ryu, C.M., Farag, M.A., Hu, C.H., et al., 2004. Bacterial volatiles induce systemic resistance in Arabidopsis. Plant Physiol. 134, 1017–1026.

Saharan, B., 2011. Plant growth promoting rhizobacteria: a critical review. Life Sci. Med. Res. 21, 1–30.

Saravanakumar, D., Thomas, A., Banwarie, N., 2019. Antagonistic potential of lipopeptide producing *Bacillus amyloliquefaciens* against major vegetable pathogens. Eur. J. Plant Pathol 154, 319–335.

Sharma, S., Chen, C., Navathe, S., Chand, R., Pandey, S., 2019. A halotolerant growth promoting rhizobacteria triggers induced systemic resistance in plants and defends against fungal infection. Sci. Rep. 9, 4054. Available from: https://doi.org/10.1038/s41598-019-40930-x.

Sharma, A., Jansen, R., Johri, B.N., Wray, V., 2007. Molecular and structural characterization of rhamnolipids of rhizobacteria that reduce post-emergence damping-off disease in Chile and tomato in Central Himalayan region. J. Nat. Prod. 187 (4), 321–335.

Sharma, A., Johri, B.N., Sharma, A.K., Glick, B.R., 2003. Plant growth promoting bacterium *Pseudomonas* sp. Strain GRP3 influences iron acquisition in mung bean (*Vigna radiata* L. *Wilczek*). Soil Biol. Biochem. 35, 887–894.

Singh, R., Sinha, A.P., 2005. Management of rice sheath blight by *Pseudomonas fluorescens* and grain yield. Ann. Plant Protect. Sci. 13 (2), 410–414.

Siyar, S., Inayat, N., Hussain, F., 2019. Plant growth promoting rhizobacteria and plants' improvement: a mini-review. PSM Biol. Res. 4 (1), 1–5.

Song, G.C., Riu, M., Ryu, C.-M., 2019. Beyond the two compartments petri-dish: optimizing growth promotion and induced resistance in cucumber exposed to gaseous bacterial volatiles in a miniature greenhouse system. Plant Methods 15, 9.

Stein, E., Molitor, A., Kogel, K.H., Waller, F., 2008. Systemic resistance in *Arabidopsis* conferred by the mycorrhizal fungus *Piriformospora indica* requires jasmonic acid signaling and the cytoplasmic function of NPR1. Plant Cell Physiol. 49, 1747–1751.

Stringlis, I.A., Zamioudis, C., Berendsen, R.L., Bakker, P.A.H.M., Pieterse, C.M.J., 2019. Type III secretion system of beneficial rhizobacteria *Pseudomonas simiae* WCS417 and *Pseudomonas defensor* WCS374. Front. Microbiol. 10, 1631. Available from: https://doi.org/10.3389/fmicb.2019.01631.

Surovy, M.Z., Gupta, D.R., Mahmud, N.U., Dame, Z.T., Roy, P.K., Islam, M.T., 2020. Genomics and post-genomics approaches for elucidating molecular mechanisms of plant growth-promoting *Bacilli*. Bacilli and Agrobiotechnology: Phytostimulation and Biocontrol, 2. pp. 161–200.

Tan, H., Zhou, S., Deng, Z., He, M., Cao, L., 2011. Ribosomal-sequence-directed selection for endophytic streptomyces strains antagonistic to *Ralstonia solanacearum* to control tomato bacterial wilt. Biol. Control 59, 245–254.

Thomma, B.P.H.J., Penninckx, I.A.M.A., Broekaert, W.F., Cammue, B.P.A., 2001. The complexity of disease signaling in arabidopsis. Curr. Opin. Immunol. 13, 63–68.

Ting, A.S.Y., Meon, S., Kadir, J., Radu, S., Singh, G., 2008. Endophytic microorganisms as potential growth- promoters of banana. Biocontrol 53, 541–553.

Trotel-Aziz, P., Abou-Mansour, E., Courteaux, B., Rabenoelina, F., Clément, C., Fontaine, F., et al., 2019. *Bacillus subtilis* PTA-271 counteracts Botryosphaeria dieback in Grapevine, triggering immune responses and detoxification of fungal phytotoxins. Front. Plant Sci. 10, 25. Available from: https://doi.org/10.3389/fpls.2019.00025.

Tunsagool, P., Leelasuphakul, W., Jaresitthikunchai, J., Phaonakrop, N., Roytrakul, S., Jutidamrongphan, W., 2019. Targeted transcriptional and proteomic studies explicate specific roles of *Bacillus subtilis* iturin A, fengycin, and surfactin on elicitation of defensive systems in mandarin fruit during stress. PLoS One 14 (5), e0217202. Available from: https://doi.org/10.1371/journal.pone.0217202.

Upchurch, R.G., Ramirez, M.E., 2011. Effects of temperature during soybean seed development on defense-related gene expression and fungal pathogen accumulation. Biotechnol. Lett. 33, 2397–2404.

Vacheron, J., Desbrosses, G., Bouffaud, M., 2013. Plant growth-promoting rhizobacteria and root system functioning. Front. Plant Sci. 4, 356.

Van Loon, L.C., Bakker, P.A.H.M., Pieterse, C.M.J., 1998. Systemic resistance induced by rhizosphere bacteria. Annu. Rev. Phytopathology 36, 453–483.

Van der Ent, S., Van Wees, S.C.M., Pieterse, C.M.J., 2009. Jasmonate signaling in plant interactions with resistance-inducing beneficial microbes. Phytochemistry 70, 1581–1588.

van Loon, L.C., Glick, B.R., 2004. Increased plant fitness by rhizobacteria. In: Sandermann, H. (Ed.), Molecular Ecotoxicology of Plants. Springer-Verlag, Berlin, Heidelberg, pp. 177–205.

Van Peer, R., Niemann, G.J., Schippers, B., 1991. Induced resistance and phytoalexin accumulation in biological control of *Fusarium* wilt of carnation by *Pseudomonas* sp. strain WCS417r. Phytopathology 81, 728–734.

Van Wees, S.C.M., Van der Ent, S., Pieterse, C.M.J., 2008. Plant immune responses triggered by beneficial microbes. Curr. Opin. Plant Biol. 11, 443–448.

Verhagen, B.W.M., Glazebrook, J., Zhu, T., Chang, H.-S., van Loon, L.C., Pieterse, C.M.J., 2004. The transcriptome of rhizobacteria-induced systemic resistance in Arabidopsis. Mol. Plant Microbe Interact. 17 (8), 895–908.

Verhagen, B., Trotel-Aziz, P., Jeandet, P., Baillieul, F., Aziz, A., 2011. Improved resistance against *Botrytis cinerea* by grapevine associated bacteria that induce a prime oxidative burst and phytoalexin production. Phytopathology 101, 768–777.

Vurukonda, S.S.K.P., Vardharajula, S., Shrivastava, M., SkZ, A., 2016. Enhancement of drought stress tolerance in crops by plant growth promoting rhizobacteria. Microbiol. Res. 184, 13–24.

Waewthongrak, W., Leelasuphakul, W., McCollum, G., 2014. Cyclic Lipopeptides from *Bacillus subtilis* ABS-S14 elicit defense-related gene expression in citrus fruit. PLoS One 9, e109386. Available from: https://doi.org/10.1371/journal.pone.0109386. PMID: 25329301.

Walters, D., Heil, M., 2007. Costs and trade-offs associated with induced resistance. Physiol. Mol. Plant Pathol. 71, 3–17. Available from: https://doi.org/10.1016/j.pmpp.2007.09.008.

Walters, R.R., Ratsep, J., Havis, N.D., 2013. Controlling crop diseases using induced resistance: challenges for the future. J. Exp. Bot. 5, 1–19.

Wang, K.L.C., Li, H., Ecker, J.R., 2002. Ethylene biosynthesis and signaling networks. Plant Cell. 14, S131–S151. Available from: https://doi.org/10.1105/tpc.001768. PMID: 12045274.

Wang, Y.Q., Ohara, Y., Nakayashiki, H., Tosa, Y., Mayama, S., 2005. Microarray analysis of the gene expression profile induced by the endophytic plant growth–promoting rhizobacteria, Pseudomonas Fluoresc*ens* FPT9601-T5 Arabidopsis. Mol. Plant Microbe Interact. 18, 385–396.

Welbaum, G., Sturz, A.V., Dong, Z., Nowak, J., 2004. Fertilizing soil microorganisms to improve productivity of agroecosystems. Crit. Rev. Plant Sci. 23, 175–193.

Weller, D.M., Mavrodi, D.V., Van Pelt, J.A., Pieterse, C.M.J., Van Loon, L.C., Bakker, P.A.H.M., 2012. Induced systemic resistance (ISR) in *Arabidopsis thaliana* against *Pseudomonas syringae* pv. *tomato* by 2,4-diacetylphloroglucinol-producing *Pseudomonas fluorescens*. Phytopathology 102, 403–412.

Whipps, J.M., 2001. Microbial interactions and biocontrol in the rhizosphere. J. Exp. Bot. 52, 487–511.

Widnyana, I.K., Sumantra, I.K., Pandawani, N.P., Eka Pasmidi Ariati, P., 2019. Induction of total phenol and salicylic acid with *Pseudomonas* spp. suspension as indications increasing systemic resistance on vanilla plant. Bulg. J. Agric. Sci. 25 (3).

Wu, G., Liu, Y., Xu, Y., Zhang, G., Shen, Q., Zhang, R., 2018. Exploring elicitors of the beneficial rhizobacterium *Bacillus amyloliquefaciens* SQR9 to induce plant systemic resistance and their interactions with plant signaling pathways. Mol. Plant Microbe Interact. Available from: https://doi.org/10.1094/MPMI-11-17-0273-R.

Wu, L., Wu, H.J., Chen, L., Yu, X.F., Borriss, R., Gao, X.W., 2015. Difficidin and bacilysin from *Bacillus amyloliquefaciens* FZB42 have antibacterial activity against *Xanthomonas oryzae* rice pathogens. Sci. Rep. 5, 12975. Available from: https://doi.org/10.1038/srep12975.

Wu, Y., Zhou, J., Li, C., Ma, Y., 2019. Antifungal and plant growth promotion activity of volatile organic compounds produced by *Bacillus amyloliquefaciens*. Microbiol. Open 8, e813.

Xie, S., Wu, H., Chen, L., et al., 2015. Transcriptome profiling of *Bacillus Subtilis* OKB105 in response to rice seedlings. BMC Microbiol. 15 (1), 21.

Yuan, J., Raza, W., Shen, Q.R., Huang, Q.W., 2012. Antifungal activity of *Bacillus amyloliquefaciens* NJN-6 volatile compounds against *Fusarium oxysporum* f. sp. cubense. Appl. Environ. Microbiol. 78, 5942–5944. Available from: https://doi.org/10.1128/AEM.01357-12.

Zohara, F., Akanda, M.A.M., Paul, N.C., Rahman, M., Islam, M.T., 2016. Inhibitory effects of *Pseudomonas* spp. On plant pathogen *Phytophthora capsici* in vitro and in planta. Biocat. Agric. Biotechnol. 5, 69–77.

CHAPTER

6

RNA interference as a promising strategy for plant disease management

Ganeshamoorthy Hariharan, Nishanthi Sivasubramaniam and Kandeeparoopan Prasannath

Department of Agricultural Biology, Faculty of Agriculture, Eastern University, Chenkalady, Sri Lanka

6.1 Introduction

Plant diseases result from multiplex interconnections between biotic and abiotic components comprising susceptible plant hosts, virulent pathogens, and environmental conditions. A diverse group of plant pathogens like fungi, bacteria, viruses, nematodes, viroids, and mollicutes (phytoplasma) are reported as major phytopathogens. They cause several plant diseases resulting in changes in plant physiology (photosynthesis, transpiration, translocation, nutrient assimilation, etc.), morphology (discoloration), and plant density that lead to substantial economic collapse and reduce food security all over the world (Agrios, 2005). Crop damages due to biotic stress, especially by those plant pathogens have been a major threat in agricultural crop production.

Agricultural commodities are often affected by phytopathogens at any time of their life cycle viz. seed, seedling, vegetative, flowering (reproductive), or during harvesting and post-harvesting periods. Phytopathogens drastically reduce actual crop yield, productivity and quality, and abiding plant pathogenic impact causes adverse effects on soil health that creates chronic stress situation to the plant, which further comforts pathogens to cause severe and drastic phytopathogenic infestations. Phytopathogenic diseases have been a foremost factor affecting food production and human societal advancement over millennia (Palmgren et al., 2015). A current investigation revealed that yield losses incurred due to 137 pathogens and pests allied with five major crops viz. wheat, rice, maize, potato, and soybean in 67 countries over the main producing regions, worldwide as 21.5%, 30.0%, 22.5%, 17.2%, and 21.4%, respectively (Savary et al., 2019), and plant diseases solely cost the global economy around $220 billion (Food and Agriculture Organization of the United Nations, 2020). Thus failures in

Food Security and Plant Disease Management
DOI: https://doi.org/10.1016/B978-0-12-821843-3.00008-8

precise plant disease diagnosis and control may lead to a heavy toll in plant production and related agricultural commodities.

Major challenges to manage plant diseases are due to growing demands for total, safe, and different foodstuffs to care the flourishing worldwide population and its improving living standards. Shrinking production potential in agriculture is due to competition for land in fertile areas, deteriorating of agroecosystems and diminishing of natural resources, and greater threat of disease epidemics resulting from agricultural intensification and monocultures (Dunchun et al., 2016). Various endeavors have been made in the area of disease management to offset phytopathogenic diseases. Present management measures to protect crops and enhance agricultural output heavily depend on synthetic chemicals, which are hazardous to the environment, decrease the populations of beneficial soil microorganisms and nontarget organisms (Wang et al., 2019) and exert lethal effects on human health (Dubrovina and Kiselev, 2019). Therefore agriculture research began to focus on alternate harmless methods.

Molecular biotechnological approaches came in the scenario to overcome the burning issues regarding phytopathogenic disease management. Accurate plant disease diagnosis and detection methods including fluorescence in situ hybridization (Kliot et al., 2014), enzyme-linked immunosorbent assay (López et al., 2003; Fang and Ramasamy, 2015), polymerase chain reaction, DNA barcoding, next-generation sequencing (Aslam et al., 2017) are used in phytopathogenic management. Transgenic crops have also been produced to successfully combat against countless diseases of economically important plants (Fagwalawa et al., 2013; Saharan et al., 2016). Genetically engineered food crops with improved disease resistance to several pathogens have been permitted for commercial production (Foster et al., 2009; Ye and Li, 2010). Molecular genome editing technologies, such as transcription activator-like effector nucleases (Li et al., 2012a), zinc-finger nucleases (Wang et al., 2014), and CRISPR/Cas9 (clustered regularly interspaced short palindromic repeats) (Ji et al., 2015; Ali et al., 2016; Macovei et al., 2018) have also been extensively used in engineering disease resistance in crops to manage several plant pathogenic diseases (Mushtaq et al., 2019).

The advent of molecular techniques in genetic engineering offers an entirely innovative tactic being pursued to render plant resistant to phytopathogens. A promising alternative strategy for managing plant pathogens is delivery of double-stranded RNA (dsRNA) for gene silencing, also known as RNA interference (RNAi), which posttranscriptionally down regulates the expression of plant pathogen genes and other negative regulators of resistance pathways (Puyam and Kaur, 2020). It is an emerging molecular therapeutic tool to lighten the intrinsic threats linked with traditional transgenic technologies against phytopathogens. This chapter emphasizes the potential of RNAi mechanism as a promising tool for managing major phytopathogens in agriculture. In addition to discussing the efficiency of RNAi in plant protection against several pathogens, the pros and cons of different strategies to induce RNAi in current technology are outlined, and future directions of this application are discussed to improve the aspects in plant disease management.

6.2 RNA interference: Definition

RNAi is a biological activity, in which small RNA (sRNA) molecules silence gene expression sequence-specifically at transcriptional or posttranscriptional level, either by directing inhibitory chromatin modifications or by reducing the stability or translation potential of the targeted mRNA (Liu et al., 2020). RNAi is a natural machinery for gene silencing, which is stimulated by the presence of short interfering RNA (siRNA). RNAi-mediated resistance is a catalytic path, which is elicited by dsRNA endogenously. This stimulation is a natural phenomenon when a dsRNA virus infects a cell, or dsRNAs are deliberately delivered to persuade user-directed damage to the cognate transcript (Farrell, 2017).

6.3 A brief evolutionary story of RNA interference

In the early 1990s, an initial phenomenon related to RNAi was testified, and it was assumed that introduced chimeric petunia chalcone synthase genes potentially cosuppressed the endogenous genes (Napoli et al., 1990). The same phenomenon of cosuppression of an endogenous gene was also detected in the fungus *Neurospora crassa* in which the occurrence of exogenous sequences triggered a severe declension of the expression of the endogenous *al-1* or *al-3* genes and the mechanism was named as "quelling" (Romano and Macino, 1992). Guo and Kemphues (1995) reported the first usage of RNA silencing strategy in a study on nematode *Caenorhabditis elegans* and they observed that the application of sense or antisense RNA caused degradation of the *par-1* gene. The molecular mechanism of RNAi was first identified in *C. elegans* (Fire et al., 1991) by identifying dsRNAs as the causative agent for posttranscriptional gene silencing in *C. elegans*, a phenomenon they termed "RNA interference." Fire et al. (1998) also indicated that *C. elegans* genes could be passed in the germline and potent silencing signal could also be passed through the sperm or the egg for up to several generations through antisense gene silencing technique (Grishok et al., 2000).

6.4 Core components in RNA interference mechanism

Dicer-like proteins (DCLs), RNA-dependent RNA polymerase proteins (RDRs), and Argonaute proteins (AGOs) are the three catalytic evolutionarily conserved core components (Castel and Martienssen, 2013; Jeseničnik et al., 2019) that described for RNAi in protists, fungi, plants, and animals (Cerutti and Casas-Mollano, 2006; Chen, 2009; Kim et al., 2009). A brief description of principle elements and the functions of their domains are given below (Table 6.1).

6.4.1 Dicer-like proteins

DCLs are endoribonucleases belonging to the RNase III family that has a DExD-box Helicase-C (similar to RNA helicase) domain, Piwi–Argonaute–Zwille (PAZ) domain, the domain of unidentified role 283 (DUF283), two RNase III domains at N′ terminal and a dsRNA binding domain (dsRBD) at the C-terminal (Carmell and Hannon, 2004; MacRae et al., 2006; Margis et al., 2006; Mateos et al., 2010). DCLs are evolutionarily conserved in a variety of eukaryotes (Kadotani et al., 2004), and they perform the initiation step of the

TABLE 6.1 Core components of RNAi and the functions of their domains.

Components of RNAi	Domains	Functions	References
Dicer-like proteins (DCLs)	DExD-box Helicase-C	Essential for the competent construction of DCLs-dependent sRNAs	Carmell and Hannon (2004)
	Piwi–Argonaute–Zwille	The phosphate-binding pocket of PAW recognizes the monophosphate of pre-miRNAs and is essential for dsRNAs cleavage	Park et al. (2011), Tian et al. (2014), Kandasamy and Fukunaga (2016)
	DUF283	It aids hybridization among the nucleic acid strands, which are complementary in nature	Kurzynska-Kokorniak et al. (2016)
	RNase III	It specifically cleaves dsRNA, thereby leaving the 2 nt overhang at 3′ end of the product	Nicholson (2014), Aguado and tenOever 2018
	dsRBD	The C-terminal regulates the protein nucleocytoplasmic circulation, and further, it can bind to dsRNAs	Banerjee and Barraud (2014), Barraud et al. (2014)
Argonaute proteins (AGOs)	PAZ	It anchors the 3′ end of the guide strand of an sRNA duplex	Ma et al. (2004)
	MID	The C-terminal has MID and PIWI domains, and a binding pocket at the joint of these two domains that anchors the 5′ end of sRNAs	Parker (2010)
	PIWI	It functions similar to RNase H that cleaves target mRNA	Simon et al. (2011)
	N-terminal	Domain fragment (DUF1785) was just selected as being vital for RNA duplex relaxing	Derrien et al. (2018)
RNA-dependent RNA polymerase proteins (RDRs)	RNA-dependent RNA polymerase catalytic domain	Recognizes and processes aberrant transcripts triggering the process, and in amplification of the silencing response	Marker et al. (2010), Nicolas and Garre (2016)

RNAi mechanism by triggering both transcriptional and posttranscriptional gene silencing, in which dsRNAs are sliced into sRNAs of 21–24 nt in length (Carmell and Hannon, 2004).

6.4.2 Argonaute proteins

Initial structural understandings into AGOs originated from crystal structures of domains of eukaryotic or prokaryotic AGOs (Jinek and Doudna, 2009). AGOs are 100 kDa in molecular weight and are characterized by four conserved domains such as PAZ,

Middle (MID) and P-element induced wimpy (PIWI), and N-terminal domains (Zhang et al., 2014; Frank et al., 2012). They are extremely specific sRNA-binding units and are taken into account to be the vital elements of the RNA-induced silencing complex (RISC) in silencing pathways (Vaucheret, 2008; Mallory et al., 2009).

6.4.3 RNA-dependent RNA polymerase proteins

RDRs (also known as RdRPs) are believed as the first known element of plants sRNAs biogenesis pathways (Voinnet, 2008), and all RDRs consists of a catalytic RDR domain (Iyer et al., 2003; Nicolas and Garre, 2016). RDRs play a significant role in the amplification of silencing, where it changes single-stranded RNAs (ssRNAs) to dsRNAs. DCLs process the dsRNAs, resulting in another cycle of gene silencing (Muhammad et al., 2019).

6.5 RNA interference mechanism

RNAi is an endogenous posttranscriptional silencing mechanism that requires an activator (commonly called as trigger), that is, dsRNA as a precursor. The trigger can arise naturally in the event of cell infection by the intentional introduction of dsRNA or by a dsRNA virus to stimulate user-driven destroy of the cognate transcript. In plants, dsRNAs could be generated through the following ways; replication of RNA viruses, synthesizing by endogenous plant RdRP (Luo and Chen, 2007), from natural sense and antisense transcript or the formation of hairpin RNAs (hpRNAs) from ssRNAs (Rosa et al., 2018) and the exogenously applied dsRNA (Dubrovina and Kiselev, 2019).

The first step in RNAi involves a recognizing and cleaving of these dsRNA molecules by a type III endoribonuclease called Dicer or DCLs. This ATP-dependent digestion of longer dsRNA (Praveen et al., 2012) creates siRNA of 21 to 23 bp staggered duplex fragments (Hamilton and Baulcombe, 1999; Fukudome and Fukuhara, 2016) with distinguishing overhangs of 3′ dinucleotide on both strands (Dorsett and Tuschl, 2004). However, it should also be noted that RNAi-related pathways in plants give rise to diverse forms of sRNAs, including micro RNAs (miRNAs), siRNAs, transacting siRNAs, and phased siRNAs (Willmann et al., 2011). Among them, miRNAs and siRNA have been most investigated and utilized molecules for translational and RNAi mechanism–related applications (Rosa et al., 2018). At the end of the first step, hpRNAs and long dsRNAs that are created by several procedures are the primary candidates in RNAi stimulation. These molecules are then recognized and converted into siRNAs, miRNAs, or other sRNAs, by DCLs (Axtell, 2013; Borges and Martienssen, 2015). A schematic diagram of the RNAi mechanism that presents major molecules for silencing is shown in Fig. 6.1. In the next step, the formed siRNA duplexes (regardless of the source) incorporate into the effector complex called RISC, where AGOs play a vital role (Khraiwesh et al., 2012). An ATP-dependent helicase of RISC unwinds the siRNA duplex (Zamore et al., 2000). RISC distinguishes antisense strand from the sense strand of the duplexes (Elbashir et al., 2001a). The sense strand is degraded, and the antisense strand of the duplex siRNA molecule is recruited as guide strand in silencing mechanism that designates to target and identify specific cognate

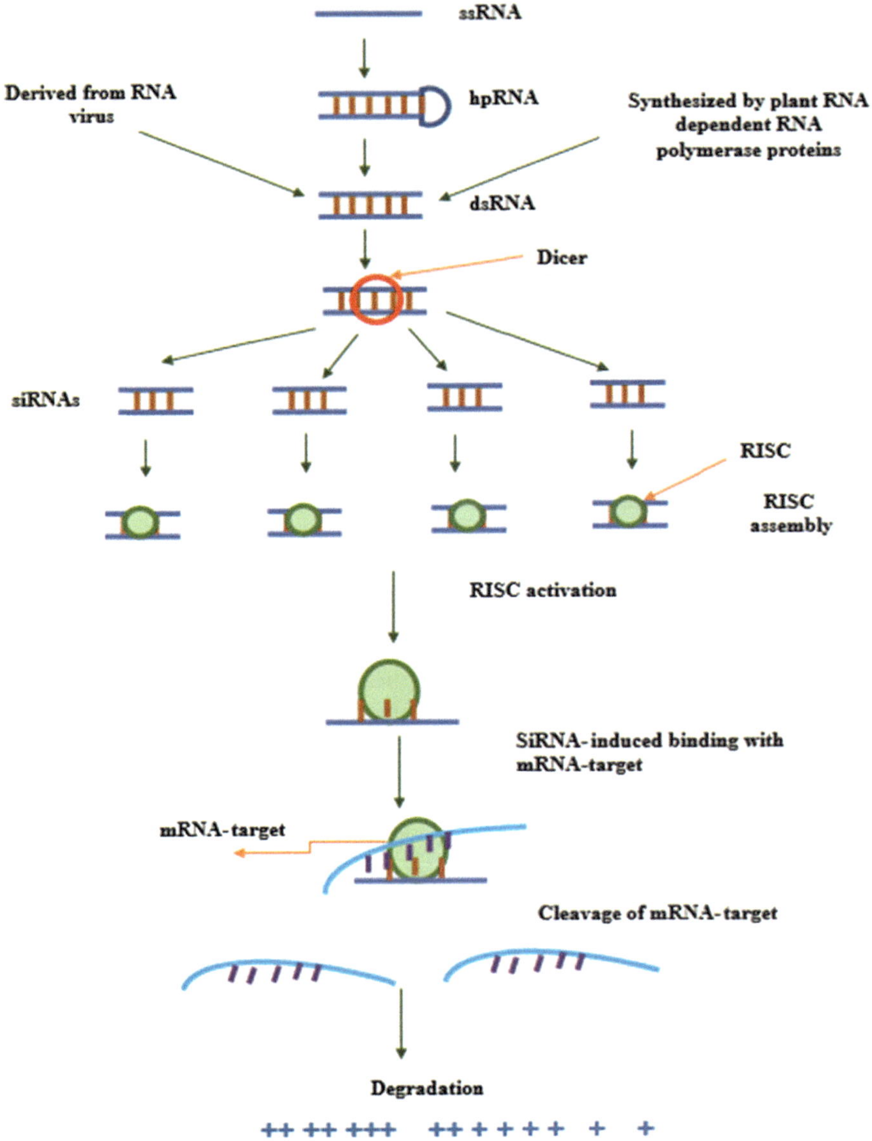

FIGURE 6.1 Schematic diagram of RNAi mechanism.

RNAs target sequence by complementary base pairing (Jinek and Doudna, 2009; Siomi and Siomi, 2009). It leads to RISC cleavage/sequence-specific degradation (by both siRNAs and miRNAs) or translational repression (by miRNAs) of the mRNA target (Farrell, 2017), and results in no protein synthesis (Brodersen, and Voinnet, 2006). RISC slices mRNA into 10–11 nucleotides (Baulcombe, 2004; Meister and Tuschl, 2004) without using ATP (Elbashir et al., 2001b) that yields one fragment missing its polyA tail and the other missing its 5′ 7-methylguanosine cap, leaving each strand susceptible that promotes further degradation by RNA surveillance machinery. Interestingly, an amplification step within RNAi mechanism has been identified in plants (Carthew and Sontheimer, 2009), where RdRP uses antisense strand to make additional dsRNAs from ssRNAs and these dsRNAs are processed one more time by DCLs, resulting in another round of RNA silencing (Tang et al., 2003; Willmann et al., 2011). Furthermore, in plants, specific siRNA can also provoke RNA-directed DNA methylation in which DNA methyltransferase DOMAINS REARRANGED METHYLTRANSFERASE 2 are recruited to start de novo methylation (Mette et al., 2000; Zhong et al., 2014; Xie and Yu, 2015), which leads to diverse epigenetic processes, such as transposon suppression and gene imprinting (Ito, 2011).

6.6 Methods to induce RNA interference in plants against phytopathogenic diseases

Induction of RNAi in plants to combat phytopathogenic diseases requires a proper method of delivery of active molecules that trigger the RNAi mechanism. Transformative (transgenic) and nontransformative approaches have been practiced to initiate the mechanism against selected pathogens (Cagliari et al., 2019). Conventional RNAi approaches to induce RNAi are primarily dependent on the use of genetically modified (GM) crops expressing dsRNAs, but it needs generations of a crop that takes few years and has interruptions for field practice due to transformability, regulations, genetic consistency, and public acceptance of GM plants (Paces et al., 2017). A recent strategy to trigger RNAi mechanism is the exogenous spray of in vitro-produced or bacterially formed long dsRNAs, siRNAs, or hpRNAs. They are directly delivered onto the plants (Dubrovina and Kiselev, 2019) by sprays (Dalakouras et al., 2016; Niehl et al., 2018; Song et al., 2018), mechanical inoculation (Yin et al., 2010; Lau et al., 2014; Konakalla et al., 2016; Kaldis et al., 2018), loading onto clay nanosheets (Mitter et al., 2017), use of materials encouraging RNA adhesion (Zhang et al., 2019b), dispersion by brushes or pipette, introducing by seed soaking or root drenching (Jiang et al., 2014), and injection (Tzean et al., 2019). RNAi could also be triggered not only by exogenously supplied dsRNA (Whangbo and Hunter, 2008) but also by hpRNA or long dsRNA transcribed from either a transgene or an endogenous gene (Carthew and Sontheimer, 2009). Since endogenously generated and exogenously supplied dsRNAs activate RNAi in a sequence-specific manner, this strategy is a promising avenue for providing low risk and environmentally safe plant protection. Following is the discussion on major transformative and nontransformative methods inducing RNAi that are used to control phytopathogens in plants.

6.6.1 Virus-induced gene silencing

Gene functional studies in terms of forward and reverse genetics have been carried out by virus-induced gene silencing (VIGS) techniques in plants (Ramegowda et al., 2014), which uses viral vectors conveying a target gene fragment to produce dsRNA, which activates

RNA-mediated (Unver and Budak, 2009) antiviral defense mechanism (Lu et al., 2003; Bekele et al., 2019) in gene silencing. VIGS is one of the much-recognized tools of choice for transient induction of gene silencing in phytopathogenic management that encompasses the formation of engineered viruses having sequences corresponding to the target host genes to be silenced. Infection turns onto the development of viral dsRNA (Senthil-Kumar and Mysore, 2011), a transitional stage in viral replication that triggers the antiviral RNA silencing pathway, following in downregulation of the host gene transcript (Padmanabhan and Dinesh-Kumar, 2009). Numerous RNA and DNA viruses including Tobacco mosaic virus (TMV), Tobacco rattle virus (TRV), and Potato virus X (PVX) have been adapted and benefited as vectors for gene expression and gene silencing (MacFarlane and Popovich, 2000; Pogue et al., 2002; Martín-Hernández and Baulcombe, 2008). TRV-based VIGS has been widely used in many crops (Liu et al., 2002; Senthil-Kumar et al., 2007) and efficiently used in tomato crop to discover the candidate genes for their participation in Verticillium wilt resistance (Fradin et al., 2009). Hein et al. (2005) implemented *Mla13*-mediated RNAi resistant technology against *Blumeria graminis* f. sp. *hordei*, the causal agent of barley powdery mildew. For VIGS experiments, the RNA-based Barley stripe mosaic virus (BSMV) vector was deployed to knock down *Pds*, *Sgt1*, *Rar1*, and *Hsp90* genes in barley (Lacomme et al., 2003). Direct infection of recombinant vector TMV caused RNAi in fungus *Colletotrichum acutatum* (strain C71) by VIGS (Mascia et al., 2014). Another critical issue is all RNA virus-acquired expression vectors cannot be considered as silencing vectors since a few RNA viruses possess proteins of antisilencing effect which directly interlinked with the silencing machinery of the host (Kumagai et al., 1995; Palmer and Rybicki, 2001). Therefore choosing appropriate vectors for the right execution is crucial. Several studies reported that VIGS could only last for about 3 weeks in the meantime capability declines after a month thereby plant can recover from silencing ability as a whole or partially (Hiriart et al., 2003; Ryu et al., 2004). Conversely, recent studies indicated that certain VIGS vectors could sustain the silencing of genes for many months by appropriately adjusting plant growth conditions that support viral duplication (Fu et al., 2005; Tuttle et al., 2008) and can carry to succeeding generation (Senthil-Kumar and Mysore, 2011), acting like GM crops. VIGS has many benefits which includes short-term suppression of the gene and requires the cloning of short sequence fragments of the desired gene to be knocked down (Unver and Budak, 2009). Further, VIGS vectors are improved to silence more than one host plant and can target both monocots and dicots. BSMV–VIGS can silence wheat and barley genes (Unver and Budak, 2009).

6.6.2 Host-induced gene silencing

Host-induced gene silencing (HIGS) is a new concept of RNAi, in which hosts synthesize sRNAs that activate silencing of genes of the pathogens during the invasion (Nowara et al., 2010; Baulcombe, 2015; Govindarajulu et al., 2015). This mechanism silences the entering pathogen precisely (Qi et al., 2019) so that it has emerged to be a successful approach in managing plant diseases prior to the establishment. HIGS is based on the instructions coded on the inverted-repeat sequence of the host genome. The gene sequence homologous to the pathogen gene combines into the plant genome and expresses the siRNA targeted gene, which makes them silence the invaded pathogen. In this regard, numerous HIGS vectors have been constructed (Sanju, et al., 2015) to drive the long dsRNA and hpRNAs through either an inverted repeated gene or an inverted promoter sequence into the host plant genome (Qi et al., 2019). HIGS has appeared as an effective and novel approach for the modification of resistance to phytopathogens by silencing genes, which are vital for these pathogens. In preharvest plant disease management, RNAi-based HIGS of key genes responsible for the phytopathogen's growth, virulence, and/or toxin production may accommodate as a substitute (Majumdar et al., 2017). HIGS strategy aims one or few of the essential phytopathogen genes that are required for growth and development or pathogenicity or host plants genes required for infection. Targets for HIGS involve genes associated with inhibition of the disease development in the host in several ways. This might be included genes those responsible for primary or secondary metabolism or genes which essential for the synthesis of structural components; chitin, ergosterol, and other necessary components for growth regulation (Yin and Hulbert, 2015). Fusarium head blight resistance in spike and leaf gained in transgenic wheat by HIGS technique. It was done by construction of an RNAi hairpin structure against β-1,3-glucan synthase gene *FcGls1* of *Fusarium culmorum* or a triple combination of FcGls1 with two additional, preexamined target genes (Chen et al., 2016). Effector genes; *Avrk1* and *Avra10* related to the haustorium formation of powdery mildew fungi suppressed by the HIGS approach (Godfrey et al., 2010). Other investigations have also suggested the functionality of HIGS in suppressing wheat rust disease caused by fungus *Puccinia striiformis* f. sp. *tritici* (Yin et al., 2011), *Phytophthora infestans* in potato (Jahan *et al.*, 2015), white mold by *Sclerotinia sclerotiorum* (Andrade *et al.*, 2016), Wheat streak mosaic virus (WSMV) in wheat using artificial miRNA (Fahim et al., 2012) and *Triticum mosaic virus* (TriMV) in wheat by constructing a hairpin sequence of the TriMV coat protein gene (Shoup Rupp, 2016). HIGS-mediated RNAi has some restrictions; available transformation protocols in many crop species are minimal, the unpredictability of engineered RNA silencing characters and communal anxiety for the creation of transgenic crop (Wang and Jin, 2017). Furthermore, it necessitates generations of transgenic plants, which may cause considerable delay in the implementation of strategies, in terms of the genetic stability and transformability of the plant species (Koch et al., 2016). Moreover, the efficacy of HIGS is dependent on sufficient supplements and the active transport of siRNA from one organism to another organism. Hence, necrotrophic fungi cannot be controlled by HIGS as these fungi receive nutrition from dead hosts, which cannot be a source of adequate quantity of siRNAs (Qi et al., 2019).

6.6.3 Microbombardment

Microbombardment also known as particle gun bombardment or biolistic process is based on bombarding siRNA, dsRNA, DNA that encodes hairpin constructs, or hpRNAs, as well as, sense or antisense RNA that stimulates the RNA resistant–mediated pathway to control phytopathogens (Mann et al., 2008). Pooggin et al. (2003) reported that DNA construct of genes such as *AC1* and *AV2* bombarded in black gram (*Vigna mungo*) against Vigna mungo yellow mosaic virus (VMYMV) revealed complete recovery from VMYMV infection, recommending the efficacy of RNA silencing tactic by developing resistance to DNA viruses. Synthetic siRNAs delivered through biolistic method into plants produced silencing of green fluorescent protein expression (Wani et al., 2010). Exogenously constructed dsRNA derived from the viral coat protein gene was applied through biolistic process and induction leading to defense against Pea seed borne mosaic virus (PSbMV) was examined by Safarova et al. (2014), where they detected a significant short-term reduction in the PSbMV concentration in experimental pea plants. Microbombardment tactic could be cost-effective and will need less skill, but requires to be consistent for implementation of this approach broadly (Puyam et al., 2017).

6.6.4 Agroinoculation

Agroinoculation also known as agroinfiltration, involves the injection of *Agrobacterium tumefaciens* possessing similar DNA constructs into the intracellular spaces of leaves for eliciting RNA silencing (Hilly and Liu, 2007). The goal of this technique is to activate a temporary RNA silencing mechanism to transmit RNA silencing molecules directly to crop tissues (Duan et al., 2012). In plants, cytoplasmic RNAi can be triggered effectively by agroinoculation, analogous to a technique for transient expression of T-DNA vectors after transmission by *A. tumefaciens* (Usharani et al., 2005, Karthikeyan et al., 2013). A gene silencing vector utilizing the characters of tasiRNA was designed to aim RNAi suppressor proteins; AC2 and AC4 of the geminivirus, Tomato leaf curl New Delhi virus (ToLCNDV). The vector comprising fragments of ToLCNDV *AC2* and *AC4* genes brought by agroinfiltration produced profuse quantities of *AC2* and *AC4* specific siRNA in both tomato and tobacco plants. When agroinfiltrated plants were inoculated with ToLCNDV, most plants exhibited a lack of symptoms and a low accumulation of viral DNA (Singh et al., 2015). In leaves of the grapevine plant, transient expression of genes was found by Agroinfiltration. In detail, RNAi constructs to express dsRNA equivalent to the defense-related gene *VvPGIP1* encoding a polygalacturonase-inhibiting protein (PGIP) were transiently expressed (Bertazzon et al., 2012). The direct transfer of such constructs for dsRNA may offer fast silencing and allows advanced studies on gene function in grapevine (Helm et al., 2011). This approach has also been successfully used to trigger *iaaM* genes, and *ipt* is responsible for the induction of disease resistance to crown gall in apple (Dunoyer et al., 2007), OsGEN-1-green fluorescent fusion protein in rice (Moritoh et al., 2005) and *hpGUS* gene in different transgenic lettuce lines (Wroblewski et al., 2007).

6.6.5 Spray-induced gene silencing

Spray-induced gene silencing (SIGS) is a revolutionary, nontransforming, and exogenous use of dsRNA or hpRNA to cause RNAi resistance to phytopathogens (Song et al., 2018). In this approach, the dsRNAs pointing a plant pathogen's gene are sprayed onto plant surfaces. Phytopathogen directly receives the dsRNAs and triggers the RNAi machinery, and/or the host plant takes dsRNAs up first, activates the plant RNAi machinery, and then dsRNAs or siRNAs are transferred into pathogen cell and stimulate the pathogen RNAi machinery (Song et al., 2018). Thus this method knocks down pathogen gene without presenting transmissible modifications into the plant genetic material (Koch et al., 2016;

Wang and Jin, 2017; Cagliari et al., 2019). In maize, viral resistance has been attained upon applying of crude extract from *Escherichia coli* HT115 (DE3) expressing dsRNA to sugarcane mosaic virus coat protein (Gan et al., 2010). Spray application of a dsRNA (791 nt *CYP3*-dsRNA) to target fungal cytochrome P450 *lanosterol C-14α-demethylases* (essential to biosynthesis of fungal ergosterol) prevents *Fusarium graminearum* growth and development (Koch et al., 2013). Unexpectedly, competent SIGS of fungal infection in the distal tissue involved a transfer of *CYP3*-dsRNA via the plant vascular system and processing into siRNAs by *FgDCL-1* after being received by the pathogen and CYP51 can persistently be present for 168 h at the local or nonsprayed distal segments (Koch et al., 2016). *Botrytis cinerea* genome encodes *DCL* genes *Bc-DCL1* and *Bc-DCL2* (*Bc-DCL1/2*) with respect to pathogenicity and growth (Weiberg et al., 2013). External application of sRNAs or dsRNAs that targets *Botrytis* DCLs on the surface of tomato, grape, lettuce, strawberry, onion, rose flowers, and *Arabidopsis* leaves, silences *Bc-DCL1/2* and reduces growth and pathogenicity of the fungus, exemplifying bidirectional cross-kingdom RNAi and sRNA trafficking between plants and fungi (Wang et al., 2016). So far, the mechanism of SIGS is not explored, and secondary siRNA amplification restricted the application of SIGS (Song et al., 2018). Several companies attempt in spray application of dsRNAs to control plant infectious diseases that aboveground pathogens cause (Regalado, 2015; Miguel and Scott, 2016). This novel SIGS method can significantly reduce the dangerous effects of synthetic chemicals on human and environment that are applied in conventional plant protection. Unlike structure-based synthetic fungicides, dsRNAs are sequence-based; thus SIGS application is an eco-friendly and sustainable method (Cai et al., 2019). Externally applied, in vitro-produced or crude nucleic acid extracts purified from dsRNAs or hpRNAs presenting bacterial strains can be considered as GMO-free approaches in plant disease management (Dubrovina and Kiselev, 2019). On the other hand, quick destroy of naked sRNAs and high expense to synthesize enough artificial sRNA molecules are the two barriers for SIGS approaches to induce RNA silencing against phytopathogens. Protecting sRNAs using clay particles or loading them in a synthetic vesicle could help address the first drawback (Schmitt et al., 2016; Mitter et al., 2017). The production cost to synthesize sRNAs is recently dropping from over $100,000 per gram a few years ago to $2 per gram currently (Le Page, 2017). Furthermore, direct delivery of sRNA molecules to battle against phytopathogenic diseases requires less time and is less complicated than GMO plants.

6.7 Management of phytopathogenic viruses using RNA interference technology

Fifty percentage of the microbes accountable for emerging crop diseases are viruses (Anderson et al., 2004; Bernardo et al., 2018) and the global economic loss cost more than $30 billion annually (Sastry and Zitter, 2014). Therefore management of virus diseases remains to be agronomic and scientific challenge, which requires advanced inventive approaches to satisfy the demands of a rising global population (Gaffar and Koch, 2019). Thus effective control mechanisms of plant pathogenic viruses infecting economically important crops are one of the key components in the formulation of sustainable crop disease management strategies (Tenllado et al., 2004). RNA-mediated resistance (a result of RNAi) emerges to be a more accessible approach to engineer highly resistant crops against viral diseases (Mansoor et al., 2006). Thus far, RNA silencing technology has been effectively exploited in antiviral breeding for more than 30 years, to target over 60 species of economically significant plant viruses, and many crops engineered to stably express sRNAs aiming countless viruses have been permitted for commercial cultivation (Zhang et al., 2019a).

Identification of suitable gene target is a critical step in developing an effective RNAi strategy. Numerous genes are appropriate as antiviral targets in both HIGS and exogenously delivered RNA molecules. RNAi strategies to manage viral diseases mainly rely on targeting one or multiple viral genes viz. replication-associated protein gene (Asad et al., 2003; Owor et al., 2011), genes that encode viral replicases (Tenllado et al., 2003; Kreuze et al., 2008; Pavli et al., 2010; Shekhawat et al., 2012; Ntui et al., 2014), transcriptional silencing of viral promoters (Seemanpillai et al., 2003; Vanderschuren et al., 2007), coat proteins (Missiou et al., 2004; Batuman et al., 2006; Kertbundit et al., 2007; Yin et al., 2009; Thu et al., 2016; Wagaba et al., 2017), viral RNAi suppressors (Varallyay et al., 2010; Law et al., 2013; Konakalla et al., 2016; Du et al., 2014; Sahana et al., 2014), and other genes (Hu et al., 2011; Shimizu et al., 2011a; Shimizu et al., 2011b). RNA silencing to combat against plant-viral diseases has focused on several plant families, namely Fabaceae, Cucurbitaceae, Poaceae, Solanaceae, Euphorbiaceae, Brassicaceae and tropical fruit crops especially citrus crops intensively (Gaffar and Koch, 2019). Sense or antisense viral sequences, virus-derived hpRNAs in transgene-mediated resistance, artificial miRNA-mediated resistance, resistance by mimicking the intact secondary structure of endogenous miRNA precursors using artificial miRNAs (amiRNAs) and transient RNA silencing-mediated resistance have been generated and effectively used in plants against several viral diseases (Duan et al., 2012). Due to the enlightened understanding of the RNA silencing mechanism and its part in antiviral immunity has been utilized to develop crops with viral resistance (Ding and Voinnet, 2007). The utilization of RNAi machinery has resulted in developing disease resistance against several viral pathogens in diverse crop-viral systems by exploiting different delivery approaches of dsRNAs to induce silencing (Table 6.2).

RNAi technology uses an innovative and alternative approach to combat against plant pathogenic viruses by exploiting the concept of parallel or simultaneous silencing of various viral invader's gene targets using single chimeric sequence to produce resistance to multiple viral pathogens (Gaffar and Koch, 2019). A single chimeric gene construct having three *CP* gene fragments of Cucumber green mottle mosaic virus (CGMMV), Watermelon mosaic virus (WMV), and Cucumber mosaic virus (CMV) was introduced to *A. tumefaciens* carrying a silencer DNA from the partial *N* gene of Watermelon silver mottle virus (WSMoV) to produce transgenic watermelon lines (Lin et al., 2012). The developed transgenic watermelon lines exhibited viral immunity to CGMMV, WMV, and CMV alone or mixed infection upon resistance evaluation (Lin et al., 2012). Fusing viral CP coding sequences derived from PVX, Potato virus S (PVS), and Potato virus Y (PVY) resulted in parallel immunity against infections caused by the tested RNA viruses in transgenic potato lines of cv. Desiree (Hameed et al., 2017). Developing broad-spectrum virus resistance by simultaneous silencing approach was also heretofore stated by the transgenic expression of fused-tandem constructs, and virus-derived dsRNAs (Arif et al., 2012; Chung and Palukaitis, 2011).

However, viruses also can respond against plant defense mechanism and make them less immune to infection. To counteract it, many viruses have genes that encode protein called viral suppressor of RNA silencing (VSR) (Burgyán and Havelda, 2011). First VSR identified

TABLE 6.2 RNAi mechanism in various plant-viral systems.

Target pathogen	Host	Delivery strategy	Target gene	Results	References
Rice stripe virus (RSV)	Rice	HIGS	Nucleocapsid protein (*NCP*) gene	Massive abundance of siRNA-derived NCP caused RSV resistance	Li et al. (2016)
Rice grassy stunt virus (RGSV)	Rice	HIGS	Nucleocapsid protein *pC5* and movement protein *pC6*	Plants exhibited high resistance to RGSV and inhibited the proliferation of RGSV	Shimizu et al. (2013)
Wheat streak mosaic virus (WSMV)	Wheat	Biolistic particle bombardment	Nuclear inclusion protein "a" (*Nla*) gene	Immune to WSMV	Fahim et al. (2010)
Maize streak virus (MSV)	Maize	Biolistic particle bombardment	Replication-associated protein gene (*rep*), viral long intergenic region (*LIR*)	Hairpin replication construct (pHPrepDI662) entirely subdued MSV replication, and the LIR hairpin construct was unsuccessful	Owor et al. (2011)
Barley yellow dwarf virus-PAV (BYDV-PAV)	Barley	Agroinfiltration	BYDV-polymerase (*BYDVpol*)	Strong immune to BYDV-PAV infection	Wang et al. (2000)
Maize dwarf mosaic virus (MDMV)	Maize	Agroinfiltration	Coat protein (*CP*) gene	Resistance to MDMV facilitated by RNA silencing was proportional to the length of the inverted-repeat sequence, copy number of T-DNA integration, and the repeatability of integration sites	Zhang et al. (2011)
Tomato yellow leaf curl virus (TYLCV)	Tomato. cv. Campbell-28	HIGS	*TYLCV C1*	Preventing replicase activity caused immunity to TYLCV infection	Fuentes et al. (2016)
Tomato leaf curl virus (ToLCV)	Tomato cv. Pusa	The upper surface of two carborundum-dusted leaves rubbed with solution	The dsRNA fragments made for AC1/AC4, AV1/AV2 (overlapping regions of ToLCV), AC1/AC4_AV1/AV2 (fusion construct), CMV-2b_ToLCV-AV1/AV2 (hybrid construct)	Hybrid constructs of AC1/AC4, AV1/AV2, AC1/AC4_AV1/AV2, and CMV-2b_ToLCV-AV1/AV2 showed 45%, 60%, 50%, and 55% defense against ToLCV, respectively	Namgial et al. (2019)
Cucumber mosaic virus (CMV)	Tobacco (*Nicotiana tabacum* cv. *xanthi*)	Mechanical inoculation via carborundum	The dsRNA fragments made for CMV-2b and hybrid construct CMV-2b_ToLCV-AV1/AV2	dsRNA construct of CMV-2b_ToLCV-*AV1/AV2* revealed 33.3% protection against CMV, while CMV-2b conferred 40% protection	Namgial et al. (2019)
African cassava mosaic virus (ACMV)	Cassava	Biolistic particle bombardment	Replication-associated protein coding sequence (*Rep/AC1*)	Transgenic cassava lines with high levels of AC1-homologous small RNAs had ACMV resistance with increasing viral load	Vanderschuren et al. (2009)

(*Continued*)

TABLE 6.2 (Continued)

Target pathogen	Host	Delivery strategy	Target gene	Results	References
Sweet potato chlorotic stunt virus (SPCSV)	Sweet potato	Blotted onto sterile filter paper	Replicase RdRp-encoding regions of SPCSV to target *RdRp*	10 of the 20 studied transgenic plants showed very mild or no symptoms following infection and accumulation of SPCSV was substantially decreased	Kreuze et al. (2008)
Bean golden mosaic virus (BGMV)	*Phaseolus vulgaris*	Posttranscriptional gene silencing	Replicase coding sequence *ACI*	No infected transgenic lines	Aragão and Faria (2009)
Mungbean yellow mosaic India virus (MYMIV)	Cowpea	HIGS	*AC2, AC4*	AC2-hpRNA and AC2 + AC4-hpRNA constructs expressing transgenic lines exhibited 100% resistance to MYMIV infection	Kumar et al. (2017)
Soybean mosaic virus (SMV)	*Glycine max*	HIGS	Helper component-proteinase (*HC-Pro*)	Stronger viral resistance	Kim et al. (2016)
Tobacco mosaic virus (TMV)	*N. tabacum* cv. *xanthi*	Exogenous (by rubbing)	P126 (TMV silencing suppressor) coat protein	Induced greater resistance against TMV	Konakalla et al. (2016)
Potato virus Y (PVY)	Potato	Agroinfiltration	Helper component-proteinase (*HC-Pro*)	Antiviral immunity	Arif et al. (2012)
Papaya ringspot virus (PRSW)	Papaya	Posttranscriptional gene silencing	Coat protein (*CP*) gene	Complete resistance to PRSW was noted in transgenic papaya	Ruanjan et al. (2007)
Banana bunchy top virus (BBTV)	Banana	HIGS	Replication initiation gene	Strong resistance to BBTV infection	Shekhawat et al. (2012)
Citrus psorosis virus (CPsV)	Sweet orange	Posttranscriptional gene silencing	Coat protein (*CP*), *54K* or *24K* genes	Resistance to CPsV	Reyes et al. (2011)
Grapevine fanleaf virus (GFLV)	Grapevine	Agroinfiltration	Coat protein (*CP*) gene	Active processing of pre-amiRNAs by plant machinery	Jelly et al. (2012)
Olive mild mosaic virus (OMMV)	*Nicotiana benthamiana*	Agroinfiltration	Coat protein (*CP*) and *P6* genes	Reduction in viral accumulation and attenuation	Varanda et al. (2018)

was the helper component-proteinase (HC-Pro) of potyviruses (Anandalakshmi et al., 1998) and many VSRs have been identified in various plant viruses (Lakatos et al., 2006; Karran and Sanfacon, 2014; Csorba et al., 2015; Kenesi et al., 2017), diverse within and across kingdoms and can shut off RNA silencing mechanism at various stages. Suppressors including the *2b* gene of CMV (Qu et al., 2007), *p25* of PVX (Ai et al., 2011), *p19* of Tomato bush stunt virus (Wei-xian et al., 2007), *p6* of *Cauliflower mosaic virus* (Laird et al., 2013), and *p20, p23, p25* and *CP* of Citrus tristeza virus (Batuman et al., 2006) are some critical examples for suppressors. These VRS are frequently associated with viral pathogenicity, mediate synergism amidst crop viruses, and contribute to the induction of severe plant diseases. Results from certain well-examined virus suppressors showed that suppressors obstruct with systemic signaling for silencing. Targeting such suppressors can help in further utilization of RNA silencing against plant virus management.

6.8 Management of plant-bacterial diseases by adopting RNA interference technology

Bacteria cause severe diseases and yield losses in plants worldwide. At present, the world faces antibiotic resistance in bacteria. This drastically looms the demand for antibiotics. Hence, a novel option is needed to rectify this problem in cultivation of horticultural crops those which previously used antibiotics to control the bacterial diseases. RNA-mediated resistance could be an effective strategy for controlling bacterial diseases in plants. This section will elaborate on the research works carried out so far on RNA silencing mechanism used in the management of phytopathogenic bacterial diseases. Although RNAi method is considered as a customized defense mechanism against viruses, it is also used as an important strategy to overcome the bacterial diseases in plants through promoting plant disease resistance (Peláez and Sanchez, 2013). Many phytobacterial diseases are caused when type III effector genes of the pathogen and disease-susceptibility genes of the host are interlinked. Unfortunately, such susceptibility genes are mostly unidentified (Yang et al., 2006). Further, in bacteria, the RNAi mechanism is absent. However, it has an RNA-based regulative mechanism. In which, 100-nt size cis and trans antisense transcripts cause transcript decay by suppressing the translation (Good and Stach, 2011).

Generally, cis-encoded asRNA is more compatible with target mRNA than that of trans-coded asRNA in bacteria, and Hfq chaperones are needed to enhance the binding in trans-asRNA. Exogenous application of asRNA is limited in use since it is a single-stranded RNA molecule so that it could be easily interfered by the nuclease activity. To shun this problem, asRNA combined with peptide nucleic acids and/or phosphorothioate morpholino oligomers could be utilized (Good and Stach, 2011). Further, these peptide nucleic acids or phosphorothioate morpholino oligomers are conjugated with cell cationic entry peptides on the bacterial cell to enable their intracellular transport (Mellbye et al., 2009). A recent investigation performed by Dalakouras et al. (2018) reported that the success rate of asRNA as an external application is significant in xylem-restricted bacteria; *Xylella fastidiosa, Clavibacter xyli, and Pseudomonas syzygii.*

In addition to the above discussed direct RNAi-based technique to control bacterial phytopathogens, there are indirect RNAi-based strategies also tested. Escobar et al. (2001) have worked with a soil-borne plant pathogenic bacterium called *A. tumefaciens* which causes crown gall diseases in a wide range of perennial crops and imposes significant economic losses. They observed that oncogenes; *iaaM* and *ipt* of this bacterium get transferred horizontally when infects the plant. Subsequently, these oncogenes integrate into the plant genome and induce tumor formation. Hence, they developed transgenic tomato and *Arabidopsis* lines by inserting hpRNA molecules which could suppress the oncogenes of the *Agrobacterium*. The developed transformants are highly resistant to crown gall disease. The transgenic plant contained modified extra genes of hpRNA sequence, which knock down the expression of the bacterial oncogenes corresponding to the gall formation upon infection (Dunoyer et al., 2007). The conservation of high level of cross-strain DNA sequence in the *iaaM* and *ipt* oncogenes recommends that this technique can deliver an extensive level of resistance to crown gall tumorigenesis (Escobar et al., 2001).

In legumes and solanaceae crops, many of miRNA constructs target genes of plant innate immune receptors (Zhai et al., 2011; Li et al., 2012b). They provided a more in-depth understanding into bacterial infection in crops that repress miRNA-mediated silencing of R genes where that miRNA can either function as up or down regulators (either by suppression of down regulators or up regulation of the immune responses) of the bacterial infestation (Jiang et al., 2009; Zhai et al., 2011; Li et al., 2012b). Table 6.3 displays some additional reports of RNAi in different plant-bacterial systems.

TABLE 6.3 RNAi mechanism in different plant-bacterial systems.

Target pathogen	Disease	Crop	Delivery strategy	Target gene	Result	References
Agrobacterium tumefaciens	Crown gall	Tomato	Transgenic plants (hairpin-loop structure)	*iaaM and ipt*	30% of the transformants showed cosuppression of hydrophobin synthesis	Escobar et al. (2001), Escobar and Dandekar (2003), Dunoyer et al. (2007)
Xanthomonas oryzae pv. *oryzae*	Bacterial blight	Rice	Transgenic plants (hairpin-loop structure)	*Os8N3*	Plants that are resistant to PXO99A infection	Yang et al. (2006)
Pseudomonas syringae	Bacterial speck	Tomato	Agroinfiltration	*PPRL*	Increment in RPS2-mediated race-specific resistance against *P. syringae* was noted while induction of nat-siRNAATGB2	Katiyar-Agarwal et al. (2006)
Xanthomonas citri subsp. *Citri*	Bacterial canker	Citrus	*A. tumefaciens* EHA105, via heat shock	*CsBZIP40*	Reduction in the expression of *CsBZIP40* as low transcriptional levels of gene	Li et al. (2019)
Erwinia amylovora	Fire blight	Apple	*A. tumefaciens*-mediated transformation	*HIPM*	The transgenic lines grown as vegetative shoots for 2 years under controlled environmental conditions showed no detectable toxic effects due to fire blight	Malnoy et al. (2008)

6.9 Management of plant fungal diseases using RNA interference technology

Perennial crop yield losses due to phytopathogenic fungi and oomycetes recently account up to 20% worldwide, and further cost incurs because of postharvest diseases caused by fungi (Fisher et al., 2018). Besides, biotrophic and necrotrophic fungi obscure agricultural production, instigating worldwide decline in crop yield up to 30%–40% in-field and various postharvest stages (Bebber and Gurr, 2015; Myers et al., 2017). Application of topical sprays or achievement of stable and robust resistance via transgenic expression of dsRNAs in infected plant tissues are used in RNAi technology to offer robust and multicrop protection against an array of fungal diseases. RNA silencing–based disease control machinery uses advanced RNA sequencing amenities to detect fungal gene targets, thereby, it remains to be an alternative to synthetic fungicides for crop protection (McLoughlin et al., 2018). This phytopathogenic fungal management chiefly aimed at reducing pathogen virulence, pathogenicity, restricting mycelial growth, delaying disease symptom development, and reducing or ceasing the production of toxic compounds. Virulence effecter genes; *Avra10, BEC1016, Ave1* (Pliego et al., 2013; Song and Thomma, 2018), F-box protein required for pathogenicity1 gene; *frp1* (Hu et al., 2015), chitin synthase genes; *Chs* and *Chs3b* (Cheng et al., 2015; Andrade et al., 2016), β-1,3-glucan synthase gene; *gl1* (Chen et al., 2016), transcription factor gene; *velvet* (Ghag et al., 2014) and aflatoxin biosynthesis gene; *aflC* (Thakare et al., 2017) are recently investigated targets in RNAi field to switch fungal plant pathogens. Reports of RNAi-based therapeutic strategies to succeed phytofungal diseases are incrementing, and the salient features in the technology deliver innovative routes to upsurge species-specificity and/or effectiveness opposed to fungi for which existing fungicides have been useless. Therefore exploitation of RNAi machinery has resulted in identifying different RNA silencing approaches in diverse crop-fungal systems to manage phytopathogenic fungal diseases (Table 6.4).

An aggressive gray mold pathogen *B. cinerea* (Bc) derived sRNAs (Bc-sRNAs; actually they are sRNA effectors and are regularly formed by *Bc-DCL1* and *Bc-DCL2*) can migrate to host plants, that is, *Arabidopsis* and tomato, and overpower the host immunity genes via hijacking their RNAi mechanism (Weiberg et al., 2013). *B. cinerea* can also receive externally delivered dsRNAs and sRNAs that hinder its growth, providing a classic case of bidirectional transkingdom RNAi (Wang et al., 2016). The authors further coined that superficial sprays of purified dsRNAs or sRNAs that trigger *Bc-DCL1* and *Bc-DCL2* genes significantly inhibit gray mold diseases in fruits (grape, strawberry, and tomato), vegetables (iceberg lettuce), and flowers (rose). Since postharvest diseases are the foremost source of commercial losses within the fresh produce supply chain and most of the infections are caused by diverse of fungal pathogens, namely, *Alternaria, Aspergillus, Botrytis, Colletotrichum, Diplodia, Monilinia, Mucor, Penicillium, Phomopsis, Rhizopus*, and *Sclerotinia* (Barkai-Golan, 2005; Sharma et al., 2009), this technique would be a breakthrough to control many postharvest fungal diseases in fruits, vegetables, and flowers in the near future. However, the ability of pathogens to uptake exogenously delivered RNAs, practical usage, cost and longevity of exogenously synthesized RNAs, and consumer safety of such products should be addressed for further development.

TABLE 6.4 RNAi mechanism in various crop-fungal systems.

Target pathogen	Disease	Host	Delivery strategy	Target gene	Results	References
Magnaporthe oryzae	Rice blast	Rice	HIGS	*M. oryzae* transcription factor *MoAP1*	Inhibition of fungal hyphae growth, formation of abnormal spores, and reduction in pathogenicity	Xiao-Yi et al. (2019)
Puccinia triticina	Leaf rust	Wheat	HIGS	Pt MAP-kinase (*PtMAPK1*) cyclophilin (*PtCYC1*)	Depletion in target fungal transcripts, reduction in fungal biomass accumulation in highly resistant crops	Panwar et al. (2018)
Fusarium graminearum	Fusarium head blight (FHB), Fusarium seedling blight (FSB)	Wheat	HIGS	*Chs3b*	Transgenic plants showed strong, stable, and consistent resistance to FHB and FSB, restricted mycelial growth and efficient downregulation of *Chs3b* in the colonizing *F. graminearum* on wheat seedlings and spikes	Cheng et al. (2015)
F. graminearum	Fusarium head blight	Barley	SIGS	*FgAGO* and *FgDCL* genes	Silencing fungal RNAi pathway genes, efficiently increased plant disease resistance	Werner et al. (2019)
F. graminearum	Fusarium head blight	Barley	SIGS	*CYP51A, CYP51B, and CYP51C*	Inhibition of *F. graminearum* mycelial growth	Koch et al. (2013)
Fusarium oxysporum	Fusarium wilt	Arabidopsis	Agroinfiltration	*F. oxysporum* Wilt 2 (FOW2), oxophytodienoate-10,11-gene (OPR), F-box protein required for pathogenicity 1 (FRP1)	Transgenic lines viz FRP1 and OPR exhibited improved resistance to *F. oxysporum* with hindered disease symptom development	Hu et al. (2015)
F. oxysporum f. sp. *cubens*	Fusarium wilt	*Musa* sp.	Agroinfiltration	Velvet and Fusarium transcription factor 1	Conidiophores count reduced by 7- to 25-fold and 70%–85% increment in resistance to Fusarium wilt	Ghag et al. (2014)

Fusarium culmorum	Fusarium head blight	Barley	SIGS	*FcCYP51A, FcCYP51B, FcCYP51C*	Growth retardation and destruction to fungus, RNAi using CYP51-dsRNAs displayed frail aberrant morphologies of hyphae including undulation, minor conglomeration, and more branching with defective septation	Koch et al. (2018)
Aspergillus flavus	Aspergillus ear rot	Maize	HIGS	*amy1*	Reduction in fungal growth and aflatoxin production in maize kernels	Gilbert et al. (2018)
A. flavus	Aspergillus toxin	Peanut	Agroinfiltration	*aflS, aflR, aflC, pes1, aflep*	~4- to 20-fold expression of the hpRNA expression in the RNAi lines, ~100% reduction in aflatoxin B1 and B2 in the seeds	Arias et al. (2015)
Colletotrichum gloeosporioides	Anthracnose	Chilli, Tomato	HIGS	*C. gloeosporioides* conidial morphology 1 gene (*CgCOM1*)	High suppression in conidial germination, germ tube development, appressoria formation, and mycelial growth resulting in reduced infection of plant tissues	Mahto et al. (2020)
Verticillium dahliae	Verticillium wilt	Tomato, Arabidopsis	HIGS	*Ave1*, *Sge1* and *NLP1*	Suppression of Verticillium wilt disease	Song and Thomma (2018)
Phytophthora infestans	Late blight	Potato	Electrophoretically delivered	Sesquiterpene cyclase gene (*PVS3p*), β-glucuronidase (*GUS*)	PVS-silenced transgenic potato tubers showed no accumulation of phytoalexins after inoculation with *P. infestans*	Yoshioka et al. (2019)
P. infestans	Late blight	Potato	HIGS	Syntaxin-related 1 (StSYR1) and soluble *N*-ethylmaleimide-sensitive factor adapter protein 33 (StSNAP33)	Improvement in resistance status and reduction in the growth of *P. infestans* on StSYR1-RNAi plants suggest involvement of syntaxins in secretory defense responses of potato, development of callose-containing papillae	Eschen-Lippold et al. (2012)

(Continued)

TABLE 6.4 (Continued)

Target pathogen	Disease	Host	Delivery strategy	Target gene	Results	References
Bremia lactucae	Downey mildew	Lettuce	HIGS	*ELF1, HAM34, CES, ACT, TUB, NLP2,* and *NLP3*	Growth reduction and inhibition of sporulation of *B. lactucae*	Govindarajulu et al. (2015)
V. dahliae	Verticillium wilt	Cotton	HIGS	*VdH1*	Inhibits virulence gene expression in *V. dahlia*	Zhang et al. (2016)
Blumeria graminis	Powdery mildew	Wheat, Barley	HIGS	Effector gene *Avra10*	Targeting ds or antisense RNA of fungal transcripts decreased fungal development	Nowara et al., 2010

Moreover, phytopathogenic fungi that belong to *Aspergillus*, *Fusarium*, and *Penicillium* genera are a foremost menace to food and feed crops due to production of mycotoxins, namely, aflatoxins, citrinin, 4-deoxynivalenol, fumonisins, fusarin C, patulin, and various toxic metabolic intermediates that considerably reduce the quality and quantity of agricultural produce (Ismaiel and Papenbrock, 2015; Majumdar et al., 2017). RNAi technology has also been explored for the control of mycotoxin contamination in crops and may enhance food safety by controlling the proliferation of plant pathogenic and mycotoxin-generating fungi (Majumdar et al., 2017; Machado et al., 2018).

6.10 Conclusions and future directions

RNA silencing offers robust, efficient, and more selective ways in plant protection, and this novel approach proved effective to create resistance against several phytopathogens that cause deleterious effects on crop production and threat world food security. It is evident from the above findings that RNAi is an effective management tool to control viral, bacterial, and fungal diseases in various plants tested by targeting most appropriate genes. A suitable way of delivery to trigger RNA silencing machinery is utmost to switch the mechanism inside pathogen−host plant systems. The phytopathogens damages can be controlled by the delivery of dsRNAs constructs (both from exogenous nontransformative and transformative methods) that target to silence the plant pathogen structural genes, and virulence and pathogenicity-related genes. Transgenic expression of dsRNA is achieved by HIGS approach that requires genetic alteration of host plants, generation of transgenic crops, which may take a longer duration to release for commercial cultivation. There are several concerns about transformability, genetic stability, and unpredictability of RNAi transgenic crops that have to move through regulations for the recommendations. Another innovative and eco-friendly tactic called SIGS, which uses nontransformative delivery of dsRNAs, is in vitro-synthesized and exogenously applied on plant surfaces. SIGS has been considered as a GMO-free approach in plant disease management. However, the consequence of exogenously applied RNAs, stability, pathways, and molecular mechanisms inside the plant systems and fate of those delivered dsRNAs within food and ecosystems should further evidently and critically be evaluated for the betterments. On the other hand, a variety of pathogenic RNA silencing suppressor proteins can encumber RNAi mechanism in agriculturally important crop plants and thereby making the plants prone to diseases. The suppressor molecules have also been targeted and examined as a tool for phytopathogen control mechanism, and their harmful effects have been stunned by RNAi. Numerous suppressor molecules in RNAi have been identified in many viruses, oomycetes, fungi, and very few in bacterial pathogens and still more to be identified in future too. Identification of novel suppressors in different pathogens could be a route to find proper RNAi strategies to cope with virulent plant pathogens.

Furthermore, RNA silencing is a promising tool in plant disease management that exploits negatively regulated gene expression at posttranscription level to combat phytopathogens that cause severe damages to plants. This novel breakthrough will be of greater acceptance to a broader audience if it is more focused on designing dsDNAs to target the suitable genes that can reduce pathogenicity and virulence of plant pathogens without any off-target effects.

References

Agrios, G.N., 2005. Plant Pathology, 5th Ed. Academic Press, New York.

Aguado, L.C., tenOever, B.R., 2018. RNase III nucleases and the evolution of antiviral systems. BioEssays 40, 1700173.

Ai, T., Zhang, L., Gao, Z., Zhu, C.X., Guo, X., 2011. Highly efficient virus resistance mediated by artificial microRNAs that target the suppressor of PVX and PVY in plants. Plant Biol. 13, 304–316.

Ali, Z., Ali, S., Tashkandi, M., Zaidi, S.S.A., Mahfouz, M.M., 2016. CRISPR/Cas9-mediated immunity to geminiviruses: differential interference and evasion. Sci. Rep. 6 (1), 1–13.

Anandalakshmi, R., Pruss, G.J., Ge, X., Marathe, R., Mallory, A.C., Smith, T.H., et al., 1998. A viral suppressor of gene silencing in plants. Proc. Natl. Acad. Sci. USA 95, 13079–13084.

Anderson, P.K., Cunningham, A.A., Patel, N.G., Morales, F.J., Epstein, P.R., Daszak, P., 2004. Emerging infectious diseases of plants: pathogen pollution, climate change and agrotechnology drivers. Trends Ecol. Evol. 19, 535–544.

Andrade, C.M., Tinoco, M.L.P., Rieth, A.F., Maia, F.C.O., Aragao, F.J.L., 2016. Host-induced gene silencing in the necrotrophic fungal pathogen *Sclerotinia sclerotiorum*. Plant Pathol. 65, 626–632.

Aragão, F., Faria, J., 2009. First transgenic geminivirus-resistant plant in the field. Nat. Biotechnol. 27, 1086–1088.

Arias, R.S., Dang, P.M., Sobolev, V.S., 2015. RNAi-mediated control of aflatoxins in peanut: method to analyze mycotoxin production and transgene expression in the peanut/*Aspergillus* pathosystem. J. Vis. Exp. 106, e53398.

Arif, M., Azhar, U., Arshad, M., Zafar, Y., Mansoor, S., Asad, S., 2012. Engineering broad-spectrum resistance against RNA viruses in potato. Transgenic Res. 21, 303–311.

Asad, S., Haris, W.A., Bashir, A., Zafar, Y., Malik, K.A., Malik, N.N., et al., 2003. Transgenic tobacco expressing geminiviral RNAs are resistant to the serious viral pathogen causing cotton leaf curl disease. Arch. Virol. 148 (12), 2341–2352.

Aslam, S., Tahir, A., Aslam, M.F., Alam, M.W., Shedayi, A.A., Sadia, S., 2017. Recent advances in molecular techniques for the identification of phytopathogenic fungi – a mini review. J. Plant Interact. 12 (1), 493–504.

Axtell, M.J., 2013. Classification and comparison of small RNAs from plants. Annu. Rev. Plant Biol. 64, 137–159.

Banerjee, S., Barraud, P., 2014. Functions of double-stranded RNA-binding domains in nucleocytoplasmic transport. RNA Biol. 11, 1226–1232.

Barkai-Golan, R., 2005. *Postharvest Diseases of Fruits and Vegetables*. A Division of Reed. Elsevier India Pvt. Limited, New Delhi.

Barraud, P., Banerjee, S., Mohamed, W.I., Jantsch, M.F., Allain, F.H.T., 2014. A bimodular nuclear localization signal assembled via an extended double-stranded RNA-binding domain acts as an RNA-sensing signal for transportin 1. Proc. Natl Acad. Sci. USA 111, 1852–1861.

Batuman, O., Mawassi, M., Bar-Joseph, M., 2006. Transgenes consisting of a dsRNA of an RNAi suppressor plus the 30 UTR provide resistance to *Citrus tristeza virus* sequences in *Nicotiana benthamiana* but not in citrus. Virus Genes 33, 319–327.

Baulcombe, D., 2004. RNA silencing in plants. Nature 431, 356–363.

Baulcombe, D.C., 2015. VIGS, HIGS and FIGS: small RNA silencing in the interactions of viruses or filamentous organisms with their plant hosts. Curr. Opin. Plant Biol. 26, 141–146.

Bebber, D.P., Gurr, S.J., 2015. Crop-destroying fungal and oomycete pathogens challenge food security. Fungal Genet. Biol. 74, 62–64.

Bekele, D., Tesfaye, K., Fikre, A., 2019. Applications of virus induced gene silencing (VIGS) in plant functional genomics studies. J. Plant Biochem. Physiol. 7 (1), 1–7.

Bernardo, P., Charles-Dominique, T., Barakat, M., Ortet, P., Fernandez, E., Filloux, D., et al., 2018. Geometagenomics illuminates the impact of agriculture on the distribution and prevalence of plant viruses at the ecosystem scale. ISME J. 12, 173–184.

Bertazzon, N., Raiola, A., Castiglioni, C., Gardiman, M., Angelini, E., Borgo, M., et al., 2012. Transient silencing of the grapevine gene *VvPGIP1* by agroinfiltration with a construct for RNA interference. Plant Cell Rep. 31, 133–143.

Borges, F., Martienssen, R.A., 2015. The expanding world of small RNAs in plants. Nat. Rev. Mol. Cell Biol. 16, 727–741.

Brodersen, P., Voinnet, O., 2006. The diversity of RNA silencing pathways in plants. Trends Genet. 22, 268–280.

Burgyán, J., Havelda, Z., 2011. Viral suppressors of RNA silencing. Trends Plant Sci. 16, 265–272.

Cagliari, D., Dias, N.P., Galdeano, D.M., Dos Santos, E.Á., Smagghe, G., Zotti, M.J., 2019. Management of pest insects and plant diseases by non-transformative RNAi. Front. Plant Sci. 10, 1319.

Cai, Q., He, B., Weiberg, A., Buck, A.H., Jin, H., 2019. Small RNAs and extracellular vesicles: new mechanisms of cross-species communication and innovative tools for disease control. PLoS Pathog. 15 (12), e1008090.

Carmell, M., Hannon, G., 2004. RNase III enzymes and the initiation of gene silencing. Nat. Struct. Mol. Biol. 11, 214–218.

Carthew, R.W., Sontheimer, E.J., 2009. Origins and mechanisms of miRNAs and siRNAs. Cell 136, 642–655.

Castel, S.E., Martienssen, R.A., 2013. RNA interference in the nucleus: roles for small RNAs in transcription, epigenetics and beyond. Nat. Rev. Genet. 14, 100–112.

Cerutti, H., Casas-Mollano, J.A., 2006. On the origin and functions of RNA-mediated silencing: from protists to man. Curr. Genet. 50, 81–99.

Chen, X., 2009. Small RNAs and their roles in plant development. Annu. Rev. Cell Dev. Biol. 25, 21–44.

Chen, W., Kastner, C., Nowara, D., Oliveira-Garcia, E., Rutten, T., Zhao, Y., et al., 2016. Host-induced silencing of *Fusarium culmorum* genes protects wheat from infection. J. Exp. Bot. 67 (17), 4979–4991.

Cheng, W., Song, X., Li, H., Cao, L., Sun, K., Qiu, X., et al., 2015. Host-induced gene silencing of an essential chitin synthase gene confers durable resistance to Fusarium head blight and seedling blight in wheat. Plant Biotechnol. J. 13, 1335–1345.

Chung, B.N., Palukaitis, P., 2011. Resistance to multiple viruses in transgenic tobacco expressing fused, tandem repeat, virus-derived double-stranded RNAs. Virus Genes 43, 454–464.

Csorba, T., Kontra, L., Burgyan, J., 2015. Viral silencing suppressors: tools forged to fine-tune host-pathogen coexistence. Virology 479-480, 85–103.

Dalakouras, A., Jarausch, W., Buchholz, G., Bassler, A., Braun, M., Manthey, T., et al., 2018. Delivery of hairpin RNAs and small RNAs into woody and herbaceous plants by trunk injection and petiole absorption. Front. Plant Sci. 9, 1253.

Dalakouras, A., Wassenegger, M., McMillan, J.N., Cardoza, V., Maegele, I., Dadami, E., et al., 2016. Induction of silencing in plants by high-pressure spraying of in vitro-synthesized small RNAs. Front. Plant Sci. 7, 1327.

Derrien, B., Clavel, M., Baumberger, N., Iki, T., Sarazin, A., Hacquard, T., et al., 2018. A Suppressor screen for AGO1 degradation by the viral F-Box P0 protein uncovers a role for AGO DUF1785 in sRNA duplex unwinding. Plant Cell 30 (6), 1353–1374.

Ding, S.W., Voinnet, O., 2007. Antiviral immunity directed by small RNAs. Cell 130 (3), 413–426.

Dorsett, Y., Tuschl, T., 2004. siRNAs: applications in functional genomics and potential as therapeutics. Nat. Rev. Drug Discov. 3, 318–329.

Du, Z., Chen, A., Chen, W., Liao, Q., Zhang, H., Bao, Y.M.J., et al., 2014. Can nuclear cytoplasmic partitioning of cucumber mosaic virus protein 2b determines the balance between its roles as a virulence determinant and an RNA-silencing suppressor. J. Virol. 88, 5228–5241.

Duan, C.G., Wang, C.H., Guo, H.S., 2012. Application of RNA silencing to plant disease resistance. Silence 3 (1), 5.

Dubrovina, A.S., Kiselev, K.V., 2019. Exogenous RNAs for gene regulation and plant resistance. Int. J. Mol. Sci. 20 (9), 2282.

Dun-chun, H., Jia-sui, Z., Lian-hui, X., 2016. Problems, challenges and future of plant disease management: from an ecological point of view. J. Integr. Agric. 15 (4), 705–715.

Dunoyer, P., Himber, C., Ruiz-Ferrer, V., Alioua, A., Voinnet, O., 2007. Intra- and Intercellular RNA interference in *Arabidopsis thaliana* requires components of the microRNA and heterochromatic silencing path–ways. Nat. Genet. 39, 848–856.

Elbashir, S.M., Lendeckel, W., Tuschl, T., 2001a. RNA interference is mediated by 21- and 22-nucleotide RNAs. Genes. Dev. 15, 188–200.

Elbashir, S.M., Martinez, J., Patkaniowska, A., Lendeckel, W., Tuschl, T., 2001b. Functional anatomy of siRNAs for mediating efficient RNAi in *Drosophila melanogaster* embryo lysate. EMBO J. 20, 6877–6888.

Eschen-Lippold, L., Landgraf, R., Smolka, U., Schulze, S., Heilmann, M., Heilmann, I., et al., 2012. Activation of defense against *Phytophthora infestans* in potato by down-regulation of syntaxin gene expression. New Phytol. 193, 985–996.

Escobar, M.A., Civerolo, E.L., Summerfelt, K.R., Dandekar, A.M., 2001. RNAi-mediated oncogene silencing confers resistance to crown gall tumorigenesis. Proc. Natl. Acad. Sci USA 98, 13437–14237.

Escobar, M.A., Dandekar, A.M., 2003. *Agrobacterium tumefaciens* as an agent of disease. Trends Plant Sci. 8, 380–386.

Fagwalawa, L.D., Kutama, A.S., Yakasai, M.T., 2013. Current issues in plant disease control: biotechnology and plant disease. BAJOPAS 6 (2), 121–126.

Fahim, M., Ayala-Navarrete, L., Millar, A.A., Larkin, P.J., 2010. Hairpin RNA derived from viral *NIa* gene confers immunity to wheat streak mosaic virus infection in transgenic wheat plants. Plant Biotechnol. J. 8, 821–834.

Fahim, M., Millar, A.A., Wood, C.C., Larkin, P.J., 2012. Resistance to *Wheat streak mosaic virus* generated by expression of an artificial polycistronic microRNA in wheat. Plant Biotechnol. J. 10, 150–163.

Fang, Y., Ramasamy, R.P., 2015. Current and prospective methods for plant disease detection. Biosensors 4, 537–561.

Farrell, R.E., 2017. Chapter 11 - RNA interference and RNA editing. In: Farrell, R.E. (Ed.), RNA Methodologies. Academic Press, United States, pp. 345–376.

Fire, A., Albertson, D., Harrison, S.W., Moerman, D.G., 1991. Production of antisense RNA leads to effective and specific inhibition of gene expression in *C. elegans* muscle. Development 113, 503–514.

Fire, A., Xu, S., Montgomery, M.K., Kostas, S.A., Driver, S.E., Mello, C.C., 1998. Potent and specific genetic interference by double-stranded RNA in *Caenorhabditis elegans*. Nature 391, 806–811.

Fisher, M.C., Hawkins, N.J., Sanglard, D., Gurr, S.J., 2018. Worldwide emergence of resistance to antifungal drugs challenges human health and food security. Science 360, 739–742.

Food and Agriculture Organization of the United Nations. (2020). <http://www.fao.org/news/story/en/item/1187738/icode/> (accessed 05.01.20).

Foster, S.J., Park, T.H., Pel, M., Brigneti, G., Sliwka, J., Jagger, L., et al., 2009. Rpi-vnt1.1, a Tm-2(2) homolog from *Solanum venturii*, confers resistance to potato late blight. Mol. Plant Microbe Interact. 22, 589–600.

Fradin, E.F., Zhang, Z., Ayala, J.C.J., Castroverde, C.D., Nazar, R.N., Robb, J., et al., 2009. Genetic dissection of Verticillium wilt resistance mediated by tomato *Ve1*. Plant Physiol. 150, 320–332.

Frank, F., Hauver, J., Sonenberg, N., Nagar, B., 2012. Arabidopsis Argonaute MID domains use their nucleotide specificity loop to sort small RNAs. EMBO J. 31 (17), 3588–3595.

Fu, D.Q., Zhu, B.Z., Zhu, H.L., Jiang, W.B., Luo, Y.B., 2005. Virus-induced gene silencing in tomato fruit. Plant J. 43, 299–308.

Fuentes, A., Carlos, N., Ruiz, Y., et al., 2016. Field trial and molecular characterization of RNAi-transgenic tomato plants that exhibit resistance to tomato yellow leaf curl geminivirus. Mol. Plant Microbe Interact. 29 (3), 197–209.

Fukudome, A., Fukuhara, T., 2016. Plant dicer-like proteins: double-stranded RNA-cleaving enzymes for small RNA biogenesis. J. Plant Res. 130 (1), 33–44.

Gaffar, F.Y., Koch, A., 2019. Catch me if you can! RNA silencing-based improvement of antiviral plant immunity. Viruses 11 (7), 673.

Gan, D., Zhang, J., Jiang, H., Jiang, T., Zhu, S., Cheng, B., 2010. Bacterially expressed dsRNA protects maize against SCMV infection. Plant Cell Rep. 29, 1261–1268.

Ghag, S.B., Shekhawat, U.K.S., Ganapathi, T.R., 2014. Host-induced post-transcriptional hairpin RNA-mediated gene silencing of vital fungal genes confers efficient resistance against Fusarium wilt in banana. Plant Biotechnol. J. 12, 541–553.

Gilbert, M.K., Majumdar, R.M., Rajasekaran, K., Chen, Z.Y., Wei, Q., Sickler, C.M., et al., 2018. RNA interference-based silencing of the alpha-amylase (amy1) gene in *Aspergillus flavus* decreases fungal growth and aflatoxin production in maize kernels. Planta 247, 1465–1473.

Godfrey, D., Böhlenius, H., Pedersen, C., Zhang, Z., Emmersen, J., Thordal-Christensen, H., 2010. Powdery mildew fungal effector candidates share N-terminal Y/F/WxC-motif. BMC Genomics 11, 317.

Good, L., Stach, J.E.M., 2011. Synthetic RNA silencing in bacteria-antimicrobial discovery and resistance breaking. Front. Microbiol. 2 (185), 1–11.

Govindarajulu, M., Epstein, L., Wroblewski, T., Michelmore, R.W., 2015. Host-induced gene silencing inhibits the biotrophic pathogen causing downy mildew of lettuce. Plant Biotechnol. J. 13, 875–883.

Grishok, A., Tabara, H., Mello, C.C., 2000. Genetic requirements for inheritance of RNAi in *C. elegans*. Science 287, 2494–2497.

Guo, S., Kemphues, K.J., 1995. par-1, a gene required for establishing polarity in *C. elegans* embryos, encodes a putative Ser/Thr kinase that is asymmetrically distributed. Cell 81, 611–620.

Hameed, A., Tahir, M.N., Asad, S., Bilal, R., Van Eck, J., Jander, G., et al., 2017. RNAi-mediated simultaneous resistance against three RNA viruses in potato. Mol. Biotechnol. 59, 73–83.
Hamilton, A.J., Baulcombe, D.C., 1999. A species of small antisense RNA in posttranscriptional gene silencing in plants. Science 286, 950–952.
Hein, I., Pacak, M.B., Hrubikova, K., Williamson, S., Dinesen, M., Soenderby, I.E., et al., 2005. Virus induced gene silencing-based functional characterization of genes associated with powdery mildew resistance in barley. Plant Physiol. 138 (4), 2155–2164.
Helm, J.M., Dadami, E., Kalantidis, K., 2011. Local RNA silencing mediated by agroinfiltration. Methods Mol. Biol. 744, 97–108.
Hilly, J.M., Liu, Z., 2007. An overview of small RNAs. In: Bassett, C.L. (Ed.), Regulation of Gene Expression in Plants. Springer, pp. 123–147.
Hiriart, J.B., Aro, E.M., Lehto, K., 2003. Dynamics of the VIGS-mediated chimeric silencing of the *Nicotiana benthamiana ChlH* gene and of the *Tobacco mosaic virus* vector. Mol. Plant Microbe. Interact. 16, 99–106.
Hu, Q., Niu, Y., Zhang, K., Liu, Y., Zhou, X., 2011. Virus derived transgenes expressing hairpin RNA give immunity to *Tobacco mosaic virus* and *Cucumber mosaic virus*. Virol. J. 8, 41.
Hu, Z.L., Parekh, U., Maruta, N., Trusov, Y., Botella, J.R., 2015. Down-regulation of *Fusarium oxysporum* endogenous genes by host-delivered RNA interference enhances disease resistance. Front. Chem. 3, 1–10.
Ismaiel, A., Papenbrock, J., 2015. Mycotoxins: producing fungi and mechanisms of phytotoxicity. Agriculture 5 (3), 492–537.
Ito, K., 2011. Small RNAs and transposon silencing in plants. Dev. Growth Differ. 54 (1), 100–107.
Iyer, L.M., Koonin, E.V., Aravind, L., 2003. Evolutionary connection between the catalytic subunits of DNA-dependent RNA polymerases and eukaryotic RNA-dependent RNA polymerases and the origin of RNA polymerases. BMC Struct. Biol. 3, 1–23.
Jahan, S.N., Asman, A.K., Corcoran, P., Fogelqvist, J., Vetukuri, R.R., Dixelius, C., 2015. Plant-mediated gene silencing restricts growth of the potato late blight pathogen *Phytophthora infestans*. J. Exp. Bot. 66, 2785–2794.
Jelly, N.S., Schellenbaum, P., Walter, B., Maillot, P., 2012. Transient expression of artificial microRNAs targeting *Grapevine fanleaf virus* and evidence for RNA silencing in grapevine somatic embryos. Transgenic Res. 21, 1319–1327.
Jeseničnik, T., Štajner, N., Radišek, S., et al., 2019. RNA interference core components identified and characterised in *Verticillium nonalfalfae*, a vascular wilt pathogenic plant fungi of hops. Sci. Rep. 9, 8651.
Ji, X., Zhang, H., Zhang, Y., Wang, Y., Gao, C., 2015. Establishing a CRISPRCas-like immune system conferring DNA virus resistance in plants. Nat. Plants 1, 15144.
Jiang, L., Ding, L., He, B., Shenm, J., Xu, Z., Yin, M., et al., 2014. Systemic gene silencing in plants triggered by fluorescent nanoparticle-delivered double stranded RNA. Nanoscale 6, 9965–9969.
Jiang, C.J., Shimono, M., Maeda, S., Inoue, H., Mori, M., Hasegawa, M., et al., 2009. Suppression of the rice fatty-acid desaturase gene OsSSI2 enhances resistance to blast and leaf blight diseases in rice. Mol. Plant Microbe Interact. 22, 820–829.
Jinek, M., Doudna, J.A., 2009. A three-dimensional view of the molecular machinery of RNA interference. Nature 457, 405–412.
Kadotani, N., Nakayashiki, H., Tosa, Y., Mayama, S., 2004. One of the two dicer-like proteins in the filamentous fungi *Magnaporthe oryzae* genome is responsible for hairpin RNA-triggered RNA silencing and related small interfering RNA accumulation. J. Biol. Chem. 279, 44467–44474.
Kaldis, A., Berbati, M., Melita, O., Reppa, C., Holeva, M., Otten, P., et al., 2018. Exogenously applied dsRNA molecules deriving from the *Zucchini yellow mosaic virus* (ZYMV) genome move systemically and protect cucurbits against ZYMV. Mol. Plant Pathol. 19, 883–895.
Kandasamy, S.K., Fukunaga, R., 2016. Phosphate-binding pocket in Dicer-2 PAZ domain for high-fidelity siRNA production. Proc. Natl. Acad. Sci. USA 113, 14031–14036.
Karran, R.A., Sanfacon, H., 2014. *Tomato ringspot virus* coat protein binds to ARGONAUTE 1 and suppresses the translation repression of a reporter gene. Mol. Plant Microbe Interact. 27, 933–943.
Karthikeyan, A., Deivamani, M., Shobhana, V.G., Sudha, M., Anandhan, T., 2013. RNA interference: evolutions and applications in plant disease management. Arch. Phytopathol. Plant Protect. 46 (12), 1430–1441.
Katiyar-Agarwal, S., Morgan, R., Dahlbeck, D., Borsani, O., Villegas, A., Zhu, J., et al., 2006. A pathogen-inducible endogenous siRNA in plant immunity. Proc. Natl. Acad. Sci. USA 103 (47), 18002–18007.

Kenesi, E., Carbonell, A., Lózsa, R., Vértessy, B., Lakatos, L., 2017. A viral suppressor of RNA silencing inhibits ARGONAUTE 1 function by precluding target RNA binding to pre-assembled RISC. Nucleic Acids Res. 45 (13), 7736–7750.

Kertbundit, S., Pongtanom, N., Ruanjan, P., Chantasingh, D., Tanwanchai, A., Panyim, S., et al., 2007. Resistance of transgenic papaya plants to papaya ringspot virus. Biol. Plant 51, 333–339.

Khraiwesh, B., Zhu, J.K., Zhu, J., 2012. Role of miRNAs and siRNAs in biotic and abiotic stress responses of plants. Biochim. Biophys. Acta 1819, 137–148.

Kim, V.N., Han, J., Siomi, M.C., 2009. Biogenesis of small RNAs in animals. Nat. Rev. Mol. Cell Biol. 10, 126–139.

Kim, H.J., Kim, M.J., Pak, J.H., Im, H.H., Lee, D.H., Kim, K.H., et al., 2016. RNAi-mediated *Soybean mosaic virus* (SMV) resistance of a Korean Soybean cultivar. Plant Biotechnol. Rep. 10, 257.

Kliot, A., Kontsedalov, S., Lebedev, G., Brumin, M., Cathrin, P.B., Marubayashi, J.M., et al., 2014. Fluorescence in situ hybridizations (FISH) for the localization of viruses and endosymbiotic bacteria in plant and insect tissues. J. Vis. Exp. 84, e51030.

Koch, A., Biedenkopf, D., Furch, A., Weber, L., Rossbach, O., Abdellatef, E., et al., 2016. An RNAi-based control of *Fusarium graminearum* infection through spraying of long dsRNAs involves a plant passage and is controlled by the fungal silencing machinery. PLoS Pathog. 12, e1005901.

Koch, A., Kumar, N., Weber, L., Keller, H., Imani, J., Kogel, K.H., 2013. Host-induced gene silencing of cytochrome P450 lanosterol C14 -demethylase-encoding genes confers strong resistance to *Fusarium* species. Proc. Natl. Acad. Sci. USA 110, 19324–19329.

Koch, A., Stein, E., Kogel, K., 2018. RNA-based disease control as a complementary measure to fight Fusarium fungi through silencing of the azole target cytochrome *P450* lanosterol C-14 α-demethylase. Eur. J. Plant Pathol. 152, 1003–1010.

Konakalla, N.C., Kaldis, A., Berbati, M., Masarapu, H., Voloudakis, A.E., 2016. Exogenous application of double-stranded RNA molecules from TMV p126 and CP genes confers resistance against TMV in tobacco. Planta 244, 961–969.

Kreuze, J.F., Klein, I.S., Lazaro, M.U., Chuquiyuri, W.J., Morgan, G.L., Mejía, P.G., et al., 2008. RNA silencing-mediated resistance to a crinivirus (Closteroviridae) in cultivated sweet potato (*Ipomoea batatas* L.) and development of sweet potato virus disease following co-infection with a potyvirus. Mol. Plant Pathol. 9, 589–598.

Kumagai, M.H., Donson, J., Della-Cioppa, G., Harvey, D., Hanley, K., Grill, L.K., 1995. Cytoplasmic inhibition of carotenoid biosynthesis with virus-derived RNA. Proc. Natl Acad. Sci. USA 92 (5), 1679–1683.

Kumar, S., Tanti, B., Patil, B.L., Mukherjee, S.K., Sahoo, L., 2017. RNAi-derived transgenic resistance to *Mungbean yellow mosaic India virus* in cowpea. PLOS One 12 (10), e0186786.

Kurzynska-Kokorniak, A., Pokornowska, M., Koralewska, N., Hoffmann, W., Bienkowska-Szewczyk, K., Figlerowicz, M., 2016. Revealing a new activity of the human Dicer DUF283 domain *in vitro*. Sci. Rep. 6, 23989.

Lacomme, C., Hrubikova, K., Hein, I., 2003. Enhancement of virus-induced gene silencing through viral-based production of inverted-repeats. Plant J. 34, 543–553.

Laird, J., McInally, C., Carr, C., et al., 2013. Identification of the domains of *Cauliflower mosaic virus* protein P6 responsible for suppression of RNA silencing and salicylic acid signalling. J. Gen. Virol. 12, 2777–2789.

Lakatos, L., Csorba, T., Pantaleo, V., Chapman, E.J., Carrington, J.C., Liu, Y.P., et al., 2006. Small RNA binding is a common strategy to suppress RNA silencing by several viral suppressors. EMBO J. 25, 2768–2780.

Lau, S., Mazumdari, P., Hee, T., Song, A., Othma, R., Harikrishna, J., 2014. Crude extracts of bacterially-expressed dsRNA protect orchid plants against *Cymbidium mosaic virus* during transplantation from *in vitro* culture. J. Hortic. Sci. Biotechnol. 89, 569–576.

Law, S.M., Zhang, B.W., Brooks, C.L., 2013. pH-sensitive residues in the P19 RNA silencing suppressor protein from *Carnation Italian ringspot virus* affect siRNA binding stability. Protein Sci. 22, 595–604.

Le Page, M. (2017). Gene-silencing spray lets us modify plants without changing DNA new scientist. <https://www.newscientist.com/article/2117460-genesilencing- spray-lets-us-modify-plants-without-changingdna/> (accessed 12.03.20).

Li, L., Guo, C., Wang, B., Zhou, T., Lei, Y., Dai, Y.H., et al., 2016. RNAi-mediated transgenic rice resistance to *Rice stripe virus*. J. Integr. Agric. 15 (11), 2539–2549.

Li, Q., Jia, R., Dou, W., Qi, J., Qin, X., Fu, Y., et al., 2019. CsBZIP40, a BZIP transcription factor in sweet orange, plays a positive regulatory role in citrus bacterial canker response and tolerance. PLoS One 14 (10), e0223498.

Li, T.B., Liu, M.H., Spalding, D., Weeks, P., Yang, B., 2012a. High-efficiency TALEN-based gene editing produces disease-resistant rice. Nat. Biotechnol. 30 (5), 390–392.

Li, X., Wang, X., Zhang, S., Liu, D., Duan, Y., Dong, W., 2012b. Identification of soybean microRNAs involved in soybean cyst nematode infection by deep sequencing. PLoS One 7 (6), 1–10.

Lin, C.Y., Ku, H.M., Chiang, Y.H., Ho, H.Y., Yu, T.A., Jan, F.J., 2012. Development of transgenic watermelon resistant to *Cucumber mosaic virus* and *Watermelon mosaic virus* by using a single chimeric transgene construct. Transgenic Res. 21, 983–993.

Liu, S., Jaouannet, M., Dempsey, D.A., Imani, J., Coustau, C., Kogel, K.H., 2020. RNA-based technologies for insect control in plant production. Biotechnol. Adv 39, 107463.

Liu, Y., Schiff, M., Dinesh-Kumar, S., 2002. Virus-induced gene silencing in tomato. Plant J. 31, 777–786.

López, M.M., Bertolini, E., Olmos, A., Caruso, P., Corris, M.T., Llop, P., et al., 2003. Innovative tools for detection of plant pathogenic viruses and bacteria. Int. Microbiol. 6, 233–243.

Lu, R., Martin-Hernandez, A.M., Peart, J.R., Malcuit, I., 2003. Virus-induced gene silencing in plants. Methods 30 (4), 296–303.

Luo, Z., Chen, Z., 2007. Improperly terminated, unpolyadenylated mRNA of sense transgenes is targeted by RDR6-mediated RNA silencing in *Arabidopsis*. Plant Cell 19, 943–958.

Ma, J.B., Ye, K., Patel, D.J., 2004. Structural basis for overhang specific small interfering RNA recognition by the PAZ domain. Nature 429, 318–322.

MacFarlane, S.A., Popovich, A.H., 2000. Efficient expression of foreign proteins in roots from tobravirus vectors. Virology 267, 29–35.

Machado, A.K., Brown, N.A., Urban, M., Kanyuka, K., Hammond-Kosack, K.E., 2018. RNAi as an emerging approach to control Fusarium head blight disease and mycotoxin contamination in cereals. Pest Manag. Sci. 74 (4), 790–799.

Macovei, A., Sevilla, N.R., Cantos, C., Jonson, G.B., Slamet-Loedin, I., Cermak, T., et al., 2018. Novel alleles of rice eIF4G generated by CRISPR/Cas9-targeted mutagenesis confer resistance to *Rice tungro spherical virus*. Plant Biotechnol. J. 16, 1918–1927.

MacRae, I.J., Zhou, K., Li, F., Repic, A., Brooks, A.N., Cande, W.Z., et al., 2006. Structural basis for double-stranded RNA processing by dicer. Science 311, 195–198.

Mahto, B.K., Singh, A., Pareek, M., Manchikatla, V., Rajam, M.V., Dhar-Ray, S., et al., 2020. Host-induced silencing of the *Colletotrichum gloeosporioides* conidial 2 morphology 1 gene (*CgCOM1*) confers resistance against Anthracnose disease in chilli and tomato. bioRxiv. Available from: https://doi.org/10.1101/2020.01.03.893677.

Majumdar, R., Rajasekaran, K., Cary, J.W., 2017. RNA interference (RNAi) as a potential tool for control of mycotoxin contamination in crop plants: concepts and considerations. Front. Plant Sci. 8, 200.

Mallory, A.C., Hinze, A., Tucker, M.R., Bouché, N., Gasciolli, V., Elmayan, T., et al., 2009. Redundant and specific roles of the ARGONAUTE proteins AGO1 and ZLL in development and small RNA-directed gene silencing. PLoS Genet. 5, e1000646.

Malnoy, M., Borejsza-Wysocka, E.E., Pascal-Omeñaca, L., Aldwinckle, H.S., Oh, C.S., Beer, S.V., 2008. Silencing of HIPM, the apple protein that interacts with hrpN of *Erwinia amylovora*. Acta Hortic. 793, 261–264.

Mann, S.K., Kashyap, P.L., Sanghera, G.S., Singh, G., Singh, S., 2008. RNA interference: an eco-friendly tool for plant disease management. Transgen. Plant J. 2 (2), 110–126.

Mansoor, S., Amin, I., Hussain, M., Zafar, Y., Briddon, R.W., 2006. Engineering novel traits in plants through RNA interference. Trends Plant Sci. 11, 559–565.

Margis, R., Fusaro, A.F., Smith, N.A., Curtin, S.J., Watson, J.M., Finnegan, E.J., et al., 2006. The evolution and diversification of dicers in plants. FEBS Lett. 580, 2442–2450.

Marker, S., Le Mouël, A., Meyer, E., Simon, M., 2010. Distinct RNA-dependent RNA polymerases are required for RNAi triggered by double-stranded RNA versus truncated transgenes in *Paramecium tetraurelia*. Nucleic Acids Res. 38 (12), 4092–4107.

Martín-Hernández, A.M., Baulcombe, D.C., 2008. Tobacco rattle virus 16-kilodalton protein encodes a suppressor of RNA silencing that allows transient viral entry in meristems. J. Virol. 82, 4064–4071.

Mascia, T., Nigro, F., Abdallah, A., et al., 2014. Gene silencing and gene expression in phytopathogenic fungi using a plant virus vector. Proc. Natl. Acad. Sci. USA 111 (11), 4291–4296.

Mateos, J.L., Bologna, N.G., Chorostecki, U., Palatnik, J.F., 2010. Identification of microRNA processing determinants by random mutagenesis of *Arabidopsis* MIR172a precursor. Curr. Biol. 20 (1), 49–54.

McLoughlin, A.G., Philip, L., Walker, P.L., Wytinck, N., Sullivan, D.S., Whyard, S., et al., 2018. Developing new RNA interference technologies to control fungal pathogens. Can. J. Plant Pathol. 40 (3), 325–335.

Meister, G., Tuschl, T., 2004. Mechanisms of gene silencing by double-stranded RNA. Nature 431, 343–349.

Mellbye, B.L., Puckett, S.E., Tilley, L.D., Iversen, P.L., Geller, B.L., 2009. Variations in amino acid composition of antisense peptide-phosphorodiamidate morpholino oligomer affect potency against *Escherichia coli in vitro* and *in vivo*. Antimicrob. Agents Chemother. 53, 525–530.

Mette, M.F., Aufsatz, W., van der Winden, J., Matzke, M.A., Matzke, A.J., 2000. Transcriptional silencing and promoter methylation triggered by double-stranded RNA. EMBO J. 19 (19), 5194–5201.

Miguel, K.S., Scott, J.G., 2016. The next generation of insecticides: dsRNA is stable as a foliar-applied insecticide. Pest. Manag. Sci. 72, 801–809.

Missiou, A., Kalantidis, K., Boutla, A., Tzortzakaki, S., Tabler, M., Tsagris, M., 2004. Generation of transgenic potato plants highly resistant to *Potato virus Y* (PVY) through RNA silencing. Mol. Breed. 14, 185–197.

Mitter, N., Worrall, E.A., Robinson, K.E., Li, P., Jain, R.G., Taochy, C., et al., 2017. Clay nanosheets for topical delivery of RNAi for sustained protection against plant viruses. Nat. Plants 3, 1620.

Moritoh, S., Miki, D., Akiyama, M., Kawahara, M., Izawa, T., Maki, H., et al., 2005. RNAi–mediated silencing of OsGEN–L (OsGEN–like), a new member of the RAD2/XPG nuclease family, causes male sterility by defect of microspore development in rice. Plant Cell Physiol. 46, 699–715.

Muhammad, T., Zhang, F., Zhang, Y., Liang, Y., 2019. RNA interference: anatural immune system of plants to counteract biotic stressors. Cells 8 (1), 38.

Mushtaq, M., Sakina, A., Wani, S.H., Shikari, A.B., Tripathi, P., Zaid, A., et al., 2019. Harnessing genome editing techniques to engineer disease resistance in plants. Front. Plant Sci. 7 (10), 550.

Myers, S.S., Smith, M.R., Guth, S., Golden, C.D., Vaitla, B., Mueller, N.D., et al., 2017. Climate change and global food systems: potential impacts on food security and undernutrition. Annu. Rev. Public Health 38, 259–277.

Namgial, T., Kaldis, A., Chakraborty, S., Voloudakis, A., 2019. Topical application of double-stranded RNA molecules containing sequences of *Tomato leaf curl virus* and *Cucumber mosaic virus* confers protection against the cognate viruses. Physiol. Mol. Plant Pathol. 108, 101432.

Napoli, C., Lemieux, C., Jorgensen, R., 1990. Introduction of a chimeric chalcone synthase gene into petunia results in reversible co-suppression of homologous genes in trans. Plant Cell 2, 279–289.

Nicholson, A.W., 2014. Ribonuclease III mechanisms of double-stranded RNA cleavage. Wiley Interdiscip. Rev. RNA 5, 31–48.

Nicolas, F.E., Garre, V., 2016. RNA interference in fungi: retention and loss. Microbiol. Spectr. 4.

Niehl, A., Soininen, M., Poranen, M.M., Heinlein, M., 2018. Synthetic biology approach for plant protection using dsRNA. Plant Biotechnol. J. 16, 1679–1687.

Nowara, D., Gay, A., Lacomme, C., Shaw, J., Ridout, C., Douchkov, D., et al., 2010. HIGS: host-induced gene silencing in the obligate biotrophic fungal pathogen *Blumeria graminis*. Plant Cell 22, 3130–3141.

Ntui, V.O., Kong, K., Azadi, P., Khan, R.S., Chin, D.P., Igawa, T., et al., 2014. RNAi-mediated resistance to *Cucumber mosaic virus* (CMV) in genetically engineered tomato. AJPS 5, 554–572.

Owor, B.E., Martin, D.P., Rybicki, E.P., Thomson, J.A., Bezuidenhout, M.E., Lakay, F.M., et al., 2011. A rep-based hairpin inhibits replication of diverse *Maize streak virus* isolates in a transient assay. J. Gen. Virol. 92 (10), 2458–2465.

Paces, J., Nic, M., Novotny, T., Svoboda, P., 2017. Literature review of baseline information to support the risk assessment of RNAi-based GM plants. EFSA Support. Publ. 314.

Padmanabhan, M., Dinesh-Kumar, S.P., 2009. Virus-induced gene silencing as a tool for delivery of dsRNA into plants. Cold Spring Harb. Protoc. Available from: https://doi.org/10.1101/pdb.prot5139.

Palmer, K.E., Rybicki, E.P., 2001. Investigation of the potential of *Maize streak virus* to act as an infectious gene vector in maize plants. Arch. Virol. 146, 1089–1104.

Palmgren, M.G., Edenbrandt, A.K., Vedel, S.E., Andersen, M.M., Landes, X., Østerberg, J.T., et al., 2015. Are we ready for back-to-nature crop breeding? Trends Plant Sci. 20, 155–164.

Panwar, V., Jordan, M., McCallum, B., Bakkeren, G., 2018. Host-induced silencing of essential genes in *Puccinia triticina* through transgenic expression of RNAi sequences reduces severity of leaf rust infection in wheat. Plant Biotechnol. J. 16, 1013–1023.

Park, J.E., Heo, I., Tian, Y., Simanshu, D.K., Chang, H., Jee, D., et al., 2011. Dicer recognizes the 5′ end of RNA for efficient and accurate processing. Nature 475, 201–205.

Parker, J.S., 2010. How to slice: snapshots of argonaute in action. Silence 1, 3.

Pavli, O.I., Panopoulos, N.J., Goldbach, R., Skaracis, G.N., 2010. BNYVV-derived dsRNA confers resistance to rhizomania disease of sugar beet as evidenced by a novel transgenic hairy root approach. Transgenic Res. 19, 915–922.

Peláez, P., Sanchez, F., 2013. Small RNAs in plant defense responses during viral and bacterial interactions: similarities and differences. Front. Plant Sci. 4, 343.

Pliego, C., Nowara, D., Bonciani, G., Gheorghe, D.M., Xu, R., Surana, P., et al., 2013. Host-induced gene silencing in barley powdery mildew reveals a class of ribonuclease-like effectors. Mol. Plant Microbe Interact. 26, 633–642.

Pogue, G.P., Lindbo, J.A., Garger, S.J., Fitzmaurice, W.P., 2002. Making an ally from an enemy: plant virology and the new agriculture. Annu. Rev. Phytopathol. 40, 45–74.

Pooggin, M., Shivaprasad, P.V., Veluthambi, K., Hohn, T., 2003. RNAi targeting of DNA virus in plants. Nat. Biotechnol. 21 (2), 131–132.

Praveen, G., Deepmala, G., Monika, M., Kumar, V., Jyoti, B., Sudesh, K.Y., 2012. MicroRNAs and their role in plants during abiotic stresses. Environmental Adaptations and Stress Tolerance of Plants in the Era of Climate Change. pp. 265–278.

Puyam, A., Kaur, K., 2020. Exploiting RNA interference mechanism in plants for disease resistance. In: Ul Haq, I., Ijaz, S. (Eds.), Plant Disease Management Strategies for Sustainable Agriculture through Traditional and Modern Approaches. Springer, Cham, pp. 217–236.

Puyam, A., Sharma, S., Kashyap, P.L., 2017. RNA interference-a novel approach for plant disease management. J. Appl. Nat. Sci. 9 (3), 1612–1618.

Qi, T., Guo, J., Peng, H., Liu, P., Kang, Z., Guo, J., 2019. Host-induced gene silencing: a powerful strategy to control diseases of wheat and barley. Int. J. Mol. Sci. 20 (1), 206.

Qu, J., Ye, J., Fang, R., 2007. Artificial microRNA-mediated virus resistance in plants. J. Virol. 81, 6690–6699.

Ramegowda, V., Mysore, K.S., Senthil-Kumar, M., 2014. Virus–induced gene silencing is a versatile tool for unraveling the functional relevance of multiple abiotic-stress-responsive genes in crop plants. Front. Plant Sci. 5, 1–12.

Regalado, A. (2015). The next great GMO debate. MIT Technology Review. <https://www.technologyreview.com/s/540136/the-next-great-gmo-debate/> (accessed 28 February 2020).

Reyes, C.A., De Francesco, A., Peña, E.J., Costa, N., Plata, M.I., Sendin, L., et al., 2011. Resistance to *Citrus psorosis virus* in transgenic sweet orange plants is triggered by coat protein-RNA silencing. J. Biotechnol. 151, 151–158.

Romano, N., Macino, G., 1992. Quelling: transient inactivation of gene expression in *Neurospora crassa* by transformation with homologous sequences. Mol. Microbiol. 6 (22), 3343–3353.

Rosa, C., Kuo, Y.W., Wuriyanghan, H., Falk, B.W., 2018. RNA interference mechanisms and applications in plant pathology. Annu. Rev. Phytopathol. 56, 581–610.

Ruanjan, P., Kertbundit, S., Juricek, M., 2007. Posttranscriptional gene silencing is involved in resistance of transgenic papayas to *Papaya ringspot virus*. Biologia Plant 51, 517–520.

Ryu, C.M., An, A., Kang, L., Mysore, K.S., 2004. Agrodrench: a novel and effective agroinoculation method for virus-induced gene silencing in roots and diverse Solanaceous species. Plant J. 40, 322–331.

Safarova, D., Brazda, P., Navratil, M., 2014. Effect of artificial dsRNA on infection of pea plants by pea seed-borne mosaic virus. Czech J. Genet. Plant Breed. 50, 105–108.

Sahana, N., Kaur, H., Jain, R.K., Palukaitis, P., Canto, T., Praveen, S., 2014. The asparagine residue in the FRNK box of potyviral helper-component protease is critical for its small RNA binding and subcellular localization. J. Gen. Virol. 95, 1167–1177.

Saharan, V., Jain, D., Pareek, S., Pal, A., Kumaraswamy, R.V., Jakhar, S.K., et al., 2016. Viral, fungal and bacterial disease resistance in transgenic plants. In: Al-Khayri, J., Jain, S., Johnson, D. (Eds.), Advances in Plant Breeding Strategies: Agronomic, Abiotic and Biotic Stress Traits. Springer, Cham, pp. 627–656.

Sanju, S., Siddappa, S., Thakur, A., Shukla, P.K., Srivastava, N., Pattanayak, D., et al., 2015. Host mediated gene silencing of a single effector gene from the potato pathogen *Phytophthora infestans* imparts partial resistance to late blight disease. Funct. Integr. Genomics 15, 697–706.

Sastry, S.K., Zitter, T.A., 2014. Management of virus and viroid diseases of crops in the tropics. *Plant Virus and Viroid Diseases in the Tropics, Epidemiology and Management*. Springer, Dordrecht, pp. 149–480.

Savary, S., Willocquet, L., Pethybridge, S.J., et al., 2019. The global burden of pathogens and pests on major food crops. Nat. Ecol. Evol. 3, 430–439.

Schmitt, C., Lippert, A.H., Bonakdar, N., Sandoghdar, V., Voll, L.M., 2016. Compartmentalization and transport in synthetic vesicles. Front. Bioeng. Biotechnol. 4, 19.

Seemanpillai, M., Dry, I., Randles, J., Rezaian, A., 2003. Transcriptional silencing of geminiviral promoter-driven transgenes following homologous virus infection. Mol. Plant Microbe Interact. 16, 429–438.

Senthil-Kumar, M., Hema, R., Anand, A., Kang, L., Udayakumar, M., Mysore, K.S., 2007. A systematic study to determine the extent of gene silencing in *Nicotiana benthamiana* and other Solanaceae species when heterologous gene sequences are used for virus-induced gene silencing. N. Phytol. 176, 782–791.

Senthil-Kumar, M., Mysore, K.S., 2011. Virus-induced gene silencing can persist for more than 2 years and also be transmitted to progeny seedlings in *Nicotiana benthamiana* and tomato. Plant Biotechnol. J. 9, 797–806.

Sharma, R.R., Singh, D., Singh, R., 2009. Biological control of postharvest diseases of fruits and vegetables by microbial antagonists: a review. Biol. Control 50, 205–221.

Shekhawat, U.K., Ganapathi, T.R., Hadapad, A.B., 2012. Transgenic banana plants expressing small interfering RNAs targeted against viral replication initiation gene display high-level resistance to banana bunchy top virus infection. J. Gen. Virol. 93 (8), 1804–1813.

Shimizu, T., Nakazono-Nagaoka, E., Akita, F., Uehara-Ichiki, T., Sasaya, T., Omura, T., 2011a. Immunity to *Rice black streaked dwarf virus*, a plant reovirus, can be achieved in rice plants by RNA silencing against the gene for the viroplasm component protein. Virus Res. 160, 400–403.

Shimizu, T., Nakazono-Nagaoka, E., Uehara-Ichiki, T., Sasaya, T., Omura, T., 2011b. Targeting specific genes for RNA interference is crucial to the development of strong resistance to *Rice stripe virus*. Plant Biotechnol. J. 9, 503–512.

Shimizu, T., Ogamino, T., Hiraguri, A., Nakazono-Nagaoka, E., Uehara-Ichiki, T., Nakajima, M., et al., 2013. Strong resistance against *Rice grassy stunt virus* is induced in transgenic rice plants expressing double-stranded RNA of the viral genes for nucleocapsid or movement proteins as targets for RNA interference. Phytopathology 103, 513–519.

Shoup Rupp, J.L., 2016. RNAi-mediated, stable resistance to Triticum mosaic virus in wheat. Crop Sci. 56, 1602–1610.

Simon, B., Kirkpatrick, J.P., Eckhardt, S., Reuter, M., Rocha, E.A., Andrade-Navarro, M.A., et al., 2011. Recognition of 2′-o-methylated 3′-end of piRNA by the PAZ domain of a Piwi protein. Structure 19, 172–180.

Singh, A., Taneja, J., Dasgupta, I., Mukherjee, S.K., 2015. Development of plants resistant to tomato geminiviruses using artificial trans-acting small interfering RNA. Mol. Plant Pathol. 16 (7), 724–734.

Siomi, H., Siomi, M.C., 2009. On the road to reading the RNA-interference code. Nature 457, 396–404.

Song, X.S., Gu, K.X., Duan, X.X., Xiao, X.M., Hou, Y.P., Duan, Y.B., et al., 2018. Secondary amplification of siRNA machinery limits the application of spray-induced gene silencing. Mol. Plant Pathol. 19, 2543–2560.

Song, Y., Thomma, B., 2018. Host-induced gene silencing compromises Verticillium wilt in tomato and Arabidopsis. Mol. Plant Pathol. 19 (1), 77–89.

Tang, G., Reinhart, B.J., Bartel, D.P., Zamore, P.D., 2003. A biochemical framework for RNA silencing in plants. Genes. Dev. 17, 49–63.

Tenllado, F., Llave, C., Díaz-Ruíz, J., 2004. RNA interference as a new biotechnological tool for the control of virus diseases in plants. Virus Res. 102, 85–96.

Tenllado, F., Martínez-García, B., Vargas, M., Díaz-Ruíz, J.R., 2003. Crude extracts of bacterially expressed dsRNA can be used to protect plants against virus infections. BMC Biotechnol. 3, 3.

Thakare, D., Zhang, J.W., Wing, R.A., Cotty, P.J., Schmidt, M.A., 2017. Aflatoxin-free transgenic maize using host-induced gene silencing. Sci. Adv. 3, e1602382.

Thu, L.T.M., Thuy, V.T.X., Duc, L.H., Son, L.V., Ha, C.H., Mau, C.H., 2016. RNAi-mediated resistance to SMV and BYMV in transgenic tobacco. Crop Breed. Appl. Biot. 16, 213–218.

Tian, Y., Simanshu, D.K., Ma, J.B., Park, J.E., Heo, I., Kim, V.N., et al., 2014. A phosphate-binding pocket within the platform-PAZ-connector helix cassette of human dicer. Mol. Cell 53, 606–616.

Tuttle, J.R., Idris, A.M., Brown, J.K., Haigler, C.H., Robertson, D., 2008. Geminivirus-mediated gene silencing from *Cotton leaf crumple virus* is enhanced by low temperature in cotton. Plant Physiol. 148, 41–50.

Tzean, Y., Lee, M., Jan, H., Chiu, Y., Tu, T., Hou, B., et al., 2019. Cucumber mosaic virus-induced gene silencing in banana. Sci. Rep. 9, 11553.

Unver, T., Budak, H., 2009. Virus-induced gene silencing, a post transcriptional gene silencing method. Int. J. Plant. Genomics 2009, 198680.

Usharani, K., Surendranath, B., Haq, Q., Malathi, V.G., 2005. Infectivity analysis of a soybean isolate of *Mungbean yellow mosaic India virus* by agroinoculation. J. Gen. Plant Pathol. 71, 230–237.

Vanderschuren, H., Akbergenov, R., Pooggin, M.M., Hohn, T., Gruissem, W., Zhang, P., 2007. Transgenic cassava resistance to *African cassava mosaic virus* is enhanced by viral DNA-A bidirectional promoter-derived siRNAs. Plant Mol. Biol. 64, 549–557.

Vanderschuren, H., Alder, A., Zhang, P., Griussem, W., 2009. Dose-dependent RNAi-mediated geminivirus resistance in the tropical root crop cassava. Plant Mol. Biol. 70, 265–272.

Varallyay, A., Valoczi, A., Agyi, J., Burgyan, J., Havelda, Z., 2010. Plant virus-mediated induction of miR168 is associated with repression of ARGONAUTE1 accumulation. EMBO J. 29, 3507–3519.

Varanda, C.M., Materatski, P., Campos, M.D., Clara, M., Nolasco, G., Félix, M., 2018. *Olive Mild Mosaic Virus* coat protein and P6 are suppressors of RNA silencing, and their silencing confers resistance against OMMV. Viruses 10 (8), 416.

Vaucheret, H., 2008. Plant argonautes. Trends Plant Sci. 13, 350–358.

Voinnet, O., 2008. Use, tolerance and avoidance of amplified RNA silencing by plants. Trends Plant Sci. 13, 317–328.

Wagaba, H., Beyene, G., Aleu, J., Odipio, J., Okao-Okuja, G., Chauhan, R.D., et al., 2017. Field level RNAi-mediated resistance to Cassava brown streak disease across multiple cropping cycles and diverse East African agro-ecological locations. Front. Plant Sci. 7, 2060.

Wang, M., Abbot, D.C., Waterhouse, P.M., 2000. A single copy of a virus-derived transgene encoding hairpin RNA gives immunity to *Barley yellow dwarf virus*. Mol. Plant Pathol. 1 (6), 347–356.

Wang, Y., Cheng, X., Shan, Q., Zhang, Y., Liu, J., Gao, C., et al., 2014. Simultaneous editing of three homoeoalleles in hexaploid bread wheat confers heritable resistance to powdery mildew. Nat. Biotechnol. 32, 947–951.

Wang, M., Jin, H., 2017. Spray-induced gene silencing: a powerful innovative strategy for crop protection. Trends Microbiol. 25 (1), 4–6.

Wang, M., Weiberg, A., Lin, F.M., Thomma, B.P., Huang, H.D., Jin, H., 2016. Bidirectional cross-kingdom RNAi and fungal uptake of external RNAs confer plant protection. Nat. Plants 2, 16151.

Wang, L., Xia, X., Zhang, W., Wang, J., Zhu, L., Wang, J., et al., 2019. Separate and joint eco-toxicological effects of sulfadimidine and copper on soil microbial biomasses and ammoxidation microorganisms abundances. Chemosphere 228, 556–564.

Wani, S.H., Sanghera, G.S., Singh, N.B., 2010. Biotechnology and plant disease control–role of RNA interference. Am. J. Plant Sci. 1, 55–68.

Weiberg, A., Wang, M., Lin, F.M., Zhao, H., Zhang, Z., Kaloshian, I., et al., 2013. Fungal small RNAs suppress plant immunity by hijacking host RNA interference pathways. Science 342 (6154), 118–123.

Wei-xian, C., Juan, C., Zhen-zhen, Z., et al., 2007. P19 of *Tomato bushy stunt virus* suppresses RNA silencing induced by short hairpin RNA in mammal cells. Virol. Sin. 22, 199–206.

Werner, B.T., Gaffar, F.Y., Schuemann, J., Biedenkopf, D., Koch, A., 2019. RNA-spray-mediated silencing of *Fusarium graminearum* AGO and DCL genes improve barley disease resistance. bioRxiv. Available from: https://doi.org/10.1101/821868.

Whangbo, J.S., Hunter, C.P., 2008. Environmental RNA interference. Trends Genet. 24, 297–305.

Willmann, M.R., Endres, M.W., Cook, R.T., Gregory, B.D., 2011. The functions of RNA-dependent RNA polymerases in arabidopsis. Arabidopsis Book 9, e0146.

Wroblewski, T., Piskurewicz, U., Tomczak, A., Ochoa, O., Michelmore, R.W., 2007. Silencing of the major family of NBS–LRR–encoding genes in lettuce results in the loss of multiple resistance specificities. Plant J. 51 (5), 803–818.

Xiao-Yi, G., Yan, L., Jing, F., Hong, X., Fu-Xian, X., Jun, S., et al., 2019. Host-induced gene silencing of MoAP1 confers broad-spectrum resistance to *Magnaporthe oryzae*. Front. Plant Sci. 10, 433.

Xie, M., Yu, B., 2015. siRNA-directed DNA methylation in plants. Curr. Genomics 16 (1), 23–31.

Yang, B., Sugio, A., White, F.F., 2006. *Os8N3* is a host disease-susceptibility gene for bacterial blight of rice. Proc. Natl. Acad. Sci. USA 103 (27), 10503–10508.

Ye, C., Li, H., 2010. 20 Years of transgenic research in China for resistance to *Papaya ringspot virus*. Transgenic Plant J. 4, 58–63.

Yin, C., Hulbert, S., 2015. Host induced gene silencing (HIGS), a promising strategy for developing disease resistant crops. Gene Technol. 4, 3.

Yin, C.T., Jurgenson, J.E., Hulbert, S.H., 2011. Development of a host-induced RNAi system in the wheat stripe rust fungus *Puccinia striiformis* f. sp. *Tritici*. Mol. Plant Microbe Interact. 24, 554–561.

Yin, G., Sun, Z., Liu, N., Zhang, L., Song, Y., Zhu, C., et al., 2009. Production of double-stranded RNA for interference with TMV infection utilizing a bacterial prokaryotic expression system. Appl. Microbiol. Biotechnol. 84, 323–333.

Yin, G.H., Sun, Z.N., Song, Y.Z., An, H.L., Zhu, C.X., Wen, F.J., 2010. Bacterially expressed double-stranded RNAs against hot-spot sequences of *Tobacco mosaic virus* or *Potato virus Y* genome have different ability to protect tobacco from viral infection. Appl. Biochem. Biotechnol. 162, 1901–1914.

Yoshioka, M., Adachi, A., Sato, Y., Doke, N., Kondo, T., Yoshioka, H., 2019. RNAi of the sesquiterpene cyclase gene for phytoalexin production impairs pre- and post-invasive resistance to potato blight pathogens. Mol. Plant Pathol. 20, 907–922.

Zamore, P., Tuschl, T., Sharp, P., Bartel, D., 2000. RNAi: double-stranded RNA directs the ATP-dependent cleavage of mRNA at 21 to 23 nucleotide intervals. Cell 101, 25–33.

Zhai, J., Jeong, D.H., DePaoli, E., Park, S., Rosen, B.D., Li, Y., 2011. Micro RNAs as master regulators of the plant NB-LRR defense gene family via the production of phased, trans- acting siRNAs. Genes Dev. 25, 2540–2553.

Zhang, H., Demirer, G.S., Zhang, H., Ye, T., Goh, N.S., Aditham, A.J., et al., 2019b. DNA nanostructures coordinate gene silencing in mature plants. Proc. Natl. Acad. Sci. USA 116, 7543–7548.

Zhang, B., Li, W., Zhang, J., Wang, L., Wu, J., 2019a. Roles of small RNAs in virus-plant interactions. Viruses 11 (9), 827.

Zhang, X., Niu, D., Carbonell, A., Wang, A., Lee, A., Tun, V., et al., 2014. ARGONAUTE PIWI domain and microRNA duplex structure regulate small RNA sorting in arabidopsis. Nat. Commun. 5, 5468.

Zhang, Z.Y., Yang, L., Zhou, S.F., Wang, H.G., Li, W.C., Fu, F.L., 2011. Improvement of resistance to *Maize dwarf mosaic virus* mediated by transgenic RNA interference. J. Biotechnol. 153, 181–187.

Zhang, T., Zhao, Y., Zhao, J., et al., 2016. Cotton plants export microRNAs to inhibit virulence gene expression in a fungal pathogen. Nat. Plants 2, 16153.

Zhong, X., Du, J., Hale, C.J., Gallego-Bartolome, J., Feng, S., Vashisht, A.A., et al., 2014. Molecular mechanism of action of plant DRM de novo DNA methyltransferases. Cell 157 (5), 1050–1060.

CHAPTER

7

Safflower disease—a sustainable protection against *Alternaria carthami* L.

Sudheeran Pradeep Kumar

Department of Postharvest Science, Institute of Postharvest and Food Sciences, Agriculture Research Organisation, The Volcani Center, Rishon LeZion, Israel

7.1 Introduction

Safflower (*Carthamus tinctorius* L.) has been grown for centuries, primarily for its colorful petals to use as a food coloring and flavoring agent for vegetable oils and also for preparing textile dye in the Far East, Central, and Northern Asia and European Caucasian (Gecgel et al., 2007). Safflower is most commonly known as "kusum" (India), derived from the Sanskrit, "kusumbha," and as "honghua" (red flower) in China. Its use as a less costly substitute for saffron is indicated by the names false saffron, thistle saffron, and dyer's saffron (Weiss, 1983). Safflower (*C. tinctorius* L.), which belongs to the family Compositae, is cultivated in several parts of the world due to its adaptability to different environmental conditions. The world production of safflower seeds in 2013 amounted to 718,161 tons. With a production value of 174,900 tons, Kazakhstan was the leading producer of this crop in the world in the same year. The production of safflower in Kazakhstan accounted for 24% of the total global safflower production. India and the United States ranked second and third in safflower production in 2013 with production values of 109,000 and 95,360 tons, respectively (Sen Nag, 2017). Safflower production in India is mostly confined to rain-fed conditions during winter. It is highly branched, herbaceous, thistle-like annual or winter annual, usually with many long sharp spines on the leaves.

The main constituents of the safflower are carthamin and carthamidin, and other constituents are safflor yellow, arctigenin, tacheloside, *N*-feruloyl tryptamine, *N*-feruloylserotonin, steroids, flavonoids, and polyacetylenes. Carthamin is responsible to produce water-insoluble red dye and carthamidin to produce water-soluble yellow color dye (Kizil et al., 2008). The selection for high oil content in modern cultivars has reduced the pericarp thickness. Safflower varieties released for commercial production in India, in general, possess low oil content of 28%–32%, except HUS-305, NARI-6, and nonspiny hybrid NARI-NH-1, each of which contains 35% oil.

Food Security and Plant Disease Management
DOI: https://doi.org/10.1016/B978-0-12-821843-3.00018-0

The quality of any oil is determined by the fatty acid composition of the oil, and the oils rich in poly or monounsaturated fatty acids are considered good, as they help in reducing the cholesterol level in the blood (Knowles, 1989). Standard safflower oil contains about 6%–8% palmitic acid, 2%–3% stearic acid, 16%–20% oleic acid, and 71%–75% linoleic acid. It was reported that the dried flowers of safflower (*C. tinctorius*) had been used in traditional Asian medicine for thousands of years (Gao et al., 2000). Florets were widely used to color and flavor soups and rice as well as cloth, potions, and unguents. In addition to these properties, safflower petals are also used for production of food and fabric colorants (Srinivas et al., 1999).

Safflower dyes were particularly important to the carpet-weaving industries of Eastern Europe, the Middle East, and the Indian subcontinent. Carthamin dye was used extensively to color cloth until the 19th century when cheaper aniline dyes became available. True saffron is perhaps the world's most costly spice and safflower is a common adulterant or substitute. The addition of safflower florets to foods is a widespread and ancient tradition. Health concerns regarding synthetic food colorings may increase demand for safflower-derived food coloring. China produces carthamin dye for use in food, particularly at a large factory in Kunming in Yunnan Province. Cosmetic rouge can be made from carthamin dye mixed with French chalk and Japanese cosmetics (beni) (Weiss, 1983) and lipsticks include safflower coloring (Smith, 1996). Until this century, when cheaper aniline dyes became available, safflower was mainly grown for dye. The water-soluble yellow dye, carthamidin, and a water-insoluble red dye, carthamin that is readily soluble in alkali, can be obtained from safflower florets (Weiss, 1983). Yellow florets contain little or no red dye (Smith, 1996).

7.2 Safflower disease resistance

Safflower is attacked by many diseases caused by fungi, bacteria, viruses, or physiological disorders due to abiotic stresses. Kolte (2018) reported that safflower is recorded to be infested around the world by 57 pathogens, including 40 fungi, 2 bacteria, 14 viruses, and 1 mycoplasma. Foliar diseases have been particularly severe in areas where rainfall occurs between the late bud stage and near maturity. Most serious and widespread is leaf blight caused by *Alternaria carthami*.

Other foliar diseases of more localized concern include those caused by *Botrytis cinerea, Cercospora carthami, Pseudomonas syringae, Puccinia carthami,* and *Ramularia carthami.* Safflower rust (*P. carthami*) is widespread in all areas of commercial production. The rust fungus has a complex life cycle and produces different types of spores, depending on the stage of the fungus' life cycle present (Doohan, 2005). Black teliospores appear at the end of the crop growth cycle and can infest seeds or persist in the soil. *Phytophthora* root rots (*Phytophthora cryptogea* and *Phytophthora drechsleri*) are widespread and are the most serious diseases of irrigated safflower. Many races of the causal fungi are endemic in California soils and attack the roots and lower stem under favorable environmental conditions (Sastry and Chattopadhyay, 2005). Plants are susceptible at all stages of growth but the visible symptoms become more apparent from flowering onward. Infected plants become light-colored, wilt, and die. In the early stages of infection, roots may show reddening of tissue. Infected roots and lower stems later become darkly discolored.

Phytophthora can be devastating, particularly when crops are irrigated late in the season and plants have developed symptoms of moisture stress prior to irrigation (Bryla and Linderman, 2007). It may also occur in nonirrigated fields if heavy rains occur, especially late in the spring.

Safflower (*C. tinctorius*) leave-spots composed of alternating light and dark brown, concentric rings of 1 cm in diameter, frequently coalescing into large, irregular lesions and assuming a brittle texture winch is apt to lead to the breaking of the leaf blade. The stems and petioles suffer less severe damage, but infected flower buds shrivel without opening the extension of the dark brown spots from the base of the calyx to other parts of the inflorescence. The optimum temperature for growth was found to lie between 25°C and 30°C. The severity of disease *Phytophthora* root rot may vary depending on whether the crop is surface-irrigated, subirrigated, or not irrigated (Reddy, 2014) is an important seed-borne pathogen of safflower (*C. tinctorius* L.) in Western Canada, the United States, Australia, and India. Mortensen et al. (1983) found that *A. carthami* was pathogenic on safflower at all growth stages in Montana, causing up to 50% seed rot and seedling blight in susceptible cultivars. A seed produced in Alberta, Manitoba, Saskatchewan, and Montana is often heavily contaminated with this pathogen, resulting in reduced germination and seedling vigor (Mortensen et al., 1983). Infection by fungal pathogens results in a change of host plant physiology and in mechanical and biochemical disruptions. Intercellular penetration of tissues results in direct effects on the middle lamellae between associated plant cells followed by invasion of plant cells, membranes, and organelles, increased nutrient leakage and host cell death, which results in visible disease symptoms (Elad, 1997). Chemical control measures for diseases and pests are expensive and ineffective.

Moreover, the increasing use of chemicals in agriculture affects the quality of the products as well as presenting problems to ecosystems. Most of the resistant lines identified earlier were able to maintain the same level of resistance over time. An understanding of the inheritance of resistance in the host and of virulence and nonvirulence of different physiological races of pathogens is needed.

7.3 Scientific evidence for sustainable plant protection

Chitinase is a type of pathogenesis-related (PR) proteins and ubiquitous enzyme of bacteria, fungi, animals, and plants. Chitinases are hydrolytic enzymes produced by plants in response to fungal infection (Punja and Zhang, 1993). They are proposed to have a role in plant defense by degrading the fungal cell walls. They hydrolyze the β-1,4-linkage between *N*-acetylglucosamine residues of chitin, a structural polysaccharide of the cell wall of many fungi and of the exoskeleton of invertebrates (Kaczmarek et al., 2019). It is one of the most abundant naturally occurring polysaccharides and has attracted tremendous attention in the field of agriculture, pharmacology, and biotechnology (Antranikian et al., 2005; Muzzarelli et al., 2005). Chitinases are used to treat fungal infections (Liu et al., 2002). Chitin also has multiple applications mainly in the bioremediation of environmental pollutants such as Pb^{2+}, Zn^{2+}, and Bi^{2+} (Rae and Gibb, 2005). The development of research in plant defense mechanisms has led to a rapid and steady interest in chitinase since they were the first pathogen-induced proteins whose function was identified. Their substrate is present in the cell wall of many fungi, as well as in insects and nematodes, which are major pathogens of crop plants. Several studies have been made on transgenic plants integrated with chitinase genes. Transgenic tobacco and canola that have been engineered with be an endochitinase gene were shown to exhibit resistance to *Rhizoctonia solani* (Broglie et al., 1991). Transgenic rice integrated with rice endochitinase driven by the 35S promoter showed enhanced resistance to sheath blight (Lin et al., 1995). The activation of the antifungal activity of chitinase, in vitro and in transgenic plants, may be defensive against fungal pathogens (Yamamoto et al., 2000). Transgenic plants and biological control agents offer many opportunities to manipulate genes from a variety of sources for the enhancement of host plant resistance to pathogens, respectively, against insects and other types of pests (Herrera-Estrella and Chet, 1999), whereas transgenic tobacco harboring rice endochitinase gene also possessed increased resistance against powdery mildew, *Erysiphe cichoracearum* (Nishizawa et al., 1993). Chitinase enzyme possesses added value in food and pharmaceutical industries, antibacterial agents, elicitors, lysozyme inducers, and immune enhancers (Kato et al., 2003; Muzzarelli et al., 2005). Furthermore, antioxidant and in situ chitinase gel activity in safflower (*C. tinctorius* L.) were studied by Kumar and Kumari (2013).

Due to the massive biological applications, the characterization and synthesis of chitinase are highly anticipated in the biotechnology industry.

7.4 Safflower biotechnology

Safflower has attracted very little attention as far as tissue culture and genetic transformation are concerned. Initial efforts in safflower were directed to develop suitable culture conditions for whole plant regeneration. It has been demonstrated that regeneration frequencies are very high, and regeneration is possible through embryogenesis and organogenesis pathways. The mode of regeneration, in general, is through direct or indirect organogenesis (Sujatha and Suganya, 1996; Walia et al., 2005). The tissue culture of safflower has been the subject of several studies focusing on establishment of in vitro propagation. It offers a method to increase valuable genotypes rapidly and expedite release of large number of plantlets. It is well advanced and regularly used for the conservation, multiplication, and distribution of elite germplasm. The alternative method to select or detect resistance is the use of in vitro selection methods. In comparison with in vivo selection that is commonly applied in plant breeding programs, in vitro selection methods have several advantages. It is possible to screen a large number of individuals within a short time under controlled conditions in the lab and, therefore help to advance breeding progress. It is possible to exclude interactions with the environment, which might cover certain traits. Instead of whole plants, organs, small amounts of tissues, or even single cells can be tested (Ahmed and Sagi, 1993). In vitro selection allows the specific use of genetic variability which is induced from the in vitro culture of plant cells or tissue. Somaclonal variation as a source for inducing genetic variability has been demonstrated in safflower (Seeta et al., 2000). Somaclones generated from cotyledon explants of safflower genotype Manjira showed an enormous variation for both qualitative and quantitative traits and were found to breed true. It is the ever-ready tool for specialization in hybridization either by sexual or asexual means. It is a clean and rapid way in genetic engineering by which the materials can be grown for identification and manipulation of genes or transfer of characters from one plant to another.

Genetic improvement in safflower through conventional breeding is limited to the development of new cultivars with widely varying oil content and quality (Knowles, 1989). Genetic improvement of this crop for agronomical attributes is constrained by the modest levels of variability available in the cultivar germplasm. The development of in vitro shoot regeneration techniques is essential for introgression of desirable traits from alien sources. In earlier studies of safflower, there are few reports demonstrating anther culture (Prasad et al., 1991), in vitro regeneration (Nikam and Shitole, 1999; Vijaya Kumar and Ranjitha Kumari, 2005; Radhika et al., 2006) and somatic embryogenesis have only been briefly documented (Mandal et al., 2001; Mandal and Dutta Gupta, 2003).

The addition of toxic compounds to the culture media or the application of abiotic stress results in defined selection pressure. Together with the in vitro selection for disease resistance, there are a number of other possible applications for in vitro selections such as selection for herbicide tolerance, selection for salt tolerance, selection for tolerance to metals, selection for tolerance to high or low temperatures, and selection for tolerance to water stress (Haines, 1993). Evaluation of NaCl tolerance in safflower callus cultures by the repeated transfer of selected clones to the NaCl-rich medium enabled identification of salt-tolerant cell lines (Nikam and Shitole, 1997). Induction of high variability for qualitative and quantitative traits through tissue culture indicated that tissue culture can be suitably utilized to detect spontaneous somaclonal variants with improved tolerance to abiotic stresses. In vitro selection for resistance to a pathogen can be realized when in vitro cultures are exposed to toxins produced by the pathogen, to synthetic toxin analogs, to a pathogen filtrate, to extracts of the pathogen, or to the pathogen itself (Daub, 1986). Numerous studies have demonstrated the utility of tissue culture–based testing of agrochemicals, microbial toxins, or allelochemicals on plant tissues. These studies have also contributed useful data on physiological and biochemical processes affected by induced stress (Smeda and Weller, 1991). Kneusel and Matern (1996) attempted to use genetic engineering techniques for safflower to introduce resistance to leaf blight caused by *Alternaria* species. This group used molecular methods to identify the macrolide brefeldin A, as the phytotoxin from *A. carthami*, which suppresses the plant's defense response and is thus identified as a virulence factor of the fungal pathogen.

Agrobacterium-mediated transformation of safflower and the efficient recovery of transgenic plants via grafting (Belide et al., 2011; Lijiao and Meili, 2013). *Agrobacterium*-based vectors have been successfully employed to transfer genes into a number of dicotyledonous plants (Gelvin, 2003). *Agrobacterium*-mediated genetic transformation is still the most widely used method of producing transgenic plants (Dunwell et al., 2000). Indirect somatic embryogenesis from transgenic immature leaf of safflower *C. tinctorius* was reported by Pradeep Kumar and Ranjitha Kumari (2017). Besides being cheaper and simpler than most direct gene transfer methods, it allows little rearrangement of transgenes and efficient integration of the transgene into the plant genome (Ingelbrecht et al., 1991). However, the most limiting factor for efficient transformation is the absence of high-yielding regeneration protocols. Transformation of plant cells results from transfer of a segment of the tumor-inducing (Ti) plasmid (T-DNA) and its subsequent integration into plant chromosomal DNA (Tzfira and Citovsky, 2006). It has been suggested that *Agrobacterium tumefaciens*-mediated transformation may offer a better alternative than the direct method for delivery of transgenes into plants. A better understanding of this process will facilitate manipulation of the plant host, both to increase transformation efficiency and to achieve the important strategic objective of targeted T-DNA integration into plant chromosomes (Gelvin, 2003; Tzfira and Citovsky, 2006). This gene delivery system results in a greater proportion of stable, low-copy number transgene events than the Biolistic gun does and offers the possibility of transferring larger DNA segments into recipient cells (Shibata and Liu, 2000). The successful reports on plant regeneration of safflower through *A. tumefaciens*-mediated transformation were reported limitedly in few cultivars (Rohini and Sankara Rao, 2000). Another important and interesting use of safflower seed has recently emerged by means of its genetic modification to produce high-value proteins as pharmaceuticals and industrial enzymes. SemBioSys, a Calgary-based (Canada) company, transforms safflower tissue genetically in order to get the proteins of interest to accumulate in the seed of the mature transgenic plant (Mündel, 2004). The process of transformation of safflower tissues follows the patented Stratosome Biologics system that facilitates the genetic attachment of target proteins of interest to oleosin, the primary protein coating the oil-containing vesicles (oil bodies) of the seed. Such attachment permits the target protein to be purified along with the oil body fraction, which upon centrifugation floats to the surface of ground seeds/water slurry (van Rooijen et al., 1992). The purification process of the Stratosome system makes it more efficient than the other transgenic systems. The attachment of proteins to the oil bodies of safflower in the Stratosome Biologics system is expected to stabilize intracellular accumulation of foreign proteins, and also provides a useful attachment matrix and delivers benefits for useful applications. The commercialization of genetically modified safflower will further increase the acreage and production of this crop in the world. Therefore genetic modification of safflower would be of enormous importance in improving productivity, production, and remuneration per unit area from the crop, which in turn would certainly help in increasing safflower area in the world.

7.5 In vitro regeneration through tissue culture

Safflower has attracted very little attention as far as tissue culture is concerned. Recent advances made in the field of tissue culture have brought about new emerging technologies for crop improvement. It has the potential to produce numerous genetically identical plants selected for special attributes. The mode of regeneration, in general, is through direct or indirect organogenesis (Walia et al., 2005). All these factors drive the tempo of in vitro organogenesis and regeneration of complete plants by integrating the individual organs (Gantait and Mandal, 2010). Further, emphasized on the synergistic obligation of both cytokinins and auxins for initiation of cell division and growth in plant tissue cultures (Kothari et al., 2010) where an array of experiments have fundamentally recognized the continuation of antagonistic as well as additive interactions involving these two types of plant growth regulators. The tissue culture of safflower has been the subject of several studies focusing on establishment of in vitro propagation. It offers a method to increase valuable genotypes rapidly and expedites release of large number of plantlets. Rapid regeneration of plants directly from explants presents an effective strategy to avoid or substantially reduce somaclonal variation as it minimizes culture duration and eliminates or minimizes callus formation in culture (Lakshmanan et al., 2006). Genetic improvement of this crop for agronomical attributes is constrained by the modest levels of variability available in the cultivar germplasm. Plant regeneration from different plant species is influenced by plant growth regulators including the ratios of cytokinin and auxin (Rout et al., 2001; Liao et al., 2004). The suitable ratio of cytokinin and auxin for plant regeneration of safflower depends on the used cultivars. George and Rao (1982) reported the induction of shoots from cotyledons on Murashige and Skoog (1962) salts as the basal medium supplemented with BA (6-benzyl adenine), but regeneration frequency was low and complete plants were rarely obtained. Direct shoot regeneration and inflorescence induction have been achieved from cotyledon on MS medium containing BA or kinetin in combinations with NAA (α-naphthalene acetic acid) (Tejovathi and Anwar, 1987). In addition, primary seedling explants were used on MS medium consisting of BA and NAA for direct as well as indirect organogenesis of safflower cultivars (Sujatha and Suganya, 1996; Nikam and Shitole, 1999). Cotyledonary explants cultured on MS medium supplemented with BA and kinetin induced shoot organogenesis (Walia et al., 2005). In earlier reports Vijaya Kumar and Ranjitha Kumari (2005) also confirmed that primary seedling explants (cotyledonary and stem node) on MS salts and B5 vitamins medium comprised of BA and kinetin were found to be the most effective for shoot organogenesis. In many cases, TDZ (Thidiazuron) (1-phenyl-3-(1,2,3-thiadiazol-5-yl)urea), a substituted phenylurea compound, was shown to possess potent activity as a cytokinin in the regulation of shoot organogenesis in several plant species (Mithila et al., 2001). In safflower, shoot regeneration frequencies on medium with TDZ/NAA were high and shoot buds were induced on a wide range of concentrations (Radhika et al., 2006). A major problem associated with in vitro culture is the occurrence of somaclonal variation amongst subclones of one parental line, arising as a direct consequence of in vitro culture of plant cells, tissues, or organs (Gould, 1986). Recently, the addition of amino acids to the tissue culture medium promoted high frequency of plant regeneration in many plant species including *Artemisia vulgaris* L. (Kumar and Kumari, 2010). However, there is only one report being described for safflower plant regeneration in amino acids supplemented medium (Walia et al., 2005). The objective of the investigation reported here was to develop a rapid and reproducible in vitro regeneration system from different explants of safflower for mass propagation of selected elite clones. All the explants on cytokinins BA, KIN (kinetin), and TDZ alone or in combination with CPPU *N*-(2-chloro-4-pyridyl-*N*-phenylurea) supplemented medium were tested to improve adventitious and multiple shoot regeneration of safflower cv. NARI-6. This technique provides a rapid reliable system for the production of large number of genetically uniform disease-free plantlets. The development of an efficient regeneration protocol is also a prerequisite for tissue culture studies for raising plants of this valuable oilseed crop plant (Fig. 7.1).

7.6 Plants resistant to *Alternaria carthami* via indirect organogenesis

For indirect organogenesis 7- to 10-day-old leaf, hypocotyls, and immature leaf explants were sectioned into 0.5–1.5 cm long segments, were superficially wounded by making scraps with a scalpel. They were precultured on MS basal with 3.0% sucrose and 0.8%

FIGURE 7.1 In vitro seed germination of safflower plant. (A) Safflower seeds. (B) In vitro seed germination in MS medium. (C) Fully germinated plant in test tube containing MS medium.

agar along with hygromycin (70 mg/L) in medium supplemented with (5.0 μM TDZ) and different compositions of amino acids such as proline, glutamine, and aspartic acid for 2–5 days. After preculture, all the explants were infected for 20 minutes with 30 mL of a bacterial suspension in MS basal medium (pH 5.6). Infected explants were placed on cocultivation medium and incubated at 25°C in the dark for 2–7 days. The pH of the cocultivation medium (MS basal salts, 3.0% sucrose, and 100 μM acetosyringone) was adjusted to 5.4, 5.5, 5.6, 5.7, and 5.8. After cocultivation, the explants were rinsed thoroughly in sterile distilled water containing 60 mg/L cefotaxime for 10 seconds to remove excess growth of bacteria on the surface of explants. Then the explants were transferred onto plant regeneration medium containing MS basal with 3.0% sucrose and 0.8% agar with TDZ (0.5 μM) alone or combination of amino acids proline, glutamine, and aspartic acid, ranges from 1.5 to 7.0 μM along with hygromycin (70 mg/L) at pH 5.7. All of the cultures were maintained at 25°C under continuous irradiation with a white fluorescent tube (15 $\mu mol/m^2$ s) for 12-hour photoperiod.

7.6.1 Callus induction

For callus initiation, the leaf and hypocotyls explants were placed in contact with the solidified MS salts with 3.0% sucrose with hygromycin (70 mg/L) in medium supplemented with 5.0 μM of TDZ in combinations with NAA (4.0 μM) and aspartic acid (5.0 μM) for leaf explant (Fig. 7.2A). Whereas hypocotyls explants showed callus induction with 5.0 μM of TDZ in combinations with NAA (3.0 μM) and glutamine (4.5 μM). The resulting callus was allowed to grow up to 45 days. In our study, individual hormones and amino acids

FIGURE 7.2 Indirect organogenesis from transgenic leaf explant of safflower. (A) Callus initiation from leaf explant. (B) Microshoot initiation from callus. (C) Shoot proliferation from callus. (D) Rooting from shoots. (E) Hardened in growth chamber. (F) Hardened in pot for field condition.

TABLE 7.1 Comparative effect of different amino acids along with TDZ (5.0 μM) on transgenic callus induction from leaf and hypocotyls explants of safflower.

MS medium composition (μM)		Callusing response (%)	
		Leaf	Hypocotyl
NAA + proline			
1.0	1.5	3.6 ± 1.1 cd	3.2 ± 1.0 cd
2.0	3.0	11.0 ± 3.1 c	10.5 ± 1.1 c
3.0	4.5	15.2 ± 1.9 b	16.5 ± 1.2 b
4.0	6.0	17.7 ± 2.4 bc	14.3 ± 2.4 b
5.0	7.0	12.5 ± 1.3 c	11.7 ± 0.7 bc
NAA + glutamine			
1.0	1.5	8.9 ± 2.1 cd	8.3 ± 0.3 c
2.0	3.0	19 ± 0.3 bc	17.3 ± 1.0 bc
3.0	**4.5**	25 ± 0.6 ab	**58.2 ± 1.5 a**
4.0	6.0	16 ± 2.0 b	15.7 ± 4.0 ab
5.0	7.0	11 ± 0.6 c	14.4 ± 6.0 b
NAA + aspartic acid			
1.0	1.5	6.7 ± 1.4 cd	7.1 ± 0.3 c
2.0	3.0	19.9 ± 1.1 bc	14.4 ± 0.8 b
3.0	4.5	33.5 ± 6.0 ab	21 ± 1.9 ab
4.0	**5.0**	**76.7 ± 1.9 a**	18 ± 2.9 ab
5.0	7.0	22.4 ± 4.3 b	9.6 ± 1.0 ab

Bold represents more significant value. Means with common letters within a column are not significantly different at $P \leq .05$, according to Duncan's Multiple Range Test (DMRT).

(NAA, TDZ and aspartic acid, glutamine) induced white friable calli for leaf and yellow-green nodular in hypocotyl explants. However, conspicuous callus mass was obtained with the medium containing NAA and glutamine (Fig. 7.2B). Whereas, leaf explants showed highest response with 76.7% (Table 7.1) per leaf and hypocotyls explants showed 58.2% (Table 7.1) for transgenic callusing response per explants. Further the explants were subcultured in shoot elongation medium for further plant regeneration.

7.6.2 Shoot elongation

All types of calli were turned organogenic after 7–14 days of culture on plant regeneration medium. The explants sources cultured on MS salts with 3.0% sucrose and 0.8% agar

TABLE 7.2 Comparative effect of different amino acids along with TDZ (5.0 μM) on transgenic adventitious shoot regeneration from leaf and hypocotyls explants of safflower.

MS medium composition (μM)	Mean no. of shoots/explant culture	
	Leaf	Hypocotyls
Proline		
1.5	4.6 ± 1.0 cd	5.2 ± 1.0 cd
3.0	10 ± 3.1 c	9.9 ± 1.5 c
4.5	18 ± 1.9 b	17.5 ± 1.2 b
6.0	15 ± 2.4 bc	14 ± 2.4 b
7.0	12 ± 1.3 c	12.8 ± 0.7 bc
Glutamine		
1.5	7.9 ± 2.2 cd	6.9 ± 0.3 c
3.0	17 ± 0.4 bc	13.3 ± 1.7 bc
4.5	23 ± 0.6 ab	**47.9 ± 1.5 a**
6.0	16 ± 2.3 b	15.7 ± 4.0 ab
7.0	13 ± 0.5 c	14.4 ± 6.0 b
Aspartic acid		
1.5	6.7 ± 1.4 cd	7.1 ± 0.3 c
3.0	16.9 ± 1.1 bc	14.4 ± 0.8 b
4.5	23.5 ± 6.0 ab	21 ± 1.9 ab
5.0	**34.7 ± 1.9 a**	18 ± 2.9 ab
7.0	20.4 ± 4.3 b	9.6 ± 1.0 ab

Bold represents more significant value. Means with common letters within a column are not significantly different at $P \leq .05$, according to Duncan's Multiple Range Test (DMRT).

(w/v) along with hygromycin (70 mg/L) in medium supplemented with plant growth regulators and their interactions were significant for organogenesis. The shoot organogenesis was observed from leaf and hypocotyl explants derived calli after 65 days of culture supplemented medium with TDZ (5.0 μM) and aspartic acid (5.0 μM) for leaf explants (Fig. 7.2D). In this case, TDZ (5.0 μM) and aspartic acid (5.0 μM) containing medium produced 34.7% (Table 7.2) shoot organogenesis per leaf explants and hypocotyls explants supplemented with TDZ (5.0 μM) in combination with glutamine (4.5 μM) for shoot proliferation (Fig. 7.2D). However, maximum of 47.9% shoot organogenesis per hypocotyls explants were achieved (Table 7.2). A comparatively low frequency of shoot organogenesis was noticed in TDZ (0.5 μM) and proline (1.5 μM) supplemented medium (Table 7.2) in leaf explants.

7.6.3 Root induction and acclimatization

Transgenic shoots produced more callus with low-frequency rooting when elongated shoots were placed on rooting medium supplemented with NAA. However, the use of half-strength MS salts with 3.0% (w/v) sucrose and 0.8% agar (w/v) along with hygromycin (70 mg/L) in medium supplemented with NAA and IBA or $AgNO_3$ produced more extensive root formation and induction of adventitious roots with less amount of callus in green healthy shoots. Leaf explants supplemented with NAA (1.5 μM) and IBA (0.5 μM) showed 56.0% of rooting frequency (Table 7.3; Fig. 7.2E), whereas hypocotyl explants in medium supplemented with NAA (2.0 μM) and $AgNO_3$ (1.5 μM) showed the highest frequency of 76.0% rooting (Table 7.3; Fig. 7.2D). The fully regenerated plants were acclimatized and successfully transferred to plant growth chamber (Fig. 7.2E and F) after 4–7 weeks of acclimatization under greenhouse conditions. There was no detectable variation on morphology or growth behavior among the acclimatized plants.

TABLE 7.3 Effect of growth regulators on rooting from transgenic shoot tip, leaf and, hypocotyl explants of safflower (*Carthamus tinctorius* L.) cv. NARI-6.

1/2 MS medium composition (μM)	Rooting frequency (%)		
	Shoot tip	Leaf	Hypocotyl
NAA			
0.5	28.3 ± 1.5 cd	22 ± 0.7 c	10 ± 1.8 cd
1.0	39.0 ± 0.6 c	30.5 ± 1.2c	17 ± 1.7 c
1.5	42.2 ± 1.0 c	38 ± 0.7 bc	25 ± 1.5 bc
2.0	**64.5 ± 1.5 bc**	40 ± 0.4 bc	31 ± 2.5 b
2.5	40.0 ± 2.0 c	35 ± 1.0 c	27 ± 0.5 bc
NAA + IBA			
0.5 0.3	42.0 ± 1.4 c	44.5 ± 1.2 bc	14 ± 1.2 c
1.0 0.4	52.3 ± 1.6 b	57.7 ± 1.0 b	22 ± 0.6 bc
1.5 0.5	60.1 ± 2.4 ab	**56.0 ± 0.4 ab**	55 ± 0.8 b
2.0 0.6	55.5 ± 1.7 ab	59.4 ± 1.4 ab	40 ± 1.4 ab
2.5 0.7	40.8 ± 1.9 bc	38.3 ± 2.5 bc	31 ± 2.9 b
NAA + $AgNO_3$			
0.5 0.5	46.6 ± 0.3 bc	44.0 ± 0.9 b	19.6 ± 1.2 c
1.0 0.5	50.0 ± 0.5 b	51.0 ± 1.7 ab	27.2 ± 1.9 bc
1.5 1.0	47.0 ± 1.6 ab	57.7 ± 1.3 ab	38.0 ± 0.4 b
2.0 1.5	55.0 ± 1.0 a	59.1 ± .0 a	**76.0 ± 1.0 a**
2.5 1.5	51.0 ± 0.4 b	49.3 ± 2.0 b	34.0 ± 0.8 b

Bold represents more significant value. Means with common letters within a column are not significantly different at $P \leq .05$, according to Duncan's Multiple Range Test (DMRT).

7.7 Genetic improvement and transformation of safflower

Genetic engineering has become an important tool for crop improvement since it provides means of the introduction of genes from diverse sources without changing the phenotype and agronomic performance of the transformed plant. *A. tumefaciens*-mediated transformations produced transgenic safflower by using the cultivar centennial (Motamedi et al., 2011). Efficient callus formation from cotyledon, stem, and leaf explants was observed. Transformation and integration of transgenes were confirmed with the application of β-glucuronidase assay and DNA hybridization in kanamycin-resistant calluses and β-glucuronidase assay in regenerated shoots.

Chitinases are a diverse family of proteins occurring in plants. Their function varies considerably, with certain chitinases having been associated with development. The majority are PR proteins that have been shown to play a role during plant–pathogen interactions (Sels et al., 2008). This has led to many investigations on the use of chitinases in providing transgenic disease resistance. These studies are usually done using a constitutive expression system. This, however, stands in contrast with the natural defense system where PR gene expression is usually only upregulated when the plant is exposed to abiotic and biotic stress factors. The constitutive expression is therefore not ideal as it increases 'cost' penalties due to the energy being spent expressing the gene. The introduction of disease-resistance genes into plants to provide increased resistance in plant–pathogen interactions has been investigated intensively (Boyd et al., 2013). With the advent of gene-mining techniques such as microarrays, new genes are continuously being discovered that can be potentially utilized.

Fungal plant pathogens can be classified as necrotrophic, biotrophic, or hemitrophic (Glazebrook, 2005). Necrotrophic pathogens act by killing and destroying plant tissue through toxins and cell wall–degrading enzymes. Biotrophic pathogens, on the other hand, can only complete their life cycle in a living host. These pathogens seldom cause the death of their hosts but they can impact severely the growth. An example of a biotrophic fungal pathogen affecting tobacco is *E. cichoracearum* causing powdery mildew (Micali et al., 2008). Hemitrophic pathogens usually have a biotrophic phase in the early infection cycle, which end up becoming necrotrophic (Koeck et al., 2011). The first evidence of *A. tumefaciens* transformation of the T-DNA fragment into the genome of the host plant was found in 1977 (Chilton et al., 1977). This leads scientists to contemplate the usefulness of *Agrobacterium* as a vector for recombinant gene transformation and the first plant was successfully transformed (Zambryski, 1983). Since then research has shown that *Agrobacterium* is not only limited to plant transformation but can also transform yeast, fungi, and human cells (Lacroix et al., 2006). The design of an expression vector is dependent on a variety of genetic elements that will allow its proper functioning in an organism. One such element is a promoter that regulates gene transcription. These promoter elements can be either constitutive or inducibly expressed. The use of constitutive promoters is well established in the engineering of disease resistance in plants and can allow constitutively active transcription of the gene. Commonly used promoters in dicotyledonous plants include the CaMV35S promoter (Wang and Oard, 2003) derived from the Cauliflower Mosaic Virus, and in monocotyledons the ubiquitin promoter derived from maize (Toki et al., 1992). In practice, however, the promoters can show expression differences in different tissues and also under different stimuli. This is due to the transformation processes where foreign DNA is randomly integrated into the genome with the integration site, copynumber, and configuration of the gene as well as environmental and metabolic factors all being able to influence the transgene expression (Cervera et al., 2000; Fladung, 1999). The use of constitutive promoters to express disease-resistance genes can also result in plant abnormalities such as reduced size (Chen and Chang, 2002), altered morphology (Li et al., 2004), and symptom development in the absence of pathogens (Lu et al., 2004). The strong constitutive expression of a gene can also increase fitness costs as resources and energy reserves are diverted in the continuous synthesis during gene expression (Pasonen et al., 2004). Inducible promoter's alternatives to constitutive promoters are inducible promoters. The use of inducible promoters is still being investigated but there are several options that can be explored to activate disease resistance. One of the most relevant options is pathogen-inducible promoters.

The successful introduction and expression of genes form an essential component in molecular research on plants. The most important criteria include stable integration and consistent transformation. *Agrobacterium tumefaciens* strain LBA 4404 containing binary vector pCAMBIA-Bar-Ubi-Chi11 with selectable markers phosphinothricin acetyltransferase gene (*bar*) and the hygromycin phosphotransferase gene (*hpt*) and kanamycin (Kan) used as a bacterial selectable marker for safflower transformation (Becker et al., 1992; Kalivoda et al., 2011). Transformation efficiency was tested by incubation of plasmid DNA with competent bacteria prepared using the transformation protocol first described by Chung et al. (1989) with incubation on ice for 30 minutes. We did not find a clear influence of dimethyl sulfoxide (Sigma, 99.9%) on transformation efficiency. The dependence of transformation efficiency on the pH of total soluble *sugars* was significant. At pH values <4 or >8, no transformants were obtained (unpublished data). This was determined to be the direct result of a loss of cell viability. Transformation of plants with *Agrobacterium* works on the basis that bacteria are eliminated with antibiotics during the regeneration stage. Transgenic plants to be released should not contain residual *Agrobacterium* as the possibility of gene escape, via the bacteria, exists. Molecular probes are often used in determining the presence of the "integrated" T-DNA in the plant genome and the presence of residual *Agrobacterium* harboring the binary vector may give false positives (Lee and Gelvin, 2008). The results of this study, using three model transformation systems, indicate the difficulty in eliminating *Agrobacterium* from cultures following transformation. The concentrations of antibiotics used in the case of carbenicillin and cefotaxime were the highest of values (500 mg/L) evident in the literature for transformation, yet they failed to eliminate contamination. In fact, the levels of *Agrobacterium*, in explants, increased and eventually became visible on the plates. Colony-forming units of bacteria in shoots declined from 20 to 24 weeks, but still a considerable percentage (24%) of the shoot material was contaminated.

According to previous reports varying the concentration and molecular weight of PEG produced dramatic results. When the concentration of PEG was $>20\%$ or $<5\%$, the transformation was not observed. The highest transformation efficiencies were obtained with 10% PEG 8000. Therefore transformation efficiency is highly dependent on the concentration of PEG in the total soluble *sugars* mixture. In our report using 10%, PEG 3350 is optimal in our transformation system, whereas Klebe et al. (1983) found that 25% PEG 1000 yielded the maximum number of transformants. A second difference is that Klebe et al. (1983) found that cells harvested in late exponential phase (OD_{600} 0.6–0.7) yielded the highest transformation efficiencies. On the other hand, we find that cells from the early exponential phase (OD_{600} 0.3–0.4) yield more transformants per microgram of plasmid DNA than those later in the growth stage. Thus there are both similarities and differences between the two PEG-mediated procedures. We also find that the transformation efficiencies obtained in our protocol are consistently higher, by 1–2 orders of magnitude than those obtained by Klebe et al. (1983). There are several factors that may contribute to our higher transformation efficiencies. One is the inclusion of dimethyl sulfoxide in our buffer system, which has been shown by others (Hanahan, 1983) to enhance transformation efficiency. Another factor is that we do not wash the bacteria after treatment with PEG, a step that could result in decreased numbers of transformants in the procedure (Klebe et al., 1983). Also, we use low concentrations (10%) of high-molecular-weight PEG rather than high concentrations (25%) of low-molecular-weight PEG. Thus the main advantages of our protocol are the significantly higher transformation efficiencies and the dual function of the transformation solution (transformation and storage). To increase efficiency we tested prolonged incubation at higher temperatures. We found highest transformation rates with incubation on ice for 10 minutes, followed by 10 minutes at room temperature and again 10 minutes on ice.

In our study, the establishment of cultural conditions for the production of 3-ketoglycosides was consistently good when pH of the medium was maintained near neutrality. At pH of 7.0–7.5, the sugars are probably transported across the cell membrane most rapidly (Fukui and Hochster, 1965). Also, 3-ketoglycosides are unstable at a pH of less than 6.0 or higher than 7.5 (Fukui and Hochster, 1963). We found that pH was adequately maintained at about 7.0 in a medium that contained 0.5% $CaCO_3$, 0.6% KH_2PO_4, and 0.09% urea. The ability to use sucrose is most likely a benefit when the bacterium is in a plant, but of dubious use when the bacterium is in the soil, where sucrose is not commonly encountered. The possession of two independent pathways might act as a buffer against loss of sucrose metabolism genes (which might occur in the absence of selection), and so bacteria coming from the soil environment into the plant would be more likely to have retained the ability to use sucrose, and perhaps would be more able to colonize the plant tissue and initiate pathogenesis. Alternatively, it may be that *Agrobacterium* routinely encounters a wide variety of glycosides, and so has evolved a generalized mechanism for metabolizing them. *Agrobacterium* is known to possess highly sensitive chemotaxis toward sucrose (Loake et al., 1988). Our interest in the practical use of glucose or its derivative prompted the study of conditions required for efficient production of 3-ketoglycosides (Fig. 7.3).

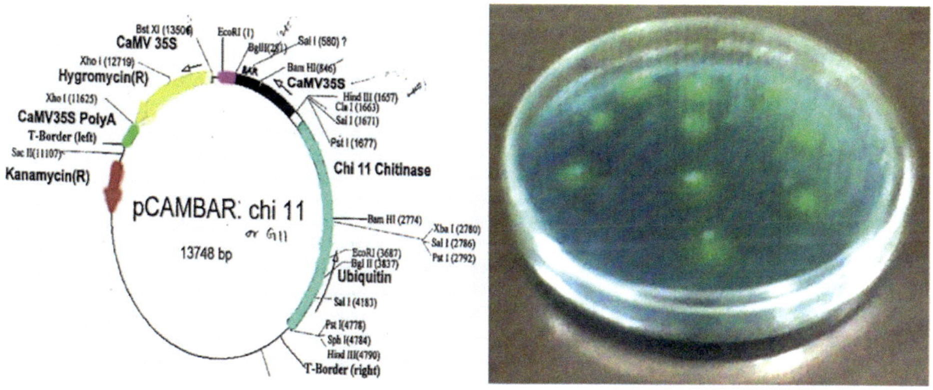

Vector map of the construct **3-Ketoglycoside test**

FIGURE 7.3 Chitinase gene construct (pCAMBIA-Bar-Ubi-Chi11) and viability test for *Agrobacterium tumefaciens* by 3-ketoglycoside test.

7.8 Field testing

Field resistance of safflower plants has been shown to correlate closely with the laboratory assays employed in this study. However, it will be necessary to assess the performance of the transgenic plants under different field conditions, when exposed to a range of *A. carthami* spores. To shed more light on the involvement of chitinases in plant growth and development further experiments are required. This protocol can now allow safflower to expand as an industrial crop platform and also offers the chance to modernize this ancient crop to benefit from advances in plant molecular biology.

7.8.1 Analyses of fungal culture filtrates

The fungal spores were initially cultured in Petri dishes containing PDA (potato dextrose agar) solid medium and were incubated at 25°C. Fungal colonies (mycelium with spores) that appeared 7 days after plating on PDA media were transferred to PDA containing antibiotics selection medium (100 μg/mL hygromycin-B). After 10 days, colony growth and morphology were observed (Fig. 7.4A). Mycelial mats with spores were collected from the Petri dishes and inoculated directly into Erlenmeyer flasks, containing MS basal liquid medium for culture filtrate production (Fig. 7.4B). All the cultures were maintained in 125 rpm under dark for continuous agitation. After 70 days of incubation, the dark brown colored culture filtrate was collected from each flask by filtering through a Whatman No.1 filter paper disc to remove spores and was centrifuged at 3000 rpm. After centrifugation, the *fungal culture* filtrate was again filtered through a Whatman No.1 filter paper disc and was stored in brown bottles at −20°C in refrigerator for further use. Spores were filtered through cheesecloth to remove most of the mycelia. The filtered spore suspension was centrifuged at $2000 \times g$ for 5 minutes and resuspended in deionized water. After the final wash, supernatant was discarded and spores were resuspended in water containing 0.1% Tween-20. Pathogenicity was assayed by spraying the inoculum of conidial suspension (2×10^5 conidial spores/mL) to the leaves of control and transgenic plants. Percent survival of hardened plants from control and transgenic plants was recorded (Fig. 7.4C and D).

FIGURE 7.4 Preparation of fungal culture filtrates and assay for pathogenicity and vegetative growth. (A) *Alternaria carthami* fungal colony growth in petri dishes. (B) Fungal culture filtrate in liquid culture. (C) Control plant after pathogenicity. (D) Transgenic plant with *Chi*11 gene after pathogenicity.

FIGURE 7.5 Analysis of *Alternaria carthami* leaf spot disease in safflower. (A) Control leaf with number of leaf spot lesion. (B) Transgenic leaf without leaf spot lesion.

7.8.2 Analysis of disease resistance

The regenerated plants from control and transgenic treated were taken for disease-resistance analysis. The survival of control and transgenic plants were significantly affected by *A. carthami* fungal spores (2×10^5 spores/mL). The appearance of leaf spots and without leaf spots was noticed on the leaves of nontransgenic and transgenic plants, respectively (Fig. 7.5A and B). *A. carthami* fungal spores sprayed on transgenic safflower plants showed better growth with decreasing numbers of brown spot lesions found.

This study has clearly demonstrated the major benefits to be gained from the introduction of disease defense genes into safflower and provides a baseline for the ongoing improvement of this high-value oilseed crop. The use of improved gene constructs, promoters and the identification of additional defense genes are likely to result in the production of commercially viable disease-resistant safflower cultivars in the near future.

7.9 Conclusion and future perspective

The intensification of agriculture has to a large extent met the demand of feeding a growing population, but the increase in agricultural production has reached a plateau in many countries. The increase in yields can partly be attributed to improved plant pest and disease management, including better understanding of the disease-causing organisms and the use of a range of different control measures. These challenges have already resulted in increased risks of establishment, spread and propagation of plant pests and pathogens. One of the greatest challenges is the development of resistance in pathogens toward pesticides, making the products less efficient or even inefficient.

Development of appropriate markers will support studies on origin and evolution of the genus *Carthamus*, map-based cloning, diversity analysis, and marker-assisted breeding programs. The availability of extensive expressed sequence tag resources within the Asteraceae members will facilitate comparative mapping studies and in development of simple sequence repeat markers. Several useful traits such as dwarfness, early maturity, resistance to biotic and abiotic stresses, high oleic (UC-1) types for enhancement of oil stability, high linoleic types, and tocopherol variants are available in the cultivar germplasm and wild species that could be mapped to accelerate the breeding programs. Unlike several other crops, development of mapping population through doubled haploid technique would be relatively easy in safflower owing to the propensity of safflower anthers for shoot regeneration. Micropropagation and in vitro conservation protocols could be developed for maintenance and multiplication of wild *Carthamus* species. Shoot regeneration frequencies are very high and need to be converted to genetic transformation frequencies for development of transgenics aimed at input traits and high-value end products. Viable linkages need to be established between institutions involved in germplasm conservation, plant breeding activities, and biotechnology for development and sharing of genetic and genomic resources and accelerating breeding programs.

References

Ahmed, K.Z., Sagi, F., 1993. Use of somaclonal variation and in vitro selection for induction of plant disease-resistance: prospects and limitations. Acta Phytopathol. Entomol. Hung. 28 (2–4), 143–159.

Antranikian, G., Vorigas, C.E., Bertoldo, C., 2005. Extreme environments as a resources for microorganisms and novel biocatalysts. Adv. Biochem. Eng. Biotechnol. 96, 219–262.

Becker, D., Kemper, E., Schell, J., Masterson, R., 1992. New plant binary vectors with selectable markers located proximal to the left T-DNA border. Plant Mol. Biol. 20 (6), 1195–1197.

Belide, S., Hac, L., Singh, S.P., Green, A.G., Wood, C.C., 2011. *Agrobacterium* mediated transformation of safflower and the efficient recovery of transgenic plants via grafting. Plant Methods 7 (1), 12.

Broglie, K., Chet, I., Holliday, M., Cressman, R., Biddle, P., Knowlton, S., et al., 1991. Transgenic plants with enhanced resistance to the fungal pathogen *Rhizoctonia solani*. Science 254, 1194–1197.

Bryla, D.R., Linderman, R.G., 2007. Implications of irrigation method and amount of water application on *Phytophthora* and *Pythium* infection and severity of root rot in highbush blueberry. Hort. Sci. 42 (6), 1463–1467.

Boyd, L.A., Ridout, C., O'Sullivan, D.M., Leach, J.E., Leung, H., 2013. Plant–pathogen interactions: disease resistance in modern agriculture. Trends Genet. 29 (4), 233–240.

Cervera, M., Orgeta, C., Navarro, A., Navarra, L., Pena, L., 2000. Generation of citrus plants with the tolerance-to-salinity gene HAL2 from yeast. J. Hort. Sci. Biotechnol. 75, 26–30.

Chen, J.T., Chang, W.C., 2002. Effects of tissue culture conditions and explant characteristics on direct somatic embryogenesis in *Oncidium* 'Gower Ramsey'. Plant Cell Tissue Organ. Cult. 69, 41–44.

Chung, C.T., Niemela, S.L., Miller, R.H., 1989. One-step preparation of competent *Escherichia coli*: transformation and storage of bacterial cells in the same solution. Proc. Natl. Acad. Sci. U.S.A. 86, 2172–2175.

Chilton, M., Drummond, M., Merlo, D., 1977. Stable incorporation of plasmid DNA into higher plant cells: the molecular basis of crown gall tumorigenesis. Cell 11, 263–271.

Daub, M.E., 1986. Tissue culture and the selection of resistance to pathogens. Annu. Rev. Phytopathol. 24, 159–186.

Doohan, F., 2005. Fungal pathogens of plants. Fungi: Biology and Applications. John Wiley, Chichester, pp. 232–263.

Dunwell, J.M., Khuri, S., Gane, P.J., 2000. Microbial relatives of the seed storage proteins of higher plants: conservation of structure and diversification of function during evolution of the cupin super family. Microbiol. Mol. Biol. Rev. 64, 153–179.

Elad, Y., 1997. Responses of plants to infection by *Botrytis cinerea* and novel means involved in reducing their susceptibility to infection. Biol. Rev. 72, 381–422.

Fladung, M., 1999. Gene stability in transgenic aspen (Populus), flanking DNA sequences and T-DNA structure. Mol. Genet. Genomics 260, 574–581.

Fukui, S., Hochster, R.M., 1963. Conversion of disaccharides to the corresponding glycoside-3-ulose by intact cells of *Agrobacterium tumefaciens*. Can. J. Biochem. Physiol. 41, 2363–2371.

Fukui, S., Hochster, R.M., 1965. On the active transport of sucrose and of 3-ketosucrose in *Agrobacterium tumefaciens*. Can. J. Biochem. 43, 1129–1141.

Gantait, S., Mandal, N., 2010. Tissue culture of *Anthurium andreanum*: a significant review and future prospective. Int. J. Bot. 6, 207–219.

Gao, W.Y., Fan, L., Paek, K.Y., 2000. Yellow and red pigment production by cell cultures of *Carthamus tinctorius* in a bioreactor. Plant Cell Tissue Organ. Cult. 60, 95–100.

Gecgel, U., Demirci, M., Esendal, E., Tasan, M., 2007. Fatty acid composition of the oil from developing seeds of different varieties of safflower (*Carthamus tinctorius* L.). J. Am. Oil Chemists' Soc. 84 (1), 47–54.

Gelvin, S.B., 2003. *Agrobacterium*-mediated plant transformation: the biology behind the "gene-jockeying" tool. Microbiol. Mol. Biol. Rev. 67, 16–37.

George, L., Rao, P.S., 1982. *In vitro* manipulation of safflower (*Carthamus tinctorius* L.) through tissue culture. Proc. Ind. Natl. Sci. Acad. B48, 791–794.

Glazebrook, J., 2005. Contrasting mechanisms of defense against biotrophic and necrotrophic pathogens. Annu. Rev. Phytopathol. 43, 205–227.

Gould, A.R., 1986. Factors controlling generation of variability *in vitro*. In: Vasil, I.K. (Ed.), Cell Culture and Somatic Cell Genetics in Plants. 3. Plant Regeneration and Genetic Variability. Academic, Orlando, pp. 549–567.

Haines, R.J., 1993. *In vitro* selection in plant breeding. In: Proceedings of the Tenth Australian Plant Breeding Conference, pp. 210 – 219.

Hanahan, D., 1983. Studies on transformation of *Escherichia coli* with plasmids. J. Mol. Biol. 166 (4), 557–580.

Herrera-Estrella, A., Chet, I., 1999. Chitinases in biological control. EXS 87, 171–184.

Ingelbrecht, I., Breyne, P., Vancomperonolle, A., van Montagu, J.M., Depicker, A., 1991. Transcriptional interferences in transgenic plants. Gene 109, 239–242.

Kaczmarek, M.B., Struszczyk-Swita, K., Li, X., Szczesna-Antczak, M., Daroch, M., 2019. Enzymatic modifications of chitin, chitosan and chito oligosaccharides. Front. Bioeng. Biotechnol. 7, 243.

Kalivoda, E.J., Horzempa, J., Stella, N.A., Sadaf, A., Kowalski, R.P., Nau, G.J., et al., 2011. New vector tools with a hygromycin resistance marker for use with opportunistic pathogens. Mol. Biotechnol. 48 (1), 7–14.

Kato, Y., Onishi, H., Machida, Y., 2003. Application of chitin and chitosan derivatives in the pharmaceutical field. Curr. Pharm. Biotechnol. 4, 303–309.

Klebe, R.J., Harriss, J.V., Sharp, Z.D., Douglas, M.G., 1983. A general method for polyethylene-glycol-induced genetic transformation of bacteria and yeast. Gene 25, 333–341.

Kizil, S., Çakmak, Ö., Kirici, S., Inan, M., 2008. A comprehensive study on safflower (*Carthamus tinctorius* L.) in semi-arid conditions. Biotechnol. Biotechnol. Equip. 22 (4), 947–953.

Kneusel, R.E., Matern, U., 1996. In vitro induction of resistance to *Alternaria* leaf blight disease in *Carthamus tinctorius* L.(safflower). Somaclonal Variation in Crop Improvement II. Springer, Berlin, Heidelberg, pp. 250–266.

Knowles, P.F., 1989. Safflower in Oil Crops of the World. McGraw Hill Publishing Company, New York, pp. 365–374.

Kolte, S.J., 2018. Diseases of Annual Edible Oilseed Crops: Volume III: Sunflower, Safflower, and Niger Seed Diseases. CRC Press.

Koeck, M., Hardham, A.R., Dodds, P.N., 2011. The role of effectors of biotrophic and hemibiotrophic fungi in infection. Cell. Microbiol. 13 (12), 1849–1857.

Kothari, S.L., Joshi, A., Kachhwaha, S., Ochoa-Alejo, N., 2010. Chilli peppers-a review on tissue culture and transgenesis. Biotechnol. Adv. 28, 35–48.

Kumar, S.P., Kumari, B.R., 2010. Effect of amino acids and growth regulators on indirect organogenesis in *Artemisia vulgaris* L. Asian J. Biotechnol. 2, 37–45.

Kumar, S.P., Kumari, B.R., 2013. Antioxidant and in situ chitinase gel activity in safflower (*Carthamus tinctorius* L). World Appl. Sci. J. 24 (1), 46–52.

Lacroix, B., Tzfira, T., Vainstein, A., Citovsky, V., 2006. A case of promiscuity: *Agrobacterium*'s endless hunt for new partners. Trends Genet. 22, 29–37.

Lakshmanan, P., Geijskes, R.J., Wang, L., Elliott, A., Grof, C.P., Berding, N., et al., 2006. Developmental and hormonal regulation of direct shoot organogenesis and somatic embryogenesis in sugarcane (*Saccharum* spp. interspecific hybrids) leaf culture. Plant Cell Rep. 25 (10), 1007–1015.

Lee, L.Y., Gelvin, S.B., 2008. T-DNA binary vectors and systems. Plant Physiol. 146 (2), 325–332.

Li, X., Wang, X.D., Zhao, X., Dutt, Y., 2004. Improvement of cotton fiber quality by transforming the acsA and acsB genes into *Gossypium hirsutum* L. by means of vacuum infiltration. Plant Cell Rep. 22, 691–697.

Liao, Z., Chen, M., Tan, F., Sun, X., Tang, K., 2004. Micropropagation of endangered Chinese aloe. Plant Cell Tissue Organ. Cult. 76, 83–86.

Lijiao, F.A.N., Meili, G.U.O., 2013. Progress of safflower (*Carthamus tinctorius* L.) regeneration through tissue culture. J. Med. Coll. PLA 28 (5), 289–301.

Lin, W., Anuratha, C.S., Datta, K., Potrykus, I., Muthukrishnan, S., Datta, S.K., 1995. Genetic engineering of rice for resistance to sheath blight. Nat. Biotechnol. 13, 686–691.

Liu, S., Graham, J.E., Bigelow, L., Morse II, P.D., Wilkinson, B.J., 2002. Identification of *Listeria monocytogenes* genes expressed in response to growth at low temperature. Appl. Environ. Microbiol. 68 (4), 1697.

Loake, G., Ashby, A., Shaw, C., 1988. Attraction of *Agrobacterium tumefaciens* C58C1 towards sugars involves a highly sensitive chemotaxis system. J. Gen. Microbiol. 134, 1427–1432.

Lu, P.D., Harding, H.P., Ron, D., 2004. Translation reinitiating at alternative open reading frames regulates gene expression in an integrated stress response. J. Cell Biol. 167 (1), 27–33.

Mandal, A.K.A., Dutta Gupta, S., 2003. Somatic embryogenesis of safflower: influence of auxin and ontogeny of somatic embryos. Plant Cell Tissue Organ. Cult. 72, 27–31.

Mandal, A.K.A., Dutta Gupta, S., Chatterji, A.K., 2001. Factors affecting somatic embryogenesis from cotyledonary explants of safflower. Biol. Plant 44, 503–507.

Micali, C., Göllner, K., Humphry, M., Consonni, C., Panstruga, R., 2008. The powdery mildew disease of Arabidopsis: a paradigm for the interaction between plants and biotrophic fungi. The Arabidopsis Book. American Society of Plant Biologists, 6.

Mithila, J., Murch, S.J., Krishnaraj, S., Saxena, P.K., 2001. Recent advances in Pelargonium in vitro regeneration systems. Plant Cell Tissue Organ. Cult. 67, 1–9.

Mortensen, K., Bergman, J.W., Burns, E.E., 1983. Importance of *Alternaria carthami* and *A. alternaria* in causing leaf spot diseases in safflower. Plant Dis. 67, 1187–1190.

Motamedi, J., Zebarjadi, A., Kahrizi, D., Salmanian, A.H., 2011. *In vitro* propagation and Agrobacterium-mediated transformation of safflower (*Carthamus tinctorius* L.) using a bacterial mutated aroA gene. Australian J. Crop Sci. 5 (4), 479.

Mündel, H.H., 2004. Safflower Production on the Canadian Prairies: Revisited in 2004. Lethbridge Research Station, Agriculture and Agri-Food Canada, Lethbridge.

Murashige, T., Skoog, F., 1962. A revised medium for rapid growth and bioassay with tobacco tissue culture. Physiol. Plant 15, 473–497.

Muzzarelli, R.A., Guerrieri, M., Goteri, G., Muzzarelli, C., Armeni, T., Ghiselli, R., et al., 2005. The biocompatibility of dibutyryl chitin in the context of wound dressings. Biomaterials 26, 5844–5854.

Nikam, T.D., Shitole, M.G., 1997. Sodium chloride tolerance in (*Carthamus tinctorius* L.). A A-1 callus cultures. In: Corleto, A., Mundel, H.H. (Eds.) Proceedings of the 4th International Safflower Conference: Safflower: A Multipurpose Species with Unexploited Potential and World Adaptability, Adriatica, Editrice, Bari, Italy, June 2–7, 1997, pp. 175–178.

Nikam, T.D., Shitole, M.G., 1999. *In vitro* culture of safflower (*Carthamus tinctorius* L.) cv. Bhima: initiation, growth optimization and organogenesis. Plant Cell Tissue Organ. Cult. 55, 15–22.

Nishizawa, Y., Nishio, K., Nakazono, K., Soma, M., Nakajima, E., Ugaki, M., et al., 1993. Enhanced resistance to blast (*Magna porthegrisea*) in transgenic Japonica rice by constitutive expression of rice chitinase. Mol. Gen. Genet. 241, 1–10.

Pasonen, H.L., Seppanen, S.K., Degefu, Y., Rytkonen, A., Von Weissenberg, K., Pappinen, A., 2004. Field performance of chitinase transgenic silver birches (*Betula pendula*): resistance to fungal diseases. Theor. Appl. Genet. 109, 562–570.

Pradeep Kumar, S., Ranjitha Kumari, B.D., 2017. Indirect somatic embryogenesis from transgenic immature leaf of safflower *Carthamus tinctorius* (Mohler, Roth, Schmidt & Boudreaux, 1967). Braz. J. Biol. Sci. 4 (8), 247–258.

Prasad, B.R., Khadeer, M.A., Seeta, S.P., Anwar, S.Y., 1991. *In vitro* induction of androgenic haploids in safflower (*Carthamus tinctorius* L.). Plant Cell Rep. 10, 48–51.

Punja, Z.K., Zhang, Y.Y., 1993. Plant chitinases and their roles in resistance to fungal diseases. J. Nematol. 25, 526–540.

Radhika, K., Sujatha, M., Nageshwar Rao, T., 2006. Thidiazuron stimulates adventitious shoot regeneration in different safflower explants. Biologia Plant 50, 174–179.

Rae, I.B., Gibb, S.W., 2005. Removal of metals from aqueous solutions using natural chitinous materials. Water Sci. Technol. 47, 189–196.

Reddy, P.P., 2014. Plant Growth-Promoting Rhizobacteria for Horticultural Crop Protection. Springer India.

Rohini, V.K., Sankara Rao, K., 2000. Embryo transformation, a practical approach for realizing transgenic plants of safflower (*Carthamus tinctorius* L.). Ann. Bot. 86, 1043–1049.

Rout, G.R., Reddy, G.M., Das, P., 2001. Studies on in vitro clonal propagation of *Paulownia tomentosa* Steud and evaluation of genetic fidelity through RAPD marker. Silvae Genet. 50, 208–212.

Sastry, R.K., Chattopadhyay, C., 2005. Diseases of safflower. In: Saharan, G.S., Naresh Mehta, Sangwan, M.S. (Eds.), Diseases of Oilseed Crops. Indus Publishing Co, New Delhi, pp. 236–268.

Seeta, P., Talat, K., Anwar, S.Y., 2000. Somaclonal variation: an alternative source of genetic variability in safflower. J. Cytol. Genet. 1, 127–135.

Sels, J., Mathys, J., De Coninck, B.M., Cammue, B.P., De Bolle, M.F., 2008. Plant pathogenesis-related (PR) proteins: a focus on PR peptides. Plant Physiol. Biochem. 46 (11), 941–950.

Sen Nag, O., 2017. World's Top Safflower Producing Countries. World Atlas.

Shibata, D., Liu, Y.G., 2000. *Agrobacterium*-mediated plant transformation with large DNA fragments. Trends Plant Sci. 5, 354–357.

Smeda, R.J., Weller, S.C., 1991. Plant cell and tissue culture techniques for weed science research. Weed Sci. 39, 497–504.

Smith, J.R., 1996. Safflower. AOCS Press, Champaign, IL, p. 624.

Srinivas, C.V.S., Praveena, B., Nagaraj, G., 1999. Safflower petals: a source of gamma linolenic acid. Plant Foods Hum. Nutr. 54, 89–92.

Sujatha, M., Suganya, A., 1996. *In vitro* organogenic comparison of different seedling tissues of safflower (*Carthamus tinctorius* L.). Sesame Safflower N. Lett. 11, 85–90.

Tejovathi, G., Anwar, S.Y., 1987. Plant regeneration from Cotyledonary cultures of safflower (*Carthamus tinctorius* L.). In: Reddy, G.M. (Ed.), Plant Cell and Tissue Culture of Economically Important Plants. Osmania University, Hyderabad, pp. 347–353.

Toki, S., Takamatsu, S., Nojiri, C., Ooba, S., Anzai, H., Iwata, M., et al., 1992. Expression of a maize ubiquitin gene promoter-bar chimeric gene in transgenic rice plants. Plant Physiol. 100, 1503–1507.

Tzfira, T., Citovsky, V., 2006. *Agrobacterium*-mediated genetic transformation of plants: biology and biotechnology. Curr. Opin. Biotechnol. 17, 147–154.

van Rooijen, G.J., Terning, L.I., Moloney, M.M., 1992. Nucleotide sequence of an *Arabidopsis thaliana* oleosin gene. Plant Mol. Biol. 18, 1177–1179.

Vijaya Kumar, J., Ranjitha Kumari, B.D., 2005. Effect of phytohormones on multiple shoot induction in cv. NARI-6 of safflower (*Carthamus tinctorius* L.). J. Plant Biotechnol. 7, 149–153.

Walia, N., Kaur, A., Babbar, S.B., 2005. *In vitro* regeneration of a high oil-yielding variety of safflower (*Carthamus tinctorius* var HUS-305). J. Plant Biochem. Biotechnol. 14, 65–68.

Wang, J., Oard, J.H., 2003. Rice ubiquitin promoters: deletion analysis and potential usefulness in plant transformation systems. Plant Cell Rep. 22 (2), 129–134.

Weiss, E.A., 1983. Oilseed crops. Chapter 6. Safflower. Longman Group Limited, Longman House, London, pp. 216–281.

Yamamoto, T., Iketan, H., Ieki, H., Nishizawa, Y., Notsuka, K., Hibi, T., et al., 2000. Transgenic grapevine plants expressing a rice chitinase with enhanced resistance to fungal pathogens. Plant Cell Rep. 19, 639–646.

Zambryski, Z.A., 1983. Ti plasmid vector for the introduction of DNA into plant cells without alteration of their normal regeneration capacity. EMBO J. 2, 2143–2150.

Further reading

Barnwal, M.K., Kotasthane, A., Magculia, N., Mukherjee, P.K., Savary, S., Sharma, A.K., et al., 2013. A review on crop losses, epidemiology and disease management of rice brown spot to identify research priorities and knowledge gaps. Eur. J. Plant Pathol. 136 (3), 443–457.

CHAPTER

8

Mycotoxin-associated food safety concerns of agriculture crops

Nitesh S. Litoriya[1] *and Arpan Modi*[2]

[1]AINP on Pesticide Residues, Anand Agricultural University, Anand, India [2]Department of Fruit Tree Sciences, Institute of Plant Sciences, Agricultural Research Organization, Volcani Center, Rishon LeZion, Israel

8.1 Introduction

Filamentous fungi of the phylum Ascomycota (molds) produce low-molecular-weight toxic secondary metabolites widely known as mycotoxins. The word is a combination of the Greek word *mykes* and the Latin word *toxicum* meanings of which imply fungus and toxic, respectively. The most common mycotoxins producing fungi are different species from *Fusarium, Alternaria, Aspergillus, Claviceps,* and *Penicillium.* Commonly occurring mycotoxins are aflatoxins, ochratoxins, fumonisins, trichothecenes, zearalenone, fusaproliferin, moniliformin, beauvericin, NX-2 toxin, enniatins, ergot alkaloids (EAs), altenuene, alternariol, alternariol methyl ether, altertoxin, tenuazonic acid, and patulin, although they don't have any characteristic odor and also any significant effect on the organoleptic properties of the food (Agriopoulou et al., 2020). The term mycotoxin was coined in 1962 when 100,000 turkey poults died due to consumption of peanut meal contaminated with aflatoxin produced by *Aspergillus flavus* (Blount, 1961).

By targets, fungal secondary metabolites are classified as antibiotics like penicillin (toxic to bacteria), phytotoxins (toxic to plants), and mycotoxins (toxic to vertebrates and other animals at lower concentrations). The mycotoxins listed earlier may be produced by more than one species of fungi and at the same time, one fungus may produce more than one type of mycotoxins. Moreover, when infected, foods may have more than one type of mycotoxins (Zain, 2011). Foods from developing and tropical countries are more susceptible to have the mycotoxin contamination than developed and temperate countries due to favorable climatic conditions (Ayofemi, 2020).

Food Security and Plant Disease Management
DOI: https://doi.org/10.1016/B978-0-12-821843-3.00011-8

8.2 Occurrence of mycotoxins

Generally, mycotoxins infect the food crops either at preharvest season or during postharvest handling, which are called as field fungi and storage fungi, respectively. Depending on plants life cycle, subclasses of field fungi also exist. This means that starting from plant's growth, senescence, and postharvest period, they are prone to get infected with different class of fungi. Some fungal species such as *Fusarium graminearum* infect healthy plants and act as plant pathogens, whereas, some (*Fusarium moniliforme* and *A. flavus*) grow on senescent or drought stressed plants (Klich, 1987). Moreover, a class of fungi colonize or infect the plants before harvest and contaminate the commodity with mycotoxin after harvest such as *Penicillium verrucosum* and *A. flavus* (Tola and Kebede, 2016). Health problems caused by various mycotoxins are cancer, immunosuppression (Bondy and Pestka, 2000), and diminished growth patterns in children during weaning period (Gong et al., 2004). Affected food crops are corn, soybean, peanuts, sorghum, wheat, and rice.

Food processing practices are not completely capable to eliminate these mycotoxins because most of them are stable against heat, physical, and chemical treatments. In addition to that, contamination of feed by mycotoxins will be more dangerous as the situation may affect both the consuming animals and animal-derived products such as egg and milk. During postharvest management of cereals, temperature and humidity play major role in occurrence and establishment of mycotoxin contamination, which can sustain for longer period. Mycotoxins can appear in the diet from both food (direct) and feed (through animal products) crops. Various well-established food agencies have proposed guidelines for the maximum permissible limits of various mycotoxins in food crops and animal feed. According to USFDA guidelines these limits for the aflatoxins (B1, B2, G1, G2), aflatoxin M1, fumonisins (B1, B2, B3), deoxynivalenol (DON), and patulin are 20, 0.5, 2000–4000, 1000, and 50 μg/kg, respectively. Limits for the same mycotoxins as described by EU (EC, 2006) are 2–15, 0.025–0.05, 200–1000, 200–500, and 10–50 μg/kg, respectively (Alshannaq and Yu, 2017). Details of various mycotoxins (aflatoxin, citrinin, EAs, fumonisins, patulin, trichothecenes, zearalenone) are discussed in the following subsections.

8.2.1 Aflatoxins

They are mainly produced by *A. flavus* and *Aspergillus parasiticus*. These fungi are commonly found in soil and other organic matter (Alshannaq and Yu, 2017). Major Aflatoxins produced by these fungi are B1 ($C_{17}H_{12}O_6$), B2 ($C_{17}H_{14}O_6$), G1 ($C_{17}H_{12}O_7$), and G2 ($C_{17}H_{14}O_7$) where B and G represent their nature of fluorescent light production (either blue or green) when illuminated under UV light as well as their relative mobility values (R_f values) when they are subjected to thin-layer chromatography (TLC). The numbers 1 and 2 given in the end of aflatoxin represent their slightly different chemical structure. B1 and G1 have additional double bond between eighth and ninth carbon (Zain, 2011). Aflatoxin B1 is the most studied and potent carcinogenic mycotoxin among other aflatoxins. Due to this reason, it represents all the aflatoxins as well (Goto et al., 1996). Apart from B and G forms of aflatoxins, M form also exists, which represents metabolites or derivatives of

B1 and B2 and mostly found in human and animal milk (Bennett et al., 2007). Chemically, they are difuranocoumarin derivatives synthesized in many strains of *A. flavus* and *A. parasiticus* by a polyketide pathway. Their molecular weights, melting points, UV absorptions, and fluorescence emissions are 312–330 g/mol, 237°C–289°C, 223–363 nm, and 425/450 nm, respectively. Their chemical properties revealed that they are colorless to pale yellow crystals, poorly soluble in water, but soluble in polar solvents such as chloroform, dimethyl sulfoxide, and methanol; however, under extreme pH conditions they are unstable. They have lactone ring backbone that can be opened under alkaline conditions leading to degradation. However, upon acidification, the reaction is reversible. Ammonia treatment at high temperature leads to complete degradation of aflatoxin and the reaction is irreversible (Kumar, 2018). Apart from these two, other aflatoxin producing species have also been isolated viz., *Aspergillus bombycis, Aspergillus ochraceoroseus, Aspergillus nomius*, and *Aspergillus pseudotamarii* (Klich et al., 2000; Peterson et al., 2001). In case of feed-based contamination of aflatoxins in food chain, a cow can transform aflatoxin B1 into its hydroxylated derivative known as aflatoxin M1. When diet, containing aflatoxin B1, is consumed and transported to mammary cells, the natural antioxidants present in the epithelial cells combat with B1 in order to minimize its toxic effect via action of drug-metabolizing enzymes (Ghadiri et al., 2019; Van Egmond, 1989). Various diseases caused by aflatoxin consumption in humans and animals are known as aflatoxicoses. According to severity of the aflatoxicoses, they are grouped as acute (which results in death) or chronic (which results in cancer, diminished immune system, and other less harmful pathological conditions) as described by Benkerroum et al. (2020).

8.2.2 Citrinin

Citrinin was discovered from *Penicillium citrinum* prior to 1939, but nowadays, reported in several species of *Penicillium* and *Aspergillus* (Manabe, 2001). In Japan, citrinin was known to be associated with yellow rice disease. It contributes toward the development of porcine nephropathy and it has also been reported to be neurotoxic in all the animal species tested (Bennett and Klich, 2003). It is mainly produced by several species of *Monascus, Penicillium*, and *Aspergillus*. Crops, affected by these fungi, are cereals, nuts, fruits, bulbs, and root vegetables. It is a polyketide mycotoxin with the molecular weight of 250.25 g/mol and melting point of 175°C. In methanol, it forms yellow crystals with maximum absorption of UV light (Doughari, 2015) whereas, under acidic, alkaline, or high temperature, loss of structure can be seen due to formation of chelate complexes (Poupko et al., 1997; Vrabcheva et al., 2000).

8.2.3 Ergot alkaloids

Common pathogen of various grass species, that is, *Claviceps* spp., produces various indole alkaloids that are derived from tetracycline ergoline ring system. EAs are classified into two groups viz., amine alkaloids (lysergic acid, ergonovine and methylergonovine, methysergide, lergotrile) and amino acid alkaloids (ergotamine, bromocriptine), which impart wide range of physiological effects and diseases (Hodgson, 2012). Disease caused by EAs in humans and animals is known as ergotism (Flieger et al., 2019). In humans, the risk of poisoning by epidemic is lower due to awareness toward the cause, whereas, veterinary animals (cattle, sheep, pigs, and chickens) are at the highest risk of ergotism (Hodgson, 2012). Common clinical symptoms found in ergotism are gangrene, abortion, suppression of lactation, ataxia, hypersensitivity, and convulsions (Bennett and Klich, 2003). These compounds proved as devastating nature of infection during mid-20th century. However, due to improvisation in agronomical, harvest, and postharvest management, their contamination can be completely avoided in the present era of sustainable agriculture (Arroyo-Manzanares et al., 2016).

8.2.4 Fumonisins

They are mainly synthesized by various species of *Fusarium* and *Alternaria alternata* f. sp. *lycopersici.* These species of fungi are mostly residing inside the plant as endophytes and generally do not impart any plant diseases. However, under favorable environmental conditions such as humidity or insect damage they can cause seedling blight, stalk rot, and ear rot (Nelson et al., 1993). To the animals, they interfere with sphingolipid metabolism and cause leukoencephalomalacia (in equines), pulmonary edema and hydrothorax (in swine), and carcinogenic and hepatotoxic effects and apoptosis of the liver (in rats). In humans also, they are reported to cause esophageal cancer (Bucci et al., 1996; Gelderblom et al., 1988; Harrison et al., 1990; Pozzi et al., 2001; Sydenham et al., 1991).

The most abundant fumonisin among all is fumonisin B1 and through this most affected crops are corn and sorghum. There are two types of fumonisins viz., cryptic fumonisins and free fumonisins. In which cryptic fumonisins are generally difficult to detect and form different structures with other primary and secondary metabolites, whereas, free fumonisins are easily detectable with various chromatographic techniques such as TLC, high-performance liquid chromatography (HPLC), and gas–liquid chromatography. Apart from these techniques, they can also be detected through immunoassays (Braun and Wink, 2018).

8.2.5 Patulin

Various species of *Aspergillus, Byssochlamys,* and *Penicillium* are responsible for the production of patulin. Patulins were originally recognized as antimicrobial agents after they were tested to be effective against various gram-positive and gram-negative bacteria. Later, their toxic effects were recognized. Fruit (mostly apples) and vegetable crops are primary target for patulin. Chemically, they are water-soluble polyketide lactones. They are heat stable with melting point of 110°C. Molecular weight of patulin is 154.12 g/mol. Adverse health effects due to patulin are nausea, vomiting, ulcers, intestinal hemorrhages, and lesion in the duodenum. Biochemical studies revealed that patulin inhibits the activity of various enzymes such as ATPase, lysosomal enzymes, and RNA polymerases. Recent studies revealed that the compound showed neurological, gastrointestinal, and immunological disorders. Similar adverse effects on health of various animals such as mice, hamsters, chickens, and monkeys were also observed (Saleh and Goktepe, 2019).

8.2.6 Trichothecenes

Trichothecenes are produced by various species of *Myrothecium, Fusarium, Trichoderma,* and *Stachybotrys*. In cereals (wheat and corn), they cause Fusarium Head Blight disease, however, legume crops are also equally susceptible as cereals for the infection of these fungi. Chemically they are sesquiterpene derivatives. Most toxic among all the trichothecenes is T-2 mycotoxin. Other trichothecenes are HT-2 toxin, 4,15-diacetoxy-scripenol (DAS), nivalenol, DON, verrucarin A, roridin A, and satratoxin H. They are nonvolatile, low molecular weight (250–550 g/mol), water insoluble, and soluble in polar solvents, such as chloroform, ethanol, acetone, ethyl acetate, dimethyl sulfoxide (DMSO), and propylene glycol (Cole and Cox, 1981). In animals, they cause emaciation, necrosis in the lymphoid tissue, subacute catarrhal gastritis, gastroenteritis lesions, intestinal hemorrhages, impaired immunity, weight loss, and thinning of eggshells (Adhikari et al., 2017) whereas, in humans, they cause nausea, vomiting, diarrhea, headache, abdominal pain, dizziness, and fever (Foroud et al., 2019).

8.2.7 Zearalenone

Fusarium spp. are known to produce substantial amount of zearalenone. Toxicological effects of these compounds are comparatively lower than other mycotoxins. The word zearalenone (zea-ral-en-one) is derived from the species name of fungus (*Gibberella* *zeae*), resorcylic acid lactone, "en" from the "-ene" representing C1 to C2 double bond, and "one" from the ket**one** group of the compound (Urry, 1966). They are synthesized from the polyketide pathway in fungus. Zearalenone-producing fungi are commonly found as contaminants in cereal grains, particularly in maize. However, in animals especially in pigs, hypoestrogenism, disrupted conception, abortion, and other problems related to impaired reproduction (Bennett and Klich, 2003). It is heat stable and thus its removal is difficult in the food chain (Zhang et al., 2018). Its solubility in water is 0.002 g/100 mL and it is highly stable. It is highly soluble in methanol and ethanol. Under alkaline condition and UV radiation, it undergoes degradation. In animals, it undergoes either reduction to produce less toxic/degraded compounds or conjugation (Rogowska et al., 2019).

As mentioned earlier, mycotoxin-producing fungi can infect the food crops during the life cycle of the crop (preharvest) or during postharvest management that includes storage, transport, and processing. Both the preharvest and postharvest contamination and containment depend on the stability of the mycotoxins and suitability of the environment for the mycotoxin-producing fungi.

8.3 Analysis and monitoring of mycotoxins

Since the occurrence, containment, and transmission of mycotoxins require crucial management practices before and after harvest, a stringent criterion should also be fixed to monitor their dynamics across the food chain. They are effective and associated with health issues even at very low concentrations. Thus detection of these mycotoxins requires analytical methods with greatest sensitivity and robustness. Common steps for the analysis of any mycotoxins are sampling, homogenization, extraction, purification, detection, and quantification. To set the criteria for complete set of analytical procedures, certain parameters should be taken into consideration. These include multiple target of interest, varied chemistry, multiple matrices, timing of testing, geographical and financial limitations, speed of testing, sampling, limits of detection and/or quantification, and nature of required measurements. As mentioned earlier, one mycotoxin may be produced by different species of fungi as well as one fungal species can also produce more than one type of mycotoxin. This phenomenon leads to the situation where single commodity may have more than one mycotoxin and thus scope of mycotoxin detection and quantification widens with reference to multiple target of interests. Depending on various chemical, physical, and environmental factors, a single molecule of mycotoxin may exhibit different forms or isomers. These forms are not always easy to isolate and quantitate. Extraction procedure greatly depends on the matrix of the sample due to the variation in carbohydrate, protein, lipid, and other primary and secondary metabolites. Time of testing is equally important as other parameters. Analysis from preharvest samples needs more attention than postharvest. Analytical procedures require highly skilled workers and costly equipment; therefore, it is also necessary to design and execute existing protocols where such conditions can be fulfilled maximally. Speed of the test is crucial before storage or during preharvest time. Depending completely on statistical methods and management practices, sampling type must be selected. Limit of detection and quantification are also set according to sample type, geographical location, type and final use of produce (Turner et al., 2015).

Guidelines for the handling of mycotoxin in food and feed can be found in Codex Alimentarius. According to their guidelines, mycotoxins were defined as contaminants unevenly distributed in the food and feed. Therefore appropriate sampling procedure must be followed. For various mycotoxins in different commodities, the maximum level (in μg/kg) has been defined. Table 8.1 represents these values. As these toxins are unevenly distributed and unintentionally present in the food and feed, the protocols for the detection of different mycotoxins vary from commodity to commodity or sample to sample. Protocols for the sampling in peanuts, tree nuts, and dried figs for detection of total aflatoxins have been described in detail (Codex Alimentarius Commission, 2013). Apart from these guidelines, several researchers also reported the variance of the batch depends on the variance of the sampling. Since the distribution of toxins is heterogenous, it is difficult to select the representative sample, values of which signify the entire batch. Whitaker et al. (1976) showed that collection of higher sized samples would result in low standard deviation, and sample preparation variance was reduced by increased size of subsamples for aflatoxin estimation. For aflatoxin M1 in milk, minimum 4.5 kg of sample from the single batch should be taken from ≥ 10 randomly selected portions (FDA, 2005).

TABLE 8.1 Various mycotoxins and their maximum level in different commodities.

Mycotoxin	Commodity/product	Maximum level (μg/kg)	Portion of commodity to which maximum level is applicable				
			A	B	C	D	E
Aflatoxins (total)	Almonds (ready-to-eat)	10		✓			
	Almonds (for processing)	15		✓			
	Brazil nuts (ready-to-eat)	10	✓				
	Brazil nuts (for processing)	15	✓				
	Hazelnuts (ready-to-eat)	10		✓			
	Hazelnuts (for processing)	15		✓			
	Pistachios (ready-to-eat)	10		✓			
	Pistachios (for processing)	15		✓			
	Dried figs (ready-to-eat)	10	✓				
	Peanuts (for processing)	15		✓	✓		
Aflatoxin M1	Milk	0.5	✓				
Deoxynivalenol	Cereal-based food for infants and young children	200				✓	
	Flour, meal, semolina, and flakes made from wheat, maize, or barley	1000					
	Cereal grains (further processing)	2000					
Fumonisin	Raw maize grain	4000	✓				
	Maize flour and maize meal	2000	✓				
Ochratoxin A	Wheat (raw common, raw durum, raw spelt, and raw emmer)	5	✓				
	Barley (raw)	5	✓				
	Rye (raw)	5	✓				
Patulin	Apple juice (ready-to-drink)	50	✓				✓
	Apple juice (used as ingredients)	50	✓				✓

Portion of commodity represented as A (whole commodity); B (whole commodity after removal of shell); C (seed or kernels); D (dry matter basis); E (reconstituted to the original juice concentrations).
Data from Codex Alimentarius Commission, 2013.

For fumonisins, 10 out of 24 batches were selected, and from each batch 10 samples were randomly selected, out of which 2 portions (25 g each) were taken for the analysis as per standard procedure mentioned in A.O.A.C. (Sydenham et al., 1996). Sampling plan for ochratoxin A (OTA) has been set as aflatoxins. Freese et al. (2000) studied the variability associated with barley for DON content. Six batches, comprised of 225 kg samples, were selected. Three subsamples (0.1, 0.8, and 7 kg) were selected in a lot and 16 lots were prepared. The result showed that the variability associated with the sampling procedure and analytical procedures were more significant than the sampling variance for all the sample sizes.

Protocols for the quantitative estimation of different mycotoxins are also available in different commodities and food samples. Analysis of mycotoxins in food and feed is carried out using common analytical techniques such as gas chromatography (GC) and HPLC as these techniques give advantages in terms of good performance, accuracy, sensitivity, precision, robustness, and reproducibility (Trucksess and Pohland, 2000). Sforza et al. (2006) reviewed various chromatographic techniques employed for the qualitative and quantitative analysis of mycotoxins in food samples. All the mycotoxins can be estimated by GC and/or liquid chromatography (LC) coupled with mass spectrometry (MS). Apart from these chromatographic techniques, enzyme-linked immunosorbent assay (ELISA) based detection of the same mycotoxin is also used. These techniques showed less accuracy as compared to chromatographic analysis. ELISA-based techniques comprised of detection of monoclonal or polyclonal antibodies generated in an organism after exposure to these toxins (Trucksess and Pohland, 2000). However, these chromatography techniques, nowadays, are employed for the detection and quantification of various mycotoxins in a sample even in single run (Woo et al., 2019; Bessaire et al., 2019). Since the food products are prone to get contaminated with different mycotoxins and the detection of these mycotoxin at individual level would be time and money consuming, standard protocols must be set to to detect multiple mycotoxins in a single run. A total of 72 internationally recognized methods are available for the detection and quantification of multiple mycotoxins in various commodities and food samples. However, very few studies have been reported so far for the cooccurrence of multiple mycotoxins in food samples and these 72 methods are more efficient in detection of multiple mycotoxins than individual ones (Bessaire et al., 2019; Smith et al., 2016). The method for the detection and estimation of many mycotoxins in food sample comprises of three parts viz., sampling, extraction, and purification, and analysis. Guidelines for the sampling from various commodities have been discussed already. For the extraction, suitable solvent systems are optimized first. According to complexity of the matrix of sample, and partition coefficient of the solvent, mycotoxins are extracted in different quantities in the solvent. This is known as recovery percentage that can be calculated by adding known quantity of mycotoxin(s) in uninfected sample (spiking). This sample will then undergo extraction and analytical procedures. This is called as method validation, for which several other parameters such as linearity studies, limit of detection, limit of quantitation, matrix effect, recovery, precision, trueness, and measurement of uncertainty are also calculated. Analytical part comprises optimization of LC/MS parameters such as source of ionization, column selection, run time, solvent selection, flow rate, and solvent running mode (isocratic or gradient). Using this practice, Woo et al. (2019) separated and analyzed 20 different mycotoxins in fermented paste of soybean.

8.4 Prevention and food safety measures

As many of the mycotoxins are heat stable and can retain inside the food for prolonged period, the first and foremost strategy to get rid of the mycotoxin is the prevention. Prevention can be targeted at the time of crop cultivation. As a preventive measure, recommendations from different organizations are released for different crop plants. Time of harvesting is a critical factor in this aspect. Preventive measures start from the selection or development of variety that is resistant to mold infection and thereby preventing the production of mycotoxins. Breeding strategies, particularly in wheat and corn, are available for the development of disease-resistant lines. Using this strategy, Germany covered 25% of total wheat cultivation area with resistant varieties. However, in breeding terminology, there is a cycle for variety with resistant gene and infectious organism, which means after few years the pathogen develops the resistance gene and thus, breeder must search for new gene to be incorporated through breeding. Alternative to this approach, a biotechnologist can find the gene that can degrade the mycotoxins produced by the fungi and incorporate it in the genome of the crop plant through genetic engineering. The transgenic plant then shows the resistance against respective mycotoxins. However, the procedure requires strict regulation and approval of various committees assigned for the release of genetically modified plants (Halasz et al., 2009). Moving toward highly reliable and practical approach, inhibition of fungal spores or inhibition of the mycotoxin synthesis may prove the best remedy for prevention. Most widely used chemicals for the inhibition of fungal growth are dichlorvos (an organophosphate insecticide) and caffeine. They are reported to inhibit the production of aflatoxins in variety of cereals and animal feeds (Trenholm et al., 1989). Apart from such substances, some surfactants were also used to inhibit the growth of *A. flavus* and thereby to stop the production of aflatoxins (Rodriguez and Mahoney, 1994). However, the Food and Agriculture Organization (FAO) has already published literature describing strategies for the prevention of growth of mycotoxigenic fungi. These include development of resistant varieties, employment of good agricultural practices (GAPs), controlling moisture content of plant seeds after harvesting, low temperature storage, use of fungicides (Table 8.2) and preservatives, use of insecticides to control

TABLE 8.2 Use of various fungicides for prevention of mycotoxin in cereal crops.

Fungicides	Target fungi	Mycotoxin	Crop	References
Metconazole, Propiconazole, Tebuconazole, Difenoconazole, Boscalid, Isopyrazam, Fludioxonil, Thiophanate-methyl, Folpet	*Fusarium graminearum, Fusarium proliferatum, Fusarium verticillioides, Aspergillus flavus*	Deoxynivalenol, nivalenol, fumonisins, zearalenone	Maize	Masiello et al. (2019)
Trifloxystrobin, Prothioconazole, Epoxiconazole, Pyraclostrobin, Mancozeb	*F. graminearum, Aspergillus* spp., *Penicillium* spp.	Deoxynivalenol, zearalenone, aflatoxin B1, B2, G1, G2, fumonisin B1, and ochratoxin A	Wheat	Da Luz et al. (2017)
Azoxystrobin	*Fusarium* spp.	Ochratoxin A, aflatoxin, deoxynivalenol	Rice	Giorni et al. (2019)

insect attack, good management practices, and the hazard analysis and critical control point system (Luo et al., 2018).

8.4.1 Physical methods

Before processing of food grains, the material may bring into several physical processes that eliminate the risk of mycotoxin contamination. Unprocessed cereal grains comprise 3%–6% of dirt, admixture, and broken kernels. Mostly broken kernels are the site of mycotoxin accumulation due to their suitability as host for most of the molds (Karlovsky et al., 2016). First physical process for the removal of the mycotoxin is sorting of grain biomass. More than 100 years ago, sorting techniques have been used since end of the 19th century. Sorting technique also has the long history of applications. Initially, based on the particle weight and size, sorting machines were used. Afterward, in the 1960s, optical sensor–based removal of kernel was established. This principle of sorting is also used presently with high-throughput capacity (Karlovsky et al., 2016). Damaged kernels are the most susceptible to aflatoxin contamination and thus, physical separation of kernels will help the control of the toxins (Kabak et al., 2006). Using same method of separation, EAs were also removed from rye (Miedaner and Geiger, 2015; Young et al., 1983). However, the process of sorting using optical sensor for the removal of mycotoxins has limitation when asymptomatic spread of the disease prevails. In such conditions, occurrence of symptoms and fungal infection (for fumonisin or aflatoxins) has very weak correlation. This means that the infected food seems healthy and thus may escape from the screening procedures (Pearson et al., 2004; Mutiga et al., 2014). Another such method based on physical properties is cleaning by sieving. The technique is used for EAs and fumonisins in wheat and corn, respectively (Murphy et al., 1993). The technique of sieve cleaning may also remove the sample by 34% (barley), 55% (wheat), and 69% (maize) in order to get significant reduction in mycotoxins. Considering physical properties of mold-damaged kernels and nondamaged ones, they can be separated using density segregation. In rye grains, flotation in NaCl solution resulted in efficient separation of damaged kernels. Likewise, flotation of samples on water and 30% sucrose solution also resulted in removal of DON by 53% and 68% in maize and wheat, respectively (Huff and Hagler, 1985). Shetty and Bhat (1999) used saturated brine solution for the removal of fumonisin from maize kernels and achieved 86% of reduced toxin with 20% loss of sample. Some of the mycotoxins (like DON) are water soluble and can be easily removed from the contaminated samples by washing with distilled water. Similarly, zearalenone (ZEA) is highly soluble in alkaline solutions and thus, sodium carbonate solution is used to remove ZEA from infected food grains. Up to 87% reduction was observed in various cereals contaminated with ZEA using 1 N solution of sodium carbonate (Trenholm et al., 1989). In the aqueous solution–based methods, solubility and partition coefficient are considered before treatments. Moreover, these methods require further step of drying and even partially failing of which may result in significant loss of food grain. Particularly for the aflatoxin in maize, dehulling (removal of outer layer of grain) showed 93% reduction of contaminant (Siwela et al., 2005). The process of milling and its effect in the removal of DON and ZEA (with their derivatives) were well studied in wheat grains (Schwake-

Anduschus et al., 2015). They showed the removal of these mycotoxins and their derivatives took place in different fractions of the milling process. The study revealed their existence in different fractions of the grain that can be removed with the process of milling. Wet milling is also another type of physical treatment, which results in the fractions such as germ (germ oil), starch, and gluten. Among them starch fraction is less prone to get infected with mycotoxins. In the process of wet milling, almost half of the mycotoxins get removed from corn to steep water and from the other half, nearly 30% remain in the fiber fraction and remaining mycotoxins are distributed in gluten, germ, and starch fractions. Water-insoluble or poorly soluble mycotoxins such as fumonisins show very low percentage in steep water fraction during wet milling. On the other hand, two-third of the total T-2 toxins were removed by steep water (Bennett et al., 1996; Collins and Rosen, 1981). Although most mycotoxins are heat stable in nature, heat treatment has been widely used in combination with time and in presence of other chemicals. Heat treatment is used for the significant reduction of aflatoxins in variety of food grains such as peanut, maize, and soybean. Because most of the mycotoxins are stable at higher temperature, they are not completely degraded by heat treatment, indeed, they are converted into other derivatives. For example, corn fumonisins are converted into *N*-(1-deoxy-D-fructos-1-yl)-fumonisin B1 (Hahn et al., 2015). Likewise, citrinin and EAs were also found to be partially degraded and less toxic after extrusion (Merkel et al., 2012). Irradiation is also used for the remediation of fungal and partial detoxification of the mycotoxins produced by them. Two types of irradiation treatments exist viz., nonionizing (solar, UV, microwave) and ionizing (gamma) irradiation. Using this technique, Herzallah et al. (2008) showed photodegradation of aflatoxins in cereal grains in combination with time of exposure. They found that toxin levels were reduced by 42.3% and 75.5% after treatment of direct sunlight for 3 and 30 hours, respectively. Similarly, they also observed the reduction in total mycotoxins in sunlight-treated, microwave-treated, and gamma-treated samples. Apart from cereals, the technique is widely used for almonds, peanuts, pistachios, date palm, fruit juices, and other food and beverages (Ghanem et al., 2008; Vita et al., 2014; Ouf et al., 2015; Ten Bosch et al., 2016). Originally used in medical engineering and material manufacturing, a cold plasma technique was used by Schlüter et al. (2013) , specific for the foods that are heat sensitive. Here, they defined plasma as gas containing electrons, ions, and neutral particles. In a cold plasma or nonthermal plasma, the gases used are molecular or inert gas such as nitrogen, argon, or helium. These gases are passed through electric field to accelerate and trigger chemical reaction via electron impact. The entire process is carried out at either low pressure or atmospheric pressure for food sector. Using air gases and sulfur hexafluoride, a low-pressure cold plasma was applied to control the growth of *A. parasiticus* in nut samples. The result showed 50% reduction of total aflatoxins (AFB1, AFB2, AFG1, and AFG2) in shelled and unshelled nut samples (hazelnuts, peanuts, and pistachio nuts) when they were subjected to plasma treatment for 20 minutes (Basaran et al., 2008). Atmospheric pressure cold plasma treatment (with argon gas) was also checked by Ouf et al. (2015) in contaminating date palm samples. Nine-minute exposure of the treatment showed complete removal of the fungal spores and significant reduction in OTA and fumonisin B2 content of the tested samples.

Apart from the methods listed earlier, steeping in combination with the addition of sulfur oxides promotes lactic acid production, which makes the treatment partially chemical. This method was also used to remove aflatoxin, fumonisin, and ochratoxin A (Canela et al., 1996; Pujol et al., 1999; Lefyedi and Taylor, 2006). Another such treatment comprises of use of activated charcoal and bentonite, surface on which, the toxins are adsorbed. They are termed as mycotoxin binders. However, very limited research has been conducted for this method (De Nijs et al., 2012; Wang et al., 2005).

8.4.2 Chemical methods

As compared to physical methods, the chemical methods require stringent criteria starting from protocol establishment to the final application. The strict rules for any country are decided because chemicals not only react with the contaminants but also with the food products. Therefore regulatory authority defines and issues guidelines for the use of chemical in the process of contaminants removal from the food and feed. For example, European Commission (EC) has published a guideline for the use of various compounds for the decontamination of feed (EC, 2015). The guideline comprised nine articles that define scope, application, acceptability criteria for physical detoxification process, acceptability criteria for chemical detoxification process, acceptability criteria for (micro)biological detoxification process, establishments, emergency situations, transitional measures, and entry into force. After screening of these chemicals to be used, detoxification process can be carried out (within the limitation of the guidelines). Chemical processes are sometimes also carried out in combination with the physical processes mentioned in previous section. This may be useful in efficient degradation of mycotoxins in the samples. The mode of application of chemical agents can be mixing, packing, fumigation, or immersion. Mycotoxins are generally resistant to mild acids and thus treatments with strong acid help to reduce the biological impacts of aflatoxins (AFB1 and AFG1). They are converted in their hemiacetal forms AFB_{2a} and AFG_{2a} after treatment with strong acids. A diluted solution of HCl (pH 2.0) showed 19% reduction of AFB1 content within 24 hours (Doyle et al., 1982). Like strong acid treatments, citric acid, acetic acid, and lactic acid were also used for the decontamination of aflatoxins. The result revealed that lactic acid was suitable among them, which also inhibited mold growth and thus may be used as preservative (Aiko et al., 2016). In case of treatment with the bases, most of the mycotoxins are unstable in the alkali solutions. However, the process is reversible and thus the toxin may reestablish its original integrity, if the process of degradation is not completed. Bases used for the treatments are sodium hydroxide, calcium hydroxide, sodium carbonate, sodium phosphate, ethanolamine, ethylene diamine, and methylamine (Karlovsky et al., 2016). Most effective method of degradation is the treatment with ammonia. It makes the lactone ring of aflatoxin to open and forms two new products viz., aflatoxin D1 and D2 (4-hydroxy-6-methoxy-3a,8a-dihydrofuro [2,3] benzofuran) which are less toxic (Cucullu et al., 1976). In maize, the process of ammoniation showed 75% reduction in aflatoxin levels and complete decomposition of ochratoxin A, in maize, wheat, and barley (Park et al., 1988; Chelkowski et al., 1981). However, after its recommendation as decontamination process in oilseed cake for feed, no country has approved this method for the removal of mycotoxins. Treatment with oxidizing agents such as ozone (O_3), hydrogen peroxide (H_2O_2), ammonium persulfate, and sodium hypochlorite shows degradation of terminal furan ring by addition of double bond, which leads to the opening of the lactone ring. Among all the oxidizing agents, ozone is the most preferred one for the removal of AFB_1, AFB_2, AFG_1, AFG_2, DON, OTA, and ZEA (Proctor et al., 2004; Inan et al., 2007; Young et al., 1986; McKenzie et al., 1997). Mycotoxins can also be subjected to degradation on exposure of reducing agents such as sodium bisulfite ($NaHSO_3$). It has been proven as potent reducing agents for the removal of aflatoxin in maize (Doyle et al., 1982), aflatoxin in dried figs (Altug et al., 1990), and DON in maize (Young et al., 1987). Apart from these chemicals, some naturally occurring secondary metabolites present in spices and medicinal plants are also useful to serve the same purpose. They are generally used in the form of extract of whole plant or any plant part. Extract of carom, *Ocimum tenuiflorum*, and *Adhatoda vasica* has found to be effective in the detoxification of aflatoxin from food (Velazhahan et al., 2010; Panda and Mehta, 2013).

8.4.3 Biochemical or biological method

Biological agents or biochemicals obtained from the organisms have potential application in number of industries including food. Under biochemical methods, certain enzymes are used for the detoxification of mycotoxins. Enzymes are known for their substrate specificity and thus may have less scope in the removal of toxins, but some enzymes such as laccase and peroxidases have wide range of substrate and can act on many mycotoxins (Vanhoutte et al., 2016). Enzyme obtained from *Gliocladium roseum* showed effective degradation of zearalenone (ZEA) into zearalanol. ZEA-degrading enzyme, ZEA lactonohydrolase, catalyzes the hydrolysis of the ZEA at the ester bond of the lactone ring. The process is followed by spontaneous decarboxylation in order to complete the degradation process (Takahashi-Ando et al., 2004; Matthies et al., 2001). The enzyme laccase from *Trametes versicolor* showed 87.34% degradation of AFB_1 (Alberts et al., 2009). Manganese peroxidase obtained from *Phanerochaete sordida* (YK-624) showed oxidation of AFB1 to AFB1-8,9-epoxide (Wang et al., 2005). It was also reported that cell-free extract of *Phoma* sp. when incubated with AFB1 in boiling water bath for 10 minutes, a significant degradation of aflatoxin was observed. This finding suggested that the enzyme was heat stable and can be used extensively (Ji et al., 2016). Like other mycotoxins, patulin (PAT) also undergoes detoxification by the action of enzymes used in juice production. These enzymes include pectinases, arabanases, glucoamylases, and other enzymes (Karlovsky et al., 2016). Black yeast *Exophiala spinifera* was efficient enough to produce fumonisin-detoxifying enzymes (Blackwell et al., 1999), gene of which was also transferred in maize (Duvick, 2001) to produce transgenic maize showing resistant to fumonisins toxification. A species of *Sphingomonas* also found to produce fumonisin-detoxifying enzymes hydrolyzing ester bonds present in polyketide residue fumonisin. Application of the enzyme fumonisin carboxylesterase, obtained from fermentation process of *Pichia pastoris*, in pigs showed reduced fumonisin content in infected pigs (Grenier et al., 2012). *Clonostachys rosea* NRRL 1859 can produce ZEA-specific hydrolases (El-Sharkawy and Abul-Hajj, 1988; Takahashi-Ando et al., 2004; Karlovsky et al., 2016) cleaving lactone ring of the zearalenone. Luckily for the degradation of ochratoxin A, many enzymes are available because it is an amide, which can easily undergo hydrolysis by carboxypeptidase and chymotrypsin (Abrunhosa et al., 2010). Stander et al. (2000) also showed its hydrolysis by the action of lipases.

In the baking and brewing industries, the enzymes are key candidates to make the final products. However, mycotoxins may hinder the activity of such enzymes and thus, the use of a stable and highly resistant enzyme is always in demand. Natural enzymes may sometimes fail to show such stability and resistance. Therefore a concept of addition of recombinant enzyme in the processing was suggested. Hong et al. (2014) used recombinant asparaginase to prevent the formation of acrylamide in bread. The recombinant enzyme was hyperthermostable and was prepared to facilitate its handling, storage, and transportation. Similarly, in the making of cheese, rennet is used. Alternative to this aspartic protease chymosin showed better performance as compared to traditional one. In malting and brewing industries, DON is a common contaminant. A recombinant version of all the enzymes used in these industries was made comprising of subunits that detoxify DON. Examples of such enzymes include amylases, glucanases, protease, and other enzymes used in beer production (Karlovsky et al., 2016).

Although there are many demonstrations available for the use of such enzymes for the degradation or detoxification of mycotoxins in food industries, a lack of the proof of concept makes this practice in a controversy. Use of such enzymes for the mycotoxin remediation at commercial scale certainly require strict regulatory norms in order to minimize the immunogenic responses as these enzymes are basically protein in nature.

Coming to the biological aspects of mycotoxin removal, several microorganisms are used for the remediation of mycotoxins in food samples. It may also be referred as biocontrol strategy. However, like enzymes, this technique is also used in the food processing industries. Fermentation is widely used in food processing. During fermentation process, the product of interest is transformed from basic metabolites of food through the action of microbes-secreted enzymes. Microbial cells also produce other enzymes that can be traced in the fermenting medium, even after cellular lysis. Activities from the secreted enzymes also reduce the toxicogenic effects of mycotoxins. However, for the processing aid with the scope of mycotoxin removal by this method, no microbes have been approved so far (Karlovsky et al., 2016). Since the last three decades, it has been observed that aflatoxin M1 found mostly in milk products, was decreased significantly after the treatment of lactic acid bacteria. Two reasons for this phenomenon have been proposed. One is the adsorption of mycotoxins on lactobacteria and the other reason could be the detoxification of aflatoxin M1 by enzymes secreted by bacteria (Pierides et al., 2000; Haskard et al., 2001). Commonly used yeast *Saccharomyces cerevisiae* in wine industries also showed its potency in the detoxification of PAT and OTA (Moss and Long, 2002; Petruzzi et al., 2014). Another approach for the control of mycotoxic fungi at the preharvest stage is the use of atoxigenic fungal species. It is widely used for the removal of aflatoxin in crops such as peanuts, pistachios, maize, and cottonseed in the Mediterranean basin. Lactic acid bacteria, atoxigenic *Aspergillus*, and other fungal strains, *Aureobasidium pullulans*, *Bacillus subtilis*, *Bacillus amyloliquefaciens*, *Kluyveromyces* sp., *Pseudomonas syringae*, *Candida sake*, *Pantoea agglomerans*, and some *Bacillus* spp. have been used for commonly occurring mycotoxins in food crops (Tsitsigiannis et al., 2012). Mechanism of action in the use of atoxigenic strains of microbes is simple. These strains compete for the survival and as a result, significant decline in the growth of toxigenic fungi can be observed. Another mechanism comprises of release of substances by atoxigenic strains, which are toxic to toxigenic strains (Medina

TABLE 8.3 List of atoxigenic strains of microorganisms used in the control of mycotoxin-producing fungus in food crops.

Biocontrol agent	Toxigenic fungi	Crops/food product	Mycotoxin	Reference
Aspergillus flavus (atoxigenic)	*A. flavus*	Cotton, maize	AFB_1	Bandyopadhyay et al. (2016)
Clonostachys rosea 016	*Fusarium verticillioides*	Maize	AFB_1	Lyn et al. (2009)
Bacillus subtilis, Brevibacillus sp., *C. rosea*	*Fusarium* spp., *Aspergillus* spp.	Peanuts	Aflatoxins	Chulze et al. (2015)
		Cereals	DON; fumonisin	
		Grapes	OTA	
Pichia anomala	*Penicillium verrucosum*	Wheat	OTA	Mokiou and Magan (2008)
Penicillium chrysogenum (isolated antifungal protein)	*A. flavus, Penicillium restrictum*	Dry-ripened meat products	–	Delgado et al. (2015)

et al., 2017). Some of the atoxigenic strains of different microbes used for the control of toxigenic strains are listed in Table 8.3.

8.5 Conclusion

Mycotoxins are chemical substances produced by several fungal species such as *Fusarium, Aspergillus,* and *Penicillium*. These chemical substances are aflatoxin, zearalenone, and ochratoxins. Widely studied mycotoxins are aflatoxins, zearalenone, ochratoxin A, patulin, fumonisins, and trichothecenes. They are less soluble in water and highly soluble in organic solvents. Most of these mycotoxins contain ring structures and exhibit fluorescence upon exposure to UV light. These mycotoxins are stable at higher temperature and may undergo reversible reaction after partial degradation treatments. They impart serious health issues to humans and farm animals. Major health issues caused by mycotoxins are respiratory diseases, immunological disorders, diabetes, cancer, neurological disorders, and cardiovascular diseases. These fungi can infect food crops amid preharvest, postharvest, handling, and storage practices. Use of GAPs in food crops (cereals, fruits, dry fruits, and spices) is preferred method for the prevention and removal of these contaminants in the beginning. In the para-agricultural industries, such as dairy, food processing, wine making, brewery, and baking, occurrence and toxic effects of these mycotoxins are also reported. Retention of these fungi and their toxic products can be seen from field to plate if proper monitoring and remediation are not employed. The most important part of the food safety is the monitoring and estimation of these mycotoxins from farm to consumer levels due to their distribution in products is heterogenous. Several food safety–related organizations have already published guidelines for sampling and subsequent analysis of mycotoxins from various commodities and food samples. Many physical, chemical, and biological (biochemical) methods are reported for the efficient degradation of mycotoxins and/or removal of pathogenic fungi from food samples. Chemical and biological methods require strict regulatory operations before, during, and after treatments. Among these methods, biocontrol strategies are highly efficient and most preferred. In the context of food safety, monitoring of mycotoxins in combinations with remediation will remove some of the barriers coming in the way of attaining prosperous human and animal health.

References

Abrunhosa, L., Paterson, R.R., Venâncio, A., 2010. Biodegradation of ochratoxin A for food and feed decontamination. Toxins 2 (5), 1078–1099.

Adhikari, M., Negi, B., Kaushik, N., Adhikari, A., Al-Khedhairy, A.A., Kaushik, N.K., et al., 2017. T-2 mycotoxin: toxicological effects and decontamination strategies. Oncotarget 8 (20), 33933.

Agriopoulou, S., Stamatelopoulou, E., Varzakas, T., 2020. Advances in occurrence, importance, and mycotoxin control strategies: prevention and detoxification in foods. Foods 9 (2), 137.

Aiko, V., Edamana, P., Mehta, A., 2016. Decomposition and detoxification of aflatoxin B1 by lactic acid. J. Sci. Food Agric. 96 (6), 1959–1966.

Alberts, J.F., Gelderblom, W.C.A., Botha, A., Van Zyl, W.H., 2009. Degradation of aflatoxin B1 by fungal laccase enzymes. Int. J. Food Microbiol. 135 (1), 47–52.

Alshannaq, A., Yu, J.H., 2017. Occurrence, toxicity, and analysis of major mycotoxins in food. Int. J. Environ. Res. Public Health 14 (6), 632.

Altug, T.O.M.R.I.S., Yousef, A.E., Marth, E.H., 1990. Degradation of aflatoxin B1 in dried figs by sodium bisulfite with or without heat, ultraviolet energy or hydrogen peroxide. J. Food Prot. 53 (7), 581–582.

Arroyo-Manzanares, N., Gámiz-Gracia, L., García-Campaña, A.M., Diana Di Mavungu, J., De Saeger, S., 2016. Ergot alkaloids: chemistry, biosynthesis, bioactivity, and methods of analysis. In: Mérillon, J.-M., Ramawat, K.G. (Eds.), Fungal Metabolites. Springer, Cham. pp. 887–929.

Ayofemi Olalekan Adeyeye, S., 2020. Aflatoxigenic fungi and mycotoxins in food: a review. Crit. Rev. Food Sci. Nutr. 60 (5), 709–721.

Bandyopadhyay, R., Ortega-Beltran, A., Akande, A., Mutegi, C., Atehnkeng, J., Kaptoge, L., et al., 2016. Biological control of aflatoxins in Africa: current status and potential challenges in the face of climate change. World Mycotoxin J. 9 (5), 771–789.

Basaran, P., Basaran-Akgul, N., Oksuz, L., 2008. Elimination of *Aspergillus parasiticus* from nut surface with low pressure cold plasma (LPCP) treatment. Food Microbiol. 25 (4), 626–632.

Benkerroum, N., 2020. Chronic and acute toxicities of aflatoxins: mechanisms of action. Int. J. Environ. Res. Public. Health 17 (2), 423.

Bennett, G.A., Richard, J.L., Eckhoff, S.R., 1996. Distribution of fumonisins in food and feed products prepared from contaminated corn. Fumonisins in Food. Springer, Boston, MA, pp. 317–322.

Bennett, J.W., Kale, S., Yu, J., 2007. Aflatoxins: background, toxicology, and molecular biology. Foodborne Diseases. Humana Press, pp. 355–373.

Bennett, J., Klich, M., 2003. chotoxins. C lin. Microbiol. Rev. 16, 497–516.

Bessaire, T., Mujahid, C., Mottier, P., Desmarchelier, A., 2019. Multiple mycotoxins determination in food by LC-MS/MS: an International Collaborative Study. Toxins 11 (11), 658.

Blackwell, B.A., Gilliam, J.T., Savard, M.E., David Miller, J., Duvick, J.P., 1999. Oxidative deamination of hydrolyzed fumonisin B1 (AP1) by cultures of *Exophiala spinifera*. Nat. Toxins 7 (1), 31–38.

Blount, W.P., 1961. Turkey "X" disease. Turkeys 9 (2), 52–55.

Bondy, G.S., Pestka, J.J., 2000. Immunomodulation by fungal toxins. J. Toxicol. Environ. Health Part B: Crit. Rev. 3 (2), 109–143.

Braun, M.S., Wink, M., 2018. Exposure, occurrence, and chemistry of fumonisins and their cryptic derivatives. Compr. Rev. Food Sci. Food Saf. 17 (3), 769–791.

Bucci, T.J., Hansen, D.K., Laborde, J.B., 1996. Leukoencephalomalacia and hemorrhage in the brain of rabbits gavaged with mycotoxin fumonisin B1. Nat. Toxins 4 (1), 51–52.

Canela, R., Pujol, R., Sala, N., Sanchis, V., 1996. Fate of fumonisins B1 and B2 in steeped corn kernels. Food Addit. Contam. 13 (5), 511–517.

Chelkowski, J., Goliński, P., Godlewska, B., Radomyska, W., Szebiotko, K., Wiewiorowska, M., 1981. Mycotoxins in cereal grain. Part IV. Inactivation of ochratoxin A and other mycotoxins during ammoniation. Food/Nahrung 25 (7), 631–637.

Chulze, S.N., Palazzini, J.M., Torres, A.M., Barros, G., Ponsone, M.L., Geisen, R., et al., 2015. Biological control as a strategy to reduce the impact of mycotoxins in peanuts, grapes and cereals in Argentina. Food Addit. Contam. Part A 32 (4), 471–479.

Codex Alimentarius Commission, 2013. Codex General Standard for Contaminants and Toxins in Food and Feed. CODEX STAN 193, 1995. Food and Agriculture Organization, Rome.

Cole, R.J., Cox, R.H., 1981. Handbook of Toxic Fungal Metabolites. Academic Press.

Collins, G.J., Rosen, J.D., 1981. Distribution of T-2 toxin in wet-milled corn products. J. Food Sci. 46 (3), 877–879.

Cucullu, A.F., Lee, L.S., Pons Jr, W.A., Stanley, J.B., 1976. Ammoniation of aflatoxin B1. Isolation and characterization of a product with molecular weight 206. J. Agric. Food Chem. 24 (2), 408–410.

Da Luz, S.R., Pazdiora, P.C., Dallagnol, L.J., Dors, G.C., Chaves, F.C., 2017. Mycotoxin and fungicide residues in wheat grains from fungicide-treated plants measured by a validated LC-MS method. Food Chem. 220, 510–516.

De Nijs, M., van Egmond, H.P., Nauta, M., Rombouts, F.M., Notermans, S., 2012. Mycotoxin binders in feed: food safety aspects. In: Report 2011.019 of RIKILT, Institute of Food Safety. Wageningen University (University & Research Centre), Wageningen.

Delgado, J., Acosta, R., Rodríguez-Martín, A., Bermúdez, E., Núñez, F., Asensio, M.A., 2015. Growth inhibition and stability of PgAFP from *Penicillium chrysogenum* against fungi common on dry-ripened meat products. Int. J. Food Microbiol. 205, 23–29.

Vita, D.S., Rosa, P., Giuseppe, A., CGA, 2014. Effect of gamma irradiation on aflatoxins and ochratoxin a reduction in almond samples. J. Food Res. 3 (4).

Doughari, J.H., 2015. The occurrence, properties and significance of citrinin mycotoxin. J. Plant Pathol. Microbiol. 6 (11), 2.

Doyle, M.P., Applebaum, R.S., Brackett, R.E., Marth, E.H., 1982. Physical, chemical and biological degradation of mycotoxins in foods and agricultural commodities. J. Food Prot. 45 (10), 964–971.

Duvick, J., 2001. Prospects for reducing fumonisin contamination of maize through genetic modification. Environ. Health Perspect. 109 (Suppl. 2), 337–342.

European Commission, 2006. Commission Regulation (EC) No 401/2006 of 23 February 2006 laying down the methods of sampling and analysis for the official control of the levels of mycotoxins in foodstuffs. Off. J. Eur. Union. 70, 12–34.

EC, European Commission, 2015. Regulation 2015/786 defining acceptability criteria for detoxification processes applied to products intended for animal feed. Off. J. Eur. Union 125, 10–14.

El-Sharkawy, S., Abul-Hajj, Y.J., 1988. Microbial cleavage of zearalenone. Xenobiotica 18 (4), 365–371.

FDA, S., 2005. 527.400 Whole milk, low fat milk, skim milk–aflatoxin M1 (CPG 7106.10). FDA/ORA Compliance Policy Guides.

Flieger, M., Stodůlková, E., Wyka, S.A., Černý, J., Grobárová, V., Píchová, K., et al., 2019. Ergochromes: heretofore neglected side of ergot toxicity. Toxins 11 (8), 439.

Foroud, N.A., Baines, D., Gagkaeva, T.Y., Thakor, N., Badea, A., Steiner, B., et al., 2019. Trichothecenes in cereal grains–an update. Toxins 11 (11), 634.

Freese, L., Friedrich, R., Kendall, D., Tanner, S., 2000. Variability of deoxynivalenol measurements in barley. J. AOAC Int. 83 (5), 1259–1263.

Gelderblom, W.C., Jaskiewicz, K., Marasas, W.F., Thiel, P.G., Horak, R.M., Vleggaar, R., et al., 1988. Fumonisins--novel mycotoxins with cancer-promoting activity produced by *Fusarium moniliforme*. Appl. Environ. Microbiol. 54 (7), 1806–1811.

Ghadiri, S., Spalenza, V., Dellafiora, L., Badino, P., Barbarossa, A., Dall'Asta, C., et al., 2019. Modulation of aflatoxin B1 cytotoxicity and aflatoxin M1 synthesis by natural antioxidants in a bovine mammary epithelial cell line. Toxicol. In Vitro 57, 174–183.

Ghanem, I., Orfi, M., Shamma, M., 2008. Effect of gamma radiation on the inactivation of aflatoxin B1 in food and feed crops. Braz. J. Microbiol. 39 (4), 787–791.

Giorni, P., Rolla, U., Romani, M., Mulazzi, A., Bertuzzi, T., 2019. Efficacy of azoxystrobin on mycotoxins and related fungi in Italian paddy rice. Toxins 11 (6), 310.

Gong, Y., Hounsa, A., Egal, S., Turner, P.C., Sutcliffe, A.E., Hall, A.J., et al., 2004. Postweaning exposure to aflatoxin results in impaired child growth: a longitudinal study in Benin, West Africa. Environ. Health Perspect. 112 (13), 1334–1338.

Goto, T., Wicklow, D.T., Ito, Y., 1996. Aflatoxin and cyclopiazonic acid production by a sclerotium-producing *Aspergillus tamarii* strain. Appl. Environ. Microbiol. 62 (11), 4036–4038.

Grenier, B., Bracarense, A.P.F., Schwartz, H.E., Trumel, C., Cossalter, A.M., Schatzmayr, G., et al., 2012. The low intestinal and hepatic toxicity of hydrolyzed fumonisin B1 correlates with its inability to alter the metabolism of sphingolipids. Biochem. Pharmacol. 83 (10), 1465–1473.

Hahn, I., Nagl, V., Schwartz-Zimmermann, H.E., Varga, E., Schwarz, C., Slavik, V., et al., 2015. Effects of orally administered fumonisin B1 (FB1), partially hydrolysed FB1, hydrolysed FB1 and N-(1-deoxy-D-fructos-1-yl) FB1 on the sphingolipid metabolism in rats. Food Chem. Toxicol. 76, 11–18.

Halasz, A., Lasztity, R., Abonyi, T., Bata, A., 2009. Decontamination of mycotoxin-containing food and feed by biodegradation. Food Rev. Int. 25 (4), 284–298.

Harrison, L.R., Colvin, B.M., Greene, J.T., Newman, L.E., Cole Jr, J.R., 1990. Pulmonary edema and hydrothorax in swine produced by fumonisin B1, a toxic metabolite of *Fusarium moniliforme*. J. Vet. Diagn. Invest. 2 (3), 217–221.

Haskard, C.A., El-Nezami, H.S., Kankaanpää, P.E., Salminen, S., Ahokas, J.T., 2001. Surface binding of aflatoxin B1 by lactic acid bacteria. Appl. Environ. Microbiol. 67 (7), 3086–3091.

Herzallah, S., Alshawabkeh, K., Fataftah, A.A., 2008. Aflatoxin decontamination of artificially contaminated feeds by sunlight, γ-radiation, and microwave heating. J. Appl. Poult. Res. 17 (4), 515–521.

Hodgson, E., 2012. Toxins and venoms, Progress in Molecular Biology and Translational Science, Vol. 112. Academic Press, pp. 373–415.

Hong, S.J., Lee, Y.H., Khan, A.R., Ullah, I., Lee, C., Park, C.K., et al., 2014. Cloning, expression, and characterization of thermophilic L-asparaginase from *Thermococcus kodakarensis* KOD1. J. Basic Microbiol. 54 (6), 500–508.

Hsieh, D.P.H., 1989. Potential human health hazards of mycotoxins. Bioact. Mol.

Huff, W.E., Hagler Jr, W.M., 1985. Density segregation of corn and wheat naturally contaminated with aflatoxin, deoxynivalenol and zearalenone. J. Food Prot. 48 (5), 416–420.

Inan, F., Pala, M., Doymaz, I., 2007. Use of ozone in detoxification of aflatoxin B1 in red pepper. J. Stored Prod. Res. 43 (4), 425–429.

Ji, C., Fan, Y., Zhao, L., 2016. Review on biological degradation of mycotoxins. Anim. Nutr. 2 (3), 127–133.

Kabak, B., Dobson, A.D., Var, I.I.L., 2006. Strategies to prevent mycotoxin contamination of food and animal feed: a review. Crit. Rev. Food Sci. Nutr. 46 (8), 593–619.

Karlovsky, P., Suman, M., Berthiller, F., De Meester, J., Eisenbrand, G., Perrin, I., et al., 2016. Impact of food processing and detoxification treatments on mycotoxin contamination. Mycotoxin Res. 32 (4), 179–205.

Klich, M.A., 1987. Relation of plant water potential at flowering to subsequent cottonseed infection by *Aspergillus flavus*. Phytopathology 77 (5), 739–741.

Klich, M.A., Mullaney, E.J., Daly, C.B., Cary, J.W., 2000. Molecular and physiological aspects of aflatoxin and sterigmatocystin biosynthesis by *Aspergillus tamarii* and *A. ochraceoroseus*. Appl. Microbiol. Biotechnol. 53 (5), 605–609.

Kumar, V.V., 2018. Aflatoxins: properties, toxicity and detoxification. Nutr. Food Sci. Int. J. 6, 555696.

Lefyedi, M.L., Taylor, J.R., 2006. Effect of dilute alkaline steeping on the microbial contamination, toxicity and diastatic power of sorghum malt. J. Inst. Brew. 112 (2), 108–116.

Luo, Y., Liu, X., Li, J., 2018. Updating techniques on controlling mycotoxins-a review. Food Control. 89, 123–132.

Lyn, M.E., Abbas, H.K., Zablotowicz, R.M., Johnson, B.J., 2009. Delivery systems for biological control agents to manage aflatoxin contamination of pre-harvest maize. Food Addit. Contam. Part A 26 (3), 381–387.

Manabe, M., 2001. Fermented foods and mycotoxins. Mycotoxins 51, 25–28.

Masiello, M., Somma, S., Ghionna, V., Logrieco, A.F., Moretti, A., 2019. In vitro and in field response of different fungicides against *Aspergillus flavus* and *Fusarium* species causing ear rot disease of maize. Toxins 11 (1), 11.

Matthies, I., Woerfel, G., Karlovsky, P., 2001. Induction of a zearalenone degrading enzyme caused by the substrate and its derivatives. Mycotoxin Res. 17 (1), 28–31.

McKenzie, K.S., Sarr, A.B., Mayura, K., Bailey, R.H., Miller, D.R., Rogers, T.D., et al., 1997. Oxidative degradation and detoxification of mycotoxins using a novel source of ozone. Food Chem. Toxicol. 35 (8), 807–820.

Medina, A., Mohale, S., Samsudin, N.I.P., Rodriguez-Sixtos, A., Rodriguez, A., Magan, N., 2017. Biocontrol of mycotoxins: dynamics and mechanisms of action. Curr. Opin. Food Sci. 17, 41–48.

Merkel, S., Dib, B., Maul, R., Köppen, R., Koch, M., Nehls, I., 2012. Degradation and epimerization of ergot alkaloids after baking and in vitro digestion. Anal. Bioanal. Chem. 404 (8), 2489–2497.

Miedaner, T., Geiger, H.H., 2015. Biology, genetics, and management of ergot (*Claviceps* spp.) in rye, sorghum, and pearl millet. Toxins 7 (3), 659–678.

Mokiou, S., Magan, N., 2008. Physiological manipulation and formulation of the biocontrol yeast *Pichia anomala* for control of *Penicillium verrucosum* and ochratoxin A contamination of moist grain. Biocontrol Sci. Technol. 18 (10), 1063–1073.

Moss, M.O., Long, M.T., 2002. Fate of patulin in the presence of the yeast *Saccharomyces cerevisiae*. Food Addit. Contam. 19 (4), 387–399.

Murphy, P.A., Rice, L.G., Ross, P.F., 1993. Fumonisin B1, B2, and B3 content of Iowa, Wisconsin, and Illinois corn and corn screenings. J. Agric. Food Chem. 41 (2), 263–266.

Mutiga, S.K., Were, V., Hoffmann, V., Harvey, J.W., Milgroom, M.G., Nelson, R.J., 2014. Extent and drivers of mycotoxin contamination: inferences from a survey of Kenyan maize mills. Phytopathology 104 (11), 1221–1231.

Nelson, P.E., Desjardins, A.E., Plattner, R.D., 1993. Fumonisins, mycotoxins produced by *Fusarium* species: biology, chemistry, and significance. Annu. Rev. Phytopathol. 31 (1), 233–252.

Ouf, S.A., Basher, A.H., Mohamed, A.A.H., 2015. Inhibitory effect of double atmospheric pressure argon cold plasma on spores and mycotoxin production of *Aspergillus niger* contaminating date palm fruits. J. Sci. Food Agric. 95 (15), 3204–3210.

Panda, P., Mehta, A., 2013. Aflatoxin detoxification potential of O cimum Tenuiflorum. J. Food Saf. 33 (3), 265–272.

Park, D.L., Lee, L.S., Price, R.L., Pohland, A.E., 1988. Review of the decontamination of aflatoxins by ammoniation: current status and regulation. J. Assoc. Off. Anal. Chem. 71 (4), 685–703.

Pearson, T.C., Wicklow, D.T., Pasikatan, M.C., 2004. Reduction of aflatoxin and fumonisin contamination in yellow corn by high-speed dual-wavelength sorting. Cereal Chem. 81 (4), 490–498.

Peterson, S.W., Ito, Y., Horn, B.W., Goto, T., 2001. *Aspergillus bombycis*, a new aflatoxigenic species and genetic variation in its sibling species. A. nomius. Mycologia 93 (4), 689–703.

Petruzzi, L., Bevilacqua, A., Baiano, A., Beneduce, L., Corbo, M.R., Sinigaglia, M., 2014. In vitro removal of ochratoxin A by two strains of *Saccharomyces cerevisiae* and their performances under fermentative and stressing conditions. J. Appl. Microbiol. 116 (1), 60–70.

Pierides, M., El-Nezami, H., Peltonen, K., Salminen, S., Ahokas, J., 2000. Ability of dairy strains of lactic acid bacteria to bind aflatoxin M1 in a food model. J. Food Prot. 63 (5), 645–650.

Poupko, R., Luz, Z., Destro, R., 1997. Carbon-13 NMR of citrinin in the solid state and in solutions. J. Phys. Chem. A 101 (28), 5097–5102.

Pozzi, C.R., Corrêa, B., Xavier, J.G., Direito, G.M., Orsi, R.B., Matarazzo, S.V., 2001. Effects of prolonged oral administration of fumonisin B 1 and aflatoxin B 1 in rats. Mycopathologia 151 (1), 21–27.

Proctor, A.D., Ahmedna, M., Kumar, J.V., Goktepe, I., 2004. Degradation of aflatoxins in peanut kernels/flour by gaseous ozonation and mild heat treatment. Food Addit. Contam. 21 (8), 786–793.

Pujol, R., Torres, M., Sanchis, V., Canela, R., 1999. Fate of fumonisin B1 in corn kernel steeping water containing SO_2. J. Agric. Food Chem. 47 (1), 276–278.

Rodriguez, S.B., Mahoney, N.E., 1994. Inhibition of aflatoxin production by surfactants. Appl. Environ. Microbiol. 60 (1), 106–110.

Rogowska, A., Pomastowski, P., Sagandykova, G., Buszewski, B., 2019. Zearalenone and its metabolites: effect on human health, metabolism and neutralisation methods. Toxicon 15 (162), 46–56.

Saleh, I., Goktepe, I., 2019. The characteristics, occurrence, and toxicological effects of patulin. Food Chem. Toxicol. 129, 301–311.

Schlüter, O., Ehlbeck, J., Hertel, C., Habermeyer, M., Roth, A., Engel, K.H., et al., 2013. Opinion on the use of plasma processes for treatment of foods. Mol. Nutr. Food Res. 57 (5), 920–927.

Schwake-Anduschus, C., Proske, M., Sciurba, E., Muenzing, K., Koch, M., Maul, R., 2015. Distribution of deoxynivalenol, zearalenone, and their respective modified analogues in milling fractions of naturally contaminated wheat grains. World Mycotoxin J. 8 (4), 433–443.

Sforza, S., Dall'Asta, C., Marchelli, R., 2006. Recent advances in mycotoxin determination in food and feed by hyphenated chromatographic techniques/mass spectrometry. Mass. Spectrometry Rev., 25 (1), 54–76.

Shetty, P.H., Bhat, R.V., 1999. A physical method for segregation of fumonisin-contaminated maize. Food Chem. 66 (3), 371–374.

Siwela, A.H., Siwela, M., Matindi, G., Dube, S., Nziramasanga, N., 2005. Decontamination of aflatoxin-contaminated maize by dehulling. J. Sci. Food Agric. 85 (15), 2535–2538.

Smith, M.C., Madec, S., Coton, E., Hymery, N., 2016. Natural co-occurrence of mycotoxins in foods and feeds and their in vitro combined toxicological effects. Toxins 8 (4), 94.

Stander, M.A., Bornscheuer, U.T., Henke, E., Steyn, P.S., 2000. Screening of commercial hydrolases for the degradation of ochratoxin A. J. Agric. Food Chem. 48 (11), 5736–5739.

Sydenham, E.W., Shephard, G.S., Thiel, P.G., Marasas, W.F., Stockenstrom, S., 1991. Fumonisin contamination of commercial corn-based human foodstuffs. J. Agric. Food Chem. 39 (11), 2014–2018.

Sydenham, E.W., Shephard, G.S., Thiel, P.G., Stockenström, S., Snijman, P.W., Van Schalkwyk, D.J., et al., 1996. Liquid chromatographic determination of fumonisins B1, B2, and B3 in corn: AOAC–IUPAC collaborative study. J. AOAC Int. 79 (3), 688–696.

Takahashi-Ando, N., Ohsato, S., Shibata, T., Hamamoto, H., Yamaguchi, I., Kimura, M., 2004. Metabolism of zearalenone by genetically modified organisms expressing the detoxification gene from *Clonostachys rosea*. Appl. Environ. Microbiol. 70 (6), 3239–3245.

Ten Bosch, L., Avramidis, G., Döll, K., Wienecke, S., Karlovsky, P., Viöl, W., 2016. Degradation of Fusarium mycotoxins deoxynivalenol and zearalenone by cold atmospheric pressure plasma, 36th Mycotoxin Workshop, Germany.

Tola, M., Kebede, B., 2016. Occurrence, importance and control of mycotoxins: a review. Cogent Food Agric. 2 (1), 1191103.

Trenholm, H.L., Prelusky, D.B., Young, J.C., Miller, J.D., 1989. A practical guide to the prevention of Fusarium mycotoxins in grain and animal feedstuffs. Arch. Environ. Contam. Toxicol. 18 (3), 443–451.

Trucksess, M.W., Pohland, A.E. (Eds.), 2000. Mycotoxin Protocols, Vol. 157. Springer Science & Business Media.

Tsitsigiannis, D.I., Dimakopoulou, M., Antoniou, P.P., Tjamos, E.C., 2012. Biological control strategies of mycotoxigenic fungi and associated mycotoxins in Mediterranean basin crops. Phytopathol. Mediterr. 158–174.

Turner, N.W., Bramhmbhatt, H., Szabo-Vezse, M., Poma, A., Coker, R., Piletsky, S.A., 2015. Analytical methods for determination of mycotoxins: an update (2009–2014). Anal. Chim. Acta 901, 12–33.

Urry, M.H., 1966. The structure of zearalenone. Tetrahedron Lett. 27, 3109–3114.

Van Egmond, H.E., 1989. Aflatoxin M1: occurrence, toxicity, regulation. Mycotoxins Dairy. Products 1989, 11–55.

Vanhoutte, I., Audenaert, K., De Gelder, L., 2016. Biodegradation of mycotoxins: tales from known and unexplored worlds. Front. Microbiol. 7, 561.

Velazhahan, R., Vijayanandraj, S., Vijayasamundeeswari, A., Paranidharan, V., Samiyappan, R., Iwamoto, T., et al., 2010. Detoxification of aflatoxins by seed extracts of the medicinal plant, Trachyspermum ammi (L.) Sprague ex Turrill–structural analysis and biological toxicity of degradation product of aflatoxin G1. Food Control. 21 (5), 719–725.

Vrabcheva, T., Usleber, E., Dietrich, R., Märtlbauer, E., 2000. Co-occurrence of ochratoxin A and citrinin in cereals from Bulgarian villages with a history of Balkan endemic nephropathy. J. Agric. Food Chem. 48 (6), 2483–2488.

Wang, J.S., Luo, H., Billam, M., Wang, Z., Guan, H., Tang, L., et al., 2005. Short-term safety evaluation of processed calcium montmorillonite clay (NovaSil) in humans. Food Addit. Contam. 22 (3), 270–279.

Whitaker, T.B., Whitten, M.E., Monroe, R.J., 1976. Variability associated with testing cottonseed for aflatoxin. J. Am. Oil Chemists' Soc. 53 (7), 502–505.

Woo, S.Y., Ryu, S.Y., Tian, F., Lee, S.Y., Park, S.B., Chun, H.S., 2019. Simultaneous determination of twenty mycotoxins in the Korean soybean paste doenjang by LC-MS/MS with immunoaffinity cleanup. Toxins 11 (10), 594.

Young, J.C., Chen, Z.J., Marquardt, R.R., 1983. Reduction in alkaloid content of ergot sclerotia by chemical and physical treatment. J. Agric. Food Chem. 31 (2), 413–415.

Young, J.C., Subryan, L.M., Potts, D., McLaren, M.E., Gobran, F.H., 1986. Reduction in levels of deoxynivalenol in contaminated wheat by chemical and physical treatment. J. Agric. Food Chem. 34 (3), 461–465.

Young, J.C., Trenholm, H.L., Friend, D.W., Prelusky, D.B., 1987. Detoxification of deoxynivalenol with sodium bisulfite and evaluation of the effects when pure mycotoxin or contaminated corn was treated and given to pigs. J. Agric. Food Chem. 35 (2), 259–261.

Zain, M.E., 2011. Impact of mycotoxins on humans and animals. J. Saudi Chem. Soc. 15 (2), 129–144.

Zhang, G.L., Feng, Y.L., Song, J.L., Zhou, X.S., 2018. Zearalenone: a mycotoxin with different toxic effect in domestic and laboratory animals' granulosa cells. Front. Genet. 9, 667.

CHAPTER

9

CRISPR/Cas in food security and plant disease management

Sopan Ganpatrao Wagh[1], *Manoj Baliram Pohare*[2] *and Ravindra Ramrao Kale*[3]

[1]Department of Bio-Science and Technology, Agri-biotech College, VNMK University, Aurangabad, India [2]Department of Infection, Immunity and Cardiovascular Diseases, University of Sheffield, Sheffield, United Kingdom [3]Crop improvement division, ICAR-IIRR, Hyderabad, India

9.1 Introduction

Most of the food comes from a limited number of crops and the total number is slowly decreasing causes narrowing of diversity that contribute as a possible challenge to food safety (Khoury et al., 2014). Food security means when everyone gets sufficient, secure, and nutritious food for life. Various factors such as limited land availability for agricultural expansion, increasing graph of population, the impact of climate change, increased environmental pressures like new diseases and insect pests, drought, soil salinity, and nutrient deficiency adversely affect crop production and challenge global food security (Ma et al., 2018; Kamoun et al., 2019; Islam, 2019). In the coming years, there will be a huge increase in population and more food production will be needed to fulfill food requirement. The increase in genetic potential can be accomplished by defining genotypes for resistance to different stresses and other desirable traits by the use of selective biotechnology techniques and tools. Initial crop improvement has started with the application of traditional and mutational plant breeding with increased agricultural productivity but then also it had some limitations in terms of labor requirement, time needed, and barriers of crossing with distant species (Pohare et al., 2012, 2013). Genetically modified (GM) crops have helped overcome the limitations of the conventional breeding approach and have resulted in increased yields within less time period. Market acceptance concerns and the high cost of legalization cause restricted use of GM crops to few countries (Cole et al., 2018). Through careful implementation and scientific controls, advanced plant breeding technology such as genome editing,

DOI: https://doi.org/10.1016/B978-0-12-821843-3.00020-9

will make a major contribution to global food demand, among which clustered regularly interspaced short palindromic repeats/CRISPR-associated protein(CRISPR/Cas) has evolved as one of the key methods for the crop genome editing in range of crops (Shan et al., 2013; Liang et al., 2014; Jacobs et al., 2015; Kumar and Jain, 2015; Svitashev et al., 2015; Li et al., 2016, 2018; Langner et al., 2018; Haque et al., 2018; Zhang et al., 2018; Chen et al., 2019a,b; Kuang et al., 2020; Ali et al., 2020). Because of simplicity, reliability, and less cost of the CRISPR method, DNA testing using CRISPR/Cas for crop disease diagnosis and nongenetically modified organism (GMO) administration would be easy to implement (Zhang et al., 2020). If CRISPR/Cas emerges as an equally strong, if not better, method for achieving efficiency than what is currently being used, it could also play a role in tackling the challenge of food security. It is therefore clear that precise crop optimization in terms of yield, nutritional quality, and plant health using genome editing will be a critical strategy for tackling current and future agricultural challenges (Ma et al., 2018).

Among the various techniques of gene editing, gene targeting notably by CRISPR/Cas has used much consequential excitement in crop improvement research. In prokaryotes, the CRISPR is used to precisely recognize and destroy identical viral DNA fragments from viruses that has previously infected the prokaryote, using Cas protein Viz., Cas9. Cas9 uses RNA-guided mechanism that can be useful for developing successful technologies for genome editing and gene expression studies (Pennisi, 2013). CRISPR/Cas in prokaryotes has contributed to a revolution in precise genome editing with an affordable, fast, effective, and rapidly flourishing technique of high precision and accuracy (Islam, 2019). Originally, CRISPR/Cas system interference was expected to be RNA-mediated and protein-dependent while later it was known that it causes sequence-specific DNA cleavage as initial mode of action (Makarova et al., 2006; Garneau et al., 2010). This chapter offers a detailed overview of the theory and updates the application of the CRISPR/Cas for plant genetic modification to develop climate smart food crop varieties against various stresses and for nutritional improvement to ensure global food and nutritional stability.

9.2 Insights of CRISPR/Cas system invention

The CRISPR/Cas research got going in 1987 with the discovery of repetitive DNA segment in *Escherichia coli* (Ishino et al., 1987) and latter in other groups of bacteria and archaea (Mojica and Rodriguez-Valera, 2016). CRISPR/Cas provides protection to the bacteria and archaea against the invading viruses (Gasiunas et al., 2012). CRISPR/Cas system consist of two components, first one is 29 bp short identical palindromic DNA repeats, which are regularly interspaced and second one is protospacer, which is nothing but a 32 bp hypervariable spacer DNA that is unique and matches up exactly with the bacteria infecting viruses (Hsu et al., 2014). In addition to these two components, other genes that are associated with CRISPR are called as *cas* genes that produces Cas proteins which are either nucleases or helicases involved in the process of DNA manipulations (Hsu et al., 2014). Based on the assembly of different components, this system is mainly grouped into two categories that are class I and II. Former requires a complex of several proteins to enact RNA-guided target cleavage while latter uses only one RNA-guided endonuclease to enact RNA-guided cleavage of invading bacteriophage (Brouns et al., 2008). These two main CRISPR groups are further divided into six subcategories, depending on their Cas counterpart and the extent of their interference with the foreign genetic material.

After the recognition of foreign genetic material especially of bacteriophages, bacterial cell transcribes repetitive DNA to produce CRISPR-RNA (crRNA) along with *cas* genes, which then produces Cas protein from it after the process of translation. The crRNA consist of two components that is a spacer at the 5′ end, which is a short RNA segment that matches perfectly to a sequence from bacteriophages, and a repetitive DNA sequence at the 3′ end. In addition to these two components, crRNA also contains transactivating RNA (tracrRNA), which provides the binding site for the Cas proteins. This crRNA is then go on to combines with Cas proteins to produce crRNA–effector complexes and this assembly will then destroys the foreign genetic material once it matches exactly with spacer sequence in crRNA. Therefore the infection is essentially ended, even before it is started to spread in bacteria. Usually, a 2–5 bp short-conserved sequence is situated either immediate upstream or downstream of the target sequence of foreign genetic material, which is also called as protospacer adjacent motif (PAM sequence; 5′-NGG-3′ for Cas9). This PAM sequence has an important role to play in target DNA selection. In case of novel bacteriophage infection, bacterial cells do not have the complementary spacer DNA sequence with them that will match foreign genetic material. Therefore under such circumstances bacterial cell produces class I type of Cas proteins, which degrades the foreign genetic material piece by piece and insert a copy of it into CRISPR cascade as another spacer DNA sequence. So for the next time whenever this bacteriophage infects again, this spacer DNA will be used along with Cas proteins to degrade the genetic material of foreign invasion (Deveau et al., 2008, 2010; Bhaya et al., 2011; Pohare et al., 2019; Wagh and Pohare, 2019) (Fig. 9.1). Doudna and Charpentier used the mechanism of RNA-led DNA interference to develop a very potent RNA-guided DNA targeting tool for genome editing (Koonin et al., 2017).

9.3 Origin and applications of CRISPR/Cas technology

The Cas9 endonuclease from bacteria, *Streptococcus pyogenes* (Class II category) has emerged as a very potent platform for genome engineering as it requires only single effector Cas protein which is easy to handle as compared to class I CRISPR systems where multiprotein effector complex is required for genome modifications (GMs). Cas9 can be simply manipulated to new targets simply with modification of its single guide RNA (sgRNA) sequence that becomes plug and play tool for GMs. Notably, sgRNA consist of two components that is tracrRNA, which is a Cas9 binding domain, and crRNA, which binds to target sequence specifically (Jinek et al., 2012; Doudna and Charpentier, 2014).

Cas9 mediated GMs involve two steps namely DNA double strand breaks (DSB) that is followed by DNA repair. DSB on a target site generated by Cas9 (in PAM's presence) under guidance of sgRNA initiates cellular repair pathways like nonhomologous end joining (NHEJ) or homology-directed repair (HDR) (Deltcheva et al., 2011; Gasiunas et al., 2012; Jinek et al., 2012; Cong et al., 2013; Chen et al., 2014). In case of NHEJ, a random insertion or deletion of a single nucleotide at the DSB site leads to a nonfunctional gene (INDELS or gene knockout) as a result of frame shift and thus proving to be a good alternative to the other gene silencing techniques (Cong et al., 2013; Jinek et al., 2013;

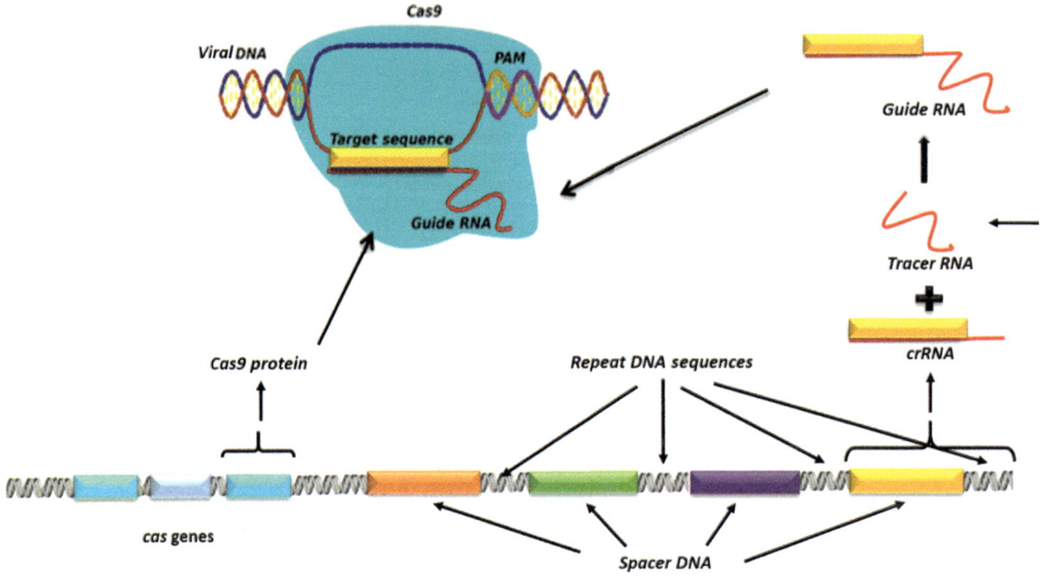

FIGURE 9.1 **CRISPR/Cas9 cascade in bacteria:** viral infection introduces viral genetic material into bacteria that triggers CRISPR/Cas cascade in bacteria. In case of novel infection, new spacer DNA specific to this infection is introduced into CRISPR cascade. Upon reinfection by this virus leads to transcription of spacer DNAs along with the repeats to generate crRNAs. The transcription of tracrRNA occurs separately that then combines with crRNA to generate sgRNA. The tracrRNA hold crRNA in position in sgRNA–Cas effector complex that then leads to DSB at a site on target DNA that matches exactly to the crRNA near PAM sequence. This immune system of bacteria is hijacked by Doudna and Charpentier to develop a CRISPR system which is a very potent RNA-guided DNA-targeting tool for genome editing (Koonin et al., 2017; Pohare et al., 2019; Wagh and Pohare, 2019; Wagh, 2020).

Unniyampurath et al., 2016; Chen et al., 2019a,b). While HDR can be applied to make required mutations or gene knock-in at the DSB site through homologous recombination directed by a donor DNA repair template, leading to targeted gene modifications, insertion, or deletion (Cristea et al., 2013; Maresca et al., 2013). Therefore the CRISPR/Cas system provides a robust platform for precise genome editing, including gene knock-in or knockout and its modifications (Pohare et al., 2019; Wagh and Pohare, 2019) (Fig. 9.2). Basic components of this system to generate site-specific DSB at DNA, includes a crRNA, tracrRNA, the presence of an adjacent PAM sequence and Cas9 protein (Deltcheva et al., 2011; Gasiunas et al., 2012; Cong et al., 2013). Multiplexing stage to generate multiple mutations can be achieved by using coexpression of Cas9 and several RNA guides (Cong et al., 2013). dCas9 is a modified form of Cas9 obtained by mutating both the RuvC and HNH nucleolytic active sites and can be directed to bind promoter sequences or fusion of dCas9 to an activation domain to subdue or initiate gene expression in cells (Gasiunas et al., 2012; Qi et al., 2013; Bikard et al., 2013). Adding certain functional domains to dCas9, such as the modifying domains KRAB and Kox1 chromatin, and green fluorescent protein (GFP) and other functional groups would generate scope for this technology (Larson et al., 2013; Gilbert et al., 2013; Malina et al., 2013; Chen et al., 2013). Recently, a

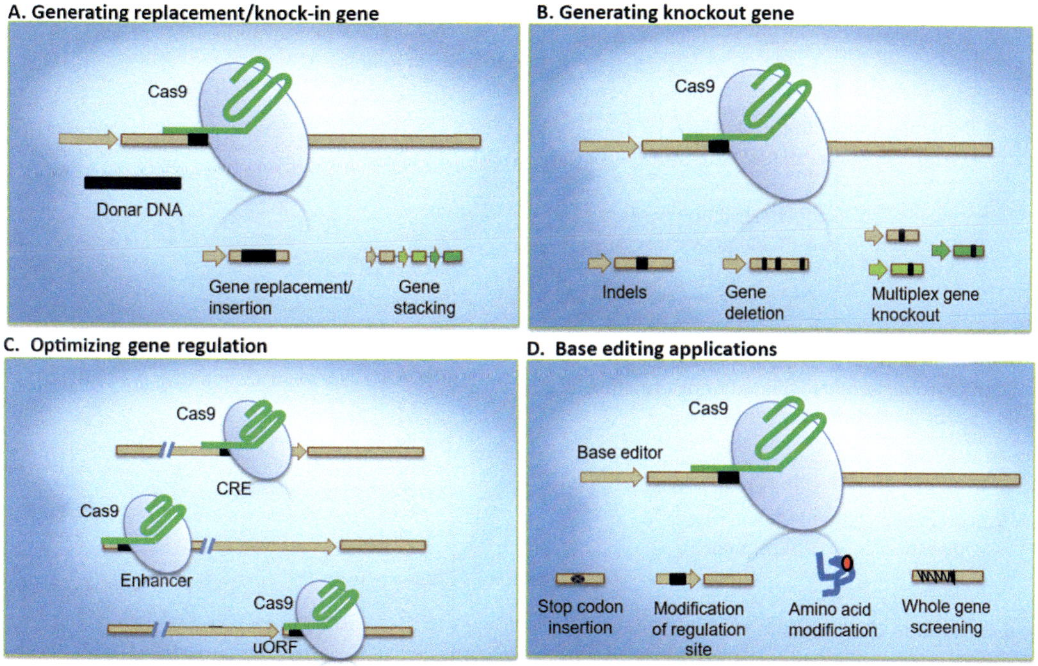

FIGURE 9.2 **Potential applications of CRISPR/Cas technology in plant breeding:** (A) Gene knock-in is used in generating gene stacking or gene replacement in recipient elite cultivars. (B) Gene knockout developed by this methodology can be used to create INDELS, gene deletions, and multiplex gene knockout. (C) Gene regulation is another possibility of CRISPR applications by modifying the regulatory sequences such as promoter or enhancer region. (D) Base editing can be used for amino acid substitution, stop codon insertion, and gene regulation in crop trait improvement. CRE, cis-regulatory element; uORF, upstream open reading frame (Wagh and Pohare, 2019).

new category of Cas protein (Cas12a or Cpf1) has been applied for genome editing in plants (Zetsche et al., 2015; Zaidi et al., 2017; Tang et al., 2017). Additional benefit of CRISPR/Cpf1 is that it requires only a single crRNA and not tracrRNA, which makes the size of sgRNA even smaller. Cpf1 recognizes different PAM sequence (5′-TTTN-3′) than that of Cas9 (5′-NGG-3′) (Zetsche et al., 2015).

Despite its many benefits, it has some limitations such as activity influenced by protein efficiency, protein size, PAM presence, and off-target effects (Barrangou and Doudna, 2016). New Cas9s like the latest nanoCas9s and other CRISPR systems like Cas12, Cas13 can be used to resolve these current limitations (Crawley et al., 2018; Chen et al., 2018; Yan et al., 2019). Nowadays cloning has become plug and play thing (Pohare and Akita, 2017) and novel CRISPR-associated genes with different characteristics has been discovered regularly and applied in genome editing research. Genome-edited products shall not be subject to the limitations imposed on genetically modified organisms, except in those cases where the mutation involves a large sequence of insertion into the genome (Georges and Ray, 2017). A large number of plants altered by CRISPR/Cas technology have been given free pass for cultivation in the United States as a GMO, and several nations, are prepared to adopt the same.

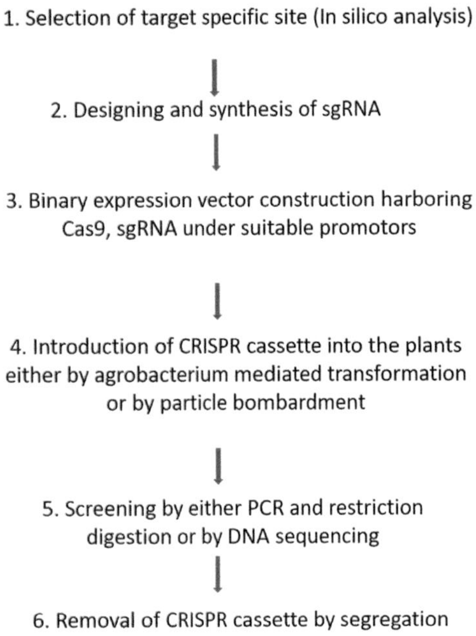

FIGURE 9.3 Flow of the application of CRISPR/Cas9 system in plant genome editing.

Typically, following steps are followed in plant gene/genome editing if one wants a gene knockout or knock-in (Fig. 9.3). In the first step, it is necessary to locate around 20 bp gene sequence that is specific to the gene of interest and is in the vicinity of PAM sequence (Jiang et al., 2013). In the second step, sgRNA is developed which contain this 20 bp gene specific sequence and tracrRNA sequence, which will bind to Cas9 (Jiang et al., 2013; Doudna and Charpentier, 2014). After successful expression of Cas9–sgRNA effector complex in plants it will bind to almost all the PAM sequence but cleavage will occur only at the site where sgRNA matches sufficient homology (Kleinstiver et al., 2016). Therefore designing highly specific sgRNA is vital for the success of gene editing and to avoid off targeting (Kleinstiver et al., 2016; Doench et al., 2016). If one wants a gene knock-in, site-specific sequence mutagenesis or gene corrections then it is essential to provide donor DNA repair template (single- or double-stranded oligonucleotide or plasmid) as well along with Cas9 and sgRNA (Li et al., 2016; Sun et al., 2016). In the third step, these cassettes are then introduced into plants usually through agrobacterium-mediated transformation (Fan et al., 2015; Zhou et al., 2014) and generated transgenic plants are identified by sequencing or by restriction digestion (Soda et al., 2018).

9.4 CRISPR/Cas in editing plants for abiotic stresses

Abiotic stress greatly affects crop production and is a quantitative trait that governed by more than one gene. Cas9 blends classical mutagens permanently with RNA interference (RNAi) programmability and can be used as a powerful tool in abiotic stress tolerance crop breeding program (Mushtaq et al., 2018). It is used for multiabiotic stress tolerance induced by modifying function of OsNCED3 gene that control abscisic acid (ABA) biosynthesis in rice (Huang et al., 2018). Ali et al. (2020) used Cas9-VirD2 chimeric protein complex, which combines the functions of Cas9, that is, targeted and unique DNA DSBs, and VirD2 relaxase, which takes the repair template close to the DSB. Here they used this complex successfully for specific allele modification to produce herbicide-resistant rice plants and for plant architecture and can be implemented effectively in agronomic trait gene selection, gene stacking or genome engineering, and crop improvement. In case of rice an endogenous gene OsALS1 has mutated to create novel rice, germplasms using a base-editing-mediated gene evolution technique. These substitute mutations conferred tolerance to the herbicide and the mutation was transferred to commercial rice cultivar Nangeng (Kuang et al., 2020). Same approach of base editing used in wheat to generate transgene-free wheat harboring herbicide tolerance (Zhang et al., 2019). CRISPR/Cas successfully applied to perform selective modification of the stress-responsive transcription factor genes in wheat that shows enormous potential for precise wheat genome engineering (Kim et al., 2018). ARGOS8 is a negative regulator for ethylene responses and decreases the sensitivity to ethylene resulting in improved yield of grain under conditions of drought stress. An advanced breeding technology allowed by CRISPR/Cas employed to generate new variants of ARGOS8 where field study showed that in stress conditions, the ARGOS8 variants produced more grain yield under drought condition compared to the wild-type that demonstrates CRISPR/Cas9 system's usefulness for drought-tolerant breeding (Shi et al., 2017).

9.5 Editing plants for disease resistance

Plant pathogens are major yield restricting factors, which cause losses in global farm productivity of 20%–40%, which poses significant challenges to food production that remain the world's largest challenge in agriculture (Savary et al., 2012). Over last 15 years, every aspect of plant science has revolutionized the implementation of the principle of GE over cultivating plants. The creation of effective and reproducible GE tools in plants would have a major influence on both applied and basic research in plants (Qi et al., 2013). Development in the editing of genomes has opened new ways to improve crop resistance. Four site-specific endonuclease systems have been used extensively to date are CRISPR/Cas9; zinc-finger nucleases, ZNF, TALEN, and meganucleases for genome editing in crops (Baltes et al., 2015; Zaidi et al., 2016; Mushtaq et al., 2018). The CRISPR/Cas method has been used over the last few years to tackle numerous arising problems, which include improvement of tolerance to biotic stresses (Arora and Narula, 2017). CRISPR/Cas approach has been investigated primarily to improve resistance to fungal and bacterial diseases accompanied by infection with viruses. The same has been discussed below that shows CRISPR/Cas technology's strength to develop resistance to these pathogens categories.

9.6 CRISPR/Cas for virus resistance

Plant viruses are a major concern and difficult to control for many commercially important staple and specialty crops, as viruses are rapidly developing and vector pests are generally involved. RNAi induced with double-stranded RNA has recently been used as an effective mechanism for virus resistance in plants (Ding and Voinnet, 2007). It has a single stranded, circular DNA genome that is repeated inside host nucleus to the intermediate double-stranded DNA (dsDNA) (Roossinck et al., 2015). This intermediary of dsDNA makes it a unique sequencing target for endonucleases. The artificial zinc-finger protein was meant targeting the replication base of the *beet extreme curly top virus* (BSCTV) by disabling the protein binds for viral replication (Rep). Arabidopsis has been demonstrated enhanced BSCTV tolerance and no signs of viral infection showed up to 80% of these transgenic plants (Sera, 2005). TALENs were used to grow *tomato yellow leaf curl China virus* (TYLCCNV), *tobacco curly shoot virus* (TbCSV), and *tomato leaf curl Yunnan virus* (TLCYnV) resistant transgenic tobacco that target the Rep genes (Cheng et al., 2015). The numerous researchers who investigated CRISPR-edited virus resistance plants recently targeted genomes of ssDNA geminivirus (Ji et al., 2015; Baltes et al., 2015; Ali et al., 2016) (Table 9.1). Large plant virus family like a *Geminiviridae* that causes crop loss among several major families worldwide and *Begomovirus* is the most important economic class of *geminiviruses* (Zaidi et al., 2016). *Begomoviruses* target dicotyledonous plants growing sweet potato and tobacco are also mainly associated with infected plant phloem (Gilbertson et al., 2015). Recently, multiple research indicated that CRISPR/Cas9 has been used for the establishment of geminivirus tolerance in plants. Delivery of coat protein (CP), intergenic region (IR), or Cas9-expressing tobacco gRNAs used to reduced DNA viruses accumulation (Ali et al., 2015, 2016). However, these investigations have shown that multiple geminiviruses could be targeted simultaneously with a conserved sequence (TAATATTAC) of gRNA in the IR region (Ali et al., 2015). The symptoms of gRNA targeting to Rep, CP, or IR were frequently diminished or removed substantially by the symptoms of dwarf yellow bean (BeYDV) and BSCTV (Baltes et al., 2015; Ji et al., 2015).

Viruses are dependent on plant machinery for a translation since viruses do not contain ribosomes in their virions. In plants, the multiconstituency translation complex along with eIF4E and its isoforms helps to recruit ribosome from 50 mRNA nontranslate regions (UTRs) (Denver et al., 2016). Arabidopsis mutant genetic screening with improved resistance to *turnip mosaic virus* (TuMV) suggested that the loss of functional mutations in the *eIF(iso)4E* gene resulted in resistance to TuMV (Lellis et al., 2002). In addition, Arabidopsis *eIF(iso)4E* knockout with CRISPR/Cas9 resulted in resistance to TuMV without losing the integrity of plant (Pyott et al., 2016). Two CRISPR/Cas9 (Chandrasekaran et al., 2016) gene sites were resistant to *papaya ring spot mosaic virus*-W, *zucchini yellow mosaic virus*, and *cucumber vein yellowing virus*. *Rice tungro spherical virus* resistance is accorded on the new alleles of rice eIF4 G created by CRISPR/Cas9-targeting (Roossinck et al., 2015). Direct antiviral use of CRISPR/Cas9 in barley to establish effective *wheat dwarf virus* (WDV) resistance. Accordingly, only CRISPR/Cas9-produced dual mutants ncbp-1/ncbp-2 symptoms showed delayed and attenuated following CBSV infection (*cassava brown streak virus*) (Gomez et al., 2019). In a first study (Table 9.1) showing RNA viruses resistance (Zhang et al., 2018) of plants known as *Nicotiana benthamiana* and arabidopsis, FnCas9 and RNA-targeting sgRNAs similar to *cucumber mosaic viruses* (CMVs) and *tobacco mosaic viruses* (TMVs) were expressed in transgenic plants reported a 40%–80% reduction in CMV and TMV accumulation relative to control plants.

TABLE 9.1 Successful examples of plant–pathogen interactions modified by CRISPR/Cas are summarized.

Host	Disease	Pathogen	Targeted gene(s) for editing	References
Viruses				
Rice	Viral (DNA)	*RTSV*	eIF4G-host factor for eIF4G translation of RNA viruses	Roossinck et al. (2015)
Barley	Viral (DNA)	*Wheat dwarf virus (WDV)*	Rep, MP, LIR essential for viral development and reproduction	Kis et al. (2019)
Cassava	Brown streak (RNA)	*CBSV*	nCBP-1 and nCBP-2/exon (plant)	Gomez et al. (2019)
Tomato	Viral (DNA)	*Bean yellow dwarf virus*	Carotenoid isomerase (CRTISO) and phytoene synthase 1 (PSY1)	Dahan-Meir et al. (2018)
Banana	Viral (DNA)	*eBSV*	Three target sites in viral genome (microbe)	Tripathi et al. (2019)
Arabidopsis, tobacco	Viral (RNA)	*TMV*	ORFCP, ORF1a, 3′-UTR (microbe)	Zhang et al. (2018)
Tobacco	Viral (RNA)	*TuMV*	Proteinase silencing suppressor (HC-Pro), *TuMV*-GFP, helper component, coat protein genes (microbe)	Aman et al. (2018)
Tobacco	Viral (DNA)	*CLCuKoV, TYLCV, TYLCSV, MeMV, BCTV*	IR, coat protein, and Rep (microbe)	Ali et al. (2016)
Cotton	Viral replication	*Cotton leaf curl disease (CLCuD)*	Replication associated protein (Rep) and noncoding intergenic regions (IR), α-Satellite Rep, and β-Sat IR	Iqbal et al. (2016)
Cotton	Leaf curl	*Leaf curl Kokhran virus (CLCuKoV)*	Coat protein (CP), RCRII domain of replication associated protein (Rep)	Ali et al. (2016)
Arabidopsis	Viral (DNA)	*CYVV*	eIF4E1 gene (plant); interact directly with viral protein and help with PmCDA1 viral replication	Bastet et al. (2019)
Bacteria				
Rice	Bacterial blight rice	*Xanthomonas oryzae pv. oryzae*	OsSWEET11, OsSWEET14 (rice bacterial blight susceptibility genes)	Jiang et al. (2013)
Tomato	Capsici bacterial speck, blight, and spot of tomato	*Pseudomonas syringae pv., Xanthomonas spp., Phytophthora*	SlDMR6-1 (plant); knockout of DMR6 enhances the levels of salicylic acid causing secondary metabolites and PR genes CRISPR/Cas9	Thomazella et al. (2016)
Tomato	Bacterial speck DC3000	*Pseudomonas syringae pv.*	Tomato SlJAZ2 (plant); directly associate with bacteria-producing coronatin that helps colonize the leaves	Ortigosa et al. (2019)
Citrus	Citrus canker	*Xanthomonas citri*	CsLOB1 (plant); pathogen induced susceptibility gene	Jia et al. (2017); Peng et al. (2017)

(Continued)

TABLE 9.1 (Continued)

Host	Disease	Pathogen	Targeted gene(s) for editing	References
Apple	Fire blight	*Erwinia amylovora*	Interact with disease pathogen-specific gene DIPM-1, 2, and 4 (plant)	Malnoy et al. (2016)
Fungus				
Rice	Rice blast	*Magnaporthe oryzae*	OsERF922 (plant); negative blast-fungus regulator	Wang et al. (2016)
Rice	Rice blast	*Magnaporthe grisea*	Rice OsMPK5 (plant); a negative rice defense regulator	Xie and Yang (2013)
Rice	Rice blast	*M. oryzae*	OsSEC3A (plant); participate in the complex exocyst and interact with proteins for defense	Ma et al. (2018)
Rice	False smut	*Ustilaginoidea virens*	USTA ustiloxin and UvSLT2 MAP kinase (microbe); to study the gene function	Liang et al. (2018)
Wheat	Powdery mildew	*Blumeria graminis f* sp. *tritici*	MLO-A1, B1, and D1 (plant); confer susceptibility to fungi	Wang et al. (2014)
Wheat	Powdery mildew	*B. graminis f.* sp. *tritici*	Three homologs of TaEDR1 (plant); plays a negative role in the immune system	Zhang et al. (2017)
Maize	Corn smut	*Ustilago maydis*	bW2 and bE1 (microbe); to evaluate the CRISPR system	Schuster et al. (2016)
Soybean	Damping-off	*Phytophthora sojae*	Avr4/6 (microbe); plays a negative role in plant immunity	Fang and Tyler (2015)
Canola	Blackleg disease	*Leptosphaeria maculans*	Histidine kinase (microbe); to study resistance mechanism against a pesticide (iprodione)	Idnurm et al. (2017)
Sunflower	Black molds	*Alternaria alternata*	Phosphate decarboxylase pyrG, polyketide-synthase, pksA, and 1,3,8-THN reductase, brm2 (microbe); to establish a CRISPR system	Wenderoth et al. (2017)
Tomato	Powdery mildew	*Oidium neolycopersici*	SlMlo1 (plant); confer susceptibility to fungi	Nekrasov et al. (2017)
Tomato, legumes, cotton	Wilt	*Fusarium oxysporum*	FoSso1 and FoSso2 (microbe); for endogenous tagging of target genes	Wang and Coleman (2019)
Vegetables, soybean	Powdery mildew, damping-off	*Phytophthora capsici and P. sojae*	Oxysterol binding protein-related protein 1 (microbe); to analyze mechanism of resistance to a pesticide (oxathiapiprolin)	Miao et al. (2018)
Grape	Powdery mildew	*Uncinula necator*	MLO-7 (plant); confer susceptibility to a fungal pathogen	Malnoy et al. (2016)
Grape	Gray mold	*Botrytis cinerea*	WRKY52 (plant); transcription factor involved in response to biotic stress	Wang et al. (2018)

(*Continued*)

TABLE 9.1 (Continued)

Host	Disease	Pathogen	Targeted gene(s) for editing	References
Cacao	Black pod disease	*Phytophthora tropicalis*	Nonexpressor of athogenesis-related3 (TcNPR3) gene (plant)	Fister et al. (2018)
Vegetables	White mold	*Sclerotinia sclerotiorum*	Oxalate biosynthesis gene Ssoah1 (microbe); mutations to study the pathogenicity	Li et al. (2018)

In addition, the resistance of the sgRNA–FnCas9 generation T6 was very stable and still active. Crucially, Zhang et al. (2018) noted that FnCas9 endonuclease activity was not necessary to interfere with the genome of CMV, even if RNA-binding activities were required, this meant that a CRISPR interference similar to the catalytically inactive SpCas9 protein programmed to suppress gene expression could be regarded as this particular FnCas9 function (Larson et al., 2013). Similar analysis of the case13a was carried out. Aman et al. (2018) used RNA-guided ribonuclease to control the RNA genome of TuMV (Table 9.1).

The majority of plant viruses have genomes containing RNA. Therefore CRISPR/Cas mechanisms that usually bind dsDNA cannot specifically target viral RNA sequences. The development of new CRISPR/Cas systems, particularly those aiming at RNA, provides new opportunities for virus-resistant plant development. FnCas9 was found to bind human ssRNA to engineered gRNA and was then driven by its binding and inhibition (Price et al., 2015). Following the above strategy, FnCas9 and CMV and sgRNA-targeted TMV showed reduced accumulation of the virus follow the above strategy. Curiously, FnCas9 required binding RNA, not cleavage activity for viral inhibition (Zhang et al., 2018). CRISPR/Cas13a (C2c2) seems to be a much better way of editing a single-stranded RNA. Tobacco resistance to TuMV has been caused by the acquisition of the viral RNA genome by the CRISPR/Cas13a (C2c2) tar-4 (Aman et al., 2018). Additionally, agronomic parameters in modified rice plants have been improved after RTSV inoculation compared to nonmodified rice plants under glasshouse conditions. The loss of a major host factor for the viral cycle is a form of resistance that should last longer than that of dominant R genes because viruses have a lower selective pressure that prevents their development to inhibit the host defense systems. One potential drawback of the knockout technique is that it may have a detrimental impact on plant vigor, promoting the development of resistance-breaking virus variants, as seen in nature (Abdul-Razzak et al., 2009).

9.7 Bacterial pathogen resistance via CRISPR/Cas

Just a few hundred of the world's bacterial species are responsible for crop destruction, sometimes displaying signs of infection (Schloss and Handelsman, 2004). Bacteria that are phytopathological are hard to avoid to control particularly after outbreaks due to the extremely complex and rapid replication of bacterial pathogens which can also propagate in several ways (Kerr, 2016). Nonetheless, many host plant groups, including some S genes, have been established as involved in these complex processes because of their marked impact on the elucidation of molecular host mechanisms—bacterial pathogenic interactions. Phytopathogenic bacteria can, as they are the cause of the disease in tomatoes, be classified as culturally specific, such as *Clavibakter michiganesis*; polyphages specific, as *Ralstonia solanacearum* causes the disease among many monocotype and dicot species; as well as kingdom cruisers, such as *Dickeya dadantii*, which can affect animals and plants. Pathogenetic bacteria are known as cultivars. This gene disruption made the *Xanthomonas oryzae* effector unable to bind *OsSWEET14* and result in disease resistance in a previous study of the *OsSWEET14* mutagenesis promoter using the TALEN approach (Li et al., 2012). CRISPR/Cas/sgRNA based rice gene modification constructs targeting bacterial susceptibility gene promoter region *OsSWEET14* and *OsSWEET11* (Jiang et al., 2013). Two recent studies have documented the use of CRISPR/Cas9 in the cultivation of CBC-resistant citrus plants. *Xanthomonas citri*. subsp. citri. is a CBC agent and is the most common commercial crop disease. In the first study, Jia et al. (2016), the promoter of Lateral Organ Boundaries 1 (*CsLOB1*), developed cancer-mutants in Duncan grapefruit. Mutated lines showed decrease in normal symptoms of canker 4 days after Xcc inoculation and could not detect more phenotypic changes. Nor has PCR-detected any possible off-mutations in other genes of the LOB family. Citrus canker is a debilitating disease brought about by *X. citri*. ssp. *Terium bac*. citri. *CsLOB1*, a member of the lateral organ boundary (LBD) family of transcription factors, has been previously identified as an Xcc S gene (Hu et al., 2014). The promoter *CsLOB1* generates an EBE that the Xcc effector PthA4 detects; which stimulates the *CsLOB1* expression to accelerate cancer growth. One study targeted the promoter *CsLOB1* EBE, while another research focused on the CsLOB1 coding region via CRISPR/Cas9, both experiments demonstrated resistance to Xcc (Jia et al., 2017; Peng et al., 2017). The mutation of one gene in Arabidopsis, DMR6 (downy mildew resistance 6), resulted in increased levels of salicylic acid and tolerance to many pathogenic plants (Zelimaker et al., 2015). Interestingly, in response to *P. syringae* pv tomato and *Phytophthora capsici* infection, the SlDMR6-1 orthologue tomato is also highly regulated and the null SlDMR6-1 mutants generated by the genome editing showed resistance to this pathogen (de Toledo et al., 2016). These findings together indicate that removal of DMR6 can be a promising strategy for conferring plant resistance to widespread diseases.

9.8 Fungal pathogen resistance via CRISPR/Cas

Fungal infections, which lead to substantial reduction of production and quality of crops lead to major global economic losses, are main factors that are responsible for the most serious crop diseases. They may invade new hosts with ease, divide and produce gene-mediated R resistance fungicide resistance with different lifestyles and a high level of genomic versatility (Doehlemann et al., 2017). Giraud et al. (2010) reported that fungi cause about 30% of new diseases, requiring new ways of improving their control. Thus a biological control strategy based on the competitiveness of spatial and nutrient products and/or the promotion by attenuate or avirulent pathogens of the plant defense against virulent pathogens may be considered a viable alternative (Ghorbanpour et al., 2018). This challenge is being tackled by gene editing of the host S genes. Powdery mildew has an impact on many plants worldwide. The best, economic, and environmentally sustainable method for managing powdery mildew is the breeding resistant cultivar. The main method of growing resistant cultivars is the introduction to elite varieties of the resistance hybridization (via R genes). Since most of these resistance genes are race specific, their resistance is gradually weakening as new breeds of pathogen emerge (Jorgensen, 1992). Therefore a broad range and long-lasting resistance of wheat varieties is highly desirable. Barley Mlo mutants (Lyngkjaer et al., 2000) have been discovered to be a significant achievement for broad-spectrum and long longevity against powdery mildew. *Mlo* orthologists have been found to be evolutionarily preserved in dicots and monocots following the *Mlo* Barley Gene cloning in 1997. Since allohexaploid bread wheat has three barley *Mlo* orthologists (TaMlo-A1, B1, and D1). With available knowledge of plant–microbe interaction, several strategies to improve the resistance of fungi in plant species have been established. The approach has the advantage that once inserted into the organism; the target gene can be modified by altering the spacer sequence of sgRNA and use different sgRNA to silence more than one gene (Hsu et al., 2014). However, the more important advantage is that the use of transient plasmids, which can replicate only under antibiotic pressure permits marker-free removals (Zhang et al., 2016; Wang et al., 2018). Using CRISPR/Cas also offers a timesaving approach to genomic functional studies and to identify better new fungal genotypes that can be used as potential competitors in the field of plant pathogens and/or in plant defenses. One possible scenario for applying silenced mutants to CRISPR/Cas9 could be Fusarium head blight (FHB), among the most harmful cereal-grain crop diseases caused worldwide by various *Fusarium* spp., with *Fusarium graminearum* and *Fusarium culmorum* being the most common agents and aggressive. The reduction in grain quality in FHB is mainly due to trichothecens accumulation coded by the fungal test genes cluster, which are extremely poisonous to humans and animals, while yield loss is due to the sterility of contaminated flowers. Wang et al. (2014) have used TALEN and CRISPR to modify these *Mlo* wheat genes and find resistance to *Blumeria graminis f.* spp. of the modified plants. In the development of Taedr1 wheat plants all three homologs of the EDR1 wheat were targeted simultaneously, the CRISPR/Cas9 system was used. As mutant plants in Arabidopsis edr1 have bacterial and oomycetes resistance, it can be fairly hypothesized that Taedr1 plants can also resist other wheat pathogens and are not resistant to Bgt without mildew deaths (Zhang et al., 2017). *Magnaporth oryzae* rice blast is one of the world's most destructive rice diseases (Dean et al., 2012). In rice adaptation to several biotic and abiotic strains, ethylene reaction factors (ERFs) of the superfamily APETELA2/ERF (AP2/ERF) are a main factor (Mizoi et al., 2012). Not only ABA and salt, but also *M. oryzae* are used to induce the expression *OserF922*. *OsERF922* (RNA) contributes to an increase in *M. oryzae* resistance, indicating that *OsERF822* is a negative rice blast resistance regulator (Liu et al., 2012). *OsERF922* has been designed to be modified using CRISPR/Cas9 knockout mutants from *OsERF922* (Wang et al., 2016). *OsRDR6* contributes to the increase in *M. oryzae, X. oryzae*, RNMV, and CMV resistance, indicating that OsRDR6 is a negative regulator of rice blast resistance (Wagh et al., 2016a,b). Such nonmutations displayed increased rice resistance without affecting other major agricultural characteristics. One example of positive CRISPR/Cas9 grapevine defense is the WRKY52 transcription factor targeted by four gRNAs (Wang et al., 2018). About 21% had biallelic mutations and were more resistant to *Botrytis cinerea* than monoallelic mutants. The CRISPR/Cas9 strategy for woody plantations showed no major changes to phenotypes between the wild-type and the biallelic mutant crops. Another method to speed up genome editing on tree crops of slow production is the use of transient leaf and diseases monitoring as demonstrated in *Theobroma cacao*.

The mechanisms by which these fungi may antagonize pathophytic funguses are another strategy, for example, the release of a broad range of degrading enzymes from the cell wall and secondary metabolites like antibiotics (Khalid, 2017). Genetic metabolic pathways which are responsible for the biosynthetic way of the secreted proteins and secondary compounds can lead to increased biocontrol activities of new fungal strains. The silent of the Ace1 gene contributes to an increase in the production of antibiotics and other secondary metabolites, leading to a substantial increase in the ability to act as a biocontrol agent for *Fusarium oxysporum*, as opposed to *Rhizoctonia solani* (Fang and Chen, 2018). Under this strategy, unknown fungal cluster activation can be inducted by the use of CRISPR/Cas9 and new secondary metabolites that may interfere with plants or phytopathogens can be identified. It may mean the release of new interesting biocontrol strains on the planet, stopping transgenes from being introduced into the ecosystem. The results show the clear and advantaged use of the CRISPR/Cas9 method to boost crop resistance to fungal diseases.

9.9 Future potentials and challenges of CRISPR/Cas technology

CRIPSR/Cas is playing an important role in advanced crop breeding and other functional genomic studies. Genetic improvement is helpful to minimize chemicals use for crop protection offers a promising alternative without a direct effect on living things and the environment (Pohare et al., 2020). Efficiency of genome editing is more due to (1) sufficient advancement of information of the molecular mechanisms that enable gene need to be edited, (2) the alteration of one single gene can result in disease resistance, as it can be preserved by inactivation of susceptibility genes a theoretically less challenging issue. This is analogous to modifying metabolic pathways, which can also have an all or nothing effect when altering a single gene, but distinct from abiotic stress tolerance, in which multiple genes involved. Genome editing prevents genetic load that potentially shows negative impacts on the agricultural output, but it also enables multiple resistances to be generated at the same time in a single generation. Cas-based genome editing also plays crucial role in human health for somatic as well germline cell editing for gene therapy. The variation of CRISPR/Cas9 that use nickase Cas9 helps to do base editing precisely which has more importance in basic research. In case of agriculture, this system is useful to carry out crop breeding for various stresses resistance and to improve nutritional quality of food crops where it introduces traits in single generation without disrupting a desired genetic background. CRISPR/Cas can be used effectively in all living organism for basic and applied studies such as full knockout gene regulation, microRNA knockdown screening, conditional allele creation, study of spatial and temporal gene expression, selection of activation/repression effector domains, epigenome editing. Due to their numerous attractive features such as simplicity, efficiency, high specificity, and multiplexing convenience, the CRISPR/Cas revolutionizes the way for crop breeding. Beside all benefits, CRISPR/Cas has some challenges. First one is to prove that the commitments of edited crop made in controlled condition can be kept in natural conditions through proof of concept thus for proper characterization of the agricultural fitness of edited crops, field tests are also required. Most of published concept include laboratory varieties that can easily be regenerated after editing have been implemented, although these also have limited agricultural values. Many CRISPR/Cas-inactivated genes are likely to play a different part in plant physiology than targeting the pathogen for disease resistance. In order to convince a society which is often hostile to this technology, this challenge must be taken seriously. Another challenge is to address today's technological limitations on selective mutagenesis and actual genome alteration in crop plants. However, true GE by HDR is still hindered by very low plant performance, even though in many of the animal species it has become routine recently. The HDR DNA template should be inserted along with the effector nuclease into the cell, which allows a custom designed sequence to be inserted into the genome. Genome editing considered as safe technology due to its precise nature for crop improvement but still it has some concerns for biosafety because of off-target effects and to minimize this effect need to improve efficiency and specificity of this system. In case of agriculture, this system is useful to carry out crop breeding for various biotic and abiotic stress resistance as well as to improve nutritional quality of food crops where it introduces traits in single generation without disrupting a desired genetic background. Additionally, it plays an important role in the development of new biopesticides (Pohare et al., 2020), gaining insights into the plant–microbe interactions (Shelake et al., 2019a) and in developing modern transgene-free mutagenesis tools (Shelake et al., 2019b). Though CRISPR/Cas genome editing is growing with speed, it needs to use properly in such a way that it will be safe and beneficial to society as well as environment. CRISPR/Cas genome editing is likely to face similar problem like GM crops depending on how regulators view this technology, and how they provide it with equal regulatory consideration. Instead, its safety CRISPR/Cas system is under regulation and product from it deregulated to few countries.

Acknowledgments

The authors are grateful to Rahul Mahadev Shelake, Division of Applied Life Science (BK21 Plus Program), Plant Molecular Biology and Biotechnology Research Center, Gyeongsang National University, Jinju 52828, Korea for his critical comments and English corrections. Authors are also thankful to Pallavi Kisan Patil, Research Associate, University of Sheffield, United Kingdom for her critical English corrections.

References

Abdul-Razzak, A., Guiraud, T., Peypelut, M., Walter, J., Houvenaghel, M.C., Candresse, T., et al., 2009. Involvement of the cylindrical inclusion (CI) protein in the overcoming of an eIF4E-mediated resistance against Lettuce mosaic potyvirus. Mol. Plant. Pathol. 10, 109–113.

Ali, Z., Abulfaraj, A., Idris, A., Ali, S., Tashkandi, M., Mahfouz, M.M., 2015. CRISPR/Cas9-mediated viral interference in plants. Genome Biol. 16, 238.

Ali, Z., Ali, S., Tashkandi, M., Shan, S., Zaidi, A., Mahfouz, M.M., 2016. CRISPR/Cas9-mediated immunity to geminiviruses: differential interference and evasion. Sci. Rep. 6, 26912.

Ali, Z., Shami, A., Sedeek, K., Kamel, R., Alhabsi, A., Tehseen, M., et al., 2020. Fusion of the Cas9 endonuclease and the VirD2 relaxase facilitates homology-directed repair for precise genome engineering in rice. Commun. Biol. 3, 44.

Aman, R., Ali, Z., Butt, H., Mahas, A., Aljedaani, F., Khan, M.Z., et al., 2018. RNA virus interference via CRISPR/Cas13a system in plants. Genome Biol. 19, 1.

Arora, L., Narula, A., 2017. Gene editing and crop improvement using CRISPR-Cas9 system. Front. Plant. Sci. 8, 1932.

Baltes, N.J., Hummel, A.W., Konecna, E., Cegan, R., Bruns, A.N., Bisaro, D.M., et al., 2015. Conferring resistance to geminiviruses with the CRISPR–Cas prokaryotic immune system. Nat. Plants 1, 15145.

Barrangou, R., Doudna, J.A., 2016. Applications of CRISPR technologies in research and beyond. Nat. Biotechnol. 34, 933–941.

Bastet, A., Zafirov, D., Giovinazzo, N., Guyon-Debast, A., Nogué, F., Robaglia, C., et al., 2019. Mimicking natural polymorphism in eIF4E by CRISPR-Cas9 base editing is associated with resistance to potyviruses. Plant. Biotechnol. J. 17, 1736–1750.

Bhaya, D., Davison, M., Barrangou, R., 2011. CRISPR/Cas systems in bacteria and archaea: Versatile small RNAs for adaptive defence and regulation. Annu. Rev. Genet. 45, 273–297.

Bikard, D., Jiang, W., Samai, P., Hochschild, A., Zhang, F., Marraffini, L.A., 2013. Programmable repression and activation of bacterial gene expression using an engineered CRISPR-Cas system. Nucleic Acids Res. 41 (15), 7429–7437.

Brouns, S.J., Jore, M.M., Lundgren, M., Westra, E.R., Slijkhuis, R.J., Snijders, A.P., et al., 2008. Small CRISPR RNAs guide antiviral defensedefence in prokaryotes. Science 321, 960–964.

Chandrasekaran, J., Brumin, M., Wolf, D., Leibman, D., Klap, C., Pearlsman, M., et al., 2016. Development of broad virus resistance in non-transgenic cucumber using CRISPR/Cas9 technology. Mol. Plant Pathol. 17, 1140–1153.

Chen, H., Choi, J., Bailey, S., 2014. Cut site selection by the two-nuclease domains of the Cas9 RNA-guided endonuclease. J. Biol. Chem. 289 (19), 13284–13294.

Chen, B., Gilbert, L.A., Cimini, B.A., Schnitzbauer, J., Zhang, W., Li, G.W., et al., 2013. Dynamic imaging of genomic loci in living human cells by an optimized CRISPR/Cas system. Cell 155 (7), 1479–1491.

Chen, H., Ino, M., Shimono, M., Wagh, S.G., Kobayashi, K., Yaeno, T., et al., 2019a. A single amino acid substitution in the intervening region of 129K protein of cucumber green mottle mosaic virus resulted in attenuated symptoms. Phytopathology 110, 146–152.

Chen, J.S., Ma, E., Harrington, L.B., Costa, M.D., Tian, X., Palefsky, J.M., et al., 2018. CRISPR-Cas12a target binding unleashes indiscriminate singlestranded DNase activity. Science 360 (6387), 436–439.

Chen, K., Wang, Y., Zhang, R., Zhang, H., Gao, C., 2019b. CRISPR/Cas genome editing and precision plant breeding in agriculture. Annu. Rev. Plant. Biol. 70, 667–697.

Cheng, X., Li, F., Cai, J., Chen, W., Zhao, N., Sun, Y., et al., 2015. Artificial TALE as a convenient protein platform for engineering broad-spectrum resistance to begomoviruses. Virus 7, 4772–4782.

Cole, M.B., Augustin, M.A., Robertson, M.J., Manners, J.M., 2018. The science of food security. NPJ Sci. Food. 2 (1).

Cong, L., Ran, F.A., Cox, D., Lin, S., Barretto, R., Habib, N., et al., 2013. Multiplex genome engineering using CRISPR/Cas systems. Science 339, 819–823.

Crawley, A.B., Henriksen, E.D., Stout, E., Brandt, K., Barrangou, R., 2018. Characterizing the activity of abundant, diverse and active CRISPR-Cas systems in *lactobacilli*. Sci. Rep. 8, 11544.

Cristea, S., Freyvert, Y., Santiago, Y., Holmes, M.C., Urnov, F.D., Gregory, P.D., et al., 2013. *In vivo* cleavage of transgene donors promotes nuclease mediated targeted integration. Biotechnol. Bioeng. 110 (3), 871–880.

Dahan-Meir, T., Filler-Hayut, S., Melamed-Bessudo, C., Bocobza, S., Czosnek, H., Aharoni, A., et al., 2018. Efficient in planta gene targeting in tomato using geminiviral replicons and the CRISPR/Cas9 system. Plant. J. 95 (1), 5–16.

Dean, R., Van Kan, J.A., Pretorius, Z.A., Hammond-Kosack, K.E., Di Pietro, A., Spanu, P.D., et al., 2012. The top10 fungal pathogens in molecular plant pathology. Mol. Plant. Pathol. 13, 414–430.

Deltcheva, E., Chylinski, K., Sharma, C.M., Gonzales, K., Chao, Y., Pirzada, Z.A., et al., 2011. CRISPR RNA maturation by trans-encoded small RNA and host factor RNase III. Nature 471 (7340), 602–607.

de Toledo, D.P., Brail, Q., Dahlbeck, D., Staskawicz, B.J., 2016. CRISPR-Cas9 mediated mutagenesis of a DMR6 ortholog in tomato confers broad-spectrum disease resistance. BioRXIV 064824.

Denver, T.E., Kinzy, T.G., Pavitt, G.D., 2016. Mechanism and Regulation of Protein Synthesis in *Saccharomyces cerevisiae*. Genetics 203 (1), 65–107.

Deveau, H., Barrangou, R., Garneau, J.E., Labonté, J., Fremaux, C., Boyaval, P., et al., 2008. Phage response to CRISPR-encoded resistance in *Streptococcus thermophilus*. J. Bacteriol. 190, 1390–1400.

Deveau, H., Garneau, J.E., Moineau, S., 2010. CRISPR/Cas system and its role in phage-bacteria interactions. Annu. Rev. Microbiol. 64, 475–493.

Ding, S.W., Voinnet, O., 2007. Antiviral immunity directed bysmall RNAs. Cell 130, 413–426.

Doehlemann, G., Okmen, B., Zhu, W., Sharon, A., 2017. Plant pathogenic fungi. In: Heitman, J., Howlett, B., Crous, P., Stukenbrock, E., James, T., Gow, N. (Eds.), The Fungal Kingdom. ASM press, Washington, DC, pp. 703–726.

Doench, J.G., Fusi, N., Sullender, M., Hegde, M., Vaimberg, E.W., Donovan, K.F., et al., 2016. Optimized sgRNA design to maximize activity and minimize off-target effects of CRISPR-Cas9. Nat. Biotechnol. 34, 184–191.

Doudna, J.A., Charpentier, E., 2014. The new frontier of genome engineering with CRISPR/Cas9. Science 346 (6213), 1258096.

Fan, D., Liu, T., Li, C., Jiao, B., Li, S., Hou, Y., et al., 2015. Efficient CRISPR/Cas9-mediated targeted mutagenesis in populus in the first generation. Sci. Rep. 5, 12217.

Fang, C., Chen, X., 2018. Potential biocontrol efficacy of *Trichoderma atroviride* with cellulase expression regulator ace1 gene knockout. Biotechnol. 8, 302.

Fang, Y., Tyler, B.M., 2015. Efficient disruption and replacement of an effector gene in the oomycetePhytophthora sojaeusing CRISPR/Cas9. Mol. Plant Pathol. 17 (1), 127–139.

Fister, A.S., Landherr, L., Maximova, S.N., Guiltinan, M.J., 2018. Transient expression of CRISPR/Cas9 machinery targeting *TcNPR3* enhances defense response in *Theobroma cacao*. Front. Plant. Sci. 9, 268.

Garneau, J.E., Dupuis, M.È., Villion, M., Romero, D.A., Barrangou, R., Boyaval, P., et al., 2010. The CRISPR/Cas bacterial immune system cleaves bacteriophage and plasmid DNA. Nature 468, 67–71.

Gasiunas, G., Barrangou, R., Horvath, P., Siksnys, V., 2012. Cas9-crRNA ribonucleoprotein complex mediates specific DNA cleavage for adaptive immunity in bacteria. PNAS 109, E2579–E2586.

Georges, F., Ray, H., 2017. Genome editing of crops: A renewed opportunity for food security. GM Crop Food 8 (1), 1–12.

Ghorbanpour, M., Omidvari, M., Abbaszadeh-Dahaji, P., Omidvar, R., Kariman, K., 2018. Mechanisms underlying the protective effects of beneficial fungi against plant diseases. Biol. Cont. 117, 147–157.

Gilbert, L.A., Larson, M.H., Morsut, L., Liu, Z., Brar, G.A., Torres, S.E., et al., 2013. CRISPR-mediated modular RNA-guided regulation of transcription in eukaryotes. Cell 154 (2), 442–451.

Gilbertson, R.L., Batuman, O., Webster, C.G., Adkins, S., 2015. Role of the insect supervectors *Bemisia tabaci* and *Frankliniella occidentalis* in the emergence and global spread of plant viruses. Annu. Rev. Virol. 2, 67–93.

Giraud, T., Gladieux, P., Gavrilets, S., 2010. Linking the emergence of fungal plant diseases with ecological speciation. Trends Ecol. Evol. 25, 387–395.

Gomez, M.A., Lin, Z.D., Moll, T., Chauhan, R.D., Hayden, L., Renninger, K., et al., 2019. Simultaneous CRISPR/Cas9- mediated editing of cassava eIF4E isoforms nCBP-1 and nCBP-2 reduces cassava brown streak disease symptom severity and incidence. Plant. Biotechnol. J. 17 (2), 421–434.

Haque, E., Taniguchi, H., Hassan, M.M., Bhowmik, P., Karim, M.R., Śmiech, M., et al., 2018. Application of CRISPR/Cas9 genome editing technology for the improvement of crops cultivated in tropical climates: recent progress, prospects, and challenges. Front. Plant. Sci. 9, 617.

Hsu, P.D., Lander, E.S., Zhang, F., 2014. Development and applications of CRISPR/Cas9 for genome engineering. Cell 157 (6), 1262–1278.

Hu, Y., Zhang, J., Jia, H., Sosso, D., Li, T., Frommer, W.B., et al., 2014. Lateral organ boundaries 1 is a disease susceptibility gene for citrus bacterial canker disease. PNAS 111, E521–E529.

Huang, Y., Guo, Y., Liu, Y., Zhang, F., Wang, Z., Wang, H., et al., 2018. 9-cis-Epoxycarotenoid dioxygenase 3 regulates plant growth and enhances multi-abiotic stress tolerance in rice. Front. Plant. Sci. 9, 162.

Idnurm, A., Urquhart, A.S., Vummadi, D.R., Chang, S., Van DeWouw, A.P., López-Ruiz, F.J., 2017. Spontaneous and CRISPR/Cas9-induced mutation of the osmosensor histidine kinase of the canola pathogen *Leptosphaeria maculans*. Fungal Biol. Biotechnol. 4, 12.

Iqbal, Z., Sattar, M.N., Shafiq, M., 2016. CRISPR/Cas9: a tool to circumscribe cotton leaf curl sisease. Front. Plant. Sci. 7, 475.

Ishino, Y., Shinagawa, H., Makino, K., Amemura, M., Nakata, A., 1987. Nucleotide sequence of the iap gene, responsible for alkaline phosphatase isozyme conversion in *Escherichia coli* and identification of the gene product. J. Bacteriol. 169, 5429–5433.

Islam, T., 2019. CRISPR-Cas technology in modifying food crops. CAB. Rev. 14 (050).

Jacobs, T.B., LaFayette, P.R., Schmitz, R.J., Parrott, W.A., 2015. Targeted genome modifications in soybean with CRISPR/Cas9. BMC Biotechnol. 15, 16.

Ji, X., Zhang, H., Zhang, Y., Wang, Y., Gao, C., 2015. Establishing a CRISPR–Cas-like immune system conferring DNA virus resistance in plants. Nat. Plants. 1, 15144.

Jia, H., Orbovic, V., Jones, J.B., Wang, N., 2016. Modification of the PthA4 effector binding elements in Type I CsLOB1 promoter using Cas9/sgRNA to produce transgenic Duncan grapefruit alleviating XccOpthA4: dCsLOB1.3 infection. Plant Biotechnol. J. 14, 1291–1301.

Jia, H., Zhang, Y., Orbovi'c, V., Xu, J., White, F.F., Jones, J.B., et al., 2017. Genome editing of the disease susceptibility gene CsLOB1 in citrus confers resistance to citrus canker. Plant Biotechnol. J. 15, 817–823.

Jiang, L., Yu, X., Qi, X., Yu, Q., Deng, S., Bai, B., et al., 2013. Multigene engineering of starch biosynthesis in maize endosperm increases the total starch content and the proportion of amylose. Transgenic Res. 22, 1133–1142.

Jinek, M., Chylinski, K., Fonfara, I., Hauer, M., Doudna, J.A., Charpentier, E.A., 2012. A programmable dual-RNA-guided DNA endonuclease in adaptive bacterial immunity. Science. 337, 816–821.

Jinek, M., East, A., Cheng, A., Lin, S., Ma, E., Doudna, J., 2013. RNA-programmed genome editing in human cells. eLife 2, e00471.

Jorgensen, I.H., 1992. Discovery, characterization and exploitation of Mlo powdery mildew resistance in barley. Euphytica 63, 141–152.

Kamoun, S., Talbot, N.J., Islam, M.T., 2019. Plant health emergencies demand open science: tackling a cereal killer on the run. PLoS Biology. 17 (6), e3000302.

Kerr, A., 2016. Biological control of Crown Gall. Australas. Plant. Pathol. 45, 15–18.

Khalid, S.A., 2017. Trichoderma as biological control weapon against soil borne plant pathogens. Afr. J. Biotech. 16, 2299–2306.

Khoury, C.K., Bjorkman, A.D., Dempewolf, H., Villegasa, J.R., Guarinof, L., Jarvisa, A., et al., 2014. Increasing homogeneity in global food supplies and the implications for food security. PNAS 111 (11), 4001–4006.

Kim, D., Alptekin, B., Budak, H., 2018. CRISPR/Cas9 genome editing in wheat. Funct. Integr. Genomics 18 (1), 31–41.

Kis, A., Hamar, É., Tholt, G., Bán, R., Havelda, Z., 2019. Creating highly efficient resistance against *wheat dwarf virus* in barley by employing CRISPR/Cas9 system. Plant. Biotechnol. J. 17 (6), 1004–1006.

Kleinstiver, B.P., Pattanayak, V., Prew, M.S., Tsai, S.Q., Nguyen, N.T., Zheng, Z., et al., 2016. High-fidelityHigh fidelity CRISPR–Cas9 nucleases with no detectable genome-wide off-target effects. Nature 529, 490–495.

Koonin, E.V., Makarova, K.S., Zhang, F., 2017. Diversity classification and evolution of CRISPR/Cas systems. Curr. Opin. Microbiol. 37, 67–78.

Kuang, Y., Li, S., Ren, B., Yan, F., Spetz, C., Li, X., et al., 2020. Base-editing-mediated artificial evolution of OsALS1 *In Planta* to develop novel herbicide-tolerant rice germplasms. Mol. Plant. 13, 565–572.

Kumar, V., Jain, M., 2015. The CRISPR-Cas system for plant genome editing: advances and opportunities. J. Exp. Bot. 66, 47–57.

Langner, T., Kamoun, S., Belhaj, K., 2018. CRISPR crops: plant genome editing toward disease resistance. Annu. Rev. Phytopathol. 56, 479–512.

Larson, M.H., Gilbert, L.A., Wang, X., Lim, W.A., Weissman, J.S., Qi, L.S., 2013. CRISPR interference (CRISPRi) for sequence-specific control of gene expression. Nat. Protoc. 8, 2180–2196.

Lellis, A.D., Kasschau, K.D., Whitham, S.A., Carrington, J.C., 2002. Loss-of-susceptibility mutants of Arabidopsis thaliana reveal an essential role for eIF(iso) 4E during potyvirus infection. Curr. Biol. 12, 1046–1051.

Li, M., Li, X., Zhou, Z., Wu, P., Fang, M., Pan, X., et al., 2016. Reassessment of the four yield-related genes Gn1a, DEP1, GS3, and IPA1 in rice using a CRISPR/Cas9 system. Front. Plant. Sci. 7, 377.

Li, T., Liu, B., Spalding, M.H., Weeks, D.P., Yang, B., 2012. High-efficiency TALEN-based gene editing produces disease-resistant rice. Nat. Biotechnol. 30, 390–392.

Li, T., Yang, X., Yu, Y., Si, X., Zhai, X., Zhang, H., et al., 2018. Domestication of wild tomato is accelerated by genome editing. Nat. Biotechnol. 36, 1160–1163.

Liang, Y., Han, Y., Wang, C., Jiang, C., Xu, J.-R., 2018. Targeted deletion of the USTA and UvSLT2 Genes effiE_ciently in *Ustilaginoidea virens* with the CRISPR-Cas9 System. Front. Plant. Sci. 9, 1–11.

Liang, Z., Zhang, K., Chen, K., Gao, C., 2014. Targeted mutagenesis in *Zea mays* using TALENs and the CRISPR/Cas system. J. Genet. Genomics 41, 63–68.

Liu, D., Chen, X., Liu, J., Ye, J., Guo, Z., 2012. The rice ERF transcription factor OsERF922 negatively regulates resistance to *Magnaporthe oryzae* and salt tolerance. J. Exp. Bot. 63, 3899–3911.

Lyngkjaer, M.F., Newton, A.C., Atzema, J.L., Baker, S.J., 2000. The marley mlo-gene: an important powdery mildew resistance source. Agronomie 20, 745–756.

Ma, X., Mau, M., Sharbel, T.F., 2018. Genome editing for global food security. Trends Biotechnol. 36, 2.

Makarova, K.S., Grishin, N.V., Shabalina, S.A., Wolf, Y.I., Koonin, E.V., 2006. A putative RNA-interference-based immune system in prokaryotes: computational analysis of the predicted enzymatic machinery, functional analogies with eukaryotic RNAi, and hypothetical mechanisms of action. Biol. Direct 1 (7), 1–26.

Malina, A., Mills, J.R., Cencic, R., Yan, Y., Fraser, J., Schippers, L.M., et al., 2013. Repurposing CRISPR/Cas9 for in situ functional assays. Genes Dev. 27, 2602–2614.

Malnoy, M., Viola, R., Jung, M.-H., Koo, O.-J., Kim, S., Kim, J.-S., et al., 2016. DNA-free genetically edited grapevine and apple protoplast using CRISPR/Cas9 ribonucleoproteins. Front. Plant. Sci. 7, 1473.

Maresca, M., Lin, V.G., Guo, N., Yang, Y., 2013. Obligate ligation-gated recombination (ObLiGaRe): custom-designed nuclease-mediated targeted integration through nonhomologous end joining. Genome Res. 23, 539–546.

Miao, J., Chi, Y., Lin, D., Tyler, B.M., Liu, X.L., 2018. Mutations in ORP1 conferring oxathiapiprolin resistance confirmed by genome editing using CRISPR/Cas9 in *Phytophthora capsici* and *P. sojae*. Phytopathology 108, 1412–1419.

Mizoi, J., Shinozaki, K., Yamaguchi-Shinozaki, K., 2012. AP2/ERF family transcription factors in plant stress responses. Biochim. Biophys. Acta 1819, 86–96.

Mojica, F.J., Rodriguez-Valera, F., 2016. The discovery of CRISPR in archaea and bacteria. FEBS J. 283, 3162–3169.

Mushtaq, M., Bhat, J.A., Mir, Z.A., Sakina, A., Ali, S., Singh, A.K., et al., 2018. CRISPR/Cas approach: a new way of looking at plant–abiotic interactions. J. Plant. Physiol. 224 (225), 156–162.

Nekrasov, V., Wang, C., Win, J., Lanz, C., Weigel, D., Kamoun, S., 2017. Rapid generation of a transgene-free powdery mildew resistant tomato by genome deletion. Sci. Rep. 7, 482.

Ortigosa, A., Gimenez-Ibanez, S., Leonhardt, N., Solano, R., 2019. Design of a bacterial speck resistant tomato byCRISPR/Cas9-mediated editing of SlJAZ2. Plant. Biotechnol. J. 17, 665–673.

Peng, A., Chen, S., Lei, T., Xu, L., He, Y., Wu, L., et al., 2017. Engineering canker-resistant plants through CRISPR/Cas9-targeted editing of the susceptibility gene CsLOB1 promoter in citrus. Plant Biotechnol. J. 15, 1509–1519.

Pennisi, E., 2013. The CRISPR craze. Science 341, 833–836.

Pohare, M.B., Akita, M., 2017. A rapid and simple, recombination-based cloning method in *Escherichia coli*. Biosci. Biotechnol. Res. Asia 14 (1), 43–52.

Pohare, M.B., Batule, B.S., Bhor, S.A., Shahakar, S.B., Kelatkar, S.K., Varandani, S.P., 2013. Effect of gamma radiations on the morphological characters in in vitro regenerated *Polianthestuberosa*. Indian J. Horticul. 3, 95–97.

Pohare, M.B., Rathod, H.P., Shahakar, S.B., Kelatkar, S.K., Suryawanshi, P.P., 2012. Effects of UV Radiations on Morphological Characters in *In vitro* regenerated *Polianthestuberosa*. Res. J. Agric. Sci. 3, 1307–1308.

Pohare, M.B., Sharma, M., Wagh, S.G., 2019. CRISPR/Cas9 genome editing and its medical potential. In: Kumar, S. (Ed.), Advances in Biotechnology and Biosciences, 3. NavNik Publications, Pusa, India, pp. 69–90.

Pohare, M.B., Wagh, S.G., Udayasuriyan, V., 2020. *Bacillus thuringiensis* as potential biocontrol agent for sustainable agriculture. In: Yadav, A.N. (Ed.), Current Trends in Microbial Biotechnology for Sustainable Agriculture. Springer Nature, Switzerland (In press).

Price, A.A., Sampson, T.R., Ratner, H.K., Grakoui, A., Weiss, D.S., 2015. Cas9-mediated targeting of viral RNA in eukaryotic cells. PNAS. 112, 6164–6169.

Pyott, D.E., Sheehan, E., Molnar, A., 2016. Engineering of CRISPR/Cas9-mediated potyvirus resistance in transgene-free arabidopsis plants. Mol. Plant. Pathol. 4, 1–13.

Qi, L.S., Larson, M.H., Gilbert, L.A., Doudna, J.A., Weissman, J.S., Arkin, A.P., et al., 2013. Repurposing CRISPR as an RNA-guided platform for sequence specific control of gene expression. Cell 152 (5), 1173–1183.

Roossinck, M.J., Martin, D.P., Roumagnac, P., 2015. Plant virus metagenomics: advances in virus discovery. Phytopathology 105, 716–727.

Savary, S., Ficke, A., Aubertot, J.N., Hollier, C., 2012. Crop losses due to diseases and their implications for global food production losses and food security. Food Secur. 4, 519–537.

Schloss, P.D., Handelsman, J., 2004. Status of the microbial census. Microbiol. Mol. Biol. Rev. 68, 686–691.

Schuster, M., Schweizer, G., Reissmann, S., Kahmann, R., 2016. Genome editing in *Ustilago maydis* using the CRISPR-Cas system. Fungal Genet. Biol. 89, 3–9.

Sera, T., 2005. Inhibition of virus DNA replication by artificial zinc finger proteins. J. Virol. 79, 2614–2619.

Shan, Q., Wang, Y., Li, J., Zhang, Y., Chen, K., Liang, Z., et al., 2013. Targeted genome modification of crop plants using a CRISPR-Cas system. Nat. Biotechnol. 31, 686–688.

Shelake, R.M., Pramanik, D., Kim, J.Y., 2019a. Exploration of plant-microbe interactions for sustainable agriculture in CRISPR era. Microorganisms 7 (8), 269.

Shelake, R.M., Pramanik, D., Kim, J.Y., 2019b. Evolution of plant mutagenesis tools: a shifting paradigm from random to targeted genome editing. Plant Biotechnol. Rep. 13, 423–445.

Shi, J., Gao, H., Wang, H., Lafitte, H.R., Archibald, R.L., Yang, M., et al., 2017. ARGOS8 variants generated by CRISPR-Cas9 improve maize grain yield under field drought stress conditions. Plant Biotechnol. J. 15, 207–216.

Soda, N., Verma, L., Giri, J., 2018. CRISPR-Cas9 based plant genome editing: Significance, opportunities and recent advances, Plant Physiol. Bioch. 131, 2–11.

Sun, Y., Zhang, X., Wu, C., He, Y., Ma, Y., Hou, H., et al., 2016. Engineering herbicide-resistant rice plants through CRISPR/Cas9-mediated homologous recombination of *acetolactate synthase*. Mol. Plant 9, 628–631.

Svitashev, S., Young, J.K., Schwartz, C., Gao, H., Falco, S.C., Cigan, A.M., 2015. Targeted mutagenesis, precise gene editing, and site-specific gene insertion in maize using Cas9 and guide RNA. Plant Physiol. 169, 931–945.

Tang, X., Lowder, L.G., Zhang, T., Malzahn, A.A., Zheng, X., Voytas, D.F., et al., 2017. A CRISPR–Cpf1 system for efficient genome editing and transcriptional repression in plants. Nat. Plants 3, 17018.

Thomazella, D.P.D.T., Brail, Q., Dahlbeck, D., Staskawicz, B.J., 2016. CRISPR-Cas9 mediated mutagenesis of a DMR6 ortholog in tomato confers broad-spectrum disease resistance. bioRxiv.

Tripathi, J.N., Ntui, V.O., Ron, M., Muiruri, S.K., Britt, A., Tripathi, L., 2019. CRISPR/Cas9 editing of endogenous banana streak virus in the B genome of Musa spp. overcomes a major challenge in banana breeding. Commun. Biol. 2, 46.

Unniyampurath, U., Pilankatta, R., Krishnan, M., 2016. RNA interference in the age of CRISPR: Will CRISPR interfere with RNAi? Int. J. Mol. Sci. 17, 291.

Wagh, S.G., 2020. Comparative updates on host induced gene silencing and CRISPR/Cas9 utilization for improved disease resistance in crops, research. J. Biotechnol. (In press).

Wagh, S.G., Alam, M.M., Kobayashi, K., Yaeno, T., Yamaoka, N., Toriba, T., et al., 2016a. Analysis of rice RNA-dependent RNA polymerase 6 (OsRDR6) gene in response to viral, bacterial and fungal pathogens. J. Gen. Plant. Pathol. 82, 12–17.

Wagh, S.G., Kobayashi, K., Yaeno, T., Yamaoka, N., Masuta, C., Nishiguchi, M., 2016b. Rice necrosis mosaic virus, a fungal transmitted bymovirus: complete nucleotide sequence of the genomic RNAs and subgrouping of bymoviruses. J. Gen. Plant Pathol. 82, 38–43.

Wagh, S.G., Pohare, M.B., 2019. Current and future prospects of plant breeding with CRISPR/Cas. Curr. J. Appl. Sci. Technol. 38 (3), 1–17.

Wang, Y., Cheng, X., Shan, Q., Zhang, Y., Liu, J., Gao, C., et al., 2014. Simultaneous editing of three homoeoalleles in hexaploid bread wheat confers heritable resistance to powdery mildew. Nat. Biotechnol. 32, 947–952.

Wang, Q., Coleman, J.J., 2019. CRISPR/Cas9-mediated endogenous gene tagging in *Fusarium oxysporum*. Fungal Genet. Biol. 126, 17–24.

Wang, W., Pan, Q., He, F., Akhunova, A., Chao, S., Trick, H., et al., 2018. Transgenerational CRISPR-Cas9 activity facilitates multiplex gene editing in allopolyploid wheat. CRISPR J. 1, 65–74.

Wang, F., Wang, C., Liu, P., Lei, C., Hao, W., Gao, Y., 2016. Enhanced rice blast resistance by CRISPR/Cas9-targeted mutagenesis of the ERF transcription factor gene OsERF922. PLoS One 11, e0154027.

Wenderoth, M., Pinecker, C., Voß, B., Fischer, R., 2017. Establishment of CRISPR/Cas9 in *Alternaria alternata*. Fungal Genet. Biol. 101, 55–60.

Xie, K., Yang, Y., 2013. RNA-guided genome editing in plants using a CRISPR-Cas system. Mol. Plant 6, 1975–1983.

Yan, F., Wang, W., Zhang, J., 2019. CRISPR-Cas12 and Cas13: the lesser-known siblings of CRISPR-Cas9. Cell Biol. Toxicol. 35, 489–492.

Zaidi, S.S.E.A., Mahfouz, M.M., Mansoor, S., 2017. CRISPR-Cpf1: a new tool for plant genome editing. Trends Plant Sci. 22, 550–553.

Zaidi, S.S., Tashkandi, M., Mansoor, S., Mahfouz, M.M., 2016. Engineering plant immunity: using CRISPR/Cas9 to generate virus resistance. Front. Plant. Sci. 7, 1673.

Zelimaker, T., Ludwig, N.R., Elberse, J., Seidl, M.F., Berke, L., Van Doorn, A., et al., 2015. DOWNY MILDEW RESISTANT 6 and DMR 6-LIKE OXYGENASE 1 are partially redundant but distinct suppressors of immunity in Arabidopsis. Plant J. 81, 210–222.

Zetsche, B., Gootenberg, J.S., Abudayyeh, O.O., Slaymaker, I.M., Makarova, K.S., Essletzbichler, P., et al., 2015. Cpf1 is a single RNA-guided endonuclease of a class 2 CRISPR-Cas system. Cell 163, 759–771.

Zhang, Y., Bai, Y., Wu, G., Zou, S., Chen, Y., Gao, C., et al., 2017. Simultaneous modification of three homologs of TaEDR1 by genome editing enhances powdery mildew resistance in wheat. Plant J. 91, 714–724.

Zhang, Y., Liang, Z., Zong, Y., Wang, Y., Liu, J., Chen, K., et al., 2016. Efficient and transgene-free genome editing in wheat through transient expression of CRISPR/Cas9 DNA or RNA. Nat. Commun. 7, 12617.

Zhang, R., Liu, J., Chai, Z., Chen, S., Bai, Y., Zong, Y., et al., 2019. Generation of herbicide tolerance traits and a new selectable marker in wheat using base editing. Nat. Plants 5, 480–485.

Zhang, Y.M., Zhang, Y., Xie, K., 2020. Evaluation of CRISPR/Cas12a-based DNA detection for fast pathogen diagnosis and GMO test in rice. Mol. Breed. 40 (11).

Zhang, T., Zheng, Q., Yi, X., An, H., Zhao, Y., Ma, S., et al., 2018. Establishing RNA virus resistance in plants by harnessing CRISPR immune system. Plant Biotechnol. J. 16, 1415–1423.

Zhou, H., Liu, B., Weeks, D.P., Spalding, M.H., Yang, B., 2014. Large chromosomal deletions and heritable small genetic changes induced by CRISPR/Cas9 in rice. Nucleic Acids Res. 42, 10903–10914.

CHAPTER

10

An overview of nanotechnology in plant disease management, food safety, and sustainable agriculture

Andleeb Zehra[1], *Ashutosh Rai*[2], *Sushil Kumar Singh*[3], *Mohd Aamir*[1,2], *Waqar Akhtar Ansari*[2] *and Ram S. Upadhyay*[1]

[1]Laboratory of Mycopathology and Microbial Technology, Centre of Advanced Study in Botany, Institute of Science, Banaras Hindu University, Varanasi, India [2]Division of Crop Improvement and Biotechnology, Indian Institute of Vegetable Research, Varanasi, India [3]Department of Agricultural Biotechnology, Assam Agricultural University, Jorhat, India

10.1 Introduction

World population is growing rapidly and has been projected to reach 10 billion by 2050 (Conforti, 2017) and would require to increase the food production by 50%. In contrast, the agricultural resources are rapidly declining due to overexploitation. At present, a staggering amount of resources are used to produce the required food through agricultural system. One data suggest that global annual production exceeds 3 billion tonnes and requires 2.7 trillion cubic meters of water, 187 million tonnes of fertilizer, 2 quadrillion British thermal units (BTU) of energy, and approximately 4 million tonnes of pesticides (Kah et al., 2019). The ongoing changes in climatic conditions due to increased anthropogenic activities are likely to disrupt the food production cycles by raising the average daily temperature or extension of drought events, more specifically occurring in agricultural sensitive zones, and therefore, open the door for various abiotic as well as biotic stresses. The actual challenges faced by farmers, plant pathologist, and other agricultural physiologist are more formidable. Moreover, the current trends for agricultural production are insufficient to support the future growth. According to reports, at present, about 815 million people are suffering from undernourishment due to nutritional deficiency and unavailability of nutrient-rich foods, and if the agricultural productivity for procurement

Food Security and Plant Disease Management
DOI: https://doi.org/10.1016/B978-0-12-821843-3.00009-X

of food crops follows the similar trend and practices, no doubt in nearby future, we will have an additional 2 billion people in the undernourished category. In this way, we can understand that, to meet the additional demand output from the agricultural sector by sustainable means, we need to focus on other alternatives rather than depending on single sources. The food production through sustainable means requires low dependences on agricultural inputs for optimizing the yield. In this context, nanotechnology has gained momentum during the recent years (Parisi et al., 2015; Kah et al., 2019).

Chemical fertilizers are essential agrochemical inputs required by plants for better growth and productivity. However, most of the applied fertilizers used in agricultural fields are rendered unavailable due to multiple factors including degradation by photolysis, leaching, decomposition, and hydrolysis safe (Elemike et al., 2019; Iqbal, 2019). Further, extensive application of chemical fertilizers has both environmental and ecological concerns and nutrients loss through leaching and volatilization has drastic impact on climate change and environmental pollution (Iqbal, 2019). To achieve agricultural sustainability for crop production, we need to develop other alternative resources for providing nutrients and modifying prevalent nutrient sources. With the advent of nanotechnology, it has now become possible to develop concentrated sources of plant nutrients with higher utilization efficiency, absorption capacity, and higher applicability with minimum losses. These nanofertilizers are composed of thin encapsulating plant nutrients into thin-coated nanomaterials that deliver nutrients in the form of nanosized emulsions. Nanofertilizers components may include metal or nonmetals or their combinations including nanoparticles composed of silica, iron, ZnS/ZnCdSe, ZnO, FeO, TiO_2, Al_2O_3, quantum dots (QDs), nanorods, and quantum shells such as core--shell QDs, InP/ZnS core–shell QDs, and Mn/ZnSe QDs, gold nanorods (Prasad et al., 2017). Further, higher nutrient uptake efficiency is accomplished and supported by stomatal openings, nanopores thus facilitating their deep penetration inside leaves for enhancing the nutrient uptake. These nanofertilizers delivered to plants in a controlled manner for delayed release of active nutrients; thus, it helps not only to increase the efficiency of nutrient uptake but also prevents the eutrophication, water pollution, and greenhouse gas. In this way, higher nutrient use efficiency (NUE) and minimal losses of nutrients through nanofertilizers result in higher productivity. However, development of these nanoformulation requires extensive skill, effective legislation as well as associated with risk management, and therefore, function as limiting factors besides their adoption as plant nutrient sources.

Nanotechnology has an eminent potential to improve the agricultural problem and issues, particularly, reducing the economic losses occurring due to rapidly occurring climatic changes. With the advent of nanotechnology, it has been now possible to enhance the efficiencies of traditionally used agricultural inputs and supplements, in a more sophisticated, reliable, and easily accessible way, for improving food productivity and nutritional security. The application of nanomaterials in agriculture sector such as genetic engineering, plant breeding, genetic transformation, precision farming, and food processing industries has diversified scopes (Batsmanova et al., 2013; Sekhon, 2014; Prasad et al., 2017; Cunningham et al., 2018; Elemike et al., 2019). Further, plant disease management through nanotools could be useful in providing protection to plants against pathogens or disease management either through targeted and highly regulated disposition of active ingredients at required sites as well as monitoring through efficient and rapid action disease diagnostic tools (Sathiyabama, 2019). For example, nanotechnology-derived tools have higher efficiency for targeted and precise delivery to the desired location with minimum essential requirements. In this way, reduces on average cost, decreasing the undesirable wastages of agricultural resources and limiting offsite interactions, and having explicit and increased effects (Prasad et al., 2017). Further, sustainable agriculture using nanomaterials allows for a quick and smarter delivery mode of active ingredients compared to conventionally used agrochemicals and fertilizers, and therefore, providing a better, convenient, and promising management technologies for essential resources (Zhao et al., 2018). With the help of nanofertilizers and nanopesticides, one can solve some pertinent objectives such as minimizing the nutrient loss through optimized nutrient management, reduction of pesticides, disease control, and management of various abiotic and biotic stresses to increase the shelf life and quality management of food products, and therefore, avoiding the issue pertaining to food security and scarcity. The use of nanotechnology and many nano-enabled tools such as nanoparticles, nanocapsules, nanocomposites, and nano-enabled biosensors could be deployed in better mode in different sectors such as increasing mineral nutrient absorption by crop plants, management of plant diseases as well as highly regulated and accurate delivery of plants active ingredients. Nanotechnology although has multidisciplinary applications in various fields including agricultural biotechnology, food industries, and disease diagnosis, and today, we have a broad range of nano-enabled tools for promotion of plant growth, disease management, and supply of quality enriched foods. However, practical application of such tools, particularly, in agricultural sector and food supply is associated with major issues, and therefore, we must answer and address these issues before their exploitation and application at commercial level.

In this chapter we have provided an overview of potential nanotechnological tools and techniques used in plant disease management, food safety, and sustainable agriculture to highlight the importance of nanotechnology in providing better services to mankind over other conventionally used methodologies.

10.2 Implications of nanotechnological tools in sustainable agriculture

Agriculture is a basic process of cultivating plants and livestock required for the survival of human being. It significantly affects our natural resources and environment. Agricultural production through sustainable means could provide long-term food supply for future growing population without harming our natural environment and depletion of natural resources. In this way, agricultural sustainability provides sufficient foods with clean and green environment.

Agricultural sustainability through implementation of nanotechnology could be useful in increasing the agricultural productivity through balanced nutrient uptake, supplying better quality of water, and timely bounded targeted application of crop-specific pesticides for improvement of soil quality and health (Gruère, 2012; Mukhopadhyay, 2014; Prasad et al., 2017). Nanotechnology has diverse application in sustainable agriculture such as control monitoring of plant growth, plant protection against various diseases, delivery of targeted chemicals, and improve quality of food crops. It has been found that application of nanofertilizers in place of chemical fertilizers has added advantages with respect to soil health and productivity. Nanofertilizers improve the soil texture and hydrophobicity without disturbing its microbiota in comparison to agrochemicals. The novel agrochemical formulation and delivery mechanism of nanofertilizer and nanopesticide ameliorated the crop productivity and sustainability in agriculture. It also helps to reduce the application of toxic pesticide and protect agriculture food crops for feeding humans and animals. The variability in toxicity of nanoparticle depends on their shape, size, structure, aggregation properties, and composition (Ion et al., 2010). It has been found that nano-enabled chemical fertilizers regulate the expression of key genes involved in plant growth promotion and increase productivity. Nanotechnology regulates the use of fertilizer by monitoring the release of specific nutrient based on the requirement of crop in the form of soluble fertilizer and nanotubes. The use of nanomaterial in fertilizers and pesticides enhances their accessible surface area, and therefore, improves the applicability and performance of used/bound fertilizers. Nano-based fertilizer or pesticide offers significant advantages over increasing the plant growth as well as the population density of microbes, and therefore, help in the establishment of a proper soil microenvironment. The various fields where nanotechnology finds applications have been summarized (Fig. 10.1). The properties of nanomaterial differ in larger scale at atomic and nanoscale level utilized by many

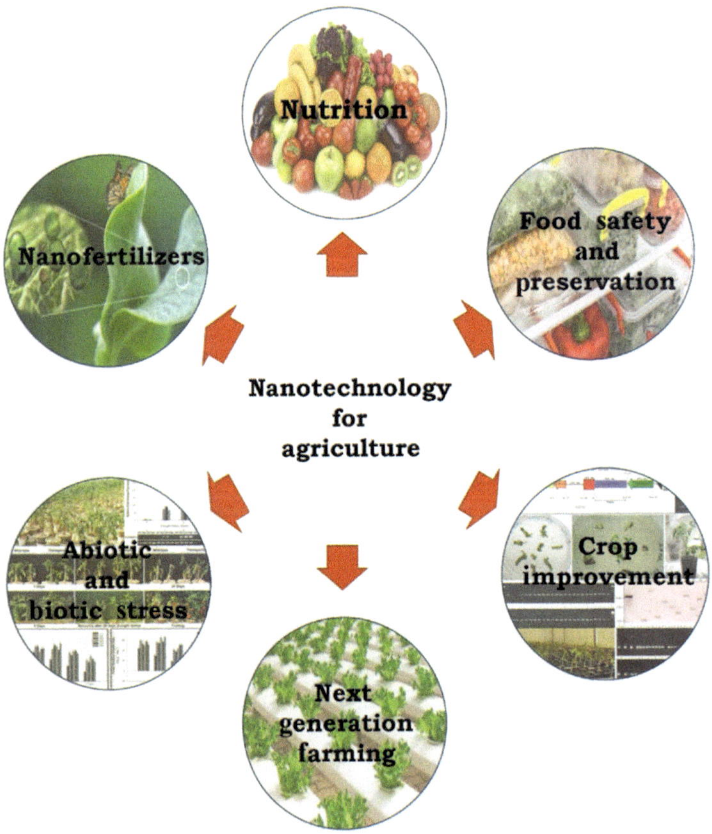

FIGURE 10.1 A simplified overview of nanotechnology in various aspects of sustainable agriculture.

scientists for innovation and development of novel nanomaterial with new properties that paved way for widespread applications, Hence, nanotechnology provides the tools and technique for exploring the properties of nanoparticles to formulate a precise and efficient compound for resource-efficient better agriculture and healthy environment (Fig. 10.2). Here we have focused on various developments and advancements made in the area of nanopesticides, nanofertilizers, nanobiosensors, and contaminated soil remediation through nanotechnology.

10.2.1 Nanotechnology in crop production

Nanotechnology is a promising control, coordinating the information on different essential science privileged insights and evidences at nanoscale level (Misra et al., 2016). Nanotechnology has been ascending as an imaginative intelligent field introducing the new time of nano-horticulture to reform the harvest improvement and sustenance application approaches through usage of various novel devices and things for improving yield profitability (Shang et al., 2019). The use of nanoparticles as micronutrient supplements for plants is given in Table 10.1. The ultra-little size of the nanoparticles just as nearness of huge number of surface molecules improves the reactivity for different reactions (Axelos and Van de Voorde, 2017). Due to increased bioactivity and bioavailability, nanoparticles have great potential in agriculture (Singh et al., 2015). Recently, sharp increase in use of nanopesticides, nanofertilizers, and nanosensors in agriculture has been reported (Boddula et al., 2019). Different nanoparticles impact on the yield of various crops is listed in Table 10.2.

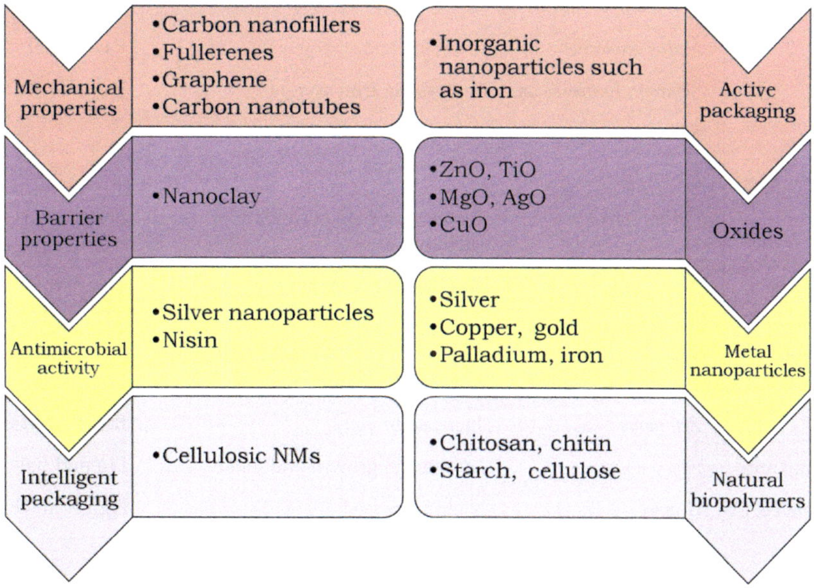

FIGURE 10.2 Overview of some important nanoparticles with properties.

TABLE 10.1 Use of nanoparticles as micronutrient supplements for plants.

Nutrient	Nanoparticle type	Reference
Copper, zinc	Cu, Zn NPs	Monreal et al. (2016)
Nitrogen	Nanozeolite	Lateef et al. (2016)
Iron	Fe_2O_3 NPs	Jahangirian et al. (2020)
	Fe_3O_4	Wang et al. (2019)
	γ-Fe_2O_3 NPs	Wang et al. (2019)
	α-Fe_2O_3	Li et al. (2018)
Iron, zinc	FeO, ZnO NPs	Elizabath et al. (2017)
	ZnO NPs	Singh et al. (2013)
Zinc and iron	ZnO and Fe_2O_3	Jahangirian et al. (2020)
Zinc, iron, and copper	Nano-ZnO, nano-FeO, and nano-ZnFeCu oxide	Rana et al. (2019)

NPs, *Nanoparticles.*

TABLE 10.2 Effect of different nanoparticles for improving the growth and quality of crops.

Metal nanoparticles	Plants	Beneficial effects	Reference
Chelated Ti solution	*Brassica oleracea*	Increase in yield	Pais (1983)
	Capsicum annuum		Pais (1992)
	Malus domestica		Pais (1983)
	Solanum lycopersicum	Increase in fruit weight	Pais (1983)
	Ribes uva-crispa	Increase in yield	
	Prunus persica		
Titanite	*Phleum pretense*	Increase in seed germination, yield increase	Radkowski et al. (2015)
Zinc nanofertilizer	*Pennisetum americanum*	Increase in grain yield	Tarafdar et al. (2014)
Silver nanoparticles	*Triticum aestivum*	Growth and quality of crop improved	Razzaq et al. (2016)
	Zea mays	Increase in yield	Berahmand et al. (2012)
Nanohydroxyapatite	*Glycine max*	Increase in growth and yield	Liu and Lal (2014)
Nanocrystalline Fe, Co, and Cu	*G. max*	Yield increased	Ngo et al. (2014)
Ti-ascorbate	*Pisum sativum*	Uptake of essential elements increased	István et al. (1991)

10.2.2 Nanofertilizers: Potential types and applications

In agriculture utilization of nanotechnology primarily spins around nutrient enrichment of plants mediated by advancement of productive implementing methodology carrying micro- and macronutrient (Shang et al., 2019). Nano fertilizers have an incredible significance in agriculture because of different properties including their higher solubility, stability, lesser toxicity, regulated and effective release of nutrients, and therefore, play an essential role in improving the crop productivity which ultimately assists in procurement and supplementation of food and derived products to world growing population (Raliya et al., 2017; Shang et al., 2019; Iqbal, 2019). Moreover, the efficiency of plants for uptake of nanofertilizers could be improved by creating new micropores, and utilizing root exudates and molecular transporters through the ionic channels (Rico et al., 2011). The following properties of nanofertilizers make them better choice as fertilizers for enhancing crop productivity over other chemically used conventional fertilizers (Iqbal, 2019).

1. Nanofertilizers are more efficacious in terms of nutrient absorption and utilization, which allow lesser losses in the form of leaching and volatilization.
2. Nanofertilizers have added advantages over conventional fertilizers. Since plant cell wall has smaller pore sizes (up to 20 nm) usages of nanofertilizers have eminent potential for higher nutrient uptake, and therefore, enhanced NUE. Higher nutrient uptake facilitated by nanofertilizers is caused by smaller surface area of nanomaterials, which increases nutrient–surface interaction.
3. Nanofertilizers feed crop plants gradually in controlled manner compared to rapid and spontaneous release of nutrients from chemical fertilizers.
4. Nanofertilizers have higher diffusion and solubility, imparting superiority to nanofertilizers over other conventional synthetic fertilizers.

Use of nanofertilizers over conventional fertilizers is more preferred as they can reduce the requirement of agrochemicals, protect the crops against abiotic and biotic stresses, and are environmentally safe (Elemike et al., 2019). However, intake and application of nanofertilizers through the plants depend on its proper formulation and its delivery method (Solanki et al., 2015).

The various delivery methods include hydroponics, foliar application, or application through soil. For hydroponics delivery methods essential parameters include volume of nutrient solution, pH, and oxygen demands and limited by high moisture or fungal contamination. Likewise, nanofertilizer application through soil needs information on soil texture, salinity, and pH and is frequently used. In contrast, foliar mode of delivery is generally preferred for transport of trace elements and mineral nutrients (Taiz and Zeiger, 2010). The size range of nanoparticles used as nano fertilizers lies in the range of 1–100 nm, and this small size is favored and facilitated by molecular transporters (nanometric pores) or other nanostructured cuticle pores, which promotes an easy and rapid uptake of plant nutrients, and thereby, play an essential role in improving the plant metabolism (Rico et al., 2011; Guo et al., 2018). However, nutrient uptake efficiency delivered through nanofertilizers depends on several factors including physiology of plants, chemical composition of nanomaterials, interaction of nanomaterials with environment or its components (both biotic and abiotic) (Pitambara et al., 2019). Several nanofertilizers playing important role in crop improvement program have been discussed.

10.2.3 Nitrogen-based nanofertilizers

Nitrogen, the nutrient most required by crop plants, is an important constituent of many important components of plant cells. The most common mechanism through which N loss in soil is ammonia volatilization, nitrate leaching, and denitrification. The huge amount of conventional fertilizer applied in the crop field is lost and flow out to the water bodies (Wilson et al., 2008). Loss of mineral nutrients through leaching and runoff to surface and ground water have environmental concerns. Further, use of conventional fertilizers is resulting in overdosing of chemical fertilizers leading to eutrophication of water bodies (algal growth on the surface of water bodies due to nutrients). Nanofertilizers are far more superior than conventional fertilizers (Adisa et al., 2019) and regulate the steady and controlled supply of nutrients. Urea-modified hydroxyl-apatite (HA) [Ca_{10} $(PO_4)_6(OH)_2$], a nitrogen-based nanofertilizer, also supply phosphorus other than nitrogen (Preetha and Balakrishnan, 2017). The large surface area of hydroxyapatite and its better interaction with urea results in slow and controlled release of N from urea and provides long-term (60 days) and continuum release of N compared to other conventionally deployed nitrogenous fertilizers (Kottegoda et al., 2011). In one study, Giroto et al. (2017) reported that nanocomposites of urea and hydroxyapatite assist in controlled release of nitrogen, lower volatilization of ammonia, and sustained release of P following the 4 weeks after incubation.

10.2.4 Phosphate-based nanofertilizers

Phosphorus is also one of the important nutrients other than nitrogen. In crops, development of reproductive structures at their early stages is highly dependent on phosphorus (Preetha and Balakrishnan, 2017). In one study, it was reported that application of phosphatic nanofertilizers compared to P-based agrochemicals resulted in higher seed yield (20%) and increased growth rate (32%) in soybean (*Glycine max* L.) (Liu and Lal, 2014). Phosphate-based nanofertilizers are also having critical role in heavy metals remediation. For example, vivianite and apatite are the two phosphorus-based nanofertilizers, which can decrease the leaching of phosphate into the soil (Liu and Lal, 2012). It has been reported that phosphate-based nanoformulated products have higher efficiency for P release (40–50 days) compared to conventional fertilizers (10–12 days) (Preetha and Balakrishnan, 2017).

10.2.5 Iron-based nanofertilizers

Deficiency of iron is very common in most of the plants. They are unable to take most of the iron from the soil because it is present in the insoluble form. Iron is an essential metal element as it is a cofactor of several important enzymes (Raliya, 2013). One of the important aspects of Fe-based nanofertilizers is that its supplementation to the crops results in increased chlorophyll content. For example, increased chlorophyll content in soybean was favored by supplementation of superparamagnetic iron oxide (Ghafariyan et al., 2013). Likewise, Fe nanoparticles when used as nanofertilizer on foliage as 0.5 g/L solution resulted in 10% rise in chlorophyll content in cowpea *Vigna unguiculata* (L. Walp.) (Karami et al., 2014). Iron chelate nanofertilizers are highly stable and having controlled release of nutrients. Plants are also protected from the early aging and senescence by using the iron chelate nanofertilizers because it does not contain ethylene-based structure. It also increases the chlorophyll synthesis in plants (Polyakov et al., 2019). Iron-based nanofertilizers offer a characteristic, minimal effort, and ecological alternative to the conventional iron fertilizers.

10.2.6 Zinc-based nanofertilizers

Zinc is also an essential macronutrient required by crop plants. Its deficiency affects the productivity and crop plants quality. Regular fertilizers are not very soluble or they are converted into their oxide phase. Zinc-based nanofertilizers are far superior than regular fertilizers because of high solubility in the soil. Nanoparticles based on ZnO and sized ZnO elements were layered on urea and monoammonium phosphate and their solubility and dissolution kinetics were compared. It was reported that the monoammonium phosphate showed higher solubility and greater dissolution rate (Milani et al., 2012). Zinc-based nanofertilizer was developed for the pearl millet crop growth and yield improvement (Tarafdar et al., 2014).

10.2.7 Silver-based nanofertilizers

Silver nanoparticles have many unique properties and silver being a nutrient for the plants can be used to increase the nutrient uptake efficiency of the plants (Anand and Bhagat, 2019). The nanosilver provides an improved uptake of nutrients from the soil than the bulk one. Therefore it can have applications in the crop production as well as crop protection. They can be used as nanofertilizers. It has been reported that silver nanoparticles could be used to enhance seed germination potential in many plants (Anand and Bhagat, 2019).

10.2.8 Other metal- and nonmetal-based nanofertilizers

The aluminum-coated nanoparticle in the form of double hydroxide as aluminum-silicate or aluminum-zinc has higher efficiency of monitored excretions of nutrients, which performs as growth regulators of plant. The combination of SiO_2 and TiO_2 composite enhances the uptake of nutrients and water in soybean by improving the metabolic activity of nitrate reductase. However, TiO_2 nanoparticle alone can also increase the growth of maize crop; it helps to regulate the growth of plants due to its optoelectronic properties that increase the light absorption and improve photosynthesis (Lu et al., 2015). Similarly, use of other nanoparticle such as kaolin, maghemite (Fe_2O_3), nanoporous zeolite, polymeric biocompatible and carbon nanotubes enhances plant growth and germination of seeds by absorbing nitrogen, phosphorus and potassium, water other micronutrients from the soil (Khodakovskaya et al., 2009; Manikandan and Subramanian, 2014). Chitosan nanoparticle and hydroxyapatite (HA) nanofertilizers have been used for the growth of crop plants in a controlled mode (Corradini et al., 2010; Kottegoda et al., 2011). Role of mesoporous silica nanoparticles is additionally applicable in gradual and monitored release of urea in soil and water by entrapping it (Wanyika et al., 2012).

10.3 Nanopesticides

Nanopesticides are a new field of agriculture technology utilizing nanomaterials for smart delivery of pesticides in protecting plants against diseases and damages caused by pests and pathogens. Nanopesticides are formed by the combination of numerous surfactants, polymers, and metal nanoparticles of nanoscale dimensions (Sekhon, 2014). Nanoencapsulated pesticide formulation has been deployed with an intention to reduce the dosages of pesticides, and avoid their exposures and risks to human health and therefore is environmentally safe for crop protection (Nuruzzaman et al., 2016). Further, application of nanopesticide is associated with conservation of both water and energy as they are required in lesser amount and less frequently than other agrochemicals used as pesticides. In this way, nanopesticide usages over other agrochemicals are economically safe as they enhance crop productivity by increasing the pesticide use efficiency, reducing the wastages and labor cost. The major concept and aim behind the development of nanopesticides are to increase the solubility of poorly dissolved active ingredients, preventing their premature degradation (stability) and their disposal to the target sites accurately (Shah et al., 2016). The slow degradation and controlled release of active ingredients in presence of suitable nanoparticles or other nanomaterials can control pests effectively over longer period of time (Chhipa, 2017). The application of conventional solutes or pesticides in the form of nanoformulations could be transported in dissolved or other colloidal states through unique mechanism and reflecting different behavior than their original bulk molecules (Kumpiene et al., 2008). Further, the solubility and mobility of the active ingredients formulated as nanopesticides are favored and increased by soil-inhabiting microorganisms and therefore, causing lesser drastic impacts on environment than other harmful pesticides (Kah and Hofmann, 2014). For example, nanoformulations of insecticide imidacloprid (IMI) (50% nano—SDS/Ag/TiO_2) against *Martianus dermestoides* were more photodegradable and effective (lethal concentration LC50: 9.86 mg/L) than their conventionally used forms [95% IMI (LC50: 13.45 mg/L)] (Guan et al., 2008). Nanoformulated derivative of permethrin resulted in its increased efficacy and uptake against *Aedes aegypti*. Similar results were obtained with polyethylene glycol–based nanoformulations of carbofuran and acephate (Pankaj et al., 2012; Pradhan et al., 2013). Similarly, it was found that nanoformulations developed from herbicides atrazine and paraquat were more effective against their target weeds compared to their original and pure forms. Further, these nanoformulations had lesser drastic effects over nontarget plants, particularly, onion (*Allium cepa* L.) as revealed and confirmed through cytotoxicity and genotoxicity assays (Grillo et al., 2014; Pereira et al., 2014). Likewise, nanoformulations composed of carbofuran, thiamethoxam, and imidacloprid show higher penetration and efficacy (Chen and Yada, 2011; Adak et al., 2012; Yata et al., 2018). Moreover, glyphosate, acephate, pheromones, and essential oil extract from neem have shown higher efficiency in the form of nanoemulsions compared to their bulk molecular form (Jiang et al., 2012; Anjali et al., 2012; Bhagat et al., 2013).

10.4 Nanotechnology in disease diagnosis and plant disease management

According to one estimate about an average 20%–40% global losses occur only due to pest and pathogens (Flood, 2010). Currently, plant disease is managed through an efficient use of fungicides and insecticides. These chemicals although have provided rapid response, easy availability, and reliability in their application. However, chemical pesticides also affect nontarget organism, and may lead to disease resistance and resurgence of the pest population (Stephenson, 2003). Further, pesticides have no long-lasting effects and 90% of used pesticides are lost during or after application (Stephenson, 2003; Ghormade et al., 2011). In this way, we need to advance cost-effective, efficient pesticides that are less toxic to the environment (Worrall et al., 2018). Nanotechnology performs significant role in disease diagnosis and plant disease management. For example, nanoparticles could be employed directly as such or as transporters for currently known pesticides or further active ingredients including double-stranded RNA (dsRNA) for their use as spray application, drenching/soaking onto seeds and over other tissues (root and leaves) (Worrall et al., 2018). Detection of infected material as well as pathogen is very important for the success of any sanitation practice (Elmer and White, 2018). In this area of diagnostics, nanotechnology offers significant advances through faster and more-sensitive pathogen probes. Rapid detection of diverse plant pathogens such as bacteria, virus, fungus, and nematodes can be done by the use of nanoparticles as a diagnostic tool (Yao et al., 2009). In this context, superparamagnetic iron oxide nanoparticles have been recently used as a diagnostic tool for pathogen detection in plant pathology (Ahmadov et al., 2014). Magnetic nanoparticles can also be used as diagnostic tool as it attaches the biological tissue or DNA. Superparamagnetic iron oxide nanoparticles are different from QDs. Magnetic nanoparticles including superparamagnetic iron oxide and quantum dots (QDs) can be used as a disease diagnostic tools as both of these found to be attached with hyphae. However, QDs have been reported to be getting absorbed on the biological surface compared to a magnetic nanoparticle that shows only surface attachment (Rispail et al., 2014). Quantification of *Meloidogyne hapla* from mineral soils was performed by adding the superparamagnetic iron oxide nanoparticles to the extraction lysate to maximize the DNA yield (Gorny et al., 2018). Furthermore, biosensors are also an important diagnostic tool in the field of plant pathology because it produces an electric signal when comes in contact with a pathogen (Elmer and White, 2018). As the density of pathogen increases, electric signal also increases and this interaction is converted into a digital output. Biosensors consist of carbon nanomaterials and metal oxide nanoparticles. However, with the advancement of nanotechnology, biosensors are to be prepared with different type of nanomaterials with fewer technical problems (Fang and Ramasamy, 2015). Khaledian et al. (2017) used a biosensor consists of nano-Au with single-stranded oligonucleotides for the detection of genomic DNA of *Ralstonia solanacearum* in the soil. Different studies have reported the potential of antibody-based biosensor for the detection of plant pathogens (Chartuprayoon et al., 2013; Lin et al., 2014a). Detection of different plant viruses has been also made possible by the formation of nano-based biosensors (Perdikaris et al., 2011).

Control of epidemic disease is conceivable effectively by different accessible nanodiagnostic units, and such innovation can give a structure of appropriate integrated disease management framework that unfavorably influences crop pathogen as well as increases crop production (Fang and Ramasamy, 2015). Plant disease resistance through nanoparticles may involve the use of metallic oxides, metalloids, and nonmetals as either bactericide/fungicide or nanofertilizer (Datnoff et al., 2007). Moreover, single or multiwall nanotubes, carbon nanomaterials, fullerenes (buckyball), graphene oxide sheet (reduced or oxidized form) have been successfully employed as antimicrobial agents for delivering antimicrobials as well as genetic material (Elmer and White, 2018). Likewise, liposomes and dendrimers have been frequently used as delivery vehicle for genetic material or antimicrobial products. In contrast, nanobiosensors, QDs, and nanoshells have been deployed us diagnostics, research tool in plant pathology (Elmer and White, 2018). In this context, silver nanoparticles have been deployed as antifungal agent against *Rhizoctonia solani*, *Alternaria alternata*, *Macrophomina phaseolina*, *C. lunata*, *Sclerotinia sclerotiorum*, and *Botrytis cinerea* (Rafique et al., 2017). Further, nanoparticles of Ag were first used in management of plant diseases (Richards, 1981). For example, a water-soluble polymer made with nano-Ag and nano-Si at 0.3 μg/mL was deployed against *Podosphaera xanthii* in cucumber and *Sphaerotheca pannosa*, a powdery mildew fungus at 10 μg/mL in roses. In one study, nano-Ag formulations biosynthesized from plant extracts were used to manage postharvest control of *Colletotrichum musae* (Jagana et al., 2017). Further, silver nanoparticles were used for plant defense against yellow mosaic virus in faba bean sprayed 24 hours postinoculation and resulted in better disease control compared to nanoparticle sprayed before or at the time of infection (Elbeshehy et al., 2015).

The synergistic impact of AgNPs and biosurfactant resulted from *Bacillus subtilis* was tested by Joanna et al. (2018) against different fungal pathogens and demonstrated that the supplementing biosurfactant improves the antifungal activity of nanoparticles. AgNPs synthesized from the metabolites of *Trichoderma harzianum* showed antimicrobial activity against a white mold fungus *S. sclerotiorum* (Guilger et al., 2017). Krishnaraj et al. (2012) reported the inhibition in growth of phytopathogenic fungi, such as *C. lunata*, *R. solani*, *A. alternata*, *M. phaseolina*, and *B. cinerea*, by using the AgNPs. Likewise, nanocopper formulations including nano-CuO, Cu_2O, and Cu/Cu_2O composites were compared with standard Cu-based fungicides, such as Kocide Opti 30 WG, Cuprofix Disperss R, and Ridomil Gold Plus, for evaluating their efficiency in controlling tomato blight disease caused by *Phytophthora infestans* (Giannousi et al., 2013). Cu-based nanoparticles (CuO) when used to leaves of young seedlings at 500 μg/mL provided better disease control caused by *Verticillium* and *Fusarium* wilt of tomato (Elmer and White, 2018). Furthermore, in many studies, nanoformulations of Zn have been found to exhibit inhibitory role against different bacteria (Indhumathy and Mala, 2013; Graham et al., 2016) as well as fungal plant pathogens including *B. cinerea*, *A. alternata*, *Mucor plumbeus*, *Fusarium oxysporum*, *R. solani*, *Penicillium expansum*, *S. sclerotiorum*, and *Rhizopus stolonifer* (He et al., 2011; Wani and Shah, 2012; Sardella et al., 2017). Further, antimicrobial activity of nanoparticles based on MgO was applied to increase tomato systemic resistance against *R. solanacearum* (Huang et al., 2005; Imada et al., 2016). In this regard, Paret et al. deployed nanotechnology combined with photocatalyst technology to develop a novel antimicrobial light-activated TiO_2/Zn nanoparticle composite for suppressing diseases developed by bacterial leaf spot pathogen *Xanthomonas* spp. in rose (Paret et al., 2013). Apart from Ag, Cu, and Zn, other metals, such as gold (Au) and titanium-based nanomaterials, more specifically, have been used as fertilizers rather than plant disease management.

10.5 Application of nanotechnology for food safety and preservation

10.5.1 Nanotechnological tools for controlling postharvest pest and pathogens

The storage of sweet and leafy horticultural products is mostly damaged by postharvest pathogens infected by pathogenic bacteria or fungi that disturb quality of postharvest product and lower the market price. Due to the presence of saccharide, organic compound and low pH fruits are highly prone to spoilage by multiple pathogens, while vegetable having comparatively higher pH is only infected by bacterial pathogens (Ruffo Roberto et al., 2019). Many physicochemical and organic methods are available for controlling such economic loss. However, an alternative novel method of nanotechnology has proved to be effective against toxic chemical fungicides and bactericide for monitoring destruction of fruits and vegetables immediately after harvest (Singh and Rattanpal, 2014). The unique properties of nanomaterial make it possible to an alternative for sustainable horticulture in protection of postharvest loss as presented in Table 10.3.

The secondary coating and encapsulation have shown positive results for enhancing shelf life of fruit and also act as value addition of products (Rajkumar et al., 2006). For example, chitosan nanoparticle in combination with cinnamon essential oil proved to be an effective measure for controlling fungal disease of cucumber caused by *Phytophthora drechsleri*, and enhanced the quality and shelf life of cucumbers for about 3 weeks (Mohammadi et al., 2015). The fungal growth can also be repressed by metallic nanomaterials such as copper, silver, zinc, and sulfur by penetrating into the fungal hypha and produce unstable oxygen containing reactive chemical species, that is, hydrogen peroxide, oxygen radicals, and super peroxides, which disturbs the development of formations of conidia and may lead to cell death (Esteban-Tejeda et al., 2009; Patra et al., 2013). The nano-chitosan has substantiated the effectiveness for ameliorating the quality, texture, and shelf life of peach fruit for about a month (Gad et al., 2016). A nanocomposite of chitosan with antimicrobial properties and pectin with properties of blockage improve the shelf life of fruit and protect against fungal spoilage of fruit (Medeiros et al., 2012).

The antimicrobial property of novel nano-chitosan coating provides an effective procedure for extending the shelf life of horticultural crops such as mangoes, strawberry, papaya, sweet cherries, tomatoes, apples, bananas, bell peppers, and citrus fruits (El-Ghaouth et al., 2000; Vargas et al., 2006; De Capdeville et al., 2002; Bautista-Baños et al., 2003; Romanazzi et al., 2003; Li and Yu, 2001). The silver and zinc oxide nanocomposites display effective antimicrobial activity and improve the storage capacity of orange fruit juice for about 28 days (Emamifar et al., 2011). Similarly, silver nanoparticle is effective against citrus fruits fungal pathogen viz., *Alternaria alternata*, *Penicillium digitatum*, and *Alternaria citri* (Abdelmalek and Salaheldin, 2016). Silver nanoparticles are highly effective for antimicrobial activity by combining with sulfhydryl group of pathogens' enzyme present in the plasma membrane causing alteration in cell membrane permeability and hinder synthesis of ATP leading to retard growth of storage pest (Semeykina and Skulachev, 1990; Yadav et al., 2015). Nanotechnology

TABLE 10.3 Applications of nanomaterials in food processing and preservation.

Function	Nanoparticles	Reference
Edible coatings	Edible coating of gelatin having cellulose nanoparticles	Fakhouri et al. (2014)
	Chitosan/nanosilica coatings	Shi et al. (2013)
	Chitosan film with nano-SiO_2	Yu et al. (2012)
	Alginate/lysozyme nanolaminate coatings	Medeiros et al. (2014)
Encapsulation of vitamins	Vitamin B_2 (riboflavin) in polyamines	Bernardos et al. (2008)
	trans-β-Carotene in mesoporous silica particles	Clifford et al. (2008)
	Vitamin B_3 (niacin) in mesoporous	Kapoor et al. (2010)
	Silica hollow microspheres	
	Ascorbic acid (vitamin C) in mesoporous silica particles	Rashidi et al. (2013)
Encapsulation of phytochemicals	Curcumin encapsulated in mesoporous silica particles	Clifford et al. (2008)
	Resveratrol in nanoporous silica-based spherical mesoporous MCM-41	Popova et al. (2014)
	Wine polyphenols in SBA-15	Cotea et al. (2012)
Liposomes	Cationic lipid incorporated liposomes modified with an acid-labile polymer hyperbranched poly(glycidol)	Yoshizaki et al. (2014)
Hydrogels	Protein hydrogels	Qiu and Park (2001)
Polymeric micelles	PEO-*b*-PCL [poly (ethylene glycol) *block*-poly(caprolactone)] polymeric micelles	Molavi et al. (2008)
Inorganic nanoparticles	Mesoporous silica nanoparticles	Tang et al. (2012)
Nanoemulsions	β-Carotene-based nanoemulsion	Kong et al. (2011)

has modernized basic knowledge of postharvest storage of tropical, subtropical, and temperate fruits and provides alternative methods for efficient and safe antimicrobial compound applied at minimum concentration for higher duration against storage pest (Aloui and Khwaldia, 2016).

10.5.2 Active and intelligent food packaging

The high cost value of agricultural food product depends on the packaging of food materials, which determines the storage and transportation of food product excluding spoilage from chemicals, pathogens, moisture, oxygen, and light with the purpose of enriching shelf life and preserve quality as well as safety of foods. Extensive study and application in the field of nanotechnology have revolutionized many food industries with its potential to develop novel intelligent packaging solution to determine the identity of packaged food and provide information about foods safety and high quality of product to the consumer (Mihindukulasuriya and Lim, 2014). Conventional food packaging materials are replaced by various nanotechnology methods such as nanocomposites act as barrier; nanoparticles act as antimicrobial agent; nanosystems for efficient delivery; nanosensors for detecting the level of pathogenic microorganism and chemicals associated to spoilage in packaged food; nanoindicators for determining time, temperature, oxygen level, freshness of food; optochemical sensors for determining quality of products by sensing gas analytes; and nanopolymers replacing conventional food packaging materials; besides this, nanomaterials also provide various other applications such as biopolymer, encapsulation, nanosolution, and emulsion (Kerry et al., 2006; Bouwmeester et al., 2007; Bratovčić et al., 2015; Majidi et al., 2013). Use of nanoencapsulation helps to mix the active components with foodstuff; regulate the flow of active agents; preserve food tastes, flavor, and odors; reduce moisture; maintain proper temperature; and enhance packaging of food (Ubbink and Kruger, 2006). Similarly, nanocapsulation of nanolaminates is used to cover meat, fish, fruits, vegetables, and bakery products and enhances shelf life of food by blocking extreme thermal and mechanical shock. Nanocomposites composed of synthetic polymers associated with nanoparticles aid to boost the freshness of food materials, act as a hindrance for any microbial infection, and block gas exchange in carbonated beverage. This may be the reason many industrial beverages' manufacturers use nanocomposite for preventing any type of leakage in packed bottles, showing additional properties of eco-friendly and biodegradable nature (Pandey et al., 2013). Nanopacking of green tea enhances the quality and preservation of nutrition, which helps to restore amino acid, vitamin C, and polyphenols (Hu and Fu, 2003). Nanoparticle plays a major role in packaging of food materials that serve to perform antimicrobial properties, maintain color, texture, and stability. Many nanoparticles such as inorganic, silver, metal oxide, chitosan, copper, titanium oxide, and zinc oxide perform antibacterial activity in low concentration being applied in packaging of many food items viz., fruits, cheeses, soups, root vegetable, breads, sauces, meats, and herbs (Bradley et al., 2011; Tan et al., 2013; Bumbudsanpharoke and Ko, 2015). This smart food packaging system would benefit many industrial stakeholders, regulatory body of food and consumers (Duncan, 2011).

10.5.3 Improved mechanical and heat resistance

The improvement of thermal and mechanical properties of food packaging materials is the fundamental characteristic of food industries. The nanocomposites made of nanomaterials with polymer reinforcement block the flow of gases and moisture from the packed food materials enhances the storage capacity of food materials in packaged food. Utilization of thin layer of packaging materials depends on the expansion of mechanical properties of nanomaterials, which requires very less amount of raw materials and reduces the cost of recycling. Improvements in heat resistance properties of nanomaterials enhance the protection from high thermal condition and provide slow spread of fire. The mechanical properties of nanomaterials depend on tensile strength, decrease friction coefficients, rigidity, isotropy, water resistance, agglomeration state, dispersion ratio, flame resistance, ratio of longitudinal and transverse lengths, hydrophilicity, gas permeability, increasing elastic modulus, etc. (Sheng et al., 2004; Youssef, 2013). Also, structural perfection of materials improves the mechanical properties of nanomaterials (Herring and Galt, 1952). The thermal properties of nanomaterials depend on nanofiller ratio and flammable properties depend on the components of nanocomposites, composed of reinforcement of polystyrene with suitable ratio of multiwalled carbon nanotubes (MWCNTs) (Cipiriano et al., 2007). These MWCNTs alter thermal properties based on characteristic of viscoelastic properties such as reinforcement of polymer and the lack of tangent (Natali et al., 2016). An extensive study and application of effective high aspect ratio and higher dispersion rate in polymer of carbon nanofiller improve the thermal conductivity of nanocomposites. The strengthening of polystyrene with MWCNTs in nanocomposites enhances the flammability properties and heat release rate of nanocomposite. MWCNTs also create stiffer nanocomposite, which decrease the rate of mass loss (Zhao et al., 2018; Evgin et al., 2016). The reinforced polymer with MWCNTs established the improvement of nanocomposite materials by adding viscoelastic properties (Natali et al., 2016). Similarly, the presence of nanographite in polymer-based nanocomposite alters the characteristic by enhancing wear resistance and lowering friction coefficient (Jia et al., 2015). Similarly, nanoclays in polymer matrices improve thermal properties by enhancing the heat storage and heat distortion temperature (Fornes and Paul, 2003). Aiming to improve such properties in nanomaterials has high mechanical and thermal potential for its utility in food packaging industry.

10.5.4 Enhancement of food shelf life, security, and observing quality of packaged foods

Food shelf life can be regarded as the period of time during which a food product could be stored and therefore maintaining its suitable quality based on different view such as safety and organoleptic parameter, depending on several essential factors (storage, packaging, processing, etc.). Foods have the perishable properties and the quality of food deteriorates through many factors during processing and supply chains. With the intension of maintaining the food quality or expiry of food the intrinsic and extrinsic factors must be considered as an important part to take care of it. The kinetic models play a vital role to understand the data obtained in the basis experimental design. These kinetic models can subsequently be utilized for assumption, process organization, maximization, and simulation of diverse food processing functions. The optimization of storage period of food materials requires the insights and exploration of the food system, allows the overall set-up of the following food system. Shelf life insights are mathematical model that explains the connection between the packaging materials of food with the environment. These models are depending on various crucial factors (degradation properties) and are important in calculating expiry of stored food, in packages designing, and achieving valuable insights about the foods under unusual conditions along the supply chain. Multiple barriers are well-thought-out to prolong the storage period of food. During packaging gas barrier extends the expiry of food materials by shielding quality of food and buyer security. The technology advancement in food processing might direct to the growth of innovative equipment of gas barrier by substituting the packaging materials of glass and metal. We need to develop some innovative technologies and discover better gas materials that can improve demand for food supply polymers viz., nylons, ethylene vinyl alcohol, and polyethylene terephthalate in a different form eloquently improve the barrier. Therefore novel innovation in the field of nanotechnology required further enhancement in the functionality of barrier. Nanotechnology in combination with polymer chemistry facilitates multiple properties such as multilayer function and elegant combination, which give rise to promising high-barrier food packaging innovations ideas. During nanopackaging, property of food wrapping substances with antimicrobial activities and barrier functions plays an important role against aroma and gases (Rhim et al., 2013). The example of antimicrobial activity of nanocomposite utilized for food packaging materials is shown in Table 10.4. The efficient utilization of food packaging depends on nanotechnology, which has exemplified its application in terms of several aspects in the leading countries. Nanopackaging of food materials has gained its importance in multiple food industries throughout the world (Echegoyen, 2015).

In recent decades, the use of nanomaterials or nanocomposites has gained popularities in food market. The nanomaterial-based packaging of food items is grouped into two different types: (1) Improved packaging, in which clay nanocomposites and nanopolymers matrix improve gas barrier of food packaging when combining with nanomaterials. For example, CO_2 and O_2 barrier properties have been improved in nanocomposites and utilized for food packaging materials. (2) Active packaging, in which direct involvement of nanomaterial with food materials improves freshness quality and storage period of food, hence, provide consumers safety (Duncan, 2011; Carbone et al., 2016). Structural factors such as exfoliation of food, volume division, phase ratio, platelet direction angle, and size

TABLE 10.4 Nanoantimicrobial against food spoilage microorganisms.

Polymer matrix	Nanoparticles	Test microorganisms	Reference
Agar	Silver nanoparticles	*Listeria monocytogenes*	Arfat et al. (2017)
	Copper nanoparticles	*Salmonella typhimurium*	
Polylactic acid/polyethylene glycol	Zinc oxide nanoparticle	*L. monocytogenes*	Ahmed et al. (2017)
	Silver nanoparticles	*S. typhimurium*	
	Copper nanoparticles		
Starch/polyvinyl alcohol	Silver nanoparticles	*Listeria innocua*	Cano et al. (2016)
		Escherichia coli	
		Aspergillus niger	
		Penicillium expansum	
Fish skin gelatin	Silver–copper nanoparticles	*L. monocytogenes*	Arfat et al. (2017)
		S. typhimurium	
Low-density polyethylene	Zinc oxide nanoparticle	*Bacillus subtilis*	Esmailzadeh et al. (2016)
		Enterobacter aerogenes	

of intercalation play a vital role in improvement of nanocomposites barrier properties (Bhunia et al., 2012).

Further, the aspect ratio and volume fraction properties of food packaging nanoparticle compounds have been improved by exfoliate organization that affects nanocomposites barriers (Bharadwaj, 2001; Sablani, 2015). Various studies have revealed mixing of nanoparticles with base polymer–enhanced barrier properties of packaging materials, such as polyamide-6 (PA6), EVOH, and poly(ethylene terephthalate) (Cabedo et al., 2004; Frounchi and Dourbash, 2009; Fasihi and Abolghasemi, 2012). Other examples of nanotechnology products involved are Durethan (LANXESS Deutschland GmbH) for covering fruit juice paper bottle and Aegis (Honeywell Polymers) for poly(ethylene terephthalate) bottles storing alcoholic beverages (Duncan, 2011; Sablani, 2015). Therefore, for the development of nano-based technology, there must be awareness program among the people for utilizing nanotechnology in the field of food safety. However, tremendous challenges are required to bring success for availability of secure food materials around the world.

10.5.5 Improvement of texture, flavor, taste, and consistency of processed foods

Nanotechnology has latent functions in all phase's food security viz., processing, packaging, distributing, and supplying. Manufacturing of processed foods such as nonalcoholic beverage, ice cream, chocolate, and potato chips, canned foods are promoted as healthy foods by decreasing content of fat, carbohydrate or calorie or by fortification of protein, vitamin, and fiber contents. At present, various nano-based food materials deployed ingredients and additives made up of nanomaterials are available for sale in the food market. Companies such as Nestle and Unilever are selling ice cream with nanoparticles emulsion that helps to reduce fat content with balanced taste and texture (Alfadul and Elneshwy, 2010).

In food processing units, nano-based technology can be used in two diverse methodologies, "bottom up" and "top down." The top-down method is accomplished mainly by the resources of processing equipment that helps in grinding and milling of ready to eat food. Dry-milling increases the shelf life of products by inhibiting the growth of microorganism and reduces the cost of packaging by reducing bulk weight of processed product (Rahimmalek and Goli, 2013). Processed wheat achieves fine-size wheat flour that can combine with other ingredients to enhance water holding capacity, provide suitable texture to the wheat products such as biscuits and other cookies. Due to its high applicability, it has also been used to process green tea powder with higher antioxidant, vitamins, and polyphenol properties. Optimum period of drying with required temperature reduced the size of leaf up to 1000 nm, which enhances the color and antioxidant properties of tea (Roshanak et al., 2016). In the milk industry, process of homogenization reduces the fat content without a compromise of other nutrimental content of the milk, converts liquid to powdered milk by high pressure and shearing force utilizing nanomaterial (Brody et al., 2008). In contrast, auto-anabolic process of biology is taken up by the bottom-up nanofood technology. The synthesis of food product takes place by adding nanometer size range compounds, direct assembly of molecules, and artificial synthesis.

The nanoemulsion technique is developed by homogenization of raw food materials under high pressure of inert gas. It helps to produce less fat food product. Due to homogenizer, size of the emulsion droplets reduces, hence reduces the use of stabilizers in food product. Using nanoemulsion, fortified ingredient can be incorporated into the food product. Nanotechnology has also revolutionized the meat industry, development of fat less meat or salt less meat without compromising flavor, test and nutrition of meat, and reduces the risk of cardiac diseases. Similarly, many nanoencapsulated active compounds are available in the market for enhancement and preservation of color, test, flavor, and other sensory quality of meat product.

Many nanocarriers are being utilized for enhancing the bioavailability of nutrient by encapsulating with micelle or liposomes. Liposome is a phospholipid bilayer spherical structure, in which antimicrobial compound can be inserted and delivered into food product, which enhances the storage period of food whereas micelles represent aggregation of surfactant molecules dispersed in liquid colloids and consist of outer polar head group and inner hydrophobic nonpolar groups structure in which nonpolar food supplements are embedded inside it for delivery of vitamins, antimicrobial compounds, preservatives, fat, etc. Many functional foods are developed using bioactive nanomaterials encapsulated into the protected covering. Such type of nanoencapsulation in the food industries is used for regulated release of active ingredients that protect color, texture, order, test of food materials; regulated release of antimicrobial compound that protects it from both biotic and abiotic factors; and enhances the shelf life of functional food. Nanoencapsulation of functional food nowadays becomes consumer friendly in order to control the test and nutrition according to time and need. Nanosize of bioactive compound ameliorates the bioavailability of compound due to high solubility rate that can even cross the lipid membrane of intestine and provides nutrition to each corner of body tissue (Singh, 2016).

Acknowledgments

AZ is thankful to the Head, Department of Botany, Banaras Hindu University for providing necessary facilities. Financial assistance from UGC-MANF is gratefully acknowledged.

Conflict of interest

The authors declare that the research was conducted in the absence of any commercial or financial relationships that could be construed as a potential conflict of interest.

Author contributions

AZ and MA conceived the idea for chapter writing. AR provided necessary table and figures. SKS helped in writing some sections. WAA helped in writing and editing the manuscript. AZ, MA, AR, SKS, and WAA all contributed equally in preparing the final version of the manuscript. RSU supervised the whole work. All authors read and approved the final version of the manuscript for publication.

References

Abdelmalek, G.A.M., Salaheldin, T.A., 2016. Silver nanoparticles as a potent fungicide for citrus phytopathogenic fungi. J. Nanomed. Res. 3, 00065.

Adak, T., Kumar, J., Dey, D., Shakil, N.A., Walia, S., 2012. Residue and bio-efficacy evaluation of controlled release formulations of imidacloprid against pests in soybean (*Glycine max*). J. Environ. Sci. Health Part B 47 (3), 226–231.

Adisa, I.O., Pullagurala, V.L.R., Peralta-Videa, J.R., Dimkpa, C.O., Elmer, W.H., Gardea-Torresdey, J.L., et al., 2019. Recent advances in nano-enabled fertilizers and pesticides: a critical review of mechanisms of action. Environ. Sci. Nano 6 (7), 2002–2030.

Ahmadov, I.S., Ramazanov, M.A., Sienkiewicz, A., Forro, L., 2014. Uptake and intracellular trafficking of superparamagnetic iron oxide nanoparticles (spions) in plants. Dig. J. Nanomater. Biostruct. 9 (3), 1149–1157.

Ahmed, J., Mulla, M., Arfat, Y.A., 2017. Application of high-pressure processing and polylactide/cinnamon oil packaging on chicken sample for inactivation and inhibition of *Listeria monocytogenes* and *Salmonella typhimurium*, and post-processing film properties. Food Control 78, 160–168.

Alfadul, S.M., Elneshwy, A.A., 2010. Use of nanotechnology in food processing, packaging and safety–review, Afr. J. Food Agric. Nutr. Develop. 10 (6).

Aloui, H., Khwaldia, K., 2016. Natural antimicrobial edible coatings for microbial safety and food quality enhancement. Compr. Rev. Food Sci. Food Saf. 15, 1080–1103.

Anand, R., Bhagat, M., 2019. Silver nanoparticles (AgNPs): as nanopesticides and nanofertilizers. MOJ Biol. Med. 4, 19–20.

Anjali, C.H., Sharma, Y., Mukherjee, A., Chandrasekaran, N., 2012. Neem oil (*Azadirachta indica*) nanoemulsion as potent larvicidal agent against *Culex quinquefasciatus*. Pest Manage. Sci. 68 (2), 158–163.

Arfat, Y.A., Ahmed, J., Hiremath, N., Auras, R., Joseph, A., 2017. Thermo-mechanical, rheological, structural and antimicrobial properties of bionanocomposite films based on fish skin gelatin and silver-copper nanoparticles. Food Hydrocoll. 62, 191–202.

Axelos, M.A., Van de Voorde, M. (Eds.), 2017. Nanotechnology in Agriculture and Food Science. John Wiley & Sons.

Batsmanova, L.M., Gonchar, L.M., Taran, N.Y., Okanenko, A.A., 2013. Using a Colloidal Solution of Metal Nanoparticles as Micronutrient Fertiliser for Cereals. Doctoral dissertation, Sumy State University.

Bautista-Baños, S., Hernández-López, M., Bosquez-Molina, E., Wilson, C., 2003. Effects of chitosan and plant extracts on growth of *Colletotrichum gloeosporioides*, anthracnose levels and quality of papaya fruit. Crop Prot. 22, 1087–1092.

Berahmand, A.A., Panahi, A.G., Sahabi, H., Feizi, H., Moghaddam, P.R., Shahtahmassebi, N., et al., 2012. Effects silver nanoparticles and magnetic field on growth of fodder maize (*Zea mays* L.). Biol. Trace Elem. Res. 149 (3), 419–424.

Bernardos, A., Aznar, E., Coll, C., Martínez-Mañez, R., Barat, J.M., Marcos, M.D., et al., 2008. Controlled release of vitamin b2 using mesoporous materials functionalized with amine-bearing gate-like scaffoldings. J. Control. Release 131 (3), 181–189.

Bhagat, D., Samanta, S.K., Bhattacharya, S., 2013. Efficient management of fruit pests by pheromone nanogels. Sci. Rep. 3, 1294.

Bharadwaj, R.K., 2001. Modeling the barrier properties of polymer-layered silicate nanocomposites. Macromolecules 34 (26), 9189–9192.

Bhunia, K., Dhawan, S., Sablani, S.S., 2012. Modeling the oxygen diffusion of nanocomposite-based food packaging films. J. Food Sci. 77 (7), N29–N38.

Boddula, R., Trivedi, U., Pothu, R., Rajput, M.S., Saran, A., 2019. Nanopesticides and nanosensors in agriculture. Plant Nanobionics. Springer, Cham, pp. 165–181.

Bouwmeester, H., Dekkers, S., Noordam, M., Hagens, W., Bulder, A., de Heer, C., et al., 2007. Health impact of nanotechnologies in food production. RIKILT/RIVM Report 2007.014.

Bradley, E.L., Castle, L., Chaudhry, Q., 2011. Applications of nanomaterials in food packaging with a consideration of opportunities for developing countries. Trends Food Sci. Technol. 22, 603–610. Available from: https://doi.org/10.1016/j.tifs.2011.01.002.

Bratovčić, A., Odobašić, A., Catić, S., Šestan, I., 2015. Application of polymer nanocomposite materials in food packaging. Croat. J. Food Sci. Technol. 7, 86–94.

Brody, A.L., Bugusu, B., Han, J.H., Sand, C.K., McHugh, T.H., 2008. Innovative food packaging solutions. J. Food Sci 73 (8), R107–R116.

Bumbudsanpharoke, N., Ko, S., 2015. Nano-food packaging: an overview of market, migration research, and safety regulations. J. Food Sci. 80, R910–R923.

Cabedo, L.S., Giménez, E., Lagaron, J.M., Gavara, R., Saura, J.J., 2004. Development of EVOH-kaolinite nanocomposites. Polymer 45, 5233–5238.

Cano, A., Cháfer, M., Chiralt, A., González-Martínez, C., 2016. Development and characterization of active films based on starch-PVA, containing silver nanoparticles. Food Packag. Shelf Life 10, 16–24.

Carbone, M., Donia, D.T., Sabatella, G., et al., 2016. Silver nanoparticles in polymeric matrices for fresh food packaging. J. King Saud. Univ. Sci. 28 (4), 273–279.

Chartuprayoon, N., Rheem, Y., Ng, J.C., Nam, J., Chen, W., Myung, N.V., 2013. Polypyrrole nanoribbon based chemiresistive immunosensors for viral plant pathogen detection. Anal. Methods 5 (14), 3497–3502.

Chen, H., Yada, R., 2011. Nanotechnologies in agriculture: new tools for sustainable development. Trends Food Sci. Technol. 22 (11), 585–594.

Chhipa, H., 2017. Nanofertilizers and nanopesticides for agriculture. Environ. Chem. Lett. 15 (1), 15–22.

Cipiriano, B.H., Kashiwagi, T., Raghavan, S.R., Yang, Y., Grulke, E.A., Yamamoto, K., et al., 2007. Effects of aspect ratio of MWNT on the flammability properties of polymer nanocomposites. Polymer 48 (20), 6086–6096.

Clifford, N.W., Iyer, K.S., Raston, C.L., 2008. Encapsulation and controlled release of nutraceuticals using mesoporous silica capsules. J. Mater. Chem. 18 (2), 162–165.

Conforti, P., 2017. Looking Ahead in World Food and Agriculture: Perspectives to 2050. Food and Agriculture Organization of the United Nations (FAO).

Corradini, E., De Moura, M.R., Mattoso, L.H.C., 2010. A preliminary study of the incorparation of NPK fertilizer into chitosan nanoparticles. Express Polym. Lett. 4 (8).

Cotea, V., Luchian, C., Bilba, N., Niculaua, M., 2012. Mesoporous silica sba-15, a new adsorbent for bioactive polyphenols from red wine. Anal. Chim. Acta 732, 180–185.

Cunningham, F.J., Goh, N.S., Demirer, G.S., Matos, J.L., Landry, M.P., 2018. Nanoparticle-mediated delivery towards advancing plant genetic engineering. Trends Biotechnol. 36 (9), 882–897.

Datnoff, L.E., Elmer, W.H., Huber, D.M., 2007. Mineral nutrition and plant disease. American Phytopathological Society. APS Press, St. Paul.

De Capdeville, G., Wilson, C.L., Beer, S.V., Aist, J.R., 2002. Alternative disease control agents induce resistance to blue mold in harvested 'Red Delicious' apple fruit. Phytopathology 92, 900–908.

Duncan, T.V., 2011. Applications of nanotechnology in food packaging and food safety: barrier materials, antimicrobials and sensors. J. Colloid Interface Sci. 363 (1), 1–24.

Echegoyen, Y., 2015. Nano-developments for food packaging and labeling applications. Nanotechnologies in Food and Agriculture. Springer, Cham, pp. 141–166.

Elbeshehy, E.K., Elazzazy, A.M., Aggelis, G., 2015. Silver nanoparticles synthesis mediated by new isolates of *Bacillus* spp., nanoparticle characterization and their activity against Bean Yellow Mosaic Virus and human pathogens. Front. Microbiol. 6, 453.

Elemike, E.E., Uzoh, I.M., Onwudiwe, D.C., Babalola, O.O., 2019. The role of nanotechnology in the fortification of plant nutrients and improvement of crop production. Appl. Sci. 9 (3), 499.

El-Ghaouth, A., Smilanick, J.L., Wilson, C.L., 2000. Enhancement of the performance of *Candida saitoana* by the addition of glycol chitosan for the control of postharvest decay of apple and citrus fruit. Postharvest Biol. Technol. 19, 103–110.

Elizabath, A., Bahadur, V., Misra, P., Prasad, V.M., Thomas, T., 2017. Effect of different concentrations of iron oxide and zinc oxide nanoparticles on growth and yield of carrot (*Daucus carota* L.). J. Pharm. Phytochem. 6 (4), 1266–1269.

Elmer, W., White, J.C., 2018. The future of nanotechnology in plant pathology. Annu. Rev. Phytopathol. 56, 111–133.

Emamifar, A., Kadivar, M., Shahedi, M., Soleimanian-Zad, S., 2011. Effect of nanocomposite packaging containing Ag and ZnO on inactivation of *Lactobacillus plantarum* in orange juice. Food Control 22 (3–4), 408–413.

Esmailzadeh, H., Sangpour, P., Shahraz, F., Hejazi, J., Khaksar, R., 2016. Effect of nanocomposite packaging containing ZnO on growth of *Bacillus subtilis* and *Enterobacter aerogenes*. Mater. Sci. Eng. C 58, 1058–1063.

Esteban-Tejeda, L., Malpartida, F., Esteban-Cubillo, A., Pecharromán, C., Moya, J.S., 2009. The antibacterial and antifungal activity of a soda-lime glass containing silver nanoparticles. Nanotechnology 20 (8), 085103.

Evgin, T., Koca, H.D., Horny, N., Turgut, A., Tavman, I.H., Chirtoc, M., et al., 2016. Effect of aspect ratio on thermal conductivity of high-density polyethylene/multi-walled carbon nanotubes nanocomposites. Composites Part A Appl. Sci. Manuf. 82, 208–213.

Fakhouri, F., et al., 2014. Effect of a gelatin-based edible coating containing cellulose nanocrystals (CNC) on the quality and nutrient retention of fresh strawberries during storage, IOP Conference Series: *Materials Science and Engineering*, vol. 64. IOP Publishing, p 012024.

Fang, Y., Ramasamy, R.P., 2015. Current and prospective methods for plant disease detection. Biosensors 5 (3), 537–561.

Fasihi, M., Abolghasemi, M.R., 2012. Oxygen barrier and mechanical properties of masterbatch-based PA6/nanoclay composite films. J. Appl. Polym. Sci. 125 (S1), E2–E8.

Flood, J., 2010. The importance of plant health to food security. Food Secur. 2 (3), 215–231.

Fornes, T.D., Paul, D.R., 2003. Modeling properties of nylon 6/clay nanocomposites using composite theories. Polymer 44 (17), 4993–5013.

Frounchi, M., Dourbash, A., 2009. Oxygen barrier properties of poly (ethylene terephthalate) nanocomposite films. Macromol. Mater. Eng. 294 (1), 68–74.

Gad, M.M., Zagzog, O.A., Hemeda, O.M., 2016. Development of nano-chitosan edible coating for peach fruits cv. Desert Red. Int. J. Environ. 5 (4), 43–55.

Ghafariyan, M.H., Malakouti, M.J., Dadpour, M.R., Stroeve, P., Mahmoudi, M., 2013. Effects of magnetite nanoparticles on soybean chlorophyll. Environ. Sci. Technol. 47 (18), 10645–10652.

Ghormade, V., Deshpande, M.V., Paknikar, K.M., 2011. Perspectives for nano-biotechnology enabled protection and nutrition of plants. Biotechnol. Adv. 29 (6), 792–803.

Giannousi, K., Avramidis, I., Dendrinou-Samara, C., 2013. Synthesis, characterization and evaluation of copper-based nanoparticles as agrochemicals against *Phytophthora infestans*. RSC Adv. 3 (44), 21743–21752.

Giroto, A.S., Guimarães, G.G., Foschini, M., Ribeiro, C., 2017. Role of slow-release nanocomposite fertilizers on nitrogen and phosphate availability in soil. Sci. Rep. 7, 46032.

Gorny, A.M., Hay, F.S., Wang, X., Pethybridge, S.J., 2018. Isolation of nematode DNA from 100 g of soil using Fe_3O_4 super paramagnetic nanoparticles. Nematology 20 (3), 271–283.

Graham, J.H., Johnson, E.G., Myers, M.E., Young, M., Rajasekaran, P., Das, S., et al., 2016. Potential of nano-formulated zinc oxide for control of citrus canker on grapefruit trees. Plant Dis. 100 (12), 2442–2447.

Grillo, R., Pereira, A.E., Nishisaka, C.S., de Lima, R., Oehlke, K., Greiner, R., et al., 2014. Chitosan/tripolyphosphate nanoparticles loaded with paraquat herbicide: an environmentally safer alternative for weed control. J. Hazard. Mater. 278, 163–171.

Gruère, G.P., 2012. Implications of nanotechnology growth in food and agriculture in OECD countries. Food Policy 37, 191–198. Available from: https://doi.org/10.1016/j.jhazmat.2014.05.079.

Guan, Y.F., Pearce, R.C., Melechko, A.V., Hensley, D.K., Simpson, M.L., Rack, P.D., 2008. Pulsed laser dewetting of nickel catalyst for carbon nanofiber growth. Nanotechnology 19 (23), 235604.

Guilger, M., Pasquoto-Stigliani, T., Bilesky-Jose, N., Grillo, R., Abhilash, P.C., Fraceto, L.F., et al., 2017. Biogenic silver nanoparticles based on *Trichoderma harzianum*: synthesis, characterization, toxicity evaluation and biological activity. Sci. Rep. 7, 44421.

Guo, H., White, J.C., Wang, Z., Xing, B., 2018. Nano-enabled fertilizers to control the release and use efficiency of nutrients. Curr. Opin. Environ. Sci. Health 6, 77–83.

He, X., Wu, X., Gao, C., Wang, K., Lin, S., Huang, W., et al., 2011. Synthesis and self-assembly of a hydrophilic, thermo-responsive poly (ethylene oxide) monomethyl ether-block-poly (acrylic acid)-block-poly (N-isopropylacrylamide) copolymer to form micelles for drug delivery, Reac. Functional Polym 71 (5), 544–552.

Herring, C., Galt, J.K., 1952. Properties of very small metal specimens, Phys. Rev. 85, 1060–1061.

Hu, A.W., Fu, Z.H., 2003. Nanotechnology and its application in packaging and packaging machinery. Packag. Eng. 24, 22–24.

Huang, L., Li, D.Q., Lin, Y.J., Wei, M., Evans, D.G., Duan, X., 2005. Controllable preparation of Nano-MgO and investigation of its bactericidal properties. J. Inorg. Biochem. 99 (5), 986–993.

Imada, K., Sakai, S., Kajihara, H., Tanaka, S., Ito, S., 2016. Magnesium oxide nanoparticles induce systemic resistance in tomato against bacterial wilt disease. Plant Pathol. 65 (4), 551–560.

Indhumathy, M., Mala, R., 2013. Photocatalytic activity of zinc sulphate nano material on phytopathogens. Int. J. Agric. Environ. Biotechnol. 6 (4S), 737.

Ion, A.C., Ion, I., Culetu, A., 2010. Carbon-based nanomaterials: environmental applications. Univ. Politehn. Bucharest 38, 129–132.

Iqbal, M.A., 2019. Nano-fertilizers for sustainable crop production under changing climate: a global perspective. Sustainable Crop Production. IntechOpen.

István, P., Fehér, M., Bokori, J., Nagy, B., 1991. Physilogically beneficial effects of titanium. Water Air Soil Pollut. 57 (1), 675–680.

Jagana, D., Hegde, Y.R., Lella, R., 2017. Green nanoparticles: a novel approach for the management of banana anthracnose caused by *Colletotrichum musae*. Int. J. Curr. Microbiol. Appl. Sci. 6 (10), 1749–1756.

Jahangirian, H., Rafiee-Moghaddam, R., Jahangirian, N., Nikpey, B., Jahangirian, S., Bassous, N., et al., 2020. Green synthesis of zeolite/Fe_2O_3 nanocomposites: toxicity & cell proliferation assays and application as a smart iron nanofertilizer. Int. J. Nanomedicine 15, 1005.

Jia, Z., Hao, C., Yan, Y., Yang, Y., 2015. Effects of nanoscale expanded graphite on the wear and frictional behaviors of polyimide-based composites. Wear 338–339, 282–287.

Jiang, L.C., Basri, M., Omar, D., Rahman, M.B.A., Salleh, A.B., Rahman, R.N.Z.R.A., et al., 2012. Green nanoemulsion intervention for water-soluble glyphosate isopropylamine (IPA) formulations in controlling *Eleusine indica*. Pest. Biochem. Physiol. 102 (1), 19–29.

Joanna, C., Marcin, L., Ewa, K., Grażyna, P., 2018. A nonspecific synergistic effect of biogenic silver nanoparticles and biosurfactant towards environmental bacteria and fungi. Ecotoxicology 27 (3), 352–359.

Kah, M., Hofmann, T., 2014. Nanopesticide research: current trends and future priorities. Environ. Int. 63, 224–235.

Kah, M., Navarro, D., Kookana, R.S., Kirby, J.K., Santra, S., Ozcan, A., et al., 2019. Impact of (nano) formulations on the distribution and wash-off of copper pesticides and fertilisers applied on citrus leaves. Environ. Chem. 16 (6), 401–410.

Kapoor, M.P., Vinu, A., Fujii, W., Kimura, T., Yang, Q., Kasama, Y., et al., 2010. Self-assembly of mesoporous silicas hollow microspheres via food grade emulsifiers for delivery systems. Microporous Mesoporous Mater. 128 (1–3), 187–193.

Karami, M., Bahabadi, M.A., Delfani, S., Ghozatloo, A., 2014. A new application of carbon nanotubes nanofluid as working fluid of low-temperature direct absorption solar collector. Solar Energy Mater. Solar Cells 121, 114–118.

Kerry, J.P., O'grady, M.N., Hogan, S.A., 2006. Past, current and potential utilization of active and intelligent packaging systems for meat and muscle-based products: a review. Meat Sci. 74 (1), 113–130.

Khaledian, S., Nikkhah, M., Shams-bakhsh, M., Hoseinzadeh, S., 2017. A sensitive biosensor based on gold nanoparticles to detect *Ralstonia solanacearum* in soil. J. General Plant Pathol. 83 (4), 231–239.

Khodakovskaya, M., Dervishi, E., Mahmood, M., Xu, Y., Li, Z., Biris, A.S., 2009. Carbon nanotubes are able to penetrate plant seed coat and dramatically affect seed germination and plant growth. ACS Nano 3 (10), 3221–3227.

Kong, M., Chen, X.G., Kweon, D.K., Park, H.J., 2011. Investigations on skin permeation of hyaluronic acid based nanoemulsion as transdermal carrier. Carbohydr. Polym. 86 (2), 837–843.

Kottegoda, N., Munaweera, I., Madusanka, N., Karunaratne, V., 2011. A green slow-release fertilizer composition based on urea-modified hydroxyapatite nanoparticles encapsulated wood. Curr. Sci. 73–78.

Krishnaraj, C., Ramachandran, R., Mohan, K., Kalaichelvan, P.T., 2012. Optimization for rapid synthesis of silver nanoparticles and its effect on phytopathogenic fungi. Spectrochim. Acta Part A Mol. Biomol. Spectrosc. 93, 95–99.

Kumpiene, J., Lagerkvist, A., Maurice, C., 2008. Stabilization of As, Cr, Cu, Pb and Zn in soil using amendments—a review. Waste Manag 28, 215–225.

Lateef, A., Nazir, R., Jamil, N., Alam, S., Shah, R., Khan, M.N., et al., 2016. Synthesis and characterization of zeolite based nano–composite: an environment friendly slow release fertilizer. Microporous Mesoporous Mater. 232, 174–183.

Li, H., Yu, T., 2001. Effect of chitosan on incidence of brown rot, quality and physiological attributes of postharvest peach fruit. J. Sci. Food Agric. 81 (2), 269–274.

Li, J., Hu, J., Xiao, L., Wang, Y., Wang, X., 2018. Interaction mechanisms between α-Fe_2O_3, γ-Fe_2O_3 and Fe_3O_4 nanoparticles and citrus maxima seedlings. Sci. Total Environ. 625, 677–685.

Lin, P., Lin, S., Wang, P.C., Sridhar, R., 2014a. Techniques for physicochemical characterization of nanomaterials. Biotechnol. Adv. 32, 711–726.

Liu, R., Lal, R., 2012. Nanoenhanced materials for reclamation of mine lands and other degraded soils: a review. J. Nanotechnol. 2012.

Liu, R., Lal, R., 2014. Synthetic apatite nanoparticles as a phosphorus fertilizer for soybean (*Glycine max*). Sci. Rep. 4, 5686.

Lu, H., Yuan, W., Zhou, J., Chong, P.L.G., 2015. Glucose synthesis in a protein-based artificial photosynthesis system. Appl. Biochem. Biotechnol. 177 (1), 105–117.

Majidi, M.R., Baj, R.F.B., Naseri, A., 2013. Carbon nanotube–ionic liquid (CNT–IL) nanocomposite modified sol-gel derived carbon-ceramic electrode for simultaneous determination of sunset yellow and tartrazine in food samples. Food Anal. Methods 6 (5), 1388–1397.

Manikandan, A., Subramanian, K.S., 2014. Fabrication and characterization of nanoporous zeolite-based N fertilizer. Afr. J. Agric. Res. 9 (2), 276–284.

Medeiros, B.G. de S., Pinheiro, A.C., Carneiro-da-Cunha, M.G., Vicente, A.A., 2012. Development and characterization of a nanomultilayer coating of pectin and chitosan – evaluation of its gas barrier properties and application on 'Tommy Atkins' mangoes. J. Food Eng. 110, 457–464.

Medeiros, B.Gd.S., Souza, M.P., Pinheiro, A.C., et al., 2014. Physical characterization of an alginate/lysozyme nanolaminate coating and its evaluation on 'coalho' cheese shelf life. Food Bioprocess Technol. 7 (4), 1088–1098.

Mihindukulasuriya, S.D.F., Lim, L.T., 2014. Nanotechnology development in food packaging: a review. Trends Food Sci. Technol. 40, 149–167.

Milani, N., McLaughlin, M.J., Stacey, S.P., Kirby, J.K., Hettiarachchi, G.M., Beak, D.G., et al., 2012. Dissolution kinetics of macronutrient fertilizers coated with manufactured zinc oxide nanoparticles. J. Agric. Food Chem. 60 (16), 3991–3998.

Misra, P., Shukla, P.K., Pramanik, K., Gautam, S., Kole, C., 2016. Nanotechnology for crop improvement. Plant Nanotechnology. Springer, Cham, pp. 219–256.

Mohammadi, A., Hashemi, M., Hosseini, S.M., 2015. Chitosan nanoparticles loaded with *Cinnamomum zeylanicum* essential oil enhances the shelf life of cucumber during cold storage. Postharvest Biol. Technol. 110, 203–213.

Molavi, O., Ma, Z., Mahmud, A., Alshamsan, A., Samuel, J., Lai, R., et al., 2008. Polymeric micelles for the solubilization and delivery of STAT3 inhibitor cucurbitacins in solid tumors. Int. J. Pharm. 347 (1–2), 118–127.

Monreal, C., DeRosa, M., Mallubhotla, S., Bindraban, P., Dimkpa, C., 2016. Nanotechnologies for increasing the crop use efficiency of fertilizer-micronutrients. Biol. Fertil. Soils 52 (3), 423–437.

Mukhopadhyay, S.S., 2014. Nanotechnology in agriculture: prospects and constraints. Nanotechnol. Sci. Appl. 7, 63–71.

Natali, M., Passeri, D., Reggente, M., Tamburri, E., Terranova, M.L., Rossi, M., 2016. Contact resonance atomic force microscopy for viscoelastic characterization of polymer-based nanocomposites at variable temperature. AIP Conf. Proc. 1749, 020008.

Ngo, Q.B., Dao, T.H., Nguyen, H.C., Tran, X.T., Van Nguyen, T., Khuu, T.D., et al., 2014. Effects of nanocrystalline powders (Fe, Co and Cu) on the germination, growth, crop yield and product quality of soybean (Vietnamese species DT-51). Adv. Nat. Sci. Nanosci. Nanotechnol. 5 (1), 015016.

Nuruzzaman, M.D., Rahman, M.M., Liu, Y., Naidu, R., 2016. Nanoencapsulation, nano-guard for pesticides: a new window for safe application. J. Agric. Food Chem. 64 (7), 1447–1483.

Pais, I., 1983. The biological importance of titanium. J. Plant Nutr. 6 (1), 3–131.

Pais, I., 1992. Criteria of essentiality, beneficiality and toxicity of chemical elements. Acta Alimentaria (Budapest) 21 (2), 145–152.

Pandey, S., Zaidib, M.G.H., Gururani, S.K., 2013. Recent developments in clay-polymer nanocomposites. Sci. J. Rev. 2 (11), 296–328.

Pankaj, Shakil, N.A., Kumar, J., Singh, M., Singh, K., 2012. Bioefficacy evaluation of con-trolled release formulations based on amphiphilic nano-polymer of carbofuran against Meloidogyne incognita infecting tomato. J. Environ. Sci. Health B 47, 520–528.

Paret, M.L., Vallad, G.E., Averett, D.R., Jones, J.B., Olson, S.M., 2013. Photocatalysis: effect of light-activated nanoscale formulations of TiO_2 on *Xanthomonas perforans* and control of bacterial spot of tomato. Phytopathology 103 (3), 228–236.

Parisi, C., Vigani, M., Rodríguez-Cerezo, E., 2015. Agricultural nanotechnologies: what are the current possibilities? Nano Today 10 (2), 124–127.

Patra, P., Choudhury, S.R., Mandal, S., Basu, A., Goswami, A., Gogoi, R., et al., 2013. Effect sulfur and ZnO nanoparticles on stress physiology and plant (*Vigna radiata*) nutrition. Advanced Nanomaterials and Nanotechnology. Springer, pp. 301–309.

Perdikaris, A., Vassilakos, N., Yiakoumettis, I., Kektsidou, O., Kintzios, S., 2011. Development of a portable, high throughput biosensor system for rapid plant virus detection. J. Virol. Methods 177 (1), 94–99.

Pereira, A.E., Grillo, R., Mello, N.F., Rosa, A.H., Fraceto, L.F., 2014. Application of poly (epsilon-caprolactone) nanoparticles containing atrazine herbicide as an alternative technique to control weeds and reduce damage to the environment. J. Hazard. Mater. 268, 207–215.

Pitambara, Archana, Shukla, Y.M., 2019. Nanofertilizers: a recent approach in crop production. In: Nanotechnology for Agriculture: Crop Production & Protection, Springer Nature, Singapore, 25.

Polyakov, A.Y., Lebedev, V.A., Volkov, D.S., Pankratov, D.A., Veligzhanin, A.A., Perminova, I.V., et al., 2019. Eco-friendly iron-humic nanofertilizers synthesis for the prevention of iron chlorosis in soybean (*Glycine max*) grown in calcareous soil. Front. Plant Sci. 10, 413.

Popova, M., Szegedi, A., Mavrodinova, V., Tušar, N.N., Mihály, J., Klébert, S., et al., 2014. Preparation of resveratrol-loaded nanoporous silica materials with different structures. J. Solid-State Chem. 219, 37–42.

Pradhan, S., Roy, I., Lodh, G., Patra, P., Choudhury, S.R., Samanta, A., Goswami, A., 2013. Entomotoxicity and biosafety assessment of PEGylated acephate nanoparticles: a bi-ologically safe alternative to neurotoxic pesticides. J. Environ. Sci. Health B 48, 559–569.

Prasad, R., Bhattacharyya, A., Nguyen, Q.D., 2017. Nanotechnology in sustainable agriculture: recent developments, challenges, and perspectives. Front. Microbiol. 8, 1014.

Preetha, P.S., Balakrishnan, N., 2017. A review of nano fertilizers and their use and functions in soil. Int. J. Curr. Microbiol. Appl. Sci. 6 (12), 3117–3133.

Qiu, Y., Park, K., 2001. Environment-sensitive hydrogels for drug delivery. Adv. Drug Deliv. Rev. 53 (3), 321–339.

Radkowski, A., Radkowska, I., Lemek, T., 2015. Effects of foliar application of titanium on seed yield in timothy (*Phleum pratense* L.). Ecol. Chem. Eng. S 22 (4), 691–701.

Rafique, M., Sadaf, I., Rafique, M.S., Tahir, M.B., 2017. A review on green synthesis of silver nanoparticles and their applications. Artif. Cells Nanomed. Biotechnol. 45, 1272–1291.

Rahimmalek, M., Goli, S.M., 2013. Evaluation of six drying treatments with respect to essential oil yield, composition and color characteristics of Thymys daenensis subsp. daenensis. Celak leaves. Ind. Crops Prod. 42, 613–619.

Rajkumar, R.S., Kandeepan, G., Susitha, K., 2006. Applications of nano-technology in poultry meat industry: a vision to 2020. In: Poultry Research Priorities to 2020. Proceedings of National Seminar. Contributory Papers. Central Avian Research Institute, Izatnagar, India, 2–3 November 2006 (pp. 43–48). Central Avian Research Institute, Indian Council of Agricultural Research.

Raliya, R., 2013. Rapid, low-cost, and ecofriendly approach for iron nanoparticle synthesis using *Aspergillus oryzae* TFR9. J. Nanopart. 2013.

Raliya, R., Saharan, V., Dimkpa, C., Biswas, P., 2017. Nanofertilizer for precision and sustainable agriculture: current state and future perspectives. J. Agric. Food Chem. 66 (26), 6487–6503.

Rana, K., Kumari, M., Mishra, A., Pudake, R.N., 2019. Engineered nanoparticles for increasing micronutrient use efficiency. Nanoscience for Sustainable Agriculture. Springer, pp. 25–49.

Rashidi, L., Vasheghani-Farahani, E., Rostami, K., Gangi, F., Fallahpour, M., 2013. Mesoporous silica nanoparticles as a nanocarrier for delivery of vitamin c. Iran. J. Biotechnol. 11 (4), 209–213.

Razzaq, A., Ammara, R., Jhanzab, H.M., Mahmood, T., Hafeez, A., Hussain, S., 2016. A novel nanomaterial to enhance growth and yield of wheat. J. Nanosci. Technol. 2 (1), 55–58.

Rhim, J.-W., Park, H.-M., Ha, C.-S., 2013. Bio-nanocomposites for food packaging applications. Prog. Polym. Sci. 38, 1629–1652.

Richards, R.M., 1981. Antimicrobial action of silver nitrate. Microbios 31 (124), 83–91.

Rico, C.M., Majumdar, S., Duarte-Gardea, M., Peralta-Videa, J.R., Gardea-Torresdey, J.L., 2011. Interaction of nanoparticles with edible plants and their possible implications in the food chain. J. Agric. Food Chem. 59 (8), 3485–3498.

Rispail, N., De Matteis, L., Santos, R., Miguel, A.S., Custardoy, L., Testillano, P.S., et al., 2014. Quantum dot and superparamagnetic nanoparticle interaction with pathogenic fungi: internalization and toxicity profile. ACS Appl. Mater. Interfaces 6 (12), 9100–9110.

Romanazzi, G., Nigrov, F., Ippolito, A., 2003. Short hypobaric treatments potentiate the effect of chitosan in reducing storage decay of sweet cherries. Postharvest Biol. Technol.

Ruffo Roberto, S., Youssef, K., Hashim, A.F., Ippolito, A., 2019. Nanomaterials as alternative control means against postharvest diseases in fruit crops. Nanomaterials 9 (12), 1752.

Roshanak, S., Rahimmalek, M., Goli, S.A.H., 2016. Evaluation of seven different drying treatments in respect to total flavonoid, phenolic, vitamin C content, chlorophyll, antioxidant activity and color of green tea (Camellia sinensis or C. assamica) leaves. J. Food Sci. Technol. 53 (1), 721–729.

Sablani, S.S., 2015. Polymer nanocomposites for food packaging applications. In: Hernández-Sánchez, H., Gutiérrez-López, F.G. (Eds.), Food Nanoscience and Nanotechnology. Springer International Publishing, Cham, pp. 205–212.

Sardella, D., Gatt, R., Valdramidis, V.P., 2017. Physiological effects and mode of action of ZnO nanoparticles against postharvest fungal contaminants. Food Res. Int. 101, 274–279.

Sathiyabama, M., 2019. Biopolymeric nanoparticles as a nanocide for crop protection. Nanoscience for Sustainable Agriculture. Springer, Cham, pp. 139–152, Sci. 74, 113–130.

Sekhon, B.S., 2014. Nanotechnology in agri-food production: an overview. Nanotechnol. Sci. Appl. 7, 31.

Semeykina, A.L., Skulachev, V.P., 1990. Sub micromolar Ag^+ increases passive Na^+ permeability and inhibits the respiration-supported formation of Na^+ gradient in *Bacillus* FTU vesicles. FEBS Lett. 269 (1), 69–72.

Shah, M.A., Wani, S.H., Khan, A.A., 2016. Nanotechnology and insecticidal formulations. J. Food Bioeng. Nanoprocess. 1, 285–310.

Shang, Y., Hasan, M., Ahammed, G.J., Li, M., Yin, H., Zhou, J., 2019. Applications of nanotechnology in plant growth and crop protection: a review. Molecules 24 (14), 2558.

Sheng, N., Boyce, M.C., Parks, D.M., Rutledge, G.C., Abes, J.I., Cohen, R.E., 2004. Multiscale micromechanical modeling of polymer/clay nanocomposites and the effective clay particle. Polymer 45 (2), 487–506.

Singh, H., 2016. Nanotechnology applications in functional foods; opportunities and challenges. Prev. Nutr. Food Sci. 21, 1–8.

Shi, S., Wang, W., Liu, L., Wu, S., Wei, Y., Li, W., 2013. Effect of chitosan/nano-silica coating on the physicochemical characteristics of longan fruit under ambient temperature. J. Food Eng. 118 (1), 125–131.

Singh, G., Rattanpal, H., 2014. Use of nanotechnology in horticulture: a review. Int. J. Agric. Sci. Vet. Med. 2, 34–42.

Singh, N., Amist, N., Yadav, K., Singh, D., Pandey, J., Singh, S., 2013. Zinc oxide nanoparticles as fertilizer for the germination, growth and metabolism of vegetable crops. J. Nanoeng. Nanomanuf. 3 (4), 353–364.

Singh, S., Singh, B.K., Yadav, S.M., Gupta, A.K., 2015. Applications of nanotechnology in agricultural and their role in disease management. Res. J. Nanosci. Nanotechnol. 5 (1), 1–5.

Solanki, P., Bhargava, A., Chhipa, H., Jain, N., Panwar, J., 2015. Nano-fertilizers and their smart delivery system. Nanotechnologies in Food and Agriculture. Springer, Cham, pp. 81–101.

Stephenson, G.R., 2003. Pesticide use and world food production: risks and benefits. In: Environmental Fate and Effects of Pesticides. ACS Publications. pp. 261–270.

Taiz, L., Zeiger, E., 2010. Plant Physiology, fifth ed. Sinauer Associates Inc, Massachusetts, p. 781.

Tan, H., Ma, R., Lin, C., Liu, Z., Tang, T., 2013. Quaternized chitosan as an antimicrobial agent: antimicrobial activity, mechanism of action and biomedical applications in orthopedics. Int. J. Mol. Sci. 14, 1854–1869. Available from: https://doi.org/10.3390/ijms14011854.

Tang, F., Li, L., Chen, D., 2012. Mesoporous silica nanoparticles: synthesis, biocompatibility and drug delivery. Adv. Mater. 24 (12), 1504–1534.

Tarafdar, J.C., Raliya, R., Mahawar, H., Rathore, I., 2014. Development of zinc nanofertilizer to enhance crop production in pearl millet (*Pennisetum americanum*). Agric. Res. 3 (3), 257–262.

Ubbink, J., Kruger, J., 2006. Physical approaches for the delivery of active ingredients in foods. Trends Food Sci. Technol. 17, 244–254.

Vargas, M., Albors, A., Chiralt, A., González-Martínez, C., 2006. Quality of cold-stored strawberries as affected by chitosan–oleic acid edible coatings. Post. Biol. Technol. 41, 164–171.

Wang, Y., Wang, S., Xu, M., Xiao, L., Dai, Z., Li, J. (2019). The impacts of γ-Fe2O3 and Fe3O4 nanoparticles on the physiology and fruit quality of muskmelon (Cucumis melo) plants. Environ. Pollut., 249, 1011–1018.

Wani, A.H., Shah, M.A., 2012. A unique and profound effect of MgO and ZnO nanoparticles on some plant pathogenic fungi. J. Appl. Pharm. Sci. 2 (3), 4.

Wanyika, H., Gatebe, E., Kioni, P., Tang, Z., Gao, Y., 2012. Mesoporous silica nanoparticles carrier for urea: potential applications in agrochemical delivery systems. J. Nanosci. Nanotechnol. 12 (3), 2221–2228.

Wilson, M.A., Tran, N.H., Milev, A.S., Kannangara, G.S.K., Volk, H., Lu, G.H.M., 2008. Nanomaterials in soils. Geoderma 146, 291–302.

Worrall, E.A., Hamid, A., Mody, K.T., Mitter, N., Pappu, H.R., 2018. Nanotechnology for plant disease management. Agronomy 8 (12), 285.

Yadav, A., Kon, K., Kratosova, G., Duran, N., Ingle, A.P., Rai, M., 2015. Fungi as an efficient mycosystem for the synthesis of metal nanoparticles: progress and key aspects of research. Biotechnol. Lett. 37, 2099–2120.

Yao, K.S., Li, S.J., Tzeng, K.C., Cheng, T.C., Chang, C.Y., Chiu, C.Y., et al., 2009. Fluorescence silica nanoprobe as a biomarker for rapid detection of plant pathogens, Advanced Materials Research, vol. 79. Trans Tech Publications Ltd, pp. 513–516.

Yata, V.K., Tiwari, B.C., Ahmad, I., 2018. Nanoscience in food and agriculture: research, industries and patents. Environ. Chem. Lett. 16 (1), 79–84.

Yoshizaki, Y., Yuba, E., Sakaguchi, N., Koiwai, K., Harada, A., Kono, K., 2014. Potentiation of pH-sensitive polymer-modified liposomes with cationic lipid inclusion as antigen delivery carriers for cancer immunotherapy. Biomaterials 35 (28), 8186–8196.

Youssef, A.M., 2013. Polymer nanocomposites as a new trend for packaging applications. Polym. Plast. Technol. Eng. 52, 635–660.

Yu, S.R., Liu, Y., Li, W., Liu, J.A., Yuan, D.S., 2012. The running-in tribological behavior of nano-SiO_2/Ni composite coatings. Composites Part B Eng. 43 (3), 1070–1076.

Zhao, F., Zhou, X., Shi, Y., Qian, X., Alexander, M., Zhao, X., et al., 2018. Highly efficient solar vapour generation via hierarchically nanostructured gels. Nat. Nanotechnol. 13 (6).

CHAPTER

11

Plant growth promoting bacteria as biocontrol agents against diseases of cereal crops

Divya Singh[1], *Sandeep Kumar Singh*[2], *Vipin Kumar Singh*[2], *Sougata Ghosh*[3], *Hariom Verma*[2] and *Ajay Kumar*[4]

[1]Central Sericulture Research and Training Institute, Mysore, India [2]Department of Botany, Centre of Advanced Study, Institute of Science, Banaras Hindu University, Varanasi, India [3]Department of Microbiology, School of Science, RK University, Rajkot, India [4]Department of Postharvest Science, Agricultural Research Organization (ARO), Volcani Center, Rishon LeZion, Israel

11.1 Introduction

Since ancient times, plant diseases are represented as an important limiting factor for food production and social development (Palmgren et al., 2015; He et al., 2016). It has been estimated that food commodity/crop with market value approximately 40 billion dollars are lost annually due to plant diseases caused by bacteria, fungi, nematodes (FAO, 2017). To manage plant diseases, pests, or pathogens farmers generally apply huge amounts of chemicals including nematicides, bactericides, and fungicides. These chemical pesticides residues have long persistence into the environment, which adversely affect the health of humans, plant, soil as well as the survivality of native microflora (Kumar et al., 2017; Singh et al., 2020). Moreover, continuous use of pesticides exerts deleterious effects on agricultural soil texture, soil quality, productivity of plants and induces resistance development in pests. The hazardous effects of chemical pesticides and the long persistence of degrading products pushed researchers and growers to explore new and sustainable methods of disease management, a major challenge because of the pathogens diversity and broad range host specificity (Lugtenberg and Kamilova, 2009).

In this context "biocontrol" is considered as a sustainable method for disease management in plants. The term "biocontrol" or "biological control" (BC) implies the use of microbial

DOI: https://doi.org/10.1016/B978-0-12-821843-3.00016-7

antagonists to suppress the growth of pest and pathogens. The organism responsible for such suppression is also referred as "biological control agent" (BCA) (Baker and Cook, 1974; Pal and Gardener, 2006). BC is not only effective against plant disease but still represents an attractive, sustainable approach because of deleterious effects of agrochemicals and save the time and expense in breeding disease resistant varieties. Although these biocontrol agents broadly be utilized as antagonistic agents against various pathogens attack during postharvest storage (Zhimo et al., 2020).

As BCAs, plant growth promoting rhizobacteria (PGPR) are one of the most effective and eco-friendly choices for disease management (Raupach and Kloepper, 2000; Compant et al., 2005; Singh et al., 2019). There are certain advantages of PGPR over conventional chemical pesticides in terms of nontoxicity, lower side effects as well as multifaceted modes of action (Pérez-Montaño et al., 2014). Some pathogens are polyphagous, and crops may be affected by a wide range of pests, including nematodes, fungi, bacteria, and viruses, therefore search for PGPR effective against variety of pests is highly desirable. Among different crops, cereals are frequent target of multiple pests, which results in huge losses of products during growth or even in postharvest (Whipps, 2001). The present chapter has attempted to discuss the current status of cereals, their diseases, PGPB as BCA, mode of action of PGPB, and its implications in management of relevant cereals crop diseases.

11.2 The status of cereal crops and diseases

Cereals produce edible grain seed that are made up of endosperm rich in carbohydrates. They belong to the monocotyledon family Poaceae. Cereals and cereal-based foods are rich sources of carbohydrates, proteins, vitamins, lipids, and minerals, and are the daily need for most of the populations in the world (Adebayo Oluwakemi and Omodele, 2015).

According to FAO (2014) (http://faostat3.fao.org), the three top cereal crops of the world, in terms of production million tonnes (MT) are maize (*Zea mays*), rice (*Oryza sativa*), and wheat (*Triticum aestivum* L.). According to Rosegrant et al. (2009), the production of maize in 2011 was approximately more than 883 MT, which should be doubled by 2050 in the developing countries. In the case of rice, global production was approximately 723 MT. Rice is considered as a strategic crop for world food security as pointed out by the FAO. The production of wheat in 2011 was approximately 704 MT, and a projected increase by 60% by the year 2050 has been estimated to fulfill the demand for wheat in developing countries. Changes in climatic conditions will negatively affect production by 29% at that time. The world population is estimated to increase from 5.8 billion to 9 billion by 2050 (Dixon et al., 2009). To assure food availability for such a huge population, farmers and researchers have to focus on sustainable methods for cereal production and postharvest storage.

Cereals are affected by soil-borne diseases most commonly than any other crop and the attack of pathogens cause diseases or even death of host plants, which results into huge yield losses (Gohel et al., 2006). The rising socioeconomic importance of cereals has called for a major focus on biocontrol strategies as a possible mean to control plant diseases for the sustainable management of crop productivity. In this context PGPR shows an important role in

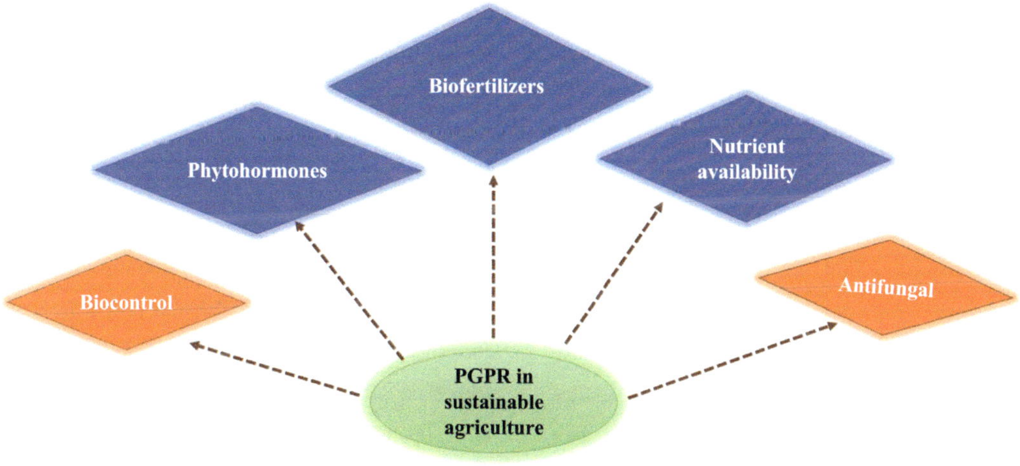

FIGURE 11.1 Schematic diagram showing perspectives of PGPR in sustainable agriculture.

the sustainable agriculture (Fig. 11.1), the mechanisms of PGPR include regulating hormonal and nutritional balance, inducing resistance against plant pathogens, and solubilizing nutrients for easy uptake by plants. Some of the most common diseases caused by pathogens in wheat are *Alternaria* leaf blight (*Alternaria triticina*), damping-off (*Arthrinium sacchari*), downy mildew (crazy top) (*Sclerophthora macrospora*), ergot (*Claviceps purpurea*), *Fusarium* head blight (*Fusarium acuminatum*), bacterial leaf blight (*Pseudomonas syringae*), and bacterial mosaic (*Clavibacter michiganensis*).

Being one of the most consumable cereals throughout the world, rice is sensitive to many diseases. Currently, more than 40 diseases are identified in rice, the most common of which include blight or brown spots (*Helminthosporium oryzae*), bunt of rice (*Tilletia barclayana*), bakanae disease (*Fusarium moniliforme*), bacterial blight (*Xanthomonas oryzae*), and stem rot (*Sclerotium oryzae*).

The pathogens affect the growth of plants during different development stages such as at seedling stage, flowering, seed setting, etc. The blight, root, and foot rot, as well as head blight (Fernandez and Jefferson, 2004) has been documented in every part of the world where cereals are cultivated. The pathogens not only affect yields but also contaminate the texture of food by release of mycotoxins (Leonard and Bushnell, 2003; Jennings et al., 2004). To mitigate these problems, effective control by PGPR strains has been widely used (Whipps, 2001; Lucy et al., 2004). Many bacterial genera such as *Pseudomonas, Bacillus, Klebsiella*, etc. have been reported as BCAs for cereals (Table 11.1).

11.3 Plant growth promoting bacteria as biological control agent

The rhizosphere is the hotspot of microbial populations and related interactions. It is the thin layer of soil surrounding a root, which is influenced by root exudates (Kumar et al., 2016; Lugtenberg and Kamilova, 2009; Kumar et al., 2015a). The bacterial species

TABLE 11.1 Biocontrol potential of some PGPR against various diseases of cereal.

PGPRs	Cereal crops	Disease causing agents	References
Pseudomonas fluorescens, Chryseobacterium balustinum	Rice	Blast disease	Lucas et al. (2009)
P. fluorescens	Rice	Leaf folder pest	Saravanakumar et al. (2007)
		Sheath rot disease	Saravanakumar et al. (2009)
Bacillus amyloliquefaciens, Bacillus subtilis	Rice	*Rhizoctonia solani*	Kumar et al. (2011)
Pseudomonas aeruginosa	Rice	Leaf blight	Yasmin et al. (2017)
B. subtilis	Rice	Rice sheath blight	Luo et al. (2015)
Bacillus methylotrophicus	Rice	Rice Blast	Shan et al. (2013)
Fluorescent Pseudomonads	Rice	Fungal pathogens	Reddy et al. (2010)
Delftia tsuruhatensis	Rice	*Xanthomonas oryzae Rhizoctonia solani*	Han et al. (2005)
Bacillus sp.	Rice	Sheath blight and bacterial panicle blight	Shrestha et al. (2016)
P. fluorescens	Barley	*Fusarium graminearum*	Henkes et al. (2011)
Actinomycetes	Wheat	*Fusarium* wilt disease	Laid et al. (2016)
P. fluorescens	Wheat	*Microdochium nivale*	Amein et al. (2008)
Bacillus megaterium TV-6D, Brevibacillus choshinensis TV-53D, Bacillus pumilus TV-3C	Wheat	Wheat root rot disease	Gökçe and Kotan (2016)
Acinetobacter, Pseudomonas, Staphylococcus, Bacillus, Enterobacter, Pantoea, Alcaligenes	Wheat	*Fusarium oxysporum, Fusarium culmorum, F. solani*	Egamberdieva et al. (2008)
Bacillus sp., *P. fluorescens*	Wheat	*G. graminis, R. solani, R. oryzae*	Cook et al. (2002)
Bacillus cereus	Wheat	Wheat sharp eyespot	Xu et al. (2014)
Bacillus strains, *Cryptococcus* strain	Wheat, barley	*Gibberella zeae, Fusarium* head blight	Khan et al. (2006)
B. subtilis	Maize	*Fusarium verticilloides*	Cavaglieri et al. (2005)
Trichoderma	Maize	Maize sheath blight	Zhang and Wen (2005)
B. amyloliquefaciens, Microbacterium oleovorans	Maize	*F. verticilloides*	Pereira et al. (2007)
Pseudomonas chlororaphis	Sorghum	*Macrophomina phaseolina*	Das et al. (2008)
Bacillus cereus, Brevibacterium laterosporus, P. fluorescens	Sorghum	*Pythium ultimum*	Idris et al. (2008)
P. fluorescens	Rye	*F. culmorum*	Kurek and Jaroszuk-Scisel (2003)

present in the rhizosphere, the rhizospheric bacteria, are directly or indirectly involved in growth promotion and disease management (Kumar et al., 2015b; Singh et al., 2019) (Fig. 11.2).

PGPR strains have some specific properties known for direct involvement in growth promotion. They include activities enhancing the availability of nutrients such as phosphorous, iron, and ammonia after converting insoluble forms to corresponding soluble forms via several mechanisms such as phosphate solubilization, siderophore and ammonia production, and synthesis of phytohormones such as indole-3-acetic acid (Kumar et al., 2014, 2016). They are also indirectly involved in growth promotion through biocontrol of phytopathogens via various methods such as the production of bioactive compounds, antibiotics, hydrogen cyanide (HCN) and cell wall–degrading enzymes (Singh et al., 2017a,b).

PGPB includes various groups of bacteria in which most common genera include: *Pseudomonas*, *Bacillus*, *Azotobacter*, *Azospirillum*, *Serratia*, *Klebsiella*, *Burkholderia* and have been broadly used for growth promotion, disease management as well as abiotic stress management (Berg and Smalla, 2009; Kumar et al., 2018). Currently many PGPR strains have been formulated as BCAs against diseases caused by nematodes, fungi, bacteria, or even viruses. Their effective potential depends on several factors such as method of application or inoculation, their physiological state, the concentration or volume of treatments, the presence of nutrients, etc. (Knudsen et al., 1997; Spadaro and Gullino, 2005; Fravel, 2005; Levenfors et al., 2008). Fig. 11.3 shows the overview of PGPB as biocontrol agent and their different aspects on cereal growth and developments.

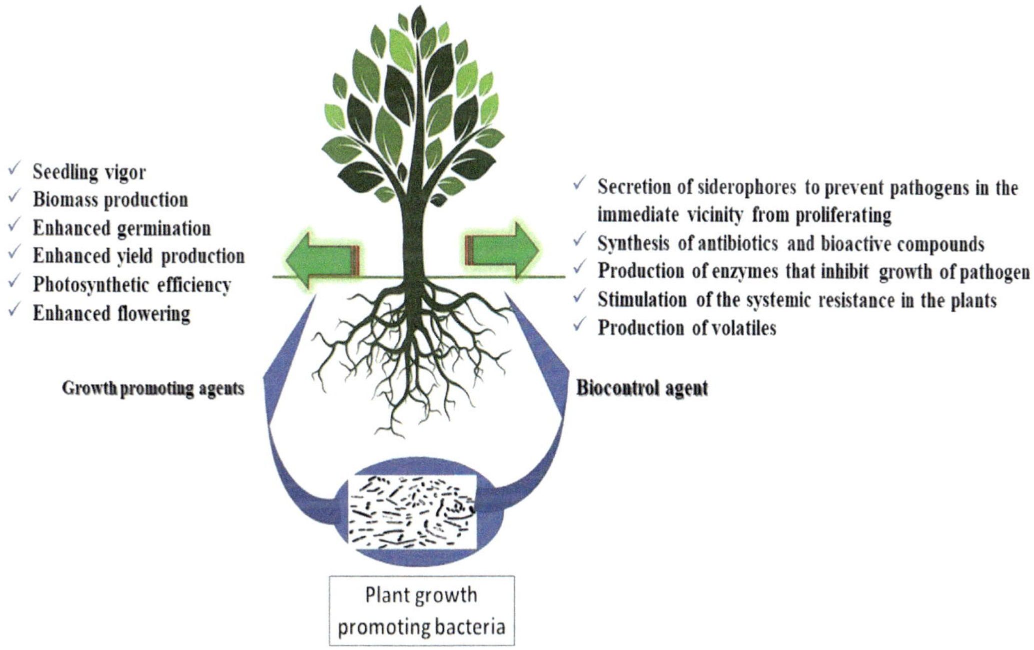

FIGURE 11.2 Application of plant growth promoting bacteria as growth promoting and biocontrol agents and their impact on cereal crops.

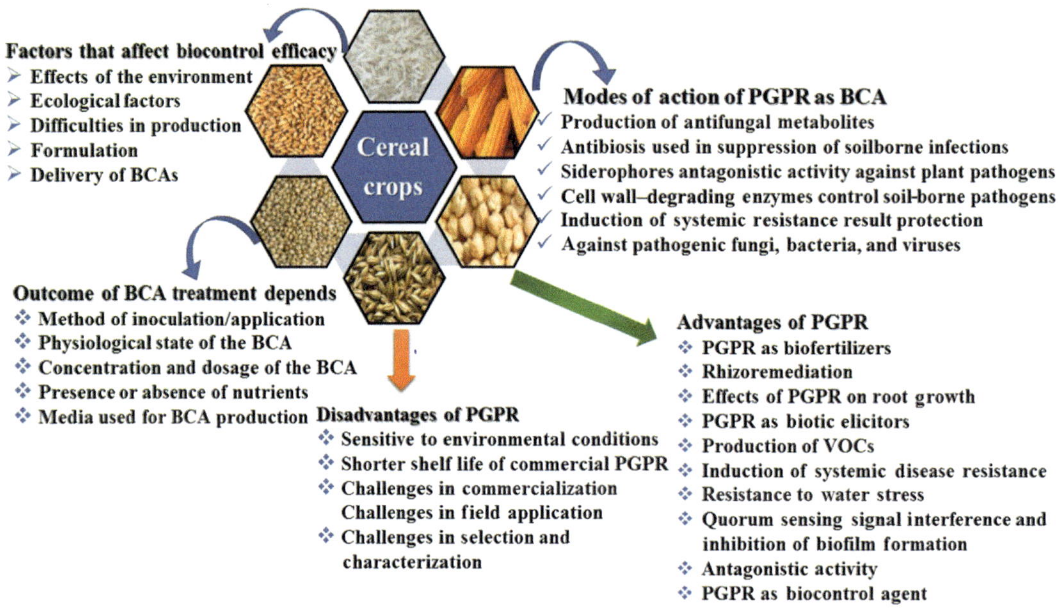

FIGURE 11.3 Overview of PGPB as biocontrol agents and their multidimensional approach on cereal crops.

11.4 Action mechanism of plant growth promoting bacteria

The mechanisms of PGPR facilitating the improvement of plant development consist of nitrogen fixation, solubilization of phosphate and mineralization of additional nutrients, production of siderophores, phytohormones (auxin-indole acetic acid, abscisic acid, ET, gibberellic acid, and cytokinin), ACC-deaminase, and induced systematic resistant. However, for effective control of plant pest or pathogens, microbial strains secrete various bioactive compounds, antibiotics, volatiles, siderophores, cell wall–degrading enzymes are being produced by microbial strains.

The details of microbial synthesized products having effective role in disease management have been discussed below.

11.4.1 Antibiotics production

Production of antibiotics is a key mechanism of PGPR action against pathogens. Antibiotics may inhibit physiological activities of microbes even at low concentration. A variety of low-molecular weight antibiotic compounds have been isolated and identified from different strains of PGPR belonging to genera *Bacillus*, *Pseudomonas*, *Trichoderma*, *Streptomyces*, and *Lysobacter*. Some of the most commonly reported antibiotics synthesized by PGPR are presented in Table 11.1. The most frequently reported antibiotics are phenazine-1-carboxylic acid, 2,4-diacetylphloroglucinol (DAPG), phenazine-1-carboxamide, pyrrolnitrin, pyoluterin, oomycin A, oligomycin A, kanosamine, butyrolactones, zwittermycin-A, aerugine, cepaciamide A, pseudomonic acid, azomycin, cepafungins, karalicin, sublancin, subtilosin A, subtilin, chlorotetain, bacilysin, mycobacillin, bacillaene, xanthobaccin, tropolone, rhizocticins, difficidin, surfactin, plipastatin, mupirocin, iturins compounds (mycosubtilins, iturins, and bacillomycins), and cyclic lipopeptides (tensin, viscosinamide, and amphisin). These compounds possess cytotoxic, phytotoxic, antioxidant, antitumor, antimicrobial, antiviral, and antihelminthic activities. Antibiotic compounds are categorized into volatile and nonvolatile antibiotics. Volatile antibiotics include HCN, benzothiazole, dimethyl trisulfide, cyclohexanol, 2-ethyl 1-hexanol, *n*-decanal, and nonanal. Nonvolatile antibiotics consist of DAPG, mupirocin, pyoluteorin, pyrrolnitrin, phenazine derivatives, zwittermicin-A, lipopeptides, and cyclic lipopeptides (Fernando et al., 2005; Lee et al., 2013; Goswami et al., 2016).

Antibiotic production is regulated by both endogenous and exogenous factors. Endogenous factors affect the synthesis of antibiotics are mainly GacS/GacA regulatory system, regulation based on LuxI and LuxR proteins, cell-density dependent regulation, and sigma factor–based regulatory systems (Fernando et al., 2005). However, exogenous factors include both abiotic and biotic factors such as pH, water content, temperature, soil composition, metabolic by-products, etc. (Milner et al., 1996; Duffy and Defago, 1999; Wegst-Uhrich et al., 2014). The addition of nutrients, carbon sources, minerals, and fertilizers also affects antibiotics production. Environmental conditions that stimulate the production of one antibiotic may suppress the synthesis of another compound. For example, glucose supplementation favors DAPG synthesis, but suppresses pyoluteorin synthesis in *Pseudomonas fluorescens* CHA0 (Duffy and Defago, 1999) and kanosamine synthesis in *Bacillus cereus* (Milner et al., 1996).

Additionally, biotic conditions also influence antibiotics synthesis. The availability of several metabolite by-products in the surrounding environment may affect production of antibiotics. The exudates or substrates released by plant shoots or roots can alter the biosynthesis of antibiotics. Furthermore, the availability of pyoluteorin and salicylate was found to affect the synthesis of DAPG in *P. fluorescens* CHA0 (Schnider-Keel et al., 2000). The composition and quantity of antibiotics produced by microbes are also remarkably influenced by the host plant. Milner et al. (1996) reported that biosynthesis and amounts of kanosamine was regulated by seedlings of alfalfa (*Medicago sativa*).

Antibiosis is an effectual mechanism of PGPR in repression of phytopathogens as observed in several crops. The fungal pathogens are inhibited by one or various groups of antibiotics produced by PGPR strains acting as BCAs (Haas and Défago, 2005). Antibiotics synthesized by PGPR biocontrol strains were found to repress soilborne infections in several crops such as barley, maize, rice, wheat, and chickpea (Raaijmakers et al., 2002). Some antibiotics also suppress pathogens by injuring membrane and inhibiting zoospore formation (de Souza et al., 2003). Others, such as phenazines, repress pathogens by inhibiting electron transport and damaging their macromolecules (Haas and Défago, 2005).

The application of consortia containing *P. fluorescens* and *Pseudomonas chlororaphis* on seeds suppressed damping-off in tomato and chilli (Kavitha et al., 2003). Aerugine application reduced the incidence of anthracnose and *Phytophthora* disease on cucumber and pepper, respectively (Lee et al., 2003). Spraying with isolate PA23 (*P. chlororaphis*) controlled *Sclerotinia sclerotiorum* infection in canola flowers (Duke et al., 2017). *P. chlororaphis* (PcPA23) and *Bacillus subtilis* (BsCBE4) is also demonstrated to reduce rhizome rot disease caused by *Pythium aphanidermatum* (Kavitha et al., 2012).

11.4.2 Volatile compounds

Microbial volatile organic compounds (mVOCs) are synthesized by soil microorganisms as secondary metabolites. They are defined as low-molecular weight (<300 g/mol) molecules, with a low boiling point and high vapor pressure that are produced by catabolic pathways such as glycolysis, lipolysis, and proteolysis. More than 300 different mVOCs have been identified to date, including 50 aromatic compounds, 75 fatty acid derivatives, 30 sulfur compounds, 74 nitrogen-containing compounds, 96 terpenoids, and 18 compounds containing halogen, tellurium, selenium, or other metals (Schulz and Dickschat, 2007). mVOCs can function over a wide area through diffusion in aqueous solutions and atmospheric infiltration. Therefore they exert physiological effects on pathogens without physical contact. In recent years, volatile organic compounds (VOCs) produced by BCAs has become important in biological control because of their eco-friendly and long-term effectiveness when used against target pathogens. Several VOCs emitted by BCAs have been studied for nematicidal, antimicrobial, plant growth promotion activities, and induced systemic resistance (ISR) (Bhattacharya and Jha, 2012; Raza et al., 2013; Audrain et al., 2015). Some species of *Pseudomonas* and *Bacillus* protected *Nicotiana* sp. and *Arabidopsis thaliana* from *Pectobacterium carotovorum* and *Pseudomonas syringae* by synthesizing short chain compounds that activated ISR via the ET pathway (Radruppa et al., 2010). Similarly, short chain organic compounds synthesized by *Bacillus* sp. protected *Cucumis sativus* from *P. syringae* through activation of ISR via the jasmonic acid (JA) pathway (Song and Ryu, 2013). VOCs produced by *Pseudomonas, Bacillus, Burkholderia, Streptomyces,* and *Serratia* sp. were capable to reduce growth and sporulation of pathogenic fungi in plants (Kai and Piechulla, 2009; Groenhagen et al., 2013).

11.4.3 Siderophores

Siderophores are low-molecular weight compounds (<10 kDa) synthesized by microorganisms for iron sequestration and biocontrol of plant diseases. They comprise pyoverdins, catechols, hydroxamates, and rhizobactin. Pyoverdine is a major class of siderophores synthesized by fluorescent strains of *Pseudomonas putida, P. syringae,* and *Pseudomonas aeruginosa* (Bouizgarne, 2013). *Pseudomonas fluorescens* NCIM5096 and *P. putida* NCIM2847 mainly produce hydroxamate-type siderophores (Sayyed et al., 2005).

The release of siderophores creates iron-depriving conditions for several fungal pathogens by chelating most of available iron ions existing in rhizospheric region. Thus they prevent the propagation of fungal pathogens (Kloepper et al., 1980; O'Sullivan and O'Gara, 1992; Mathiyazhagan et al., 2004). Iron deficiency inhibits growth and sporulation of *Fusarium oxysporum, Pythium ultimum,* etc. causing root rot and wilt diseases. Coinoculation of *Ralstonia* and *Bradyrhizobium* with *Pseudomonas* strains is noticed to suppress root rot disease (Sahu and Sindhu, 2011). Siderophore synthesizing *P. fluorescens* CHA0 (Couillerot et al., 2009), *P. syringae* pv. *syringae* strain 22d/93 (Wensing et al., 2010), and *P. putida* WCS strains (Weller, 2007) can be used as BCAs against soil-borne diseases. Battu and Reddy (2009) reported suppression of the fungal pathogens *Rhizoctonia solani* and *Pyricularia oryzae* through siderophore-mediated process, under laboratory conditions. *Pythium* induced root rot disease of wheat was inhibited by siderophores produced by *Pseudomonas* strain B324 (Becker and Cook, 1988). Moreover, some of the siderophores such as pyoveridin and pyocyanin are also required for induction of systemic resistance (Bouizgarne, 2013). Thus siderophore production is an important mechanism of plant protection implemented by some bacterial strains against root pathogens.

11.4.4 Cell wall–degrading enzymes

The production of cell wall–degrading enzyme is a key mechanism executed by BCAs against soil-borne pathogens. Cell wall–degrading enzymes (chitinase, cellulose, β-1,3-glucanase, protease, and lipase) produced by one or more PGPR strains can degrade cell wall of fungal hyphae and thus inhibit their growth and pathogenicity. Chitinase deteriorates chitin, a polymer of β-1,4-*N*-acetyl-glucoseamine, an integral component of fungal cell wall that is insoluble in water (Husson et al., 2017). The β-1,3-glucanase, a cell wall carbohydrate produced by *Paenibacillus* and *Streptomyces* strains, destroys the cell wall of *F. oxysporum* (Compant et al., 2005; Vaddepalli et al., 2017). Similarly, β-1,3-glucanase of *Burkholderia cepacia* degrades cell walls of *P. ultimum, R. solani,* and *Sclerotium rolfsii* (Compant et al., 2005). Chitinolytic activity has been reported in several BCAs including *Bacillus thuringiensis, B. subtilis, Bacillus licheniformis, Bacillus circulans,* and *B. cereus* (Sadfi et al., 2001). Some Gram-negative bacteria such as *Enterobacter agglomerans, P. fluorescens, P. aeruginosa,* and *Serratia marcescens* also possess a chitinolytic activity (Neiendam-Nielsen and Sørensen, 1999).

Cell wall–degrading enzymes synthesized by rhizobacteria influence the structural stability of the pathogens cell wall (Budi et al., 2000). Chitinases, β-1,3-glucanases, and peroxidases are component of pathogenesis-related (PR) proteins. The activation of these enzymes can induce ISR in plants (Yedidia et al., 1999). Kim et al. (2015) reported that *Bacillus* sp. JS stimulated up-regulation of PR-3 and PR-2 genes that encode chitinase and β-1,3-glucanase, respectively.

11.4.5 Induced systemic resistance

Plants have two types of defense mechanisms: ISR and systemic acquired resistance (SAR). SAR is switched on after pathogen's attack. Primary infection of pathogens activates SAR-based defense mechanisms in the host plants. The plant identifies the pathogen on a molecular basis and detoxifies pathogenic effects by modifying gene expression and synthesis of metabolites and hormones (Sunkar et al., 2012). SAR is executed through salicylic acid (SA) that is generally synthesized after infection and changes expression of PR protein-associated genes. On the other hand, ISR develops resistance in plants against future pathogenic attacks (Jain et al., 2013). It is induced by PGPR colonization (van Loon et al., 1998; Kloepper and Beauchamp, 1992) and executed through the activation of ethylene (ET) and JA pathways (Nawrocka and Malolepsza, 2013). Earlier studies indicated that PGPR stimulate ISR either by fortifying the cell wall structure or by modifying the host reactions. This phenomenon induces synthesis of defense signals such as proteinase, chitinase, and peroxidase inhibitors (Ramamoorthy et al., 2001; Nandakumar et al., 2001; Silva et al., 2004). Molecular studies suggested that SAR and ISR are linked through the NPR1gene (Ramamoorthy et al., 2001).

PGPR such as *P. fluorescens* WCS417r elicit ISR in various plant species through JA/ET signaling pathways, altering the expression of NPR1. Other PGPR strains such as *Pseudomonas protegens* CHA0, *P. fluorescens* Q2–87, and *Serratia marcescens* 90–166 induce ISR in *Arabidopsis* (Iavicoli et al., 2003; Ryu et al., 2004). PGPRs also induce systemic resistance through the expression of defense enzymes such as glucanase, chitinase, peroxidase, polyphenyl oxidase, and phenylalanine ammonia-lyase (Bharathi et al., 2004). In some cases, PGPR intervene in JA/ET signaling, overlapping the SA signaling pathway (Visca et al., 1993; Mur et al., 2006).

PGPR-induced ISR has been reported in several plant species (cucumber, bean, radish, tomato, and tobacco) that inhibit a wide range of phytopathogens including fungi, viruses, and bacteria (van Loon et al., 1998). ISR elicitors include flagella and lipopolysaccharides (cell wall components) as well as siderophores and antibiotics (metabolites) (Bakker et al., 2003; Lavicoli et al., 2003; Van loon and Bakker, 2005). Mostly, ISR-triggering PGPR reported to date are Gram-negative bacteria species associated with *Pseudomonas*.

Over the past decades, *Bacillus* spp., such as *Bacillus amyloliquefaciens*, *B. cereus*, *Bacillus mycoides*, *Bacillus pasteurii*, *Bacillus pumilus*, *B. subtilis*, and *Bacillus sphaericus*, have attracted much attention in relation to the ISR route, against several phytopathogens. Huang et al. (2012) reported that *B. cereus* produced dimethyl disulfide having an ISR-eliciting activity against fungal pathogens. It has been reported that *B. amyloliquefaciens* SQR9 synthesizes elicitors that act synergistically to stimulate ISR activity against diverse pathogens by acting on various genes involved in signaling pathways (Wu et al., 2018). VOCs may play important role in ISR signaling pathways. *B. amyloliquefaciens* IN937a and *B. subtilis* GB03 strains synthesize 2,3-butanediol that stimulates plant defense reactions (Ryu et al., 2004). In addition, surfactins and fengycins produced by *B. subtilis* can also control many plant diseases (Ongena et al., 2007).

Besides, endophytic bacteria may also persuade ISR activity: *P. fluorescens* 89B-61 mediated ISR acting against anthracnose in cucumber, and *B. pumilus* INR7, reduced the frequency of diseases such as angular leaf spot and cucurbit wilt by inhibiting the pathogen's developmental stages. *B. pumilus* INR7 also promotes resistance to *Colletotrichum gloeosporioides*, cucumber mosaic virus (CMV), *Ralstonia solanacearum*, *S. rolfsii*, and *Rhizoctonia solani* (in pepper and tomato) (Wei et al., 1996; Enebak and Carey, 2000; Zehnder et al., 2001; Murphy et al., 2000). Similarly, Harish et al. (2009) reported that endophytes stimulated ISR responses against banana bunchy top virus.

11.5 Biocontrol of soil-borne diseases in cereal crops

PGPR are very safe and effective choice for application as BCAs against various plants diseases. Generally it is found that, under laboratory and greenhouse conditions, many of them act better than in field conditions (Knudsen et al., 1997). The inconsistency of PGPR under field conditions is a major drawback of biological control (Gerhardson, 2002). This may be due to the lack of knowledge regarding formulations or effective applications in various environmental conditions (Emmert and Handelsman, 1999). The biocontrol efficacy of PGPR depends upon environmental conditions such as soil temperature, moisture, or clay content and pH (Callan et al., 1990). Many authors reported various PGPR strains as BCAs in cereal disease management. *Pseudomonas* and *Bacillus* spp. are the most commonly included important BCAs to manage root- and soil-borne pathogens, as reported in many crops such as chickpea, rice, wheat, maize, etc. (Grover et al., 2010; Dashti et al., 2012; Pérez-Montaño et al., 2014).

In the case of rice, Shan et al. (2013) used a strain of *Bacillus methylotrophicus* to control growth of *Magnaporthe oryzae*, the causal agent of rice blast. During assessment of its activity, they found significant biological control over the pathogen after inoculation. Phenaminomethylacetic acid was also observed as an active substance under fermentation conditions, with strong inhibitory capacity against the pathogen. In another study Luo et al. (2015) investigated the antagonistic activity of *B. subtilis* against *R. solani*, the causable agents for rice sheath, to elucidate the mechanism of plant leaves colonization. *B. subtilis* exhibited better biological control actions than of other bacterial mutants. Moreover, surfactin and bacillomycin L secretes of *Bacillus* strains contributed differently but synergistically to control the rice sheath blight, through the biofilm formation, antifungal activities and effective colonization. Zhang et al. (2012) utilized strains of *Bacillus* sp., *Burkholderia* sp., and *Streptomyces katrae* to assess their colonization ability and biological control effects against rice sheath blight, in detached leave, seedling and tillering stage. The results indicated that both *Bacillus* sp. and *Burkholderia* strains could colonize rice leaves at a high density. The biological control efficacy in detached leaves was very high for *Bacillus* and *Burkholderia* spp. strains, whereas lower (by approx. 50%) for *S. katrae*. All the three strains showed control (by 59.9%, 53.7%, and 38.4%, respectively) of rice sheath blight at the seedling stage. However, all showed unsatisfactory effect at tillering.

Delftia tsuruhatensis, a novel PGPR isolated from the rhizoplane of rice, has antagonistic activity against various plant pathogens. In rice, strains were used to control the growth of some common rice pathogens such as *Xanthomonas oryzae* pv. *oryzae*, *R. solani*, and *Pyricularia oryzae*. Moreover, the strains showed a high nitrogen-fixing activity in N-free Döbereiner culture medium (Han et al., 2005)

Most common wheat diseases include root or foot rot, seedling or head blight, and sharp eyespot. *Fusarium* spp. are necrotrophic pathogens of winter wheat that infect crops in all cultivated regions (Wachowska et al., 2013). They reduce wheat grain yields, and may also colonize stored grains. It has been estimated that fungal pathogens of wheat alone cause crop losses up to 29% (Oerke and Dehne, 2004), whereas in United States, the crown rot of wheat and barley, caused by *Fusarium* sp., are responsible of yield losses up to 35%, in commercial fields (Smiley et al., 2005).

Wheat sharp eyespot is also a disastrous disease caused by *Rhizoctonia cerealis* that infects the sheath or stem of wheat plants (Chen et al., 2008). Xu et al. (2014) reported an effective biological control of wheat sharp using an endophytic strain of *B. cereus*. However, this species may show virulence in humans because of toxin production (Kotiranta et al., 2000). In another study, Dang et al. (2014) also reported the potential of *B. amyloliquefaciens* inhibiting mycelial growth and germination of spores of *Rhizoctonia cerealis* and *Bipolaris maydis*.

Maize is an important crop, with an increasing area of cultivation in temperate regions, because of the increasing demand for food and livestock feed. There is a growing interest in the introduction of specific PGPR into the rhizosphere of this crop. Some rhizobacteria were selected for inoculation in maize, once their plant growth promoting potential and disease suppression were established (Nacamulli et al., 1997; Chiarini et al., 1998; Bevivino et al., 1998, 2000). *Bacillus mojavensis* is an endophytic bacterium patented for control of fungal diseases in maize and other plants. Culture extracts and filtrates from this bacterium were antagonistic to the pathogenic and mycotoxigenic fungus *Fusarium verticillioides* (Snook et al., 2009).

F. verticillioides is one of the most commonly reported soil-borne fungal pathogens of maize (Abbas et al., 1988; Bacon and Hinton, 1996) and the most prevalent *Fusarium* isolated in freshly harvested maize in Argentina. This microorganism not only causes severe reductions in yields and quality, but also produces secondary metabolites such as fumonisins and in particular fumonisin B_1, which affects human and animal health (Chu and Li, 1994; Harrison et al., 1990; Marasas et al., 1988). *F. verticillioides* can be found in maize fields at different stages of maize ear development (Kedera et al., 1992)

Several commercial products are currently available for biological control of diseases in cereals and a variety of other crops (Gardener and Fravel, 2002). New products are constantly entering in the market. Although there are several PGPR-based products available as soil inoculants for cereals, most of them are marketed as biofertilizers and not as BCA (Ryder et al., 1999)

11.6 Conclusions

In the current scenario, sustainable agriculture is the ultimate need for the whole world population because of lower marginal increases in yields, insurgence of resistance in pests and pathogens, adverse effects of chemical pesticides, and fertilizers on texture and productivity of soil and plants, as well as negative impacts on human health and the environment. Currently, huge amount of cereals are lost annually during growth (preharvest) or storage (at the postharvest stage). In this context, use of PGPR as BCAs appears as a safe, easy, and sustainable alternative. Increased knowledge and development of locally suitable application methods are, however, required for their extensive exploitation aiming at replacement of one or more synthetic pesticides. If applied as BCAs, some PGPR display a number of advantages over conventional control approaches. With the exception of human pathogenic species, they appear safe and less toxic because of exploitation of naturally occurring microorganism already present in the environment. Moreover, formulation and applications methods appear relatively easy, once a strain has been identified and characterized in the laboratory. Diverse PGPR action modes, such as the production of antibiotics, volatiles, siderophores, cell wall–degrading enzymes, and ISR appear as holding more benefits than conventional chemical approaches. Although, application of conventional pesticides represents the very important baseline in intensive agriculture, continuous use adversely affects food crops and soil ecosystems. In conclusion, the development of PGPR as suitable alternatives can contribute greatly to sustain food safety in many circumstances through local implementation of eco-friendly management practices to control cereal diseases.

References

Abbas, H.K., Mirocha, C.J., Meronuk, R.A., Pokorny, J.D., Gould, S.L., Kommedahl, T., 1988. Mycotoxins and *Fusarium* spp. associated with infected ears of corn in Minnesota. Appl. Env. Microbiol. 54, 1930–1933.

Adebayo Oluwakemi, O., Omodele, I., 2015. The current status of cereal (maize, rice and sorghum) crops cultivation in Africa: need for integration of advances in transgenic for sustainable crop production. IJAPR 3, 3.

Amein, T., Omer, Z., Welch, C., 2008. Application and evaluation of *Pseudomonas* strains for biocontrol of wheat seedling blight. Crop. Prot. 27, 532–536.

Audrain, B., Farag, M.A., Ryu, C.M., Ghigo, J.M., 2015. Role of bacterial volatile compounds in bacterial biology. FEMS Microbiol. Rev. 39, 222–233.

Bacon, C.W., Hinton, D.M., 1996. Symptomless endophytic colonization of maize by *Fusarium moniliforme*. Can. J. Bot. 74, 1195–1202.

Baker, K.F., Cook, R.J., 1974. Biological Control of Plant Pathogens. WH Freeman and Company.

Bakker, P.A.H.M., Ran, L.X., Pieterse, C.M.J., Van Loon, L.C., 2003. Understanding the involvement of induced systemic resistance in rhizobacteria-mediated biocontrol of plant diseases. Can. J. Plant. Pathol. 25, 5–9.

Battu, P.R., Reddy, M.S., 2009. Siderophore-mediated antibiosis of rhizobacterial fluorescent *Pseudomonads* against rice fungal pathogens. Int. J. Pharmtech Res. 1, 227–229.

Becker, J.O., Cook, R.J., 1988. Role of siderophores in suppression of *Pythium* species and production of increased-growth response of wheat by fluorescent pseudomonads. Phytopathology. 78, 778–782.

Berg, G., Smalla, K., 2009. Plant species and soil type cooperatively shape the structure and function of microbial communities in the rhizosphere. FEMS Microbiol. Ecol. 68, 1–13.

Bevivino, A., Dalmastri, C., Tabacchioni, S., Chiarini, L., 2000. Efficacy of *Burkholderia cepacia* MCI 7 in disease suppression and growth promotion of maize. Biol. Fert. Soils 31, 225–231.

Bevivino, A., Sarrocco, S., Dalmastri, C., Tabacchioni, S., Cantale, C., Chiarini, L., 1998. Characterization of a free-living maize-rhizosphere population of *Burkholderia cepacia*: effect of seed treatment on disease suppression and growth promotion of maize. FEMS Microbiol. Ecol. 27, 225–237.

Bharathi, R., Vivekananthan, R., Harish, S., Ramanathan, A., Samiyappan, R., 2004. Rhizobacteria-based bioformulations for the management of fruit rot infection in chillies. Crop. Prot. 23, 835–843.

Bhattacharya, P.N., Jha, D.K., 2012. Plant growth-promoting rhizobacteria (PGPR): emergence in agriculture. World J. Microbiol. Biotechnol. 28, 1327–1350.

Bouizgarne, B., 2013. Bacteria for plant growth promotion and disease management. In: Maheshwari, D. (Ed.), Bacteria in Agrobiology: Disease Management. Springer, Berlin, Heidelberg, pp. 15–47.

Budi, S.W., van Tuinen, D., Arnould, C., Dumas-Gaudot, E., Gianinazzi-Pearson, V., Gianinazzi, S., 2000. Hydrolytic enzyme activity of *Paenibacillus* sp. Strain B2 and effects of the antagonistic bacterium on cell integrity of two soil-borne pathogenic fungi. Appl. Soil. Ecol. 15, 191–199.

Callan, N.W., Mathre, D., Miller, J.B., 1990. Bio-priming seed treatment for biological control of *Pythium ultimum* pre emergence damping-off in sh-2 sweet corn. Plant. Dis. 74, 368–372.

Cavaglieri, L., Orlando, J.R.M.I., Rodriguez, M.I., Chulze, S., Etcheverry, M., 2005. Biocontrol of *Bacillus subtilis* against *Fusarium verticillioides* in vitro and at the maize root level. Res. Microbiol. 156, 748–754.

Chen, L., Zhang, Z., Liang, H., Liu, H., Du, L., Xu, H., et al., 2008. Overexpression of TiERF1 enhances resistance to sharp eyespot in transgenic wheat. J. Exp. Bot. 59, 4195–4204.

Chiarini, L., Bevivino, A., Tabacchioni, S., Dalmastri, C., 1998. Inoculation of *Burkholderia cepacia*, *Pseudomonas fluorescens* and *Enterobacter* sp. on *Sorghum bicolour*: root colonization and plant growth promotion of dual strain inocula. Soil. Biol. Biochem. 30, 81–87.

Chu, F.S., Li, G.Y., 1994. Simultaneous occurrence of fumonisin B1 and other mycotoxins in moldy maize collected from People's Republic of China in regions with high incidences of esophageal cancer. Appl. Env. Microbiol. 60, 847–852.

Compant, S., Duffy, B., Nowak, J., Clément, C., Barka, E.A., 2005. Use of plant growth-promoting bacteria for biocontrol of plant diseases: principles, mechanisms of action, and future prospects. Appl. Env. Microbiol. 71, 4951–4959.

Cook, J.R., Weller, D.M., El-Banna, A.Y., Vakoch, D., Zhang, H., 2002. Yield responses of direct-seeded wheat to rhizobacteria and fungicide seed treatments. Plant. Dis. 86, 780–784.

Couillerot, O., Prigent-Combaret, C., Caballero-Mellado, J., Moënne-Loccoz, Y., 2009. *Pseudomonas fluorescens* and closely-related fluorescent pseudomonads as biocontrol agents of soil-borne phytopathogens. Lett. Appl. Microbiol. 48, 505–512.

Dang, N.X., Popova, A.V., Hundertmark, M., Hincha, D.K., 2014. Functional characterization of selected LEA proteins from *Arabidopsis thaliana* in yeast and in vitro. Planta 240, 325–336.

Das, I.K., Indira, S., Annapurna, A., Seetharama, N., 2008. Biocontrol of charcoal rot in sorghum by fluorescent pseudomonads associated with the rhizosphere. Crop. Prot. 27, 1407–1414.

Dashti, N.H., Ali, N.Y., Cherian, V.M., Montasser, M.S., 2012. Application of plant growth-promoting rhizobacteria (PGPR) in combination with a mild strain of cucumber mosaic virus (CMV) associated with viral satellite RNAs to enhance growth and protection against a virulent strain of CMV in tomato. Can. J. Plant. Pathol. 34, 177–186.

de Souza, J.T., Weller, D.M., Raaijmakers, J.M., 2003. Frequency, diversity and activity of 2, 4-diacetylphloroglucinol producing fluorescent *Pseudomonas* spp. in Dutch take-all decline soils. Phytopathol 93, 54–63.

Dixon, J., Braun, H.J., Kosina, P., Crouch, J.H., 2009. Wheat Facts and Futures 2009. Cimmyt.

Duffy, B.K., Defago, G., 1999. Environmental factors modulating antibiotic and siderophore biosynthesis by *Pseudomonas fluorescens* biocontrol strains. Appl. Env. Microbiol. 65, 2429–2438.

Duke, K.A., Becker, M.G., Girard, I.J., Millar, J.L., Fernando, W.G.D., Belmonte, M.F., et al., 2017. The biocontrol agent *Pseudomonas chlororaphis* PA23 primes *Brassica napus* defenses through distinct gene networks. BMC Genomics. 18, 467.

Egamberdieva, D., Kamilova, F., Validov, S., Gafurova, L., Kucharova, Z., Lugtenberg, B., 2008. High incidence of plant growth-stimulating bacteria associated with the rhizosphere of wheat grown on salinated soil in Uzbekistan. Env. Microbiol. 10, 1–9.

Emmert, E.A., Handelsman, J., 1999. Biocontrol of plant disease: a (Gram-) positive perspective. FEMS Microbiol. Lett. 171, 1–9.

Enebak, S.A., Carey, W.A., 2000. Evidence for induced systemic protection to fusiform rust in loblolly pine by plant growth-promoting rhizobacteria. Plant. Dis. 84, 306–308.

FAO (Food and Agriculture Organization), 2014. FAOSTAT Agricultural Production 2013. FAO, Rome. <http://faostat3.fao.org/browse/Q/QC/E>.

FAO, IFAD, UNICEF, WFP, 2017. The state of food security and nutrition in the world. Food and Agriculture Organization of the United Nations, Rome.

Fernandez, M.R., Jefferson, P.G., 2004. Fungal populations in roots and crowns of common and durum wheat in Saskatchewan. Can. J. Plant. Pathol. 26, 325–334.

Fernando, W.G.D., Nakkeeran, S., Zhang, Y., 2005. Biosynthesis of antibiotics by PGPR and its relation in biocontrol of plant diseases. In: Siddiqui, Z.A. (Ed.), PGPR: Biocontrol and Biofertilization. Springer, Dordrecht, pp. 67–109.

Fravel, D.R., 2005. Commercialization and implementation of biocontrol. Annu. Rev. Phytopathol. 43, 337–359.

Gardener, B.B.M., Fravel, D.R., 2002. Biological control of plant pathogens: research, commercialization, and application in the USA. Plant Health Progress 3 (1), 17.

Gerhardson, B., 2002. Biological substitutes for pesticides. Trends Biotechnol. 20, 338–343.

Gohel, V., Singh, A., Vimal, M., Ashwini, P., Chhatpar, H.S., 2006. Bioprospecting and antifungal potential of chitinolytic microorganisms. Afr. J. Biotechnol. 5, 54–72.

Gökçe, A.Y., Kotan, R., 2016. Investigation of biological control possibilities of wheat root rot disease caused by *Bipolaris sorokiniana* (Sacc.) using PGPR and bio-control bacteria in controlled condition. Bitki Koruma Bül. 56, 49–75.

Goswami, D., Thakker, J.N., Dhandhukia, P.C., 2016. Portraying mechanics of plant growth promoting rhizobacteria (PGPR): a review. Cogent Food Agric. 2, 1127500.

Groenhagen, U., Baumgartner, R., Bailly, A., Gardiner, A., Eberl, L., Schulz, S., et al., 2013. Production of bioactive volatiles by different *Burkholderia ambifaria* strains. J. Chem. Ecol. 39, 892–906.

Grover, M., Nain, L., Singh, S.B., Saxena, A.K., 2010. Molecular and biochemical approaches for characterization of antifungal trait of a potent biocontrol agent *Bacillus subtilis* RP24. Curr. Microbiol. 60, 99–106.

Haas, D., Défago, G., 2005. Biological control of soil-borne pathogens by fluorescent pseudomonads. Nat. Rev. Microbiol. 3, 307–319.

Han, J., Sun, L., Dong, X., Cai, Z., Sun, X., Yang, H., et al., 2005. Characterization of a novel plant growth-promoting bacteria strain *Delftia tsuruhatensis* HR4 both as a diazotroph and a potential biocontrol agent against various plant pathogens. Syst. Appl. Microbiol. 28, 66–76.

Harish, S., Kavino, M., Kumar, N., Samiyappan, R., 2009. Differential expression of pathogenesis-related proteins and defense enzymes in banana: interaction between endophytic bacteria Banana bunchy top virus and *Pentalonia nigronervosa*. Biocontrol Sci. Tech. 19, 843–857.

Harrison, L.R., Colvin, B.M., Greene, I.T., Newman, L.E., Cole, J.R., 1990. Pulmonary edema and hydrothorax in swine produced by fumonisin B1, a toxic metabolite of *Fusarium moniliforme*. J. Vet. Diagn. Invest. 2, 217–221.

He, D.C., Zhan, J.S., Xie, L.H., 2016. Problems, challenges and future of plant disease management: from an ecological point of view. J. Integr. Agric. 15, 705–715.

Henkes, G.J., Jousset, A., Bonkowski, M., Thorpe, M.R., Scheu, S., Lanoue, A., et al., 2011. *Pseudomonas fluorescens* CHA0 maintains carbon delivery to *Fusarium graminearum*-infected roots and prevents reduction in biomass of barley shoots through systemic interactions. J. Exp. Bot. 62, 4337–4344.

Huang, C.J., Tsay, J.F., Chang, S.Y., Yang, H.P., Wu, W.S., Chen, C.Y., 2012. Dimethyl disulfide is an induced systemic resistance elicitor produced by *Bacillus cereus* C1L. Pest. Manag. Sci. 68, 1306–1310.

Husson, E., Hadad, C., Huet, G., Laclef, S., Lesur, D., Lambertyn, V., et al., 2017. The effect of room temperature ionic liquids on the selective biocatalytic hydrolysis of chitin via sequential or simultaneous strategies. Green. Chem. 19, 4122–4131.

Iavicoli, A., Boutet, E., Buchala, A., Métraux, J.P., 2003. Induced systemic resistance in *Arabidopsis thaliana* in response to root inoculation with *Pseudomonas fluorescens* CHAo. Mol. Plant. Microbe Interact. 16, 851–858.

Idris, H.A., Labuschagne, N., Korsten, L., 2008. Suppression of *Pythium ultimum* root rot of sorghum by rhizobacterial isolates from Ethiopia and South Africa. Biol. Control. 45, 72–84.

Jain, S., Vaishnav, A., Kasotia, A., Kumari, S., Gaur, R.K., Choudhary, D.K., 2013. Bacteria-induced systemic resistance and growth promotion in *Glycine max* L. Merrill upon challenge inoculation with *Fusarium oxysporum*. Proc. Natl Acad. Sci. India B: Biol. Sci 83, 561–567.

Jennings, P., Coates, M.E., Walsh, K., Turner, J.A., Nicholson, P., 2004. Determination of deoxynivalenol-and nivalenol-producing chemotypes of *Fusarium graminearum* isolated from wheat crops in England and Wales. Plant. Pathol. 53, 643–652.

Kai, M., Piechulla, B., 2009. Plant growth promotion due to rhizobacterial volatiles-an effect of CO_2? FEBS Lett. 583, 3473–3477.

Kavitha, K., Nakkeeran, S., Chandrasekar, G., 2012. Rhizobacterial-mediated induction of defense enzymes to enhance the resistance of turmeric (*Curcuma longa* L) to *Pythium aphanidermatum* causing rhizome rot. Arch. Phytopath Plant. Prot. 45, 199–219.

Kavitha, K., Nakkeeran, S., Chandrasekar, G., Fernando, W.G.D., Mathiyazhagan, S., Renukadevi, P., et al., 2003. Role of antifungal antibiotics, siderophores and IAA production in biocontrol of Pythium aphanidermatum inciting damping off in tomato by Pseudomonas chlororaphis and Bacillus subtilis. Proceedings of the 6th International Workshop on PGPR, Organised by IISR, Calicut. pp. 493–497.

Kedera, C.J., Leslie, J.F., Clattin, L.E., 1992. Systemic infection of corn by *Fusarium moniliforme*. Phytopathol 84, 603–607.

Khan, M.R., Fischer, S., Egan, D., Doohan, F.M., 2006. Biological control of *Fusarium* seedling blight disease of wheat and barley. Phytopathology. 96, 386–394.

Kim, J.S., Lee, J., Lee, C.H., Woo, S.Y., Kang, H., Seo, S.G., et al., 2015. Activation of pathogenesis-related genes by the rhizobacterium, *Bacillus* sp. JS, which induces systemic resistance in tobacco plants. Plant. Pathol. J. 31, 195–201.

Kloepper, J.W., Beauchamp, C.J., 1992. A review of issues related to measuring colonization of plant roots by bacteria. Can. J. Microbiol. 38, 1219–1232.

Kloepper, J.W., Leong, J., Teintze, M., Schroth, M.N., 1980. *Pseudomonas* siderophores: A mechanism explaining disease suppressive soils. Curr. Microbiol. 4, 317–320.

Knudsen, I.M., Hockenhull, J., Jensen, D.F., Gerhardson, B., Hökeberg, M., Tahvonen, R., et al., 1997. Selection of biological control agents for controlling soil and seed-borne diseases in the field. Eur. J. Plant. Pathol. 103, 775–784.

Kotiranta, A., Lounatmaa, K., Haapasalo, M., 2000. Epidemiology and pathogenesis of *Bacillus cereus* infections. Microb. Infect. 2, 189–198.

Kumar, K.V.K., Reddy, M.S., Kloepper, J.W., Lawrence, K.S., Yellareddygari, S.K.R., Zhou, X.G., et al., 2011. Screening and selection of elite plant growth promoting rhizobacteria (PGPR) for suppression of *Rhizoctonia solani* and enhancement of rice seedling vigor. J. Pure Appl. Microbiol. 5, 641–651.

Kumar, A., Singh, R., Giri, D.D., Singh, P.K., Pandey, K.D., 2014. Effect of *Azotobactor chroococcum* CL13 inoculation on growth and curcumin content of turmeric (*Curcuma longa* L). Int. J. Curr. Microbiol. Appl. Sci. 3, 275–283.

Kumar, A., Singh, M., Singh, P.P., Singh, S.K., Singh, P.K., Pandey, K.D., 2016. Isolation of plant growth promoting rhizobacteria and their impact of growth and curcumin content in *Curcuma longa* L. Biocatal. Agric. Biotechnol. 8, 1–7.

Kumar, A., Singh, V.K., Tripathi, V., Singh, P.P., Singh, A.K., 2018. Plant growth-promoting rhizobacteria (PGPR): perspective in agriculture under biotic and abiotic stress. In: Prasad, R., Gill, S.S., Tuteja, N. (Eds.), Crop Improvement Through Microbial Biotechnology. Elsevier, pp. 333–342.

Kumar, A., Vandana, R.S., Singh, M., Pandey, K.D., 2015b. Plant growth promoting rhizobacteria (PGPR): A promising approach for disease management. In: Singh, J.S., Singh, D.P. (Eds.), Microbes and Environmental Management. Studium Press, New Delhi, pp. 195–209.

Kumar, A., Vandana, R.S., Yadav, V., Giri, D.D., Singh, P.K., Pandey, K.D., 2015a. Rhizosphere and their role in plant microbe interaction. In: Chaudhary, K.K., Dhar, D.W. (Eds.), Microbes Soil and Their Agricultural Prospects. Nova Publisher, New York, pp. 83–97.

Kumar, A., Verma, H., Singh, V.K., Singh, P.P., Singh, S.K., Ansari, W.A., et al., 2017. Role of *Pseudomonas* sp. in sustainable agriculture and disease management. In: Meena, V., Mishra, P., Bisht, J., Pattanayak, A. (Eds.), Agriculturally Important Microbes for Sustainable Agriculture. Springer, Singapore, pp. 195–215.

Kurek, E., Jaroszuk-Scisel, J., 2003. Rye (Secale cereale) growth promotion by *Pseudomonas fluorescens* strains and their interactions with *Fusarium culmorum* under various soil conditions. Biol. Control. 26, 48–56.

Laid, B., Kamel, K., Mouloud, G., Manel, S., Walid, S., Amar, B., et al., 2016. Effects of plant growth promoting rhizobacteria (PGPR) on in vitro bread wheat (*Triticum aestivum* L.) growth parameters and biological control mechanisms. Adv. Microbiol. 6, 677–690.

Lavicoli, A., Boutet, E., Buchala, A., Métraux, J.P., 2003. Induced systemic resistance in *Arabidopsisthaliana* in response to root inoculation with *Pseudomonas fluorescens* CHA0. Mol. Plant Microbe Interact. 16, 851–858.

Lee, J.Y., Moon, S.S., Hwang, B.K., 2003. Isolation and antifungal and antioomycete activities of aerugine produced by *Pseudomonas fluorescens* Strain MM-B16. Appl. Env. Microbiol. 69, 2023–2031.

Lee, K.J., Oh, B.T., Seralathan, K.K., 2013. Advances in plant growth promoting rhizobacteria for biological control of plant diseases. In: Maheshwari, D. (Ed.), Bacteria in Agrobiology: Disease Management. Springer, Berlin, Heidelberg, pp. 1–13.

Leonard, K.J., Bushnell, W.R., 2003. Fusarium head blight of wheat and barley. American Phytopathological Society (APS Press).

Levenfors, J.P., Eberhard, T.H., Levenfors, J.J., Gerhardson, B., Hokeberg, M., 2008. Biological control of snow mould (*Microdochium nivale*) in winter cereals by *Pseudomonas brassicacearum*, MA250. BioControl 53, 651–665.

Lucas, J.A., Solano, B.R., Montes, F., Ojeda, J., Megias, M., Manero, F.J.G., 2009. Use of two PGPR strains in the integrated management of blast disease in rice (*Oryza sativa*) in Southern Spain. Field Crops Res. 114, 404–410.

Lucy, M., Reed, E., Glick, B.R., 2004. Application of free living plant growth-promoting rhizobacteria. Antonie Van Leeuwenhoek 86, 1–25.

Lugtenberg, B., Kamilova, F., 2009. Plant-growth-promoting rhizobacteria. Annu. Rev. Microbiol. 63, 541–556.

Luo, C., Zhou, H., Zou, J., Wang, X., Zhang, R., Xiang, Y., et al., 2015. Bacillomycin L and surfactin contribute synergistically to the phenotypic features of *Bacillus subtilis* 916 and the biocontrol of rice sheath blight induced by *Rhizoctonia solani*. Appl. Microbiol. Biotechnol. 99, 1897–1910.

Marasas, W.F.O., Kellerman, T.S., Gelderblom, W.C.A., Coetzer, J.A.W., Thiel, P.G., Der Lugt, J.J., 1988. Leukoencephalomalacia in a horse induced by fumonisin B1 isolated from *Fusarium moniliforme*. Onderstepoort J. Vet. Res. 55, 197–203.

Mathiyazhagan, S., Kavitha, K., Nakkeeran, S., Chandrasekar, G., Manian, K., Renukadevi, P., et al., 2004. PGPR mediated management of stem blight of *Phllanthus amarus* (Schum and Thonn) caused by *Corynespora cassiicola* (Berk and Curt) Wei. Arch. Phytopathol. Plant. Prot. 37, 183–199.

Milner, J.L., Silo-Suh, L., Lee, J.C., He, H., Clardy, J., Handelsman, J., 1996. Production of kanosamine by *Bacillus cereus* UW85. Appl. Env. Microbiol. 62, 3061–3065.

Mur, L.A.J., Kenton, P., Atzom, R., Miersch, O., Wasternack, C., 2006. The outcomes of concentration-specific interactions between salicylate and jasmonate signaling include synergy, antagonism, and oxidative stress leading to cell death. Plant Physiol 140, 249–262.

Murphy, J.F., Zehnder, G.W., Schuster, D.J., Sikora, E.J., Polstan, J.E., Kloepper, J.W., 2000. Plant growth-promoting rhizobacteria mediated protection in tomato against tomato mottle virus. Plant. Dis. 84, 779–784.

Nacamulli, C., Bevivino, A., Dalmastri, C., Tabacchioni, S., Ciarini, L., 1997. Perturbation of maize rhizosphere microflora following seed bacterization with *Burkholderia cepacia* MCI 7. FEMS Microbiol. Ecol. 23, 183–193.

Nandakumar, R., Babu, S., Viswanathan, R., Sheela, J., Raguchander, T., Samiyappan, R., 2001. A ner bioformulation containing plant growth promoting rhizobacterial mixture for the management of sheath blight and enhanced grain yield in rice. BioControl 46, 493–510.

Nawrocka, K., Malolepsza, U., 2013. Diversity in plant systemic resistance induced by *Trichoderma*. Biol. Control 67, 149–156.

Neiendam-Nielsen, M., Sørensen, J., 1999. Chitinolytic activity of *Pseudomonas fluorescens* isolates from barley and sugar beet rhizosphere. FEMS Microbiol. Ecol. 30, 217–227.

O'Sullivan, D.J., O'Gara, F., 1992. Traits of fluorescent *Pseudomonas* spp involved in suppression of plant root pathogens. Microbiol. Rev. 56, 662–676.

Oerke, E.C., Dehne, H.W., 2004. Safeguarding production-losses in major crops and the role of crop protection. Crop. Prot. 23, 275–285.

Ongena, M., Jourdan, E., Adam, A., Paquot, M., Brans, A., Joris, B., et al., 2007. Surfactin and fengycin lipopeptides of Bacillus subtilis as elicitors of induced systemic resistance in plants. Env. Microbiol. 9, 1084–1090.

Pal, K.K., Gardener, B.M., 2006. Biological control of plant pathogens. Plant Health Instr. 2, 1117–1142.

Palmgren, M.G., Edenbrandt, A.K., Vedel, S.E., Andersen, M.M., Landes, X., Østerberg, J.T., et al., 2015. Are we ready for back-to-nature crop breeding? Trends Plant Sci. 20, 155–164.

Pereira, P., Nesci, A., Etcheverry, M., 2007. Effects of biocontrol agents on *Fusarium verticillioides* count and fumonisin content in the maize agroecosystem: impact on rhizospheric bacterial and fungal groups. Biol. Control. 42, 281–287.

Pérez-Montaño, F., Alías-Villegas, C., Bellogín, R.A., Del Cerro, P., Espuny, M.R., Jiménez-Guerrero, I., et al., 2014. Plant growth promotion in cereal and leguminous agricultural important plants: from microorganism capacities to crop production. Microbiol. Res. 169, 325–336.

Raaijmakers, J.M., Vlami, M., de Souza, J.T., 2002. Antibiotic production by bacterial biocontrol agents. Antonie Van Leeuwenhoek 81, 537–547.

Radruppa, T., Biedrzycki, M.L., Kunjeti, S.G., Donofrio, N.M., Czymmek, K.J., Paré, P.W., et al., 2010. The rhizobacterial elicitor acetoin induces resistance in *Arabidopsis thaliana*. Commun. Integr. Biol. 3, 130–138.

Ramamoorthy, V., Viswanathan, R., Raguchandar, J., Prakasham, T., Samiyappan, R., 2001. Induction of systemic resistance by plant growth promoting rhizobacteria in crop plants against pest and diseases. Crop Prot. 20, 1–11.

Raupach, G.S., Kloepper, J.W., 2000. Biocontrol of cucumber diseases in the field by plant growth-promoting rhizobacteria with and without methyl bromide fumigation. Plant disease 84 (10), 1073–1075.

Raza, W., Faheem, M., Yousaf, S., Rajer, F.U., Yamin, M., 2013. Volatile and nonvolatile antifungal compounds produced by *Trichoderma harzianum* SQR-T037 suppressed the growth of *Fusarium oxysporum* f. sp. Niveum. Sci. Lett. 1, 21–24.

Reddy, B.P., Rani, J., Reddy, M.S., Kumar, K.V.K., 2010. Isolation of siderophore-producing strains of rhizobacterial fluorescent Pseudomonads and their biocontrol against rice fungal pathogens. Int. J. Appl. Biol. Pharm. Technol. 1, 133–137.

Rosegrant, M.R., Ringler, C., Sulser, T.B., Ewing, M., Palazzo, A., Zhu, T., et al., 2009. Agriculture and food security under global change: Prospects for 2025/2050. International Food Policy Research Institute, Washington, DC, pp. 145–178.

Ryder, M.H., Nong, Y.Z., Terrace, T.E., Rovira, A.D., Hua, T.W., Correll, R.L., 1999. Use of strains of *Bacillus* isolated in China to suppress take-all and *Rhizoctonia* root rot, and promote seedling growth of glasshousegrown wheat in Australian soils. Soil. Biol. Biochem. 31, 19–29.

Ryu, C.M., Farag, M.A., Hu, C.H., Reddy, M.S., Kloepper, J.W., Pare, P.W., 2004. Bacterial volatiles induce systemic resistance in *Arabidopsis*. Plant. Physiol. 134, 1017–1026.

Sadfi, N., Cherif, M., Fliss, I., Boudabbous, A., Antoum, H., 2001. Evaluation of bacterial isolates from salty soils and *Bacillus thuringiensis* strains for the biocontrol of *Fusarium* dry rot of potato tubers. J. Plant. Pathol. 83, 101–117.

Sahu, G., Sindhu, S., 2011. Disease control and plant growth promotion of green gram by siderophore producing *Pseudomonas* sp. Res. J. Microbiol. 6, 735–749.

Saravanakumar, D., Lavanya, N., Muthumeena, K., Raguchander, T., Samiyappan, R., 2009. Fluorescent pseudomonad mixtures mediate disease resistance in rice plants against sheath rot (*Sarocladium oryzae*) disease. BioControl 54, 273–286.

Saravanakumar, D., Vijayakumar, C., Kumar, N., Samiyappan, R., 2007. PGPR-induced defense responses in the tea plant against blister blight disease. Crop Prot. 26, 556–565.

Sayyed, R.Z., Badgujar, M.D., Sonawane, H.M., Mhaske, M.M., Chincholkar, S.B., 2005. Production of microbial iron chelators (siderophores) by *Fluorescent paeudomonads*. Indian. J. Biotechnol. 4, 484–490.

Schnider-Keel, U., Seematter, A., Maurhofer, M., Blumer, C., Duffy, B., Gigot-Bonnefoy, C., et al., 2000. Autoinduction of 2, 4-diacetylphloroglucinol biosynthesis in the biocontrol agent *Pseudomonas fluorescens* CHA0 and repression by the bacterial metabolites salicylate and pyoluteorin. J. Bacteriol. 182, 1215–1225.

Schulz, S., Dickschat, J.S., 2007. Bacterial volatiles: the smell of small organisms. Nat. Prod. Rep. 24, 814–842.

Shan, H., Zhao, M., Chen, D., Cheng, J., Li, J., Feng, Z., et al., 2013. Biocontrol of rice blast by the phenaminomethylacetic acid producer of *Bacillus methylotrophicus* strain BC79. Crop Prot. 44, 29–37.

Shrestha, B.K., Karki, H.S., Groth, D.E., Jungkhun, N., Ham, J.H., 2016. Biological control activities of rice-associated Bacillus sp. strains against sheath blight and bacterial panicle blight of rice. PLoS one. 11, 1.

Silva, H.S.A., da Silva Romeiro, R., Macagnan, D., de Almeida Halfeld-Vieira, B., Pereira, M.C.B., Mounteer, A., 2004. Rhizobacterial induction of systemic resistance in tomato plants: non-specific protection and increase in enzyme activities. Biol. Control 29, 288–295.

Singh, M., Kumar, A., Singh, R., Pandey, K.D., 2017b. Endophytic bacteria: a new source of bioactive compounds. 3 Biotech. 7, 315.

Singh, M., Singh, D., Gupta, A., Pandey, K.D., Singh, P.K., Kumar, A., 2019. Plant growth promoting rhizobacteria: application in biofertilizers and biocontrol of phytopathogens. PGPR Amelioration in Sustainable Agriculture. Woodhead Publishing, pp. 41–66.

Singh, V.K., Singh, A.K., Kumar, A., 2017a. Disease management of tomato through PGPB: current trends and future perspective. 3 Biotech. 7, 255.

Singh, V.K., Singh, S.K., Singh, P.K., Verma, H., Pandey, K.D., Singh, P.K., et al., 2020. Impact of pesticides applications on the growth and function of cyanobacteria. Advances in Cyanobacterial Biology. Academic Press, pp. 151–162.

Smiley, R.W., Gourlie, J.A., Easley, S.A., Patterson, L.M., Whittaker, R.G., 2005. Crop damage estimates for crown rot of wheat and barley in the Pacific Northwest. Plant. Dis. 89, 595–604.

Snook, M.E., Mitchell, T., Hinton, D.M., Bacon, C.W., 2009. Isolation and characterization of Leu7-surfactin from the endophytic bacterium Bacillus mojavensis RRC 101, a biocontrol agent for Fusarium verticillioides. J. Agric. Food Chem. 57, 4287–4292.

Song, G.C., Ryu, C.M., 2013. Two volatile organic compounds trigger plant self-defense against a bacterial pathogen and a sucking insect in cucumber under open field conditions. Int. J. Mol. Sci. 14, 9803–9819.

Spadaro, D., Gullino, M.L., 2005. Improving the efficacy of biocontrol against soilborne pathogens. Crop. Prot. 24, 601–613.

Sunkar, R., Li, Y.F., Jagadeeswaran, G., 2012. Functions of microRNAs in plant stress responses. Trends Plant. Sci. 17, 196–203.

Vaddepalli, P., Fulton, L., Wieland, J., Wassmer, K., Schaeffer, M., Ranf, S., et al., 2017. The cell wall-localiszed atypical β-1,3 glucanase ZERZAUST controls tissue morphogenesis in *Arabidopsis thaliana*. Development 144, 2259–2269.

van Loon, L.C., Bakker, P.A., Pieterse, C.M., 1998. Systemic resistance induced by rhizosphere bacteria. Annu. Rev. Phytopathol. 36, 453–483.

Van Loon, L.C., Bakker, P.A.H.M., 2005. Induced systemic resistance as a mechanism of disease suppression by rhizobacteria. In: PGPR: Biocontrol and biofertilization. Springer, Dordrecht, pp. 39–66.

Visca, C., Ciervo, A., Sanfilippo, V., Orsi, N., 1993. Regulated salicylate synthesis by *Pseudomonas* spp. Gen. Microbiol. 139, 1995–2001.

Wachowska, U., Kucharska, K., Jędryczka, M., Łobik, N., 2013. Microorganisms as biological control agents against *Fusarium* pathogens in winter wheat. Pol. J. Env. Stud. 22, 591–597.

Wegst-Uhrich, S.R., Navarro, D.A.G., Zimmerman, L., Aga, D.S., 2014. Assessing antibiotic sorption in soil: a literature review and new case studies on sulfonamides and macrolides. Chem. Cent. J. 8, 5.

Wei, G., Kloepper, J.W., Tuzun, S., 1996. Induced systemic resistance to cucumber diseases and increased plant growth by plant growth-promoting rhizobacteria under field conditions. Phytopathology 86, 221–224.

Weller, D.M., 2007. *Pseudomonas* biocontrol agents of soil borne pathogens: looking back over 30 years. Phytopathology 97, 250–256.

Wensing, A., Braun, S.D., Büttner, P., Expert, D., 2010. Impact of siderophore production by *Pseudomonas syringae* pv. *syringae* 22d/93 on epiphytic fitness and biocontrol activity against *Pseudomonas syringae* pv. *glycinea* 1a/96. Appl. Env. Microbiol. 76, 2704–2711.

Whipps, J.M., 2001. Microbial interactions and biocontrol in the rhizosphere. J. Exp. Bot. 52, 487–511.

Wu, G., Liu, Y., Xu, Y., Zhang, G., Shen, Q., Zhang, R., 2018. Exploring elicitors of the beneficial rhizobacterium *Bacillus amyloliquefaciens* SQR9 to induce plant systemic resistance and their interactions with plant signaling pathways. Mol. Plant Microbe Interact. 31, 560–567.

Xu, Y.B., Chen, M., Zhang, Y., Wang, M., Wang, Y., Huang, Q.B., et al., 2014. The phosphotransferase system gene ptsI in the endophytic bacterium *Bacillus cereus* is required for biofilm formation, colonization, and biocontrol against wheat sharp eyespot. FEMS Microbiol. Lett. 354, 142–152.

Yasmin, S., Hafeez, F.Y., Mirza, M.S., Rasul, M., Arshad, H.M., Zubair, M., et al., 2017. Biocontrol of bacterial leaf blight of rice and profiling of secondary metabolites produced by rhizospheric *Pseudomonas aeruginosa* BRp3. Front. Microbiol. 8, 1895.

Yedidia, I., Benhamou, N., Chet, I., 1999. Induction of defense responses in cucumber plants (*Cucumis sativus* L.) by the biocontrol agent *Trichoderma harzianum*. Appl. Env. Microbiol. 65, 1061–1070.

Zehnder, G.W., Murphy, J.F., Sikora, E.J., Kloepper, J.W., 2001. Application of rhizobacteria for induced resistance. Eur. J. Plant. Pathol. 107, 39–50.

Zhang, D.T., Peng, Z.K., Cao, Q.Q., Yang, M., Zheng, L., Zhou, E.X., 2012. Colonization of three antagonistic strains on rice plant and their biocontrol effects on rice sheath blight. J. Northwest. A&F Univ. 40, 97–102.

Zhang, G., Wen, C., 2005. Biocontrol of maize sheath blight with *Trichoderma* spp. Acta Phytophylacica Sin. 32, 353–356.

Zhimo, V.Y., Biasi, A., Kumar, A., Feygenberg, O., Salim, S., Vero, S., et al., 2020. Yeasts and bacterial consortia from kefir Grains are effective biocontrol agents of postharvest diseases of fruits. Microorganisms 8 (3), 42.

Further reading

Savary, S., Ficke, A., Aubertot, J.N., Hollier, C., 2012. Crop losses due to diseases and their implications for global food production losses and food security. Food Secur. 4, 519–537.

CHAPTER

12

Commercial production and formulation of microbial biocontrol agents

Simranjeet Singh[1,2,3,4,*], *Vijay Kumar*[5,*], *Shivika Datta*[6], *Satyender Singh*[4], *Daljeet Singh Dhanjal*[1], *Vaishali Dhaka*[1,2], *Kankan Sharma*[1] *and Joginder Singh*[1,2]

[1]Department of Biotechnology, Lovely Professional University, Phagwara, India [2]Climate Mitigation and Sustainable Agriculture Research lab (CMaSAR), Division of Research and Development, Lovely Professional University, Phagwara, India [3]Punjab Biotechnology Incubator (PBTI), S.A.S. Nagar, India [4]Regional Advance Water Testing Lab (RAWTL), Department of Water Supply and Sanitation, S.A.S. Nagar, India [5]Regional Ayurveda Research Institute for Drug Development, Gwalior, India [6]Department of Zoology, Doaba College Jalandhar, Jalandhar, India

12.1 Introduction

Food security has become the primary concern worldwide, as plant disease are decreasing their yield and making the yield insufficient as per the demand (Strange and Scott, 2005). The incurring of various pathogens in crops has severely affected the yield of cereals and fruits. Multiple methods like the use of resistant plants and rotation of crop are used to control the plant pathogens (Shuping and Eloff, 2017). Even chemical pesticides are used to regulate these pathogens and are the most popular method. But, excessive use of these chemical pesticides has led to significant environmental and health problems (Nicolopoulou-Stamati et al., 2016). Hence, now environmentally friendly approaches are being explored to control damage induced by plant pathogens. They are using natural

[*] Equal contributors are Simranjeet Singh and Vijay Kumar.

DOI: https://doi.org/10.1016/B978-0-12-821843-3.00014-3

antagonistic microbes, also known as biocontrol agents (BCA), which is an eco-friendly and, in a few cases, the only option to control these plant pathogens (Köhl et al., 2019). Other than controlling these plant diseases, these BCA also aid in plant growth. Addition to that, improve tolerance against stress, develop resistance in plant and help in the acquisition of nutrients (Andersen et al., 2018).

Based on the effect and mechanism, these BCA are used as biofertilizers, biopesticides, and in some cases, as plant strengtheners (Berg, 2009). Even, the global biopesticides market has gained the net worth of about 83.7 billion USD (Thakore, 2006). In general, BCA can be defined as the process in which antagonistic nonpathogenic microbes are used to suppress the growth of pathogens responsible for the disease development in an eco-friendly manner (O'Brien, 2017). Whereas, the US National Research Council defines BCA as the use of natural or genetically modified microbes or genes or gene products to reduce the harmful effects of plant pathogens (Heydari and Pessarakli, 2010). The two main principles of biological control agents are:

1. It is the phenomena of using the natural controller to regulate the population of harmful plant pathogens.
2. These BCAs does not restrict the pests but reduce the number of pest species, and substantially maintain the number of pest and natural enemies at lower density in the natural system.

The BCAs are used in both field and greenhouse to suppress the pathogens infecting various cereals, fruits, flowers, legumes, and ornamental plants (Lecomte et al., 2016). Broadly, BCA can also be stated as an approach in which activities of natural enemies are used to regulate the growth of plant pathogens. Mainly, BCA involves the use of only one antagonist against the specific pathogen in particular crop (Ab Rahman et al., 2018). The term "antagonism" was first coined by William Roberts in 1874 while studying the antagonistic activity of the microbes (Foster and Raoult, 1974). Later, in 1914 C.F. Von introduced the term biological control. In 1926 Sanford reported about the antagonistic activity of microbes residing the manure against *Streptomyces scabies* responsible for causing potato scab disease (Lin et al., 2018). Later, in 1932 Weindling reported about the antagonistic activity of *Trichoderma lignorum* which was effective in controlling and suppressing the growth of *Rhizoctonia solani* responsible for the onset of various soil diseases (Naglot et al., 2015). In 1980 Kloepper reported by the production of siderophores by antagonistic microbes as a controlling mechanism to reduce the growth of *Erwinia carotovora* (Shanmugaiah et al., 2015). Whereas, in 1993 Howell reported about two strains of *Trichoderma virens*, that is, P and Q. The P strain of *T. virens* was found to be ineffective to serve as BCA against disease related to seed in cotton, and it does not induce the resistance in cotton plant. On the other hand, strain Q was found to be an effective and potential strain to serve as BCA against seedling disease in cotton as it triggers the overexpression of phytoalexins in the cotton plant (Howell and Puckhaber, 2005). These studies highlight the real potential of BCAs against plant disease and make it an eco-friendly and effective method to circumvent the health issues linked with constant use of pesticides, control plant pathogens, and to preserve our natural habitat (Spadaro and Gullino, 2005).

This chapter intends to highlight the mechanisms employed by BCAs to control the phytopathogens, comprehend the potential of various microbes like bacteria, fungi, and yeast to suppress the growth of phytopathogens and highlight the potential of these BCAs as plant growth–promoting bacteria. Moreover, the existing status of these BCA, their composition, mass production, registered and commercially available BCAs will also be discussed.

12.2 Mode of action of biocontrol agents

There are two types of modes of action of biocontrol agents, namely direct and indirect (Fig. 12.2; Table 12.1). Direct modes of action involve antibiosis, parasitism, and competition, whereas indirect interactions are those in which plant itself responds to the antagonist presence resultant into the plant growth promotion or induced resistance. In competition mode, pathogen and antagonist compete for space or nutrients while in antibiosis antagonists produce the metabolites harmful to the pathogen. When the pathogen gets infected by the biocontrol agent, that phenomenon is known as parasitism (Köhl et al., 2019). Competition may happen at different levels for specific infection sites on seeds and roots. When the iron supply is limiting in the environment, the production of siderophores (iron-chelating agent) by various bacterial species has been proved as a mode of action of biocontrol agent in soil. Pyoverdines, *Pseudomonas* species, and pseudobactins are the main siderophores producing biocontrols (Lemanceau et al., 1993).

For the living organisms, iron is a vital nutrient element for the growth of plant and insufficiency of its availability gives rise to a furious competition (Loper and Henkels, 1997). The protection of plants from the phytopathogens using biocontrol agents through competing for niches and nutrients to survive is an important mechanism (Duffy, 2001). *Enterobacter cloacae* shows the competitive mechanism for the catabolism of nutrients by suppressing the *Pythium ultimum* (Kageyama and Nelson, 2003; van Dijk and Nelson, 2000). Different effects of biocontrol agents on phytopathogens and plants are well represented in Fig. 12.1.

To prevent the growth of the soil-borne pathogen, cell wall–degrading enzymes productions are the main characteristic of biocontrol agents (Singh et al., 2017). Degradation of β-1,3-glucan and chitin, which is present in the fungal cell walls by β-1,3-glucanase and chitinase, respectively which ultimately kills the pathogen (Burns and Dick., 2002; Goswami et al., 2016). In *Trichoderma*, competition is an effective mechanism of action for carbon nutrient, but it is also shown by some other strains of *Fusarium oxysporum* (Alabouvette et al., 2009; Sarrocco et al., 2009). In *Trichoderma* species, mycoparasitism is another mechanism, which is involved in the antagonism mode of action (Zeilinger et al., 1999). A necrotrophic bacterium which is plant pathogenic, kill and invade the tissue as well as utilize the nutrients available to the host plant. The advantage of the pathogens, which are necrotrophic is that they first colonize directly just after the incited death of the cell (Köhl et al., 2019). It was also found that PGPR inhibits the *Phytophthora capsici*, which is a notorious phytopathogen (Islam et al., 2016). When there is the interaction between the two microorganisms in which one is affecting the other in terms of gaining nutrients that competitive direct interaction is known as parasitism. For example, if the plant pathogen is the host and as well as act as a parasite that particular interaction is hyperparasitism, which is mostly observed in the fungi. In the case of the predatory bacterium, that is, *Bdellovibrio bacteriovorus*, which uses the nutrient present in the cytoplasm of

TABLE 12.1 Mode of action of biocontrol agents in association with development, risk assessment, and application of microbial biological control agents.

Mode of action	Risk of resistance	Risk of acute toxicity	Risk by environmental fate	Risk of metabolites	Risk of phytotoxicity	Environmental risk	Pathogen specificity	Dependency on plant physiology	Dependency on environmental conditions	Analytical method	Method for screening
Antimicrobial metabolites produced *in situ*	L	L	L	L	L	L	Specific-broad	L	M	SS	Simple bioassay
Antimicrobial metabolites in product	M	RAR	RAR	RAR	RAR	RAR	Broad	L	L	MS	In vitro assay
Assembled consortia combining different modes of action	L	L	L	L	L	L	Broad	L	L	MSS	In silico design associated with complex bioassay
Competition	L	VL	VL	VL	VL	VL	Broad	L	H	SS	Simple bioassay
Hyperparasitism	L	VL	VL	VL	VL	VL	Pathogen specific interactions	L	H	SS	Simple bioassay
Helper strains[a]	L	L	L	L	L	L	Depends on MBCA	R	R	SS	Complex bioassay
Induced resistance	L	VL	VL	VL	VL	VL	Specific -broad	H	L	SS	Complex bioassay conducted on plant
Modulation of indigenous microbiota	L	L	L	L	L	L	Broad	L	MD	MC	Site-specific complex bioassay

[a]*Used in combinations with MBCAs.*

H, *high;* L, *low;* M, *moderate;* MBCA, *microbial biological control agents;* MC, *microbiome characterization;* MD, *medium;* MS, *metabolite specific;* MSS, *multiple strain specific;* R, *reduced;* RAR, *risk assessment relevant;* SS, *strain specific;* VL, *very low.*

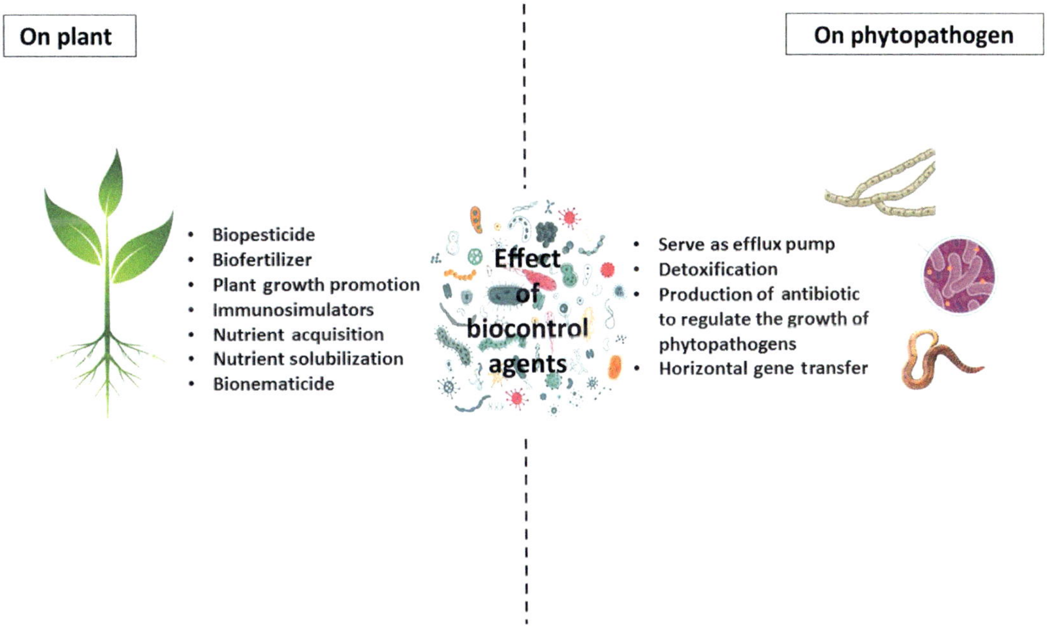

FIGURE 12.1 Different effects of biocontrol agents on phytopathogens and plants.

the Gram-negative bacteria (McNeely et al., 2017). The fungal species an approximate 30 in total showed the hyperparasitism against the rust pathogens (Zheng et al., 2017). The various strains of *Sclerotinia sclerotiorum* produce sclerotia of several thicknesses of their melanized cortical tissue. But, the morphological traits did not match with vulnerability to hyperparasitism by *Coniothyrium minitans* (Nicot et al., 2018). Hyperparasitism resistance has not been reported by a biocontrol agent. However, the resting structures viz. endospores, melanized sclerotia, and chlamydospores can be developed in the pathogen with high resistance against the hyperparasitism (Bardin et al., 2015) (Fig. 12.2).

12.3 Antibiosis

In 1889–90 Vuillemin coined the term antibiosis to explain antagonistic effects among the microorganisms). The condition in which two organisms shows the antagonistic effect; for example, one organism is harmful to another is known as antibiosis, for example, protozoal–bacterial antibiosis (Bentley and Bennett, 2003). The bacteria release various types of antibiotics during stressful conditions (Glick, 2012). Antibiotics are oligopeptide of low molecular weight, which shows deleterious effects at low concentration on numerous microorganisms (Maksimov et al., 2011). This antagonistic impact is named as antibiosis. To control the soil-borne infections by PGPR in different types of crops, antibiotics act as an effective approach (Handelsman and Stabb, 1996). *Pythium* species which is pathogenic, antibiotics disrupt their membrane by preventing the formation of zoospores

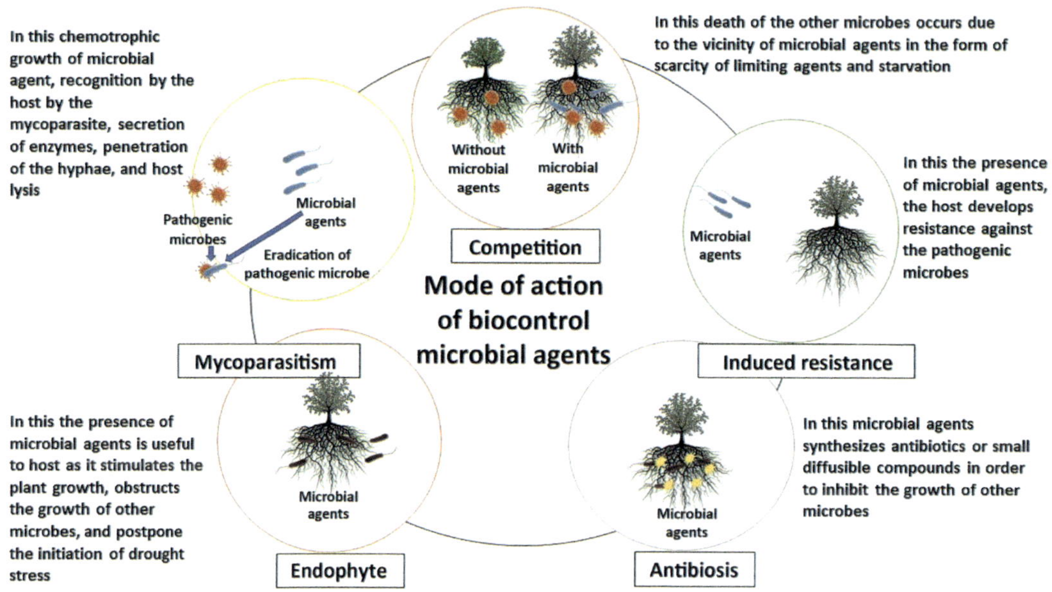

FIGURE 12.2 Mode of action of biocontrol agents.

(de Souza et al., 2003). *Pseudomonas fluorescens* strains produce the antibiotic phenazine-1-carboxylic acid, which suppresses take-all in the wheat seeds (Weller, 2007). Other than *Pseudomonas*, *Bacillus* species has an efficiency to act as biocontrol agents in several agriculture crops (Ashwini and Srividya, 2014). By means of parasitism or competition, antibiotics play a vital role in the control of various pests and act as a biocontrol agent (Raguchander et al., 2011). Hydrogen cyanide produced by some rhizobacteria kills the organisms, which are deleterious to the plant growth and helps in creating a favorable environment in the rhizospheric region (Tian et al., 2007). To reduce the growth of nematodes, Rizvi et al. (2012) obtained *P. fluorescens* as an effective bacteria fused with neem cake. To inhibit fungal pathogens, most PGPR biocontrol agents produce different types of antibiotics (Haas and Défago, 2005). Phenazines antibiotic inhibits the electron transport in organism suffering from a disease, which acts by disrupting the lipid layer and macromolecules. Pereira et al. (2009) reported that *Bacillus amyloliquefaciens* has the ability to suppress the *Fusarium verticillioides*, which causes the loss of yield and root rot. Pusey et al. (2011) evaluated the antibiosis against *Erwinia amylovora* on the stigma of apple flower using E325ad1 and E325ad2 antibiosis deficient mutants. For the control of *Sclerotinia homoeocarpa* on crawling twisted grass, PCA from *Pseudomonas aureofaciens* has been utilized as an immediate field treatment (Powell et al., 2000). Antinematode, antibacterial, and antifungal activities, their antibiosis action are observed by the production of metabolites nonspecific or specific (De La Fuente et al., 2002). The microorganism producing an antibiotic should have its own resistance mechanism against that particular antibiotic so that they can prevent themselves from suicide during the production of their toxin (Slininger et al., 2003). When the dose of a particular antibiotic is given constantly to a population of microorganism, those microbes which developed the natural resistance can survive leading to the antibiotic resistance into them, on the other hand, those microorganisms will die which do not contain the resistance factor (Fernando et al., 2005).

12.4 List of antibiotics produced by some biocontrol agents

1. *Pseudomonas* species—DAPG (2,4-diacetylphloroglucinol) (Raaijmakers and Weller 2001) and Phenazines (Chin-A-Woeng et al., 1998).
2. *P. fluorescens* Hv37a—*Oomycin* A (Gutterson et al., 1986).
3. *P. fluorescens* Pf5—*Pyoluteorin* (Howell and Stipanovic, 1979).
4. *P. fluorescens* BL915—*Pyrrolnitrin* (Ligon et al., 2000) along with *Burkholderia cepacia* B37w (Burkhead et al., 1994), *Enterobacter agglomerans* (Chernin et al., 1996) and *Serratia* spp. (Kalbe et al., 1996).
5. *P. borealis* MA342—2,3-de-epoxy-2,3-didehydro-rhizoxin (*DDR*) (Hokeberg et al., 1998).
6. *P. fluorescens* DR54—*Viscosinamide* (Nielsen et al., 1998).
7. *P. aureofaciens*—*Butyrolactone* (Pascale et al., 1997).
8. *Pseudomonas* sp. AB2—*N*-Butylbenzenesulfonamide (*N-BBS*) (Kim et al., 2000).
9. *Streptomyces violaceusniger* YCED-9—*AFA* (Trejo-Estrada et al., 1998).
10. *Pantoea agglomerans* EH318—*Pantocin A and B* (Wright et al., 2001).
11. *Stenotrophomonas* SB-K88—*Xanthobaccins* (Nakayama et al., 1999).
12. *B. cepacia* BC11—*AFC-BC11* (Kang et al., 1998).
13. *B. cereus* UW85—*Kanosamine* (Milner et al., 1996).
14. *B. cereus* UW85—*Zwittermicin A* (Silo-Suh et al., 1998).
15. *Actinomycetes*—tetracyclin, streptomycin, and kanamycin (Ulloa-Ogaz et al., 2015).
16. Streptomyces—Daptomycin (Ulloa-Ogaz et al., 2015).
17. PGPR—pyoluterin, pseudomonic acid, cepafungins, rhamnolipids, antiviral antibiotic karalicin, butyrolactones, phenazine-1-carboxamide, aerugine, 2,4-diacetylphloroglucinol, antitumor antibiotics FR901463, pyrrolnitrin, and phenazine-1-carboxylic acid (Fernando et al., 2005).
18. *T. virens*—Gliotoxin (Yadav et al., 2015).
19. *Chaetomium globosum* Cg-13—chaetomin (Yadav et al., 2015).

12.5 Plant growth–promoting attributes as biocontrol agents

On the basis of plant growth, bacteria associated with plant can be divided into three different groups viz. neutral, beneficial, and deleterious groups (Dobbelaere et al., 2003). Plant growth–promoting rhizobacteria (PGPR) are known as the free-living beneficial soil bacteria (Kloepper et al., 1989). They are more focused as a biocontrol agent toward the plant diseases like *Bacillus subtilis,* which is antagonistic to the soil-borne pathogens. In two distinct ways, direct or indirect, they can affect plant growth. In direct, PGPR involves either bacterial synthesized compound, for example, phytohormones supplying to the plant or nutrients uptake from nature. When PGPR prevent the phytopathogenic organisms promoting the plant growth indicates the indirect mechanism (Glick, 1995). PGPR as a biocontrol agent, irrespective of their role in the promotion of direct growth of plant they can function by a various mechanism, for example, decrement in the ethylene levels of the plant (Glick et al., 2007) or by producing the known auxin phytohormone (Patten and Glick, 2002). To support the chance of PGPR as biocontrol is because of the antagonistic activity across numerous pathogens (Ambrosini et al., 2012). Since PGPR are naturally present rhizosphere inhabitant, which directly promote the plant growth, and they are environment-friendly and nonpathogenic. Therefore they are inclined toward the more merits as compared to the chemical compounds (Kannojia et al., 2019). Another fact is that their applications are long term. But they have one disadvantage also that they are very sensitive toward the conditions specifically to soil conditions, desiccations, and temperature, etc. as being alive microorganisms. Compare to the chemical pesticide biocontrol PGPR has a short shell life (Labuschagne et al., 2010). Interactions of soil fauna and microorganisms with PGPR have been aimed as biocontrol or ISR against the bacterial, viral, and fungal diseases along with insect and nematode pests (Antoun and Prevost, 2005). In plants, they have varied defense actions which include cell wall–degradation enzymes, volatiles, antibiosis, ISR, and biosurfactants, etc. (Saraf et al., 2014). Another significant advancement, that is, in vivo expression technology which helps in the identifying genes of *P. fluorescens* related with the colonization of root and that is explicitly present in the rhizosphere region (Bloemberg and Lugtenberg, 2001). The improvement of the root colonization ability in a noncompetent strain was achieved by transferring a recombinase site-specific gene from the competent strain, namely, *P. fluorescens* (Compant et al., 2005). Different microorganisms may improve the consistency and level of biocontrol agents by rhizospheric bacteria or fungi and viability over different atmospheric conditions (Siddiqui, 2005).

12.6 Induced systematic host resistance

For stimulating defense response in the plant, beneficial bacteria interact with them in the rhizospheric microbiome against the pathogens. This defensive ability is known as induced systemic resistance (ISR) (Pieterse et al., 2014). The PGPR were used for their ability to regulate pathogens, which are soil-borne formerly the discovery of ISR (Bakker et al., 1991). The signaling mechanism requires jasmonic acid and ethylene (Fu and Dong, 2013). In disease suppression, the primary mechanism is ISR by PGPR. To trigger this response in potato plants, rhizobacterial isolates were found effective; namely, *Micrococcus luteus* strain TRK2-2, *Flexibacteraceae* bacterium and *Pseudomonas putida* strain TRL2–3 were found effective against the late blight disease (Kim and Jeun, 2006). *Pseudomonas* species are attacking the same crop plants against the various pathogens aimed at commercial exploitation by PGPR on the induction of systemic resistance (Ramamoorthy et al., 2001). The main factors of PGPR that facilitate ISR are siderophores, salicylic acid, and lipopolysaccharides. In some host plants, *Bacillus subtilis* and *Bacillus amyloliquefaciens* caused a major reduction in diseases (Kloepper et al., 2004). Huang et al. (2012) reported that ISR eliciting activity shown by the *B. cereus* strain, which produces the dimethyl disulfide antibiotic against the varied range of fungal pathogens. To induce the ISR, *B. amyloliquefaciens* SQR9, which act as synergistically against the different phytopathogens using the genes involved in the different signaling pathway (Wu et al., 2018). Proteinase inhibitors, lipoxygenase, peroxidase, phenylalanine ammonia-lyase, ascorbate peroxidase, and superoxide dismutase are the various defense enzymes, which are associated with ISR (Mhatre et al., 2017; Pokhare et al., 2012). By producing the phenolic compounds and phytoalexin, they initiate the ISR (Viswanathan et al., 2003). By inducing this mechanism, the infection has been reduced efficiently in root-knot and potato cyst nematode (Hallmann et al., 2001). It has been accounted for a sole activity for different *Bacillus* species utilized as biocontrol agents (Kloepper et al., 2004; Ongena et al., 2007). By expanding the mechanical and physical strength of cell wall, PGPR evokes ISR in plants along with the biochemical and physiological responses of the host (Silva et al., 2004). PGPR mediated ISR bear a resemblance phenotypically with exemplary pathogen resistance, wherein noninfected part of a formerly pathogen-infected plant turns out to be more resistant toward infection. This resistance is known as systemic acquired resistance (Ross, 1961). In ISR, nonpathogenic bacteria are vital to induce resistance, whereas, in systemic acquired resistance, necrotizing pathogens are required. Additionally, salicylic acid is an essential signaling molecule; however, jasmonic acid and ethylene glycol are involved in the ISR (Kuć, 1982; Van Loon et al., 1998). Ran et al. (2005) reported the bacterium *P. fluorescens* WCS374 in Eucalyptus, which induces the systematic resistance against the disease *Ralstonia solanacearum*, which was triggered by siderophore pseudobactin. Another strain of this bacterium known WCS365 in tomato induces the ISR (Kamilova et al., 2005). This strain expresses the chemotaxis against the components of root exudate, that is, citric acid (de Weert et al., 2002). *Arachis hypogaea* which are a peanut, *B. subtilis* strain AF1 shows the systemic resistance against *Aspergillus niger*. Due to the induction in seedling of an activity lipoxygenase, results in the reduction of crown rot seedlings (Sailaja et al., 1998). A virus *Bemisia tabaci* which is found in potato plants, *B. subtilis* BEB-DN found from the rhizospheric region of the same plant found to produce the systemic response (Valenzuela-Soto et al., 2010). The strains SE34 and C-9 of *Bacillus pumilus* and *Bacillus pasteurii*, respectively elicit ISR against the *Peronospora tabacina*, which causes tobacco (blue mold) (Zhang et al., 2004). Park et al. (2006) reported that ISR was induced against potato virus X and Y by the bacterium *Bacillus vallismortis* EXTN-1 in potato plants. Hwang and Benson (2003) demonstrated that the species, as well as strains of plant pathogenic and nonpathogenic, is present in the biocontrol agents *Rhizoctania* fungal genus. Kilic-Ekici and Yuen (2003) used *Lysobacter enzymogenes* C3 strain for inducing the resistance against leaf spot disease in *Festuca arundinacea*, which is caused by *Bipolaris sorokiniana*. Since, the plants ability which induces ISR compared to the necrotrophic pathogens is more effective than hemibiotrophic or biotrophic pathogens as they do not produce a hypersensitive response (Glazebrook, 2005). The species of *bacillus* protects the rice plant from the disease leaf blight (Udayashankar et al., 2011).

12.7 Commercial products of microbial biocontrol agents

Various microbial populations have been commercialized and used as biocontrol agents in different industrial sectors. They are used either to act against phytopathogens or for producing various drugs. Some of the commercialized products biocontrol agents are documented in Table 12.2.

TABLE 12.2 Commercially marketed different types of biocontrol agents.

Biocontrol agent	Manufacturer	Product name	Targeted crop/pathogen
Agrobacterium radiobacter strain 84	AgBioChem (United States)	Galtrol	*Agrobacterium tumefaciens*
A. radiobacter strain K 1026	Biocare	Nagol	*A. tumefaciens*
Ampelomyces quisqualis isolate M-10	Ecogen (United States)	AQ10	Powdery mildew
Aspergillus flavus AF36	Circle One Global (United States)	Alfa guard	*A. flavus*
Aureobasidium pullulans	BIOMIN (Austria)	Boni Protect	*Botrytis cinerea*
Azospirillum spp.	Manidharma Biotech	Biopromoter	Banana, millets, oilseeds, paddy, and sugarcane
Bacillus subtilis strain FZB24	AbiTep	RhizoPlus	Bulbs, strawberries, vegetables
B. subtilis strain GB34	Gustafson (United States)	GB34	*Rhizoctonia, Fusarium*
B. subtilis strain GB03	Growth products (USA)	Kodiak, Companion	*Rhizoctonia, Aspergillus*
B. subtilis QST716	AgraQuest	Serenade	Lettuce, spinach, tobacco, and tomato
Bradyrhizobium japonicum	Nitragin	Soil implant	Soybean
Burkholderia cepacia	Stine Microbial Products (United States)	Deny	*Rhizoctonia, Fusarium* and *Pythium* spp.
Candida oleophila	Ecogen, Inc. (United States)	Aspire	*Botrytis* spp., *Penicillium* spp.
Cryptococcus albidus	Lallemand group (South Africa)	Yield plus	*Botrytis* spp., *Penicillium expansum*
Fusarium oxysporum	Natural Plant Protection (France)	Fusaclean	*F. oxysporum*
Gliocladium catenulatum strain JI446	Kemira Agro Oy (Finland)	Prima stop soil guard	Soil-borne pathogens
Glomus intraradices	Bioscientific Inc. (United States)	Ascend/BuRIZE	Cotton
Metschnikowia fructicola	Lesaffre-Bionext (France/ Europe)	Shemer	*Aspergillus niger, Botrytis cinerea, Fusarium, Penicillium italicum, Rhizopus, Sclerotium*, and *stolonifer*
Pseudomonas aureofaciens strain TX-1	Eco Soil System	Bio-jet, spot less	*Pythium, Rhizoctonia solani*
Pseudomonas chlororaphis 6328	Eco Soil Systems Inc. (San Diego)	AtEze	*F. oxysporum*, and *R. solani*

(Continued)

TABLE 12.2 (Continued)

Biocontrol agent	Manufacturer	Product name	Targeted crop/pathogen
Pseudomonas fluorescens strain A506	Plant Health Technologies	Frostban	Bunch rot and Fire blight
Pseudomonas syringae ESC10 and *ESC11*	EcoScience Corp. (Longwood, FL)	Bio-Save 10LP, 110	*Geotrichum candidum, Mucor piriformis*, and *Penicillium* spp.
Pseudomonas trivialis 3Re27	Sourcon Padena	Salavida	Lettuce
Pythium oligandrum	Plant Protection Institute (Slovak Republic)	Polyversum, Polygandron	*Pythium ultimum*
Serratia plymuthica HROC48	Prophyta Biologischer Pflanzenschutz	Rhizostar	Oilseed rape and strawberries
Streptomyces griseoviridis K61	Kemira Oyj (Finland)	Mycostop	*Botrytis* spp., *Phomopsis* spp., *Phythophthora* sp., and *Pythium* spp.
Trichoderma harzianum T-22	BioWorks (United States)	Planter Box, Plant shield T-22 and Root Shield	*Fusarium* spp., and *Rhizoctonia solani*
T. harzianum T-39	BioWorks (United States)	Trichodex	*Botrytis cinerea*
Trichoderma harzianum, Trichoderma polysporum, and Trichoderma viride	Wood BioInnovation (Sweden), Agrimms Biologicals (New Zealand)	Binab T, Trichodowels, Trichoject, and Trichoseal	Foliar pathogens and wilt-causing fungi

12.8 Conclusions

Biocontrol agents nowadays are replacing the chemical agents to protect the plant from phytopathogens and act as plant growth stimulators, biopesticides, and biofertilizers. They were different action mechanism and act on hyperparasitism resulting in the production of lytic enzymes, which disintegrate the membrane of plant pathogens. Biocontrol agents interact with plants resulting in the mutualistic and commensalism interaction between them. Various species of bacteria, yeast, and fungi are reported till date, which acts as biocontrol agents which contribute to sustainable agricultural practices. Some phytopathogens are vulnerable to biocontrol agents, so it is necessary to develop effective strains using recombinant technologies, which increase the biocontrol efficiency. It is also important to develop formulated biocontrol agents having a good shelf life, suitable additives, and good delivery systems, which effectiveness and commercialization of biocontrol agents.

References

Ab Rahman, S.F.S., Singh, E., Pieterse, C.M., Schenk, P.M., 2018. Emerging microbial biocontrol strategies for plant pathogens. Plant. Sci. 267, 102–111.

Alabouvette, C., Olivain, C., Migheli, Q., Steinberg, C., 2009. Microbiological control of soil-borne phytopathogenic fungi with special emphasis on wilt-inducing *Fusarium oxysporum*. N. Phytol. 184 (3), 529–544.

Ambrosini, A., Beneduzi, A., Stefanski, T., Pinheiro, F.G., Vargas, L.K., Passaglia, L.M., 2012. Screening of plant growth promoting rhizobacteria isolated from sunflower (*Helianthus annuus* L.). Plant Soil 356 (1–2), 245–264.

Andersen, E.J., Ali, S., Byamukama, E., Yen, Y., Nepal, M.P., 2018. Disease resistance mechanisms in plants. Genes 9 (7), 339.

Antoun, H., Prévost, D., 2005. Ecology of plant growth promoting rhizobacteria. PGPR: Biocontrol and biofertilization. Springer, Dordrecht, pp. 1–38.

Ashwini, N., Srividya, S., 2014. Potentiality of *Bacillus subtilis* as biocontrol agent for management of anthracnose disease of chilli caused by *Colletotrichum gloeosporioides* OGC1. 3 Biotech. 4 (2), 127–136.

Bakker, P.A., Van Peer, R.O.N., Schippers, B., 1991. Suppression of soil-borne plant pathogens by fluorescent pseudomonads: mechanisms and prospects, Developments in Agricultural and Managed Forest Ecology, Vol. 23. Elsevier, pp. 217–230.

Bardin, M., Ajouz, S., Comby, M., Lopez-Ferber, M., Graillot, B., Siegwart, M., et al., 2015. Is the efficacy of biological control against plant diseases likely to be more durable than that of chemical pesticides? Frontiers in Plant Science 6, 566.

Bentley, R., Bennett, J.W., 2003. What is an antibiotic? Revisited. Advances in applied microbiology 52 (2003), 303–332.

Berg, G., 2009. Plant–microbe interactions promoting plant growth and health: perspectives for controlled use of microorganisms in agriculture. Appl. Microbiol. Biotechnol. 84 (1), 11–18.

Bloemberg, G.V., Lugtenberg, B.J., 2001. Molecular basis of plant growth promotion and biocontrol by rhizobacteria. Curr. Opin. Plant. Biol. 4 (4), 343–350.

Burkhead, K.D., Schisler, D.A., Slininger, P.J., 1994. Pyrrolnitrin production by biological control agent *Pseudomonas cepacia* B37w in culture and in colonized wounds of potatoes. Appl. Environ. Microbiol. 60 (6), 2031–2039.

Burns, R.G., Dick, R.P., 2002. Enzymes in the environment: activity, ecology, and applications. CRC Press.

Chernin, L., Brandis, A., Ismailov, Z., Chet, I., 1996. Pyrrolnitrin production by an *Enterobacter agglomerans* strain with a broad spectrum of antagonistic activity towards fungal and bacterial phytopathogens. Curr. Microbiol. 32 (4), 208–212.

Chin-A-Woeng, T.F., Bloemberg, G.V., van der Bij, A.J., van der Drift, K.M., Schripsema, J., Kroon, B., et al., 1998. Biocontrol by phenazine-1-carboxamide-producing *Pseudomonas chlororaphis* PCL1391 of tomato root rot caused by *Fusarium oxysporum* f. sp. radicis-lycopersici. Mol. Plant Microbe Interact. 11 (11), 1069–1077.

Compant, S., Duffy, B., Nowak, J., Clément, C., Barka, E.A., 2005. Use of plant growth-promoting bacteria for biocontrol of plant diseases: principles, mechanisms of action, and future prospects. Appl. Environ. Microbiol. 71 (9), 4951–4959.

De La Fuente, L., Quagliotto, L., Bajsa, N., Fabiano, E., Altier, N., Arias, A., 2002. Inoculation with *Pseudomonas fluorescens* biocontrol strains does not affect the symbiosis between rhizobia and forage legumes. Soil. Biol. Biochem. 34 (4), 545–548.

de Souza, J.T., Weller, D.M., Raaijmakers, J.M., 2003. Frequency, diversity, and activity of 2,4-diacetylphloroglucinol-producing fluorescent Pseudomonas spp. in Dutch take-all decline soils. Phytopathology 93 (1), 54–63.

de Weert, S., Vermeiren, H., Mulders, I.H., Kuiper, I., Hendrickx, N., Bloemberg, G.V., et al., 2002. Flagella-driven chemotaxis towards exudate components is an important trait for tomato root colonization by *Pseudomonas fluorescens*. Mol. Plant Microbe Interact. 15 (11), 1173–1180.

Dobbelaere, S., Vanderleyden, J., Okon, Y., 2003. Plant growth-promoting effects of diazotrophs in the rhizosphere. Crit. Rev. Plant. Sci. 22 (2), 107–149.

Duffy, B.K., 2001. Encyclopedia of Plant Pathology. John Wiley & Sons Inc, New York, pp. 243–244.

Fernando, W.D., Nakkeeran, S., Zhang, Y., 2005. Biosynthesis of antibiotics by PGPR and its relation in biocontrol of plant diseases. PGPR: biocontrol and biofertilization. Springer, Dordrecht, pp. 67–109.

Foster, W., Raoult, A., 1974. Early descriptions of antibiosis. J. R. Coll. Gen. Pract. 24 (149), 889–894.

Fu, Z.Q., Dong, X., 2013. Systemic acquired resistance: turning local infection into global defense. Annu. Rev. Plant. Biol. 64, 839–863.

Glazebrook, J., 2005. Contrasting mechanisms of defense against biotrophic and necrotrophic pathogens. Annu. Rev. Phytopathol. 43, 205–227.

Glick, B.R., 1995. The enhancement of plant growth by free-living bacteria. Can. J. Microbiol. 41 (2), 109–117.

Glick, B.R., 2012. Plant growth-promoting bacteria: mechanisms and applications. Scientifica 2012, 1–15. Article ID 963401.

Glick, B.R., Cheng, Z., Czarny, J., Duan, J., 2007. Promotion of plant growth by ACC deaminase-producing soil bacteria. In: New perspectives and approaches in plant growth-promoting Rhizobacteria research. Springer, Dordrecht, pp. 329–339.

Goswami, D., Thakker, J.N., Dhandhukia, P.C., 2016. Portraying mechanics of plant growth promoting rhizobacteria (PGPR): a review. Cogent Food Agric. 2 (1), 1127500.

Gutterson, N.I., Layton, T.J., Ziegle, J.S., Warren, G.J., 1986. Molecular cloning of genetic determinants for inhibition of fungal growth by a fluorescent pseudomonad. J. Bacteriol. 165 (3), 696–703.

Haas, D., Défago, G., 2005. Biological control of soil-borne pathogens by fluorescent pseudomonads. Nat. Rev. Microbiol. 3 (4), 307–319.

Hallmann, J., Quadt-Hallmann, A., Miller, W.G., Sikora, R.A., Lindow, S.E., 2001. Endophytic colonization of plants by the biocontrol agent *Rhizobium etli* G12 in relation to Meloidogyne incognita infection. Phytopathology 91 (4), 415–422.

Handelsman, J., Stabb, E.V., 1996. Biocontrol of soilborne plant pathogens. Plant Cell 8 (10), 1855.

Heydari, A., Pessarakli, M., 2010. A review on biological control of fungal plant pathogens using microbial antagonists. J. Biol. Sci. 10 (4), 273–290.

Hokeberg, M., Wright, S.A.I., Svensson, M., Lundgren, L.N., & Gerhardson, B. (1998). Mutants of *Pseudomonas chlororaphis* defective in the production of an antifungal metabolite express reduced biocontrol activity. In: Abstract Proceedings ICPP98, Edinburgh, Scotland.

Howell, C.R., Puckhaber, L.S., 2005. A study of the characteristics of "P" and "Q" strains of *Trichoderma virens* to account for differences in biological control efficacy against cotton seedling diseases. Biol. Control 33 (2), 217–222.

Howell, C.R., Stipanovic, R.D., 1979. Control of *Rhizoctonia solani* on cotton seedlings with *Pseudomonas fluorescens* and with an antibiotic produced by the bacterium. Phytopathology 69 (5), 480–482.

Huang, C.J., Tsay, J.F., Chang, S.Y., Yang, H.P., Wu, W.S., Chen, C.Y., 2012. Dimethyl disulfide is an induced systemic resistance elicitor produced by Bacillus cereus C1L. Pest Manag. Sci. 68 (9), 1306–1310.

Hwang, J., Benson, D.M., 2003. Expression of induced systemic resistance in poinsettia cuttings against Rhizoctonia stem rot by treatment of stock plants with binucleate Rhizoctonia. Biol. Control 27 (1), 73–80.

Islam, S., Akanda, A.M., Prova, A., Islam, M.T., Hossain, M.M., 2016. Isolation and identification of plant growth promoting rhizobacteria from cucumber rhizosphere and their effect on plant growth promotion and disease suppression. Front. Microbiol. 6, 1360.

Kageyama, K., Nelson, E.B., 2003. Differential inactivation of seed exudate stimulation of *Pythium ultimum* sporangium germination by *Enterobacter cloacae* influences biological control efficacy on different plant species. Appl. Environ. Microbiol. 69 (2), 1114–1120.

Kalbe, C., Marten, P., Berg, G., 1996. Strains of the genus Serratia as beneficial rhizobacteria of oilseed rape with antifungal properties. Microbiol. Res. 151 (4), 433–439.

Kamilova, F., Validov, S., Azarova, T., Mulders, I., Lugtenberg, B., 2005. Enrichment for enhanced competitive plant root tip colonizers selects for a new class of biocontrol bacteria. Environ. Microbiology 7 (11), 1809–1817.

Kang, Y., Carlson, R., Tharpe, W., Schell, M.A., 1998. Characterization of genes involved in biosynthesis of a novel antibiotic from *Burkholderia cepacia* BC11 and their role in biological control of *Rhizoctonia solani*. Appl. Environ. Microbiol. 64 (10), 3939–3947.

Kannojia, P., Choudhary, K.K., Srivastava, A.K., Singh, A.K., 2019. PGPR Bioelicitors: induced systemic resistance (ISR) and proteomic perspective on biocontrol. PGPR Amelioration in Sustainable Agriculture. Woodhead Publishing, pp. 67–84.

Kilic-Ekici, O., Yuen, G.Y., 2003. Induced resistance as a mechanism of biological control by *Lysobacter enzymogenes* strain C3. Phytopathology 93 (9), 1103–1110.

Kim, H.J., Jeun, Y.C., 2006. Resistance induction and enhanced tuber production by pre-inoculation with bacterial strains in potato plants against *Phytophthora infestans*. Mycobiology 34 (2), 67–72.

Kim, K.K., Kang, J.G., Moon, S.S., Kang, K.Y., 2000. Isolation and identification of antifungal N-butylbenzenesulphonamide produced by *Pseudomonas* sp. AB2. J. Antibiot. 53 (2), 131–136.

Kloepper, J.W., Lifshitz, R., Zablotowicz, R.M., 1989. Free-living bacterial inocula for enhancing crop productivity. Trends Biotechnol. 7 (2), 39–44.

Kloepper, J.W., Ryu, C.M., Zhang, S., 2004. Induced systemic resistance and promotion of plant growth by *Bacillus* spp. Phytopathology 94 (11), 1259–1266.

Köhl, J., Kolnaar, R., Ravensberg, W.J., 2019. Mode of action of microbial biological control agents against plant diseases: Relevance Beyond Efficacy. Front. Plant. Sci. 10, 845.

Kuć, J., 1982. Induced immunity to plant disease. Bioscience 32 (11), 854–860.

Labuschagne, N., Pretorius, T., Idris, A.H., 2010. Plant growth promoting rhizobacteria as biocontrol agents against soil-borne plant diseases. Plant growth and health promoting bacteria. Springer, Berlin, Heidelberg, pp. 211–230.

Lecomte, C., Alabouvette, C., Edel-Hermann, V., Robert, F., Steinberg, C., 2016. Biological control of ornamental plant diseases caused by *Fusarium oxysporum*: a review. Biol. Control. 101, 17–30.

Lemanceau, P., Bakker, P.A., De Kogel, W.J., Alabouvette, C., Schippers, B., 1993. Antagonistic effect of nonpathogenic *Fusarium oxysporum* Fo47 and pseudobactin 358 upon pathogenic *Fusarium oxysporum* f. sp. dianthi. Appl. Environ. Microbiol. 59 (1), 74–82.

Ligon, J.M., Hill, D.S., Hammer, P.E., Torkewitz, N.R., Hofmann, D., Kempf, H.J., et al., 2000. Natural products with antifungal activity from Pseudomonas biocontrol bacteria. Pest Manag. Sci. 56 (8), 688–695.

Lin, C., Tsai, C.H., Chen, P.Y., Wu, C.Y., Chang, Y.L., Yang, Y.L., et al., 2018. Biological control of potato common scab by *Bacillus amyloliquefaciens* Ba01. PLoS One 13 (4), e0196520.

Loper, J.E., Henkels, M.D., 1997. Availability of iron to *Pseudomonas fluorescens* in rhizosphere and bulk soil evaluated with an ice nucleation reporter gene. Appl. Environ. Microbiol. 63 (1), 99–105.

Maksimov, I.V., Abizgil'Dina, R.R., Pusenkova, L.I., 2011. Plant growth promoting rhizobacteria as alternative to chemical crop protectors from pathogens. Appl. Biochem. Microbiol. 47 (4), 333–345.

McNeely, D., Chanyi, R.M., Dooley, J.S., Moore, J.E., Koval, S.F., 2017. Biocontrol of *Burkholderia cepacia* complex bacteria and bacterial phytopathogens by *Bdellovibrio bacteriovorus*. Can. J. Microbiol. 63 (4), 350–358.

Mhatre, P.H., Kumar, J., Shakil, N.A., Kumar, R., Adak, T., 2017. New formulations of salicylic acid and their bioefficacy evaluation on wheat against cereal cyst nematode. Indian J. Nematol. 47 (2), 155–165.

Milner, J.L., Silo-Suh, L.A.U.R.A., Lee, J.C., He, H., Clardy, J., Handelsman, J.O., 1996. Production of kanosamine by *Bacillus cereus* UW85. Appl. Environ. Microbiol. 62 (8), 3061–3065.

Naglot, A., Goswami, S., Rahman, I., Shrimali, D.D., Yadav, K.K., Gupta, V.K., et al., 2015. Antagonistic potential of native *Trichoderma viride* strain against potent tea fungal pathogens in North East India. Plant Pathol. J. 31 (3), 278–289.

Nakayama, T., Homma, Y., Hashidoko, Y., Mizutani, J., Tahara, S., 1999. Possible role of xanthobaccins produced by *Stenotrophomonas* sp. strain SB-K88 in suppression of sugar beet damping-off sisease. Appl. Environ. Microbiol. 65 (10), 4334–4339.

Nicolopoulou-Stamati, P., Maipas, S., Kotampasi, C., Stamatis, P., Hens, L., 2016. Chemical pesticides and human h: the urgent need for a new concept in agriculture. Front. Public Health 4, 148-148.

Nicot, P.C., Roy, C., Duffaud, M., Villeneuve, F., Bardin, M., 2018. Can sclerotium size of *Sclerotinia sclerotiorum* be used as a predictor of susceptibility to *Coniothyrium minitans*? IOBC/WPRS Bull. 133, 181–186.

Nielsen, M.N., Sørensen, J., Fels, J., Pedersen, H.C., 1998. Secondary metabolite-and endochitinase-dependent antagonism toward plant-pathogenic microfungi of *Pseudomonas fluorescens* isolates from sugar beet rhizosphere. Appl. Environ. Microbiol. 64 (10), 3563–3569.

O'Brien, P.A., 2017. Biological control of plant diseases. Australas. Plant Pathol. 46 (4), 293–304.

Ongena, M., Jourdan, E., Adam, A., Paquot, M., Brans, A., Joris, B., et al., 2007. Surfactin and fengycin lipopeptides of *Bacillus subtilis* as elicitors of induced systemic resistance in plants. Environ. Microbiol. 9 (4), 1084–1090.

Park, K., Paul, D., Ryu, K.R., Kim, E.Y., Kim, Y.K., 2006. *Bacillus vallismortis* strain EXTN-1 mediated systemic resistance against potato virus Y and X in the field. Plant. Pathol. J. 22 (4), 360.

Pascale, G., Sauriol, F., Benhamou, N., Bélanger, R.R., Paulitz, T.C., 1997. Novel butyrolactones with antifungal activity produced by *Pseudomonas aureofaciens* strain 63-28. J. Antibiot. 50 (9), 742–749.

Patten, C.L., Glick, B.R., 2002. Role of *Pseudomonas putida* indoleacetic acid in development of the host plant root system. Appl. Environ. Microbiol. 68 (8), 3795–3801.

Pereira, P., Nesci, A., Etcheverry, M.G., 2009. Efficacy of bacterial seed treatments for the control of *Fusarium verticillioides* in maize. BioControl 54 (1), 103–111.

Pieterse, C.M., Zamioudis, C., Berendsen, R.L., Weller, D.M., Van Wees, S.C., Bakker, P.A., 2014. Induced systemic resistance by beneficial microbes. Annu. Rev. Phytopathol. 52, 347–375.

Pokhare, S., Shakil, N.A., Kumar, J., Singh, K., 2012. Foliar application of chemical elicitors induces biochemical changes in wheat against the cereal cyst nematode, Heterodera *a*venae. Nematol. Medit. 40, 181–187.

Powell, J.F., Vargas Jr, J.M., Nair, M.G., Detweiler, A.R., Chandra, A., 2000. Management of dollar spot on creeping bentgrass with metabolites of *Pseudomonas aureofaciens* (TX-1). Plant Dis. 84 (1), 19–24.

Pusey, P.L., Stockwell, V.O., Reardon, C.L., Smits, T.H.M., Duffy, B., 2011. Antibiosis activity of *Pantoea agglomerans* biocontrol strain E325 against *Erwinia amylovora* on apple flower stigmas. Phytopathology 101 (10), 1234–1241.

Raaijmakers, J.M., Weller, D.M., 2001. Exploiting genotypic diversity of 2,4 diacetylphloroglucinol-producing Pseudomonas spp.: characterization of superior root-colonizing *P. fluorescens* strain Q8r1-96. Appl. Environ. Microbiol. 67 (6), 2545–2554.

Raguchander, T., Saravanakumar, D., Balasubramanian, P., 2011. Molecular approaches to improvement of biocontrol agents of plant diseases. J. Biol. Control 25 (2), 71–84.

Ramamoorthy, V., Viswanathan, R., Raguchander, T., Prakasam, V., Samiyappan, R., 2001. Induction of systemic resistance by plant growth promoting rhizobacteria in crop plants against pests and diseases. Crop. Prot. 20 (1), 1–11.

Ran, L.X., Li, Z.N., Wu, G.J., Van Loon, L.C., Bakker, P.H., 2005. Induction of systemic resistance against bacterial wilt in *Eucalyptus urophylla* by fluorescent *Pseudomonas* spp. Eur. J. plant. Pathol. 113 (1), 59–70.

Rizvi, R., Mahmood, I., Tiyagi, S.A., Khan, Z., 2012. Conjoint effect of oil-seed cakes and *Pseudomonas fluorescens* on the growth of chickpea in relation to the management of plant-parasitic nematodes. Braz. Arch. Biol. Technol. 55 (6), 801–808.

Ross, A.F., 1961. Systemic acquired resistance induced by localized virus infections in plants. Virology 14 (3), 340–358.

Sailaja, P.R., Podile, A.R., Reddanna, P., 1998. Biocontrol strain of *Bacillus subtilis* AF 1 rapidly induces lipoxygenase in groundnut (*Arachis hypogaea* L.) compared to crown rot pathogen Aspergillus niger. Eur. J. Plant Pathol. 104 (2), 125–132.

Saraf, M., Pandya, U., Thakkar, A., 2014. Role of allelochemicals in plant growth promoting rhizobacteria for biocontrol of phytopathogens. Microbiol. Res. 169 (1), 18–29.

Sarrocco, S., Guidi, L., Fambrini, S., Degl'Innocenti, E., Vannacci, G., 2009. Competition for cellulose exploitation between *Rhizoctonia solani* and two trichoderma isolates in the decomposition of wheat straw. J. Plant Pathol. 331–338.

Shanmugaiah, V., Nithya, K., Harikrishnan, H., Jayaprakashvel, M., Balasubramanian, N., 2015. Biocontrol mechanisms of siderophores against bacterial plant pathogens. Sustainable Approaches to Controlling Plant Pathogenic Bacteria. pp. 167–190.

Shuping, D.S.S., Eloff, J.N., 2017. The use of plants to protect plants and food against fungal pathogens: a review. Afr. J. Tradit. Complement. Altern. Med. 14 (4), 120–127.

Siddiqui, Z.A., 2005. PGPR: prospective biocontrol agents of plant pathogens. PGPR: biocontrol and biofertilization. Springer, Dordrecht, pp. 111–142.

Silo-Suh, L.A., Stabb, E.V., Raffel, S.J., Handelsman, J., 1998. Target range of zwittermicin A, an aminopolyol antibiotic from *Bacillus cereus*. Curr. Microbiol. 37 (1), 6–11.

Silva, H.S.A., da Silva Romeiro, R., Macagnan, D., de Almeida Halfeld-Vieira, B., Pereira, M.C.B., Mounteer, A., 2004. Rhizobacterial induction of systemic resistance in tomato plants: non-specific protection and increase in enzyme activities. Biol. Control 29 (2), 288–295.

Singh, D., Raina, T.K., Singh, J., 2017. Entomopathogenic fungi: An effective biocontrol agent for management of insect populations naturally. Journal of Pharmaceutical Sciences and Research 9 (6), 833.

Slininger, P.J., Behle, R.W., Jackson, M.A., Schisler, D.A., 2003. Discovery and development of biological agents to control crop pests. Neotrop. Entomol. 32 (2), 183–195.

Spadaro, D., Gullino, M.L., 2005. Improving the efficacy of biocontrol agents against soilborne pathogens. Crop. Prot. 24 (7), 601–613.

Strange, R.N., Scott, P.R., 2005. Plant disease: a threat to global food security. Annu. Rev. Phytopathol. 43, 83–116.

Thakore, Y., 2006. The biopesticide market for global agricultural use. Ind. Biotechnol. 2 (3), 194–208.

Tian, B.Y., Yang, J.K., Lian, L.H., Wang, C.Y., Zhang, K.Q., 2007. Role of neutral protease from *Brevibacillus laterosporus* in pathogenesis of nematode. Appl. Microbiol. Biotechnol. 74, 372–380.

Trejo-Estrada, S.R., Paszczynski, A., Crawford, D.L., 1998. Antibiotics and enzymes produced by the biocontrol agent *Streptomyces violaceusniger* YCED-9. J. Ind. Microbiol. Biotechnol. 21 (1–2), 81–90.

Udayashankar, A.C., Nayaka, S.C., Reddy, M.S., Srinivas, C., 2011. Plant growth-promoting rhizobacteria mediate induced systemic resistance in rice against bacterial leaf blight caused by *Xanthomonas oryzae* pv. oryzae. Biol. Control 59 (2), 114–122.

Ulloa-Ogaz, A.L., Muñoz-Castellanos, L.N., Nevárez-Moorillón, G.V., 2015. Biocontrol of phytopathogens: antibiotic production as mechanism of control. The Battle Against Microbial Pathogens: Basic Science, Technological Advances and Educational Programes. Formatex Research Center, Spain, pp. 305–309.

Valenzuela-Soto, J.H., Estrada-Hernández, M.G., Ibarra-Laclette, E., Délano-Frier, J.P., 2010. Inoculation of tomato plants (*Solanum lycopersicum*) with growth-promoting *Bacillus subtilis* retards whitefly Bemisia tabaci development. Planta 231 (2), 397.

van Dijk, K., Nelson, E.B., 2000. Fatty acid competition as a mechanism by which *Enterobacter cloacae* suppresses *Pythium ultimum* sporangium germination and damping-off. Appl. Environ. Microbiol. 66 (12), 5340–5347.

Van Loon, L.C., Bakker, P.A.H.M., Pieterse, C.M.J., 1998. Systemic resistance induced by rhizosphere bacteria. Annu. Rev. Phytopathol. 36 (1), 453–483.

Viswanathan, R., Rajitha, R., Sundar, A.R., Ramamoorthy, V., 2003. Isolation and identification of endophytic bacterial strains from sugarcane stalks and their in vitro antagonism against the red rot pathogen. Sugar Tech. 5 (1–2), 25–29.

Weller, D.M., 2007. Pseudomonas biocontrol agents of soilborne pathogens: looking back over 30 years. Phytopathology 97 (2), 250–256.

Wright, S.A., Zumoff, C.H., Schneider, L., Beer, S.V., 2001. *Pantoea agglomerans* strain EH318 produces two antibiotics that inhibit *Erwinia amylovorain* vitro. Appl. Environ. Microbiol. 67 (1), 284–292.

Wu, G., Liu, Y., Xu, Y., Zhang, G., Shen, Q., Zhang, R., 2018. Exploring elicitors of the beneficial rhizobacterium *Bacillus amyloliquefaciens* SQR9 to induce plant systemic resistance and their interactions with plant signaling pathways. Mol. Plant Microbe Interact., 31 (5), 560–567.

Yadav, M.K., Aravindan, S., Ghritlahre, S.K., Adak, T., Berliner, J., 2015. Induced systemic resistance and plant responses to fungal biocontrol agents. Annu. Rev. Phytopathol. 48 (1), 21–43.

Zeilinger, S., Galhaup, C., Payer, K., Woo, S.L., Mach, R.L., Fekete, C., et al., 1999. Chitinase gene expression during mycoparasitic interaction of *Trichoderma harzianum* with its host. Fungal Genet. Biol. 26 (2), 131–140.

Zhang, S., Reddy, M.S., Kloepper, J.W., 2004. Tobacco growth enhancement and blue mould disease protection by rhizobacteria: relationship between plant growth promotion and systemic disease protection by PGPR strain 90–166. Plant Soil. 262 (1–2), 277–288.

Zheng, L., Zhao, J., Liang, X., Zhan, G., Jiang, S., Kang, Z., 2017. Identification of a novel *Alternaria alternata* strain able to hyperparasitize *Puccinia striiformis* f. sp. tritici, the causal agent of wheat stripe rust. Front. Microbiol. 8, 71.

CHAPTER

13

Harnessing the potential of biostimulants and biocontrol agents for sustainable management of agricultural productivity

Vipin Kumar Singh[1], *Rishikesh Singh*[2], *Ajay Kumar*[3] *and Rahul Bhadouria*[4]

[1]Department of Botany, Centre of Advanced Study, Institute of Science, Banaras Hindu University, Varanasi, India [2]Institute of Environment and Sustainable Development, Banaras Hindu University, Varanasi, India [3]Department of Postharvest Science, Agricultural Research Organization (ARO), Volcani Center, Rishon LeZion, Israel [4]Department of Botany, University of Delhi, New Delhi, India

13.1 Introduction

Increased demand of food has necessitated the improvement in current agricultural practices to feed the rapidly rising human population. The indiscriminate utilization of synthetic fertilizers and hazardous pesticides in agroecosystem has introduced substantial amount of agrochemicals and disturbed the soil natural characteristics, posing severe threat to natural environment and human health. Contrary to this, application of diverse microbial resources and substances has promising potential in improving agricultural productivity in a sustainable manner. Plant biostimulants, in this connection, have immense importance in crop yield enhancement and are defined as natural products and microorganisms useful in improving plant growth and development (Calvo et al., 2014). Globally, plant stimulants have attracted researchers as well as industries in previous decades. Most strikingly, the mechanisms of enhancement in plant growth and development fostered by plant biostimulants are very much different as compared to synthetic fertilizers (European Biostimulants Industry Council, 2012). Further, the amount of synthetics required for plant growth is very high in

Food Security and Plant Disease Management
DOI: https://doi.org/10.1016/B978-0-12-821843-3.00006-4

comparison to biostimulants (du Jardin, 2015). Although du Jardin (2012) categorized biostimulants in eight main types, Calvo et al. (2014) have categorized in five major types as (1) microbial inoculants, (2) humic acids, (3) fulvic acid, (4) protein hydrolysate, and (5) extracts from seaweeds. Microbial inoculants are the chief foundation of biostimulants and refer to the formulations based on agriculturally important microorganisms responsible for improved plant growth, development, and yield (Sammauria et al., 2020). The judicious application of microbial inoculants can be helpful in minimizing the exploitation of currently used agrochemicals, thereby paving the way for sustainable agriculture (Singh et al., 2018). The two most commonly exploited microbial inoculants are plant growth–promoting rhizobacteria (PGPR) and arbuscular mycorrhizal fungi (AMF; Li et al., 2020a; Rouphael et al., 2015). Biostimulants, particularly PGPR, have demonstrated potential applicability in alleviating the negative consequences of drought (Asghari et al., 2020), pesticides (Nathiya et al., 2020), salinity (Sultana et al., 2020), high as well as low temperature (Mukhtar et al., 2020; Yarzábal, 2020), and heavy metal stress (Khatri et al., 2020), indicating diverse implications in abiotic stress management (Kumar et al., 2018; Goswami and Deka, 2020). In addition, biostimulants have also been documented with contribution in biotic stress management (Attia et al., 2020). The beneficial mechanism of action associated with plant biostimulant may be of direct or indirect nature involving induction of carbon and nitrogen metabolism, enhancement in nutrient acquisition, changes in root architecture, conferring improved plant processes and ability to cope up with environmental stresses (Rouphael and Colla, 2018).

Annual crop productivity losses up to 17%–30% in five economically important crops by various plant diseases and pests have challenged the target of food security (Savary et al., 2019). Currently, agricultural productivity losses are managed largely through application of conventional techniques such as deployment of disease-resistant plants (van Esse et al., 2020) and utilization of agrochemicals of hazardous nature (Belan et al., 2020). In spite of beneficial effects of agrochemicals such as fertilizers and pesticides in improving crop productivity as well as effective management of devastating plant diseases, the hazardous impacts associated with continued application including contamination of soil and aquatic ecosystem, in addition to substantial negative consequences on soil microbiological features (Carvalho, 2006; Damalas and Eleftherohorinos, 2011; Geiger et al., 2010; Kim et al., 2017; Singh et al., 2020a,b) and non-target organisms are well reported. Moreover, indiscriminate application of synthetic pesticides has resulted in the induction of resistance development in various pests (Ash, 2018; Borel, 2017) infesting agricultural crops. Contrary to this, application of biological agents and their products for controlling crop losses due to pest and pathogen attack is environmentally safe, natural, sustainable, and do not induce resistance development in target organisms. Such biological agents as well as important metabolic products thereof meant for controlling destructive plant diseases either directly or indirectly are called as "biocontrol agents" and include naturally existing antagonistic microorganisms such as bacteria, fungi, bacteriophage, and protozoans from diverse habitats (Köhl et al., 2019; Singh et al., 2019; Huss and Raman, 2020). The direct mechanism involves the inhibitory action of biocontrol agents (BCAs) via parasitic action, secretion of antibiotic substances and potential competency against infesting pests facilitated by accelerated acquisition of nutrients, and rapid recognition of pathogen attack points. The indirect action of BCAs encompasses rapid induction of some plant defense responses against infesting pests and pathogens (Vos et al., 2015) known as induced systemic resistance (ISR). ISR, being not specific in general, offers protection against varied plant diseases. Among different BCAs employed in agriculture, *Trichoderma* is the most widely deployed fungus to promote plant growth and control numbers of plant diseases because of induction of disease resistance and mycoparasitic behavior (Pimentel et al., 2020). To date, BCAs have shown effectiveness against multiple plant diseases caused by pathogens such as *Phytophthora, Fusarium, Pythium, Colletotrichum, Rhizoctonia, Macrophomina, Sclerotinia, Magnaporthe*, and *Meloidogyne*. However, application of BCAs under field conditions is challenged substantially by environmental conditions, presence of other microorganisms, host plant studied, strain of BCAs used, prevailing soil characteristics, field performance, cost effectiveness, safety, registration from concerned authorities and support from government bodies, research institutions, academicians, and industries (Velivelli et al., 2014). This chapter is an attempt to deal with various aspects of biostimulants and BCAs along with potential implications and challenges in management of cultivated ecosystems.

13.2 Biostimulants

Abiotic and biotic stress factors are important limitations restricting the productivity of agroecosystem (Yakhin et al., 2017; Tsaniklidis et al., 2020) in multiple orders of magnitude. Conventionally, the negative consequences of abiotic stressors can be minimized via introduction of water, fertilizers, and natural growth-regulating molecules, the application of biostimulant as a novel strategy for increasing agricultural productivity is much attractive and environment friendly. Biostimulants are defined as microbial resources and derived substances helpful in improving plant growth and productivity even under stressful environmental conditions (Mattarozzi et al., 2020). The improved growth and productivity of host plant through biostimulants are facilitated by activities such as nitrogen fixation, organic material decomposition, solubilization of important nutrients, and protection against biological and non-biological stress factors (Pii et al., 2015; Singh et al., 2018; Ye et al., 2020b). For easier understanding, biostimulants are classified as (1) microbial inoculants, (2) humic acids, (3) fulvic acids, (4) protein hydrolysate, and (5) extracts from seaweeds.

Among microbial inoculants, plant growth–promoting bacteria especially, rhizobacteria and AMF have been widely studied (Rouphael et al., 2015; Li et al., 2020a; Moreira et al., 2020) as biofertilizer. Microbial inoculants in free and encapsulated forms (Vassilev et al., 2020) can be applied to soil or seed in order to have beneficial effect on plant growth and development. The soil inoculation may lead to changes in microbiological community structure (Trabelsi and Mhamdi, 2013), thereby affecting agricultural productivity. Moreover, the successful employment of wide range of microbial inoculants in natural ecosystem could have possible contributions in reducing the utilization of agrochemicals such as fertilizers and hazardous pesticides, herbicides, and fungicides. The role of rhizospheric microbial inoculants with promising potential in improving arsenic accumulation characteristics of fern is recently presented (Yang et al., 2020) indicating implications of microbial resources in decontamination of cultivated ecosystem. Microbial inoculants harboring the potential to eliminate metal can be opted as a vital and sustainable tool to counteract metal toxicity–induced inhibition of plant growth and development. Role of microbial inoculants in improving soil attributes, inhibition of plant pathogens, induction of disease resistance as well as augmentation in plant growth and productivity in a sustainable manner is presented by various authors (Li et al., 2019, 2020a; Shahid et al., 2019; Moreira et al., 2020). However, outcomes under field conditions are not satisfactory (Sammauria et al., 2020) and need further investigations.

Diverse microorganisms dwelling in soil ecosystem are documented to degrade plant/animal debris and release substantial proportions of humic acids and beneficial nutrients. Humic acid, one of the important constituents of soil organic materials and chief natural carbon stock of terrestrial ecosystem (Dou et al., 2020), is recognized as biostimulant holding great promise in sustainable agricultural practices and possesses the potential to induce crop productivity (Saudy et al., 2020). The positive influences of humic acid associated with plant growth could be either of indirect nature via increased soil fertility, minimized soil compression, or of direct nature through amplification in total plant biomass (Nardi et al., 2002). Most notably, the biological effects of humic substances are considerably affected by the source and content. The beneficial impacts of humic acid in agroecosystem are presented by various researchers across the globe (Zanin et al., 2019; Nunes et al., 2019; Dehsheikh et al., 2020; Kaya et al., 2020; Rekaby et al., 2020). Humic acid as viable instruments of agricultural sustainability offers immense opportunities in managing soil nutrient characteristics.

Fulvic acid is the important constituent of humic substances and is recognized to modulate the physicochemical and biological attributes of soil environment. The beneficial impact of fulvic acid is rendered chiefly through improved acquisition of soil nutrients by host root system (Du Jardin, 2015), augmented root development, adaptation to varied environmental stress factors, and improved enzymatic and non-enzymatic defense responses (Canellas et al., 2015; Elrys et al., 2020). The beneficial impacts of fulvic acid alone and combined with other substances as biostimulant are presented by various researchers (Tarantino et al., 2018; Wang et al., 2020; Basile et al., 2020), however, more studies under natural field conditions are necessary to elucidate the advantageous effects associated with fulvic acid application as biostimulant.

Protein hydrolysates comprising of peptides and amino acids (Colla et al., 2015) and other nitrogen containing molecules serving as biostimulants are described to pose positive influence on plant growth and development in an ecologically friendly manner. Protein hydrolysates, in general, are obtained through chemical processes and enzymatic actions over agricultural and industrial products. The stimulatory impact of protein hydrolysates on plants is mediated through alterations in nitrogen absorption and regulatory actions over enzymes participating in nitrogen incorporation and tricarboxylic acid (TCA) cycle (Schiavon et al., 2008). Nevertheless, detailed biological mechanism of action of protein hydrolysate and other nitrogen-based plant biostimulants is not yet completely deciphered. The beneficial impact of amino acid–based biostimulants is recently presented (Klokić et al., 2020; Kocira et al., 2020; Di Mola et al., 2020).

Seaweed extracts as plant biostimulant (Battacharyya et al., 2015) have profound implications in sustainable management of cultivated ecosystems by improving crop productivity and alleviation of stress-induced negative consequences (Boukhari et al., 2020). The plant growth–inducing action of seaweed extract is rendered by polysaccharides, nutritional molecules, nitrogenous substances, sterols, and hormones (Craigie, 2011; Khan et al., 2009), polysaccharides, minerals, and vitamins (Battacharyya et al., 2015). As per an estimate, more than 20 different products from seaweeds have been acknowledged for biostimulant action (Khan et al., 2009). Recent studies on seaweed and extract application as plant biostimulant have displayed considerable prospects for management of cultivated ecosystems (Łangowski et al., 2019; Yao et al., 2020; Omidbakhshfard et al., 2020; Gutiérrez-Gamboa et al., 2020; Patel et al., 2020).

13.2.1 Biostimulants against abiotic stresses

Plant exposure to abiotic stress factors such as salinity, drought, heavy metals, and extreme temperature poses negative consequences on plant growth and productivity (Shinwari et al., 2020). The changing environmental conditions have favored the commencement of abiotic environmental stresses causing yield differences reaching up to 70% (Rouphael et al., 2017). The combined action of abiotic and biotic stresses has potential to decline the productivity of important crops by greater than 50% (Bray et al., 2000). Induction of oxidative stress resulting from environmental stresses modifies plant productivity through substantial alterations in plant physiological and biochemical processes, thereby reducing crop yield (Wang et al., 2003; Farooq et al., 2020; Adrees et al., 2020). Management of abiotic stressors through conventional agricultural practices and breeding techniques is not much successful, expensive, time consuming and requires the integration of ecologically sustainable approaches to mitigate the negative impacts associated. Plant biostimulants, in this regard, offer important avenues for alleviating the harmful actions of abiotic environmental factors (Van Oosten et al., 2017) on economically important crops. Role of different biostimulants, especially plant growth–promoting microbes in alleviating the negative consequences of different environmental stresses, is briefly discussed one by one.

13.2.1.1 Drought

One of the most widely investigated environmental stressors and agricultural disaster is drought that is demonstrated to pose negative impacts on productivity of multitude of important crop plants worldwide, eventually causing vulnerability to global food security. The role of biostimulants, particularly growth-inducing microorganisms in amelioration of drought-induced harmful impacts in crop could serve as an economically viable tool for sustainable management of cultivated ecosystems (Mishra et al., 2020). PGPR as microbial inoculants have shown successful outcomes in alleviation of drought stresses through modulations in hormones such as indole-3-acetic acid, gibberellin, abscisic acid, and ethylene (Singh et al., 2018) leading to considerable changes in root architecture and carbohydrate metabolism, augmented nutrient uptake, reduced water losses through leafy surfaces, and increased adaptation to environmental stress factors (Mantelin and Touraine, 2004; Bisht et al., 2019, 2020). Adaptation to drought stress in plants through association with microorganisms may involve the potential role of microbially synthesized volatile substances encompassing multiple beneficial properties. Drought tolerances in *Arabidopsis* conferred through microbially synthesized airborne medicine referred as volatile organic compound (Garbeva and Weisskopf, 2020) are advocated by Cho et al. (2008). Some of the important microbial species conferring drought stress tolerance in plants include *Bacillus*

(Woo et al., 2020), *Pseudomonas* (Chadra et al., 2020), *Acinetobacter* (Kour et al., 2020), *Streptomyces* (Abbasi et al., 2020), *Flavobacterium* (Kim et al., 2020), *Mesorhizobium* (Abdela et al., 2020), *Azospirillum* (Asghari et al., 2020), *Trichoderma* (Khoshmanzar et al., 2020), *Beauveria* (Kuzhuppillymyal-Prabhakarankutty et al., 2020), and AMF (Aalipour et al., 2020). Nevertheless, more investigations under field conditions are necessary for effective utilization of biostimulants in management of drought stress in cultivated ecosystems.

13.2.1.2 Salinity

Salinity is the challenging environmental constraint affecting non-reproductive as well as reproductive stages of plants, restricting crop productivity severely via generation of reactive oxygen species (Gupta et al., 2020a). According to studies, 45 million hectares of cultivated agricultural areas are severely influenced by high salt content and 1.5 million hectares land resources have been expected to convert into less fertile soil in coming years (Munns and Tester, 2008). The negative consequences of soil salinity are described to decline 70% yield of major food commodities including rice, wheat, maize, and barley (Etesami and Beattie, 2018). Traditional agricultural strategies and employment of salinity-resistant crops produced by breeding have largely remained unsuccessful. The active participation of biostimulants with salt-tolerant nature and growth-promoting characteristics harbors ample opportunities in sustainable management of salinity affected agroecosystems (Kushwaha et al., 2020). The growth-promoting effect of bacteria in salt-stressed plant is rendered by diversified pool of antioxidant defense system (Habib et al., 2016), bacterial synthesis of 1-aminocyclopropane-1-carboxylate (ACC) deaminase (del Carmen Orozco-Mosqueda et al., 2020), induced synthesis of hormones by host (Numan et al., 2018), generation of microbial volatile organic compounds, changes in expression of genes regulating stress responses (Baek et al., 2020), alterations in ethylene metabolism (Misra and Chauhan, 2020), and regulation of proline level and ionic balances (Chen et al., 2007; Bano and Fatima, 2009). The research progress in growth-enhancing microbial resources could offer a novel boulevard in agriculture sector for augmenting host tolerance to salinity stress with the resultant increase in crop growth and productivity. Some of the important microbes conferring salt tolerance to plants are *Piriformospora indica* and *Azotobacter chroococcum* (Arora et al., 2020), *Bacillus* (Baek et al., 2020), *Pseudomonas* (Singh et al., 2020a,b), *Serratia* (Turhan et al., 2020), mycorrhizal fungus *Glomus* (Pan et al., 2020a), *Sphingobacterium* (Vaishnav et al., 2020), *Pantoea* and *Rhizoglomus* (Moreira et al., 2020), and *Flavobacterium* (Kim et al., 2020).

13.2.1.3 Temperature

Global changes in climatic conditions leading to variations in temperature have led to devastating effects on crop productivity (Soltani et al., 2019), leading to global vulnerability to food security (Hassan et al., 2020). As per an estimate, the atmospheric temperature would increase from 1.8°C to 4.0°C as compared to present day condition by the year 2100 (IPCC, 2007). Temperature stress is known to negatively influence and perturb different components of photosynthetic activities such as transport of electrons, acceptor and receptor molecules, and disruption of enzymatic activities. Further, the heat stress is noted to provoke leaf senescence, degeneration of chlorophyll molecules (Zhang et al., 2019), galvanize the uncoupling of non-cyclic photophosphorylation (Al-Khatib and Paulsen, 1999;

Mohanty et al., 2002). In addition, commencement of temperature-guided stressful environment leads to enhanced synthesis of antioxidant molecules, implying the induction of host defense responses. The introduction of different agronomic practices and development of heat- and cold-resistant crop varieties are currently being employed in agroecosystem, however, not much success is evident so far. The deployment of biostimulants including microbial inoculants and protein hydrolysates (Kauffman et al., 2007) is a promising strategy to overcome the harmful impact of temperature variations via improvement in nutrient absorption and storage (Liu et al., 2017). Low-level expression of defense-related genes governing the synthesis of enzymes such as superoxide dismutase, ascorbate peroxidase, and catalase, augmentation in biomass, and seed dimensions after host interaction with *Pseudomonas putida* strain AKMP7 are demonstrated by Ali et al. (2011). Role of plant-derived biostimulants (Francesca et al., 2020), AMF (Liu et al., 2017), seaweed extract (Cabo et al., 2020), PGPB (Chatterjee et al., 2020; Shekhawat et al., 2020), and *Trichoderma* (Ghorbanpour et al., 2018) in providing host tolerance to temperature stress is documented by various researchers. In general, the possible mechanism underlying the mitigation of temperature stress in host by using growth-promoting microbes relies on synthesis of ACC deaminase enzyme, induced production of plant hormones, and increased biosynthesis of enzymes facilitating in elimination of reactive oxygen species.

13.2.1.4 Heavy metal stress

Increasing global food demand has forced the current day farmers to use agrochemicals consequently resulting in soil contamination (Guo et al., 2020). Heavy metals and metalloids such as chromium, copper, nickel, zinc, manganese, lead, cadmium, arsenic, and selenium released into soil environment not only provoke oxidative stress in plants (Huo et al., 2020) but also modulate soil microbial community structure substantially (Pan et al., 2020b) leading to harmful impact on plant growth and overall productivity (Rizvi and Khan, 2018). Traditional strategies to alleviate the hazardous impact of heavy metals and metalloids are expensive, employ harmful chemical substances, ecologically unsound, and time demanding. In contrast, the application of biostimulants, especially PGPR to enhance the plant growth and productivity, in addition to remediation of metals and metalloids contaminants is environmentally sustainable and quite promising for cultivated ecosystems. The growth-promoting microorganism induces crop productivity by release of siderophores with characteristic ability to sequester noxious metals (Manoj et al., 2020) as well as secretion of acidic substances of organic nature affecting the accessibility of metals in rhizospheric region, thereby simultaneously offering host tolerance to contaminants by restricting contaminant accumulation in host plants. Furthermore, plant growth promoting (PGP) microbes facilitate the volatilization and transformation in order to alleviate the toxicity of inorganic metal contaminants (Ullah et al., 2015; Pramanik et al., 2018; Ma et al., 2020). In addition, reports are also available indicating expression of genes present in growth-promoting bacteria conferring resistance and facilitating detoxification of metals in vicinity (Asad et al., 2019). Some of the microbial inoculants providing tolerance to plants against heavy metals and metalloids stress are *Klebsiella* sp. CPSB4 and *Enterobacter* sp. CPSB49 (Gupta et al., 2020b), *Stenotrophomonas maltophilia* and *Bacillus* (Shreya et al., 2020), *Serratia* (El-Esawi et al., 2020), *Paenibacillus mucilaginosus* and *Sinorhizobium meliloti* (Ju et al., 2020), *Azotobacter chroococcum* (Rizvi and Khan, 2018), *Bacillus* and *Acinetobacter*

(Rahman et al., 2020), and *Enterobacter* (Sharma et al., 2020). Although the deployment of growth-promoting microorganisms in soil is influenced by soil characteristics, prevailing environmental conditions, fertilizers applied, and host plant, successful result needs extensive studies under natural field conditions to make the process much applicable to cultivated ecosystem. Some studies have demonstrated restricted PGP activity of agriculturally important microorganisms under field condition generating suspicions among growers (Tabassum et al., 2017) regarding effectiveness.

13.2.2 Growth-promotion mechanisms in biostimulants

Biostimulants, especially microbial inoculants are promising candidate for sustainable management of cultivated ecosystems by improving crop productivity and protecting against destructive diseases. The detailed information about the growth-promotion mechanism deployed by biostimulants for enhancing crop productivity is essential for isolation of potential microbial inoculants as well as for effective employment in agricultural ecosystems. The growth-augmenting mechanism harbored by growth-promoting microbes is explored in terms of induced synthesis of phytohormones (Li et al., 2020b), secretion of siderophores (Ye et al., 2020a), release of ACC deaminase enzyme (Danish et al., 2020), biological nitrogen fixation (Matse et al., 2020), synthesis of cell wall–destabilizing enzymes (Ali et al., 2020), changes in root architecture (Moretti et al., 2020), increased nutrient acquisition by host (Ju et al., 2020), solubilization of unavailable nutrient substances (Boroumand et al., 2020), and induction of systemic resistance (Kumari and Khanna, 2020). The microorganisms possessing growth-promoting traits must be able to multiply rapidly and withstand harshness of environmental stresses such as salinity, drought, extreme temperature variations, and metal contamination. Therefore survival of growth-promoting microbes containing beneficial traits, in an agroecosystem exposed to environmental stressors, is of considerable concern (Etesami and Beattie, 2018). Moreover, the application of growth-promoting biostimulants equipped with multistress-tolerance mechanisms could serve as a better substitute to alleviate the harmful impact of abiotic and biotic stress factors in crop plants (Amna et al., 2020). Studies on antagonistic potential of growth-promoting microbes would be helpful in development of bioformulation for the management of crop losses rendered by pests and pathogens (Attia et al., 2020) together with the minimized application of hazardous agrochemicals (Srivastava et al., 2016). In essence, the exploitation of one microbial species holding multiple growth-promoting attributes may be regarded as an eco-friendly and more cost-effective biological tool as compared to consortia of different inocula with fewer desirable attributes. Nevertheless, isolation and characterization of potential biostimulants from varied soil ecosystem having most of the desirable growth-promoting traits and consistent expression of in vitro outcomes under field condition are very much daunting task (Amna et al., 2020), thereby limiting successful employment in agroecosystem. In addition, the possible interaction of agriculturally applied growth-promoting microbes with pathogenic microbes at biochemical and molecular level also needs to be investigated in order to abolish the negative consequences, if any, in agroecosystem.

13.3 Biocontrol agents

Plant diseases caused by different pests and pathogens result in huge economic losses because of substantial decline in crop productivity, eventually posing severe threat to global food security. The losses in agricultural productivity of economically important crops and products thereof under field condition or in postharvest storage condition (Salas-Méndez et al., 2019) may be carried out by insects, fungi, bacteria, viruses, parasitic plants nematodes, and protozoans (Strange and Scott, 2005). However, the more productivity losses at primary level are accounted for by the fungal pathogens. As per an estimate, destructive plant diseases account for 78% loss in productivity of fruits, 54% of vegetables, and 32% of cereals (Zhang, 2018). The currently available control measures to minimize the productivity losses are mainly dependent on expensive, environmentally hazardous, and complicated physicochemical methodologies. Previous decades have shown unprecedented rise in application of noxious agrochemicals such as pesticides rendering contamination of biosphere and harmful impact on humans and nontarget organisms. The strategies based on conventional breeding programs are also not much successful mainly due to time-consuming nature and effectiveness against only specific pests. The application of antagonistic biological agents known as "biocontrol agents" for controlling crop losses, on the other hand, offers an invaluable and economical strategy without exerting negative influence on environmental integrity. Further, the biological control measures would be helpful in minimizing the excessive employment of agrochemicals together with the alleviation of harmful impact on soil physicochemical and biological attributes. The unexpected large pools of microbial entities and metabolic products belonging to different fungi, bacteria, and bacteriophages have been tested for biological control activities in vitro and field environment under controlled conditions (Table 13.1) with potential outcomes relevant to sustainable agricultural practices. Further, many of the currently known BCAs are quite suitable even with self-perpetuating agricultural practices meant for protection of natural resources The exploitation under natural field conditions, however, is much challenging task due to variations in prevailing environmental conditions, implying further investigations for better interventions of microbial resources in management of cultivated ecosystems.

13.3.1 Mechanism of action of biological control agents

The biocontrol action of microorganisms and products thereof is based on important characteristics including induction of resistance in host via synthesis and release of elicitors, secretion of antimicrobial substances, parasitic action over target pathogens, activity loss of pathogen's enzymes, augmentation in host's nutrient uptake attributes, and host tolerance to diverse abiotic and biotic stressors (Gomes et al., 2015). The biocontrol activity represented by microorganisms can be classified as direct or indirect depending on the interaction with plant pathogen and host plant. The direct action encompasses secretion of antibiotic compounds, volatiles, enzymes participating in cellular lysis, degradation of pathogenicity associated enzymes such as peptic enzymes (Herron et al., 2000; Singh et al., 2017), competency for particular habitat and nutrient molecules, and pathogen inactivation rendered by mycoparasitism. Indirect action, on the other hand, involves considerable

TABLE 13.1 Some important biocontrol agents used for management of plant diseases.

Biocontrol agent	Pathogen	Host	Reference
Different endophytic fungi	*Rhizoctonia solani, Sclerotinia sclerotiorum, Fusarium oxysporum*	*Cucumis sativus*	Huang et al. (2020)
Pseudomonas resinovorans, Pseudomonas putida, Pseudomonas fluorescens	*F. oxysporum*	*Citrullus lanatus*	Hua et al. (2020)
Trichoderma	*Fusarium* and *Phoma*	*Lycopersicum esculentum*	López et al. (2020)
Bacillus amyloliquefaciens and *Trichoderma asperellum*	*Xanthomonas perforans*	*L. esculentum*	Chien and Huang (2020)
Trichoderma harzianum	*Botrytis cinerea, Pseudoperonospora cubensis, S. sclerotiorum,* and *Sphaerotheca fusca*	*C. sativus*	Elad (2000)
Cupriavidus	*Xanthomonas campestris*	*Raphanus sativus*	Ye et al. (2020a)
Bacillus velezensis	*Fusarium graminearum*	*Triticum aestivum*	Palazzini et al. (2020)
B. velezensis	*Verticillium dahliae*	*Olea europaea*	Azabou et al. (2020)
Streptomyces	*Ralstonia solanacearum*	*L. esculentum*	Ling et al. (2020)
Hypoxylon rubiginosum	*Hymenoscyphus fraxineus*	*Fraxinus excelsior*	Halecker et al. (2020)
Pseudomonas chlororaphis and *Clonostachys rosea*	*F. oxysporum*	*L. esculentum*	Kamou et al. (2020)
Paenibacillus polymyxa NMA1017 and *Burkholderia cenocepacia*	*R. solani* and *Pythium ultimum*	*Phaseolus vulgaris*	Chávez-Ramírez et al. (2020)
Lytic bacteriophage	*R. solanacearum*	*Musa* sp.	Ramírez et al. (2020)
Bacteriophage	*Acidovorax citrulli*	*C. lanatus*	Rahimi-Midani et al. (2020)
T. asperellum	*Stemphylium vesicarium*	*Allium cepa*	Zapata-Sarmiento et al. (2020)

modulation in physiology, biochemistry, and morphology, offering resistance to biotic and abiotic stress factors, increased nutrient availability, and acceleration of resistance phenomena in infected host (Viterbo and Chet, 2006). Such biocontrol actions may work singly or in combination with other known mechanisms.

The secretion of antibiotic substances by soil-inhabiting bacteria serving as BCA to control root rot disease in pea plants resulting from infection with *Aphanomyces euteiches* in a sustainable manner is represented recently (Godebo et al., 2020). Contribution of 22 different volatiles in growth promotion in lettuce, induced synthesis of wall-degrading enzymes and inhibition of mycelial growth facilitating disease suppression is documented by Wonglom et al. (2020). Suppression of fungal pathogen enzymes serving as pathogenicity factors including chitinase, glucanase, and cutinase by action of BCAs is illustrated (Elad, 2000), indicating sustainable application in plant disease control. In addition, the BCAs having potential competitiveness with plant pathogens for acquisition of nutrients by producing siderophore (Hua et al., 2020), causing disease suppression is well-recognized mechanism. Mycoparasitism denotes the parasitic interaction of a fungus with another fungus. The biocontrol activity based on parasitic nature of endophytic fungi to control soil-borne plant diseases is illustrated recently by Huang and coworkers (2020). The mycoparasitic action of BCAs directly inhibits the multiplication of fungal pathogens and controls the further progression of plant diseases. The mycoparasitic action of BCA, *Trichoderma* is recently deciphered to be controlled by novel class of protein identified as TBRG-1 (Dautt-Castro et al., 2020).

13.3.2 Application of biocontrol agents in disease management

Plant diseases are recorded to be responsible for severe losses in productivity as well as the quality of various crop produce, globally. Although currently employed approaches are mainly based on chemical approaches, long-term agrochemical application has badly affected environmental homeostasis, human health together with the perturbations in soil physicochemical and biological attributes (Satapute et al., 2019). In contrast, BCAs have long been registered as important tools of sustainable agriculture for controlling the agricultural crop damages caused by destructive plant diseases. Unexpectedly large microbial resources of nature, however, are still under investigation in view of selection of better microorganisms and products derived harboring potential biocontrol features. Surprisingly, a smaller fraction of microbial resources is explored by scientists working in the area of agronomy, biochemistry, and microbiology. Contribution of endophytic fungi procured from vegetables in order to manage the soil-borne fungal diseases induced by *Sclerotinia* and *Fusarium* is illustrated by Huang et al. (2020). The contribution of rhizosphere isolated fluorescent pseudomonads as potential biocontrol candidate in suppressing *Fusarium*-induced wilt disease along with augmented growth in *Citrullus lanatus* is described by Hua et al. (2020). The implication of endophytic *Trichoderma* species in restricting multiplication of phytopathogens linked with *Ilex paraguariensis* and stimulation of growth in *Lycopersicum esculentum* is presented by López et al. (2020). The observed biocontrol activity, in most of the cases, is rendered by secretion of inhibitory substances, rapid nutrient acquisition, cell wall–degrading enzymes as well as the growth-promoting behavior. To date, multiple biocontrol products have been prepared, commercialized, and being employed in field to control plant diseases (Elad, 2000; Li et al., 2020c). Nevertheless, under natural condition, the biocontrol potential of microbial resources and derived products is substantially affected by environmental factors such as pH, temperature, soil attributes, nature of pathogens, application methodologies, cost effectiveness, host characteristics at genotypic and phenotypic level, and shelf life (Abd-Elgawad and Askary, 2020; Palazzini et al., 2020). Therefore the field outcomes of known BCAs are much challenging and are of paramount concern, making its large-scale application quite doubtful among farmers. The detailed investigation of BCA-pathogen-plant nexus at physiological, biochemical, and molecular level could be of immense importance to augment the overall effectiveness of BCAs.

13.4 Challenges associated with biostimulants and biocontrol agents

Large-scale deployment of biostimulants and BCAs for improving plant growth and productivity together with the successful management of destructive plant diseases under cultivated ecosystem is considerably challenged by various factors. The important challenge, most often, lies in the unavailability of effective and commercialized products (Raymaekers et al., 2020). Recommendation for formulation and release of biostimulants and BCAs (Ehlers et al., 2020) constitutes important limiting factor constraining massive application under field condition. Furthermore, regulatory policies of different nations may not allow the introduction of biostimulants and BCAs, especially microbial ones, despite of success under natural field environment because of issues concerned with biodiversity and intellectual property rights. Another challenging factor linked with application of microbial resources as biostimulants and BCAs includes possible negative consequences on human health and environmental integrity. The communication between agriculturally important microbes being exploited as biostimulants and BCAs and pathogenic microbes dwelling in agroecosystem may exert unexpected environmental hazard. Therefore further investigations pertaining to interaction of soil applied microorganisms with host and host-associated organisms must be taken into consideration (Abd-Elgawad and Askary, 2018). The survival of deployed microbes put major challenge for field application. The microbes giving successful results for a specific soil may not necessarily be effective in similar way for another soil with significantly different characteristics. In addition, the lack of fast and durable techniques essential for evaluating exceedingly large numbers of potential BCAs hampers the pace of research and development in this direction.

13.5 Conclusion

Rapid population rise and changes in climatic conditions have exerted huge pressure on available land resources to raise the agricultural productivity simultaneously with the protection of crop productivity losses caused by plant diseases so as to minimize the risks of global food security. The current agronomic practices based on fertilizers and agrochemicals such as pesticides have facilitated the improvement in crop productivity to a great extent. However, the contamination of environment, toxicity to non-target organisms, harmful impact on humans, resistance development in pests and pathogens, and deteriorating impact on soil characteristics have necessitated the search for ecologically sound, economical, and sustainable tool for management of cultivated ecosystem. The application of biostimulants and BCAs has long been considered to improve the crop productivity. The biostimulants, especially microbial inoculants possessing growth-promoting traits have been proved effective and recognized to alleviate the negative consequences of abiotic and biotic stress factors. BCAs have demonstrated promising potential in suppressing plant diseases via various mechanisms such as antibiosis and induction of resistance, thereby improving crop productivity. Nevertheless, the deployment of biostimulants and BCAs in agroecosystem under natural environmental condition is facing various challenges rendered by changes in soil characteristics, viability of microbes used, adaptation in given environmental condition and most strikingly, regulatory issues, implying further extensive research and policy development in this direction for successful application.

References

Aalipour, H., Nikbakht, A., Etemadi, N., Rejali, F., Soleimani, M., 2020. Biochemical response and interactions between arbuscular mycorrhizal fungi and plant growth promoting rhizobacteria during establishment and stimulating growth of Arizona cypress (*Cupressus arizonica* G.) under drought stress. Sci. Hortic. 261, 108923.

Abbasi, S., Sadeghi, A., Safaie, N., 2020. *Streptomyces* alleviate drought stress in tomato plants and modulate the expression of transcription factors ERF1 and WRKY70 genes. Sci. Hortic. 265, 109206.

Abd-Elgawad, M.M., Askary, T.H., 2018. Fungal and bacterial nematicides in integrated nematode management strategies. Egypt. J. Biol. Pest Control 28 (1), 74.

Abd-Elgawad, M.M., Askary, T.H., 2020. Factors affecting success of biological agents used in controlling the plant-parasitic nematodes. Egypt. J. Biol. Pest Control 30 (1), 1–11.

Abdela, A.A., Barka, G.D., Degefu, T., 2020. Co-inoculation effect of *Mesorhizobium ciceri* and *Pseudomonas fluorescens* on physiological and biochemical responses of Kabuli chickpea (*Cicer arietinum* L.) during drought stress. Plant Physiol. Rep. 25 (2), 359–369.

Adrees, M., Khan, Z.S., Ali, S., Hafeez, M., Khalid, S., ur Rehman, M.Z., et al., 2020. Simultaneous mitigation of cadmium and drought stress in wheat by soil application of iron nanoparticles. Chemosphere 238, 124681.

Ali, S.Z., Sandhya, V., Grover, M., Linga, V.R., Bandi, V., 2011. Effect of inoculation with a thermotolerant plant growth promoting *Pseudomonas putida* strain AKMP7 on growth of wheat (*Triticum* spp.) under heat stress. J. Plant Interact. 6 (4), 239–246.

Ali, S., Hameed, S., Shahid, M., Iqbal, M., Lazarovits, G., Imran, A., 2020. Functional characterization of potential PGPR exhibiting broad-spectrum antifungal activity. Microbiol. Res. 232, 126389.

Al-Khatib, K., Paulsen, G.M., 1999. High-temperature effects on photosynthetic processes in temperate and tropical cereals. Crop Sci. 39 (1), 119–125.

Amna, Xia, Y., Farooq, M.A., Javed, M.T., Kamran, M.A., Mukhtar, T., et al., 2020. Multi-stress tolerant PGPR *Bacillus xiamenensis* PM14 activating sugarcane (*Saccharum officinarum* L.) red rot disease resistance. Plant Physiol. Biochem. 151, 640–649.

Arora, M., Saxena, P., Abdin, M.Z., Varma, A., 2020. Interaction between *Piriformospora indica* and *Azotobacter chroococcum* diminish the effect of salt stress in *Artemisia annua* L. by enhancing enzymatic and non-enzymatic antioxidants. Symbiosis 80 (1), 61–73.

Asad, S.A., Farooq, M., Afzal, A., West, H., 2019. Integrated phytobial heavy metal remediation strategies for a sustainable clean environment-a review. Chemosphere 217, 925–941.

Asghari, B., Khademian, R., Sedaghati, B., 2020. Plant growth promoting rhizobacteria (PGPR) confer drought resistance and stimulate biosynthesis of secondary metabolites in pennyroyal (*Mentha pulegium* L.) under water shortage condition. Sci. Hortic. 263, 109132.

Ash, C., 2018. Meeting resistance. Science 360 (6390), 726–727.

Attia, M.S., El-Sayyad, G.S., Abd Elkodous, M., El-Batal, A.I., 2020. The effective antagonistic potential of plant growth-promoting rhizobacteria against *Alternaria solani*-causing early blight disease in tomato plant. Sci. Hortic. 266, 109289.

Azabou, M.C., Gharbi, Y., Medhioub, I., Ennouri, K., Barham, H., Tounsi, S., et al., 2020. The endophytic strain *Bacillus velezensis* OEE1: an efficient biocontrol agent against *Verticillium* wilt of olive and a potential plant growth promoting bacteria. Biol. Control. 142, 104168.

Baek, D., Rokibuzzaman, M., Khan, A., Kim, M.C., Park, H.J., Yun, D.J., et al., 2020. Plant-growth promoting *Bacillus oryzicola* YC7007 modulates stress-response gene expression and provides protection from salt stress. Front. Plant Sci. 10, 1646.

Bano, A., Fatima, M., 2009. Salt tolerance in *Zea mays* (L). following inoculation with *Rhizobium* and *Pseudomonas*. Biol. Fertil. Soils 45 (4), 405–413.

Basile, B., Rouphael, Y., Colla, G., Soppelsa, S., Andreotti, C., 2020. Appraisal of emerging crop management opportunities in fruit trees, grapevines and berry crops facilitated by the application of biostimulants. Sci. Hortic. 267, 109330.

Battacharyya, D., Babgohari, M.Z., Rathor, P., Prithiviraj, B., 2015. Seaweed extracts as biostimulants in horticulture. Sci. Hortic. 196, 39–48.

Belan, L.L., de Jesus Junior, W.C., de Souza, A.F., Zambolim, L., Cardoso Filho, J., Barbosa, D.H.S.G., et al., 2020. Management of coffee leaf rust in *Coffea canephora* based on disease monitoring reduces fungicide use and management cost. Eur. J. Plant Pathol. 1–12.

Bisht, N., Tiwari, S., Singh, P.C., Niranjan, A., Chauhan, P.S., 2019. A multifaceted rhizobacterium *Paenibacillus lentimorbus* alleviates nutrient deficiency-induced stress in *Cicer arietinum* L. Microbiol. Res. 223, 110–119.

Bisht, N., Mishra, S.K., Chauhan, P.S., 2020. *Bacillus amyloliquefaciens* inoculation alters physiology of rice (*Oryza sativa* L. var. IR-36) through modulating carbohydrate metabolism to mitigate stress induced by nutrient starvation. Int. J. Biol. Macromol. 143, 937–951.

Borel, B., 2017. When the pesticides run out. Nature 543, 302–304.

Boroumand, N., Behbahani, M., Dini, G., 2020. Combined effects of phosphate solubilizing bacteria and nanosilica on the growth of land cress plant. J. Soil Sci. Plant Nutr. 20 (1), 232–243.

Boukhari, E.L., Barakate, M.E., Bouhia, M.Y., Lyamlouli, K., 2020. Trends in seaweed extract based biostimulants: manufacturing process and beneficial effect on soil-plant systems. Plants 9 (3), 359.

Bray, E.A., Bailey-Serres, J., Weretilnyk, E., 2000. Responses to abiotic stresses. In: Gruissem, W., Buchannan, B., Jones, R. (Eds.), Response to Abiotic Stresses. American Society of Plant Physiologists, Rockville, pp. 1156–1249.

Cabo, S., Morais, M.C., Aires, A., Carvalho, R., Pascual-Seva, N., Silva, A.P., et al., 2020. Kaolin and seaweed-based extracts can be used as middle and long-term strategy to mitigate negative effects of climate change in physiological performance of hazelnut tree. J. Agron. Crop Sci. 206 (1), 28–42.

Calvo, P., Nelson, L., Kloepper, J.W., 2014. Agricultural uses of plant biostimulants. Plant Soil 383 (1–2), 3–41.

Canellas, L.P., Olivares, F.L., Aguiar, N.O., Jones, D.L., Nebbioso, A., Mazzei, P., et al., 2015. Humic and fulvic acids as biostimulants in horticulture. Sci. Hortic. 196, 15–27.

Carvalho, F.P., 2006. Agriculture, pesticides, food security and food safety. Environ. Sci. Policy 9 (7-8), 685–692.

Chandra, D., Srivastava, R., Glick, B.R., Sharma, A.K., 2020. Rhizobacteria producing ACC deaminase mitigate water-stress response in finger millet (*Eleusine coracana* (L.) Gaertn.). 3 Biotech 10 (2), 65.

Chatterjee, P., Kanagendran, A., Samaddar, S., Pazouki, L., Sa, T.M., Niinemets, Ü., 2020. Influence of *Brevibacterium linens* RS16 on foliage photosynthetic and volatile emission characteristics upon heat stress in *Eucalyptus grandis*. Sci. Total Environ. 700, 134453.

Chávez-Ramírez, B., Kerber-Díaz, J.C., Acoltzi-Conde, M.C., Ibarra, J.A., Vásquez-Murrieta, M.-S., Estrada-de los Santos, P., 2020. Inhibition of *Rhizoctonia solani* RhCh-14 and *Pythium ultimum* PyFr-14 by *Paenibacillus polymyxa* NMA1017 and *Burkholderia cenocepacia* CACua-24: a proposal for biocontrol of phytopathogenic fungi. Microbiol. Res. 230, 126347.

Chen, M., Wei, H., Cao, J., Liu, R., Wang, Y., Zheng, C., 2007. Expression of *Bacillus subtilis* proBA genes and reduction of feedback inhibition of proline synthesis increases proline production and confers osmotolerance in transgenic Arabidopsis. BMB Rep. 40 (3), 396–403.

Chien, Y.C., Huang, C.H., 2020. Biocontrol of bacterial spot on tomato by foliar spray and growth medium application of *Bacillus amyloliquefaciens* and *Trichoderma asperellum*. Eur. J. Plant Pathol. 156, 995–1003.

Cho, S.M., Kang, B.R., Han, S.H., Anderson, A.J., Park, J.Y., Lee, Y.H., et al., 2008. 2R, 3R-butanediol, a bacterial volatile produced by *Pseudomonas chlororaphis* O6, is involved in induction of systemic tolerance to drought in *Arabidopsis thaliana*. Mol. Plant-Microbe Interact. 21 (8), 1067–1075.

Colla, G., Nardi, S., Cardarelli, M., Ertani, A., Lucini, L., Canaguier, R., et al., 2015. Protein hydrolysates as biostimulants in horticulture. Sci. Hortic. 196, 28–38.

Craigie, J.S., 2011. Seaweed extract stimuli in plant science and agriculture. J. Appl. Phycol. 23 (3), 371–393.

Damalas, C.A., Eleftherohorinos, I.G., 2011. Pesticide exposure, safety issues, and risk assessment indicators. Int. J. Environ. Res. Public Health 8 (5), 1402–1419.

Danish, S., Zafar-ul-Hye, M., Mohsin, F., Hussain, M., 2020. ACC-deaminase producing plant growth promoting rhizobacteria and biochar mitigate adverse effects of drought stress on maize growth. PLoS one 15 (4), e0230615.

Dautt-Castro, M., Estrada-Rivera, M., Olguin-Martínez, I., del Carmen Rocha-Medina, M., Islas-Osuna, M.A., Casas-Flores, S., 2020. TBRG-1 a Ras-like protein in *Trichoderma virens* involved in conidiation, development, secondary metabolism, mycoparasitism, and biocontrol unveils a new family of Ras-GTPases. Fungal Genet. Biol. 136, 103292.

del Carmen Orozco-Mosqueda, M., Glick, B.R., Santoyo, G., 2020. ACC deaminase in plant growth-promoting bacteria (PGPB): an efficient mechanism to counter salt stress in crops. Microbiol. Res. 126439.

Dehsheikh, A.B., Sourestani, M.M., Zolfaghari, M., Enayatizamir, N., 2020. Changes in soil microbial activity, essential oil quantity, and quality of Thai basil as response to biofertilizers and humic acid. J. Clean. Prod. 256, 120439.

Di Mola, I., Cozzolino, E., Ottaiano, L., Nocerino, S., Rouphael, Y., Colla, G., et al., 2020. Nitrogen use and uptake efficiency and crop performance of baby spinach (*Spinacia oleracea* L.) and lamb's lettuce (*Valerianella locusta* L.) grown under variable sub-optimal n regimes combined with plant-based biostimulant application. Agronomy 10 (2), 278.

Dou, S., Shan, J., Song, X., Cao, R., Wu, M., Li, C., et al., 2020. Are humic substances soil microbial residues or unique synthesized compounds? A perspective on their distinctiveness. Pedosphere 30 (2), 159–167.

du Jardin, P., 2012. The science of plant biostimulants-a bibliographic analysis. Contract 30-CE0455515/00-96, ad hoc Study on bio-stimulants products. Available from: http://ec.europa.eu/enterprise/sectors/chemicals/files/fertilizers/final_report_bio_2012_en.pdf.

du Jardin, P., 2015. Plant biostimulants: definition, concept, main categories and regulation. Sci. Hortic. 196, 3–14.

Ehlers, G.C., Caradus, J.R., Fowler, S.V., 2020. The regulatory process and costs to seek approval for the development and release of new biological control agents in New Zealand. BioControl 65 (1), 1–12.

El-Esawi, M.A., Elkelish, A., Soliman, M., Elansary, H.O., Zaid, A., Wani, S.H., 2020. *Serratia marcescens* BM1 enhances Cadmium stress tolerance and phytoremediation potential of soybean through modulation of osmolytes, leaf gas exchange, antioxidant machinery, and stress-responsive genes expression. Antioxidants 9 (1), 43.

Elad, Y., 2000. Biological control of foliar pathogens by means of *Trichoderma harzianum* and potential modes of action. Crop Prot. 19 (8–10), 709–714.

Elrys, A.S., Abdo, A.I., Abdel-Hamed, E.M., Desoky, E.S.M., 2020. Integrative application of licorice root extract or lipoic acid with fulvic acid improves wheat production and defenses under salt stress conditions. Ecotoxicol. Environ. Saf. 190, 110144.

Etesami, H., Beattie, G.A., 2018. Mining halophytes for plant growth-promoting halotolerant bacteria to enhance the salinity tolerance of non-halophytic crops. Front. Microbiol. 9, 148. Available from: https://doi.org/10.3389/fmicb.2018.00148.

European Biostimulants Industry Council, 2012. What are biostimulants? Available from: http://www.biostimulants.eu.

Farooq, M., Romdhane, L., Al Sulti, M.K., Rehman, A., Al-Busaidi, W.M., Lee, D.J., 2020. Morphological, physiological and biochemical aspects of osmopriming-induced drought tolerance in lentil. J. Agron. Crop Sci. 206 (2), 176–186.

Francesca, S., Arena, C., Hay Mele, B., Schettini, C., Ambrosino, P., Barone, A., et al., 2020. The use of a plant-based biostimulant improves plant performances and fruit quality in tomato plants grown at elevated temperatures. Agronomy 10 (3), 363.

Garbeva, P., Weisskopf, L., 2020. Airborne medicine: bacterial volatiles and their influence on plant health. N. Phytol. 226 (1), 32–43.

Geiger, F., Bengtsson, J., Berendse, F., Weisser, W.W., Emmerson, M., Morales, M.B., et al., 2010. Persistent negative effects of pesticides on biodiversity and biological control potential on European farmland. Basic Appl. Ecol. 11 (2), 97–105.

Ghorbanpour, A., Salimi, A., Ghanbary, M.A.T., Pirdashti, H., Dehestani, A., 2018. The effect of *Trichoderma harzianum* in mitigating low temperature stress in tomato (*Solanum lycopersicum* L.) plants. Sci. Hortic. 230, 134–141.

Godebo, A.T., Germida, J.J., Walley, F.L., 2020. Isolation, identification, and assessment of soil bacteria as biocontrol agents of pea root rot caused by *Aphanomyces euteiches*. Can. J. Soil Sci. Available from: https://doi.org/10.1139/cjss-2019-0133.

Gomes, E.V., do Nascimento Costa, M., De Paula, R.G., De Azevedo, R.R., Da Silva, F.L., Noronha, E.F., et al., 2015. The Cerato-Platanin protein Epl-1 from *Trichoderma harzianum* is involved in mycoparasitism, plant resistance induction and self cell wall protection. Sci. Rep. 5, 17998.

Goswami, M., Deka, S., 2020. Plant growth-promoting rhizobacteria—alleviators of abiotic stresses in soil: a review. Pedosphere 30 (1), 40–61.

Guo, J., Muhammad, H., Lv, X., Wei, T., Ren, X., Jia, H., et al., 2020. Prospects and applications of plant growth promoting rhizobacteria to mitigate soil metal contamination: a review. Chemosphere 246, 125823.

Gupta, A., Singh, S.K., Singh, M.K., Singh, V.K., Modi, A., Singh, P.K., et al., 2020a. Plant growth–promoting rhizobacteria and their functional role in salinity stress management. Abatement of Environmental Pollutants. Elsevier, pp. 151–160.

Gupta, P., Kumar, V., Usmani, Z., Rani, R., Chandra, A., Gupta, V.K., 2020b. Implications of plant growth promoting *Klebsiella* sp. CPSB4 and *Enterobacter* sp. CPSB49 in luxuriant growth of tomato plants under chromium stress. Chemosphere 240, 124944.

Gutiérrez-Gamboa, G., Garde-Cerdán, T., Rubio-Bretón, P., Pérez-Álvarez, E.P., 2020. Biostimulation to Tempranillo grapevines (*Vitis vinifera* L.) through a brown seaweed during two seasons: effects on grape juice and wine nitrogen compounds. Sci. Hortic. 264, 109177.

Habib, S.H., Kausar, H., Saud, H.M., 2016. Plant growth-promoting rhizobacteria enhance salinity stress tolerance in okra through ROS-scavenging enzymes. BioMed. Res. Int. 2016. Available from: https://doi.org/10.1155/2016/6284547.

Halecker, S., Wennrich, J.P., Rodrigo, S., Andrée, N., Rabsch, L., Baschien, C., et al., 2020. Fungal endophytes for biocontrol of ash dieback: the antagonistic potential of *Hypoxylon rubiginosum*. Fungal Ecol. 45, 100918.

Hassan, M.U., Chattha, M.U., Khan, I., Chattha, M.B., Barbanti, L., Aamer, M., et al., 2020. Heat stress in cultivated plants: nature, impact, mechanisms, and mitigation strategies—a review. Plant Biosystems 1–24.

Herron, S.R., Benen, J.A., Scavetta, R.D., Visser, J., Jurnak, F., 2000. Structure and function of pectic enzymes: virulence factors of plant pathogens. Proc. Natl. Acad. Sci. 97 (16), 8762–8769.

Huang, L.Q., Niu, Y.C., Su, L., Deng, H., Lyu, H., 2020. The potential of endophytic fungi isolated from cucurbit plants for biocontrol of soil borne fungal diseases of cucumber. Microbiol. Res. 231, 126369.

Hua, G.K.H., Wang, L., Chen, J., Ji, P., 2020. Biological control of *Fusarium* wilt on watermelon by fluorescent pseudomonads. Biocontrol Sci. Technol. 30 (3), 212–227.

Huo, Y., Kang, J.P., Ahn, J.C., Kim, Y.J., Piao, C.H., Yang, D.U., et al., 2020. Siderophore-producing rhizobacteria reduce heavy metal-induced oxidative stress in *Panax ginseng* Meyer. J. Ginseng Res. Available from: https://doi.org/10.1016/j.jgr.2019.12.008.

Huss, P., Raman, S., 2020. Engineered bacteriophages as programmable biocontrol agents. Curr. Opin. Biotechnol. 61, 116–121.

IPCC, 2007. Climate Change 2007: Impacts, adaptation and vulnerability. In: Parry, M.L., Canziani, O.F., Palutikof, J.P., van der Linden, P.J., Hanson, C.E. (Eds.), Contribution of Working Group II to the Fourth Assessment Report of the Intergovernmental Panel on Climate Change. Cambridge University Press, Cambridge, 976 pp.

Ju, W., Jin, X., Liu, L., Shen, G., Zhao, W., Duan, C., et al., 2020. Rhizobacteria inoculation benefits nutrient availability for phytostabilization in copper contaminated soil: drivers from bacterial community structures in rhizosphere. Appl. Soil Ecol. 150, 103450.

Kamou, N.N., Cazorla, F., Kandylas, G., Lagopodi, A.L., 2020. Induction of defense-related genes in tomato plants after treatments with the biocontrol agents *Pseudomonas chlororaphis* ToZa7 and *Clonostachys rosea* IK726. Arch. Microbiol. 202 (2), 257–267.

Kauffman, G.L., Kneivel, D.P., Watschke, T.L., 2007. Effects of a biostimulant on the heat tolerance associated with photosynthetic capacity, membrane thermostability, and polyphenol production of perennial ryegrass. Crop Sci. 47 (1), 261–267.

Kaya, C., Şenbayram, M., Akram, N.A., Ashraf, M., Alyemeni, M.N., Ahmad, P., 2020. Sulfur-enriched leonardite and humic acid soil amendments enhance tolerance to drought and phosphorus deficiency stress in maize (*Zea mays* L.). Sci. Rep. 10 (1), 1–13.

Khan, W., Rayirath, U.P., Subramanian, S., Jithesh, M.N., Rayorath, P., Hodges, D.M., et al., 2009. Seaweed extracts as biostimulants of plant growth and development. J. Plant Growth Regul. 28 (4), 386–399.

Khatri, S., Sharma, R.K., Shridhar, V., 2020. Influence of Cadmium-Tolerant and Plant growth-promoting rhizobacteria on Cadmium accumulation and growth response of wheat seedlings under mountain ecosystem. Agric. Res. 9 (1), 56–65.

Khoshmanzar, E., Aliasgharzad, N., Neyshabouri, M.R., Khoshru, B., Arzanlou, M., Lajayer, B.A., 2020. Effects of *Trichoderma* isolates on tomato growth and inducing its tolerance to water-deficit stress. Int. J. Environ. Sci. Technol. 17 (2), 869–878.

Kim, K.H., Kabir, E., Jahan, S.A., 2017. Exposure to pesticides and the associated human health effects. Sci. Total Environ. 575, 525–535.

Kim, J.E., Woo, O.G., Bae, Y., Keum, H.L., Chung, S., Sul, W.J., et al., 2020. Enhanced drought and salt stress tolerance in Arabidopsis by *Flavobacterium crocinum* HYN0056 T. J. Plant Biol. 63 (1), 63–71.

Klokić, I., Koleška, I., Hasanagić, D., Murtić, S., Bosančić, B., Todorović, V., 2020. Biostimulants' influence on tomato fruit characteristics at conventional and low-input NPK regime. Acta Agric. Scand. Sect. B—Soil Plant Sci. 1–8.

Kocira, A., Lamorska, J., Kornas, R., Nowosad, N., Tomaszewska, M., Leszczyńska, D., et al., 2020. Changes in biochemistry and yield in response to biostimulants applied in bean (*Phaseolus vulgaris* L.). Agronomy 10 (2), 189.

Köhl, J., Kolnaar, R., Ravensberg, W.J., 2019. Mode of action of microbial biological control agents against plant diseases: relevance beyond efficacy. Front. Plant Sci. 10, 845.

Kour, D., Rana, K.L., Yadav, A.N., Sheikh, I., Kumar, V., Dhaliwal, H.S., et al., 2020. Amelioration of drought stress in Foxtail millet (*Setaria italica* L.) by P-solubilizing drought-tolerant microbes with multifarious plant growth promoting attributes. Environ. Sustainability 1–12.

Kumar, A., Singh, V.K., Tripathi, V., Singh, P.P., Singh, A.K., 2018. PlantGrowth-Promoting Rhizobacteria (PGPR): perspective in agriculture under biotic and abiotic stress. Crop Improvement through Microbial Biotechnology, Elsevier. pp. 333–342.

Kumari, S., Khanna, V., 2020. Induction of systemic resistance in chickpea (*Cicer arietinum* L.) against *Fusarium oxysporum* f. sp. *ciceris* by antagonistic rhizobacteria in assistance with native mesorhizobium. Curr. Microbiol. 77 (1), 85–98.

Kushwaha, P., Kashyap, P.L., Bhardwaj, A.K., Kuppusamy, P., Srivastava, A.K., Tiwari, R.K., 2020. Bacterial endophyte mediated plant tolerance to salinity: growth responses and mechanisms of action. World J. Microbiol. Biotechnol. 36 (2), 26.

Kuzhuppillymyal-Prabhakarankutty, L., Tamez-Guerra, P., Gomez-Flores, R., Rodriguez-Padilla, M.C., Ek-Ramos, M.J., 2020. Endophytic *Beauveria bassiana* promotes drought tolerance and early flowering in corn. World J. Microbiol. Biotechnol. 36 (3), 1–10.

Łangowski, Ł., Goñi, O., Quille, P., Stephenson, P., Carmody, N., Feeney, E., et al., 2019. A plant biostimulant from the seaweed *Ascophyllum nodosum* (Sealicit) reduces podshatter and yield loss in oilseed rape through modulation of IND expression. Sci. Rep. 9 (1), 1–11.

Li, Y., Guo, Q., Li, Y., Sun, Y., Xue, Q., Lai, H., 2019. *Streptomyces pactum* Act12 controls tomato yellow leaf curl virus disease and alters rhizosphere microbial communities. Biol. Fertil. Soils 55 (2), 149–169.

Li, H., Qiu, Y., Yao, T., Ma, Y., Zhang, H., Yang, X., 2020a. Effects of PGPR microbial inoculants on the growth and soil properties of *Avena sativa*, *Medicago sativa*, and *Cucumis sativus* seedlings. Soil Tillage Res. 199, 104577.

Li, X., Sun, P., Zhang, Y., Jin, C., Guan, C., 2020b. A novel PGPR strain *Kocuria rhizophila* Y1 enhances salt stress tolerance in maize by regulating phytohormone levels, nutrient acquisition, redox potential, ion homeostasis, photosynthetic capacity and stress-responsive genes expression. Environ. Exp. Bot. 104023.

Li, J., Zhang, G., Yu, H., Huang, L., Zeng, W., Wang, Y., 2020c. Complete mitochondrial genome of the important bio-control fungus *Purpureocillium lilacinum* (Ophiocordycipitaceae, Hypocreales) and its phylogenetic analysis. Mitochondrial DNA Part B 5 (1), 240–242.

Ling, L., Han, X., Li, X., Zhang, X., Wang, H., Zhang, L., et al., 2020. A *Streptomyces* sp. NEAU-HV9: isolation, identification, and potential as a biocontrol agent against *Ralstonia solanacearum* of tomato plants. Microorganisms 8 (3), 351.

Liu, X.M., Xu, Q.L., Li, Q.Q., Zhang, H., Xiao, J.X., 2017. Physiological responses of the two blueberry cultivars to inoculation with an arbuscular mycorrhizal fungus under low-temperature stress. J. Plant Nutr. 40 (18), 2562–2570.

López, A.C., Alvarenga, A.E., Vereschuk, M.L., Barua, R.C., Zapata, P.D., Luna, M.F., et al., 2020. Trichoderma strains isolated from *Ilex paraguariensis* ST. HIL: promising biocontrol agents with chitinolytic activity and plant growth promoter on *Lycopersicum esculentum*. Arab. J. Basic Appl. Sci. 27 (1), 105–113.

Ma, H., Wei, M., Wang, Z., Hou, S., Li, X., Heng, X., 2020. Bioremediation of cadmium polluted soil using a novel cadmium immobilizing plant growth promotion strain *Bacillus* sp. TZ5 loaded on biochar. J. Hazard. Mater. 122065.

Manoj, S.R., Karthik, C., Kadirvelu, K., Arulselvi, P.I., Shanmugasundaram, T., Bruno, B., et al., 2020. Understanding the molecular mechanisms for the enhanced phytoremediation of heavy metals through plant growth promoting rhizobacteria: a review. J. Environ. Manag. 254, 109779.

Mantelin, S., Touraine, B., 2004. Plant growth-promoting bacteria and nitrate availability: impacts on root development and nitrate uptake. J. Exp. Bot. 55 (394), 27–34.

Matse, D.T., Huang, C.H., Huang, Y.M., Yen, M.Y., 2020. Effects of coinoculation of *Rhizobium* with plant growth promoting rhizobacteria on the nitrogen fixation and nutrient uptake of *Trifolium repens* in low phosphorus soil. J. Plant Nutr. 43 (5), 739–752.

Mattarozzi, M., Di Zinno, J., Montanini, B., Manfredi, M., Marengo, E., Fornasier, F., et al., 2020. Biostimulants applied to maize seeds modulate the enzymatic activity and metaproteome of the rhizosphere. Appl. Soil Ecol. 148, 103480.

Mishra, S.K., Khan, M.H., Misra, S., Dixit, V.K., Gupta, S., Tiwari, S., et al., 2020. Drought tolerant *Ochrobactrum* sp. inoculation performs multiple roles in maintaining the homeostasis in *Zea mays* L. subjected to deficit water stress. Plant Physiol. Biochem. 150, 1–14.

Misra, S., Chauhan, P.S., 2020. ACC deaminase-producing rhizosphere competent *Bacillus* spp. mitigate salt stress and promote *Zea mays* growth by modulating ethylene metabolism. 3 Biotech 10 (3), 1–14.

Mohanty, P., Vani, B., Prakash, J.S., 2002. Elevated temperature treatment induced alteration in thylakoid membrane organization and energy distribution between the two photosystems in *Pisum sativum*. Z. Naturforsch. C 57 (9–10), 836–842.

Moreira, H., Pereira, S.I., Vega, A., Castro, P.M., Marques, A.P., 2020. Synergistic effects of arbuscular mycorrhizal fungi and plant growth-promoting bacteria benefit maize growth under increasing soil salinity. J. Environ. Manag. 257, 109982.

Moretti, L.G., Crusciol, C.A., Kuramae, E.E., Bossolani, J.W., Moreira, A., Costa, N.R., et al., 2020. Effects of growth-promoting bacteria on soybean root activity, plant development, and yield. Agron. J. 112 (1), 418–428.

Mukhtar, T., Smith, D., Sultan, T., Seleiman, M.F., Alsadon, A.A., Ali, S., et al., 2020. Mitigation of heat stress in *Solanum lycopersicum* L. by ACC-deaminase and exopolysaccharide producing *Bacillus cereus*: effects on biochemical profiling. Sustainability 12 (6), 2159.

Munns, R., Tester, M., 2008. Mechanisms of salinity tolerance. Annu. Rev. Plant Biol. 59, 651–681.

Nardi, S., Pizzeghello, D., Muscolo, A., Vianello, A., 2002. Physiological effects of humic substances on higher plants. Soil Biol. Biochem. 34 (11), 1527–1536.

Nathiya, S., Janani, R., Kannan, V.R., 2020. Potential of plant growth promoting Rhizobacteria to overcome the exposure of pesticide in *Trigonella foenum-graecum* (fenugreek leaves). Biocatal. Agric. Biotechnol. 101493.

Numan, M., Bashir, S., Khan, Y., Mumtaz, R., Shinwari, Z.K., Khan, A.L., et al., 2018. Plant growth promoting bacteria as an alternative strategy for salt tolerance in plants: a review. Microbiol. Res. 209, 21–32.

Nunes, R.O., Domiciano, G.A., Alves, W.S., Melo, A.C.A., Nogueira, F.C.S., Canellas, L.P., et al., 2019. Evaluation of the effects of humic acids on maize root architecture by label-free proteomics analysis. Sci. Rep. 9 (1), 1–11.

Omidbakhshfard, M.A., Sujeeth, N., Gupta, S., Omranian, N., Guinan, K.J., Brotman, Y., et al., 2020. A biostimulant obtained from the seaweed *Ascophyllum nodosum* protects *Arabidopsis thaliana* from severe oxidative stress. Int. J. Mol. Sci. 21 (2), 474.

Palazzini, J.M., Llabot, J.M., Cantoro, R., Chiotta, M.L., Allemandi, D.A., Torres, A.M., et al., 2020. Spray-drying process as a suitable tool for the formulation of *Bacillus velezensis* RC218, a proved biocontrol agent to reduce *Fusarium* Head Blight and deoxynivalenol accumulation in wheat. Biocontrol Sci. Technol. 30, 329–338.

Pan, J., Huang, C., Peng, F., Zhang, W., Luo, J., Ma, S., et al., 2020a. Effect of arbuscular mycorrhizal fungi (AMF) and plant growth-promoting bacteria (PGPR) inoculations on *Elaeagnus angustifolia* L. in saline soil. Appl. Sci. 10 (3), 945.

Pan, X., Zhang, S., Zhong, Q., Gong, G., Wang, G., Guo, X., et al., 2020b. Effects of soil chemical properties and fractions of Pb, Cd, and Zn on bacterial and fungal communities. Sci. Total Environ. 715, 136904.

Patel, J.S., Selvaraj, V., Gunupuru, L.R., Rathor, P.K., Prithiviraj, B., 2020. Combined application of *Ascophyllum nodosum* extract and chitosan synergistically activates host-defense of peas against powdery mildew. BMC Plant Biol. 20 (1), 1–10.

Pii, Y., Mimmo, T., Tomasi, N., Terzano, R., Cesco, S., Crecchio, C., 2015. Microbial interactions in the rhizosphere: beneficial influences of plant growth-promoting rhizobacteria on nutrient acquisition process: a review. Biol. Fertil. Soils 51 (4), 403–415.

Pimentel, M.F., Arnao, E., Warner, A.J., Subedi, A., Rocha, L.F., Srour, A.Y., et al., 2020. *Trichoderma* isolates inhibit *Fusarium virguliforme* growth, reduce root rot, and induce defense-related genes on soybean seedlings. Plant Dis. Available from: https://doi.org/10.1094/PDIS-08-19-1676-RE.

Pramanik, K., Mitra, S., Sarkar, A., Maiti, T.K., 2018. Alleviation of phytotoxic effects of cadmium on rice seedlings by cadmium resistant PGPR strain *Enterobacter aerogenes* MCC 3092. J. Hazard. Mater. 351, 317–329.

Rahimi-Midani, A., Kim, J.O., Kim, J.H., Lim, J., Ryu, J.G., Kim, M.K., et al., 2020. Potential use of newly isolated bacteriophage as a biocontrol against *Acidovorax citrulli*. Arch. Microbiol. 202 (2), 377–389.

Rahman, M.S., Jamal, M.A.H.M., Biswas, P.K., Rahman, S.M., Sharma, S.P., Saha, S.K., et al., 2020. Arsenic Remediation in Bangladeshi rice varieties with enhance plant growth by unique arsenic-resistant bacterial isolates. Geomicrobiol. J. 37 (2), 130–142.

Ramírez, M., Neuman, B., Ramírez, C.A., 2020. Bacteriophages as promising agents for the biological control of moko disease (*Ralstonia solanacearum*) of banana. Biol. Control. 104238.

Raymaekers, K., Ponet, L., Holtappels, D., Berckmans, B., Cammue, B.P., 2020. Screening for novel biocontrol agents applicable in plant disease management—a review. Biol. Control. 104240.

Rekaby, S.A., Awad, M.Y., Hegab, S.A., Eissa, M.A., 2020. Effect of some organic amendments on barley plants under saline condition. J. Plant Nutr. 1–12. Available from: https://doi.org/10.1080/01904167.2020.1750645.

Rizvi, A., Khan, M.S., 2018. Heavy metal induced oxidative damage and root morphology alterations of maize (*Zea mays* L.) plants and stress mitigation by metal tolerant nitrogen fixing *Azotobacter chroococcum*. Ecotoxicol. Environ. Saf. 157, 9–20.

Rouphael, Y., Colla, G., 2018. Synergistic biostimulatory action: designing the next generation of plant biostimulants for sustainable agriculture. Front. Plant Sci. 9, 1655.

Rouphael, Y., Franken, P., Schneider, C., Schwarz, D., Giovannetti, M., Agnolucci, M., et al., 2015. Arbuscular mycorrhizal fungi act as biostimulants in horticultural crops. Sci. Hortic. 196, 91–108.

Rouphael, Y., Cardarelli, M., Bonini, P., Colla, G., 2017. Synergistic action of a microbial-based biostimulant and a plant derived-protein hydrolysate enhances lettuce tolerance to alkalinity and salinity. Front. Plant Sci. 8, 131.

Salas-Méndez, E., Vicente, A., Pinheiro, A.C., Ballesteros, L.F., Silva, P., Rodríguez-García, R., et al., 2019. Application of edible nanolaminate coatings with antimicrobial extract of Flourensia cernua to extend the shelf-life of tomato (Solanum lycopersicum L.) fruit. Postharvest Biol. Technol. 150, 19–27.

Sammauria, R., Kumawat, S., Kumawat, P., Singh, J., Jatwa, T.K., 2020. Microbial inoculants: potential tool for sustainability of agricultural production systems. Arch. Microbiol. 1–17.

Satapute, P., Kamble, M.V., Adhikari, S.S., Jogaiah, S., 2019. Influence of triazole pesticides on tillage soil microbial populations and metabolic changes. Sci. Total Environ. 651, 2334–2344.

Saudy, H.S., Hamed, M.F., Abd El-Momen, W.R., Hussein, H., 2020. Nitrogen use rationalization and boosting wheat productivity by applying packages of humic, amino acids, and microorganisms. Commun. Soil Sci. Plant Anal. 1–12.

Savary, S., Willocquet, L., Pethybridge, S.J., Esker, P., McRoberts, N., Nelson, A., 2019. The global burden of pathogens and pests on major food crops. Nat. Ecol. Evol. 3 (3), 430–439.

Schiavon, M., Ertani, A., Nardi, S., 2008. Effects of an alfalfa protein hydrolysate on the gene expression and activity of enzymes of the tricarboxylic acid (TCA) cycle and nitrogen metabolism in *Zea mays* L. J. Agric. Food Chem. 56 (24), 11800–11808.

Shahid, M., Hameed, S., Zafar, M., Tahir, M., Ijaz, M., Tariq, M., et al., 2019. *Enterobacter* sp. strain Fs-11 adapted to diverse ecological conditions and promoted sunflower achene yield, nutrient uptake, and oil contents. Braz. J. Microbiol. 50 (2), 459–469.

Sharma, R.K., Barot, K., Archana, G., 2020. Root colonization by heavy metal resistant Enterobacter and its influence on metal induced oxidative stress on *Cajanus cajan*. J. Sci. Food Agric. 100 (4), 1532–1540.

Shekhawat, K., Sheikh, A., Mariappan, K., Jalal, R., Hirt, H., 2020. *Enterobacter* sp. SA187 mediates plant thermotolerance by chromatin modification of heat stress genes. bioRxiv. Available from: https://doi.org/10.1101/2020.01.16.908756.

Shinwari, Z.K., Jan, S.A., Nakashima, K., Yamaguchi-Shinozaki, K., 2020. Genetic engineering approaches to understanding drought tolerance in plants. Plant Biotechnol. Rep. 1–12.

Shreya, D., Jinal, H.N., Kartik, V.P., Amaresan, N., 2020. Amelioration effect of chromium-tolerant bacteria on growth, physiological properties and chromium mobilization in chickpea (*Cicer arietinum*) under chromium stress. Arch. Microbiol. 1–8.

Singh, V.K., Singh, A.K., Kumar, A., 2017. Disease management of tomato through PGPB: current trends and future perspective. 3 Biotech 7 (4), 255.

Singh, V.K., Singh, A.K., Singh, P.P., Kumar, A., 2018. Interaction of plant growth promoting bacteria with tomato under abiotic stress: a review. Agric. Ecosyst. Environ. 267, 129–140.

Singh, M., Singh, D., Gupta, A., Pandey, K.D., Singh, P.K., Kumar, A., 2019. Plant growth promoting rhizobacteria: application in biofertilizers and biocontrol of phytopathogens. PGPR Amelioration in Sustainable Agriculture. Woodhead Publishing, pp. 41–66.

Singh, V.K., Singh, S.K., Singh, P.K., Verma, H., Pandey, K.D., Singh, P.K., et al., 2020a. Impact of pesticides applications on the growth and function of cyanobacteria. Advances in Cyanobacterial Biology. Academic Press, pp. 151–162.

Singh, S., Singh, U.B., Trivedi, M., Sahu, P.K., Paul, S., Paul, D., et al., 2020b. Seed biopriming with salt-tolerant endophytic *Pseudomonas geniculata*-modulated biochemical responses provide ecological fitness in Maize (*Zea mays* L.) grown in saline sodic soil. Int. J. Environ. Res. Public Health 17 (1), 253.

Soltani, A., Weraduwage, S.M., Sharkey, T.D., Lowry, D.B., 2019. Elevated temperatures cause loss of seed set in common bean (*Phaseolus vulgaris* L.) potentially through the disruption of source-sink relationships. BMC Genom. 20 (1), 312.

Srivastava, S., Bist, V., Srivastava, S., Singh, P.C., Trivedi, P.K., Asif, M.H., et al., 2016. Unraveling aspects of *Bacillus amyloliquefaciens* mediated enhanced production of rice under biotic stress of *Rhizoctonia solani*. Front. Plant Sci. 7, 587.

Strange, R.N., Scott, P.R., 2005. Plant disease: a threat to global food security. Annu. Rev. Phytopathol. 43, 83–116.

Sultana, S., Paul, S.C., Parveen, S., Alam, S., Rahman, N., Jannat, B., et al., 2020. Isolation and identification of salt-tolerant plant-growth-promoting rhizobacteria and their application for rice cultivation under salt stress. Can. J. Microbiol. 66 (2), 144–160.

Tabassum, B., Khan, A., Tariq, M., Ramzan, M., Khan, M.S.I., Shahid, N., et al., 2017. Bottlenecks in commercialisation and future prospects of PGPR. Appl. Soil Ecol. 121, 102–117.

Tarantino, A., Lops, F., Disciglio, G., Lopriore, G., 2018. Effects of plant biostimulants on fruit set, growth, yield and fruit quality attributes of 'Orange rubis®' apricot (*Prunus armeniaca* L.) cultivar in two consecutive years. Sci. Hortic. 239, 26–34.

Trabelsi, D., Mhamdi, R., 2013. Microbial inoculants and their impact on soil microbial communities: a review. BioMed. Res. Int. 2013, Article ID 863240. Available from: https://doi.org/10.1155/2013/863240.

Tsaniklidis, G., Pappi, P., Tsafouros, A., Charova, S.N., Nikoloudakis, N., Roussos, P.A., et al., 2020. Polyamine homeostasis in tomato biotic/abiotic stress cross-tolerance. Gene 727, 144230.

Turhan, E., Kiran, S., Ates, Ç., Ates, O., Kusvuran, S., Ellialtioglu, S.S., 2020. Ameliorative effects of inoculation with *Serratia marcescens* and grafting on growth of eggplant seedlings under salt stress. J. Plant Nutr. 43 (4), 594–603.

Ullah, A., Heng, S., Munis, M.F.H., Fahad, S., Yang, X., 2015. Phytoremediation of heavy metals assisted by plant growth promoting (PGP) bacteria: a review. Environ. Exp. Bot. 117, 28–40.

van Esse, H.P., Reuber, T.L., van der Does, D., 2020. Genetic modification to improve disease resistance in crops. N. Phytol. 225 (1), 70–86.

Van Oosten, M.J., Pepe, O., De Pascale, S., Silletti, S., Maggio, A., 2017. The role of biostimulants and bioeffectors as alleviators of abiotic stress in crop plants. Chem. Biol. Technol. Agric. 4 (1), 5.

Vaishnav, A., Singh, J., Singh, P., Rajput, R.S., Singh, H.B., Sarma, B.K., 2020. *Sphingobacterium* sp. BHU-AV3 induces salt tolerance in tomato by enhancing antioxidant activities and energy metabolism. Front. Microbiol. 11. Available from: https://doi.org/10.3389/fmicb.2020.00443.

Vassilev, N., Vassileva, M., Martos, V., del Moral, L.F.G., Kowalska, J., Tylkowski, B., et al., 2020. Formulation of microbial inoculants by encapsulation in natural polysaccharides: focus on beneficial properties of carrier additives and derivatives. Front. Plant Sci. 11. Available from: https://doi.org/10.3389/fpls.2020.00270.

Velivelli, S.L., De Vos, P., Kromann, P., Declerck, S., Prestwich, B.D., 2014. Biological control agents: from field to market, problems, and challenges. Trends Biotechnol. 32 (10), 493–496.

Viterbo, A.D.A., Chet, I., 2006. TasHyd1, a new hydrophobin gene from the biocontrol agent *Trichoderma asperellum*, is involved in plant root colonization. Mol. Plant Pathol. 7 (4), 249–258.

Vos, C.M., De Cremer, K., Cammue, B.P., De Coninck, B., 2015. The toolbox of *Trichoderma* spp. in the biocontrol of *Botrytis cinerea* disease. Mol. Plant Pathol. 16 (4), 400–412.

Wang, W., Vinocur, B., Altman, A., 2003. Plant responses to drought, salinity and extreme temperatures: towards genetic engineering for stress tolerance. Planta 218 (1), 1–14.

Wang, S., Tian, X., Liu, Q., 2020. The Effectiveness of foliar applications of zinc and biostimulants to increase zinc concentration and bioavailability of wheat grain. Agronomy 10 (2), 178.

Wonglom, P., Ito, S.I., Sunpapao, A., 2020. Volatile organic compounds emitted from endophytic fungus *Trichoderma asperellum* T1 mediate antifungal activity, defense response and promote plant growth in lettuce (*Lactuca sativa*). Fungal Ecology 43, 100867.

Woo, O.G., Kim, H., Kim, J.S., Keum, H.L., Lee, K.C., Sul, W.J., et al., 2020. *Bacillus subtilis* strain GOT9 confers enhanced tolerance to drought and salt stresses in *Arabidopsis thaliana* and *Brassica campestris*. Plant Physiol. Biochem. 148, 359–367.

Yakhin, O.I., Lubyanov, A.A., Yakhin, I.A., Brown, P.H., 2017. Biostimulants in plant science: a global perspective. Front. Plant Sci. 7, 2049.

Yang, C., Ho, Y.N., Makita, R., Inoue, C., Chien, M.F., 2020. A multifunctional rhizobacterial strain with wide application in different ferns facilitates arsenic phytoremediation. Sci. Total Environ. 712, 134504.

Yao, Y., Wang, X., Chen, B., Zhang, M., Ma, J., 2020. Seaweed extract improved yields, leaf photosynthesis, ripening time, and net returns of tomato (*Solanum lycopersicum* Mill.). ACS omega 5 (8), 4242–4249.

Yarzábal, L.A., 2020. Perspectives for using glacial and periglacial microorganisms for plant growth promotion at low temperatures. Appl. Microbiol. Biotechnol. 1–12.

Ye, T., Zhou, T., Li, Q., Xu, X., Fan, X., Zhang, L., et al., 2020a. *Cupriavidus* sp. HN-2, a novel quorum quenching bacterial isolate, is a potent biocontrol agent against *Xanthomonas campestris* pv. *campestris*. Microorganisms 8 (1), 45.

Ye, D., Li, T., Liu, J., Yi, Y., Zou, L., Zhang, X., et al., 2020b. Characteristics of endophytic bacteria from *Polygonum hydropiper* and their use in enhancing P-phytoextraction. Plant Soil 1–17.

Zanin, L., Tomasi, N., Cesco, S., Varanini, Z., Pinton, R., 2019. Humic substances contribute to plant iron nutrition acting as chelators and biostimulants. Front. Plant Sci. 10.

Zapata-Sarmiento, D.H., Palacios-Pala, E.F., Rodríguez-Hernández, A.A., Melchor, D.L.M., Rodríguez-Monroy, M., Sepúlveda-Jiménez, G., 2020. *Trichoderma asperellum*, a potential biological control agent of *Stemphylium vesicarium*, on onion (*Allium cepa* L.). Biol. Control. 140, 104105.

Zhang, W., 2018. Global pesticide use: profile, trend, cost/benefit and more. Proc. Int. Acad. Ecol. Environ. Sci. 8 (1), 1.

Zhang, J., Xing, J., Lu, Q., Yu, G., Xu, B., Huang, B., 2019. Transcriptional regulation of chlorophyll-catabolic genes associated with exogenous chemical effects and genotypic variations in heat-induced leaf senescence for perennial ryegrass. Environ. Exp. Bot. 167, 103858.

CHAPTER

14

Microbial formulation approaches in postharvest disease management

Madhuree Kumari[1,2,3], *Siya Kamat*[1], *Ritu Dixit*[2,4], *Shipra Pandey*[2,3], *Ved Prakash Giri*[2,4] *and Aradhana Mishra*[2,3]

[1]Indian Institute of Science, Bengaluru, India [2]National Botanical Research Institute, Lucknow, India [3]Academy of Scientific and Innovative Research (AcSIR), Ghaziabad, India [4]Department of Botany, Lucknow University, Lucknow, India

14.1 Introduction

And so, from hour to hour we ripe and ripe,
And then from hour to hour we rot and rot,
And thereby hangs a tale.

William Shakespeare

Agriculture sector is the largest source of employment for people all over the world. It is thus the major driver of poverty reduction. Mechanical innovations followed by chemical creativity have transformed agriculture to meet the increasing population growth. Hence, a challenge of agricultural sustainability chronicles all the world economies today (Schreiner, 2005; Bawa, 2019).

Horticultural perishables such as fruits, nuts, vegetables, and spices play a pertinent role in the agribusiness as well as human nutrition. However, the value of these commodities can be measured at every stage along the chain, that is, from the farm to the final buyer/during agricultural production, postharvest, and processing stages in the commodity channel (Fig. 14.1). At every stage it is packed, processed, and transported, which incurs qualitative and quantitative losses (Kader and Rolle, 2004). Qualitative losses relate to changes in appearance, texture, nutritional value, freshness, edibility, flavor, and are hence very difficult to assess than the quantitative losses.

The Indian Ministry of Food Processing Industries estimated a loss of 23 million tons of grains, 12 million tons of fruits, and 21 million tons of vegetables in 2014. This amounts to

DOI: https://doi.org/10.1016/B978-0-12-821843-3.00007-6

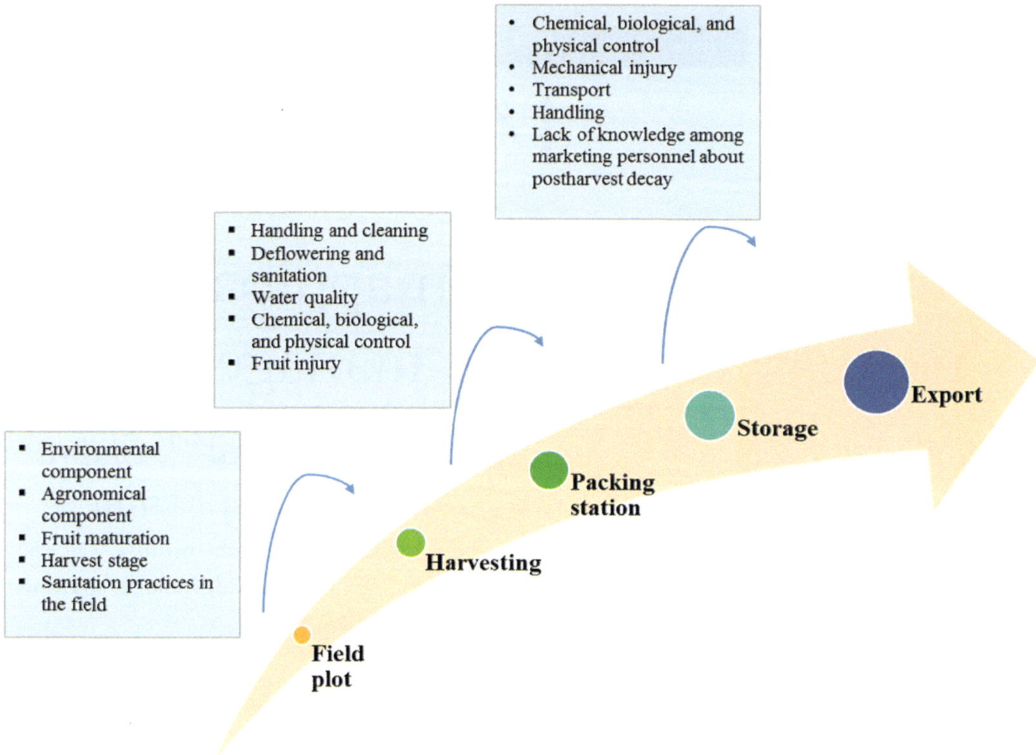

FIGURE 14.1 Diagram representing the stages during commodity channel and key factors that influence the quality and quantity of the farm produce.

10.6 billion USD (Ganesh et al., 2016). Even though highest postharvest losses were recorded in developing nations of Asia and Africa, according to the Food and Agriculture Organization of the United States (FAO Report, 1997), food loss is not restricted in developing economies. The global postharvest loss is nearly 30% for cereals, 40% for crops among other perishable items contributed by world's leading powers (Ministry of Finance, 2019). The 2011 report of FAO documented that with respect to the amounts of globally produced fruits and vegetables, 15% and 50% are lost due to postharvest decay, mainly due to fungal pathogens (FAO, 2011).

The Central Institute of Post-harvest Engineering and Technology, Ludhiana, India, reported that about 16% of fruits and vegetables valued at 6 billion USD were lost between 2012 and 2014 in the Indian agrobusiness. It was also reported that a major share of the budget was spent on cold storage facilities (https://pib.gov.in/newsite/printrelease.aspx?relid = 136922).

Hence, there is a necessity to develop better strategies to meet the demand of postharvest management.

In the present scenario, registered chemicals available in market for postharvest losses are prone to environmental and health hazards, making it prohibited in some countries (Wisniewski et al., 2016). Widespread and prolonged interaction of fungicides showed resistance toward pathogens. Therefore requirement of an alternative solution has become an utmost need of society (Marcet-Houben et al., 2012). A number of strategies have been evolved including biocontrol agents for postharvest disease management, but microbial formulation has been widely accepted at industrial and consumer levels due to their environment-friendly, cost-effective, user-friendly, and higher tolerance properties.

14.2 Postharvest disorders caused by biotic factors

Malfunction of physiological conditions of crops, fruits, and vegetables due to abiotic stresses such as temperature, relative humidity, moisture level, and pH makes a conducive environment for pathogen attack. Most postharvest pathogens invade through the cuts and wounds or when the host defense is weak. They are also selective in their host tissues and pH. Fungal pathogens prefect in an acidic environment and hence, infect citrus fruits that generally have pH less than 4. Hence, there aren't any major commercial reports on bacterial postharvest diseases. Conversely, bacterial infections are quite common in vegetables that generally have a pH of more than 5.

Postharvest pathogens can be grouped into two categories (Mincuzzi et al., 2020):

Latent pathogens: They infect the farm produce in the preharvest stage viz., during blooming or fruit development. They remain dormant until a postharvest favorable condition allows their reactivation. Example: *Botrytis* sp., *Alternaria* sp.

Wound pathogens: As the name suggests, these pathogens need an entry site in the form of a lesion, injury, or other abiotic damages to penetrate the host, in the pre- or postharvest stage. Example: *Aspergillus* sp. and *Penicillium* sp.

Common postharvest decay pathogens infecting fruits:

14.2.1 Penicillium rot

Penicillium rots are the most significant and widely reported infections in pome and citrus fruits. The key threat of this pathogen is due to their powdery spores. It usually infects the fruit from its stem end. *Penicillium digitatum* and *Penicillium italicum* are the two most common pathogens in this category. After infecting a fruit, there happens plentiful sporulation that results in a completely white fruit covered with greenish blue spores. The fungal pathogens produce abundant quantities of ethylene culminating in rapid senescence of adjacent fruits. Synthetic chemical mixtures containing limonene, ethanol, acetaldehyde, and CO_2 at particular levels could also *stimulate* the growth of *P. digitatum*.

Blue mold is a common terminology applied to refer several *Penicillium* species that cause postharvest decay. In study by Yin et al. (2017), blue rot fungi from pome fruits such as stored apples and pears were identified using a suite of molecular DNA markers. The pathogens were identified as *Penicillium expansum, Penicillium solitum, Penicillium carneum*, and *Penicillium paneum*. Every species of *Penicillium* contributing to the blue rot is different although they seem similar morphologically. For instance, *P. expansum* is more dangerous since it produces more toxins such as ochratoxins, patulin, citrinin on apples and pears than its cousin *P. solitum*. This puts forth the need to develop and implement a fast yet reliable method for differentiating between two species (Pitt et al., 1991).

Green mold that includes *P. digitatum* is widely observed in India. In contrast with blue mold, the green mold infection grows rapidly at higher temperatures (25°C–30°C) through wounds. Its mitigation begins at the preharvest stage using benomyl spray.

14.2.2 Aspergillus rot

This fungus penetrates fruits through microwound and bruises. Fruits infected with Aspergillus rot appear black. Hence, it is also called black mold. It spreads very fast at higher temperatures 30°C–35°C. This fungus is temperature specific. Cold storage (7°C–10°C) of farm produce usually helps to avoid sporulation and decay (Mincuzzi et al., 2020). Fruit decay due to *Aspergillus niger* is common in India.

14.2.3 Rhizopus rot

This troublesome parasitic pathogen follows the "nesting strategy" of infection. It infects its neighboring fruits with mycelium. Unlike Penicillium rot and Aspergillus rot, Rhizopus rot spreads rapidly at 15°C (Ladanyia and Ladaniya, 2010). One of the deadliest postharvest pathogens in this category is *Rhizopus stolonifer* and only infects mature fruits (Gonçalves et al., 2010). It is known to infect nectarine and stone fruits mainly at wound sites at harvest and packing stage. Its spores are abundant in the atmosphere (Qing and Shiping, 2000).

14.2.4 Brown rot

This latent infection is caused by *Monilinia laxa* (Aderhold & Ruhland) Honey, *Monilinia fructigena* (Aderhold & Ruhland) Honey, and *Monilinia fructicola* (Winter) Honey. This rot majorly affects peaches, nectarines, apricots, sour and sweet cherry, apples, and pears. It infects the plant at the field stage although most of the losses are in the postharvest stage (Martini and Mari, 2014).

14.2.5 Crown rot

This is a complex infection resulting from a cocktail of fungal pathogens. The microorganisms commonly contributing to crown rot are *Musicillium theobromae, Colletotrichum musae, Ceratocystis paradoxa, Lasiodiplodia theobromae, Nigrospora sphaerica, Cladosporium* spp., *Acremonium* spp., *Penicillium* spp., and *Aspergillus* spp., as well as many *Fusarium* species. This postharvest disease affects bananas. The infection begins at harvest in the peduncle region and is hence called crown rot. However, the symptoms develop gradually during shipping and ripening. The fungal complex brings about softening and darkening of the crown region and finally results in detachment of the fruit. It affects the fruit quality due to necrosis and early ripening because of ethylene produced by mycelia of the fungus *C. musae* (Lassois and de Bellaire, 2014). The postharvest decay symptoms caused by *L. theobromae* are called diplodia stem-end rot. The fungus commonly occurs on citrus fruits grown in hot, humid, and tropical regions. The economically important fruit, mango is also affected by this postharvest disease (Diskin et al., 2019). The rot develops from latent infections. Postharvest, the fungus reactivates and infects the fruit through natural openings.

14.2.6 *Botrytis cinerea* rot

The fungus *Botrytis cinerea* (gray mold) takes up the second place in the list of world's top 10 pathogens based on economic importance. It has a wide range of hosts and is hence very easy to find it in pre- or postharvest stage. A study by Williamson et al. (2007) estimated more than 200 dicotyledonous crop species to be hosts for gray mold. There is a huge difference in the global expense of gray mold control and product loss despite the control measures. The crops that are susceptible to gray mold postharvest include artichoke, apple, asparagus, bean, beet, blackberry, blueberry, broccoli, cabbage, carrot, cauliflower, celery, cucumber, currant, eggplant, endive, grape, kale, kaki, kiwi, leek, lentil, lemon, lettuce, onion, pea, peanut, pepper, pear, peach, plum, pomegranate, potato, pumpkin, raspberry, strawberry, sweet cherry, and tomato (Droby, 2007; Romanazzi et al., 2012). Infection may occur at preharvest stage and remain latent. Due to the nesting strategy of infection, the fungus causes extensive spoilage of the produce. The pathogen can sustain itself at storage temperatures as well. However, gray mold infection in grapes is considered beneficial since it contributes positively in production of rare dessert wines. Hence, it is also called "noble rot" (Tosi et al., 2012).

14.2.7 Sour rot

This disease is caused by the yeast such as fungal pathogen *Galactomyces citri-aurantii*. Most of the decay in Nagpur Mandarins during the March–April harvest season is contributed by this rot. Low temperatures below 10°C negatively regulate the fungal growth and hence the infection. The disease develops when the fruit is handled in warm areas during the packaging stage when the surrounding temperature is more than 27°C. As a result of the rapid infection, the fruit becomes soft and starts to stink since the fungus secretes tissue-degrading enzymes.

14.2.8 *Alternaria alternata* rot

This fungal pathogen has been reported all over the world and can proliferate in different environments. It causes postharvest decay in mangoes, black rot in cherry tomatoes, kiwifruit, melons, citrus fruits, figs, and many more. It can infect both fruits and vegetables. Typically, wounded, stressed, and senescent tissues are more susceptible to *Alternaria* infection.

14.2.9 *Thielaviopsis paradoxa*, *Thielaviopsis basicola* rot

These fungal pathogens are one of the major threats to agronomic produce such as fruits, crops as well as some ornamental plants. *Thielaviopsis paradoxa* infects economically significant farm produce such as pineapple, coconut, sugarcane, cocoa, banana, corn, sweet potato, and many more. The disease caused by *T. paradoxa* is commonly called black rot and is mostly reported in tropical regions (Ploetz et al., 2003). On the contrary, *Thielaviopsis basicola* infection, also called as black root rot targets ornamental foliage plants, greenhouse plants, and field crops such as peanut, tobacco, carrot, and bean. In a case study carried out by Eshel et al. (2009), it was reported that the fungus *T. basicola* could contaminate farm-produced carrots during the postharvest stage through wounds and brushing.

14.2.10 Diplodia stem-end rot

The postharvest decay is caused by fungus *L. theobromae* and commonly occurs on citrus fruits grown in hot, humid, and tropical regions. The economically important fruit, mango is also affected by this postharvest disease (Diskin et al., 2019). The rot develops from latent infections. Postharvest, the fungus reactivates and infects the fruit through natural openings.

14.3 Control strategies for postharvest diseases

A lot of conventional methods are still practiced in order to maintain the quality of harvested fruits, vegetables, and crops. These include deflowering, maintaining water quality standards and sanitation at packing stations, proper environmental controls during storage and shipping of the farm produce (Bastiaanse et al., 2010). Some control strategies mentioned in Table 14.1 have been reported to have considerable potential for mitigation of postharvest decay.

Nevertheless, commercial synthetic fungicides are the preferred choice for agricultural sector specialists and farmers to control the postharvest decay. This strategy was introduced back in 1960s since the discovery of benzimidazole derivatives (Lapeyre de Bellaire and Nolin, 1994). Eventually, many industries such as Syngenta, Bayer Crop Science, and BASF introduced into the agro-market various fungicides to control postharvest decay caused by gray mold (http://www.frac.info). Even though fungicide treatment can reduce postharvest losses, the effectiveness decreases with emergence of resistant strains of pathogens. There has also been a consistent public demand internationally to reduce the use of chemical pesticides due to their severity on the consumer's health. A high-risk pathogen, *B. cinerea* that has been often reported worldwide developed resistance to the commonly used benzimidazole fungicides. The mutants that emerged were resistant to myriads of fungicides. As a result, many countries revoked or restricted the use of certain synthetic fungicides (Myresiotis et al., 2007; Dayan et al., 2019). The Fungicide Resistant Action Committee (FRAC) has proposed guidelines that suggest a limited number of applications per year (Hermann and Stenzel, 2019). Hence, there is a considerable interest in developing alternative methods to eliminate fungicide treatment (Mari et al., 2003).

14.4 Microbes' role in controlling postharvest disease

The potential of microbes in plant growth and development is well known, but the role of microbes to inhibit growth of phytopathogen that causes postharvest disease is needed to be explored. Postharvest disease causes greater losses to crops. These diseases could be successfully controlled with application of chemical fertilizers and pesticides. But, they pose a constraint to plants and human health. It also leads to the development of resistant strains of

TABLE 14.1 Summary of commercially practiced postharvest control strategies, their benefits and drawbacks.

S. no	Control strategy	Technique	Benefits	Drawbacks	References
1.	Hot water treatment/hot water rinsing and brushing	Disinfecting fruits and vegetables with hot water 44°C–48°C and brushes for <5 min.	Inhibits fungal spore germination and subsequent growth. Pesticide-free method.	Fruit peel injury, heat damage.	Porat et al. (2000), Obagwu and Korsten (2003), Jing et al. (2010)
2.	Ultraviolet C (UV-C) irradiation (254 nm)	Low doses of UV-C (2.5 kJ/m^2).	Clean, safe, and sustainable. Pesticide-free method.	High doses (>5 kJ/m^2) may cause fruit damage. Change in metabolic profile of the fruit.	Brown et al. (2002), Stevens et al. (2004), Charles et al. (2008), Terao et al. (2015)
3.	Gaseous treatment	Application of fumigants in storage houses using "Closed Loop System" in the United States, SIROFLO, SIROCIRC in Australia, and PHYTO EXPLO in Europe. The gases used are methyl bromide, phosphine (FUME, FRISIN), sulfuryl fluoride (ProFume), propylene oxide, carbonyl sulfide, ethyl formate, hydrogen cyanide, carbon disulfide, methyl iodide, ozone, and carbon dioxide.	High speed of action, penetrative.	Toxic, ozone-depleting potential. Change in fruit quality and residues, corrosive, emergence of resistant pathogens, propylene oxide is flammable.	Navarro (2006), Venta et al. (2010), Liu (2008)
4.	Synthetic fungicides	Fungicide dip solutions are prepared in large tanks and are used on pack lines. Examples: Imazalil, thiabendazole, pyrimethanil, prochloraz, guazatine, benzothiazole.	Easy to use, high potency.	Generation of toxic waste, hazardous waste disposal is expensive, emergence of resistant pathogens. Environmental and human hazard.	Smilanick et al. (2008), FAO Report (1997), Altieri et al. (2004)
5.	Biological control: antagonistic bacteria and yeast	The common mechanisms effective against postharvest decay are competition for nutrients and space, siderophores, parasitism, formation of biofilms, and initiation of resistance. The currently registered postharvest biocontrol agents are Aspire (Ecogen, USA) containing *Candida oleophila*, strain I-182, BioSave 110 (EcoEcience, USA) containing saprophytic strain of the bacterium *Pseudomonas syringae*.	Genetically stable, effective at low concentrations, environment friendly, cost effective.	Limited spectrum of activity and efficacy, application complexity.	Mari et al. (2003), Bélanger et al. (2012), Carmona-Hernandez et al. (2019)
6.	Generally recognized as safe and natural compounds, plant volatiles	Application of aqueous extracts of various plants, chitosan, plant secondary metabolites such as limonene, camphone with fungicidal potential.	Environment friendly, cost effective, limited health hazard.	Laborious production process, continuous use may be harmful.	Leuba and Stossel (1986), Ali et al. (2016)

pathogens (Buchholz et al., 2018). Hence, there is a need for an eco-friendly treatment that could be an alternative to pesticides and chemical applications (Brzozowski et al., 2016). In some cases, antagonistic microorganisms also help in integrated disease management strategies to avoid crop quality loss. The mechanisms used by microbes for postharvest disease management involve siderophores, biological control, hydrolytic enzymes, antibiosis, biofilm formation, and induction of plant resistance (Köhl et al., 2019).

An enormous number of microorganism-based formulation including yeast, bacteria, and fungus involved in pathogen control have adopted an array of mechanism such as production of volatile organic compounds (VOCs; Di Francesco et al., 2018), antagonistic activities (Calvente et al., 1999; Castoria et al., 2001; Rajendiran et al., 2010; Wu et al., 2019). These microbial attributes induced resistance in crops and fruits by enhancing phenolic content and antioxidant enzymes (Romanazzi et al., 2016; Yang et al., 2017). Use of microbes in terms of bioformulation ultimately enhanced the plant health because interaction of plant microbe in rhizosphere promoted sustainable growth of host plant, altered soil atmosphere, and protected crop from phytopathogen (Table 14.2; Jones et al., 2019). Most commonly used fungi in bioformulation are *Metarhizium anisopliae* and *Beauveria bassiana*. The synergetic interactions of microbes with phytosanitary products cause mortality of the target pathogen, ultimately becoming a tool for deployment in integrated pest management (IPM; Bukhari et al., 2011). A *Bacillus* sp.–based formulation having remarkable antimicrobial activity against myriad of pathogens viz., *Botryosphaeria dothidea, R. stolonifer, B. cinerea, Botrytis mali, P. digitatum, Phytophthora infestans, Fusarium oxysporum* by producing secondary metabolites such as macrolactin, bacillaene, mycosubtilin, surfactin, fengycin, and iturins has been reported by many researchers (Wu et al., 2019; Cozzolino et al., 2020; Tian et al., 2020). Another advantage of *Bacillus* sp. over other formulations is endospore formation and survival in extreme conditions (higher and lower temperature, irradiation, pH, chemical pressure).

Yeast, a single cell microbial flora also plays a key role in postharvest pathogen control. Various yeasts have been reported earlier for showing proficient antagonistic activity against pathogens. *Aureobasidium pullulans* has been shown as antagonist by producing extracellular enzymes by directly interacting with fungal hyphae (Castoria et al., 2001), while some strains of *A. pullulans* produce lipopeptides (LPs). These are low-molecular-weight VOCs able to reduce brown rot disease by strengthening cuticle and membrane protein synthesis (Di Francesco et al., 2018). Likewise, according to Zheng and Chen (2009), *Cryptococcus laurentii*, a biocontrol yeast upregulated the expression of protein responsible for postharvest pathogen resistance. Yeasts are also able to elicit pathogen-triggered immunity in fruit by upregulating defense responsive genes, in response to bacterial flagellin exposure at cell surface(Zhao et al., 2020b). Various microbial formulations are being used for commercial practices for controlling broad and narrow range of postharvest pathogens (Calvo-Garrido et al., 2019), but there should be a regulation for optimization and IPM strategies for standard application.

Table 14.2 shows list of postharvest diseases, casual organism, and their antagonist microorganism that can effectively inhibit postharvest diseases.

14.5 Bioformulation

Bioformulation is a mixture of bioactive ingredients of beneficial microbial strains in a form of formulated product with inert (inactive) substances (Stamenković et al., 2018). It can be

TABLE 14.2 List of bioformulation used for postharvest loss.

S. no.	Formulation	Source	Effect	Crop	Reference
1.	Fungal formulation	*Aureobasidium pullulans* and *Candida vanderwaltii*	Fungicide against *Botrytis cinerea* and *Rhizopus stolonifer* at varied temperature	Strawberry	Lima et al. (1997)
2.	Leafy formulation	*Ocimum kilimandscharicum*	Insect pest	Maize and sorghum grains	Jembere et al. (1995)
3.	Natural product formulation	Salicylic acid + putrescine	Fruit decay	Murcott mandarin fruit	Ennab et al. (2020)
4.	Natural product formulation	25 mg/L kojic acid and 10 g/L calcium chloride (KA + $CaCl_2$)	Inhibit off-odor production	Broccoli florets	Yan et al. (2020)
5.	Bacterial formulation	*Bacillus amyloliquefaciens* (VB2) and *Bacillus subtilis* (AP)	*B. cinerea* infection after harvesting	Rose	Nakkeeran et al. (2019)
6.	Bacterial formulation	Bacillus and salicylic formulation 0^3–10^8 CFU/mL and 0.05 mM, respectively	Resistance to *Phytophthora infestans* and *Fusarium oxysporum*	Potato tuber	Lastochkina et al. (2019)
7.	Organic formulation	Methyl jasmonate	Phenolic content enhances	Plum	Karaman et al. (2013)
8.	Yeast formulation	*Rhodotorula glutinis* and *Cryptococcus laurentii*in combination with 238 mmol/L sodium bicarbonate or 15 mmol/L ammonium molybdate	Against blue mold	Jujube fruits	Wan et al. (2003)
9.	Bacterial formulation	Bacterial strain B106 1×10^8 CFU/mL	*Colletotrichum musae*	Banana	Fu et al. (2010)
11.	Yeast formulation	*R. glutinis* 1×10^8 CFU/mL	Gray and blue mold	Apple	Zhang et al. (2009)
12.	Yeast formulation	*Meyerozyma guilliermondii* with methyl jasmonate	Blue mold decay	Apple	He et al. (2020)
13.	Yeast formulation	*Pichia membranifaciens* 5×10^8 CFU/mL	Rhizopus rot	Nectarine	Qing and Shiping (2000)
14.	Bacterial formulation	*Bacillus* sp. w176	Green mold	Citrus	Tian et al. (2020)
15.	Bacterial formulation	*Pseudomonas fluorescens*	*Penicillium italicum*	Oranges	Wang et al. (2020)

(*Continued*)

TABLE 14.2 (Continued)

S. no.	Formulation	Source	Effect	Crop	Reference
16.	Yeast formulation	*Cryptococcus podzolicus* with β-glucan	Blue mold	Pears	Zhao et al. (2020a)
17.	Yeast formulation	*Sporidiobolus pararoseus* Y16 and 1 mM glycine betaine	Apple blue mold	Apple	Abdelhai et al. (2019)
18.	Bacterial formulation	*Stenotrophomonas rhizophila* 1×10^8 cells/mL	*Colletotrichum gloeosporioides*	Mango	Reyes-Perez et al. (2019)
19.	Yeast formulation	*P. membranifaciens* + 20% (w/v) trehalose, 2% (w/v) sodium glutamate, 5% (w/v) polyvinyl pyrrolidone, and 20% (w/v) skim milk (vacuum dried)	Enhanced fruit viability	*Citrus sinensis*	Zhang et al. (2020)
20.	Yeast formulation	*Hanseniaspora uvarum* with β-aminobutyric acid	Gray mold and black rot	Kiwifruit	Cheng et al. (2019)
21.	Yeast formulation	*Saccharomyces cerevisiae* EBY100	Fruit decay by *B. cinerea*	Tomato	Zhao et al. (2020b)
22.	Yeast formulation	*A. pullulans* strains L1 and L8	*Monilinia laxa*	Peach	Di Francesco et al. (2018)
23.	Bacterial formulation	*Bacillus ginsengihumi*	*Botrytis* bunch rot	Grapes	Calvo-Garrido et al. (2019)
24.	Bacterial formulation	*B. amyloliquefaciens*	Brown rot	Crops	Gotor-Vila et al. (2017)
25.	Bacterial formulation	*Burkholderia spinosa*	Banana anthracnose	Banana	Ray et al. (2011)
26.	Yeast formulation	*Pichia anomala*	Penicillium rots	Citrus	Liu et al. (2017)

defined as use of microorganism with synergistic effects as a substitute for pesticides to improve soil fertility, plant growth, and phytopathogen suppression (Arora et al., 2010; Bargaz et al., 2018). Basically, various types of formulations are used in the world but two types of bioformulations are commonly available, solid and liquid (Fig. 14.2; Mishra and Arora, 2016).

Uses of microbes in bioformulation or emulsion are secure, safe, and reliable (Bulgari et al., 2019a). Various microbes have been used in bioformulation such as *Bacillus cereus* and *Pseudomonas rhodesiae* enhanced plant vegetative growth (Kalita et al., 2015). In industry, microbes are used in emulsion for their beneficial effect such as prevention of contamination of phytopathogen. Many studies reported use of microbes in emulsion such as *Bacillus subtilis* (Felix et al., 2019), *Lactococcus lactis* subsp. *lactis* biovar *diacetylactis* (Ly et al., 2006), *Acinetobacter*, *Pseudomonas aeruginosa*, *Pseudomonas carboxydohydrogena*, and *Alcaligenes latus* (Huang et al., 2010).

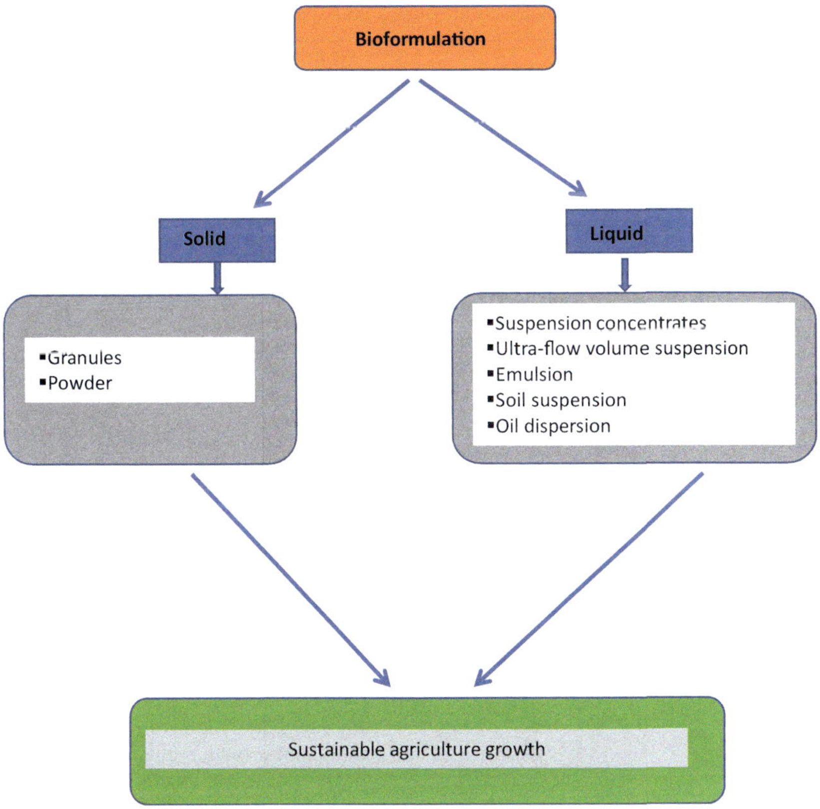

FIGURE 14.2 Types of bioformulations commonly used to control pathogens.

14.6 Volatile organic compounds or antimicrobial secondary metabolites produced by antagonists as direct pathogen control

VOCs or antimicrobial secondary metabolites produced by microbial antagonists can play a major role in showing antagonistic behavior against postharvest pathogens. Many of the VOCs have been purified from the biocontrol agents and their mechanisms against pathogens have been elucidated (Table 14.3).

14.7 Action mechanism of bioformulations

There are several mechanisms employed by the microbes involved in bioformulations to control postharvest diseases. Recent advances in techniques of microbiology, molecular biology, bioinformatics, and biochemistry have provided an insight into the multiple mechanisms employed by microbes to combat postharvest diseases. Several mechanisms including mycoparasitism, antibiosis, competition for iron and nutrients, secretion of

TABLE 14.3 List of secondary metabolites, their sources and applications.

S. no.	Secondary metabolites	Source	Application	References
1.	T22-azaphilone	*Trichoderma harzianum* 22	Biological control	Keswani et al. (2017)
2.	Genistein	*Bradyrhizobium japonicum*	Increase nodules numbers and dry matter	Muñoz et al. (2014), Morel et al. (2016)
3.	Hesperitin and naringenin	*Rhizobium leguminosarum*	Increase number of nodule and plant dry matter	Morel et al. (2016)
4.	6-Pentyl-2 Hyran-2-one (6-PP)	*T. harzianum, T. viride, T. koningii, T. atroviride*	Biological control, plant-growth promotion	Keswani et al. (2017)
5.	Butyrolactone	*Streptomyces* sp.	Morphological differentiation and secondary metabolites	Van der Meij et al. (2017)
6.	Paenilarvins	*Paenibacillus* sp.	Biocontrol	Hertlein et al. (2016)
7.	Iturin	*Bacillus amyloliquefaciens*	Biocontrol activity	Dunlap et al. (2019)
8.	Fengycin	*Bacillus subtilis*	Antifungal activity	Meena and Kanwar (2015)
9.	2,5-Diketopiperazines	*Pseudomonas aeruginosa*	Quorum sensing, plant-growth promotion	Alshaibani et al. (2017)
10.	Trichodermin	*T. harzianum*	Pharmaceutical compounds and plant-growth regulators	Contreras-Cornejo et al. (2018)

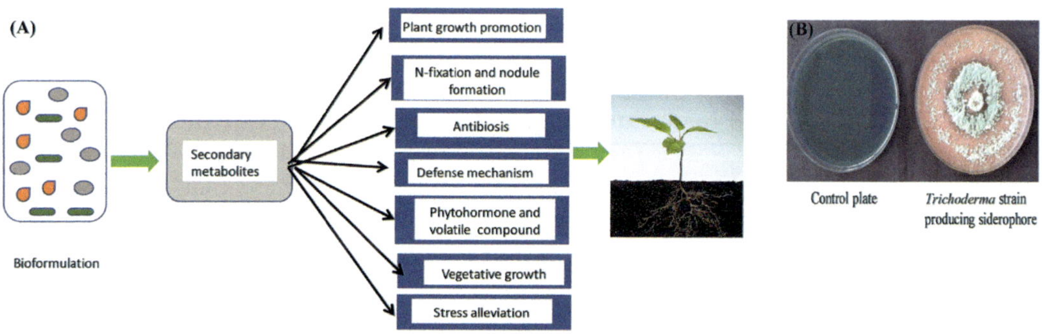

FIGURE 14.3 (A) Mode of mechanisms of bioformulations against postharvest pathogens. (B) Plate showing production of siderophore from *Trichoderma* sp.

antimicrobial compounds, and induction of induced systemic resistance (ISR) in crops play an important role during tripartite interactions of plant–pathogen and microbes for biotic stress amelioration and postharvest of crops (Fig. 14.3A). A detailed discussion with examples is provided in the following section.

14.7.1 Competition for space and nutrients

Nutrients including carbohydrates, amino acids, vitamins, and minerals are vital for growth of a pathogen and so is the space. The space and nutritional resources present in plant as host during tripartite interactions are limited (Spadaro and Droby, 2016; Dukare et al., 2017). A microbial agent that competes with the pathogen for nutrient and space and wins the battle can be an effective biocontrol agent against postharvest diseases. A number of effective biocontrol agents for control of postharvest diseases including fungi and yeast compete and exclude out their pathogenic counterpart because of their fast growth rate and ability to utilize resources effectively (Carmona-Harnandez et al., 2019).

Tryfinopoulou et al. (2020) in their studies against *Aspergillus carbonarius* in grapes/juice found yeast very efficient to inhibit the growth of *A. carbonarius* by competition for space and reducing the production of toxin ochratoxin A. Yeasts are also a successful competitor for sugar and nitrogen as observed by Spadaro and Gullino (2004), where strong inhibition of radiolabeled glucose use was observed in *B. cinerea* while competing with yeast *Sporobolomyces roseus*.

Pichia caribbica supplemented with vitamin C upregulated proteins involved in the glucose metabolism, thereby competing successfully with pathogen *P. expansum* for carbon source to inhibit blue mold decay in harvested apples (Yang et al., 2020).

14.7.2 Production of siderophores and competition for iron

Siderophores are low-molecular-weight (500–1000 Da) compounds secreted by many fungi and bacteria to chelate ferric ions from surroundings by forming stable ferric complexes and transporting them to the cells (Wilson et al., 2016). Iron is a major nutrient required by all pathogens to sustain successfully. Bioformulations producing siderophores mobilize, limit, and deplete the iron source to pathogen inducing imbalance in iron homeostasis and ultimately death of the pathogens (Glick, 2012).

Pseudomonas fluorescens is a well-known biocontrol agent for control of postharvest diseases (Wang et al., 2019; Carmona-Harnandez et al., 2019) owing to the production of siderophores. Wang et al. (2019) demonstrated efficient control of green mold caused by *P. digitatum* on oranges by application of bacteria antagonist *P. fluorescens*.

Trichoderma sp. often used as a potent postharvest disease control agents is also a prolific producer of siderophores (Fig. 14.3B; Köhl et al., 2019). Many synergistic applications of combing *Trichoderma* sp. and *P. fluorescens* have also been developed against postharvest diseases. *Metschnikowia citriensis* was shown to deplete iron and thus inhibiting the growth of *Geotrichum citri-aurantii* while controlling sour rot of postharvest citrus fruit (Kurtzman and Droby, 2001).

Yeast was also developed as an efficient postharvest disease management and a prolific producer of siderophores (Mukherjee et al., 2020). Antagonistic action of siderophores from *Rhodotorula glutinis* upon the postharvest pathogen *P. expansum* was assessed by Calvente et al. (1999).

14.7.3 Production of volatile secondary metabolites and organic compounds (VOCs)

Some of the biocontrols used in the management of postharvest diseases are able to produce antifungal and antibacterial secondary metabolites and VOCs, and other biological activities (Table 14.3; Costa et al., 2019; Kamat et al., 2020).

Trichoderma sp. is known to produce a number of antifungal secondary metabolites including peptibols, terpenoids, and alkaloids (Kumari et al., 2019). Li et al. (2018) in their studies demonstrated volatile compound–mediated recognition and inhibition between *Trichoderma* biocontrol agents and *F. oxysporum*.

Earlier, antagonistic effect of two peptibols producing isolates of *Trichoderma harzianum* was observed against postharvest pathogens of tomato (*Lycopersicon esculentum*; El-Katatny et al., 2011).

Bacillus sp. is well known for production of VOCs against fungal pathogens. The main bioactive compounds produced by *Bacillus* sp. are LPs, enzymes, 1-aminocyclopropane-1-carboxylate (ACC) deaminase, and exopolysaccharides (Lastochkina et al., 2019). Iturin A produced by *B. amyloliquefaciens* PPCB004 was found to be the main component involved in inhibition of postharvest fungal pathogen (Arrebola et al., 2010).

Yeasts have also gained an equal importance by producing antimicrobial VOCs in postharvest disease management. Several killer toxins produced by yeasts (K1, PMKT) kill the pathogen by forming membrane pores but are immune to their own class of toxins (Spadaro and Droby, 2016).

14.7.4 Mycoparasitism and antibiosis

By producing lytic enzymes, fungal biocontrol agents can directly kill the pathogenic fungus by destructing the fungal cell wall (Kumari et al., 2017a). Biocontrol agents used in postharvest disease management are able to release lytic enzymes including glucanase, chitinase, cellulase, which are able to break the fungal cell wall. Chitinase produced by many biocontrol fungi and yeasts is able to break b-1–4 linked subunits of acetylated amino sugar *N*-acetylglucosamine while glucan acts as a filling material and constitutes about 50%–60% of the total cell wall (Spadaro and Droby, 2016).

Similarly, antibiosis is the secretion of potent antibiotics by bacteria and actinomycetes, which directly kills the pathogen. *Bacillus*, *Pseudomonas*, and *Streptomyces* sp. are some of the bacterial species known to produce antibiotics (Dukare et al., 2017). Blasticidin-S was the first antibiotics used to control rice blast disease in Japan, which was isolated from *Streptomyces griseochromogenes* (Law et al., 2017). Earlier, Smilanick et al. (1993) had demonstrated role of antibiotics producing *Pseudomonas* sp. to control postharvest brown rot of nectarines and peaches.

14.7.5 Induction of disease resistance in crops and fruits

Biocontrols used during postharvest disease management are able to induce systemic resistance in the plants in general and wounded area in particular. If applied earlier, before occurrence of disease, they are able to generate a more profound effect on plant immune systems. Microbial biocontrols are known to induce priming, systemic acquired resistance, ISR, and hypersensitive reactions (Köhl et al., 2019). Enhancement of secondary metabolite production, activation of mitogen-activated protein kinase signaling pathways, production of phytoalexins, and strengthening of plant cell wall are some of the mechanisms employed by microbial biocontrols to elicit plant immune response to alleviate biotic stresses (Shoresh et al., 2010; Conrath et al., 2015).

Petriacq et al. (2018) in their review summarized the role of induced resistance in controlling fruit decay diseases. Shoresh et al. (2010) studied alleviation of ISR of plants after treatment with fungal biocontrol agents. Hase et al. (2008) studied involvement of jasmonic acid signaling in bacterial wilt disease resistance induced by biocontrol agent *Pythium oligandrum* in tomato. *P. fluorescens* strains WCS374r and WCS358r induced protection in both wild-type *Arabidopsis* and SA-nonaccumulating NahG plants without activating pathogenesis-related gene expression by inducing ISR (Van Wees et al., 1997).

14.7.6 Change in reactive oxygen species production profile in plants

Microbial biocontrols are known to interfere with reactive oxygen species (ROS) profile of plants in postharvest disease management after their application. ROS can be a double-edged sword, which is manipulated by microbial biocontrols to enhance the immunity of host plants during tripartite interactions (Kumari et al., 2017b; Dixit et al., 2016). In early plant defense response, ROS level is increased by biocontrol agents, thereby increasing the immunity of plants against pathogens while in compatible plant–pathogen interactions, biocontrols help to lower down the ROS level to cope up with the oxidative stress (Dukare et al., 2017).

Yarrowia lipolytica induced PPO, POD, PAL, APX, GLU, and CAT activities of grapes to ameliorate oxidative stress to control postharvest decay of table grapes caused by *Penicillium rubens* (Wang et al., 2019). β-Glucan induced PPO, POD, and CAT activities of pears treated with *Cryptococcus podzolicus*, which helped to overcome tomatoes infected with *Ralstonia pseudosolanacearum* (Zhao et al., 2020a). Castoria et al. (2003) demonstrated role of antagonistic yeast to make the postharvested fruit more tolerant to ROS against pathogen *R. glutinis*. Antagonistic yeast enhanced H_2O_2 accumulation in apple fruit exocarp around infected area enhancing the defense response of citrus flavedo to postharvest pathogen (Macarisin et al., 2010).

14.7.7 Quorum sensing and biofilm formation

Quorum sensing and biofilm formation can help biocontrol agents to successfully colonize and adhere to the plant surface for their better thriving in comparison to pathogens. Many biocontrol agents secrete small diffusible molecules such as quinolone, indole, and γ-amino butyrate as their mode of communication to each other (Crépin et al., 2012). This phenomenon is well described for phenazine-producing *Pseudomonas* sp. (Köhl et al., 2019).

Though this phenomenon has not been studied in much detail in tripartite interactions of plant–pathogen and microbial biocontrol during postharvest disease management, some reports suggest in favor of this. Lastochkina et al. (2019) described biofilm and surfactin components play an important role in biocontrol efficacy of *Bacillus* sp. Efficacy of antagonistic yeasts was also attributed due to quorum sensing and biofilm formation (Freimoser et al., 2019).

14.7.8 Change in "ome" of plants

Biocontrol agents can modulate the proteome, metabolome, transcriptome, lipidome, and secretome of plant and pathogens during tripartite interaction of plant–pathogen and microbial biocontrol. *Saccharomyces cerevisiae* successfully inhibited the growth of pathogen *Fusarium graminearum* by modulating proteome and transcriptome of pathogen (Zhao et al., 2019). *Trichoderma atroviride* during interaction with *B. cinerea* and *Phytophthora capsici* regulated transcription of various genes involved in cell wall degradation (Reithner et al., 2011). Hershkovitz et al. (2013) observed differential expression of more than 250 genes involved in antagonist–pathogen interaction, genes related to transmembrane, multidrug transport, and amino acid metabolism in *Metschnikowia fructicola* during interactions with *P. digitatum* and grapefruit peel.

Though many mechanisms are employed by biocontrols against pathogen during management of postharvest diseases, they are often interrelated. An in-depth study finding their correlation can provide further insight into the mechanistic aspects of postharvest biocontrol agents. Several important aspects including the microbiome composition on the plant surfaces, interaction of antagonistic microbes with epiphyte and endophyte of plant, cultivar of plant, and the disease resistance developed in pathogens also play a vital role in successful application of bioformulations to control postharvest diseases.

14.8 Commercial production of bioformulations to control postharvest diseases

Though several successful reports of antagonist biocontrols are reported to control postharvest diseases, very few successful examples are there which has crossed all the barriers from lab to land, and has reached to the stage of commercial production. The biocontrol product has to undergo a large-scale production maintaining the similar efficacy in large-scale pilot production (Droby et al., 2009) to reach to land from lab. Their biosafety and efficacy evaluations are some of the measures that need to be guaranteed before their large-scale commercial application. A good bioformulation should have the following characteristic features to sustain in the market for commercial production.

1. High efficacy against postharvest pathogens
2. Sustainable release of the active ingredient(s)
3. Nontoxic toward host
4. Increased shelf life
5. Stability and wide host range

Table 14.4 describes some of the successful examples of bioformulations produced commercially and available in market under different brand names.

TABLE 14.4 List of bioformulations available commercially across the globe.

Name of bioformulation	Microbial ingredient	Company	Effective against pathogen	Target plants	Reference
Kalisena SD and Kalisena SL		Ms Cadila Pharmaceuticals Ltd., IARI	Soilborne pathogens	Crops	Newsletter of IARI
None	FAT3 and FAT6	PDBC, Bangalore	Multiple pathogens	Chickpea, tomato	Wahab (2009)
Avogreen	*Bacillus subtilis*	S. Africa	Cercospora spots	Avocado	Korsten and Bornman (2004)
Bio-Save 10 LP	*Pseudomonas syringae*	Jet Harvest Solutions USA	*Botrytis*, *Penicillium* spp.	Citrus fruits	Korsten and Bornman (2004)
BlightBan A506	*Pseudomonas fluorescens* A506	Nufarm	Russet-inducing bacteria	Tomato, potato, peach	https://nufarm.com
Galltroll	*B. subtilis* K84		*Agrobacterium tumefaciens*, crown gal disease	Fruit, nut	Mehnaz (2016)
Candifruit	Candida sake	IRTA/Sipcam-Inagra, Spain	*Botrytis*, *Penicillium* spp.	Pome	Mehnaz (2016)
Aspire	*Candida oleophila*	Ecogen, USA	*Botrytis*, *Penicillium* spp.	Pome	Mehnaz (2016)
Shemer	*Metschnikowia fructicola*	Bayer, The Netherlands	*Botrytis*, *Penicillium*, *Rhizopus* spp.	Pome, grape, potato	Mehnaz (2016)
TrichoderMax	*Trichoderma asperellum*	Europe	Soilborne diseases	Broad host range	Mehnaz (2016)
Cedomon	*Pseudomonas chlororaphis*	Europe	*Ustilago nuda*	Barley	Mehnaz (2016)
Bioscrop BT16	*Bacillus thuringiensis*	Europe	Lepidopteran larvae and beetles	Citrus, cauliflower, deciduous fruit trees, horticultural brassicas, olives, pepper, banana, and tomato	Mehnaz (2016)
Biovaccine	*Trichoderma viride*	BioPower, Sri Lanka	*Pythium*, *Rhizoctonia* and *Fusarium* spp.	Multiple host range	Mehnaz (2016)
Biotilis	*B. subtilis*	Biopesticide	Multiple pathogens	Multiple host range	Mehnaz (2016)
BioKuprum	*Chaetomium cupreum*	Biopesticide	Rust, blight, rots	Multiple host range	Mehnaz (2016)

14.9 Hurdles during commercialization of bioformulations and measures employed to increase the transition from lab to land

Though bioformulations are an economic and eco-friendly way to combat postharvest pathogen attacks, their commercialization needs lot of efforts from researcher, government, and policy makers.

14.9.1 Stress-tolerant microbes

The major hurdle faced during commercialization of bioformulations is to maintain the efficacy of microbial component from lab to land. There are several kinds of stresses faced by the microbes in field conditions which are important to overcome to be a good bioformulant to control postharvest diseases.

Stress-tolerant microbes or synergistic application of antagonistic microbes can play an important role in maintaining the efficacy of antagonistic microbes in field conditions. Many *Trichoderma* strains have been isolated and developed, which can tolerate drought and salt while acting as an excellent biocontrol (Hirpara et al., 2017; Ndolo et al., 2019). Synergistic application of microbes has shown to help horticultural crops under abiotic stress conditions (Bulgari et al., 2019b).

14.9.2 Increasing shelf life and efficacy of antagonistic microbes

Increasing shelf life of antagonistic microbes in field and stressed condition is a major hurdle to commercialize the product. Selection of stress-tolerant microbes can be good criteria for their successful commercialization. Their genetic stability and the colloidal release of bioformulations provide them an edge for their successful commercialization (Martin et al., 2017). Edible coating of fruits and crops with polymers to control water loss and exchange of gases can also increase the shelf life of biocontrol agents (Martin et al., 2017).

The form in which biocontrol agents are prepared also affects the shelf life of antagonistic microbial agent. In general, dry powder forms of microbes have a better shelf life than the liquid form (Usall et al., 2009). However, solid formulations also have their own drawbacks, owing to their viability loss due to constant dehydration–rehydration process. Many biologically available coating materials such as milk and whey protein, maltodextrins have been reported to be used to protect microbes from dehydration (Martin et al., 2017).

Certain measures have also been employed to increase the efficacy of biocontrol agents. Combining vitamin C, salicylic acid, and lysozyme with the potent antagonistic microbes are some of the examples of strategy developed to increase the efficacy (Dukare et al., 2017).

14.9.3 Sustained release of volatile organic compounds or microbes in field

When antimicrobial secondary metabolites or VOCs are used to control postharvest diseases, the major hurdle encountered is their solubility in water and their sustained release on the plant surface. Most of the VOCs being small organic molecules face the problem of solubility in hydrophilic environment. Nanotechnology has become a common term used along with agriculture coining the new term agri-nanotechnology (Mishra et al., 2014; Kumari et al., 2017c).

Active antimicrobial ingredients encapsulated in nontoxic and degradable polymers such as sodium alginate and chitosan allow them to release slowly and confer them hydrophilicity (Soh and Lee, 2019). Chitosan has been studied for its triple actions including elicitation, coating, and antimicrobial properties to control postharvest decay in fruits and vegetables (Romanazzi et al., 2018). A combined antimicrobial nanoparticle and microbial formulation approach is also helpful for pivotal success and commercialization of the products (González-Estrada et al., 2019). Recently, Housseiny and Aboelmagd (2019) studied that nanoencapsulated naringinase produced by *Trichoderma longibrachiatum* ATCC18648 can work efficiently against citrus juice debittering.

14.9.4 Preharvest versus postharvest application of microbial formulations

Selecting the appropriate timing and mode of application is a prerequisite for successful antagonistic application of microbial bioformulations.

Some of the pathogens are carried along with fruits, vegetables during their harvest from field to storehouses to increase their susceptibility toward diseases. Preapplication of bioformulations does not only prevent the diseases but also strengthen the first line of defense in plant system (Kumari et al., 2017b). Vivekananthan et al. (2006) demonstrated in their studies that preharvest application of *P. fluorescens* and a strain of *B. subtilis* in mango field enhanced disease resistance and fruit yield. However, in some cases, preharvest application is reported to be less effective than postharvest applications (Carmona-Harnandez et al., 2019). It becomes very necessary to study the mechanism and mode of application of each bioformulation prepared and standardize them for their successful commercial applications.

14.9.5 Safety evaluation of microbial bioformulations

For any product development, it is an essential prerequisite to test their bioaccumulation and toxicity on the native microbiome and the end users. A product can only be considered as safe when it does not harm the environment, nor does it has any cytotoxic potential on healthy animal cells. Stable bioformulations derived from *Xenorhabdus stockiae* PB09 have controlled mushroom mite diseases *Luciaphorus perniciosus* Rack successfully without hampering the health of nontargeted organisms (Namsena et al., 2016). However, on a deeper screening, many biocontrol agents or their VOCs have been observed to be causative agents of human or animal diseases or some are very potent enough to interfere with the native microbiota (Keswani et al., 2019). *Bacillus gobisporus* used as an excellent antagonist against a wide range of pathogenic fungi is known as opportunistic human pathogen (Ghosh et al., 2003; Keswani et al., 2019).

To check the stability, ecotoxicity and cytotoxicity of bioformulation products are of utmost importance before their production on a large scale for their successful commercial applications.

14.9.6 Policy making, regulation, and promotion of bioformulation uses

To commercialize and promote uses of bioformulations, it is necessary that strong inputs should be provided by government, policy makers, and funding agencies.

In India, Indian Council of Agricultural Research (ICAR), Central Integrated Pest Management Centres, Council of Scientific and Industrial Research (CSIR), Department of Biotechnology (DBT), and Department of Science & Technology (DST) are putting their constant efforts to commercialize the application of antagonistic microbes. The programmes such as National Biocontrol Network Programme (NBNP) by DBT, which are running resulted in identification and mass production of more than 30 biopesticides. Continuous workshops and symposia are organized by ICAR and other agencies regarding importance and commercialization of biocontrol agents during postharvest disease management. Pilot plant set up and economical support during that process should be provided by governing bodies.

Further, there is also a need of spreading awareness about bioformulations to farmers who are the end users of the product. Awareness campaigns should periodically be organized by the government to promote uses of eco-friendly and economical bioformulation approaches for storage pest managements. Development of bioformulation technologies and the technology transfer to companies for product development should be encouraged by policy makers.

14.10 Conclusions and future aspects

Postharvest diseases put a significant pressure on the yield and quality of crops and horticulture. Biocontrols have been developed as an eco-friendly and economical strategy to combat postharvest pathogens to overcome financial and mental burden on farmers. The multiple mode of actions employed by them to mitigate the biotic stresses make them favorable antagonistic agents against postharvest pathogens. Some of the bioformulations and VOCs have been produced successfully on a large scale and commercialized, though most of them still have to travel a long path to write their successful stories. Synergistic application of bioformulations, supplementing them with proper elicitors, policy making and regulations by funding agencies, government bodies and promotion of their uses are some of the parameters that can be worked upon to increase the safe use of bioformulations. Recent development in "omics" technologies can be employed to gain a mechanistic insight into their mode of actions against postharvest pathogens. Further development on better biocontrol strategies displaying broad-spectrum activities in a cost-effective manner can provide a ray of hope to a better future for both bioformulation technologies as well as farmers.

Acknowledgment

MK thanks CSIR for providing her Senior Research Associateship under Scientists' pool scheme. SK and SP are thankful to UGC and CSIR for their Senior Research Fellowships (SRF). MK and SK are thankful to Prof. C. Jayabaskaran, Department of Biochemistry, Indian Institute of Science for his constant encouragement and support.

References

Abdelhai, M.H., Awad, F.N., Yang, Q., Mahunu, G.K., Godana, E.A., Zhang, H., 2019. Enhancement the biocontrol efficacy of *Sporidiobolus pararoseus* Y16 against apple blue mold decay by glycine betaine and its mechanism. Biol. Control. 139, 104079.

Ali, A., Bordoh, P.K., Singh, A., Siddiqui, Y., Droby, S., 2016. Post-harvest development of anthracnose in pepper (*Capsicum* spp): etiology and management strategies. J. Crop. Prot. 90, 132–141.

Alshaibani, M., Zin, N., Jalil, J., Sidik, N., Ahmad, S.J., Kamal, N., et al., 2017. Isolation, purification, and characterization of five active diketopiperazine derivatives from endophytic *Streptomyces* SUK 25 with antimicrobial and cytotoxic activities. J. Microbiol. Biotechnol. 27, 1249–1256.

Altieri, G., Di Renzo, G.C., Lanza, G., June 2004. Imazalil on-line control in post-harvest treatments of citrus fruit. In: V International Post-harvest Symposium, 682, pp. 1773–1780.

Arora, N.K., Khare, E., Maheshwari, D.K., 2010. Plant growth promoting rhizobacteria: constraints in bioformulation, commercialization, and future strategies. In: Maheshwari, D.K. (Ed.), Plant Growth and Health Promoting Bacteria, 18. Springer, pp. 97–116.

Arrebola, E., Jacobs, R., Korsten, L., 2010. Iturin A is the principal inhibitor in the biocontrol activity of *Bacillus amyloliquefaciens* PPCB004 against post-harvest fungal pathogens. J. Appl. Microbiol. 108 (2), 386–395.

Bargaz, A., Lyamlouli, K., Chtouki, M., Zeroual, Y., Dhiba, D., 2018. Soil Microbial resources for improving fertilizers efficiency in an integrated plant nutrient management system. Front. Microbiol. 9.

Bastiaanse, H., de Lapeyre de Bellaire, L., Lassois, L., Misson, C., Jijakli, M.H., 2010. Integrated control of crown rot of banana with *Candida oleophila* strain O, calcium chloride and modified atmosphere packaging. Biol. Control. 53, 100–107.

Bawa, A., 2019. Agriculture and food security in Northern Ghana. Asian J. Agric. Extension, Econ. Sociol. 31 (2), 1–7. Available from: https://doi.org/10.9734/ajaees/2019/v31i230127.

Bélanger, R.R., Labbé, C., Lefebvre, F., Teichmann, B., 2012. Mode of action of biocontrol agents: all that glitters is not gold. Can. J. Plant Pathol. 34 (4), 469–478.

Brown, J.E., Lu, T.Y., Stevens, C., Khan, V.A., Lu, J., Wilson, C.L., et al., 2002. The effect of low dose ultraviolet light-C seed treatment on induced resistance in cabbage to black rot (*Xanthomonas campestris* pv. campestris). J. Crop. Prot. 20 (10), 873–883.

Brzozowski, L., Leckie, B.M., Gardner, J., Hoffmann, M.P., Mazourek, M., 2016. Cucurbita pepo subspecies delineates striped cucumber beetle (*Acalymma vittatum*) preference. Hortic. Res. 3, 16028.

Buchholz, F., Kostić, T., Sessitsch, A., Mitter, B., 2018. The potential of plant microbiota in reducing postharvest food loss. Microb. Biotechnol. 11 (6), 971–975.

Bukhari, T., Takken, W., Koenraadt, C.J., 2011. Development of *Metarhizium anisopliae* and *Beauveria bassiana* formulations for control of malaria mosquito larvae. Parasit. Vectors 4, 23.

Bulgari, D., Montagna, M., Gobbi, E., Faoro, F., 2019a. Green technology: bacteria-based approach could lead to unsuspected microbe–plant–animal interactions. Microorganisms 7, 44.

Bulgari, R., Franzoni, G., Ferrante, A., 2019b. Biostimulants application in horticultural crops under abiotic stress conditions. Agron. J. 9 (6), 306.

Calvente, V., Benuzzi, D., de Tosetti, M.S., 1999. Antagonistic action of siderophores from *Rhodotorula glutinis* upon the postharvest pathogen *Penicillium expansum*. Int. Biodeterior. Biodegrad. 43 (4), 167–172.

Calvo-Garrido, C., Roudet, J., Aveline, N., Davidou, L., Dupin, S., Fermaud, M., 2019. Microbial antagonism toward *Botrytis* bunch rot of grapes in multiple field tests using one *Bacillus ginsengihumi* strain and formulated biological control products. Front. Plant Sci. 10, 105.

Carmona-Hernandez, S., Reyes-Pérez, J.J., Chiquito-Contreras, R.G., Rincon-Enriquez, G., Cerdan-Cabrera, C.R., Hernandez-Montiel, L.G., 2019. Biocontrol of postharvest fruit fungal diseases by bacterial antagonists: a review. Agron. J. 9 (3), 121.

Castoria, R., De Curtis, F., Lima, G., Caputo, L., Pacifico, S., De Cicco, V., 2001. *Aureobasidium pullulans* (LS-30) an antagonist of postharvest pathogens of fruits: study on its modes of action. Postharvest Biol. Technol. 22 (1), 7–17.

Castoria, R., Caputo, L., De Curtis, F., De Cicco, V., 2003. Resistance of postharvest biocontrol yeasts to oxidative stress: a possible new mechanism of action. Phytopathology 93, 564–572.

Charles, M.T., Benhamou, N., Arul, J., 2008. Physiological basis of UV-C induced resistance to *Botrytis cinerea* in tomato fruit: III. Ultrastructural modifications and their impact on fungal colonization. Postharvest Biol. Technol. 47 (1), 27–40.

Cheng, L., Nie, X., Jiang, C., Li, S., 2019. The combined use of the antagonistic yeast *Hanseniaspora uvarum* with β-aminobutyric acid for the management of postharvest diseases of kiwifruit. Biol. Control. 137, 104019.

Conrath, U., Beckers, G.J.M., Langenbach, C.J.G., Jaskiewicz, M.R., 2015. Priming for enhanced defense. Annu. Rev. Phytopathol. 53, 97–119.

Contreras-Cornejo, H., Macías-Rodríguez, L., del-Val, E., Larsen, J., 2018. The root endophytic fungus *Trichoderma atroviride* induces foliar herbivory resistance in maize plants. Appl. Soil. Ecol. 124, 45–53.

Costa, J.H., Wassano, C.I., Angolini, C.F.F., Scherlach, K., Hertweck, C., Fill, T.P., 2019. Antifungal potential of secondary metabolites involved in the interaction between citrus pathogens. Sci. Rep. 9, 18647.

Cozzolino, M.E., Distel, J.S., García, P.A., Mascotti, M.L., Ayub, M.J., Benazzi, L.M., et al., 2020. Control of postharvest fungal pathogens in pome fruits by lipopeptides from a *Bacillus* sp. isolate SL-6. Sci. Hortic. 261, 108957.

Crépin, A., Barbey, C., Beury-Cirou, A., Hélias, V., Taupin, L., Reverchon, S., et al., 2012. Quorum sensing signaling molecules produced by reference and emerging soft-rot bacteria (*Dickeya* and *Pectobacterium* spp). PLoS One 7 (4), e35176.

Dayan, F.E., Haesaert, G., Van Leeuwen, T., Holden-Dye, L., Crossthwaite, A., Nauen, R., 2019. Pesticides modes of action and resistance: a perspective from the 2019 IUPAC congress. Outlooks Pest. Manag. 30 (4), 157–163.

Di Francesco, A., Calassanzio, M., Ratti, C., Mari, M., Folchi, A., Baraldi, E., 2018. Molecular characterization of the two postharvest biological control agents *Aureobasidium pullulans* L1 and L8. Biol. Control. 123, 53–59.

Diskin, S., Sharir, T., Feygenberg, O., Maurer, D., Alkan, N., 2019. Fludioxonil–a potential alternative for postharvest disease control in mango fruit. J. Crop. Prot. 124, 104855.

Dixit, R., Agrawal, L., Gupta, S., Kumar, M., Yadav, S., Chauhan, P.S., et al., 2016. Southern blight disease of tomato control by 1-aminocyclopropane-1-carboxylate (ACC) deaminase producing *Paenibacillus lentimorbus* B-30488. Plant Signal. Behav. 11 (2), e1113363.

Droby, S., Wisniewski, M., Macarisin, D., Wilson, C., 2009. Twenty years of postharvest biocontrol research: Is it time for a new paradigm? Postharvest Biol. Technol. 52, 137–145.

Droby, S., Lichter, A., 2007. Postharvest Botritys. In: Elad, Y., Williamson, B., Tudzynski, P., Delen, N. (Eds.), Botrytis: Biology, Pathology and Control. Springer, Heidelberg, pp. 349–367.

Dukare, A.S., Paul, S., Nambi, V.E., Gupta, R.K., Singh, R., Sharma, K., et al., 2017. Exploitation of microbial antagonists for the control of postharvest diseases of fruits: a review. Crit. Rev. Food Sci. Nutr. 59 (9), 1498–1513.

Dunlap, C.A., Bowman, M.J., Rooney, A.P., 2019. Iturinic lipopeptide diversity in the *Bacillus subtilis* species group—important antifungals for plant disease biocontrol applications. Front. Microbiol. 10, 1794.

El-Katatny, M.H., El-Katatny, M.S., Fadl-Allah, E.M., Emam, A.S., 2011. Antagonistic effect of two isolates of *Trichoderma harzianum* against postharvest pathogens of tomato (*Lycopersicon esculentum*). Arch. Phytopathol. Plant Protec. 44 (7), 637–654.

Ennab, H.A., El-Shemy, M.A., Alam-Eldein, S.M., 2020. Salicylic acid and putrescine to reduce post-harvest storage problems and maintain quality of murcott mandarin fruit. Agronomy 10 (1), 115.

Eshel, D., Regev, R., Orenstein, J., Droby, S., Gan-Mor, S., 2009. Combining physical, chemical and biological methods for synergistic control of postharvest diseases: a case study of black diseases: a case study of black root rot of carrot. Postharvest Biol. Technol. 54 (1), 48–52.

FAO Report, 1997. Pesticide Residues in Food, FAO Plant Production and Protection Paper 145, Food and Agriculture Organization of the United Nations, Rome, 1998. <http://www.fao.org/docrep/W8141E/W8141E00.htm>.

FAO, 2011. Global food losses and food waste. Study conducted for the International Congress "SAVE FOOD!" at Interpack, Düsseldorf, Germany. <http://www.fao.org/3/mb060e/mb060e00.htm>.

Felix, A.K.N., Martins, J.J., Almeida, J.G.L., Giro, M.E.A., Cavalcante, K.F., Melo, V.M.M., et al., 2019. Purification and characterization of a biosurfactant produced by *Bacillus subtilis* in cashew apple juice and its application in the remediation of oil-contaminated soil. Colloids Surf. B 175, 256–263.

Freimoser, F.M., Rueda-Mejia, M.P., Tilocca, B., Migheli, Q., 2019. Biocontrol yeasts: mechanisms and applications. World J. Microb. Biotechnol. 35 (10), 154.

Fu, G., Huang, S., Ye, Y., Wu, Y., Cen, Z., Lin, S., 2010. Characterization of a bacterial biocontrol strain B106 and its efficacies on controlling banana leaf spot and post-harvest anthracnose diseases. Biol. Control. 55 (1), 1–10.

Ganesh, U., Shankar, M., Banerji, S., Mahadevan, K., Borthakur, S., Thurkal, C., et al., 2016. Reducing Post-Harvest Losses in India: Key Initiatives and Opportunities. The Rockfeller Foundation.

Ghosh, S., Penterman, J.N., Little, R.D., Chavez, R., Glick, B.R., 2003. Three newly isolated plant growth-promoting bacilli facilitate the seedling growth of canola, *Brassica campestris*. Plant Physiol. Biochem. 41 (3), 277–281.

Glick, B.R., 2012. Plant growth-promoting bacteria: mechanisms and applications. Scientifica 2012, 963401.

Gonçalves, F.P., Martins, M.C., Junior, G.J.S., Lourenço, S.A., Amorim, L., 2010. Postharvest control of brown rot and Rhizopus rot in plums and nectarines using carnauba wax. Postharvest Biol. Technol. 58 (3), 211–217.

González-Estrada, R.R., Blancas-Benitez, F.J., Moreno-Hernández, C.L., Coronado-Partida, L., Ledezma-Delgadillo, A., Gutiérrez-Martínez, P., 2019. Nanotechnology: a promising alternative for the control of post-harvest pathogens in fruits. Nanotechnology for Agriculture: Crop Production & Protection. Springer, Singapore, pp. 323–337.

Gotor-Vila, A., Teixidó, N., Casals, C., Torres, R., De Cal, A., Guijarro, B., et al., 2017. Biological control of brown rot in stone fruit using *Bacillus amyloliquefaciens* CPA-8 under field conditions. Crop. Prot. 102, 72–80.

Hase, S., Takahashi, S., Takenaka, S., Nakaho, K., Arie, T., Seo, S., et al., 2008. Involvement of jasmonic acid signalling in bacterial wilt disease resistance induced by biocontrol agent *Pythium oligandrum* in tomato. Plant Pathol. 57 (5), 870–876.

He, F., Zhao, L., Zheng, X., Abdelhai, M.H., Boateng, N.S., Zhang, X., et al., 2020. Investigating the effect of methyl jasmonate on the biocontrol activity of *Meyerozyma guilliermondii* against blue mold decay of apples and the possible mechanisms involved. Physiol. Mol. Plant P 109, 101454.

Hermann, D., Stenzel, K., 2019. FRAC mode-of-action classification and resistance risk of fungicides, Modern Crop Protection Compounds, second ed. Wiley, pp. 589–608.

Hershkovitz, V., Sela, N., Taha-Salaime, L., Liu, J., Rafael, G., Kessler, C., et al., 2013. De-novo assembly and characterization of the transcriptome of *Metschnikowia fructicola* reveals differences in gene expression following interaction with *Penicillium digitatum* and grapefruit peel. BMC Gen. 14 (1), 168.

Hertlein, G., Seiffert, M., Gensel, S., Garcia-Gonzalez, E., Ebeling, J., Skobalj, R., et al., 2016. Biological role of *Paenilarvins*, iturin-like lipopeptide secondary metabolites produced by the honeybee pathogen *Paenibacillus* larvae. PLoS One 11, e0164656.

Hirpara, D.G., Gajera, H.P., Hirapara, J.G., Golakiya, B.A., 2017. Inhibition coefficient and molecular diversity of multi stress tolerant *Trichoderma* as potential biocontrol agent against *Sclerotium rolfsii* Sacc. Infect. Genet. Evol. 55, 75–92.

Housseiny, M.M., Aboelmagd, H.I., 2019. Nano-encapsulation of naringinase produced by *Trichoderma longibrachiatum* ATCC18648 on thermally stable biopolymers for citrus juice debittering. J. Microbiol. 57 (6), 521–531.

Huang, X.F., Guan, W., Liu, J., Lu, L.J., Xu, J.C., Zhou, Q., 2010. Characterization and phylogenetic analysis of biodemulsifier-producing bacteria. Bioresour. Technol. 101, 317–323.

Jembere, B., Obeng-Ofori, D., Hassanali, A., Nyamasyo, G.N.N., 1995. Products derived from the leaves of *Ocimum kilimandscharicum* (Labiatae) as post-harvest grain protectants against the infestation of three major stored product insect pests. Bull. Entomol. Res. 85 (3), 361–367.

Jing, W., Tu, K., Shao, X.F., Su, Z.P., Zhao, Y., Wang, S., et al., 2010. Effect of postharvest short hot-water rinsing and brushing treatment on decay and quality of strawberry fruit. J. Food Qual. 33, 262–272.

Jones, P., Garcia, B.J., Furches, A., Tuskan, G.A., Jacobson, D., 2019. Plant host-associated mechanisms for microbial selection. Front. Plant Sci. 10.

Kader, A.A., Rolle, R.S., 2004. The role of post-harvest management in assuring the quality and safety of horticultural produce. Food Agric. Org. 152, 1–22.

Kalita, M., Bharadwaz, M., Dey, T., Gogoi, K., Dowarah, P., Unni, B.G., et al., 2015. Developing novel bacterial based bioformulation having PGPR properties for enhanced production of agricultural crops. Indian J. Exp. Biol. 53, 56–60.

Kamat, S., Kumari, M., Taritla, S., Jayabaskaran, C., 2020. Endophytic fungi of marine alga from Konkan coast, India-a rich source of bioactive material. Front. Mar. Sci. 7.

Karaman, S., Ozturk, B., Genc, N., Celik, S.M., 2013. Effect of pre-harvest application of methyl jasmonate on fruit quality of plum (*Prunus salicina* L indell cv. "Fortune") at harvest and during cold storage. J. Food Process. Preserv. 37 (6), 1049–1059.

Keswani, C., Bisen, K., Chitara, M.K., Sarma, B.K., Singh, H.B., 2017. Exploring the role of secondary metabolites of *Trichoderma* in tripartite interaction with plant and pathogens. In: Singh, J.S., Seneviratne, G. (Eds.), Agro-Environmental Sustainability. Springer, Cham, pp. 63–79.

Keswani, C., Prakash, O., Bharti, N., Vílchez, J.I., Sansinenea, E., Lally, R.D., et al., 2019. Re-addressing the biosafety issues of plant growth promoting rhizobacteria. Sci. Total Environ. 10, 841–852.

Köhl, J., Kolnaar, R., Ravensberg, W.J., 2019. Mode of action of microbial biological control agents against plant diseases: relevance beyond efficacy. Front. Plant Sci. 10, 845.

Korsten, L., Bornman, C.H., 2004. Biological control in Africa: can it provide a sustainable solution for control of fruit diseases. S. Afr. J. Bot. 70 (1), 128–139.

Kumari, M., Shukla, S., Pandey, S., Giri, V.P., Bhatia, A., Tripathi, T., et al., 2017a. Enhanced cellular internalization: a bactericidal mechanism more relative to biogenic nanoparticles than chemical counterparts. ACS Appl. Mater. Interfaces. Available from: http://dx.doi.org/10.1021/acsami.6b15473.

Kumari, M., Pandey, S., Bhattacharya, A., Mishra, A., Nautiyal, C.S., 2017b. Protective role of biosynthesized silver nanoparticles against early blight disease in *Solanum lycopersicum*. Plant Physiol. Biochem. 121, 216–225.

Kumari, M., Pandey, S., Mishra, S., Mishra, A., Nautiyal, C.S., 2017c. Effect of biosynthesized silver nanoparticles on native soil microflora via plant transport during plant–pathogen–nanoparticles interaction. 3 Biotech. 7 (5), 345.

Kumari, M., Giri, V.P., Pandey, S., Kumar, M., Katiyar, R., Nautiyal, C.S., et al., 2019. An insight into the mechanism of antifungal activity of biogenic nanoparticles than their chemical counterparts. Pesticide Physiol. Biochem. 157, 45–52.

Kurtzman, C.P., Droby, S., 2001. *Metschnikowia fructicola*, a new ascosporic yeast with potential for biocontrol of postharvest fruit rots. Sys. Appl. Microbiol. 24, 395–399.

Ladanyia, M., Ladaniya, M., 2010. Citrus Fruit: Biology, Technology and Evaluation. Academic press.

Lapeyre de Bellaire, L., Nolin, J., 1994. Amélioration du contrôle du chancre sur les bananes d'exportation et traitements post-récolte. Fruits 49, 179–185.

Lassois, L., de Bellaire, L.D.L., 2014. Crown rot disease of bananas. Postharvest Decay. Academic Press, pp. 103–130.

Lastochkina, O., Seifikalhor, M., Aliniaeifard, S., Baymiev, A., Pusenkova, L., Garipova, S., et al., 2019. *Bacillus* spp.: efficient biotic strategy to control postharvest diseases of fruits and vegetables. Plants (Basel, Switz.) 8 (4), 97.

Law, J.W., Ser, H.L., Khan, T.M., Chuah, L.H., Pusparajah, P., Chan, K.G., et al., 2017. The potential of *Streptomyces* as biocontrol agents against the rice blast fungus, *Magnaporthe oryzae* (*Pyricularia oryzae*). Front. Microbiol. 8, 3.

Leuba, J.L., Stossel, P., 1986. Chitosan and other polyamines: antifungal activity and interaction with biological membranes. Chitin in Nature and Technology. Springer, Boston, MA, pp. 215–222.

Li, N., Alfiky, A., Wang, W., Islam, M., Nourollahi, K., Liu, X., et al., 2018. Volatile compound – mediated recognition and inhibition between *Trichoderma* biocontrol agents and *Fusarium oxysporum*. Front. Microbiol. 9, 2614.

Lima, G., Ippolito, A., Nigro, F., Salerno, M., 1997. Effectiveness *of Aureobasidium pullulans* and *Candida oleophila* against postharvest strawberry rots. Postharvest Biol. Technol. 10 (2), 169–178.

Liu, Y.B., 2008. Low temperature phosphine fumigation for postharvest control of western flower thrips (*Thysanoptera*: *Thripidae*) on lettuce, broccoli, asparagus, and strawberry. J. Econ. Entomol. 101 (6), 1786–1791.

Liu, Y., Wang, W.H., Zhou, Y.H., Yao, S.X., Deng, L.L., Zeng, K.F., 2017. Isolation, identification and in vitro screening of Chongqing orangery yeasts for the biocontrol of *Penicillium digitatum* on citrus fruit. Biol. Control. 110, 18–24.

Ly, M.H., Naitali-Bouchez, M., Meylheuc, T., Bellon-Fontaine, M.N., Le, T.M., Belin, J.M., et al., 2006. Importance of bacterial surface properties to control the stability of emulsions. Int. J. Food Microbiol. 112 (1), 26–34.

Macarisin, D., Droby, S., Bauchan, G., Wisniewski, M., 2010. Superoxide anion and hydrogen peroxide in the yeast antagonist–fruit interaction: a new role for reactive oxygen species in postharvest biocontrol? Postharvest Biol. Technol. 58 (3), 194–202.

Marcet-Houben, M., Ballester, A.R., de la Fuente, B., Harries, E., Marcos, J.F., González-Candelas, L., et al., 2012. Genome sequence of the necrotrophic fungus *Penicillium digitatum*, the main postharvest pathogen of citrus. BMC Genomics 13 (1), 646.

Mari, M., Bertolini, P., Pratella, G.C., 2003. Non-conventional methods for the control of post-harvest pear diseases. J. Appl. Microbiol. 94 (5), 761–766.

Martin, A., Atares, L., Chiralt, A., 2017. Improving function of biocontrol agents incorporated in antifungal fruit coatings: a review. Biocontrol Sci. Technol. 27, 10.

Martini, C., Mari, M., 2014. *Monilinia fructicola, Monilinia laxa* (Monilinia rot, brown rot). Postharvest Decay. Academic Press, pp. 233–265.

Meena, K.R., Kanwar, S.S., 2015. Lipopeptides as the antifungal and antibacterial agents: applications in food safety and therapeutics. Biomed. Res. Int. Available from: http://dx.doi.org/10.1155/2015/473050.

Mehnaz, S., 2016. An overview of globally available bioformulations. Bioformulations: for Sustainable Agriculture. Springer, New Delhi, pp. 267–281.

Mincuzzi, A., Ippolito, A., Montemurro, C., Sanzani, S.M., 2020. Characterization of *Penicillium* ss and *Aspergillus* sect. nigri causing postharvest rots of pomegranate fruit in Southern Italy. Int. J. Food Microbiol. 314, 108389.

Ministry of Finance, 2019. Press Information Bureau Government of India Ministry of Finance, 186586.

Mishra, J., Arora, N.K., 2016. Bioformulations for plant growth promotion and combating phytopathogens: a sustainable approach. In: Arora, N.K., Mehnaz, S., Balestrini, R. (Eds.), Bioformulations: for Sustainable Agriculture. Springer, India, pp. 3–33.

Mishra, A., Kumari, M., Pandey, S., Chaudhary, V., Gupta, K.C., Nautiyal, C.S., 2014. Biocatalytic and antimicrobial activities of gold nanoparticles synthesized by *Trichoderma* sp. Biores. Technol. 166, 235–242.

Morel, M.A., Cagide, C., Castro-Sowinski, S., 2016. The contribution of secondary metabolites in the success of bioformulations. In: Arora, N.K., Mehnaz, S., Balestrini, R. (Eds.), Bioformulations: For Sustainable Agriculture. Springer, New Delhi, pp. 235–250.

Mukherjee, A., Verma, J.P., Gaurav, A.K., Chouhan, G.K., Patel, J.S., Hesham, A.E.L., 2020. Yeast a potential bio-agent: future for plant growth and postharvest disease management for sustainable agriculture. Appl. Microbiol. Biotechnol. 1–14.

Muñoz, N., Soria-Díaz, M.E., Manyani, H., Sánchez-Matamoros, R.C., Serrano, A.G., Megías, M., et al., 2014. Structure and biological activities of lipochito oligosaccharide nodulation signals produced by *Bradyrhizobium japonicum* USDA 138 under saline and osmotic stress. Biol. Fertil. Soils 50, 207–215.

Myresiotis, C.K., Karaoglanidis, G.S., Tzavella-Klonari, K., 2007. Resistance of *Botrytis cinerea* isolates from vegetable crops to anilinopyrimidine, phenylpyrrole, hydroxyanilide, benzimidazole, and dicarboximide fungicides. Plant Dis. 91, 407–413.

Nakkeeran, S., Surya, T., Vinodkumar, S., 2019. Antifungal potential of plant growth promoting *Bacillus* Species against blossom blight of rose. J. Plant Growth Regul. 39, 99–111. Available from: https://doi.org/10.1007/s00344-019-09966-1.

Namsena, P., Bussaman, P., Rattanasena, P., 2016. Bioformulation of *Xenorhabdus stockiae* PB09 for controlling mushroom mite, *Luciaphorus perniciosus* Rack. Bioresour. Bioprocess. 3, 19.

Navarro, S., 2006. New global challenges to the use of gaseous treatments in stored products. In: 9th International Working Conference on Stored Product Protection, pp. 495–509.

Ndolo, D., Njuguna, E., Adetunji, C.O., Harbor, C., Rowe, A., Den Breeyen, A., et al., 2019. Research and development of biopesticides: challenges and prospects. Outlooks Pest. Manag. 30 (6), 267–276.

Obagwu, J., Korsten, L., 2003. Integrated control of citrus green and blue molds using *Bacillus subtilis* in combination with sodium bicarbonate or hot water. Postharvest Biol. Technol. 28 (1), 187–194.

Pétriacq P., López A., Luna E., 2018. Fruit decay to diseases: can induced resistance and priming help?. Plants. 7 (4), 77.

Pitt, J.I., Spotts, R.A., Holmes, R.J., Cruickshank, R.H., 1991. *Penicillium solitum* revived, and its role as a pathogen of pomaceous fruit. Phytopathology 81 (10), 1108–1112.

Ploetz, R.C., Lim, T.K., Menge, J.A., Rohrbach, K.G., Michailides, T.J., 2003. Common pathogens of tropical fruit. In: Ploetz, R.C. (Ed.), Diseases of Tropical Fruit Crops. CABI Publishing, Wallingford, pp. 1–20.

Porat, R., Daus, A., Weiss, B., Cohen, L., Fallik, E., Droby, S., 2000. Reduction of postharvest decay in organic citrus fruit by a short hot water brushing treatment. Postharvest Biol. Technol. 18 (2), 151–157.

Qing, F., Shiping, T., 2000. Postharvest biological control of Rhizopus rot of nectarine fruits by *Pichia membranifaciens*. Plant Dis. 84 (11), 1212–1216.

Rajendiran, R., Jegadeeshkumar, D., Sureshkumar, B.T., Nisha, T., 2010. *In vitro* assessment of antagonistic activity of *Trichoderma viride* against post-harvest pathogens. IJAT 6 (1), 31–35.

Ray, R.C., Swain, M.R., Panda, S.H., et al., 2011. Microbial control of postharvest diseases of fruits, vegetables, roots, and tubers. In: Singh, A. (Ed.), Bioaugmentation, Biostimulation and Biocontrol, Soil Biology, vol. 28. Springer-Verlag, Berlin, Heidelberg. Available from: http://doi.org/10.1007/978-3-642-19769-7-13.

Reithner, B., Ibarra-Laclette, E., Mach, R.L., Herrera-Estrella, A., 2011. Identification of mycoparasitism-related genes in *Trichoderma atroviride*. Appl. Environ. Microbiol. 77, 4361–4370.

Reyes-Perez, J.J., Hernandez-Montiel, L.G., Vero, S., Noa-Carrazana, J.C., Quiñones-Aguilar, E.E., Rincón-Enríquez, G., 2019. Postharvest biocontrol of *Colletotrichum gloeosporioides* on mango using the marine bacterium *Stenotrophomonas rhizophila* and its possible mechanisms of action. J. Food Sci. Technol. 56 (11), 4992–4999.

Romanazzi, G., Lichter, A., Gabler, F.M., Smilanick, J.L., 2012. Recent advances on the use of natural and safe alternatives to conventional methods to control postharvest gray mold of table grapes. Postharvest Biol. Technol. 63 (1), 141–147.

Romanazzi, G., Sanzani, S.M., Bi, Y., Tian, S., Martínez, P.G., Alkan, N., 2016. Induced resistance to control postharvest decay of fruit and vegetables. Postharvest Biol. Technol. 122, 82–94.

Romanazzi, G., Feliziani, E., Sivakumar, D., 2018. Chitosan, a biopolymer with triple action on postharvest decay of fruit and vegetables: eliciting, antimicrobial and film-forming properties. Front. Microbiol. 9, 2745.

Schreiner, M., 2005. Vegetable crop management strategies to increase the quantity of phytochemicals. Eur. J. Nutr. 44 (2), 85–94.

Shoresh, M., Harman, G.E., Mastouri, F., 2010. Induced systemic resistance and plant responses to fungal biocontrol agents. Annu. Rev. Phytopathol. 48, 21–43.

Smilanick, J.L., Mansour, M.F., Gabler, F.M., Sorenson, D., 2008. Control of citrus postharvest green mold and sour rot by potassium sorbate combined with heat and fungicides. Postharvest Biol. Technol. 47 (2), 226–238.

Smilanick, J.L., Denis-Arrue, R., Bosch, J.R., Gonzalez, A.R., Henson, D., Janisiewicz, W.J., 1993. Control of postharvest brown rot of nectarines and peaches by Pseudomonas species. Crop Protection. 12 (7), 513–520.

Soh, S.H., Lee, L.Y., 2019. Microencapsulation and nanoencapsulation using supercritical fluid (SCF) techniques. Pharmaceutics 11 (1), 21.

Spadaro, D., Droby, S., 2016. Development of biocontrol products for postharvest diseases of fruit: the importance of elucidating the mechanisms of action of yeast antagonists. Trends Food Sci. Technol. 47, 39–49.

Spadaro, D., Gullino, M.L., 2004. State of the art and future prospects of the biological control of postharvest fruit diseases. Int. J. Food Microbiol. 91, 185–194.

Stamenković, S., Beškoski, V., Karabegović, I., Lazić, M., Nikolić, N., 2018. Microbial fertilizers: a comprehensive review of current findings and future perspectives. Span. J. Agric. Res. 16 (1), 1–18.

Stevens, C., Liu, J., Khan, V.A., Lu, J.Y., Kabwe, M.K., Wilson, C.L., et al., 2004. The effects of low-dose ultraviolet light-C treatment on polygalacturonase activity, delay ripening and Rhizopus soft rot development of tomatoes. J. Crop Prot. 23 (6), 551–554.

Terao, D., de Carvalho Campos, J.S., Benato, E.A., Hashimoto, J.M., 2015. Alternative strategy on control of postharvest diseases of mango (*Mangifera indica* L.) by use of low dose of ultraviolet-C irradiation. Food Eng. Rev. 7 (2), 171–175.

Tian, Z., Chen, C., Chen, K., Liu, P., Fan, Q., Zhao, J., et al., 2020. Biocontrol and the mechanisms of *Bacillus* sp. w176 against postharvest green mold in citrus. Postharvest Biol. Technol. 159, 111022.

Tosi, E., Fedrizzi, B., Azzolini, M., Finato, F., Simonato, B., Zapparoli, G., 2012. Effects of noble rot on must composition and aroma profile of Amarone wine produced by the traditional grape withering protocol. Food Chem. 130 (2), 370–375.

Tryfinopoulou, P., Chourdaki, A., Nychas, G.J.E., Panagou, E.Z., 2020. Competitive yeast action against *Aspergillus carbonarius* growth and ochratoxin A production. Int. J. Food Microbiol. 317, 108460.

Usall, J., Teixidó, N., Abadias, M., Torres, R., Cañamas, T., Viñas, I., 2009. Improving formulation of biocontrol agents manipulating production process. Postharvest Pathology. Springer, Dordrecht, pp. 149–169.

Van der Meij, A., Worsley, S.F., Hutchings, M.I., van Wezel, G.P., 2017. Chemical ecology of antibiotic production by actinomycetes. FEMS Microbiol. Rev. 41, 392–416.

Van Wees, S.C., Pieterse, C.M., Trijssenaar, A., Van't Westende, Y.A., Hartog, F., Van Loon, L.C., 1997. Differential induction of systemic resistance in Arabidopsis by biocontrol bacteria. Molecular Plant-microbe Interactions.
10 (6), 716–724.

Venta, M.B., Broche, S.S.C., Torres, I.F., Pérez, M.G., Lorenzo, E.V., Rodriguez, Y.R., et al., 2010. Ozone application for postharvest disinfection of tomatoes. Ozone-Sci. Eng. 32 (5), 361–371.

Vivekananthan, R., Ravi, M., Ramanathan, A., Kumar, N., Samiyappan, R., 2006. Pre-harvest application of a new biocontrol formulation induces resistance to post-harvest anthracnose and enhances fruit yield in mango. Phytopathol. Med. 45 (2), 126–138.

Wahab, S., 2009. Biotechnological approaches in the management of plant pests, diseases and weeds for sustainable agriculture. J. Biopest. 2 (2), 115–134.

Wan, Y.K., Tian, S.P., Qin, G.Z., 2003. Enhancement of biocontrol activity of yeasts by adding sodium bicarbonate or ammonium molybdate to control postharvest disease of jujube fruits. Lett. Appl. Microbiol. 37 (3), 249–253.

Wang, M., Zhao, L., Zhang, X., Dhanasekaran, S., Abdelhai, M.H., Yang, Q., et al., 2019. Study on biocontrol of postharvest decay of table grapes caused by *Penicillium rubens* and the possible resistance mechanisms by *Yarrowia lipolytica*. Biol. Control. 130, 110–117.

Wang, Z., Mei, X., Du, M., Chen, K., Jiang, M., Wang, K., et al., 2020. Potential modes of action of *Pseudomonas fluorescens* ZX during biocontrol of blue mold decay on postharvest citrus. J. Sci. Food Agric. 100 (2), 744–754.

Williamson, B., Tudzynski, B., Tudzynski, P., Van Kan, J.A., 2007. *Botrytis cinerea*: the cause of grey mould disease. Mol. Plant Pathol. 8 (5), 561–580.

Wilson, B.R., Bogdan, A.R., Miyazawa, M., Hashimoto, K., Tsuji, Y., 2016. Siderophores in iron metabolism: from mechanism to therapy potential. Trends Mol. Med. 22 (12), 1077–1090.

Wisniewski, M., Droby, S., Norelli, J., Liu, J., Schena, L., 2016. Alternative management technologies for postharvest disease control: the journey from simplicity to complexity. Postharvest Biol. Technol. 122, 3–10.

Wu, L., Shang, H., Gu, H., Zheng, J., 2019. Bacterial iturins mediate biocontrol activity of *Bacillus* sp. against postharvest pear fruit-rotting fungi. J. Phytopathol. 167 (9), 501–509.

Yan, Z., Shi, J., Gao, L., Wang, Q., Zuo, J., 2020. The combined treatment of broccoli florets with kojic acid and calcium chloride maintains post-harvest quality and inhibits off-odor production. Scientia Horticulturae. 262, 109019.

Yang, J., Sun, C., Zhang, Y., Fu, D., Zheng, X., Yu, T., 2017. Induced resistance in tomato fruit by γ-aminobutyric acid for the control of alternaria rot caused by *Alternaria alternata*. Food Chem. 221, 1014–1020.

Yang, Q., Diao, J., Solairaj, D., Ngea, N., Legrand, G., Zhang, H., 2020. Investigating possible mechanisms of Pichia caribbica induced with ascorbic acid against postharvest blue mold of apples. Biol. Control. 141, 104129.

Yin, G., Zhang, Y., Pennerman, K., Wu, G., Hua, S., Yu, J., et al., 2017. Characterization of blue mold *Penicillium* species isolated from stored fruits using multiple highly conserved loci. J. Fungi 3 (1), 12.

Zhang, H., Wang, L., Ma, L., Dong, Y., Jiang, S., Xu, B., et al., 2009. Biocontrol of major postharvest pathogens on apple using *Rhodotorula glutinis* and its effects on postharvest quality parameters. Biol. Control. 48 (1), 79–83.

Zhang, H., Deng, L., Yao, S., Ming, J., Zeng, K., 2020. Optimization of a vacuum-drying protectants for the biocontrol agent *Pichia membranifaciens* and its influence on viability and efficacy. Biol. Control. 142, 104155.

Zhao, L., Cheng, Y., Li, B., Gu, X., Zhang, X., Boateng, N.A.S., et al., 2019. Integration of proteome and transcriptome data reveals the mechanism involved in controlling of *Fusarium graminearum* by *Saccharomyces cerevisiae*. J. Sci. Food Agric. 99 (13), 5760–5770.

Zhao, L., Wang, Y., Li, B., Gu, X., Zhang, X., Boateng, N.A.S., et al., 2020a. Effect of β-glucan on the biocontrol efficacy of *Cryptococcus podzolicus* against postharvest decay of pears and the possible mechanisms involved. Postharvest Biol. Technol. 160, 111057.

Zhao, S., Guo, Y., Wang, Q., Luo, H., He, C., An, B., 2020b. Expression of flagellin at yeast surface increases biocontrol efficiency of yeast cells against postharvest disease of tomato caused by *Botrytis cinerea*. Postharvest Biol. Technol. 162, 111112.

Zheng, X., Chen, J., 2009. Microarray analysis of gene expression profile induced by the biocontrol yeast *Cryptococcus laurentii* in cherry tomato fruit. Gene 430 (1–2), 12–16.

CHAPTER

15

Utilization of antagonistic microbes for the eco-friendly management of fungal diseases of the harvested fruits during postharvest handling and storage

Ajinath Dukare[1], *Seema Sangwan*[2], *Hemant Maheshwari*[3], *P.N. Guru*[1], *Yogesh Khade*[4] *and Rajesh Kumar Vishwakarma*[1]

[1]ICAR-Central Institute of Post-Harvest Engineering & Technology, Ludhiana, India [2]ICAR-Indian Agricultural research Institute, New Delhi, India [3]ICAR-Indian Institute of Soybean Research, Indore, India [4]ICAR-Directorate of Onion and Garlic Research, Pune, India

15.1 Introduction

Fungal decay of postharvest fruits and vegetable causes substantial amount of wastage and reduce storage duration and produce shelf-life. According an estimate, more than one-third of harvested perishable horticultural produce are wasted (FAO, 2011). The remarkable quantity of losses is due to many fungal spoilage and decay occurring during postharvest handling and storage. Further, such wastage is often huge owing improper cold storage and transportation facilities in the developing world. Harvested fruits are more prone to fungal attack mainly due to high moisture, enough nutrients, low pH, and decreased intrinsic resistance (Droby et al., 1992). During postharvest operation, fungal spoiled fruit are also contaminated with the many pathogens such as *Alternaria*, *Fusarium*, and *Penicillium*, which are potent producer of health risk associated mycotoxins and other toxic metabolites. For example, *Penicillium expansum*, a blue mold causing fungi releases several toxins including citrinin, patulin, and chaetoglobosins that are notorious carcinogenic

Food Security and Plant Disease Management
DOI: https://doi.org/10.1016/B978-0-12-821843-3.00015-5

metabolites (Andersen et al., 2004). Traditionally, postharvest fungal disease are chiefly managed via use of synthetic fungicides and sometimes, effective cold chain management (Vitoratos et al., 2013). However, use of chemical fungicides in postharvest fungal disease control are linked with pathogen resistance to fungicides, rise of new fungal biotype, accumulation of chemical residue in produce, and associated harmful impacts on human and ecosystem health (Dukare et al., 2018). Henceforth, safer and other eco-friendly alternative method has been suggested for fruits and vegetable spoilage control.

Among various eco-friendly strategies, microbes-mediated management of fungal diseases is an emerging approach (Dukare et al., 2011). As compared to fungicides, biological control approach for disease control is not associated with any kind of adverse environmental effects (Dukare et al., 2013). In general, large numbers of beneficial microorganism are used for alleviation of different kinds of biotic as well as abiotic stresses in agricultural production system (Paul et al., 2017; Sangwan and Dukare, 2018). In postharvest scenario, numerous microbial antagonists belonging to yeasts, bacteria, and fungi taxonomic group have been explored for managing diseases of harvested tropical, subtropical, and temperate fruit crops. Some of microbial antagonist strains have been developed into liquid and dry formulations-based biocontrol products and marketed (Melin et al., 2007; Mokhtarnejad et al., 2011). With this background, present chapter gives an microbial antagonists–based postharvest biocontrol systems including ways of biocontrol actions, efficacy enhancement, formulation development, and its commercial application.

15.2 Different kinds of microbial antagonists used as postharvest biocontrol agents

In past few decades, many microorganisms with antifungal properties have been isolated from variety of plant surface and identified for controlling fungal diseases on a variety of harvested fruits (Wisniewski et al., 2016). Numerous fungal antagonistic species of yeasts, bacterial, and certain fungal genera have successfully used as postharvest diseases biocontrol agent. Among different taxonomic group, yeast represent as effective biocontrol agent due to their ability of rapid growth and fruit surface colonization, quick utilization of host nutrients, extracellular secretion of polysaccharides, and survival under diverse and stressful environments (Dukare et al., 2020). A few examples of antagonist yeast used for successful biocontrol include *Candida oleophila, Candida sake, Cryptococcus laurentii*, and *Debaryomyces hansenii* (Dukare et al., 2020). In addition, certain yeast genera such as *Kluyveromyces, Saccharomyces, Hansenula*, and *Pichia* are also potential bioagents due to their ability of extracellular production of antifungal toxins.

Bacterial antagonists are also effective biocontrol agents due to their ability to produce several antifungal and antibacterial metabolites with an effective property (Lucon et al., 2010; Yanez-Mendizabal et al., 2011). Among bacterial genera, certain strains of *Bacillus, Burkholderia, Enterobacter, Pantoea*, and *Pseudomonas* have found effective in reducing fungal disease on fruit crops. However, only notable antagonistic bacterial species have been formulated and marketed. Beside these groups, few capable antagonistic fungal species are also utilized to protect harvested produce from fungal attack and decay. Among fungi, much more attention has been paid on various species of *Trichoderma* have got a more attention as successful biological control agent of decay causing pathogens of fruits like grapes, mango, kiwifruit, pear, strawberries, and rambutan. *Trichoderma*-mediated biocontrol is mainly achieved through mechanisms of nutrients competition, release of antifungal compounds, mycoparasitism, and induced host resistance (Harman et al., 2004).

15.3 Sources of the microbial antagonist and criteria for selecting the ideal microbial antagonist

Naturally occurring and endemic to fruit surfaces are most commonly used microbial antagonists to manage postharvest diseases. The fruit surface is an exceptional source for isolating and screening microbes for developing them as effective postharvest bioagent (Janisiewicz and Korsten, 2002). Many antifungal microbes isolated from surface of several fruits and vegetables have identified as suitable fruit decay biocontrol agent (Janisiewicz et al., 2013). Similarly, plant associated surfaces such as phyllosphere, root, and soil are also natural habits, from which microbial antagonists have been isolated and identified (Liu et al., 2013).

For example, the tomato phyllosphere yeast, *Rhodotorula glutinis* (strainY-44), control incidence of *Botrytis cinerea*, an etiological agent of gray mold disease on tomato fruit (Kalogiannis et al., 2006). Apart from this, some unique habitats has also been explored for searching effective microbial antagonist of spoilage causing pathogens. To mention, *Leucosporidium scottii*, a cold tolerant yeast antagonists, isolated from Antarctic soil was successful in preventing incidence of *P. expansum* and *B. cinerea* on apple fruits (Vero et al., 2013). Similarly, marine osmotolerant yeast, *Rhodosporidium paludigenum*, inhibited the growth of *P. expansum* and *Alternaria alternata* pear and Chinese winter jujube, respectively (Wang et al., 2010).

In order to develop a potential antagonistic microbe into biocontrol products, organism should possess enviable attributes for making it as effective and suitable for managing after harvest fungal disease. The criteria for the selection of ideal postharvest microbial biocontrol agent have been given in Table 15.1. In addition to these attributes, biocontrol agent must perform over the full spectrum of environmental conditions that are conducive to pathogen growth and spoilage (Janisiewicz and Korsten, 2002). Furthermore, microbial

TABLE 15.1 Different criteria used for the selecting appropriate microbial antagonist as biocontrol agent of postharvest fruit diseases (Barkai-Golan, 2001).

Criteria for the selection of ideal postharvest microbial biocontrol agent
• Genetical stability • Efficient at low concentrations • Successful against a broad spectrum of fungal pathogens • Ability to survive and remain active for longer time under unfavorable set of environment • Easy and low-cost nutritional requirements • Cheaper to produce and formulate with longer shelf-life • Easier for application • Resistant to routinely used postharvest fungicides • Nonpathogenic toward human health and host • Companionable with other postharvest treatments • Nonproducer of toxic metabolites

antagonist must colonize, survive, and multiply in the pathogen conducive environment (Manso and Nunes, 2011). Another foremost factor is the microbial antagonist's viable population required for effective disease suppression. For developing microbial antagonist into as marketable biocontrol product, they must be efficient at reasonable cell concentrations (Nunes, 2012). Thus above all considerations should be kept in mind during successful commercialization of microbial antagonist as a biocontrol agent.

15.4 Mechanisms of actions of microbial antagonists

The tritrophic interaction comprising host–pathogen and microbial antagonist are involved in postharvest biocontrol of fungal spoilage (Fig. 15.1). Numerous modes of action including nutrients and space competition, antibiotics production, mycoparasitism via mycolytic enzymes, biocidal volatile production, and induction of host defense have reported for the biocontrol potential of microbial antagonists. In addition, some recent studies demonstrated the role of biofilm, quorum sensing, and oxidative stress mitigation in fungal disease control (Liu et al., 2013). Advances in genomic- and omics-based approaches such as transcriptomics, proteomics, metagenomics, comparative, and functional genomics helped in gaining deeper perceptive and precise insights into microbial-pathogen biocontrol interactions and mechanism of action involved in it (An et al., 2014). Mechanisms of action of microbial antagonist are discussed detail in following heads.

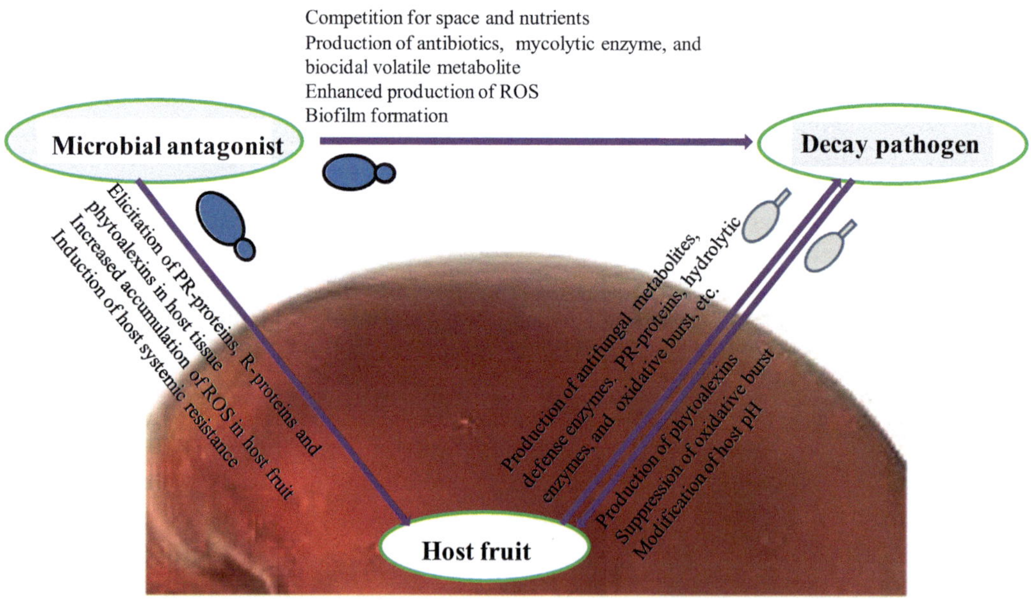

FIGURE 15.1 Schematic diagram exemplifying the mechanisms of antagonistic microbes–mediated biological control actions implicated in tritrophic interaction system, encompassing the microbial antagonists, fungal pathogen, and host fruit. *Source: Adapted after modification from Dukare, A.S., Singh, R.K., Jangra, R.K., Bhushan, B., 2020. Non-fungicides-based promising technologies for managing post-production Penicillium induced spoilage in horticultural commodities: a comprehensive review. Food Rev. Int. doi:https://doi.org/10.1080/87559129.2020.1727497.*

15.4.1 Competition for nutrients and space

The effective struggle between microbial antagonist and pathogen for the accessible limiting growth nutrients such as sugars, minerals, and vitamins and limiting space suppresses pathogen growth and development. This is most vital mode of action reported in numerous postharvest biocontrol studies for different microbial antagonists such as *Pantoea agglomerans, Serratia plymuthica, Aureobasidium pullulans, Metschnikowia pulcherrima, D. hansenii,* and *C. laurentii* (Dukare et al., 2018). However, for a competition to be an efficient pathogen biocontrol via competitive mode of action is effective when microbial antagonist is abundantly present at the right time and place and location and utilizes limited resources more proficiently than the pathogen (Spadaro and Droby, 2016).

Majority of yeast and several bacterial antagonists inhibit pathogen growth by virtue of their strong ability of successful competition for utilizing limited resources under postharvest scenarios (Zhang et al., 2010). Under competitive scenario, microbial antagonists rapidly diminish the accessible nutrients and make it unavailable for fungal growth and pathogenesis in fruit wound. The bacterium *P. agglomerans* CPA-2 prevents conidia germination at low-nutrient concentrations, which signifies role of competition process in disease control (Poppe et al., 2003). Similarly, an antagonistic yeasts attaches pathogen hyphae via nutrient competition and at same time, hinder the process of fungal pathogenesis. In general, yeast is more proficient in utilizing wide range of sugars and nitrogenous compounds as compared to the spoilage pathogen (Spadaro et al., 2010). Through the process of competition for sugars and nitrates, *Pichia guilliermondii,* an yeast bioagent control disease caused by *B. cinerea* on apple (Zhang et al., 2011) and *Colletotrichum* spp. on pepper (Chanchaichaovivat et al., 2008). The competition process in pathogen infected fruit wounds also happens for other limiting factors such as nitrogen, oxygen, vitamins, and amino acids.

In addition, effective competition between antagonist and pathogen for iron, an essential mineral required pathogen growth and development, results in pathogen biocontrol on infected fruit during storage (Saravanakumar et al., 2008). Under iron limiting conditions, bacterial or yeast antagonist produces siderophore (low-molecular-weight compounds) for competitively acquiring ferric ion present in the growth media. Based on chemical structure, siderophores are of either catecholates or hydroxymates type (Saraf et al., 2014). Microbial antagonists produced siderophore compete for iron and obstruct the fungal pathogen growth, and conidial germination. For example, bacterium *Rahnella aquatilis* prevents infection of postharvest apple pathogens (*B. cinerea* and *P. expansum)* mainly due to production of siderophore (Calvo et al., 2007). Likewise, siderophore (pulcherrimin) producing yeast antagonists such as *M. pulcherrima* and *Metschnikowia fructicola* controls the growth of *B. cinerea, A. alternata,* and *P. expansum* on apple via competitive exclusion of iron in the medium (Saravanakumar et al., 2008).

15.4.2 Antibiosis through antibiotics production

In antibiosis, microbial antagonists inhibit or kill pathogens present nearby through production of growth inhibitory compounds referred to as antibiotics. This is one of prime mode by which fungal spoilage on harvested fruits has been controlled. Antibiotics are low-molecular-weight containing compounds that are detrimental to developmental activities of other organisms (Duffy, 2003). Various bacterial genera such as *Pseudomonas, Burkholderia, Bacillus, Pantoea, Streptomyces,* and *Enterobacter,* known to produce antibiotics possess antifungal activity against fungal pathogens of postharvest diseases. The most widespread efficient antifungal antibiotics of bacteria includes lipopeptides, bacillomycin, surfactin, fengycin, and gramicidin produced by *Bacillus* spp.; pyrrolnitrin produced by *Pseudomonas* spp. and *S. plymuthica* and syringomycin produced by *Pseudomonas syringae* (Dukare et al., 2018). Such compounds prevent pathogen proliferation through inhibiting cell walls biosynthesis, changing cell membrane structures, and hindering the protein synthesis process.

The antibiotic producing various strains of Bacillus are reported as biocontrol agents' postharvest fungal spoilage (Korsten et al., 2007; Yanez-Mendizabal et al., 2011). Other antibiotic producing bacterial antagonist deployed in successful postharvest biocontrol includes *Pseudomonas cepacia* for controlling *Penicillium digitatum* in lemon, *B. cinerea,* and *P. expansum control* on apple and *P. syringae* for controlling citrus green mold (Dukare et al., 2020a). Despite of these successful examples, presently, focus is being given on non-antibiotic producing microbial antagonists for disease control due to its wider consumer acceptance and less chance of rising pathogen resistance to antimicrobial antibiotics (Di Francesco et al., 2016).

15.4.3 Mycoparasitism via lytic enzyme production

Direct fungal parasitism, also refereed as mycoparasitism, is caused by antagonistic microorganisms capable of secreting fungal cell wall–degrading extracellular hydrolytic enzymes. The process of mycoparasitism often yields in either absolute killing and/or direct lysis of fungal hyphae, cellular structural, and infection causing propagules (Dukare et al., 2020b). Chitin and glucan, an important structural component of fungal cell wall sugar are responsible for mechanical strength and cellular structural integrity. The cell walls of filamentous fungi is comprised of 20% or even more chitin and glucan content may be approximately 50%–60% of the cell wall (Seidl, 2008). In addition, protein, in the form of glycoprotein, is also an important and structural part of cell wall.

The individual or combined action of antagonist produced hydrolytic enzymes (such as chitinases, chitosanases, glucanases, and protease) contributes to the biocontrol potential of bioagent. Accordingly, several reported postharvest biocontrol agent are known to produce enough quantity of cell wall–degrading lytic enzymes. The postharvest biocontrol potential of halophilic bacteria *Bacillus* spp., against strawberry gray mold disease is due to various lytic enzymes such as chitinase, β-1,3-glucanase, cellulose, and protease by bacteria (Essghaier et al., 2009). The chitinase producing antagonistic yeast, *Metschnikowia fructicola,* was found effective in suppressing fungal decay caused by *Monilinia fructicola* on peach fruit (Banani et al., 2015). Recently, the role of exo-β-1,3-glucanase enzyme produced by biocontrol yeast, *C. oleophila* in preventing *P. expansum* induced spoilage in harvested apples was demonstrated (Urbina et al., 2016). The purified enzyme lowered pathogen spore germination and hyphal growth. Enzymatic action on fungal hyphae induces causes cytological damages, mycelia lysis and distortion, alteration in cellular membrane permeability, and finally loss of protoplasm that eventually leading to fungal death (Di Francesco et al., 2016).

15.4.4 Induction of systemic resistance

Elicitation of host defense system to infecting decay fungi, following the microbial antagonist's application to fruit surfaces is one of mechanism implicated in the post-harvest disease biocontrol (Romanazzi et al., 2016). Induction of systemic resistance (ISR) in hosts buildup various structural barriers and activates many biochemical and molecular responses such as mitogen-activated protein kinase signaling, production of reactive oxygen species, phytoalexins, PR-proteins and terpenoid, augmented production of phenolics, glycoproteins, and callose, and lignifications at the site of pathogen infection (Lloyd et al., 2011).

The induction of host defenses against many decay causing fungal pathogens following external application of microbial antagonist application have been reported. For example, peach fruits, when inoculated with yeast antagonist, *C. laurentii* along with methyl jasmonic acid enhanced the production and activities of different defense enzymes such as glucanase, chitinase, phenylalanine ammonia-lyase, and peroxidase (POD) that further lowered disease incidence caused by *P. expansum* and *Mon. fructicola* (Yao and Tian, 2005). With the help of modern techniques including genomic, transcriptomic, and proteomics, differential expression of gene and protein level in response to biocontrol agent application have also been shown in both antagonists and host fruit. For example, DNA microarray analysis of the cherry tomato following application of *C. laurentii*, revealed the up-regulation of genes involved in growth metabolism, signaling and stress response, which in turn were associated with improved host resistance toward invading fungal infection. Similarly, the increased host resistance against postharvest mandarin disease following yeast *R. paludigenum* application was found due to stimulation of defense proteins such as glucanase, POD, phenylalanine ammonia-lyase, and polyphenoloxidase in host tissue (Lu et al., 2013). Although, there is no direct correlation between ISR and pathogen growth suppression, however, with the use of molecular and proteomic tools, correlation between microbial antagonist and their role in host resistance enhancement can be established.

15.4.5 Production of antifungal volatile compounds

Among the antifungal biocidal metabolite, the volatile organic compounds (VOCs) produced by biocontrol microbes are also implicated in suppression of the growth, metabolism, and virulence ability of several spoilage causing fungi. Chemically, VOCs are complex mixture of low-molecular weight and lipophilic compounds release by yeast, fungus, or bacterium (Mari et al., 2016). The role of VOCs produced by bacterial, fungal, and yeasts antagonists in control of postharvest disease control has reported. For example, VOCs produced by bacteria *Bacillus thuringiensis* and *Bacillus pumilus* lowered the anthracnose disease symptoms in harvested mango fruits by almost 88.5% (Zheng et al., 2013). Volatile compounds producing *Bacillus subtilis* reduced *P. digitatum* infection by 30%–70% in vitro (Leelasuphakul et al., 2008).

Likewise, biological disease control potential of some yeast and fungal antagonists is also correlated with their antifungal VOCs. Several well-known biocontrol yeast and fungal species such as *Trichoderma harzianum*, *Fusarium oxysporum*, and *A. pullulans*, produce low concentrations of antifungal VOCs (Mari et al., 2012). The yeast antagonists, *A. pullulans*, capable of producing several VOCs (phenethyl alcohol, 2-methyl-1-butanol, 3-methyl-1-butanol, and

2-methyl-1-propanol) are fungistatic toward *B. cinerea, Colletotrichum acutatum*, and *Penicillium spp.* pathogens of fruit spoilage (Di Francesco et al., 2014). Electronic microscopy studies revealed the deformities such as cell membrane changes in cell permeability cell vacuolation and hyphal bulging swelling, and poor conidial germination of conidia, was observed in pathogen when it was exposed to the biocodal volatiles of microbial antagonists (Li et al., 2012). Further, antifungal VOCs producing fungal antagonists can also be deployed as biofumigants and new biocontrol agent for controlling storage fungal diseases. A few examples of microbial antagonist successfully deployed for the biological control of postharvest disease of harvested fruits are shown in Table 15.2.

TABLE 15.2 Examples of microbial antagonist successfully deployed for biological control of postharvest diseases of fruits.

Microbial antagonists	Target pathogen(s)	Host fruit(s)	References
Bacterial antagonists			
Bacillus subtilis	*Rhizopus stolonifer*	Tomato	Ma et al. (2015)
	Colletotrichum gloeosporioides	Apple	Kim and Chung (2004)
	Colletotrichum acutatum		
Bacillus atrophaeus	*C. acutatum*	Pepper	Han et al. (2015)
	C. gloeosporioides		
Pseudomonas syringae	*Fusarium pallidoroseum*	Banana	Williamson et al. (2008)
	Fusarium proliferatum		
Pantoea agglomerans	*Penicillium digitatum*	Citrus	Torres et al. (2007)
	Penicillium italicum		
Enterobacter cloacae	*Fusarium sambucinum*	Potato	Al-Mughrabi (2010)
Yeasts and fungal antagonists			
Candida oleophila	*Colletotrichum musae*	Banana	Lassois et al. (2008)
Metschnikowia fructicola	*P. digitatum*	Grapefruit	Hershkovitz et al. (2013)
Pichia guilliermondii	*Rhizopus nigricans*	Tomato	Zhao et al. (2008)
	Botrytis cinerea	Apple, kiwifruit	Sui and Liu (2014)
Rhodotorula mucilaginosa	*Penicillium expansum*	Pear	Hu et al. (2015)
Debaryomyces hansenii	*R. stolonifer*	Peach	Mandal et al. (2007)
Cryptococcus laurentii	*B. cinerea*	Strawberry	Wei et al. (2014)
Trichoderma harzianum	*C. gloeosporioides*	Rambutan	Marikar et al. (2008)
T. harzianum, Trichoderma koningii	*Lasiodiplodia theobromae*	Banana	Sangeetha et al. (2009)
Trichoderma virens	*P. expansum*	Apple	Bordbar et al. (2010)
Verticillium lecanii	*P. digitatum*	Citrus	Benhamou (2004)

15.5 Delivery methods of postharvest biocontrol agents

Usually, both before and after harvest approaches are practiced for delivering biocontrol agent in order to efficiently control the postharvest decay of fruits. The preharvest application of biocontrol agent in field conditions is crucial for controlling the field infestation of fruits caused by fungal pathogens. Field delivery of biocontrol microbes improves their biocontrol efficiency by effectively allowing sufficient interaction time between antagonists and pathogen and precolonizing fruit surface before the pathogen arrival (Ippolito et al., 2004).

Several postharvest disease biocontrol has been achieved using this approach of antagonist delivery. For instance, *C. sake* CPA-1, a yeast antagonist applied 2 days prior to harvest diminished the incidence of blue mold by 50% on the wounded apple fruits when they were cold stored for the period of 4 months (Teixidó et al., 1999). Similarly, field inoculation of biocontrol bacterium *Epicoccum nigrum* was effective in managing the fungal brown rot in the harvested peach fruits. The increased biocontrol was observed due to combined application yeast *C. sake* and bacterium *P. syringae* at field application in controlling *P. expansum* induced rots on pears and apples (Teixidó et al., 2010). Though quite successful, this strategy has limitation on its use at commercial scale owing to poor survival of microbial antagonists in the field situation. In order to improve postharvest biocontrol via field inoculation approach, microbial antagonists must be resistant toward stressful environments such as dry condition, extreme temperature, UV radiations and low-nutrient availability, prevailing in field (Nunes, 2012).

Due to these limitations, most of the biocontrol studies have employed postharvest microbial antagonist's application for successful management of fungal diseases. In this method, biocontrol microbes can be applied either as sprays or as fruit dips in an antagonist's prepared solution (Irtwange, 2006). A considerable reduction in fungal decay is achieved by bringing potential antagonist in direct contact with fungal infected wounds in the peel of harvested fruits. This approach of microbial antagonist's delivery has profound successes in biocontrol than preharvest application.

15.6 Enhancement in biocontrol performance and efficacy of microbial antagonists

Despite the microbial antagonists' potential for postharvest disease control, their individual application is usually not adequate in obtaining constant high level (more than 95%) of disease control in harvested horticultural commodities. Also, alone applications of neither physical methods (thermotherapy and UV-C illumination) nor chemical methods (food additives or generally regarded as safe compounds) (Nunes 2012) are enough to achieve almost complete postharvest pathogen suppression. This limitation can be overcome by combinational and integrative approach that comprises use of bioagent with other method of disease control in an integrated disease management.

The combinational approach overall improves efficacy of biocontrol via taking advantage of synergistic effects of each method. For example, the use of chitosan, a natural antifungal compound, along with yeast antagonistis species such as *Candida saitoana* and *C. laurentii* contributes in improving the overall biocontrol efficacy on host fruits (Meng et al., 2010). In similar manner, the biocontrol efficacy of *R. glutinis* against sweet cherry fruit decay pathogens (*P. expansum* and *A. alternata*) was augmented via its integrative application with salicyclic acid (SA). This compound did not directly inhibited pathogen growth, however, pathogen suppression was observed via induction of defense resistance in host fruits following use of SA (Qin et al., 2003). The biocontrol efficacy of bacterium, *P. agglomerans*, when applied in combination with either hot air treatment (at 33°C for 6 h) or sodium bicarbonate for the control of green and blue mold pathogens of citrus was significantly improved (Usall et al., 2008). In this way, improvement in overall biocontrol performance and efficacy has been achieved by their integrative usage with food grade chemicals (calcium chloride, glucose, ammonium molybdate, silicon, etc.), UV-C light, heat treatments, surfactants and organic salts, low doses of synthetic fungicides, and modified atmospheric storage (Arras et al., 2002; Wisniewski et al., 2016). Certain commercial biocontrol product such as "Biocoat" that mainly combines yeast antagonists and chitosan works through the integrative approach of disease control (Wisniewski et al., 2007). Therefore integrating approaches will be the key for successfully developing safe and reliable substitute of postharvest disease management.

15.7 Development of effective biocontrol formulations and their commercial applications

In general, microbial-based biocontrol formulation comprises antagonists strain (as an active ingredient), carrier material and adjuvant, to improve survival of the antagonist cells and keep them safe from various environmental stresses (Droby et al., 2016). The biocontrol formulated product, in order to be successful at commercial scale, should maintain antagonists viable cell, have a longer storage life and must be compatible with accessible application equipment and methods.

For commercial application, both dry and liquid formulations are prepared with keeping maximum biocontrol performance and shelf-life duration (Melin et al., 2006). Dry formulated biocontrol products are prepared as wettable powders, dusts, or/and granules, while liquid products are prepared in water, oils, or combinations of both. Dry formulations have advantage of longer shelf-life, lower contamination risk, and an ease of handling and application (Li and Tian, 2006). However, dehydration and rehydration mediated higher cell mortality is often limitation of this formulation. On contrary, liquid formulation overcome the limitations of dry formulation, as it has higher cell viability and relatively cheaper to produce.

Sometimes, the external inclusion of protectant in both formulations is crucial to mitigate the adverse impact several environmental stresses encountered by biocontrol strains during the process of large-scale fermentation, formulation development, and commercial product preparation. Addition of protectant helps in maintaining higher cell viability and improving overall biocontrol efficacy and performance of antagonistic organism. For example, trehalose, exogenously included protectant at the concentration of 5% or 10% in freeze dried formulation of *C. laurentii* and *R. glutinis* noticeably amplified their cell longevity (Li et al., 2008). In the same way, oxidative stress usually encountered by biocontrol organism during formulation development can be alleviated by addition of antioxidants compound, which significantly increases cell viability cell and their overall (Liu et al., 2009).

Once formulation is developed, it should have a maximum and reliable level of postharvest diseases control under wide processing and storage conditions. For this, efficacy of a biocontrol product need to be assessed in pilot, semicommercial and commercial tests in diverse packinghouses situation (Droby et al., 2009). Once found successful, next step in commercialization involves regulatory licensing and approval of a biocontrol formulation by respective government agencies. Worldwide, numerous microbial antagonists–based biocontrol products formulations available in the market (Table 15.3). These biocontrol products are registered for use on numerous horticultural produce against different decay causing fungal pathogens. Some biocontrol products are effective against different mycobial rots on different commodities. For example, Shemer, biocontrol product based on osmotolerant yeast strain of *Metschnikowia fructicola* NRRL Y-27328 has been found effective against several mycobial rots caused by *Botrytis*, *Aspergillus*, *Penicillium*, and *Rhizopus* (Blachinsky et al., 2007).

TABLE 15.3 Worldwide commercially available microbial biocontrol formulation for managing postharvest fungal spoilage/diseases in fruits.

Biocontrol products	Active ingredient	Producing firm and country	Target fruit(s)	Target fungal pathogens
Bacterial antagonists–based biocontrol products				
Biosave	*Pseudomonas syringae*	Jet Harvest Solutions, United States	Pome, citrus, strawberry, cherry, potato	Botrytis, Penicillium, Mucor
Avogreen	*Bacillus subtilis*	South Africa	Avocado	Cercospora, Colletotrichum
Pantovital	*Pantoea agglomerans*	IRTA/Sipcam-Inagra, Spain	Citrus, pome	Penicillium, Botrytis, Monilinia
Yeast antagonists–based biocontrol products				
Candifruit	*Candida sake*	IRTA/Sipcam-Inagra, Spain	Pome	Botrytis, Penicillium, Rhizopus
Aspire	*Candida oleophila*	Ecogen, United States	Pome, citrus, strawberry, stone fruit	Botrytis, Penicillium, Monilinia
Nexy	*C. oleophila*	Lesaffre, Belgium	Pome	Botrytis, Penicillium
Yield Plus	*Cryptococcus albidus*	Lallemand, South Africa	Pome, citrus	Botrytis, Penicillium, Mucor
Boni Protect	*Aureobasidium pullulans*	Bio-ferm, Austria	Pome	Penicillium, Botrytis, Monilinia
Shemer	*Metschnikowia fructicola*	Bayer/Koppert, The Netherlands	Table grape, pome, strawberry, stone fruit, sweet potato	Botrytis, Penicillium, Rhizopus, Aspergillus

Adapted after modification from Dukare, A.S., Paul, S., Nambi, V.E., Gupta, R.K., Singh, R., Sharma, K., et al., 2018. Exploitation of microbial antagonists for the control of postharvest diseases of fruits: a review. Crit. Rev. Food Sci. Nutr. 582, 1–16.

15.8 Conclusions and future prospects

The adverse impact associated with synthetic fungicide in postharvest disease control has allowed giving high priority for adopting other eco-friendly and nonfungicide-based sustainable technologies. This chapter has provided a brief overview on the use of microbial antagonists for biological and integrated approaches in postharvest diseases control. Some biofungicides are already registered and commercialized for postharvest disease management, however, it is unrealistic that alone biofungicide will control postharvest disease under all set of conditions due to lack of perfect conditions for effective growth and the development of biocontrol agent. Owing to this, biocontrol effectiveness of antagonist is relatively less than synthetic chemicals. Henceforth, biological control should be perceived as significant constituent of an integrated disease management strategy to minimize the fungicide use on produce.

The biological fungicide is expected to gain momentum as a vital component of an integrated diseases management. Before this, several challenges that need to be addressed in order to develop a commercially successful and economically viable biocontrol products include identification of microbial antagonists with a broad spectrum of antifungal activity, development of cheaper method of mass formulation, retention of cell viability and biocontrol efficacy enhancement under commercial conditions and an effective marketing and its promotion approach.

References

Al-Mughrabi, K.I., 2010. Biological control of *Fusarium* dry rot and other potato tuber diseases using *Pseudomonas fluorescens* and *Enterobacter cloacae*. Biol. Control 53, 280–284.

An, B., Chen, Y., Li, B., Qin, G., Tian, S., 2014. Ca2þ-CaM regulating viability of *Candida guilliermondii* under oxidative stress by acting on detergent resistant membrane proteins. J. Proteom. 109, 38–49.

Andersen, B., Smedsgaard, J., Frisvad, J.C., 2004. *Penicillium expansum*, consistent production of patulin, chaetoglobosins, and other secondary metabolites in culture and their natural occurrence in fruit products. J. Agric. Food Chem. 528, 2421–2428.

Arras, G., Scherm, B., Migheli, Q., 2002. Improving biocontrol activity of *Pichia guillermondii* against postharvest decay oranges in commercial packing-houses by reduced concentrations of fungicides. Biocontrol Sci. Technol. 12, 547–553.

Banani, H., Spadaro, D., Zhang, D., Matic, S., Garibaldi, A., Gullino, M.L., 2015. Postharvest application of a novel chitinase cloned from *Metschnikowia fructicola* and over expressed in *Pichia pastoris* to control brown rot of peaches. Int. J. Food Microbiol. 199, 54–61.

Barkai-Golan, R., 2001. Postharvest Diseases of Fruit and Vegetables, Development and Control, 1st edition Elsevier Sciences, Amsterdam, The Netherlands, p. 442.

Benhamou, N., 2004. Potential of the mycoparasite, *Verticillium lecanii*, to protect citrus fruit against *Penicillium digitatum*, the causal agent of green mold, a comparison with the effect of chitosan. Phytopathology 94, 693–705.

Blachinsky, D., Antonov, J., Bercovitz, A., Elad, B., Feldman, K., Husid, A., et al., 2007. Commercial applications of "Shemer" for the control of pre- and postharvest diseases. IOBC-WPRS Bull. 30, 75–78.

Bordbar, F.T., Etebarian, H.R., Sahebani, N., Rohani, H., 2010. Control of postharvest decay of apple fruit with *Trichoderma virens* isolates and induction of defense responses. J. Plant Prot. Res. 50, 146–152.

Calvo, J., Calvente, V., de Orellano, M.E., Benuzzi, D., de Tosetti, M.I.S., 2007. Biological control of postharvest spoilage caused by *Penicillium expansum* and *Botrytis cinerea* in apple by using the bacterium *Rahnella aquatilis*. Int. J. Food Microbiol. 113, 251–257.

Chanchaichaovivat, A., Panijpan, B., Ruenwongsa, P., 2008. Putative mode of action of *Pichia guilliermondii* strain R13 in controlling chilli anthracnose after harvest. Biol. Control. 47, 207–215.

Di Francesco, A., Martini, C., Mari, M., 2016. Biological control of postharvest diseases by microbial antagonists, how many mechanisms of action? Eur. J. Plant Pathol. 145, 711–717.

Di Francesco, A., Ugolini, L., Lazzeri, L., Mari, M., 2014. Production of volatile organic compounds by *Aureobasidium pullulans* as a potential mechanism of action against postharvest fruit pathogens. Biol. Control 81, 8–14.

Droby, S., Chalutz, E., Wilson, C.L., Wisniewski, M.E., 1992. Biological control of postharvest diseases, a promising alternative to the use of synthetic fungicides. Phytoparasitica. 20, 1495–1503.

Droby, S., Wisniewski, M., Macarisin, D., Wilson, C., 2009. Twenty years of postharvest biocontrol research, is it time for a new paradigm? Postharvest Biol. Technol. 52, 137–145.

Droby, S., Wisniewski, M., Teixidó, N., Spadaro, D., Jijakli, M.H., 2016. The science, development, and commercialization of postharvest biocontrol products. Postharvest Biol. Technol. 122, 22–29.

Duffy, B., 2003. Pathogen self-defense: Mechanisms to counteract microbial antagonism. Annu. Rev. Phytopathol. 41, 501–538.

Dukare, A.S., Prasanna, R., Dubey, S.C., Chaudhary, V., Nain, L., Singh, R., et al., 2011. Evaluating novel microbe amended composts as biocontrol agents in tomato. Crop. Prot. 30, 436–442.

Dukare, A.S., Paul, S., Nambi, V.E., Gupta, R.K., Singh, R., Sharma, K., et al., 2018. Exploitation of microbial antagonists for the control of postharvest diseases of fruits: a review. Crit. Rev. Food Sci. Nutr. 582, 1–16.

Dukare, A.S., Prasanna, R., Nain, L., Saxena, A.K., 2013. Optimization and evaluation of microbe fortified composts as biocontrol agents against phytopathogenic fungi. J. Microbiol. Biotechnol. Food Sci. 25, 2272–2276.

Dukare, A.S., Singh, R.K., Jangra, R.K., Bhushan, B., 2020a. Non-fungicides-based promising technologies for managing post-production Penicillium induced spoilage in horticultural commodities: a comprehensive review. Food Rev. Int. Available from: https://doi.org/10.1080/87559129.2020.1727497.

Dukare, A., Paul, S., Arambam, A., 2020b. Isolation and efficacy of native chitinolytic rhizobacteria for biocontrol activities against Fusarium wilt and plant growth promotion in pigeon pea (*Cajanus cajan* L.). Egypt J. Biol. Pest Control 30, 56. Available from: https://doi.org/10.1186/s41938-020-00256-7.

Essghaier, B., Fardeau, M.L., Cayol, J.L., Hajlaoui, M.R., Boudabous, A., Jijakli, H., et al., 2009. Biological control of grey mould in strawberry fruits by halophilic bacteria. J. Appl. Microbiol. 106, 833–846.

FAO, 2011. Global Food Losses and Food Waste-extent, Causes and Prevention. Food and Agriculture Organization, Rome, Italy, <http://www.fao.org/docrep/014/mb060e/mb060e00.pdf>.

Han, J., Chen, D., Huang, J., Li, X., Zhou, W.W., Gao, W., et al., 2015. Antifungal activity and biocontrol potential of *Paenibacillus polymyxa* HT16 against white rot pathogen *Coniella diplodiella* Speq. in table grapes. Biocontrol Sci. and. Technol. 2510, 1120–1132.

Harman, G.E., Howell, C.R., Viterbo, A., Chet, I., Lorito, M., 2004. Trichoderma species – opportunistic, virulent plant symbionts. Nat. Rev. Microbiol. 2, 43–56.

Hershkovitz, V., Sela, N., Taha-Salaime, L., Liu, J., Rafael, G., Kessler, C., et al., 2013. De-novo assemble and characterization of the transcriptome of *Metschnikowia fructicola* reveals differences in gene expression following interaction with *Penicillium digitatum* and grapefruit peel. BMC Genomics 14, 168.

Hu, H., Yan, F., Wilson, C., Shen, Q., Zheng, X., 2015. The ability of a cold-adapted *Rhodotorula mucilaginosa* strain from Tibet to control blue mold in pear fruit. Antonie van Leeuwenhoek 108, 1391–1404.

Ippolito, A., Nigro, F., Schena, L., 2004. Control of postharvest diseases of fresh fruit and vegetables by preharvest application of antagonistic microorganisms. In: Niskanen, D.R., Jain, R. (Eds.), Crop Management and Postharvest Handling of Horticultural Products, Diseases and Disorders of Fruit and Vegetables, Vol. IV. SM Science Publishers Inc, Enfield, USA, pp. 1–30.

Irtwange, S.V., 2006. Application of biological control agents in pre- and post-harvest operations. Agri. Eng. Intl. 8, Invited Overview 3.

Janisiewicz, W.J., Jurick, I.W.M., Vicob, I., Peter, K.A., Buyer, J.S., 2013. Culturable bacteria from plum fruit surfaces and their potential for controlling brown rot after harvest. Postharvest Biol. Technol. 76, 145–151.

Janisiewicz, W.J., Korsten, L., 2002. Biological control of postharvest diseases of fruit. Annu. Rev. Phytopathol. 40, 411–441.

Kalogiannis, S., Tjamos, S.E., Stergiou, A., Antoniou, P.P., Ziogas, B.N., Tjamos, E.C., 2006. Selection and evaluation of phyllosphere yeasts as biocontrol agents against grey mould of tomato. Eur. J. Plant. Pathol. 116, 69–76.

Kim, P.I., Chung, K.C., 2004. Production of an antifungal protein for control of *Colletotrichum lagenarium* by *Bacillus amyloliquefaciens* MET 0908. FEMS Microbiol. Lett. 234, 177–183.

Korsten, L., Havenga, W., Zeeman, K., Regnier, T., 2007. Mode of action of *Bacillus subtilis* as biocontrol agent of fruit diseases. Bull. OILB 30, 107–111.

Lassois, L., de Bellaire, L., Jijakli, M.H., 2008. Biological control of crown rot of bananas with *Pichia anomala* strain K and *Candida oleophila* strain O. Biol. Control 453, 410–418.

Leelasuphakul, W., Hemmanee, P., Chuenchitt, S., 2008. Growth inhibitory properties of *Bacillus subtilis* strains and their metabolites against the green mold pathogen *Penicillium digitatum* Sacc. of citrus fruit. Postharvest Biol. Technol. 48, 113–121.

Li, Q., Ning, P., Zheng, L., Huang, J., Li, G., Hsiang, T., 2012. Effects of volatile substances of *Streptomyces globisporus* JK-1 on control of *Botrytis cinerea* on tomato fruit. Biol. Control 61, 113–120.

Li, B.Q., Tian, S.P., 2006. Effects of trehalose on stress tolerance and biocontrol efficacy of *Cryptococcus laurentii*. J. Appl. Microbiol. 100, 854–861.

Li, B.Q., Zhou, Z.W., Tian, S.P., 2008. Combined effects of endo- and exogenous trehalose on stress tolerance and biocontrol efficacy of two antagonistic yeasts. Biol. Control 46, 187–193.

Liu, J., Sui, Y., Wisniewski, M., Droby, S., Liu, Y., 2013. Review, utilization of antagonistic yeasts to manage postharvest fungal diseases of fruit. Int. J. Food Microbiol.167 153–160.

Liu, J., Tian, S.P., Li, B.Q., Qin, G.Z., 2009. Enhancing viability of two biocontrol yeasts in liquid formulation by applying sugar protectant combined with antioxidant. Biol. Control. 54, 817–824.

Lloyd, A.J., William, A.J., Winder, C.L., Dunn, W.B., Heald, J.K., Cristescu, S.M., et al., 2011. Metabolomic approaches reveal that cell wall modifications play a major role in ethylene mediated resistance against *Botrytis cinerea*. Plant Pathol. J. 67, 852–868.

Lu, L., Lu, H., Wu, C., Fang, W., Yu, C., Ye, C., et al., 2013. *Rhodosporidium paludigenum* induces resistance and defense-related responses against *Penicillium digitatum* in citrus fruit. Postharvest Biol. Technol. 85, 196–202.

Lucon, C., Guzzo, S., de Jesus, C., Pascholati, S., de Goes, A., 2010. Postharvest harpin or *Bacillus thuringiensis* treatments suppress citrus black spot in 'Valencia' oranges. Crop Prot. 29, 766–772.

Ma, X., Wang, X., Cheng, J., Nie, X., Yu, X., Zhao, Y., et al., 2015. Microencapsulation of *Bacillus subtilis* B99-2 and its biocontrol efficiency against *Rhizoctonia solani* in tomato. Biol. Control 90, 31–41.

Mandal, G., Singh, D., Sharma, R.R., 2007. Effect of hot water treatment and biocontrol agent *Debaryomyces hansenii* on shelf life of peach. Indian. J. Hortic. 64 (1), 25–28.

Manso, T., Nunes, C., 2011. *Metschnikowia andauensis* as a new biocontrol agent of fruit postharvest diseases. Postharvest Biol. Technol. 61, 64–71.

Mari, M., Bautista-Baños, S., Selvakumar, D., 2016. Decay control in the postharvest system, Role of microbial and plant volatile organic compounds. Postharvest Biol. Technol. 122, 70–81.

Mari, M., Martini, C., Spadoni, A., Rouissi, W., Bertolini, P., 2012. Biocontrol of apple postharvest decay by *Aureobasidium pullulans* strains. Postharvest Biol. Technol. 73, 56–62.

Marikar, F., Sivakumar, D., Wijerathnam, R.S.W., 2008. Biological control of rambutan post-harvest anthracnose *Colleotrichum gloeosporioides* by combined treatment of *Trichoderma harzianum*-TrH40 culture filtrates and calcium salts. Food Biotechnol. 224, 326–337.

Melin, P., Håkansson, S., Eberhard, T.H., Schnürer, J., 2006. Survival of the biocontrol yeast *Pichia anomala* after long term storage in liquid formulations at different temperatures, assessed by flow cytometry. J. Appl. Microbiol. 100, 264–271.

Melin, P., Håkansson, S., Schnürer, J., 2007. Optimization and comparison of liquid and dry formulations of the biocontrol yeast *Pichia anomala*. Appl. Microbiol. Biotechnol. 73, 1008–1016.

Meng, X., Qin, G., Tian, S., 2010. Influences of preharvest spraying Cryptococcus laurentii combined with postharvest chitosan coating on postharvest diseases and quality of table grapes in storage. LWT Food Sci. Technol. 43, 596–601.

Mokhtarnejad, L., Etebarian, H.R., Fazeli, M.R., Jamalifar, H., 2011. Evaluation of different formulations of potential biocontrol yeast isolates efficacy on apple blue mold at storage condition. Arch. Phytopathol. 44, 970–980.

Nunes, C.A., 2012. Biological control of postharvest diseases of fruit. Eur. J. Plant Pathol. 133, 181–196.

Paul, S., Dukare, A.S., Bandeppa, G.S., Manjunatha, B.S., Annapurna, K., 2017. Plant growth-promoting rhizobacteria for abiotic stress alleviation in crops. In: Adhya, T., Mishra, B., Annapurna, K., Verma, D., Kumar, U. (Eds.), Advances in Soil Microbiology: Recent Trends and Future Prospects. Microorganisms for Sustainability, Vol. 4. Springer, Singapore, pp. 57–79.

Poppe, L., Vanhoutte, S., Höfte, M., 2003. Modes of action of *Pantoea agglomerans* CPA-2, an antagonist of postharvest pathogens on fruits. Eur. J. Plant Pathol. 109, 963–973.

Qin, G.Z., Tian, S.P., Xu, Y., Wan, Y.K., 2003. Enhancement of biocontrol efficacy of antagonistic yeasts by salicylic acid in sweet cherry fruit. Physiol. Mol. Plant Pathol. 62, 147–154.

Romanazzi, G., Sanzani, S.M., Bi, Y., Tian, S., Martínez, P.G., Alkan, N., 2016. Induced resistance to control postharvest decay of fruit and vegetables. Postharvest Biol. Technol. 122, 82–94.

Sangeetha, G., Usharani, S., Muthukumar, A., 2009. Biocontrol with Trichoderma species for the management of postharvest crown rot of banana. Phytopathol. Mediterr. 48, 214–225.

Sangwan, S., Dukare, A., 2018. Microbe-mediated bioremediation: an eco-friendly sustainable approach for environmental clean-up. In: Adhya, T., Lal, B., Mohapatra, B., Paul, D., Das, S. (Eds.), Advances in Soil Microbiology: Recent Trends and Future Prospects. Microorganisms for Sustainability, vol. 3. Springer, Singapore, pp. 145–163.

Saraf, M., Pandya, U., Thakkar, A., 2014. Role of allelochemicals in plant growth promoting rhizobacteria for biocontrol of phytopathogens. Microbiol. Res. 169, 18–29.

Saravanakumar, D., Ciavorella, A., Spadaro, D., Garibaldi, A., Gullino, M.L., 2008. *Metschnikowia pulcherrima* strain MACH1 out competes *Botrytis cinerea, Alternaria alternata* and *Penicillium expansum* in apples through iron depletion. Postharvest Biol. Technol. 49, 121–128.

Seidl, V., 2008. Chitinases of filamentous fungi, a large group of diverse proteins with multiple physiological functions. Fungal Biol. Rev. 22, 36–42.

Spadaro, D., Ciavorella, A., Zhang, D., Garibaldi, A., Gullino, M.L., 2010. Effect of culture media and pH on the biomass production and biocontrol efficacy of a *Metschnikowia pulcherrima* strain to be used as a biofungicide for postharvest disease control. Can. J. Microbiol. 56, 128–137.

Spadaro, D., Droby, S., 2016. Development of biocontrol products for postharvest diseases of fruit, the importance of elucidating the mechanisms of action of yeast antagonists. Trends Food Sci. Technol. 47, 39–49.

Sui, Y., Liu, J., 2014. Effect of glucose on thermotolerance and biocontrol efficacy of the antagonistic yeast *Pichia guilliermondii*. Biol. Control 74, 59–64.

Teixidó, N., Usall, J., Nunes, C., Torres, R., Abadias, M., Vinas, I., 2010. Pre-harvest strategies to control postharvest diseases in pome fruits. In: Prusky, D., Gullino, M.L. (Eds.), Post-harvest Pathology Series, Plant Pathology in the 21st Century, Vol. 2. Springer, Dordrecht, The Netherlands, pp. 89–106.

Teixidó, N., Usall, J., Viñas, I., 1999. Efficacy of preharvest and postharvest *Candida sake* biocontrol treatments to prevent blue mold on apples during storage. Int. J. Food Microbiol. 50, 203–210.

Torres, R., Nunes, C., Garcia, J.M., Abadias, M., Vinas, I., Manso, T., et al., 2007. Application of *Pantoea agglomerans* CPA-2 in combination with heated sodium bicarbonate solutions to control the major postharvest diseases affecting citrus fruit at several Mediterranean locations. Eur. J. Plant Pathol. 118 (1), 73–83.

Urbina, C.T., Prieto, V.G., Lopez, C.G., Albores, F.V., Reyes, D.B., Muniz, C.A., et al., 2016. Purification and characterization of β-1,3-glucanase from *Candida Oleophila* for the biocontrol of *Penicillium expansum*. Res. Rev. J. Bot. Sci. 51, 38–45.

Usall, J., Smilanick, J., Palou, L., Denis-Arrue, N., Teixidó, N., Torres, R., et al., 2008. Preventive and curative activity of combined treatments of sodium carbonates and *Pantoea agglomerans* CPA-2 to control postharvest green mold of citrus fruit. Postharvest Biol. Technol. 50, 1–7.

Vero, S., Garmendia, G., González, M.B., Bentancur, O., Wisniewski, M., 2013. Evaluation of yeasts obtained from Antarctic soil samples as biocontrol agents for the management of postharvest diseases of apple *Malus domestica*. FEMS Yeast Res. 13, 189–199.

Vitoratos, A., Bilalis, D., Karkanis, A., Efthimiadou, A., 2013. Antifungal activity of plant essential oils against *Botrytis cinerea, Penicillium italicum* and *Penicillium digitatum*. Not. Bot. Horti Agrobot. 411, 86–92.

Wang, Y., Wang, P., Xia, J., Yu, T., Luo, B., Wang, J., et al., 2010. Effect of water activity on stress tolerance and biocontrol activity in antagonistic yeast *Rhodosporidium paludigenum*. Int. J. Food Microbiol. 143, 103–108.

Wei, Y., Mao, S., Tu, K., 2014. Effect of preharvest spraying *Cryptococcus laurentii* on postharvest decay and quality of strawberry. Biol. Control 73, 68–74.

Williamson, S.M., Guzmán, M., Marin, D.H., Anasa, O., Jin, X., Sutton, T.B., 2008. Evaluation of *Pseudomonas syringae* strain ESC-11 for biocontrol of crown rot and anthracnose of banana. Biol. Control 46, 279–286.

Wisniewski, M., Droby, S., Norelli, J., Liu, J., Schena, L., 2016. Alternative management technologies for postharvest disease control: the journey from simplicity to complexity. Postharvest Biol. Technol. 122, 3–10.

Wisniewski, M., Wilson, C., Droby, S., Chalutz, E., El Ghaouth, A., Stevens, C., 2007. Postharvest biocontrol, the discovery of new concepts and applications. In: Vincent, C., Goettal, M., Lazarovits, G. (Eds.), Biological Control, International Case Studies. CABI Publishing, UK, pp. 262–273.

Yanez-Mendizabal, V., Usall, J., Vinas, I., Casals, C., Marın, S., Solsona, C., et al., 2011. Potential of a new strain of *Bacillus subtilis* CPA-8 to control the major postharvest diseases of fruit. Biocontrol Sci. Technol. 21, 409–426.

Yao, H.J., Tian, S.P., 2005. Effects of a biocontrol agent and methyl jasmonate on postharvest diseases of peach fruit and the possible mechanisms involved. J. Appl. Microbiol. 98, 941–950.

Zhang, D., Spadaro, D., Garibaldi, A., Gullino, M.L., 2010. Selection and evaluation of new antagonists for their efficacy against postharvest brown rot of peaches. Postharvest Biol. Technol. 55, 174–181.

Zhang, D., Spadaro, D., Garibaldi, A., Gullino, M.L., 2011. Potential biocontrol activity of a strain of *Pichia guilliermondii* against grey mould of apples and its possible modes of action. Biol. Control 57, 193–201.

Zhao, Y., Tu, K., Shao, X., Jing, W., Su, Z., 2008. Effects of the yeast *Pichia guilliermondii* against *Rhizopus nigricans* on tomato fruit. Postharvest Biol. Technol. 491, 113–120.

Zheng, M., Shi, J., Shi, J., Wang, Q., Li, Y., 2013. Antimicrobial effects of volatiles produced by two antagonistic *Bacillus strains* on the anthracnose pathogen in postharvest mangos. Biol. Control. 65, 200–206.

Further reading

Dukare, A.S., Paul, S., 2018. Effect of chitinolytic biocontrol bacterial inoculation on the soil microbiological activities and fusarium population in the rhizosphere of pigeonpea *Cajanas cajan*. Ann. Plant Prot. Sci. 261, 98–103.

CHAPTER

16

Microbiological efficacy of decontamination methodologies for fresh produce

S. Vijay Rakesh Reddy[1], *Ram Roshan Sharma*[2], *H.R. Raghavendra*[2] *and Uma Prajapati*[2]

[1]ICAR-Indian Institute of Horticultural Research, Bengaluru, Karnataka, India

[2]ICAR-Indian Agricultural Research Institute, New Delhi, India

16.1 Introduction

Fresh fruits and vegetables are the storehouse for several essential vitamins and minerals. Thus they constitute the essential component of the world's population diet [World Health Organization (WHO), 1998]. The diet enriched with fresh/fresh-cut fruits and vegetables has been shown to be protective against various chronic diseases as they possess various essential components that help in neutralizing the oxidation of cellular components by the free radicals produced through various means. Although the fresh fruits and vegetables are visibly considered as part of a healthy diet, foodborne illness relating to the consumption of produce is widely reported (Lynch et al., 2009; WHO, 2008). Since 1990 till 2005, there were about 713 outbreaks of foodborne illnesses related directly to the produce, as reported by Food Safety Project in the United States. Similarly, at least 12% of the foodborne disease outbreaks during 1990s were directly related to the fresh fruits and vegetables (Food and Drug Adulteration Act FDA, 2004). Later, between 1996 and 2004, the FDA has registered about 14 outbreaks of foodborne illness, which were traced back to the sources of fresh lettuce or tomatoes with approximately 859 confirmed cases of foodborne illness (Smith and Bhuiya, 2009).

In the past decade, there was a sudden outbreak of *Escherichia coli* O157:H7 in the United States, which was traced back to spinach, wherein about 276 cases were reported with three deaths [Centre for Disease Control and Prevention (CDC, 2006)]. Similarly, during the decade (1996–2006), the United Kingdom has reported about 3435 cases of foodborne illness extubating from 88 outbreaks and all of which were related to fresh fruits

Food Security and Plant Disease Management
DOI: https://doi.org/10.1016/B978-0-12-821843-3.00005-2

and vegetables (WHO, 2008). In 2011 the Advisory Committee on the Microbiological Safety of Food (ACMSF) has reported 531 cases of foodborne illnesses relating to fruits and vegetables in the United Kingdom (ACMSF, 2011). During 2011, there was an outbreak of *E. coli* (STEC), serotype O104:H4 in Germany, which was producing the Shiga toxin and resulted in 3785 cases of illness and 45 deaths which was later attributed to the sprouted seeds (European Food Safety Authority EFSA, 2011).

In India there were three outbreaks of Nipah Virus Encephalitis in West Bengal (2001 and 2007) with 23 confirmed cases and later in the state of Kerala during 2018, causing 18 confirmed cases and 17 deaths. Later on, the disease was traced back to fruits damaged by fruit bats (WHO, 2018).

From a global perspective, WHO has declared leafy greens to be taken with greatest safety priority during 2008. This recommendation was given on the basis of various factors viz. frequency and severity of the disease, size and scope of production, diversity of supply chain/industry, and scope for amplification of the pathogens (WHO, 2008). Some foodborne pathogens, such as *Salmonella* spp., *Shigella* spp., *E. coli*, *Campylobacter*, *Yersinia enterocolitica*, *Listeria monocytogenes*, *Staphylococcus aureus*, *Clostridium* spp., *Bacillus cereus*, *Vibrio* spp., viruses, and parasites have been identified to be inherited from fresh/raw fruits and vegetables (WHO, 1998). As the fruits and vegetables are mostly cultivated in the natural environment, they are highly susceptible to contamination from variable sources such as soil, water, wild animals, labor, tools/equipment, and transportation containers.

Although there are greater chances of contamination of fresh fruits and vegetables by foodborne pathogens, yet these are consumed on a larger scale by the world population as they are packed with vital nutrients viz. minerals and vitamins. There have been a number of Good Agricultural Practices (GAPs) guidelines developed by various agencies for reducing these risks related to foodborne illness and protect the produce industry. Codex Alimentarius, Code of Good Practice for Fresh Fruits and Vegetables (2010) is another such accepted code recognized internationally for setting out requirements for the safe production of fresh produce. Other documents setting out specific practice safety guidance for the produce industry include the Chilled Foods Association, Microbiological Guidance for Produce Suppliers to Chilled Food Manufacturers (2007) and Campden and Chorleywood Food Research Association (2004) Risks of Pathogensin Ready to Eat Fruits, Salads through the production process. More general requirements are laid down in the Global Gap Standard for Integrated Farm Management (2012), though these standards do not include specific controls for microbiological safety of produce.

A bulk of the fresh fruits and vegetables are consumed either raw or in minimally processed state without being subjected to lethal process treatments like boiling/cooking to retain their heat-sensitive nutrient components (Bhilwadikar et al., 2019). However, late scientists are working for developing technologies that could disinfect/decontaminate the fresh ready-to-eat food products. Though the preharvest precautionary strategies such as GAPs are strictly followed during growing and harvesting of the fruits and vegetables, there is still a minor risk for contamination and hence the processing industries are much relied on the produce decontamination strategies. Various methods of disinfection technologies have been developed over a period of time viz. chemical washings, sprays, irradiation treatments, hurdle technology, and other natural/biological methods

(Goodburn and Wallace, 2013). However, the food industry has relied mostly on the chemical techniques that are discussed in this chapter.

16.2 Decontamination technologies

16.2.1 Peracetic acid

For preserving the quality and controlling the microbial contamination related to food spoilage of fresh produce, cleaning and disinfection are the critical unit operations (Davidson, 1989). An initial cleaning, that is, washing plays an important role in lowering down the product temperature, removes dirt, and also improves appearance. Secondary application of disinfection methods prior to packaging can further reduce the presence and transfer of spoilage pathogenic microbes on surface of the produce, which prevents foodborne illness (Herdt and Feng, 2009).

16.2.1.1 Peracetic acid as a microbial disinfectant

Peracetic acid/peroxyacetic acid (PA) is an equivalent mixture of hydrogen peroxide and acetic acid. It was patented in 1950 for the dip or spray treatment of fruits and vegetables to reduce spoilage from bacteria and fungi (Greenspan and Margulies, 1950). US FDA has approved the use of PA for the equipment, food contact surface, and direct washing of produce as a sanitizer. It has a promising tool for food industry as it is a strong oxidizing agent, which does not release harmful by-products or leaves any residues and accepted for use in organic cultivation. There are many opportunities existing in replacement of fresh-cut produce of chlorinated chemicals with PA in microbial decontamination (Gomez-Lopez et al., 2007a).

16.2.1.2 Mode of action

The mode of action involves the transfer of electrons, therefore the stronger the oxidizer, the faster the transferring of electrons into the microorganisms causing inactivation (Appels et al., 2011). The cell membrane contains plentiful of fatty, lipid-type chemistry, for which PA has an affinity. When the PA comes in contact with fatty acid nature of the cell membrane, the oxidizing portion of PA opens a penetration point by disruption of cell wall. Then, it oxidizes proteins, enzymes, DNA, and metabolites in microbes (bacteria) causing rapid death (Kitis, 2004). The end product of the peracetic acid gives harmless compounds, that is, peroxide and acetic acid. The effectivity of this compound depends on concentration, exposure time, and target microbe.

16.2.1.3 Applications

Several research works have been carried out on the use of PA on fresh produce, which has shown it to be an efficient antimicrobial agent against pathogenic microbes in various fruits and vegetables such as apples (Wisniewsky et al., 2000; Kim et al., 2018), strawberries (Van de Velde et al., 2016), potato (Hilgren and Salverda, 2000), grated carrot (Vandekinderen et al., 2009), fresh-cut carrots (Ruiz-Cruz et al., 2007), iceberg, and romaine lettuce (Beuchat and Scouten, 2004; Bachelli et al., 2013).

Keeratipibul et al. (2011) revealed that PA at 50 mg/L reduced *E. coli* by 4.5 and 2.5 log CFU/g on tomatoes and lettuce, respectively. Reduction of 1.7 log10 over natural microbiota

TABLE 16.1 Effect of peracetic acid on the microbial load reduction of different fruits and vegetables.

Peracetic acid	Microorganism	Log10 reduction	Produce	Reference
50 mg/L	*Escherichia coli* O157:H7	4.5	Tomato	Keeratipibul et al. (2011)
50 mg/L	*E. coli* O157:H7	2.5	Lettuce	Keeratipibul et al. (2011)
80 mg/L	*E. coli* O157:H7	1.7	Cabbage	Vandekinderen et al. (2007)
80 mg/L	*Enterobacter sakazakii*	4.25	Apple, tomato, and cantaloupe	Kim et al. (2006)
80 mg/L	Lactic acid bacteria	1.1	Cut carrot	Vandekinderen et al. (2007)
–	*E. coli* O157:H7	4.4	Apple, lettuce, strawberry, and cantaloupe	Rodgers et al. (2004)
PA(80 mg/L) + UV (900 MJ/cm^2)	*Salmonella typhimurium*	1–1.49	Iceberg lettuce	Ge et al. (2013)
PA (75 mg/L) + Cold water (4°C)	*E. coli* O157:H7	>2.2	Lettuce	Ho et al. (2011)
	Listeria innocua	>2.2	Lettuce	Ho et al. (2011)
	Lactobacillus plantarum	>2.2	Lettuce	Ho et al. (2011)

of cabbage used a concentration of 80 mg/L of PA for 5 minutes (Vandekinderen et al., 2007). PA at 80 mg/L for 5 minutes treatment was used by Hilgren and Salverda (2000) to achieve 0.69–1.69 log10 reduction on mold, yeast, and coliforms bacteria over celery and cabbage (Table 16.1). Kim et al. (2006) achieved 4.25 log10 reduction of *Enterobacter sakazakii* over apples, tomatoes, cantaloupe, strawberries, and iceberg lettuce.

The advantages of using PA as an effective biocide against pathogenic fungi, bacteria, virus, and spoilage microorganisms in fresh produce are that it acts as a strong oxidizing agent, does not produce harmful by-products, and has been approved for direct food treatment.

16.2.2 Electrolyzed water

Minimally processed perishable commodities are highly susceptible to microbial damage as shredders and slicers involved in the processing might act as the primary sites of contamination (Garg et al., 1990) and there are greater chances of microbial contamination through the exposed inner tissues. There are many chemical formulations such as organic acids, ozone, sodium bisulfite, chlorine dioxide, and calcium chloride, which have shown to reduce microbial load of fresh produce.

Sodium hypochlorite is a common disinfectant. Its efficacy depends on its concentration, pH, temperature, time of exposure and growth stage of microbe (Boyette et al., 1993). The effectiveness increases with increasing concentration of available chlorine but at higher concentrations, it causes produce tainting and sodium residue on product (Ritenour and Crisosto, 1996). When sodium hypochlorite is added to water, it generates undissociated hypochlorous acid (HOCl) that is an active antimicrobial component. The acid may be dissociating into hypochlorite ions at higher concentrations and pH, or into chlorine gas at lower pH. Hence, it must be kept stable by maintaining pH in the range of 6.5–7.5 (Boyette et al., 1993) and thus electrolyzed water can be used.

16.2.2.1 Generation of electrolyzed water

Electrolyzed water contains hypochlorous acid that is generated by the reaction of Cl_2 and H_2O in an anode site. When NaCl solution of 10% is electrolyzed through a septum, it results in acid water of pH 2–3. Neutral water at pH 6.8 is generated by electrolysis of NaCl solution without a septum because HCl formed at the anode site neutralizes the NaOH at the cathode site (Hirano, 1997).

16.2.2.2 Mode of action

The antimicrobial efficacy of electrolyzed water depends on various factors such as pH, ORP (oxidation–reduction potential), and concentration of chlorine (Forghani, 2019). High ORP in electrolyzed water can cause modification of metabolic fluxes and ATP production due to electron flow change in cell. Active chlorine can disrupt membrane of microbes, inhibit oxygen uptake, decarboxylate the amino acids, disrupt protein synthesis, and unbalance metabolism of key enzymes (Kiura et al., 2002; Hricova et al., 2008).

16.2.2.3 Applications

Electrolyzed water has been tried by many research workers for plant disease control in pre- and postharvest stages. Gómez-López et al. (2007b) have revealed that application of electrolyzed water can extend the shelf life of shredded cabbage when stored under equilibrium modified atmosphere. Rico et al. (2008) have observed a log reduction of 2.2–2.4 log in mesophilic bacteria when minimally processed lettuce was treated with neutral electrolyzed water. Electrolyzed water has many benefits over other methods *viz.* (i) inexpensive, (ii) could be generated on-site, (iii) produced by electrolysis of water with dilute salts such as NaCl, KCl, or $MgCl_2$, (iv) storability for 1–2 years, (v) safer to the environment, and (vi) reduces the safety and cost issues linked with handling and application of produce without harmful effects (Table 16.2).

TABLE 16.2 Effect of different concentrations of electrolyzed water on the surface disinfection of microbes from the surface of fresh fruits and vegetables.

Electrolyzed water (ppm)	Microorganism	Log10 reduction	Produce	Reference
≤100	*Salmonella enteritidis*	3.7	Tomato	Deza et al. (2003)
≤100	*Escherichia coli* O157:H7	4.4	Tomato	Deza et al. (2003)
300	*Listeria monocytogenes*	2.1	Lettuce	Yang et al. (2003)
300	*E. coli* O157:H7	2.0	Lettuce	Yang et al. (2003)
300	*Salmonella typhimurium*	2.0	Lettuce	Yang et al. (2003)
≤100	*L. monocytogenes*	4.7	Tomato	Deza et al. (2003)
≤50	*E. coli* O157:H7	2.8	Lettuce	Park et al. (2001)
≤50	*L. monocytogenes*	2.4	Lettuce	Park et al. (2001)

16.2.3 Hydrogen peroxide

Hydrogen peroxide plays an important role in extending shelf life of horticultural commodities. It has been reported to have both bactericidal and bacteriostatic effects. It is generally used at the concentration of 1%–5% for sanitation of produce surface and decontamination of packaging materials. Hydrogen peroxide is popularly regarded as safe substance for microbial decontamination as leaving behind no toxic residues after decomposes into water and oxygen. Antimicrobial efficacy of 4%–5% H_2O_2 was comparable to effect of chlorine at 100–200 ppm (Ölmez and Kretzschmar, 2009).

H_2O_2 is a potential oxidizing agent, which forms hydrogen free radicals (OH) that would target the components of bacterial cell-like proteins, lipids, and DNA, which are essential for cell proliferation (McDonnell and Russell, 2001). The oxidizing potential of H_2O_2 is higher (1.8) than efficacy of chlorine (1.5). It helps in the degradation of large number of chemical pollutants without production of toxic by-products (Miguel et al., 2012).

16.2.3.1 *Mode of action*

Hydrogen peroxide has shown potent oxidizing property and hence it is mainly used as sanitizing and antimicrobial agent. It becomes highly toxic and undergoes reaction with lipids, nucleic acids, and protein, which leads to cellular damage. It also exhibits indirect oxidizing effects on cytotoxic oxidative species such as singlet oxygen species, hydroxyl radicals, and hydrogen peroxides, which also has role in damaging lipids membrane and DNA (Feliziani et al., 2016).

16.2.3.2 *Applications*

The main application of hydrogen peroxide involves direct immersion of the product into the solution of concentration 1%–5% as a sanitizing agent. It can also be used with combination of other sanitizers. Back et al. (2014) have reported that application of 10% fumes of H_2O_2 for lettuce leaves leads to log reduction of 3 log CFU/g of *E. coli, Salmonella typhimurium*, and *L. monocytogenes* counts after a treatment of 10 minutes. Dipping in 200 ppm of H_2O_2 of 5% concentration eliminated 2–4 log CFU/g of mesophiles from strawberry, *Listeria innocua* from the surface of red bell pepper, and coliform from watercress (Alexandre et al., 2012). Aerosolized treatment of H_2O_2 as a surface treatment to tomato resulted in undetectable levels of *E. coli* while their populations were reduced by 4.9 log CFU/g on cantaloupe rind and 1.5 log CFU/g on surface of spinach leaves (Jiang et al., 2017).

For the process of disinfection and sterilization of fruits and vegetables, hydrogen peroxide is effectively used as alternative source. It also preserves the quality attribute and extending the shelf life of perishable commodities. Compared to the effectiveness of chlorine, hydrogen peroxide presented a higher capacity to eliminate pathogenic microorganisms, which is safe and environment friendly (Table 16.3).

16.2.4 Ultrasonication

Ultrasound treatment is considered as nonthermal treatment having better potential and a suitable alternative to thermal treatment (Rastogi, 2011). Ultrasound is a form of

TABLE 16.3 Effect of different concentrations of hydrogen peroxide on the surface disinfection of microbes from the surface of fresh fruits and vegetables.

H_2O_2 (%)	Microorganism	Log10 reduction	Produce	Reference
1.0	*Escherichia coli* O157:H7	2.8	Apple	Sapers and Sites (2003)
2.5	*Listeria monocytogenes*	3.0	Melon	Ukuku et al. (2005)
5.0	Total plate count	1.8	Cantaloupe	Sapers and Sites (2003)
2.5	*E. coli* O157:H7	3.0	Melon	Ukuku et al. (2005)
5.0	*Salmonella* spp.	2.2	Cantaloupe	Ukuku (2004)
1	Total mesophiles	1.4	Strawberry	Alexandre et al. (2012)
1	*Listeria innocua*	2.12	Bell pepper	Alexandre et al. (2012)
10	*E. coli* O157:H7	3.15	Lettuce	Back et al. (2014)
10	*L. monocytogenes*	2.95	Lettuce	Back et al. (2014)

energy that is generated by sound waves of different frequencies that are too high to be detected by human ear, that is, 16 kHz (Jayasooriya et al., 2004). In the last few years, ultrasound treatment has taken wider application in most of the fields such as mineral processing, nanotechnology, nondestructive testing, surface cleaning, medical scanning, ultrasonic therapy, industrial welding, environmental decontamination, and food industry. It is also used as a processing aid in food processing such as mixing of materials, foam formation, precipitation, agglomeration, which improves efficiency of filtration, drying, and extraction techniques of solid materials (Rastogi, 2011).

16.2.4.1 Principle

The basic principle on which the ultrasound can work is the cavitation. Ultrasound is a form of energy generated by sound waves having frequency that is inaudible to human ears. When sound waves are propagated through any product, there will be a production of high amount of energy due to compression and rarefaction of the medium particles. Thus cavitation is the formation, growth, and collapse of bubbles that generate a localized mechanical and chemical energy (Gogate and Kabadi, 2009). When ultrasound waves pass through a liquid medium, formation of gas bubbles inside a liquid due to cavitation occurs (Kentish and Ashokkumar, 2011). It is the interaction among sound waves, liquid, and dissolved gas. It results in pressure change around the dissolved gas nuclei and leads to oscillations (Fig. 16.1).

Further, the dissolved gas and solvent vapor spread in and around the oscillating bubbles. Then the bubbles will get expanded in successive cycles to an unstable size and burst. Bursting of bubbles releases very high heat and pressure around the collapsing bubbles, which break the compounds in the liquid and give localized sterilization or pasteurization effect depending on the intensity of applied sound. As a result, particle dispersion and cell disruption occur (Abdullah and Chin, 2014).

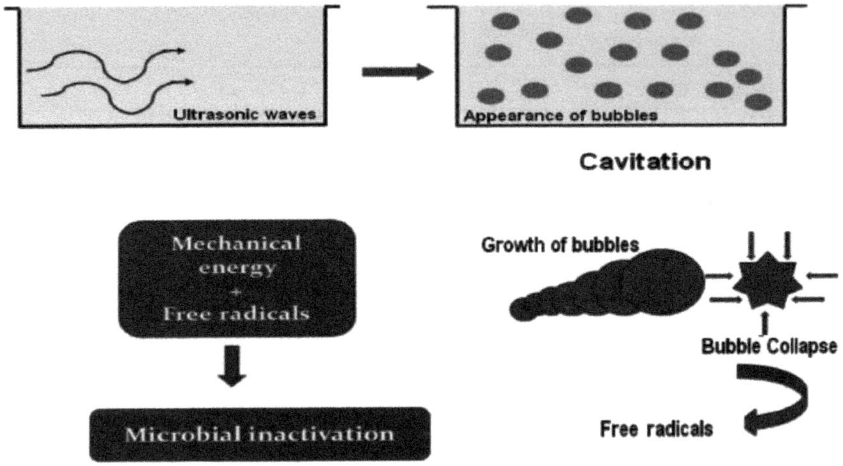

FIGURE 16.1 Ultrasonic waves and the cavitation phenomenon. *Reproduced with permission from: de São José, J. F.B., de Andrade, N.J., Ramos, A.M., Vanetti, M.C.D., Stringheta, P.C., Chaves, J.B.P., 2014. Decontamination by ultrasound application in fresh fruits and vegetables. Food Control. 45, 36–50.*

16.2.4.2 Microbial inactivation

Ultrasonication has emerged as a nondestructive technology to destroy harmful microorganisms without affecting much quality of the produce. Actually, microorganisms are directly destructed by cavitation, which is having combinations of (1) mechanical effects include the generation of turbulence, compression, rarefaction, and shearing effect; (2) chemical effects include cavitation and production of free radicals ($H+$ and OH^-). Free radicals formed in aqueous medium attack the chemical structure of cell wall of microorganism and weaken the cell wall to the point of disintegration. However, these free radicals are short lived and there is no adverse effect on human beings consuming these ultrasound-treated foods and (3) physical effects include extreme temperature and pressure in and around the food products locally, which preserve the food for longer period (Yuting et al., 2013).

16.2.4.3 Applications

The microbial decontamination by the application of ultrasound is characterized by the action of reactive oxygen species, which causes oxidation of the cell wall components and DNA damage (Birmpa et al., 2013; Hulsmans et al., 2010; Seymour et al., 2002). The use of ultrasound at 20 kHz and 106.19 W/L delayed postharvest ripening in cherry tomatoes by inhibiting ethylene production and respiration rate (Wang et al., 2003), which also maintained antioxidant activity, flavor, and firmness of the treated fruits (Table 16.4). Ultrasound treatment of 45 kHz frequency with 80%–100% amplitude and 3-minute duration increased the postharvest life of tomatoes during storage at 10°C (Pinheiro et al., 2015).

TABLE 16.4 Influence of ultrasonication on the surface microbial reduction of fresh fruits and vegetables.

Ultrasound frequency (kHz)	Microorganism	Log10 reduction	Produce	Reference
40	Total bacteria	0.71	Cherry tomato	Ding et al. (2015)
20	Yeast and mold	0.70	Cherry tomato (mature-green)	Wang et al. (2015)
37	*Listeria monocytogenes*	0.77	Cabbage	Srey et al. (2014)
37	*Escherichia coli* O157:H7	2.3	Lettuce	Birmpa et al. (2013)
40	Total bacteria	0.52	Strawberry	Ding et al. (2015)
40	Yeast and mold	0.30	Strawberry	Ding et al. (2015)
37	*E. coli* O157:H7	3.04	Strawberry	Birmpa et al. (2013)
37	*Listeria innocua*	6.12	Strawberry	Birmpa et al. (2013)
37	*Staphylococcus aureus*	2.41	Strawberry	Birmpa et al. (2013)

16.2.5 Irradiation

Food irradiation is a physical method of decontamination process that involves exposing bulk or prepackaged foodstuffs to ionizing radiation. Irradiation is also referred to as "cold pasteurization" as inactivation of microorganisms is achieved at low temperature unlike heat treatment. The use of gamma radiation for the fresh produce has been studied extensively. Using this technique, we can improve food safety and prolong the shelf life without any changes in the chemical, nutritional, and physical attributes of food. It also eliminates pests of agricultural commodities thus reducing losses and use of chemical fumigants. Several international food organizations such as FAO, WHO have recommended that an overall dosage of 10 kGy is safe and effective technology. Nonetheless, food irradiation is not as widespread as other conventional technologies due to the high costs of irradiation units and, particularly, because of a negative perception of consumers relatively to its safety (Calado et al., 2014).

16.2.5.1 Principle

The irradiation is a process of exposing food to ionizing energy, such as gamma photons emitted by ^{60}Co (or infrequently by ^{137}Cs) radioisotopes, X-rays generated by machines operated below a nominal energy of 5 MeV, and accelerated electrons generated by machines operated below a nominal energy of 10 MeV (Farkas and Mohácsi-Farkas, 2011). Only these sources can be used for food irradiation because energies emitted by these sources are much low to induce radioactivity in any exposed material.

16.2.5.2 Mode of action

When living cells are exposed to ionization radiations, there is lethal damage to DNA including strand breakage, disruption in cellular membrane, and mechanical damage to cell walls (Lado and Yousef, 2002). During the food irradiation, DNA is strongly damaged, primarily by this mechanism, as a result, microorganisms, insect gametes, and plant meristems are prevented from reproducing (Farkas, 2006). DNA damage may result from a direct action of the ionizing radiation or from an indirect action of the oxidative radicals that originated from the radiolysis of cellular water (Farkas, 2006).

The radiolysis of water takes approximately 10^{-6} seconds. Water molecules undergo sequence when it gets irradiated that forms several radiolysis products that are reactive with other chemicals. Ionization causes the splitting of water molecules into positive (H_2O^+) and negative charged radicals (e^-) at the end of the process due to various combination and cross-combination reactions (Le Caër, 2011).

16.2.5.3 Applications

From a practical point of view, there are three categories of irradiations when food is exposed to irradiation: (1) a lower dose of 1 kGy, which is used to delay ripening and sprout inhibition/insect disinfection; (2) a medium dosage of 1–10 kGy, which reduces spoilage microorganisms and non-spore-forming pathogens; and (3) a high dosage of 10–50 kGy, which is used to eliminate microorganisms up to the point of sterility for specific products (Satin, 1996). There are mainly five groups when considering the application of irradiation on produces. It includes disinfection of spices and dry vegetables, which represents 46% of all irradiated products; the inhibition of garlic and potato sprouts, which represents 22%; disinfection of fruits consists of 20%; disinfection of meat and seafood, which represents 8%; and the treatment of other products such as mushrooms, honey, health foods, which consist of remaining 4% (Kume et al., 2009).

The FDA has recommended the application of gamma radiation of fresh produce up to 4 times of Gy to control foodborne illness, extends the shelf life of produces (Table 16.5). Irradiation has been an effective measure for the decontamination of produce. It has been observed that irradiation treatment on coriander can cause log reduction of 6.7 log for *E. coli* O517:H7 at 1.05 kGy (Foley et al., 2004). Bari et al. (2005) revealed that at 1 kGy a log reduction of 4.6, 4.9, 4.1, and 5.3 for *L. monocytogenes* was observed for mung beans, broccoli sprouts, tomatoes, and cabbage, respectively.

All irradiated products need to be labeled with the RADURA-logo and a statement "treated by irradiation" or "treated with irradiation" either on the package or at the point of sale.

TABLE 16.5 Effect of irradiation exposure on the surface disinfection of fresh fruits and vegetables.

Irradiation (kGy)	Microorganism	Log10 reduction	Produce	Reference
28.7	Norovirus and Tulane virus	6	Strawberries	DiCaprio et al. (2016)
0.75	*Escherichia coli* O157:H7	4.2	Tomatoes	Mahmoud (2010)
0.75	*Salmonella*	3.7	Tomatoes	Mahmoud (2010)
1	*E. coli* O157:H7	5	Salad spinach	Jeong et al. (2010)
0.7	*Listeria monocytogenes*	5	Salad spinach	Gomes et al. (2011a)
1.5	*Salmonella* spp.	3.6	Cantaloupe	Palekar et al. (2015)

Currently more research is needed to evaluate the dose–response effects on different quality parameters of specific fresh produce, as well as to educate retailers and consumers about irradiation processing to advance commercial applications of this technology. The use of this technology as a hurl for packaged fresh produces also needs more research.

16.2.6 Ultraviolet treatment

Fresh horticultural produce is difficult to decontaminate by using conventional thermal treatment as it affects physical, nutritional, or bioactive properties (Rawson et al., 2011). Besides, chemical sanitizing agents are not much useful to afford safety to the consumer (Rowan et al., 2015) as they are harmful to environment and humans (e.g., sodium hypochlorite generates environmental and health risks associated with the formation of carcinogenic halogenated compounds; Escalona et al., 2015). Therefore, to meet these requirements, several scientists are working on the development of new technologies that are alternative to nonthermal treatments.

16.2.6.1 Ultraviolet-C radiation treatment

UV (Ultraviolet) treatment is one of the nonthermal treatments used to reduce decay of the fresh produce and also minimally processed commodities. UV radiation has been checked with UV-B (medium waves, 345–280 nm), UV-A (long waves, 400–315 nm), and most commonly with UV-C having short waves of 280–100 nm since it has more energy. UV radiation is having 100–400 nm wavelength in which UV-C at 254 nm has the highest surface disinfection and germicidal activity (Gayán et al., 2014). It has poor penetrating power in air but can generate ozone when reacts with gas at wavelength less than 200 nm.

16.2.6.2 Mode of action

Treatment of fresh produce with UV-C radiation (280–100 nm) can help in giving resistance mechanism against pathogen by damaging the bacterial DNA (Ramos et al., 2013). It produces significant damage to the cytoplasmic membrane integrity and cellular enzyme activity (Gómez et al., 2011b). However, inactivation of microbes present on food surface is more as UV-C has low penetration into solids.

16.2.6.3 Antimicrobial activity and its application

There are various reports on positive effects of UV-C radiation on quality parameters, physiological and microbiological parameters of fresh produce. Adhikari et al. (2015) reported that UV-C radiation has the efficacy to prevent spread of *Penicillium expansum* when it was inoculated on surface of strawberry, raspberries, cherries, and apples. This treatment also has been shown to reduce *E. coli* O157:H7 on surface of apples by 3 log cfu and *Salmonella* on tomatoes by 2 log cfu reduction (Yaun et al., 2004). This novel technique has high potential of pathogen reduction on uneven surfaces such as cantaloupes about 1 log cfu with dosage of 12 kJ/m^2 (Adhikari et al., 2015). Manzocco et al. (2011) reported that it shows positive effect on surface decontamination of minimally processed apricots and fresh-cut melon (Table 16.6).

TABLE 16.6 Effect of ultraviolet radiation on the microbial decontamination of fresh fruits and vegetables.

Concentration	Microorganism	Log10 reduction	Produce	Reference
23–34 J/cm^2	*Salmonella* spp.	2.8	Strawberries	Bialka et al. (2008)
23–34 J/cm^2	*Escherichia coli* O157:H7	2.1	Strawberries	Bialka et al. (2008)
150 mJ/cm^2	*Salmonella typhimurium*	2.1	Lettuce	Ge et al. (2013)
20 kHz	*Listeria monocytogenes*	0.4of	Lettuce	Yoon et al. (2013)
59–72 J/cm^2	*E. coli* O157:H7	3.9	Raspberries	Bialka et al. (2008)
59–72 J/cm^2	*Salmonella* spp.	3.4	Raspberries	Bialka et al. (2008)
20 kHz	*E. coli* O157:H7	1.5	Apple	Yoon et al. (2013)
150 mJ/cm^2	*S. typhimurium*	1.5	Green onion	Ge et al. (2013)

The decontamination of fresh and minimally processed commodities remains an unsolved technical difficulty. Hence, there are certain nonthermal physical treatments such as UV-C radiation, which are significantly useful in decontamination and surface disinfection of perishable produce, which extends the shelf life. It can also be used as an effective and safer technique as compared to thermal and chemical techniques.

16.2.7 Heat treatment/hot water treatment

Hot water treatment was first reported in 1922 on control of citrus decay (Fawcett, 1922). Prestorage heat treatments are applied for a relatively short time (in minutes) because the target pathogens are present in the first few layers under the skin of fruit or vegetables. Water is used as the most preferable medium as it is more efficient heat transferring medium than air. In this physical treatment, heat is applied to commodities in several ways such as hot water dips, vapor heat, or hot dry air (Lurie, 1998) or by hot water rinsing and brushing (HWRB) (Fallik et al., 1996). Vapor heat treatment (VHT) was developed mainly for insect control while hot dry air has been used for both fungal and insect control (Lurie, 1998). Heat treatments can also be used to inhibit ripening processes or to induce resistance to chilling injury and external skin damage during storage and thus extending storability and marketing value (Woolf, 1997; Paull and Chen, 2000; Lurie, 1998).

16.2.7.1 Mode of action

A mode of action of hot water treatment is to wash off the spores from the fruit surface. Heat treatment also induces resistance against pathogens by increasing the concentrations of phytoalexins such as scoparone and scopoletin.

16.2.7.2 Methods of treatment

Hot water dips:

Hot water dip or immersion is one of the methods of treatment used for quarantine purposes. This consists of dipping of fruit in hot water kept at 43°C–49°C for several minutes to 2 hours (Fallik, 2004). The time of exposure of hot water depends on the size of fruit. Larger the fruit, longer time is required and vice versa. This treatment provides quarantine security against Mexican fruit fly without adversely affecting fruit' market quality. Blue mold on grapefruit caused by *Penicillium* sp. has been controlled by dipping fruit in hot water for 2 minutes at 50°C. However, incidence of Hot Water Treatment (HWT)-associated damage varied between regions, harvest dates, and orchards.

Hot water rinsing and brushing:

Fresh produce is rinsed by nozzles from above with pressurized recycled hot water, while rolling on brushes made from medium–soft synthetic bristles. Fruits are first rinsed from above with nonrecycled tap water (ambient temperature) from several nozzles while being brushed for about 10 seconds in order to remove the heavy dirt, pesticides, and fungal spores that are covering the fresh harvested produce (Ilic et al., 2001; Lichter et al., 2000; Smilanick et al., 2003).

Vapor heat treatment:

VHT was developed to control fruit flies in the harvested produce. This method consists of stacking the produce, and subjecting to humidified air produced by steam. The temperature and exposure time are adjusted to kill all the stages of insect without damaging fruits. The most difficult stage for HWT or VHT is larval stage, which is further from the fruit. To control papaya fruit flies, fruits are exposed to 43°C and 40% RH for 11 hours followed by 43°C and 100% RH for 8.75 hours (Pantastico, 1975).

16.2.7.3 Applications

Immersion of citrus fruit in hot water at 50°C–53°C for 2–3 minutes can control *Penicillium* spp. or *Alternaria* spp., which causes rots in lemons, orange, or grapefruit (Couey, 1989; Schirra et al., 1997). Hot water treatments at 40°C and 41°C for 25–30 minutes reduced Colletotrichum rots in avocado. Immersion of bell peppers at 53°C for 4 minutes prior to storage was effective in reducing decay after 14 and 28 days storage at 8°C (González-Aguilar et al., 1999, 2000). The HWRB 52°C treatment enhanced resistance against artificially inoculated *Botrytis cinerea* when tomatoes were inoculated 24 hours after treatment (Fallik et al., 2002). Immersing apple fruit at 80°C and 95°C for 15 seconds produced a reduction of more than 5 log of *E. coli* O157:H7 (Fleischman et al., 2001). Dipping minimally processed green onion in hot water for 2 minutes at 55°C or 4 minutes at 52.5°C was the most effective means to control extension growth and enhance microbial disinfection.

Hot water treatments also influence the structure and composition of epicuticle waxes and covering of cracks and wounds and the formation of antifungal substances in the wax after heating. HWT and VHT are quite common in apple and mango in India. Although heat treatment provides an alternative to fungicide applications, yet acceptance of this technology by fruit growers has been hampered by high-energy costs and also the need for added labor at the peak work period during harvest time.

16.2.8 Chitosan

The quality of the harvested commodities could be significantly reduced during the storage because fresh produce is vulnerable to all kinds of spoilage pathogens. Edible coating is beneficial to the extension of shelf life of fruits and vegetables. In recent years, chitosan-based coating has shown biodegradable, nontoxic, and biocompatible properties. Chitosan-based coating with antipathogenic agents could form a thin layer of films on the surface of commodities and inhibits the growth of the microbes.

The antibacterial and antifungal properties of chitosan-based coating could be affected by many factors such as its properties, concentration of antimicrobial agents, and type of the microorganism. There are two methods to improve the properties of chitosan-based coating. One method is that chitosan is applied with organic/inorganic compounds or biological agents, and another method is that coating is used with noncoating technique such as heat treatment, gas fumigation, hypobaric treatment, and modified atmospheric packaging.

16.2.8.1 Chitosan composition

Chitosan [β-(1–4)-2-amino-2-deoxy-D-glucopyranose] is a copolymer of D-glucosamine and *N*-acetyl-D-glucosamine units derived from the deacetylation of chitin in an alkaline medium (Islam et al., 2011). Chitosan is a natural biopolymer, biodegradable, and nontoxic with several antimicrobial properties. Also, as an elicitor molecule, it activates enzymes related to the defense system of the plant against infection by microorganisms (Chandra et al., 2015; Varela et al., 2015). Because of this, chitosan has been proposed as natural alternative system for the control of postharvest microorganisms, with fungicidal characteristics and defense mechanism–inducing properties (Xoca-Orozco et al., 2017).

16.2.8.2 Benefits of chitosan

The main benefits of chitosan-based edible coating are it maintains the quality and extends the shelf life of perishable commodities. Postharvest chitosan application is a suitable alternative fungicide replacement in packaging and storage, preventing postharvest pathogen attack and increasing the postharvest life of horticultural products (Romanazzi et al., 2016; Bautista-Baños et al., 2017). It has proven as one of the best edible and biological safe preservative coating materials because of its film-forming, lack of toxicity, biodegradability, biochemical, and antimicrobial properties. It can control numerous pre- and postharvest diseases of fresh fruits and vegetables and extends shelf life by minimizing the rate of respiration and water loss. With limited increase in the concentration of chitosan coating, the beneficial effect of chitosan on postharvest life and quality of the food is enhanced.

16.2.8.3 Applications

Devlieghere et al. (2004) reported the inhibition of psychotropic bacteria in strawberries and lettuces, which increased the shelf life and preserved the organoleptic properties. Seth et al. (2015) pointed out that it can be applied/sprayed on vegetables and some fresh fruits such as tomatoes and blackberries. Li et al. (2010) reported a significant reduction in disease incidence and lesion diameter of broccoli inoculated with *Pseudomonas fluorescens*, pre- or posttreated with six different combinations of chitosan solutions. Jovanović et al. (2016) showed that population of *L. monocytogenes* ATCC-19115 inoculated in grated black radish in the presence of chitosan-gelatin (0.5%) films decreased from 5 log10 CFU/g to undetectable levels after 3 days at 4°C.

Due to numerous potent properties and applications, it has gained lot of interest as antimicrobial, antifungal, and film-forming properties that make it an efficient biodegradable and antimicrobial packaging material to improve storability of commodities. The decontamination activity of chitosan against a wide range of foodborne pathogens has made it a potential food preservative. Since chitosan application modifies the respiration pattern of fruits, further research is underway to evaluate the effect of this treatment on the fruit flavor profile.

16.2.9 Gaseous chlorine dioxide

For the fresh produce industry, chlorine dioxide has got interest as an alternative to commercial chlorine sanitizers for over the past 10 years. It has several advantages over chlorine, which include broad range of pH (Parish et al., 2003), lower reactivity with organic matter (Gordon and Rosenblatt, 2005), and higher effectiveness at lower concentrations (Huang et al., 1997). Gaseous ClO_2 has attracted much interest for its strong antimicrobial property. It is commonly applied in batch or continuous module for the sanitation of produce (Han et al., 2004).

16.2.9.1 Generation of gaseous chlorine dioxide

In batch treatment, when the fresh fruits and vegetables are kept in chamber containing sachet of ClO_2, a gas begins gradually from the sachet. In continuous method, continuous injection of ClO_2 gas between gas generator and chamber is given through a tube that provides uniform supply. Selection of methods can be influenced by many factors such as microbial load, growth, and generation technique of ClO_2 gas involving in the antimicrobial activity.

From the literature, it is clear that ClO_2 gas is traditionally generated by chemical reaction between secondary chemicals such as chlorine, hydrochloric acid, and sodium chloride (Mahmoud and Linton, 2008; Trinetta et al., 2011; Bhagat et al., 2011). Researchers found that 2%–4% of chlorine gas can also be generated from sodium chlorite.

16.2.9.2 Mode of action

When organic substances of bacterial cells react with chlorine dioxide, it leads to interruption of cellular process. Chlorine dioxide reacts directly with the proteins and lipids present in the cell membrane resulting in immediate death of the microorganism.

16.2.9.3 Applications

Du et al. (2003) revealed that treatment of 15 ppm ClO_2 gas for 6 minutes causes a 7.0 log10 reduction in *E. coli* O157:H7 inoculated on apple. Wu and Kim (2007) observed a 2.4 log10 reduction in *L. monocytogenes* on blueberry at the concentration of 10 ppm ClO_2 gas. Treatment of *Pseudomonas aeruginosa* on blueberry with 10 ppm ClO_2 for 10 minutes resulted in a 2.2 log10 reduction. Han et al. (2000) reported a reduction of 6.5 log10 CFU/g in *E. coli* that was inoculated on green peppers after treatment with 1.2 ppm for ClO_2 for 30 minutes. Commonly the produce having smooth surface shows higher level of microbial reduction as in apple, tomato, lettuce, orange, and others (Table 16.7).

TABLE 16.7 Effect of different concentration of gaseous chlorine dioxide on the microbial load reduction of different fruits and vegetables.

Chlorine dioxide (ppm)	Microorganism	Log10 reduction	Produce	Reference
70	*Salmonella* spp.	3.9	Tomato	Trinetta et al. (2010)
100	*Salmonella* spp.	>5.0	Tomato	Yuk et al. (2006)
50	*Listeria monocytogenes*	4.3	Cantaloupe	Mahmoud and Linton (2008)
50	*Escherichia coli* O157:H7	4.6	Cantaloupe	Mahmoud and Linton (2008)
5	*Salmonella* spp.	5.0	Lettuce	Mahmoud and Linton (2008)
5	*E. coli* O157:H7	5.0	Lettuce	Mahmoud and Linton (2008)
10	*E. coli* K-2	<1.0	Paprika	Matsufuji et al. (2009)
10	*E. coli* K-2	<1.0	Cucumber	Matsufuji et al. (2009)
10	*E. coli* K-2	>2.0	Celery	Matsufuji et al. (2009)
10	*L. monocytogenes*	4.2	Apple	Sy et al. (2005)
10	*Salmonella* spp.	2.3	Strawberry	Sy et al. (2005)
10	*Salmonella* spp.	0.5	Raspberry	Sy et al. (2005)
4	*L. monocytogenes*	3.6	Blueberry	Popa et al. (2007)

Gaseous chlorine dioxide is used as an alternative sanitizer and it is less reactive to organic compounds and shows 2.5 times the oxidation capacity of chlorine (Beuchat et al., 2004). But at higher concentration, it can be explosive. However, compared to gaseous form, aqueous form is advantageous for decontamination purpose. Typically, a log reduction of less than 6.0 was observed for chlorine dioxide, which was considerably better than chlorine where a log reduction of less than 2.0 was observed.

16.2.10 Sodium and calcium hypochlorite

It is well known that sanitizer plays an important role in preventing microbial count, which maintains the quality and extends shelf life of fruits and vegetables. Although water reduces contaminants, it can also transfer pathogens. Hence, washing of commodities with chemical disinfectant is important for commodities to produce hygiene produce by removing dust and dirt and avoids cross-contamination between infected and clean commodities. Among various chlorine disinfectants, hypochlorite salts (sodium and calcium hypochlorite) are widely used for postharvest disinfection. These salts are GRAS (generally recognized as safe).

Sodium hypochlorite (NaOCl) is the source of chlorine generally used at the concentration of 5.25 percent or 12.75 percent *a.i.* in the form of liquid as solid forms readily absorb water from air and release chlorine gas. It is more expensive than other forms of chlorine. Calcium hypochlorite ($CaCl_2O_2$) is the most common source of chlorine for disinfection of water at the registered formulations of 65 percent or 68 percent *a.i.* It is available in the forms of granulated powder and compressed tablets. The primary applications are intended to prevent the transmission of pathogens from adhering soil, infested fruits, and vegetable surface to non-infested surfaces such as trimming cuts or natural plant surface openings.

16.2.10.1 *Mode of action*

The antimicrobial activity of sodium and calcium hypochlorite is based on its high pH (hydroxyl ion action). The high pH interferes with the cytoplasmic membrane integrity with an irreversible enzymatic inhibition, alteration in biosynthetic cellular metabolism, and degradation of phospholipids in lipid peroxidation.

16.2.10.2 *Advantages*

Sodium and calcium hypochlorite have been an important part of properly managed horticultural commodities having advantages such as minimization of transfer and redistribution of microorganisms from adhering soil, infested fruit, or vegetable to healthy ones. These can be used for postharvest handling and cooling as disinfectant agent. Compared to other disinfectants, these are relatively inexpensive and are usually used for water disinfection. These have been proved to act as very effective microbial decontaminating sources for fungi and bacteria of fresh commodities and helps in increasing the storage life of produce.

16.2.10.3 *Postharvest applications*

Romanazzi et al. (2016) revealed that solution of 50 mg/L free chlorine without pH control and 100 mg/L maintained at pH of 7.0, prepared from sodium hypochlorite showed microbial decontamination with less damage for artichoke. Tomás-Callejas et al. (2012) found that cross-contamination of *E. coli* O157:H7 can be prevented by sodium hypochlorite in fresh-cut Red Chard baby leaves.

From the commercial perspective, sodium and calcium hypochlorite exhibit several benefits (Table 16.8). These are relatively inexpensive, readily available, and can be easily applied to control postharvest pathogens. Hypochlorite salts of chlorine are widely utilized for sanitizing and washing in packing chamber. However, care has to be taken that at higher concentration, it may cause damage especially to leafy vegetables. These salts are used under GRAS and can be safely utilized as surface disinfectant.

TABLE 16.8 Microbial efficacy of sodium hypochlorite (NaOCl) on the surface disinfection of fresh fruits and vegetables.

Sodium hypochlorite	Microorganism	Log10 reduction	Produce	Reference
100 mg/L	*Escherichia coli* O157:H7	2.5	Cabbage	Inatsu et al. (2017)
100 ppm	*Salmonella* spp.	2.35	Cabbage	Fukuyama et al. (2009)
100 ppm	*Staphylococcus aureus*	2.48	Cabbage	Fukuyama et al. (2009)
100 ppm	*E. coli* O157:H7	2.61	Chinese celery	Issa-Zacharia et al. (2011)
100 ppm	Aerobic mesophilic bacteria	2.42	Lettuce	Issa-Zacharia et al. (2011)
100 ppm	*E. coli* O157:H7	2.79	Lettuce	Issa-Zacharia et al. (2011)

16.2.11 Pulsed light

Pulsed light (PL) is a technique for surface decontamination in which short time pulses of intense broad spectrum rich in UV-C light are used. UV-C is the portion of electromagnetic spectrum having band between 200 and 280 nm. It is the technology of producing PL that multiply the power many folds. The emitted light flash has a wavelength from 200 to 1100 nm (Dunn et al., 1997). Pulsed light is also referred as high-intensity broad-spectrum PL (Roberts and Hope, 2003), pulsed UV light (Sharma and Demirci, 2003), intense pulsed light (Choi et al., 2010), and pulsed white light (Marquenie et al., 2003).

16.2.11.1 Principle of pulsed light

PL works with xenon lamps that can produce broad-spectrum white light that consists of short time pulses of UV wavelength of 200 nm to infrared wavelength of 1000 nm. The power is magnified by storing energy in a capacitor for long time and releases it over a short period of time and produces high energy per seconds. This contributes for the inactivation of both spoilage and pathogenic microbial cells.

16.2.11.2 Mode of action

High-intensity light pulse is a fast and nature-friendly technology for microbial decontamination and surface disinfection (Gomez-Lopez et al., 2007a). The UV-C light having wavelength range of 200–280 nm emitted by the lamp is most important for microbial inactivation as it causes photochemical damage of DNA, denaturation of proteins, agglutination of cytoplasmic materials, leading to disruption of cell membrane (Krishnamurthy et al., 2010; Ramos-Villarroel et al., 2012).

16.2.11.3 Applications

PL can deliver a wide spectrum of high-intensity flash 200–1100 nm that achieves reduction of 1–3 log cfu of surface bacteria (Kramer et al., 2017). The foods that are rich in carbohydrates, such as fruits and vegetables, are most suitable for PL treatment (Elmnasser et al., 2007). Application of PL can cause surface reduction of natural and inoculated microbes, that is, *Saccharomyces cerevisiae* in fresh tomatoes without losses in nutritional quality (Table 16.9).

TABLE 16.9 Effect of pulsed light on the surface microbial reduction of fresh fruits and vegetables.

Pulsed light	Microorganism	Log10 reduction	Produce	Reference
40 kJ/m^2	*Escherichia coli* O157:H7	2.2	Spinach	Agüero et al. (2016)
12 J/cm^2	*E. coli* O157:H7	5.0	Watermelon	Ramos-Villarroel et al. (2015)
12 J/cm^2	*Listeria innocua*	2.79	Watermelon	Ramos-Villarroel et al. (2015)
32.4 J/cm^2	*E. coli* O157:H7	4.3	Blueberry	Bialka and Demirci (2007)
2.68 J/cm^2	Aerobic mesophilic bacteria	1.0	Tomato	Aguiló-Aguayo et al. (2013)
53.9 J/cm^2	*E. coli* O157:H7	4.5	Raspberry	Huang and Chen (2015)
63.2 J/cm^2	*Salmonella* spp.	2.4	Strawberry	Huang and Chen (2015)

16.2.12 Electrostatic spray

Electrostatic spraying is one of the most efficient techniques of microbial decontamination and there is currently mere published work on this topic. It can benefit the commodity as a fine coating on the surface of a food with a substance. In electrostatic spray, intense electric field is applied to the liquid surface, which induces an electrostatic force to overcome surface tension and disrupt the liquid. This disruption of the droplet surface becomes a cloud of charged droplets that are attracted to the surface (Bailey, 1974).

Compared to other technologies of decontamination, efficiency of electrostatic spray is better in log reduction of microbial load. The major benefit of this technique is that it can be applied to foods with the organic acids that are less corrosive and cause less damage to processing equipments. It is organic in nature, which can be easily applied and gives more stable results (Rozin, 2005).

16.2.12.1 Applications

Ganesh et al. (2010) observed the antimicrobial activity of organic and inorganic acids that were applied conventionally and electrostatically to observe its effects on the *S. typhimurium*-inoculated spinach. They revealed that combinations of malic acid and lactic acid, and malic acid and grape seed were effective against *S. typhimurium* on spinach with log reductions of 4.3 and 3.3, respectively, when sprayed electrostatically.

Electrostatic spraying can be adopted to improve the application of chemicals and distribution during sanitization of fresh produce. It is a relatively quick, simple and the process can access most of the perishable produce surface and offer protection against pathogens. It also typically doubles the droplet deposition efficiency and increases biological efficiency.

16.2.13 Natural antimicrobials

16.2.13.1 Plant extracts

Plant extracts of natural plant origin were proved to possess a substantial scope for use in a range of food applications and a majority of them enjoy GRAS status. The antimicrobial activities of the plant extracts might be due to presence of a variety of phytochemical metabolites such as polyphenols (Table 16.10). The polyphenols are known to function by adsorbing to the cell membranes of bacteria resulting in its disruption and subsequent leakage of cellular components (Ikigai et al., 1993), and also through generation of hydroperoxides from polyphenols. The plant extracts also exhibited antifungal activity against a wide range of fungi (Davidson, 1989), antioxidant and antimutagenic activities (Boubaker et al., 2011; Cherdshewasart et al., 2009; Horn and Vargas, 2003; Negi et al., 2010), and inhibited lipid oxidation in foods (Shan et al., 2009).

16.2.13.2 Essential oils

Wide varieties of aromatic and medicinal crops have been identified and explored for their essential oils nowadays. Essential oils are defined as complex volatile compounds synthesized naturally from plant parts during the process of secondary metabolism. It has a great strength in the field of biomedicine as they effectively control a wide range of viral, bacterial, and fungal pathogens due to presence of different types of aldehydes, terpenes,

TABLE 16.10 Plant extracts as a natural source of antimicrobials.

Plant	Phytochemical	Part(s) used	Reference
Elephant foot yam (*Amorphophallus campanulatus*)	Amblyone	Root	Khan et al. (2008)
Aristolochia (*Aristolochia bracteata*)	Aristolochic acid	Root	Negi et al. (2003)
Jakaranda (*Jacaranda mimosifolia*)	Flavones, flavonoids, iridoids, triterpenes	Flowers/leaves	Rojas et al. (2006)
Achiote (*Bixa orellana*)	Alkaloids, tannins, flavonoids, terpenoids, phenolics, glycosides	Leaves/seeds	Rojas et al. (2006)
Seabuckthorn (*Hippophae rhamnoides* L.)	Flavonoids, tannins, triterpenes	Berries	Gupta et al. (2011)
	Isorhamnetin, quercetin	Leaves	Shipulina (2001)
	Phenolics	Seeds	Chauhan et al. (2007)
Kokum (*Garcinia* spp., *G. indica, G. cambogia, G. guttifera, G. mangostana, G. subelliptica, G. diocia*)	Garcinol, α-mangostin, xanthones and xanthone derivatives, xantho-chymol rubraxanthone	Fruit	Iinuma et al. (1996a,b), Mackeen et al. (2000), Selvi et al. (2003), Gopalakrishnan et al. (1997)
Pudina (*Mentha* spp., *M. piperita, M. longifolia*)	Flavonoids, menthol, terpenoids	Leaves	Gulluce et al. (2007)
Cinnamon (*Cinnamomum* spp.)	Cinnamaldehyde, flavan-3-ols, *p*-coumaric acid, eugenol, camphor, α-pinene, β-pinene, β-caryophyllene, γ-cadinene	Bark/leaves/root bark/fruit	Jayaprakasha et al. (2007)
Onion (*Allium cepa* L.)	Saponins, kaempferol, ferulic acid, β-sitosterol, myristic acid, prostaglandins, quercetin, alkenyl cysteine sulfoxide	Bulb	Kubec et al. (2000), Tang and Cronin (2007)
Garlic (*Allium sativum* L.)	Allicin, diallylisothiocyanate, allicin, diallyl disulfide, diallyl trisulfide, alkenyl cysteine sulfoxide, ajoene, methyl allyl thiosulfinate	Bulb	Naganawa et al. (1996), Horníčková et al. (2009), Yin and Cheng (2003), Kubec et al. (2000)
Neem (*Azadirachta indica* A.)	Azadirachtin, margosic acid, nimbin, nimbinin, nimbidin, nimbolide, gedunin, gallic acid, epicatechin, catechin, mahmoodin, margolone, margolonone	Bark	Biswas et al. (2002)
Pipal (*Ficus religiosa* L.)	Bergaptol and bergapten	Leaves	Swami and Bisht (1996)
Clove (*Syzygium aromaticum* L.)	Eugenol, eugeniin, acetyl eugenol, quercetic acid, gallic acid, vanillin	Bud	Shan et al. (2009)

(*Continued*)

TABLE 16.10 (Continued)

Plant	Phytochemical	Part(s) used	Reference
Oat (*Avena sativa* L.)	Phenolic acids Class I chitinases	Seed	Sørensen et al. (2010)
Pomegranate (*Punica granatum* L.)	Anthocyanins, ellagic acid, gallotannins, gallic acid, punicalagin, ellagitannins	Rind	Machado et al. (2003)
	Punicalagin and its isomers, punicalin, ellagic acid	Husk	
	Punicic acid	Seeds	
	Ellagitannins and gallotannins	Heartwood	
	Tannins (gallo- and ellagitannins)	Leaves	
Black cumin (*Nigella sativa* L.)	Glucoside, melanthin, saponin, thymoquinone	Seeds	Kumar et al. (2010)
Coorg mandarins (*Citrus reticulata* Blanco)	Polymethoxylated flavones	Rind	Jayaprakasha et al. (2002)
Grapefruit (*Citrus paradisi*)	Coumarins, flavanones, methoxyflavones	Rind	Negi and Jayaprakasha (2001)
Soapnut (*Sapindus* sp.)	Acetylated triterpenes, saponins	Fruit	Damke et al. (2011)
Grape (*Vitis vinifera* L.)	Catechin, epicatechin, rutin, myricetin, *trans*-resveratrol, gallic acid, caffeic acid, ferulic acid, ellagic acid, quercetin	Fruits	Butkhup et al. (2010)
Turmeric (*Curcuma longa* L.)	Curcumin, turmerone, ar-turmerone, curlone, *trans*-β-farnesene, α-zingiberene, β-bisabolene	Rhizome	Negi et al. (1999), Majeed et al. (1995), Srimal (1997)

phenolics, and other antimicrobial compounds. The reactivity of essential oil depends upon the nature, composition, and orientation of its functional groups.

Some essential oils have antimicrobial property that enhances the storage life of perishable produce by controlling postharvest spoilage (Table 16.11). Several reports have stated that effective usage of essential oils either by spraying or dipping can control decay of several fruits and vegetables (Tiwari et al., 1988; Smid et al., 1993; Dixit et al., 1995). For instance, an essential oil extracted from thyme (*Thymus capitatus*), that is, thymol has the ability to reduce the disease incidence of apricot gray mold and brown mold without phytototoxicity fumigation method at a lower concentration (2–4 mg/L) (Liu et al., 2002). Tripathi (2001) reported that essential oils of *Zingiber officinale*, *Ocimum canum*, and *Mentha arvensis* have been found to enhance storage life of citrus by controlling blue mold incidence. Similar reports of blue mold and green mold inhibition were reported in kinnow fruits with the application of essential oils such as lemon grass, eucalyptus, clove, and neem (Jhalegar et al., 2015).

TABLE 16.11 Efficacy of different essential oils on the microbial load reduction of different perishable commodities.

Essential oils	Microorganism	Log10 reduction	Produce	Reference
Thyme oil (5 ppm)	*Escherichia coli* O157:H7	1.0	Carrot	Tornuk et al. (2011)
Thyme oil (5 ppm)	*Salmonella typhimurium*	1.0	Apple	Tornuk et al. (2011)
Thyme oil (0.1%)	*Shigella sonnei*	5.0	Lettuce	Bagamboula et al. (2004)
Thyme oil (0.5%)	Aerobic plant count	2.0	Bell pepper	Uyttendaele et al. (2004)
Carvacrol (0.6 ppm)	*Aeromonas hydrophila*	<3.0	Mixed vegetables	de Sousa et al. (2012)
Carvacrol (0.6 ppm)	*Listeria monocytogenes*	<2.5	Mixed vegetables	de Sousa et al. (2012)
Cineol oil (20 ppm)	*A. hydrophila*	<1.5	Mixed vegetables	de Sousa et al. (2012)
Cineol oil (20 ppm)	*L. monocytogenes*	<1.0	Mixed vegetables	de Sousa et al. (2012)
Oregano oil (100 ppm)	*S. typhimurium*	3.7	Tomato	Gündüz et al. (2010)
Oregano oil (250 ppm)	Lactic acid bacteria	1.1	Carrot	Berumen-Varela et al. (2015)

16.2.14 Biological agents

Considering the safety of consumers, now there is a dramatic shift from use of synthetic fungicides to safer and eco-friendly approaches on harvested commodities. Of these approaches, use of bioagents for controlling postharvest diseases of fruits and vegetables is an emerging field (Mari et al., 2007; Sharma et al., 2003).

16.2.14.1 Modes of action

The competition for nutrients, antibiosis, induction of resistance, and direct parasitism is the main modes of action of the microbial antagonists that suppress the activity of postharvest pathogens during handling (Sharma et al., 2003).

16.2.14.2 Use of bioagents

Postharvest spoilage of fruits and vegetables has also been successfully controlled by several microbial antagonists (Van Lenteren et al., 2018). Two approaches for use of bioagents are usually followed. First approach is to manage and promote antagonistic microbes that are already settled on the fruits and vegetables itself and second approach is the artificial induction of desired antagonistic microbes against postharvest pathogens.

Natural bioagents

Natural bioagents are the microbes that are already present on the outer peel of fruits and vegetables, and have the ability to control postharvest infection. Several such agents have been identified and isolated from fruit surface of citrus and apples and again reapplied to the affected fruits as biocontrol agents. Chalutz and Wison (1990) reported that when we plated concentrated washings of citrus fruit surface, only yeast and bacteria appeared. After dilution of washings, there was reappearance of rot fungus on agar medium, which indicated that yeast and bacteria may suppress their growth.

Artificial introduced bioagents

Compared to other ways of biological control for the control of postharvest spoilage, use of artificial antagonists is more efficient and effective. Many researchers have tried various bio-control agents for controlling of postharvest diseases of fruits and vegetables with variable success (Table 16.12).

TABLE 16.12 List of microbial antagonists identified for controlling various postharvest diseases.

Microbial antagonist	Disease and its causal agent	Fruit/vegetable	Reference
Aureobasidium pullulans	Soft rot (*Monilinia laxa*)	Grape	Barkai-Golan (2001)
	Botrytis rot (*Botrytis cinerea*)	Grape	Schena et al. (2003)
	Penicillium rots (*Penicillium* spp.)	Citrus	Wilson and Chalutz (1989)
Bacillus subtilis	Anthracnose (*Colletotrichum gloeosporioides*)	Mango	Govender et al. (2005)
	Brown rot (*Lasiodiplodia theobromae*)	Peach, plum, and nectarine	Pusey and Wilson (1984)
	Green mold (*Penicillium digitatum*)	Citrus	Singh and Deverall (1984)
Bacillus pumilus	Gray mold (*B. cinerea*)	Pear	Mari et al. (1996)
Bacillus amyloliquefaciens	*B. cinerea*	Pear	Qu et al. (2016)
Candida guilliermondii	Gray mold (*B. cinerea*)	Tomato	Saligkarias et al. (2002)
	Gray mold (*B. cinerea*)	Peach and nectarine	Tian et al. (2002)
Candida oleophila	Anthracnose (*C. gloeosporioides*)	Papaya	Gamagae et al. (2003)
	Penicillium rot (*Penicillium expansum*)	Apple	El-Neshawy and Wilson (1997), Urbina et al. (2016)
	Crown rot (*Colletotrichum musae*)	Banana	Lassois et al. (2008)
Trichoderma atroviride	*Phomopsis vexans*	Brinjal	Das et al. (2014)
Wickerhamomyces anomalus	*B. cinerea*	Grapes	Parafati et al. (2015)
Paenibacillus polymyxa, B. subtilis	*C. gloeosporioides, Colletotrichum acutatum, Botryosphaeria dothidea*	Apple	Kim et al. (2016)
Pseudomonas fluorescens, B. subtilis	*L. theobromae*	Mango	Seethapathya et al. (2016)

16.3 Conclusion

There has been an exponential rise in the demand for fresh produce due to their nutritional composition and ability to prevent various degenerative diseases. Thus, at the same time, they require reliable disinfection/decontamination techniques for assuring food safety and better consumer health. Of late, there has been much research focus on developing fresh produce decontamination technologies. Various studies are being carried out to develop alternatives to chlorine, the traditional chemical method of disinfection. Researchers around the globe are actively investigating various technologies from household chemicals to more recent nonthermal processes for their efficacy in reducing the microbial loads over the fresh produce. Recent advances in food engineering have led to development and scaling-up of various modern decontamination technologies; however, these processes are often found too expensive to be commercialized. In case of chemical sanitizers, the use of large volumes of fresh produce and reuse of washing solutions risk occurrences of cross-contamination within subsequent batches, which may necessitate the use of higher volumes of sanitizers thus increasing cost and pollution. Photolytic techniques such as use of UV light/PL can be easily scaled up to the industrial level, which also does not contribute to the environment pollution, indicating that these can be an effective alternative to chemical solutions. Combination technologies are a new concept of using more than one technique to reduce the microbial loads. Not much work has been done in this sector and there is a need to conduct elaborative research for identification of complimenting technologies that can be used in combination to reduce foodborne microbial contaminations. Use of combination technologies may aid in overcoming the challenges of reduced efficacy of individual treatments; however, the effectiveness of possible combination processes in dissipating different pesticide groups and inactivating a broad spectrum of microorganisms remains unexplored. The combination of lower concentrations of chemical solutions with modern decontamination technologies (photolysis) may potentially reduce treatment duration, chemical effluents, and operational costs for industrial level processing, and result in ensuring safe supply of fresh fruits and vegetables to consumers.

References

Abdullah, N., Chin, N.L., 2014. Application of thermosonication treatment in processing and production of high quality and safe-to-drink fruit juices. Agric. Agric. Sci. Procedia 2, 320–327.

Adhikari, A., Syamaladevi, R.M., Killinger, K., Sablani, S.S., 2015. Ultraviolet-C light inactivation of *Escherichia coli* O157: H7 and *Listeria monocytogenes* on organic fruit surfaces. Int. J. Food Microbiol. 210, 136–142.

Advisory Committee on the Microbiological Safety of Food, ACMSF, 2011. Annual report 2011. Retrieved November 2012. <http://www.food.gov.uk/multimedia/pdfs/committee/acmsf-report-2011.pdf>.

Agüero, M.V., Jagus, R.J., Martín-Belloso, O., Soliva-Fortuny, R., 2016. Surface decontamination of spinach by intense pulsed light treatments: impact on quality attributes. Postharvest Biol. Technol. 121, 118–125.

Aguiló-Aguayo, I., Charles, F., Renard, C.M., Page, D., Carlin, F., 2013. Pulsed light effects on surface decontamination, physical qualities and nutritional composition of tomato fruit. Postharvest Biol. Technol. 86, 29–36.

Alexandre, E.M., Brandão, T.R., Silva, C.L., 2012. Assessment of the impact of hydrogen peroxide solutions on microbial loads and quality factors of red bell peppers, strawberries and watercress. Food Control. 27, 362–368.

Appels, L., Van Assche, A., Willems, K., Degrève, J., Van Impe, J., Dewil, R., 2011. Peracetic acid oxidation as an alternative pre-treatment for the anaerobic digestion of waste activated sludge. Bioresour. Technol. 102, 4124–4130.

Bachelli, M.L.B., Amaral, R.D.Á., Benedetti, B.C., 2013. Alternative sanitization methods for minimally processed lettuce in comparison to sodium hypochlorite. Braz. J. Microbiol. 44, 673–678.

Back, K.-H., Ha, J.-W., Kang, D.-H., 2014. Effect of hydrogen peroxide vapor treatment for inactivating *Salmonella typhimurium*, *Escherichia coli* O157: H7 and *Listeria monocytogenes* on organic fresh lettuce. Food Control. 44, 78–85.

Bagamboula, C.F., Uyttendaele, M., Debevere, J., 2004. Inhibitory effect of thyme and basil essential oils, carvacrol, thymol, estragol, linalool and p-cymene towards *Shigella sonnei* and *S. flexneri*. Food Microbiol. 21 (1), 33–42.

Bailey, A.G., 1974. Electrostatic atomization of liquids. Sci. Prog. (1933-) 61, 555–581.

Bari, M., Nakauma, M., Todoriki, S., Juneja, V.K., Isshiki, K., Kawamoto, S., 2005. Effectiveness of irradiation treatments in inactivating *Listeria monocytogenes* on fresh vegetables at refrigeration temperature. J. Food Prot. 68, 318–323.

Barkai-Golan, R., 2001. Postharvest Diseases of Fruits and Vegetables: Development and Control. Elsevier.

Bautista-Baños, S., Ventura-Aguilar, R.I., Correa-Pacheco, Z., Corona-Rangel, M.L., 2017. Chitosan: a versatile antimicrobial polysaccharide for fruit and vegetables in postharvest—a review. Rev. Chapingo Ser. Hortic. 23, 103 121.

Berumen-Varela, G., Coronado-Partida, D., Ochoa-Jiménez, V., Chacón-López, A., Gutiérrez-Martínez, P., 2015. Effect of chitosan on the induction of disease resistance against *Colletotrichum* sp. in mango (*Mangifera indica* L) cv. Tommy Atkins. Invest. Cienc. 66, 16–21.

Beuchat, L.R., Scouten, A.J., 2004. Factors affecting survival, growth, and retrieval of *Salmonella poona* on intact and wounded cantaloupe rind and in stem scar tissue. Food Microbiol. 21, 683–694.

Beuchat, L.R., Pettigrew, C.A., Tremblay, M.E., Roselle, B.J., Scouten, A.J., 2004. Lethality of chlorine, chlorine dioxide, and a commercial fruit and vegetable sanitizer to vegetative cells and spores of *Bacillus cereus* and spores of *Bacillus thuringiensis*. J. Food Prot. 67, 1702–1708.

Bhagat, A., Mahmoud, B.S., Linton, R.H., 2011. Effect of chlorine dioxide gas on *Salmonella enterica* inoculated on navel orange surfaces and its impact on the quality attributes of treated oranges. Foodborne Pathog. Dis. 8, 77–85.

Bhilwadikar, T., Pounraj, S., Manivannan, S., Rastogi, N., Negi, P., 2019. Decontamination of microorganisms and pesticides from fresh fruits and vegetables: a comprehensive review from common household processes to modern techniques. Compr. Rev. Food Sci. Food Saf. 18, 1003–1038.

Bialka, K.L., Demirci, A., 2007. Decontamination of *Escherichia coli* O157: H7 and *Salmonella enterica* on blueberries using ozone and pulsed UV-light. J. Food Sci. 72, M391–M396.

Bialka, K.L., Demirci, A., Puri, V.M., 2008. Modeling the inactivation of *Escherichia coli* O157: H7 and *Salmonella enterica* on raspberries and strawberries resulting from exposure to ozone or pulsed UV-light. J. Food Eng. 85, 444–449.

Birmpa, A., Sfika, V., Vantarakis, A., 2013. Ultraviolet light and ultrasound as non-thermal treatments for the inactivation of microorganisms in fresh ready-to-eat foods. Int. J. Food Microbiol. 167, 96–102.

Biswas, K., Chattopadhyay, I., Banerjee, R.K., Bandyopadhyay, U., 2002. Biological activities and medicinal properties of neem (*Azadirachta indica*). Curr. Sci.-Bangalore 82 (11), 1336–1345.

Boubaker, J., Mansour, H.B., Ghedira, K., Chekir-Ghedira, L., 2011. Antimutagenic and free radical scavenger effects of leaf extracts from *Accacia salicina*. Ann. Clin. Microbiol. Antimicrob. 10 (1), 37.

Boyette, M., Ritchie, D., Carballo, S., Blankenship, S., Sanders, D., 1993. Chlorination and postharvest disease control. Horttechnology 3, 395–400.

Butkhup, L., Chowtivannakul, S., Gaensakoo, R., Prathepha, P., Samappito, S., 2010. Study of the phenolic composition of Shiraz red grape cultivar (*Vitis vinifera* L.) cultivated in north-eastern Thailand and its antioxidant and antimicrobial activity. South Afr. J. Enol. Viticulture 31 (2), 89–98.

Calado, T., Venâncio, A., Abrunhosa, L., 2014. Irradiation for mold and mycotoxin control: a review. Compr. Rev. Food Sci. Food Saf. 13, 1049–1061.

Center for Disease Control and Prevention, CDC, 2006. *E. coli* outbreak from fresh spinach. <http://www.cdc.gov/ecoli/2006/september/>.

Chalutz, E., Wilson, C.L., 1990. Postharvest biocontrol of green and blue mold and sour rot of citrus fruit by *Debaryomyces hansenii*. Plant Dis. 74 (2), 134–137.

Chandra, S., Chakraborty, N., Dasgupta, A., Sarkar, J., Panda, K., Acharya, K., 2015. Chitosan nanoparticles: a positive modulator of innate immune responses in plants. Sci. Rep. 5, 15195.

Chauhan, A.S., Negi, P.S., Ramteke, R.S., 2007. Antioxidant and antibacterial activities of aqueous extract of Seabuckthorn (*Hippophae rhamnoides*) seeds. Fitoterapia 78 (7–8), 590–592.

Cherdshewasart, W., Sutjit, W., Pulcharoen, K., Chulasiri, M., 2009. The mutagenic and antimutagenic effects of the traditional phytoestrogen-rich herbs, *Pueraria mirifica* and *Pueraria lobata*. Braz. J. Med. Biol. Res. 42 (9), 816–823.

Choi, M.S., Cheigh, C.I., Jeong, E.A., Shin, J.K., Chung, M.S., 2010. Non-thermal sterilization of *Listeria monocytogenes* in infant foods by intense pulsed-light treatment. J. Food Eng. 97 (4), 504–509.

Couey, H.M., 1989. Heat treatment for control of postharvest diseases and insect pests of fruits. Hortscience 24, 198–202.

Damke, E., Tsuzuki, J.K., Cortez, D.A., Ferreira, I.C., Bertoni, T.A., Batista, M.R., et al., 2011. *In vivo* activity of *Sapindus saponaria* against azole-susceptible and-resistant human vaginal *Candida* species. BMC Complement. Altern. Med. 11 (1), 35.

Davidson, P., 1989. Methods for testing the efficacy of food antimicrobials. Food Technol. 43, 148–155.

de Sousa, J.P., de Azerêdo, G.A., de Araújo Torres, R., da Silva Vasconcelos, M.A., da Conceição, M.L., de Souza, E.L., 2012. Synergies of carvacrol and 1, 8-cineole to inhibit bacteria associated with minimally processed vegetables. Int. J. Food Microbiol. 154 (3), 145–151.

de São José, J.F.B., de Andrade, N.J., Ramos, A.M., Vanetti, M.C.D., Stringheta, P.C., Chaves, J.B.P., 2014. Decontamination by ultrasound application in fresh fruits and vegetables. Food Control. 45, 36–50.

Devlieghere, F., Vermeulen, A., Debevere, J., 2004. Chitosan: antimicrobial activity, interactions with food components and applicability as a coating on fruit and vegetables. Food Microbiol. 21, 703–714.

Deza, M., Araujo, M., Garrido, M., 2003. Inactivation of *Escherichia coli* O157: H7, *Salmonella enteritidis* and *Listeria monocytogenes* on the surface of tomatoes by neutral electrolyzed water. Lett. Appl. Microbiol. 37, 482–487.

Dicaprio, E., Phantkankum, N., Culbertson, D., Ma, Y., Hughes, J.H., Kingsley, D., et al., 2016. Inactivation of human norovirus and tulane virus in simple media and fresh whole strawberries by ionizing radiation. Int. J. Food Microbiol. 232, 43–51.

Ding, T., Ge, Z., Shi, J., Xu, Y.-T., Jones, C.L., Liu, D.-H., 2015. Impact of slightly acidic electrolyzed water (saew) and ultrasound on microbial loads and quality of fresh fruits. LWT-Food Sci. Technol. 60, 1195–1199.

Dixit, S.N., Chandra, H., Tiwari, R., Dixit, V., 1995. Development of a botanical fungicide against blue mould of mandarins. J. Stored Prod. Res. 31 (2), 165–172.

Du, J., Han, Y., Linton, R., 2003. Efficacy of chlorine dioxide gas in reducing *Escherichia coli* O157: H7 on apple surfaces. Food Microbiol. 20, 583–591.

Dunn, J., Bushnell, A., Ott, T., Clark, W., 1997. Pulsed white light food processing. Cereal Foods World 42, 510–515.

Elmnasser, N., Guillou, S., Leroi, F., Orange, N., Bakhrouf, A., Federighi, M., 2007. Pulsed-light system as a novel food decontamination technology: a review. Can. J. Microbiol. 53, 813–821.

Escalona, C.V., Hinojosa, A., Char, C.D., Villena, P., Bustamante, A., Sáenz, H., et al., 2015. Use of alternative sanitizers on minimally processed watercress harvested in two different seasons. J. Food Process. Preserv. 39 (6), 1287–1298.

European Food Safety Authority, EFSA, 2011. Shiga toxin-producing *E. coli* (STEC) o104:H4 2011 outbreaks in Europe: taking stock. EFSA J. 9 (10), 2390. Available from: https://doi.org/10.2903/j.efsa2011.2390 [22 pp.].

Fallik, E., 2004. Prestorage hot water treatments (immersion, rinsing and brushing). Postharvest Biol. Technol. 32, 125–134.

Fallik, E., Aharoni, Y., Yekutieli, O., Wiseblum, A., Regev, R., Beres, H., et al., 1996. A Method for Simultaneously Cleaning and Disinfecting Agricultural Produce. Israel Patent Application.

Fallik, E., Ilic, Z., Alkalai-Tuvia, S., Copel, A., Polevaya, Y., 2002. A short hot water rinsing and brushing reduces chilling injury and enhances resistance against *Botrytis cinerea* in fresh harvested tomato. Adv. Hortic. Sci. 3–6.

Farkas, J., 2006. Irradiation for better foods. Trends Food Sci. Technol. 17, 148–152.

Farkas, J., Mohácsi-Farkas, C., 2011. History and future of food irradiation. Trends Food Sci. Technol. 22, 121–126.

Fawcett, H., 1922. Packing house control of brown rot. Citrograph 7, 232–234.

Feliziani, E., Lichter, A., Smilanick, J.L., Ippolito, A., 2016. Disinfecting agents for controlling fruit and vegetable diseases after harvest. Postharvest Biol. Technol. 122, 53–69.

Fleischman, G., Bator, C., Merker, R., Keller, S., 2001. Hot water immersion to eliminate *Escherichia coli* O157: H7 on the surface of whole apples: thermal effects and efficacy. J. Food Prot. 64, 451–455.

Foley, D., Euper, M., Caporaso, F., Prakash, A., 2004. Irradiation and chlorination effectively reduces *Escherichia coli* O157: H7 inoculated on cilantro (*Coriandrum Sativum*) without negatively affecting quality. J. Food Prot. 67, 2092–2098.

Food and Drug Administration (FDA), 2004. Produce safety from production to consumption: 2004 action plan to minimize foodborne illness associated with fresh produce consumption. <http://www.fda.gov/Food/FoodSafety/ProductSpecificInformation/FruitsVegetablesJuices/FDAProduceSafetyActivities/ProduceSafetyActionPlan/ucm129487.htm>.

Forghani, F., 2019. Application of electrolyzed water in agriculture. Electrolyzed Water in Food: Fundamentals and Applications. Springer, Singapore, pp. 223–230.

Fukuyama, S., Watanabe, Y., Kondo, N., Nishinomiya, T., Kawamoto, S., Isshiki, K., et al., 2009. Efficiency of sodium hypochlorite and calcinated calcium in killing *Escherichia coli* O157: H7, *Salmonella* Spp., and *Staphylococcus aureus* attached to freshly shredded cabbage. Biosci. Biotechnol. Biochem. 73, 9–14.

Gamagae, S.U., Sivakumar, D., Wijeratnam, R.W., Wijesundera, R.L.C., 2003. Use of sodium bicarbonate and *Candida oleophila* to control anthracnose in papaya during storage. Crop Prot. 22 (5), 775–779.

Ganesh, V., Hettiarachchy, N.S., Ravichandran, M., Johnson, M.G., Griffis, C.L., Martin, E.M., et al., 2010. Electrostatic sprays of food-grade acids and plant extracts are more effective than conventional sprays in decontaminating *Salmonella typhimurium* on spinach. J. Food Sci. 75, M574–M579.

Garg, N., Churey, J.J., Splittstoesser, D.F., 1990. Effect of processing conditions on the microflora of fresh-cut vegetables. J. Food Prot. 53, 701–703.

Gayán, E., García-Gonzalo, D., Álvarez, I., Condón, S., 2014. Resistance of *Staphylococcus aureus* to UV-C light and combined UV-heat treatments at mild temperatures. Int. J. Food Microbiol. 172, 30–39.

Ge, C., Bohrerova, Z., Lee, J., 2013. Inactivation of internalized *Salmonella typhimurium* in lettuce and green onion using ultraviolet C irradiation and chemical sanitizers. J. Appl. Microbiol. 114, 1415–1424.

Gogate, P.R., Kabadi, A.M., 2009. A review of applications of cavitation in biochemical engineering/biotechnology. Biochem. Eng. J. 44, 60–72.

Gomes, C., Moreira, R.G., Castell-Perez, E., 2011a. Microencapsulated antimicrobial compounds as a means to enhance electron beam irradiation treatment for inactivation of pathogens on fresh spinach leaves. J. Food Sci. 76, E479–E488.

Gómez, P.L., Welti-Chanes, J., Alzamora, S.M., 2011b. Hurdle technology in fruit processing. Annu. Rev. Food Sci. Technol. 2, 447–465.

Gomez-Lopez, V.M., Ragaert, P., Debevere, J., Devlieghere, F., 2007a. Pulsed light for food decontamination: a review. Trends Food Sci. Technol. 18, 464–473.

Gómez-López, V.M., Ragaert, P., Ryckeboer, J., Jeyachchandran, V., Debevere, J., Devlieghere, F., 2007b. Shelf-life of minimally processed cabbage treated with neutral electrolysed oxidising water and stored under equilibrium modified atmosphere. Int. J. Food Microbiol. 117, 91–98.

González-Aguilar, G., Cruz, R., Baez, R., Wang, C., 1999. Storage quality of bell peppers pretreated with hot water and polyethylene packaging. J. Food Qual. 22, 287–299.

González-Aguilar, G., Gayosso, L., Cruz, R., Fortiz, J., Báez, R., Wang, C., 2000. Polyamines induced by hot water treatments reduce chilling injury and decay in pepper fruit. Postharvest Biol. Technol. 18, 19–26.

Goodburn, C., Wallace, C.A., 2013. The microbiological efficacy of decontamination methodologies for fresh produce: a review. Food Control. 32, 418–427.

Gopalakrishnan, G., Banumathi, B., Suresh, G., 1997. Evaluation of the antifungal activity of natural xanthones from *Garcinia mangostana* and their synthetic derivatives. J. Nat. Prod. 60 (5), 519–524.

Gordon, G., Rosenblatt, A.A., 2005. Chlorine dioxide: the current state of the art. Ozone: Sci. Eng. 27, 203–207.

Govender, V., Korsten, L., Sivakumar, D., 2005. Semi-commercial evaluation of *Bacillus licheniformis* to control mango postharvest diseases in South Africa. Postharvest Biol. Technol. 38 (1), 57–65.

Greenspan, F.P., Margulies, P.H., 1950. Treatment of Raw Plant Tissue. Google Patents.

Gulluce, M., Sahin, F., Sokmen, M., Ozer, H., Daferera, D., Sokmen, A., et al., 2007. Antimicrobial and antioxidant properties of the essential oils and methanol extract from *Mentha longifolia* L. ssp. *longifolia*. Food Chem. 103 (4), 1449–1456.

Gündüz, G.T., Gönül, Ş.A., Karapinar, M., 2010. Efficacy of sumac and oregano in the inactivation of *Salmonella typhimurium* on tomatoes. Int. J. Food Microbiol. 141 (1–2), 39–44.

Gupta, S.M., Gupta, A.K., Ahmed, Z., Kumar, A., 2011. Antibacterial and antifungal activity in leaf, seed extract and seed oil of seabuckthorn (*Hippophae salicifolia* D. Don) plant. J. Plant Pathol. Microbiol. 2 (2), 1–4.

Han, Y., Linton, R.H., Nielsen, S., Nelson, P., 2000. Inactivation of *Escherichia coli* O157: H7 on surface-uninjured and-injured green pepper (*Capsicum annuum* L.) by chlorine dioxide gas as demonstrated by confocal laser scanning microscopy. Food Microbiol. 17, 643–655.

Han, Y., Selby, T., Schultze, K., Nelson, P., Linton, R.H., 2004. Decontamination of strawberries using batch and continuous chlorine dioxide gas treatments. J. Food Prot. 67, 2450–2455.
Herdt, J., Feng, H., 2009. Aqueous antimicrobial treatments to improve fresh and fresh-cut produce safety. Microb. Saf. Fresh Produce 169.
Hilgren, J., Salverda, J., 2000. Antimicrobial efficacy of a peroxyacetic/octanoic acid mixture in fresh-cut-vegetable process waters. J. Food Sci. 65, 1376–1379.
Hirano, H., 1997. Characteristics of electrolyzed neutral water possibility of the practical use for food hygiene. Shokuhin Kogyo, Jpn. 40, 25–35.
Ho, K.-L.G., Luzuriaga, D.A., Rodde, K.M., Tang, S., Phan, C., 2011. Efficacy of a novel sanitizer composed of lactic acid and peroxyacetic acid against single strains of nonpathogenic *Escherichia coli* K-12, *Listeria innocua*, and *Lactobacillus plantarum* in aqueous solution and on surfaces of romaine lettuce and spinach. J. Food Prot. 74, 1468–1474.
Horn, R.C., Vargas, V.M.F., 2003. Antimutagenic activity of extracts of natural substances in the *Salmonella*/microsome assay. Mutagenesis 18 (2), 113–118.
Horníčková, J., Velíšek, J., Ovesna, J., Stavělíková, H., 2009. Distribution of S-alk (en) yl-L-cysteine sulfoxides in garlic (*Allium sativum* L.). *Czech.* J. Food Sci. 27, 232–235.
Hricova, D., Stephan, R., Zweifel, C., 2008. Electrolyzed water and its application in the food industry. J. Food Prot. 71, 1934–1947.
Huang, Y., Chen, H., 2015. Inactivation of *Escherichia coli* O157: H7, *Salmonella* and human norovirus surrogate on artificially contaminated strawberries and raspberries by water-assisted pulsed light treatment. Food Res. Int. 72, 1–7.
Huang, J., Wang, L., Ren, N., Ma, F., 1997. Disinfection effect of chlorine dioxide on bacteria in water. Water Res. 31, 607–613.
Hulsmans, A., Joris, K., Lambert, N., Rediers, H., Declerck, P., Delaedt, Y., et al., 2010. Evaluation of process parameters of ultrasonic treatment of bacterial suspensions in a pilot scale water disinfection system. Ultrason. Sonochem. 17, 1004–1009.
Iinuma, M., Tosa, H., Tanaka, T., Asai, F., Kobayashl, Y., Shimano, R., et al., 1996a. Antibacterial activity of xanthones from guttiferaeous plants against methicillin-resistant *Staphylococcus aureus*. J. Pharm. Pharmacol. 48 (8), 861–865.
Iinuma, M., Tosa, H., Tanaka, T., Kanamaru, S., Asai, F., Kobayashi, Y., et al., 1996b. Antibacterial activity of some Garcinia benzophenone derivatives against methicillin-resistant *Staphylococcus aureus*. Biol. Pharm. Bull. 19 (2), 311–314.
Ikigai, H., Nakae, T., Hara, Y., Shimamura, T., 1993. Bactericidal catechins damage the lipid bilayer. Biochim. Biophys. Acta 1147, 132–136.
Ilic, Z.P.Y., Sharon, A.K., Azica, C., Elazar, F., 2001. A short prestorage hot water rinse and brushing reduces decay development in tomato, while maintaining its quality. Trop. Agri. Res. Ext. 4 (1), 1–5.
Inatsu, Y., Weerakkody, K., Bari, M.L., Hosotani, Y., Nakamura, N., Kawasaki, S., 2017. The efficacy of combined (Naclo and organic acids) washing treatments in controlling *Escherichia coli* O157: H7, *Listeria monocytogenes* and spoilage bacteria on shredded cabbage and bean sprout. LWT-Food Sci. Technol. 85, 1–8.
Islam, M.M., Masum, S.M., Rahman, M.M., Molla, M.A.I., Shaikh, A., Roy, S., 2011. Preparation of chitosan from shrimp shell and investigation of its properties. Int. J. Basic Appl. Sci. 11, 116–130.
Issa-Zacharia, A., Kamitani, Y., Miwa, N., Muhimbula, H., Iwasaki, K., 2011. Application of slightly acidic electrolyzed water as a potential non-thermal food sanitizer for decontamination of fresh ready-to-eat vegetables and sprouts. Food Control. 22, 601–607.
Jayaprakasha, G.K., Jena, B.S., Negi, P.S., Sakariah, K.K., 2002. Evaluation of antioxidant activities and antimutagenicity of turmeric oil: a byproduct from curcumin production. Z. Naturforsch. C 57 (9–10), 828–835.
Jayaprakasha, G.K., Negi, P.S., Jena, B.S., Rao, L.J.M., 2007. Antioxidant and antimutagenic activities of *Cinnamomum zeylanicum* fruit extracts. J. Food Compos. Anal. 20 (3–4), 330–336.
Jayasooriya, S., Bhandari, B., Torley, P., D'arcy, B., 2004. Effect of high power ultrasound waves on properties of meat: a review. Int. J. Food Prop. 7, 301–319.
Jeong, S., Marks, B.P., Ryser, E.T., Moosekian, S.R., 2010. Inactivation of *Escherichia coli* O157: H7 on lettuce, using low-energy x-ray irradiation. J. Food Prot. 73, 547–551.
Jhalegar, M.J., Sharma, R.R., Singh, D., 2015. *In vitro* and *in vivo* activity of essential oils against major postharvest pathogens of Kinnow (*Citrus nobilis* × *C. deliciosa*) mandarin. J. Food Sci. Technol. 52 (4), 2229–2237.

Jiang, Y., Sokorai, K., Pyrgiotakis, G., Demokritou, P., Li, X., Mukhopadhyay, S., et al., 2017. Cold plasma-activated hydrogen peroxide aerosol inactivates *Escherichia coli* O157: H7, *Salmonella typhimurium*, and *Listeria innocua* and maintains quality of grape tomato, spinach and cantaloupe. Int. J. Food Microbiol. 249, 53–60.

Jovanović, G.D., Klaus, A.S., Nikšić, M.P., 2016. Antimicrobial activity of chitosan coatings and films against *Listeria monocytogenes* on black radish. Rev. Argent. Microbiol. 48, 128–136.

Keeratipibul, S., Phewpan, A., Lursinsap, C., 2011. Prediction of *Coliforms* and *Escherichia coli* on tomato fruits and lettuce leaves after sanitizing by using artificial neural networks. LWT-Food Sci. Technol. 44, 130–138.

Kentish, S., Ashokkumar, M., 2011. The physical and chemical effects of ultrasound. Ultrasound Technologies for Food and Bioprocessing. Springer, New York, pp. 1–12.

Khan, A., Rahman, M., Islam, M.S., 2008. Antibacterial, antifungal and cytotoxic activities of amblyone isolated from *Amorphophallus campanulatus*. Indian J. Pharmacol. 40 (1), 41.

Kim, H., Ryu, J.-H., Beuchat, L.R., 2006. Survival of *Enterobacter sakazakii* on fresh produce as affected by temperature, and effectiveness of sanitizers for its elimination. Int. J. Food Microbiol. 111, 134–143.

Kim, Y.S., Balaraju, K., Jeon, Y., 2016. Effects of rhizobacteria *Paenibacillus polymyxa* APEC136 and *Bacillus subtilis* APEC170 on biocontrol of postharvest pathogens of apple fruits. J. Zhejiang Univ.-Sci. B 17 (12), 931–940.

Kim, S.A., Park, S.H., Knueven, C., Basel, R., Ricke, S.C., 2018. A decontamination approach using a combination of bisulfate of soda and peracetic acid against *Listeria innocua* inoculated on whole apples. Food Control. 84, 106–110.

Kitis, M., 2004. Disinfection of wastewater with peracetic acid: a review. Environ. Int. 30, 47–55.

Kiura, H., Sano, K., Morimatsu, S., Nakano, T., Morita, C., Yamaguchi, M., et al., 2002. Bactericidal activity of electrolyzed acid water from solution containing sodium chloride at low concentration, in comparison with that at high concentration. J. Microbiol. Methods 49, 285–293.

Kramer, B., Wunderlich, J., Muranyi, P., 2017. Recent findings in pulsed light disinfection. J. Appl. Microbiol. 122, 830–856.

Krishnamurthy, K., Tewari, J.C., Irudayaraj, J., Demirci, A., 2010. Microscopic and spectroscopic evaluation of inactivation of *Staphylococcus aureus* by pulsed UV light and infrared heating. Food Bioprocess. Technol. 3, 93.

Kubec, R., Svobodova, M., Velíšek, J., 2000. Distribution of S-alk(en)ylcysteine sulfoxides in some *Allium* species. Identification of a new flavor precursor: S-ethylcysteine sulfoxide (ethiin). J. Agric. Food Chem. 48 (2), 428–433.

Kumar, T.S., Negi, P.S., Sankar, K.U., 2010. Antibacterial activity of "*Nigella sativa* L." seed extracts. Br. J. Pharmacol. Toxicol. 1 (2), 96–100.

Kume, T., Furuta, M., Todoriki, S., Uenoyama, N., Kobayashi, Y., 2009. Status of food irradiation in the world. Radiat. Phys. Chem. 78, 222–226.

Lado, B.H., Yousef, A.E., 2002. Alternative food-preservation technologies: efficacy and mechanisms. Microbes Infect. 4, 433–440.

Lassois, L., de Bellaire, L.D.L., Jijakli, M.H., 2008. Biological control of crown rot of bananas with *Pichia anomala* strain K and *Candida oleophila* strain O. Biol. Control. 45 (3), 410–418.

Le Caër, S., 2011. Water radiolysis: influence of oxide surfaces on H_2 production under ionizing radiation. Water 3, 235–253.

Li, B., Liu, B., Su, T., Fang, Y., Xie, G., Wang, G., et al., 2010. Effect of chitosan solution on the inhibition of *Pseudomonas fluorescens* causing bacterial head rot of broccoli. Plant Pathol. J. 26, 189–193.

Lichter, A., Dvir, O., Rot, I., Akerman, M., Regev, R., Wiesblum, A., et al., 2000. Hot water brushing: an alternative method to SO_2 fumigation for color retention of litchi fruits. Postharvest Biol. Technol. 18, 235–244.

Liu, W.T., Chu, C.L., Zhou, T., 2002. Thymol and acetic acid vapors reduce postharvest brown rot of apricots and plums. HortScience 37 (1), 151–156.

Lurie, S., 1998. Postharvest heat treatments of horticultural crops. Hort. Rev. 22, 121.

Lynch, M.F., Tauxe, R.V., Hedberg, C.W., 2009. The growing burden of foodborne outbreaks due to contaminated fresh produce: risks and opportunities. Epidemiol. Infect. 137, 307–315.

Machado, T.B., Pinto, A.V., Pinto, M.C.F.R., Leal, I.C.R., Silva, M.G., Amaral, A.C.F., et al., 2003. *In vitro* activity of Brazilian medicinal plants, naturally occurring naphthoquinones and their analogues, against methicillin-resistant *Staphylococcus aureus*. Int. J. Antimicrob. Agents 21 (3), 279–284.

Mackeen, M.M., Ali, A.M., Lajis, N.H., Kawazu, K., Hassan, Z., Amran, M., et al., 2000. Antimicrobial, antioxidant, antitumour-promoting and cytotoxic activities of different plant part extracts of *Garcinia atroviridis* Griff. ex T. Anders. J. Ethnopharmacol. 72 (3), 395–402.

Mahmoud, B.S., 2010. The effects of x-ray radiation on *Escherichia coli* O157: H7, *Listeria monocytogenes*, *Salmonella enterica* and *Shigella flexneri* inoculated on whole roma tomatoes. Food Microbiol. 27, 1057–1063.

Mahmoud, B., Linton, R., 2008. Inactivation kinetics of inoculated *Escherichia coli* O157: H7 and *Salmonella enterica* on lettuce by chlorine dioxide gas. Food Microbiol. 25, 244–252.

Majeed, M., Badmaev, V., Shivakumar, U., Rajendran, R., 1995. *Curcuminoids*. Antioxidant Phytonutrients. Nutriscience, Piscataway, NJ.

Manzocco, L., Da Pieve, S., Maifreni, M., 2011. Impact of UV-C light on safety and quality of fresh-cut melon. Innov. Food Sci. Emerg. Technol. 12, 13–17.

Mari, M., Guizzardi, M., Pratella, G.C., 1996. Biological control of graymold in pears by antagonistic bacteria. Biol. Control. 7 (1), 30–37.

Mari, M., Neri, F., Bertolini, P., 2007. Novel approaches to prevent and control postharvest diseases of fruits. Stewart Postharvest Rev. 3 (6), 1–7.

Marquenie, D., Michiels, C.W., Van Impe, J.F., Schrevens, E., Nicolaï, B.N., 2003. Pulsed white light in combination with UV-C and heat to reduce storage rot of strawberry. Postharvest Biol. Technol. 28, 455–461.

Matsufuji, H., Furukawa, S., Teranishi, K., Kawaharada, K., Chino, M., Yamagata, K., et al., 2009. Effects of non-thermal processes on the inactivation of microorganisms and antioxidants in minimally processed vegetables. Food Sci. Technol. Res. 15, 153–162.

Mcdonnell, G., Russell, A.D., 2001. Antiseptics and disinfectants: activity, action, and resistance. Clin. Microbiol. Rev. 14, 227.

Miguel, N., Ormad, M.P., Mosteo, R., Ovelleiro, J.L., 2012. Photocatalytic degradation of pesticides in natural water: effect of hydrogen peroxide. Int. J. Photoenergy 2012.

Naganawa, R., Iwata, N., Ishikawa, K., Fukuda, H., Fujino, T., Suzuki, A., 1996. Inhibition of microbial growth by ajoene, a sulfur-containing compound derived from garlic. Appl. Environ. Microbiol. 62 (11), 4238–4242.

Negi, P., Jayaprakasha, G., 2001. Antibacterial activity of grapefruit (*Citrus paradisi*) peel extracts. Eur. Food Res. Technol. 213 (6), 484–487.

Negi, P.S., Jayaprakasha, G.K., Jagan Mohan Rao, L., Sakariah, K.K., 1999. Antibacterial activity of turmeric oil: a byproduct from curcumin manufacture. J. Agric. Food Chem. 47 (10), 4297–4300.

Negi, P.S., Anandharamakrishnan, C., Jayaprakasha, G.K., 2003. Antibacterial activity of *Aristolochia bracteata* root extracts. J. Med. Food 6 (4), 401–403.

Negi, P.S., Jayaprakasha, G.K., Jena, B.S., 2010. Evaluation of antioxidant and antimutagenic activities of the extracts from the fruit rinds of *Garcinia cowa*. Int. J. Food Prop. 13 (6), 1256–1265.

Ölmez, H., Kretzschmar, U., 2009. Potential alternative disinfection methods for organic fresh-cut industry for minimizing water consumption and environmental impact. LWT-Food Sci. Technol. 42, 686–693.

Palekar, M.P., Taylor, T.M., Maxim, J.E., Castillo, A., 2015. Reduction of *Salmonella enterica* serotype poona and background microbiota on fresh-cut cantaloupe by electron beam irradiation. Int. J. Food Microbiol. 202, 66–72.

Pantastico, E.B., 1975. Postharvest Physiology, Handling and Utilization of Tropical and Subtropical Fruits and Vegetables. Avi Publishing Company, p. 634.

Parafati, L., Vitale, A., Restuccia, C., Cirvilleri, G., 2015. Biocontrol ability and action mechanism of food-isolated yeast strains against *Botrytis cinerea* causing post-harvest bunch rot of table grape. Food Microbiol. 47, 85–92.

Park, C.M., Hung, Y.C., Doyle, M., Ezeike, G., Kim, C., 2001. Pathogen reduction and quality of lettuce treated with electrolyzed oxidizing and acidified chlorinated water. J. Food Sci. 66, 1368–1372.

Paull, R.E., Chen, N.J., 2000. Heat treatment and fruit ripening. Postharvest Biol. Technol. 21, 21–37.

Pinheiro, J., Alegria, C., Abreu, M., Gonçalves, E.M., Silva, C.L., 2015. Influence of postharvest ultrasounds treatments on tomato (*Solanum lycopersicum* cv. Zinac) quality and microbial load during storage. Ultrason. Sonochem. 27, 552–559.

Popa, I., Hanson, E.J., Todd, E.C., Schilder, A.C., Ryser, E.T., 2007. Efficacy of chlorine dioxide gas sachets for enhancing the microbiological quality and safety of blueberries. J. Food Prot. 70, 2084–2088.

Pusey, P.L., Wilson, C., 1984. Postharvest biological control of stone fruit brown rot by *Bacillus subtilis*. Plant Dis. 68 (9), 753–756.

Qu, H., Zhao, L., Zhao, F., Liu, Y., Yang, Z., 2016. Biocontrol of graymold decay in pear by *Bacillus amyloliquefaciens* strain BA3 and its effect on postharvest quality parameters. Pol. J. Microbiol. 65 (2), 171–176.

Ramos, B., Miller, F., Brandão, T.R., Teixeira, P., Silva, C.L., 2013. Fresh fruits and vegetables—an overview on applied methodologies to improve its quality and safety. Innov. Food Sci. Emerg. Technol. 20, 1–15.

Ramos-Villarroel, A.Y., Aron-Maftei, N., Martín-Belloso, O., Soliva-Fortuny, R., 2012. The role of pulsed light spectral distribution in the inactivation of *Escherichia coli* and *Listeria innocua* on fresh-cut mushrooms. Food Control. 24, 206–213.

Ramos-Villarroel, A., Martín-Belloso, O., Soliva-Fortuny, R., 2015. Combined effects of malic acid dip and pulsed light treatments on the inactivation of *Listeria innocua* and *Escherichia coli* on fresh-cut produce. Food Control. 52, 112–118.

Rastogi, N.K., 2011. Opportunities and challenges in application of ultrasound in food processing. Crit. Rev. Food Sci. Nutr. 51, 705–722.

Rawson, A., Patras, A., Tiwari, B., Noci, F., Koutchma, T., Brunton, N., 2011. Effect of thermal and non thermal processing technologies on the bioactive content of exotic fruits and their products: review of recent advances. Food Res. Int. 44, 1875–1887.

Rico, D., Martín-Diana, A.B., Barry-Ryan, C., Frías, J.M., Henehan, G.T., Barat, J.M., 2008. Use of neutral electrolysed water (EW) for quality maintenance and shelf-life extension of minimally processed lettuce. Innov. Food Sci. Emerg. Technol. 9, 37–48.

Ritenour, M.A., Crisosto, C.H., 1996. Hydrocooler water sanitation in the San Joaquin Valley stone fruit industry. Cent. Val. Postharvest Newsl. 5, 15–17.

Roberts, P., Hope, A., 2003. Virus inactivation by high intensity broad spectrum pulsed light. J. Virol. Methods 110 (1), 61–65.

Rodgers, S.L., Cash, J.N., Siddiq, M., Ryser, E.T., 2004. A Comparison of different chemical sanitizers for inactivating *Escherichia coli* O157: H7 and *Listeria monocytogenes* in solution and on apples, lettuce, strawberries, and cantaloupe. J. Food Prot. 67, 721–731.

Rojas, J.J., Ochoa, V.J., Ocampo, S.A., Muñoz, J.F., 2006. Screening for antimicrobial activity of ten medicinal plants used in Colombian folkloric medicine: a possible alternative in the treatment of non-nosocomial infections. BMC Complement. Altern. Med. 6 (1), 1–6.

Romanazzi, G., Sanzani, S.M., Bi, Y., Tian, S., Martínez, P.G., Alkan, N., 2016. Induced resistance to control postharvest decay of fruit and vegetables. Postharvest Biol. Technol. 122, 82–94.

Rowan, N.J., Valdramidis, V.P., Gómez-López, V.M., 2015. A review of quantitative methods to describe efficacy of pulsed light generated inactivation data that embraces the occurrence of viable but non culturable state microorganisms. Trends Food Sci. Technol. 44, 79–92.

Rozin, P., 2005. The meaning of "natural" process more important than content. Psychol. Sci. 16 (8), 652–658.

Ruiz-Cruz, S., Acedo-Félix, E., Díaz-Cinco, M., Islas-Osuna, M.A., González-Aguilar, G.A., 2007. Efficacy of sanitizers in reducing *Escherichia coli* O157: H7, *Salmonella* Spp. and *Listeria monocytogenes* populations on fresh-cut carrots. Food Control. 18, 1383–1390.

Saligkarias, I.D., Gravanis, F.T., Epton, H.A., 2002. Biological control of *Botrytis cinerea* on tomato plants by the use of epiphytic yeasts *Candida guilliermondii* strains 101 and US 7 and *Candida oleophila* strain I-182: II. A study on mode of action. Biol. Control. 25 (2), 151–161.

Sapers, G., Sites, J., 2003. Efficacy of 1% Hydrogen Peroxide wash in decontaminating apples and cantaloupe melons. J. Food Sci. 68, 1793–1797.

Satin, M. 1996. Food Irradiation: A Guidebook. 2nd Ed. CRC Press, Boca Raton, Fla.

Schena, L., Nigro, F., Pentimone, I., Ligorio, A., Ippolito, A., 2003. Control of postharvest rots of sweet cherries and table grapes with endophytic isolates of *Aureobasidium pullulans*. Postharvest Biol. Technol. 30 (3), 209–220.

Schirra, M., Agabbio, M., D'hallewin, G., Pala, M., Ruggiu, R., 1997. Response of tarocco oranges to picking date, postharvest hot water dips, and chilling storage temperature. J. Agric. Food Chem. 45, 3216–3220.

Seethapathy, P., Gurudevan, T., Subramanian, K.S., Kuppusamy, P., 2016. Bacterial antagonists and hexanal-induced systemic resistance of mango fruits against *Lasiodiplodia theobromae* causing stem-end rot. J. Plant Interact. 11 (1), 158–166.

Selvi, A.T., Joseph, G.S., Jayaprakasha, G.K., 2003. Inhibition of growth and aflatoxin production in *Aspergillus flavus* by *Garcinia indica* extract and its antioxidant activity. Food Microbiol. 20 (4), 455–460.

Seth, D., Badwaik, L.S., Ganapathy, V., 2015. Effect of feed composition, moisture content and extrusion temperature on extrudate characteristics of yam-corn-rice based snack food. J. Food Sci. Technol. 52, 1830–1838.

Seymour, I., Burfoot, D., Smith, R., Cox, L., Lockwood, A., 2002. Ultrasound decontamination of minimally processed fruits and vegetables. Int. J. Food Sci. Technol. 37, 547–557.

Shan, B., Cai, Y.Z., Brooks, J.D., Corke, H., 2009. Antibacterial and antioxidant effects of five spice and herb extracts as natural preservatives of raw pork. J. Sci. Food Agric. 89 (11), 1879–1885.

Sharma, R., Demirci, A., 2003. Inactivation of *Escherichia coli* O157: H7 on inoculated alfalfa seeds with pulsed ultraviolet light and response surface modeling. J. Food Sci. 68, 1448–1453.

Shipulina, L.D., February 2001. A study of antiviral activity and other biological properties of hiporamin. A new antiviral drug from Seabuckthorn leaves. In: Proceeding of International Workshop on Seabuckthorn, Seabuckthorn—A Resource for Health and Environment in Twenty First Century, New Delhi, pp. 212–213.

Singh, V., Deverall, B.J., 1984. *Bacillus subtilis* as a control agent against fungal pathogens of citrus fruit. Trans. Br. Mycol. Soc. 83 (3), 487–490.

Smid, E.J., De Witte, Y., Vrees, O.D., Gorris, L.G.M., August 1993. Use of secondary plant metabolites for the control of postharvest fungal diseases on flower bulbs. In: International Symposium on Postharvest Treatment of Horticultural Crops 368, pp. 523–530.

Smilanick, J.L., Sorenson, D., Mansour, M., Aieyabei, J., Plaza, P., 2003. Impact of a brief postharvest hot water drench treatment on decay, fruit appearance, and microbe populations of California lemons and oranges. Horttechnology 13, 333–338.

Smith, D.W.C., Bhuiya, F., 2009. Outbreaks by the Numbers: Fruits and Vegetables 1990e2005. The Centre for Science in the Public Interest.

Sørensen, H.P., Madsen, L.S., Petersen, J., Andersen, J.T., Hansen, A.M., Beck, H.C., 2010. Oat (*Avena sativa*) seed extract as an antifungal food preservative through the catalytic activity of a highly abundant class I chitinase. Appl. Biochem. Biotechnol. 160 (6), 1573–1584.

Srey, S., Park, S.Y., Jahid, I.K., Ha, S.-D., 2014. Reduction effect of the selected chemical and physical treatments to reduce *L. monocytogenes* biofilms formed on lettuce and cabbage. Food Res. Int. 62, 484–491.

Srimal, R.C., 1997. Turmeric: a brief review of medicinal properties. Fitoterapia (Milano) 68 (6), 483–493.

Swami, K.D., Bisht, N.P.S., 1996. Constituents of *Ficus religiosa* and *Ficus infectoria* and their biological activity. J. Indian. Chem. Soc. 73, 11.

Sy, K.V., Mcwatters, K.H., Beuchat, L.R., 2005. Efficacy of gaseous chlorine dioxide as a sanitizer for killing salmonella, yeasts, and molds on blueberries, strawberries, and raspberries. J. Food Prot. 68, 1165–1175.

Tang, X., Cronin, D.A., 2007. The effects of brined onion extracts on lipid oxidation and sensory quality in refrigerated cooked turkey breast rolls during storage. Food Chem. 100 (2), 712–718.

Tian, S.P., Fan, Q., Xu, Y., Qin, G.Z., Liu, H.B., 2002. Effect of biocontrol antagonists applied in combination with calcium on the control of postharvest diseases in different fruit. Bull. OILB/SROP 25 (10), 193–196.

Tiwari, R., Mishra, D.N., Upadhyay, P.S., 1988. Efficacy of some plant volatiles for the control of black mould of onion caused by *Aspergillus niger* Van Tiegh during storage. Natl. Acad. Sci. Lett. 11, 345–347.

Tomás-Callejas, A., López-Gálvez, F., Sbodio, A., Artés, F., Artés-Hernández, F., Suslow, T.V., 2012. Chlorine dioxide and chlorine effectiveness to prevent *Escherichia coli* O157: H7 and *Salmonella* cross-contamination on fresh-cut Red Chard. Food Control. 23 (2), 325–332.

Tornuk, F., Cankurt, H., Ozturk, I., Sagdic, O., Bayram, O., Yetim, H., 2011. Efficacy of various plant hydrosols as natural food sanitizers in reducing *Escherichia coli* O157: H7 and *Salmonella typhimurium* on fresh cut carrots and apples. Int. J. Food Microbiol. 148 (1), 30–35.

Trinetta, V., Morgan, M., Linton, R., 2010. Use of high-concentration-short-time chlorine dioxide gas treatments for the inactivation of *Salmonella enterica* Spp. inoculated onto roma tomatoes. Food Microbiol. 27, 1009–1015.

Trinetta, V., Vaidya, N., Linton, R., Morgan, M., 2011. A comparative study on the effectiveness of chlorine dioxide gas, ozone gas and e-beam irradiation treatments for inactivation of pathogens inoculated onto tomato, cantaloupe and lettuce seeds. Int. J. Food Microbiol. 146, 203–206.

Tripathi, P., 2001. Evaluation of Some Plant Products Against Fungi Causing Post Harvest Diseases of Some Fruits (Doctoral dissertation, Ph. D. thesis). Department of Botany, Banaras Hindu University.

Ukuku, D.O., 2004. Effect of hydrogen peroxide treatment on microbial quality and appearance of whole and fresh-cut melons contaminated with *Salmonella* Spp. Int. J. Food Microbiol. 95, 137–146.

Ukuku, D.O., Bari, M., Kawamoto, S., Isshiki, K., 2005. Use of hydrogen peroxide in combination with nisin, sodium lactate and citric acid for reducing transfer of bacterial pathogens from whole melon surfaces to fresh-cut pieces. Int. J. Food Microbiol. 104, 225–233.

Urbina, C.T., Prieto, V.G., Lopez, C.G., Albores, F.V., Reyes, D.B., Muniz, C.A., et al., 2016. Purification and characterization of b-1, 3-glucanase from *Candida oleophila* for the biocontrol of *Penicillium expansum*. Res. Rev.: J. Bot. Sci. 5 (1), 38–45.

Uyttendaele, M., Neyts, K., Vanderswalmen, H., Notebaert, E., Debevere, J., 2004. Control of *Aeromonas* on minimally processed vegetables by decontamination with lactic acid, chlorinated water, or thyme essential oil solution. Int. J. Food Microbiol. 90 (3), 263–271.

Van De Velde, F., Grace, M.H., Pirovani, M.É., Lila, M.A., 2016. Impact of a new postharvest disinfection method based on peracetic acid fogging on the phenolic profile of strawberries. Postharvest Biol. Technol. 117, 197–205.

Van Lenteren, J.C., Bolckmans, K., Köhl, J., Ravensberg, W.J., Urbaneja, A., 2018. Biological control using invertebrates and microorganisms: plenty of new opportunities. BioControl 63 (1), 39–59.

Vandekinderen, I., Van Camp, J., De Meulenaer, B., Veramme, K., Denon, Q., Ragaert, P., et al., 2007. The effect of the decontamination process on the microbial and nutritional quality of fresh-cut vegetables. In: International Conference on Quality Management of Fresh Cut Produce 746, pp. 173–180.

Vandekinderen, I., Devlieghere, F., Van Camp, J., Denon, Q., Alarcon, S.S., Ragaert, P., et al., 2009. Impact of a decontamination step with peroxyacetic acid on the shelf-life, sensory quality and nutrient content of grated carrots packed under equilibrium modified atmosphere and stored at 7°C. Postharvest Biol. Technol. 54, 141–152.

Varela, G.B., Partida, L.D.C., Jiménez, V.A.O., López, M.A.C., Martínez, P.G., 2015. Efecto Del Quitosano En La Inducción De Resistencia Contra *Colletotrichum* Sp. En Mango (*Mangifera indica* L.) Cv. Tommy Atkins. Invest. Cienc. 23, 16–21.

Wang, L., Ganjyal, G., Weller, C., Jones, D., Hanna, M., 2003. Transport characteristics of single-screw extrusion cooking process. In: The Institute of Food Technologists Annual Meeting, pp. 12–16.

Wang, W., Ma, X., Zou, M., Jiang, P., Hu, W., Li, J., et al., 2015. Effects of ultrasound on spoilage microorganisms, quality, and antioxidant capacity of postharvest cherry tomatoes. J. Food Sci. 80, C2117–C2126.

Wilson, C.L., Chalutz, E., 1989. Postharvest biological control of *Penicillium* rots of citrus with antagonistic yeasts and bacteria. Sci. Hortic. 40 (2), 105–112.

Wisniewsky, M.A., Glatz, B.A., Gleason, M.L., Reitmeier, C.A., 2000. Reduction of *Escherichia coli* O157: H7 counts on whole fresh apples by treatment with sanitizers. J. Food Prot. 63, 703–708.

Woolf, A.B., 1997. Reduction of chilling injury in Storedhass' avocado fruit by 38°C water treatments. Hortscience 32, 1247–1251.

World Health Organization, Food Safety Unit, 1998. Food Safety Issues: Surface Decontamination of Fruits and Vegetables Eaten Raw: A Review. World Health Organization. WHO/FSF/FOS/98.2, Geneva.

World Health Organization, 2008. Microbiological Risk Assessment Series in Microbiological Hazards in Fresh Fruits and Vegetables. <www.fao.org/ag/AGN/agns/files/FFV_2007>.

World Health Organization, 2018. Nipah Virus-India. <https://www.who.int/csr/don/07-august-2018-nipah-virus-india/en/>.

Wu, V.C., Kim, B., 2007. Effect of a simple chlorine dioxide method for controlling five foodborne pathogens, yeasts and molds on blueberries. Food Microbiol. 24, 794–800.

Xoca-Orozco, L.-Á., Cuellar-Torres, E.A., González-Morales, S., Gutiérrez-Martínez, P., López-García, U., Herrera-Estrella, L., et al., 2017. Transcriptomic analysis of avocado hass (*Persea americana* Mill) in the interaction system fruit-chitosan-colletotrichum. Front. Plant Sci. 8, 956.

Yang, H., Swem, B., Li, Y., 2003. The effect of pH on inactivation of pathogenic bacteria on fresh-cut lettuce by dipping treatment with electrolyzed water. J. Food Sci. 68, 1013–1017.

Yaun, B.R., Sumner, S.S., Eifert, J.D., Marcy, J.E., 2004. Inhibition of pathogens on fresh produce by ultraviolet energy. Int. J. Food Microbiol. 90, 1–8.

Yin, M.C., Cheng, W.S., 2003. Antioxidant and antimicrobial effects of four garlic-derived organosulfur compounds in ground beef. Meat Sci. 63 (1), 23–28.

Yoon, J.-H., Bae, Y.-M., Jung, K.-S., Heu, S., Lee, S.-Y., 2013. Combined effect of calcium oxide and sonication to reduce foodborne pathogens on fresh produce. Food Sci. Biotechnol. 22, 275–278.

Yuk, H.G., Yoo, M.Y., Yoon, J.W., Moon, K.D., Marshall, D.L., Oh, D.H., 2006. Effect of combined ozone and organic acid treatment for control of *Escherichia coli* O157: H7 and *Listeria monocytogenes* on lettuce. J. Food Sci. 71, M83–M87.

Yuting, X., Lifen, Z., Jianjun, Z., Jie, S., Xingqian, Y., Donghong, L., 2013. Power ultrasound for the preservation of postharvest fruits and vegetables. Int. J. Agric. Biol. Eng. 6, 116–125.

CHAPTER

17

Sustainable agricultural practices using microbial strains for crop production

Sandeep Kumar Singh[1], *Hariom Verma*[1], *Vipin Kumar Singh*[1], *Sougata Ghosh*[2], *Deen Dayal Giri*[3], *Manisha Mishra*[4] and *Ajay Kumar*[5]

[1]Department of Botany, Centre of Advanced Study, Institute of Science, Banaras Hindu University, Varanasi, India [2]Department of Microbiology, School of Science, RK University, Rajkot, India [3]Department of Botany, Maharaj Singh College, Saharanpur, India [4]University Department of Botany, T.M. Bhagalpur University, Bhagalpur, India [5]Department of Postharvest Science, Agricultural Research Organization (ARO), Volcani Center, Rishon LeZion, Israel

17.1 Introduction

The food security is expected in the near future due to rapid increasing global human population, especially exponentially growth in developing countries. As per estimate of United Nations by 2050 the human population will reach to the number of 8.9 billion (Wood, 2001). To feed the increased population in the limited resources of land, water in the changing climatic scenario will be a severe challenge for the farmers, researchers, and government policymakers. However, currently a range of physical–chemical methods had been employed to enhance the productivity of crops via using chemical fertilizers to maintain the nutrient status in the soil or controlling the attack of pest or pathogens during growth, harvesting, or postharvest storage of the crops, fruits. In conventional agriculture system farmers use huge amount of chemical fertilizers to enhance growth promotion, yield enhancement of crops and also use chemical pesticides abundantly to control the growth of phytopathogens at the different stages of crop growth. However, continuous use of chemical fertilizers and pesticides has adverse consequences on texture and productivity of soil

DOI: https://doi.org/10.1016/B978-0-12-821843-3.00002-7

(environment in terms of soil and water pollution), as well as health concern of human beings (Galloway et al., 2008; Youssef and Eissa, 2014). Besides this, excessive use of chemical pesticides also affect the natural microflora or indigenous microbial community of soil, which is responsible for maintaining the natural physiochemical properties as well as productivity of the soil (McLaughlin and Mineau, 1995; Dash et al., 2018; Singh et al., 2020).

Therefore in the context of conventional agriculture practices impact, nowadays beneficial microbes including bacteria, fungi, cyanobacteria, actinomycetes have been directly or indirectly used in the different forms at the each steps of agriculture for growth promotion (Kumar et al., 2017b) as biocontrol agents (Kumar et al., 2015a; Singh et al., 2019; Zhimo et al., 2020); and various biotic and abiotic stress management (Kumar et al., 2018; Modi et al., 2020; Gupta et al., 2020). In the last few years, microbial consortia or the beneficial microbes have been also very broadly utilized as biopesticides with different names, brand, and different microbial composition including bacteria–bacteria, bacteria–fungi, bacteria–yeast (Zhimo et al., 2020) and broadly utilized as antagonistic organism against various crop plants and fruit pathogens during development, harvesting and postharvest storage (Droby et al., 2009; Singh et al., 2017a).

Sustainable agriculture is a broad concept insisting maintenance of environment health by conserving the natural resources for the upcoming generation without compromising the needs of present. However, to regulate the agriculture process in sustainable ways, different factors including physicochemical and biological properties of soil, surrounding environment, human health must be managed without affecting negatively. The physicochemical properties of the soil depend upon the natural indigenous microbial communities, which maintain the productivity of the soil. Although increased biological activity and nutrient cycling in the soil maintains texture and productivity of the soil for the long term (Kumar et al., 2017b).

17.2 Role of rhizosphere in plant–microbe interaction

Rhizosphere of the plant is a region around the root surface which is directly adhered with the soil and considered as the hot spot for various microbial communities (Antoun and Prévost, 2005). The plant root secretes a significant amount of the exudates, which are composed of diverse array of biochemical compounds including amino acids, proteins, lipids, polysaccharides flavonoids, nucleotides, which effectively act as nutrient source as well as signalling molecules to attract the microbial communities (Oku et al., 2012; Kumar et al., 2015b). It have been proven in many studies that the population of microbes is higher in rhizospheric soil compared to the bulk soil (Bahadur et al., 2017). The roots exudates effectively help in the colonization of microbes and affect the growth of plants via growth promotion and diseases management. Although some of the microbial communities enter to the plant tissue and resides as endophytic microbes without causing any harm to the host (Gupta et al., 2019; Kumar et al., 2020).

However, both the rhizospheric and endophytic microbes follow similar mechanism of action for growth promotion as well as disease management. According to an estimate about 2%–5% of studied microbial population have properties of plant growth promotion (Jha et al., 2012; Sgroy et al., 2009). The plant growth–promoting bacterial strains generally follow direct or indirect mechanism for growth-promoting attributes, in which phytohormone synthesis, phosphate solubilization, nutrient availability, nitrogen fixation, ammonia production, etc. are the common growth-promoting attributes, whereas antibiotic production, volatiles production, siderophore production, quorum sensing, ACC deaminase activity are the some common mechanism used by microbial strain to inhibit the growth of phytopathogens or act as biocontrol agents during the various stages of crop/fruit production (Singh et al., 2017a; Morrissey et al., 2004; Zakry et al., 2012). Besides these attributes nowadays, beneficial microbes extensively utilized in the mitigation of abiotic stress including salinity, draught, heavy metal stress as well as biodegradation of complex xenobiotics or environmental contaminants (Kumar et al., 2018) (Table 17.1).

17.3 Plant growth–promoting bacteria

Plant growth–promoting bacteria (PGPB) are the group of bacteria which directly or indirectly associated with the plants for their growth promotion, yield enhancement, and disease management (Kumar et al., 2016b) (Fig. 17.1). However, the word plant growth–promoting rhizobacteria (PGPR) was first introduced by Kloepper et al. (1989) for the bacterial communities which are inhabitants of the rhizospheric region of the plants and help in growth promotion of the plants. Initially PGPR term used for the *Pseudomonas,* which was able to control the phytopathogens and enhanced growth of plants. Later, it was generalized for all rhizospheric bacteria that performed function similar to *Pseudomonas* (Kapulnik et al., 1981, Haghighi et al., 2011).

17.4 Plant growth–promoting bacteria and their attributes

The action mode of a PGPB strain depends on the type of plant (García-Fraile et al., 2015) or other variable environmental conditions (Vacheron et al., 2013). Although growth-promoting attributes of beneficial microbes include various direct or indirect mechanism, which maintain the nutrient availability of the soil, signal molecules, phytohormone synthesis (Fig. 17.2). Some of the common attributes discussed in the following subheadings.

17.4.1 Nitrogen fixation

Nitrogen is the one of the most important constituents of chemical fertilizers applied in the soil to maintain the nutrient status in the soil for their optimum productivity. Although nitrogen is the most abundantly present in the atmosphere and also in the huge amount in the soil. But in the soil, nitrogen presents in a complex or unutilizable, which is not utilized in the raw form. The symbiotic nitrogen fixing bacteria such as *Rhizobium, Bradyrhizobium, Sinorhizobium,* and *Mesorhizobium* convert complex atmospheric N_2 into utilizable ammonia or nitrate form (Zahran, 2001). Although numerous free living bacteria also fix nitrogen non-symbiotically in nonleguminous plants, in which some of the most common nonsymbiotic bacteria are *Azotobacter, Acetobacter, Azospirillum, Enterobacter, Burkholderia, Diazotrophicus,*

TABLE 17.1 Plant growth–promoting bacteria (PGPB) as biofertilizer and their plant growth promotion (PGP) activity.

Sr. No.	PGPB	Region	PGP activity	Plant	References
1.	*Serratia* sp. Rh269	Rhizosphere	IAA, siderophore, phosphate solubilization	*Oryza sativa* L.	Yasmin et al. (2016)
2.	*Bacillus* sp. Rh219	Rhizosphere	Siderophore	*O. sativa* L.	Yasmin et al. (2016)
3.	*Pseudomonas* sp. E227	Rhizosphere	IAA, siderophore, HCN, phosphate solubilization	*O. sativa* L.	Yasmin et al. (2016)
4.	*Rhizobium* sp., *Azospirillum* sp.	Endophyte	IAA, N_2 fixation	*O. sativa* L.	Sev et al. (2016)
5.	*Pseudomonas fluorescens*	Rhizosphere	ACC deaminase activity	*Triticum aestivum* L.	Shaharoona et al. (2008)
6.	*Bradyrhizobium japonicum*	Rhizosphere	ACC deaminase activity	*Zea mays* L.	Shaharoona et al. (2006)
7.	*Pseudomonas thivervalensis* STF3 and *Serratia marcescens* STJ5	Rhizosphere	ACC deaminase	*Z. mays* L.	Shahzad et al. (2013)
8.	*Proteus penneri* Pp1, *Pseudomonas aeruginosa* Pa2, and *Alcaligenes faecalis* AF3	Drought stress	Phosphate solubilizing	*Z. mays* L.	Naseem and Bano (2014)
9.	*Azospirillum lipoferum* CRT1	Rhizosphere	Modifications of root development, resistance to mechanical stress	*Z. mays* L.	El Zemrany et al. (2007)
10.	*Pseudomonas monteilii* FMZR2 *Sinorhizobium meliloti* MRC31	Endophyte	Ammonia, Phosphate solubilization, HCN	*Z. mays* L.	Sandhya et al. (2017)
11.	*P. aeruginosa* FTR, *Enterobacter asburiae* MRC12 *Acitenobacter brumalii* MZ30V92	Endophyte	Ammonia, Phosphate solubilization Siderophore, HCN	*Z. mays* L.	Sandhya et al. (2017)
12.	*Bacillus licheniformis* RC04, *Paenibacillus polymyxa* RC05, *Pseudomonas putida* RC06, and *Bacillus* OSU-142	Rhizosphere	Phosphate solubilization, increased root, and shoot weight	*H. vulgare* L.	Canbolat et al. (2006)
13.	*B. licheniformis* RC02, *Rhodobacter capsulatus* RC04, *P. polymyxa* RC05, *P. putida* RC06, and *Bacillus* OSU-142	Rhizosphere	Indole acetic acid, phosphate solubilization	*H. vulgare* L.	Cakmakci et al. (2007)
14.	*Hartmannibacter diazotrophicus* $E19^T$	Rhizosphere	ACC deaminase activity	*H. vulgare* L.	Suarez et al. (2015)
15.	*Bacillus amyloliquifaciens*	Saline condition	Improving salt stress tolerance of barley	*H. vulgare* L.	Kasim et al. (2016)

(Continued)

TABLE 17.1 (Continued)

Sr. No.	PGPB	Region	PGP activity	Plant	References
16.	*Azospirillum brasilense* NO40	Saline condition	Salt stress reduced photosynthetic pigments, photosynthetic capacity, stomatal conductance, transpiration rate, accumulation of K, P, Mg, Ca, and Fe in the shoots and roots	*H. vulgare* L.	Omar et al. (2009)
17.	*P. putida* 11, *P. putida* 4, and *P. fluorescens* 169	Rhizosphere	Production of IAA, siderophore and ACC deaminase activity	*H. vulgare* L.	Baharlouei et al. (2011)
18.	*Bacillus* sp. AW1, *Providencia* sp. AW5, and *Brevundimonas diminuta* AW7	Rhizosphere	IAA, siderophore, ammonia, HCN, phosphate solubilization, ACC deaminase activity, acetylene reduction assay, and antifungal activity	*T. aestivum* L.	Rana et al. (2011)
19.	*Bacillus subtilis* SU47, and *Arthrobacter* sp. SU18	Saline condition	Salt tolerance	*T. aestivum* L.	Upadhyay et al. (2012)
20.	*P. aeruginosa*	Oxidative stress condition	Stress tolerance, IAA, siderophore	*T. aestivum* L.	Islam et al. (2014)
21.	Azospirillum sp. 245	Rhizosphere	Increased uptake of macronutrients (N, P, K, Ca, Mg, and S) and micronutrients (Fe, Mn, Zn, and Cu)	*T. aestivum* L.	Turan et al. (2010)
22.	*Bacillus* sp., *P. putida* (ECL5)	Endophyte	IAA, phosphate solubilization Siderophore	*Curcuma longa* L.	Kumar et al. (2016b)
23.	*Clavibacter michiganensis*	Endophyte	Indole acetic acid	*C. longa* L.	Kumar et al. (2016b)
24.	*Azotobacter chroococcum* CL13	Rhizosphere	IAA, HCN, ammonia production, phosphate solubilization	*C. longa* L.	Kumar et al. (2014)
25.	*Achromobacter xylosoxidans* AUM54	Endophyte	IAA, phosphate solubilization, siderophore	*Catharanthus roseus* L.	Karthikeyan et al. (2012)
26.	*A. xylosoxidans* Fd2, *Herbaspirillum seropedicae* Oci9, *Ochrobactrum rhizosphaerae* Oci13	Rhizosphere	IAA, siderophore	*Ocimum sanctum* L.	Barnawal et al. (2012)
27.	*Serratia ureilytica* Bac5	Rhizosphere	Siderophore, ACC deaminase phosphate solubilization	*O. sanctum* L.	Barnawal et al. (2012)
28.	*Pseudomonas stutzeri* CSP03 *B. subtilis* TTP02, *P. putida* PHP03	Rhizosphere	IAA, N_2 fixation, phosphate solubilization, siderophore	*Capparis spinosa* L.	El-Sayed et al. (2014)

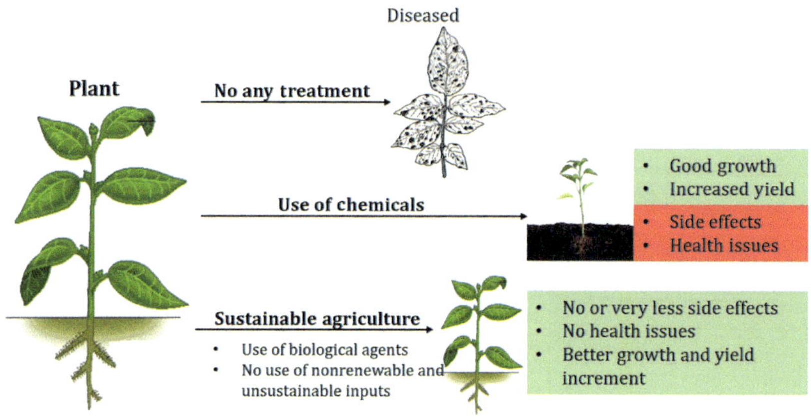

FIGURE 17.1 Overview of plant growth–promoting bacteria in sustainable agriculture.

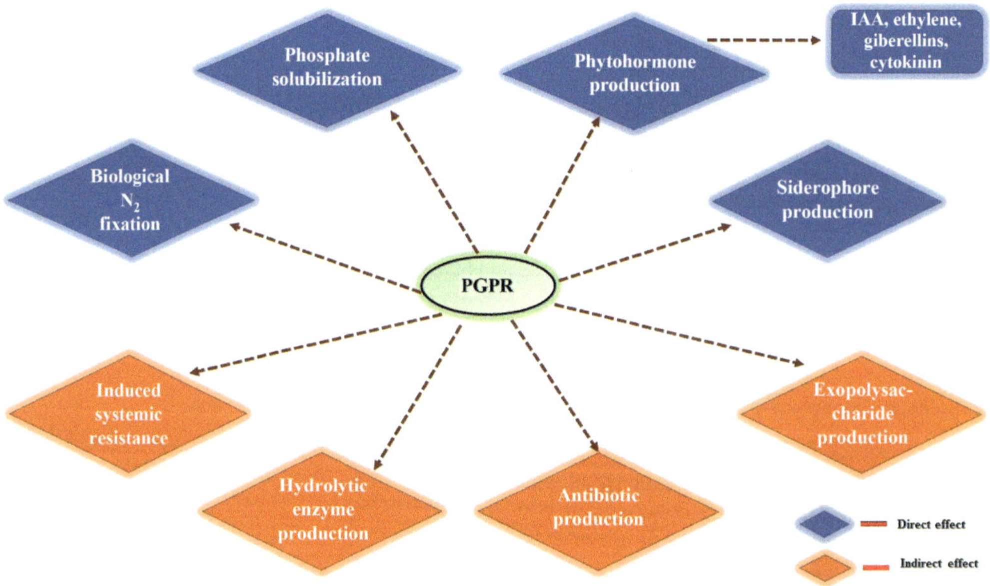

FIGURE 17.2 Action mechanism of plant growth–promoting bacteria (PGPB) in plant growth promotion.

Pseudomonas, and cyanobacteria like *Nostoc* and *Anabaena* (Bhattacharyya and Jha, 2012; Vessey, 2003). These symbiotic and nonsymbiotic bacteria help in maintaining nitrogen status in the field and help in plant growth and development.

17.4.2 Phosphate solubilization

Phosphorus is the second most important component of chemical fertilizers, required for maintenance of optimum productivity plant and soil (Anand et al., 2016). In the atmosphere phosphorous is present in the nonutilizable or insoluble rock salt $Ca_3(PO_4)_2$ and this complex form of phosphate doesn't be utilized by plants (Vassilev and Vassileva, 2003). The complex rock phosphates get transformed or solubilized into the mineral phosphates are solubilized into bioavailable anionic forms ($H_2PO_4^-$ and HPO_4^{-2}) through a range of processes such as solubilization, production of organic acids, chelation, exchange reactions, acidification, and this simpler form of phosphate can be utilized by plants or soil as nutrient source (Rodríguez and Fraga, 1999; Lavakush et al., 2014). Currently various microbial strain including *Pseudomonas, Bacillus, Azotobacter, Burkholderia, Arthrobacter, Enterobacter, Beijerinckia* have been reported for their effective phosphate solubilization property (Kumar et al., 2014, 2016b) A large proportion of inorganic P applied as chemical fertilizer get converted to unavailable forms due to their low solubility in the soil. The declining trend of pH against the increase of soluble P content was probably due to the secretion of bacterial metabolites, most probably organic acids facilitate the release of P into the medium (Glick, 2012).

17.4.3 Siderophore production

Siderophores are low-molecular weight chelating agents of 200–2000 Da that are produced by bacteria, fungi, and plants to facilitate uptake of iron (Chu et al., 2010; Hider and Kong, 2010). Iron is an essential element for the growth and development of almost all living organisms. Generally, siderophores act as biological catalyst in some common enzymatic processes, including oxygen metabolism, electron transfer, and DNA and RNA synthesis. In most aerobic microbial habitats iron is found in insoluble oxide hydrate complexes. Though present abundantly in the earth crust, but unavailable to the plant or organisms due to its complex insoluble forms (Ammari and Mengel, 2006; Singh et al., 2017a,b). Bacteria secrete siderophore that play a crucial role in iron metabolism. Siderophore in addition to growth promotion in crop plants and also helps in controlling phytopathogens and diminishing the toxic effects of heavy metal in soil.

In the nature, siderophore could have numerous chemical structures possessing electron-rich atoms able to bind metal ions (Chu et al., 2010; Verma et al., 2011; Ghavami et al., 2017). According to Hider and Kong (2010) more than 500 siderophores have been discovered and chemical structures of 270 siderophores had been elucidated. Siderophores produced by PGPB in iron-limiting soil conditions elevate the uptake of iron and make it available to the plants (Whipps, 2001; Li et al., 2016).

Different PGPR like *Pseudomonas, Bacillus, Enterobacter, Burkholderia, Klebsiella,* etc. produced siderophore for plant growth promotion (Kumar et al., 2014, 2016a). The bacterial siderophore acts as a valuable source for the maintenance of iron in the rhizospheric soil, however, these siderophores produced by one bacteria may also be utilized by other nearby bacteria for their iron metabolism. For example, *Pseudomonas putida* utilized the siderophore, produced by other bacteria to increase iron availability in the rhizosphere (Rathore, 2014). Besides the iron mobilization siderophores also used in the acquisition of nitrogenase cofactors, molybdenum, and vanadium by *Azotobacter vinelandii* (McRose et al., 2017). Currently siderophores also utilized in various biological purposes such as growth modulator, biocontrol agents, mitigation of abiotic stress as well as medicinal and industrial purposes (Venkat et al., 2017).

17.4.4 Indole acetic acid production

Microbial strains synthesized a diverse array of phytohormones, which regulate physiology and biochemistry of plants and the amount and nature of phytohormone synthesized by PGPB is strain specific. Indole acetic acid (IAA), auxin, and cytokinin are the some common phytohormones are among the common hormones synthesized by various PGPB strains, control or regulate multiple aspects including morphogenesis, growth, cell division mitigation from biotic and abiotic stresses (Waadt et al., 2015; Wani et al., 2016).

IAA is the most common growth hormones, synthesized by majority of the PGPB strains and potentially involved in division and differentiation of plant cells (Bhardwaj et al., 2014). In the study it has been mentioned that *Pseudomonas* is a potent synthesizer of IAA in comparison to other bacterial genera. However, among the *Pseudomonas* genera the synthesizing ability also varies or strain specific *P. putida* species. Among various sp of *Pseudomonas*, *P. putida* synthesizes higher amount of IAA than *Pseudomonas fluorescens* (Bharucha et al., 2013; Kumar et al., 2015a). Besides cell differentiation, microbial synthesized IAA has also been reported for different attributes of growth promotion or plant development. In a study Barea and Brown (1974) reported the *Azotobacter paspali* secreting IAA in the culture media and increased the dry weight of leaves and roots of several plant species up to a significant level following the root treatment. Increase in lateral roots, length and number was also observed in the wheat seedlings after inoculation with *Azospirillum brazilance* (Barbieri et al., 1986). Similarly, when canola seeds were inoculated with *P. putida* GR12-2 producing low levels of IAA resulted in twofold to threefold increases in the seedling length (Glick et al., 1986; Caron et al., 1995).

17.5 Fungi as plant growth–promoting agents

Like PGPB, nowadays various fungal strains have been reported as an effective biofertlizers and biocontrol agents, due to significant enhancement in the crop yield (Rai et al., 2013). Some of the growth-promoting attributes are very similar to the bacterial strain like phosphate solubilization, nitrogen fixation, phytohormone modulation for the plant growth promotion, while antibiotics, bioactive compounds, volatiles production, etc. are the common traits for plant disease management. Fungal isolates were also reported to have biocontrol functioning against several postharvest diseases and appear as a promising alternative to chemical fungicides for the management of postharvest decay of fruits, vegetables, and grains (Droby et al., 2009). There are several reports available that showed the effectivity of fungal strain as antagonistic agents against range of plant pathogens.

Sui et al. (2015) reported the recent research and overview on the antagonistic yeast species and reported that these yeast species can encounter various abiotic stresses and confers improved stress tolerance. Similarly Lopes et al. (2015) reported six isolates of *Saccharomyces cerevisiae* in controlling *Colletotrichum acutatum*, which causes the postbloom fruit drop in citrus plants. All the isolates were reported to produce antifungal compounds, which inhibit the pathogen growth. Ferraz et al. (2016) applied *Rhodotorula minuta*, *Candida azyma*, and *Aureobasidium pullulans* as the biocontrol agents to inhibit the growth of postharvest fungal pathogen *Geotrichum citri-aurantii* causal agent of sour rot disease in citrus fruits. All the three strains produced hydrolytic enzyme, which significantly inhibit the growth of pathogen.

Mushrooms, widely used medicinal and fungal food since decades, and also confers medicinal effects such as antifungal, antibacterial, antioxidant, and antiviral properties (Wani et al., 2010). In addition, they decorate the food stuffs with rich sources of phenolic compounds, flavonoids, polysaccharides, triterpenoids, dietary fiber steroids, glycopeptides, terpenes, etc. and these compounds have significant antimicrobial, antiviral, anticancer, antitumor, antiinflammatory, and immunomodulation properties (Wang and Ng, 2004; Periasamy, 2005; Carbonero et al., 2006; Iwalokun et al., 2007).

17.6 Microbial antagonistic in plant disease management

Currently plant diseases are most severe concern for the food security because of their adverse effect on quality and quantity of fruits and cereal crops. Traditionally to combat the adverse impact of plant diseases farmers utilized huge amount of chemical pesticides in the field regularly to control the growth of phytopathogen. However, regular and excessive amount of chemicals undoubtedly inhibit the growth of phytopathogens but have long-term adverse consequences on the environment, texture, and productivity of the soil and also severe health concern for the human beings because of low food quality and presence of chemical residues in the food (Guo et al., 2013; Dun-chun et al., 2016).

Currently a number of microbial genera, bacteria, fungi, and yeast strains have been applied on the cereal crops, fruits as biocontrol agents at the different stages of growth, harvesting, and postharvest storage (Droby et al., 2009; Zhimo et al., 2020). The application of microbial strains including bacteria, fungi, yeast as an antagonistic organism considered as a best suitable alternatives of chemical pesticides (Compant et al., 2005; Olanrewaju et al., 2017).

Microbial strains followed different strategies to control the growth of phytopathogens in which synthesis of bioactive compounds, antibiotics, siderophores, hydrogen cyanide (HCN), volatiles, are the some common attributes (Mishra et al., 2019). Many reports are available, which showed the effectivity of microbial strain in synthesizing a range of compounds that have significant aspect in control of phytopathogens. Pseudomonads has been reported for the synthesis of various compounds including siderophore HCN, oomycin A, phenazine, tropolone, tensin, pyrrolnitrin, pyoluteorin, cyclic lipopeptides, amphisin, and DAPG (2,4-diacetylphloroglucinol) and these compounds have effective role in plant disease management (Compant et al., 2005). Many different microbes such as *Serratia marcescens*, *Serratia plymuthica* C48, *Streptomyces* sp., *Pseudomonas stutzeri*, and *Paenibacillus* sp. have property to degrade fungal mycelia by producing enzyme chitinase. Cell wall of fungus can be degraded by β-1,3-glucanase which is produced by *Bacillus* sp., *Paenibacillus*, and *Streptomyces* (Compant et al., 2005)

However, several PGPB also mediate in the synthesis of enzymes like lipase and protease, by which they can lysis cell wall–related lipids and proteins. Similarly hydrogen cyanide has been also produced by several PGPB strain that have antifungal activity (Guo et al., 2013; Olanrewaju et al., 2017). Similarly microbial produced siderophores control the spread of pathogens by chelating the free rhizospheric iron and in the absence of free iron molecules in rhizospheric soil, pathogens can't be proliferated (Kumar et al., 2017a,b).

Induction of systemic resistance is also one of the way to impede pathogens. Mycorrhiza, *Bacillus*, *Trichoderma*, and *Pseudomonas* increase defense system of plants against various phytopathogens. Induced systemic resistance is responsible for the development of special receptors for the identification of phytopathogens. Systemic acquired resistance is alike induced systemic resistance induced through the production of salicylic acid by PGPB. Detoxification of phytopathogen virulence factor is one of the effective mechanism of biocontrol. *Fusarium* sp. and *Xanthomonas albilineans* produced toxins can be detoxified by PGPB (Compant et al., 2005). Similarly quorum sensing mechanism widely used by PGPB to control the virulence factor production. Although PGPB control the quorum sensing capacity of the pathogens through impairment of an autoinducer signals, which results in controlled expression of virulence factors (Olanrewaju et al., 2017).

17.7 Conclusion

The conventional use of chemical fertilizers and pesticides for the enhancement in agricultural productivity are easily available and very convenient in application for farmers that's why they used very frequently in the agriculture fields. Their continuous unrestricted application in agriculture field adversely affected soil texture, productivity, microflora, environment, and human health. The bacteria, fungi, and actinomycetes residing in rhizosphere or as endophyte in the plant perform beneficial function to plant without causing any harm to the host plant. These microbes perform essential function for plant life like essential mineral and nutrient acquisition, phytohormone synthesis, HCN, and NH_3 production. These microbes protect them from the biotic and abiotic stress like salinity, heavy metal. Active research in the area is need of sustainable agriculture and it is the actual challenge to scientific community to popularize the biocontrol agents from laboratory to field level for farmers use.

References

Ammari, T., Mengel, K., 2006. Total soluble Fe in soil solutions of chemically different soils. Geoderma 136, 876–885.

Anand, K., Kumari, B., Mallick, M., 2016. Phosphate solubilizing microbes: an effective and alternative approach as biofertilizers. Int. J. Pharm. Pharm. Sci. 8, 37–40.

Antoun, H., Prévost, D., 2005. Ecology of plant growth promoting rhizobacteria. PGPR: Biocontrol and Biofertilization. Springer, Dordrecht, pp. 1–38.

Bahadur, I., Maurya, B.R., Meena, V.S., Saha, M., Kumar, A., Aeron, A., 2017. Mineral release dynamics of tricalcium phosphate and waste muscovite by mineral-solubilizing rhizobacteria isolated from Indo-Gangetic plain of India. Geomicrobiol. J. 34, 454–466.

Baharlouei, J., Pazira, E., Solhi, M., 2011. Evaluation of inoculation of plant growth-promoting rhizobacteria on cadmium uptake by canola and barley. In: International Conference on Environmental Science and Technology IPCBEE, vol. 6. Singapore.

Barbieri, P., Zanelli, T., Galli, E., Zanetti, G., 1986. Wheat inoculation with *Azospirillum brasilense* Sp6 and some mutants altered in nitrogen fixation and indole-3-acetic acid production. FEMS Microbiol. Lett. 36, 87–90.

Barea, J.M., Brown, M.E., 1974. Effect of plant growth produced by *Azotobacter paspali* related to synthesis of plant growth regulating substances. J. Appl. Bacteriol. 37, 583–593.

Barnawal, D., Bharti, N., Maji, D., Chanotiya, C.S., Kalra, A., 2012. 1-Aminocyclopropane-1 carboxylic acid (ACC) deaminase-containing rhizobacteria protect *Ocimum sanctum* plants during water logging stress via reduced ethylene. Plant Physiol. Biochem. 58, 227–235.

Bhardwaj, D., Ansari, M.W., Sahoo, R.K., Tuteja, N., 2014. Biofertilizers function as key player in sustainable agriculture by improving soil fertility, plant tolerance and crop productivity. Microb. Cell Fact. 13, 66.

Bharucha, U., Patel, K., Trivedi, U.B., 2013. Optimization of indole acetic acid production by *Pseudomonas putida* UB1 and its effect as plant growth-promoting rhizobacteria on mustard (*Brassica nigra*). Agric. Res. 2, 215–221.

Bhattacharyya, P.N., Jha, D.K., 2012. Plant growth-promoting rhizobacteria (PGPR): emergence in agriculture. World J. Microbiol. Biotechnol. 28, 1327–1350.

Cakmakci, R., Dönmez, M.F., Erdoğan, Ü., 2007. The effect of plant growth promoting rhizobacteria on barley seedling growth, nutrient uptake, some soil properties, and bacterial counts. Turk. J. Agric. For. 31, 189–199.

Canbolat, M.Y., Barik, K., Çakmakçi, R., Sahin, F., 2006. Effects of mineral and biofertilizers on barley growth on compacted soil. Acta. Agric. Scand. B Soil Plant Sci. 56, 324–332.

Carbonero, E.R., Gracher, A.H.P., Smiderle, F.R., Rosado, F.R., Sassaki, G.L., Gorin, P.A.J., et al., 2006. A β-glucan from the fruit bodies of edible mushrooms *Pleurotu seryngii* and *Pleurotus ostreatoroseus*. Carbohydr. Polym. 66, 252–257.

Caron, M., Patten, C., Ghosh, S., Glick, B.R., 1995. Effects of the plant growth promoting rhizobacterium *Pseudomonas putida* GR12-2 on the physiology of canola roots. In: Proceedings Plant Growth Regulator Society of America annual meeting 22, 297–302.

Chu, B.C., Garcia-Herrero, A., Johanson, T.H., Krewulak, K.D., Lau, C.K., Peacock, R.S., et al., 2010. Siderophore uptake in bacteria and the battle for iron with the host; a bird's eye view. Biometals 23, 601–611.

Compant, S., Duffy, B., Nowak, J., Clément, C., Barka, E.A., 2005. Use of plant growth-promoting bacteria for biocontrol of plant diseases: principles, mechanisms of action, and future prospects. Appl. Env. Microbiol. 71, 4951–4959.

Dash, N.P., Kaushik, M.S., Kumar, A., Abraham, G., Singh, P.K., 2018. Toxicity of biocides to native cyanobacteria at different rice crop stages in wetland paddy field. J. Appl. Phycol. 30, 483–493.

Droby, S., Wisniewski, M., Macarisin, D., Wilson, C., 2009. Twenty years of postharvest biocontrol research: is it time for a new paradigm? Postharvest Biol. Tec. 52, 137–145.

Dun-chun, H.E., Zhan, J.S., Xie, L.H., 2016. Problems, challenges and future of plant disease management: from an ecological point of view. J. Integr. Agric. 15, 705–715.

El-Sayed, W.S., Akhkha, A., El-Naggar, M.Y., Elbadry, M., 2014. In vitro antagonistic activity, plant growth promoting traits and phylogenetic affiliation of rhizobacteria associated with wild plants grown in arid soil. Front. Microbiol. 5, 651.

El Zemrany, H., Czarnes, S., Hallett, P.D., Alamercery, S., Bally, R., Monrozier, L.J., 2007. Early changes in root characteristics of maize (*Zea mays*) following seed inoculation with the PGPR *Azospirillum lipoferum* CRT1. Plant Soil 291, 109–118.

Ferraz, L.P., da Cunha, T., da Silva, A.C., Kupper, K.C., 2016. Biocontrol ability and putative mode of action of yeasts against *Geotrichum citri-aurantii* in citrus fruit. Microbiol. Res. 188, 72–79.

Galloway, J.N., Townsend, A.R., Erisman, J.W., Bekunda, M., Cai, Z., Freney, J.R., et al., 2008. Transformation of the nitrogen cycle: recent trends, questions, and potential solutions. Science 320, 889–892.

García-Fraile, P., Menéndez, E., Rivas, R., 2015. Role of bacterial biofertilizers in agriculture and forestry. AIMS Bioeng. 2, 183–205.

Ghavami, N., Alikhani, H.A., Pourbabaei, A.A., Besharati, H., 2017. Effects of two new siderophore-producing rhizobacteria on growth and iron content of maize and canola plants. J. Plant. Nutr. 40, 736–746.

Glick, B.R., 2012. Plant growth-promoting bacteria: mechanisms and applications. Scientifica 963401.

Glick, B.R., Brooks, H.E., Pasternak, J.J., 1986. Physiological effects of plasmid DNA transformation on *Azotobacter vinelandii*. Can. J. Microb. 32, 145–148.

Guo, R.F., Yuan, G.F., Wang, Q.M., 2013. Effect of NaCl treatments on glucosinolate metabolism in broccoli sprouts. J. Zhejiang Univ. Sci. B 14, 124–131.

Gupta, A., Singh, S.K., Singh, M.K., Singh, V.K., Modi, A., Singh, P.K., et al., 2020. Plant growth–promoting rhizobacteria and their functional role in salinity stress management. Abatement of Environmental Pollutants. Elsevier, pp. 151–160.

Gupta, A., Verma, H., Singh, P.P., Singh, P., Singh, M., Mishra, V., et al., 2019. Rhizome endophytes: roles and applications in sustainable agriculture. Seed Endophytes. Springer, Cham, pp. 405–421.

Haghighi, B.J., Alizadeh, O., Firoozabadi, A.H., 2011. The role of plant growth promoting rhizobacteria (PGPR) in sustainable agriculture. Adv. Env. Bio 5, 3079–3083.

Hider, R.C., Kong, X., 2010. Chemistry and biology of siderophores. Nat. Prod. Rep. 27, 637–657.

Islam, F., Yasmeen, T., Ali, Q., Ali, S., Arif, M.S., Hussain, S., et al., 2014. Influence of *Pseudomonas aeruginosa* as PGPR on oxidative stress tolerance in wheat under Zn stress. Ecotoxicol. Env. Saf. 104, 285–293.

Iwalokun, B.A., Usen, U.A., Otunba, A.A., Olukoya, D.K., 2007. Comparative phytochemical evaluation, antimicrobial and antioxidant properties of *Pleurotus ostreatus*. Afr. J. Biotech. 6, 1732–1739.

Jha, B., Gontia, I., Hartmann, A., 2012. The roots of the halophyte *Salicornia brachiata* are a source of new halotolerant diazotrophic bacteria with plant growth-promoting potential. Plant Soil 356, 265–277.

Kapulnik, Y., Okon, Y., Kigel, J., Nur, I., Henis, Y., 1981. Effects of temperature, nitrogen fertilization, and plant age on nitrogen fixation by *Setaria italica* inoculated with *Azospirillum brasilense* (strain cd). Plant Physiol. 68, 340–343.

Karthikeyan, B., Joe, M.M., Islam, R., Sa, T., 2012. ACC deaminase containing diazotrophic endophytic bacteria ameliorate salt stress in *Catharanthus roseus* through reduced ethylene levels and induction of antioxidative defense systems. Symbiosis 56, 77–86.

Kasim, W.A., Gaafar, R.M., Abou-Ali, R.M., Omar, M.N., Hewait, H.M., 2016. Effect of biofilm forming plant growth promoting rhizobacteria on salinity tolerance in barley. Ann. Agric. Sci. 61, 217–227.

Kloepper, J.W., Lifshitz, R., Zablotowicz, R.M., 1989. Free-living bacterial inocula for enhancing crop productivity. Trends Biotech. 7, 39–44.

Kumar, A., Droby, S., Singh, V.K., Singh, S.K., White, J.F., 2020. Entry, colonization, and distribution of endophytic microorganisms in plants. Microbial Endophytes. Woodhead Publishing, pp. 1–33.

Kumar, A., Singh, R., Giri, D.D., Singh, P.K., Pandey, K.D., 2014. Effect of *Azotobacterchroococcum* CL13 inoculation on growth and curcumin content of turmeric (*Curcuma longa* L.). Int. J. Curr. Microbiol. Appl. Sci. 3, 275–283.

Kumar, A., Singh, A.K., Kaushik, M.S., Mishra, S.K., Raj, P., Singh, P.K., et al., 2017a. Interaction of turmeric (Curcuma longa L.) with beneficial microbes: a review. 3 Biotech. 7 (6), 357.

Kumar, A., Singh, V.K., Tripathi, V., Singh, P.P., Singh, A.K., 2018. Plant growth-promoting rhizobacteria (PGPR): perspective in agriculture under biotic and abiotic stress. Crop Improvement Through Microbial Biotechnology. Elsevier, pp. 333–342.

Kumar, A., Singh, R., Yadav, A., Giri, D.D., Singh, P.K., Pandey, K.D., 2016a. Isolation and characterization of bacterial endophytes of *Curcuma longa* L. 3 Biotech. 6, 60.

Kumar, A., Vandana, R.S., Singh, M., Pandey, K.D., 2015a. Plant growth promoting rhizobacteria (PGPR). A promising approach for disease management. In: Singh, J.S., Singh, D.P. (Eds.), Microbes and Environmental Management. Studium Press, New Delhi, pp. 195–209.

Kumar, A., Vandana, R.S., Singh, M., Singh, P.P., Singh, S.K., Singh, P.K., et al., 2016b. Isolation of plant growth promoting rhizobacteria and their impact on growth and curcumin content in *Curcuma longa* L. Biocatal. Agric. Biotech. 8, 1–7.

Kumar, A., Vandana, R.S., Yadav, A., Giri, D.D., Singh, P.K., Pandey, K.D., 2015b. Rhizosphere and their role in plant–microbe interaction. Microbes in soil and their agricultural prospects. Nova Science Publisher, Inc, New York, pp. 83–97.

Kumar, A., Verma, H., Singh, V.K., Singh, P.P., Singh, S.K., Ansari, W.A., et al., 2017b. Role of Pseudomonas sp. in sustainable agriculture and disease management. Agriculturally Important Microbes for Sustainable Agriculture. Springer, Singapore, pp. 195–215.

Lavakush, Y., Yadav, J., Verma, J.P., Jaiswal, D.K., Kumar, A., 2014. Evaluation of PGPR and different concentration of phosphorus level on plant growth, yield and nutrient content of rice (*Oryza sativa*). Ecol. Eng. 62, 123–128.

Li, K., Chen, W.H., Bruner, S.D., 2016. Microbial siderophore-based iron assimilation and therapeutic applications. Biometals 29, 377–388.

Lopes, M.R., Klein, M.N., Ferraz, L.P., da Silva, A.C., Kupper, K.C., 2015. *Saccharomyces cerevisiae*: A novel and efficient biological control agent for *Colletotrichum acutatum* during pre-harvest. Microbiol. Res. 175, 93–99.

McLaughlin, A., Mineau, P., 1995. The impact of agricultural practices on biodiversity. Agric. Ecosyst. Env. 55, 201–212.

McRose, D.L., Zhang, X., Kraepiel, A.M.L., Morel, F.M.M., 2017. Diversity and activity of alternative nitrogenases in sequenced genomes and coastal environments. Front. Microbiol. 8, 114.

Mishra, P., Singh, P.P., Singh, S.K., Verma, H., 2019. Sustainable agriculture and benefits of organic farming to special emphasis on PGPR. Role of Plant Growth Promoting Microorganisms in Sustainable Agriculture and Nanotechnology. Woodhead Publishing, pp. 75–87.

Modi, A., Kanani, P., Kumar, A., 2020. Fungal endophytes-induced gene expression studies in biotic and abiotic stress management. Microbial Endophytes. Woodhead Publishing, pp. 251–271.

Morrissey, J.P., Dow, J.M., Mark, G.L., O'Gara, F., 2004. Are microbes at the root of a solution to world food production? Rational exploitation of interactions between microbes and plants can help to transform agriculture. EMBO Rep. 5, 922–926.

Naseem, H., Bano, A., 2014. Role of plant growth-promoting rhizobacteria and their exopolysaccharide in drought tolerance of maize. J. Plant Interact 9, 689–701.

Oku, S., Komatsu, A., Tajima, T., Nakashimada, Y., Kato, J., 2012. Identification of chemotaxis sensory proteins for amino acids in *Pseudomonas fluorescens* Pf0-1 and their involvement in chemotaxis to tomato root exudate and root colonization. Microbes Environ. 27, 462–469.

Olanrewaju, O.S., Glick, B.R., Babalola, O.O., 2017. Mechanisms of action of plant growth promoting bacteria. World J. Microbiol. Biotech. 33, 197.

Omar, M.N.A., Osman, M.E.H., Kasim, W.A., El-Daim, I.A., 2009. Improvement of salt tolerance mechanisms of barley cultivated under salt stress using *Azospirillum brasilense*. Salinity and water stress. Springer, Dordrecht, pp. 133–147.

Periasamy, K., 2005. Novel antibacterial compounds obtained from some edible mushrooms. Int. J. Med. Mushrooms 7, 3.

Rai, A., Rai, S., Rakshit, A., 2013. Mycorrhiza-mediated phosphorus use efficiency in plants. Environ. Exp. Biol. 11, 107–117.

Rana, A., Saharan, B., Joshi, M., Prasanna, R., Kumar, K., Nain, L., 2011. Identification of multi-trait PGPR isolates and evaluating their potential as inoculants for wheat. Ann. Microbiol. 61, 893–900.

Rathore, P., 2014. A review on approaches to develop plant growth promoting rhizobacteria. Int. J. Recent Sci. Res. 5, 403–407.

Rodríguez, H., Fraga, R., 1999. Phosphate solubilizing bacteria and their role in plant growth promotion. Biotech. Adv. 17, 319–339.

Sandhya, V., Shrivastava, M., Ali, S.Z., Prasad, V.S.K., 2017. Endophytes from maize with plant growth promotion and biocontrol activity under drought stress. Russ. Agric. Sci. 43, 22–34.

Sev, T.M., Khai, A.A., Aung, A., Yu, S.S., 2016. Evaluation of endophytic bacteria from some rice varieties for plant growth promoting activities. J. Sci. Innov. Res. 5, 144–148.

Sgroy, V., Cassán, F., Masciarelli, O., Del Papa, M.F., Lagares, A., Luna, V., 2009. Isolation and characterization of endophytic plant growth-promoting (PGPB) or stress homeostasis-regulating (PSHB) bacteria associated to the halophyte *Prosopis strombulifera*. Appl. Microbiol. Biotechnol. 85, 371–381.

Shaharoona, B., Arshad, M., Zahir, Z.A., 2006. Effect of plant growth promoting rhizobacteria containing ACC-deaminase on maize (*Zea mays* L.) growth under axenic conditions and on nodulation in mung bean (*Vignaradiata* L.). Lett. Appl. Microbiol. 42, 155–159.

Shaharoona, B., Naveed, M., Arshad, M., Zahir, Z.A., 2008. Fertilizer-dependent efficiency of Pseudomonads for improving growth, yield, and nutrient use efficiency of wheat (*Triticum aestivum* L.). Appl. Microbiol. Biotechnol. 79, 147–155.

Shahzad, S.M., Arif, M.S., Riaz, M., Iqbal, Z., Ashraf, M., 2013. PGPR with varied ACC-deaminase activity induced different growth and yield response in maize (*Zea mays* L.) under fertilized conditions. Eur. J. Soil Biol. 57, 27–34.

Singh, M., Kumar, A., Singh, R., Pandey, K.D., 2017b. Endophytic bacteria: a new source of bioactive compounds. 3 Biotech. 7, 315.

Singh, M., Singh, D., Gupta, A., Pandey, K.D., Singh, P.K., Kumar, A., 2019. Plant growth promoting rhizobacteria: application in biofertilizers and biocontrol of phytopathogens. PGPR Amelioration in Sustainable Agriculture. Woodhead Publishing, pp. 41–66.

Singh, V.K., Singh, A.K., Kumar, A., 2017a. Disease management of tomato through PGPB: current trends and future perspective. 3 Biotech. 7, 255.

Singh, V.K., Singh, S.K., Singh, P.K., Verma, H., Pandey, K.D., Singh, P.K., et al., 2020. Impact of pesticides applications on the growth and function of cyanobacteria, Advances in Cyanobacterial Biology., 151–162. Academic Press.

Suarez, C., Cardinale, M., Ratering, S., Steffens, D., Jung, S., Montoya, A.M., et al., 2015. Plant growth-promoting effects of *Hartmannibacter diazotrophicus* on summer barley (*Hordeum vulgare* L.) under salt stress. Appl. Soil Ecol. 95, 23–30.

Sui, Y., Wisniewski, M., Droby, S., Liu, J., 2015. Responses of yeast biocontrol agents to environmental stress. Appl. Env. Microbiol. 81, 2968–2975.

Turan, M., Gulluce, M., Cakmakci, R., Oztas, T., Sahin, F., Gilkes, R.J., et al., 2010. The effect of PGPR strain on wheat yield and quality parameters. Proceedings of the 19th World Congress of Soil Science; Soil Solutions for a Changing World. pp. 140–143.

Upadhyay, S.K., Singh, J.S., Saxena, A.K., Singh, D.P., 2012. Impact of PGPR inoculation on growth and antioxidant status of wheat under saline conditions. Plant Biol. 14, 605–611.

Vacheron, J., Desbrosses, G., Bouffaud, M.L., Touraine, B., Moënne-Loccoz, Y., Muller, D., et al., 2013. Plant growth-promoting rhizobacteria and root system functioning. Front. Plant Sci. 4, 356.

Vassilev, N., Vassileva, M., 2003. Biotechnological solubilization of rock phosphate on media containing agro-industrial wastes. Appl. Microbiol. Biotechnol. 61, 435–440.

Venkat, K.S., Soumya, M.S., Agarwal, H., Divya, G.D., 2017. Characterization and optimization of bacterium isolated from soil samples for the production of siderophores. Resource Efficient Technol. 3, 434–439.

Verma, V.C., Singh, S.K., Prakash, S., 2011. Bio-control and plant growth promotion potential of siderophore producing endophytic Streptomyces from *Azadirachta indica* A. Juss. J. Basic Microbiol. 51, 550–556.

Vessey, J.K., 2003. Plant growth promoting rhizobacteria as biofertilizers. Plant Soil 255, 571–586.

Waadt, R., Hsu, P.K., Schroeder, J.I., 2015. Abscisic acid and other plant hormones: methods to visualize distribution and signaling. Bioessays 37, 1338–1349.

Wang, H., Ng, T.B., 2004. Eryngin, a novel antifungal peptide from fruiting bodies of the edible mushroom *Pleurotus eryngii*. Peptides 25, 1–5.

Wani, B.A., Bodha, R.H., Wani, A.H., 2010. Nutritional and medicinal importance of mushrooms. J. Med. Plants Res. 4, 2598–2604.

Wani, S.H., Kumar, V., Shriram, V., Sah, S.K., 2016. Phytohormones and their metabolic engineering for abiotic stress tolerance in crop plants. Crop J. 4, 162–176.

Whipps, J.M., 2001. Microbial interactions and biocontrol in the rhizosphere. J. Exp. Bot. 52, 487–511.

Wood, N.T., 2001. Nodulation by numbers: the role of ethylene in symbiotic nitrogen fixation. Trends Sci. 6, 501–502.

Yasmin, S., Zaka, A., Imran, A., Zahid, M.A., Yousaf, S., Rasul, G., et al., 2016. Plant growth promotion and suppression of bacterial leaf blight in rice by inoculated bacteria. PLoS One 11, e0160688.

Youssef, M.M., Eissa, M.F., 2014. Biofertilizers and their role in management of plant parasitic nematodes. A review. E3 J. Biotechnol. Pharm. Res. 5, 1–6.

Zahran, H.H., 2001. Rhizobia from wild legumes: diversity, taxonomy, ecology, nitrogen fixation and biotechnology. J. Biotechnol. 91, 143–153.

Zakry, F.A.A., Shamsuddin, Z.H., Rahim, K.A., Zakaria, Z.Z., Rahim, A.A., 2012. Inoculation of *Bacillus sphaericus* UPMB-10 to young oil palm and measurement of its uptake of fixed nitrogen using the ^{15}N isotope dilution technique. Microbiol. Env. 27, 257–262.

Zhimo, V.Y., Biasi, A., Kumar, A., Feygenberg, O., Salim, S., Vero, S., et al., 2020. Yeasts and bacterial consortia from Kefir grains are effective biocontrol agents of postharvest diseases of fruits. Microorganisms. 8, 428.

CHAPTER

18

Probiotics in edible coatings: Approaches to food security and fruits disease management

Ajay Kumar[1], V. Yeka Zhimo[1], Antonio Biasi[1], Oleg Feygenberg[1], Shoshana Salim[1], James Francis White[2], Michael Wisniewski[3] and Samir Droby[1]

[1]Department of Postharvest Science, Agricultural Research Organization (ARO), Volcani Center, Rishon LeZion, Israel [2]Department of Plant Biology, Rutgers University, New Brunswick, NJ, United States [3]Appalachian Fruit Research Station, Agricultural Research Service, United States Department of Agriculture, Kearneysville, WV, United States

18.1 Introduction

In the last few years edible coatings or edible films have been frequently employed to improve shelf life and fruit quality of fresh produce via incorporating bioactive antimicrobials or otherwise to reduce the rate of deterioration by reducing the pathogen load from the fruits/food surface during postharvest storage. This appears to be an eco-friendly and economic alternative over the traditional chemical methods to treat fruits. The "coatings" and "films" are similar, however, they may be distinguished by differences that arise during the preparation and application processes. Edible films are usually obtained in parallel to food and then applied to the surface, whereas coatings are directly prepared on the food surface (Guilbert et al., 1995; Paul, 2019; Pop et al., 2020).

The functional ingredients of edible coatings may enhance the stability as well as quality of the food by suppressing certain physiological phenomenon such as water releases through minimizing the rate of transpiration, moisture content, respiration rate, and also by enhancing nutrient source, antioxidant properties, flavor (Raghav et al., 2016). In various fresh produce types, especially fruits, whose metabolic processes continue after harvesting, the quality and other physiological activities of the fruits deteriorate over time. However,

Food Security and Plant Disease Management
DOI: https://doi.org/10.1016/B978-0-12-821843-3.00017-9

application of a biofilm or coating on the fresh produce adds to the nutrient content flavor stability and controls the proliferation of microbes associated with food pathogen or other microbes responsible for spoilage (Galie et al., 2018).

Currently postharvest deterioration of fresh produce is a global challenge and adversely affects the economy of several countries, whose export industries are based upon agricultural commodities, especially fruits and vegetables. However, quality of fresh produce depends on growth conditions and various other physiological and biochemical changes that occur after harvesting. Some agricultural commodities especially fruits are highly susceptible to pathogenic attack or wound infection during postharvest handling and transportation (Droby et al., 2009; Spadaro and Droby, 2016). It has been assumed that approximately 20%–50% of total production is lost in postharvest storage due to certain diseases (Kader, 1992). Therefore to withstand mechanical injury, pathogen attack, and enhance shelf life, fresh produce relies on various postharvest practices in the "pack house" that includes wax applications and chemical fungicide treatments. However, application of chemical fungicides and wax coatings causes health concerns after consumption of chemical residues (Palou et al., 2008). Recently, a new and alternative methodology based on formulation of edible coatings carrying probiotics have been employed to enhance shelf life, improve fruit quality as well as control of phytopathogens associated with fresh produce (Pop et al., 2020).

18.2 Edible films and coatings

Edible films and coatings have been composed of several biomaterials and act as carriers of living microorganism or probiotics, or as a medium for incorporation of functional bioactive compounds over the surface of fresh produce to enhance shelf life, maintain nutrient quality, fresh appearance, or to act as biocontrol agents during postharvest storage (Iñiguez-Moreno et al., 2020).

The chemical composition of the coatings and films is more or less same and sometimes used as synonym for each other. However, they are differentiated on the basis of functionality and purpose of utilization. Film has been defined as thin layer of materials that can be used as covers or wrappings, whereas coatings involve formation of films directly on the product surface to which they are intended to be applied (Krochta, 2002). Each of the biomolecules used in coatings have some specific functions. Polysaccharides, lipids, proteins are the major constituents of coating and biofilm. Moreover polysaccharides and proteins contained within biofilms may control gas (oxygen, ethylene, CO_2) permeability but they are not able to control the moisture; lipids within biofilms control the moisture content (Debeaufort et al., 1998). Although combined application of these molecules with some additive or surfactants enhance the viability and shelf life of fresh produce (Guimarães et al., 2018).

Selection of coating materials also depends upon their physical and chemical properties. A biomaterial having high water content should be able to control surface dehydration of fresh produce and also control oxygen movement, which reduces oxidation or free radical generation. Another application of edible coatings is to improve nutrient status, quality, and fragrance which may be reduced in the original food products (Guimarães et al., 2018).

Coatings may also have health benefits. Some biopolymers used in biocoating are easily biodegraded after consumption (Galus and Kadzińska, 2015).

18.3 Materials for edible films/coatings for probiotic applications

The composition of edible films and coatings is similar, both derived from film-forming materials and classified into composites, hydrocolloids, and lipids (Bourtoom, 2008; Rojas-Graü et al., 2009). However, some of the additional emulsifiers and plasticizers have been added in the formulation to enhance stability and mechanical support properties (Valencia-Chamorro et al., 2011). Emulsifiers have been added when the coating materials are of lipids or polysaccharides, whereas plasticizers added in protein and polysaccharide coating materials (Pop et al., 2020). Emulsifiers have been added to improve the stabilization of dispersed systems and enhance the distribution of particles in composite emulsion (Galus and Kadzińska, 2015); however, plasticizers are generally used to enhance the film flexibility and for tear resistance (Vieira et al., 2011). Chitosan, polyethylene glycol, glycerol, sorbitol, sucrose, dextran, alginate, carrageenan, and pectin are the some common polysaccharides (Bourtoom, 2008; Valencia-Chamorro et al., 2011; Ramos et al., 2012; Šuput et al., 2015; Cerqueira et al., 2011), whereas soy protein, wheat gluten, corn zein, and casein are common protein films generally used as active constituents in the edible coating (Bourtoom, 2008; Šuput et al., 2015; Pop et al., 2020). These biopolymers frequently used as physical barriers, which after application on the fruit surface efficiently delays the ripening process of fresh produce. There are several reports published which were used to produce biopolymers as edible coatings or biofilms to control pathogens or enhance functional shelf life of fresh produce (Table 18.1).

Adetunji et al. (2019) used chitosan as an edible coating to enhance the shelf life of green bell pepper. The application of chitosan significantly delayed loss of firmness or weight and vitamin C content in green bell pepper and also inhibited the growth of bacteria, yeast, and mold during 5 weeks of storage. Similarly Frazão et al. (2017) used chitosan, cassava starch, and Myrcia ovata essential oils (MYRO-174 or 175) to evaluate their antagonistic activity against eight foodborne bacteria. The application of MYRO-175 essential oil-containing coatings was shown most effective to inhibit the growth of Bacillus cereus, Bacillus subtilis, and Serratia marcescens pathogens. Coatings containing 0.5% cassava starch, 0.5% chitosan, and 1.25% MYRO essential oil, also reduced the total count of aerobic bacteria, mold, and yeast from mangabas fruit surface by reducing up to ~3 log from 4 to 12 days.

Similarly Ibrahim et al. (2017) evaluated the effectivity of chitosan with lemongrass and thyme essential oils to enhance the shelf life and quality of strawberries. In this study strawberries were coated with 1.0% chitosan with either (0.1% and 0.2%) lemongrass oil or (0.1% and 0.2%) thyme oil. Application of bioformulated edible coatings reduced the rate of weight loss, delayed change in anthocyanin content as well as microbiological growth and coatings maintained the firmness, total acidity, ascorbic acid, and quality of strawberry fruits. However, control samples had a shelf life of 2 and 8 days for storage at room temperature (24°C) and at cold storage (4°C), respectively, whereas coated samples had shelf life of 4 and 15 days, respectively, at the same storage temperatures.

TABLE 18.1 Edible coating/film and their impact after application on fruits/crops.

Coating/biofilm	Activity	Fruits	References
Chitosan–aloe vera—banana starch	Increasing shelf life	*Fragaria × ananassa*	Pinzon et al. (2020)
Sodium alginate and cyclolipopeptides (produced by *Bacillus subtilis)*	Antifungal	*Semen trigonellae*	Xu et al. (2020)
Methylcellulose and walnut flour	Shelf life	*Juglans regia*	Grosso et al. (2020)
Chitosan and lactoperoxidase	Antifungal activity	*Mangifera indica*	Mohamed et al. (2020)
Cassava starch/gelatin	Antifungal activity *(A. alternate)*	*Solanum lycopersicum*	Santa Catharina et al. (2020)
Aloe vera gel	Postharvest quality	*S. lycopersicum*	Firdous et al. (2020)
Cinnamaldehyde and citronellal	Antifungal activity	*Citrus unshiu*	OuYang et al. (2020)
Chitosan–carboxymethyl cellulose (CMC) layer-by-layer assembly	Enhance postharvest quality	*Citrus limon* L.	Chen et al. (2020)
Chitosan, carboxymethyl cellulose (CMC), Moringa leaf extracts	Reduced peteca spot incidence, reduced mass loss, ascorbic acid loss, and delayed color change of fruit	*C. limon* L.	Rikhotso et al. (2019)
Gel + fungal chitosan nanoparticles	Delay ripening	*Vitis vinifera*	Melo et al. (2018)
Chitosan/montmorillonite	Enhanced shelf Life	*Citrus reticulata*	Xu et al. (2018)
Chitosan and pomegranate peel extract (PPE)	Microbiological, chemical, and sensorial quality enhancement	*Oreochromis niloticus*	Alsaggaf et al. (2017)
Hydroxypropyl methylcellulose-beeswax	Enhance gas barrier properties and increase storability	*Prunus domestica*	Gunaydin et al. (2017)

Abu-Shama et al. (2020) evaluated edible coatings of different combinations of chitosan, gelatin, and guar gum on date fruits to evaluate their efficiency in delaying ripening and enhanced shelf life. Application of different combinations showed beneficial responses toward enhancing shelf life of date fruits in comparison to controls (untreated). However, effectivity of the treatment was found maximum after application of 0.5% guar gum and gelatin + chitosan followed by gelatin + guar gum treatment. Similarly Gao et al. (2018) evaluated the efficacy of coating formulation, chitosan (CH), cinnamaldehyde + chitosan (0.5% CI–CH), cinnamaldehyde + chitosan (1.0% CI–CH), and cinnamaldehyde + chitosan (1.5% CI–CH) on the quality maintenance of fresh mandarin. Chitosan treatment effectively reduced the postharvest decay and rate of weight loss in comparison to control. The treatment significantly controlled nutritional depletion, inhibited malondialdehyde accumulation, and enhanced antioxidant properties.

1.5% CI–CH coating did not have such impact and did not improve the storage, although it inhibited color change in the fruits and increased the accumulation of MDA in comparison to chitosan coating. However, application of 0.5% CI–CH and 1.0% CI–CH significantly controlled fruit decay as well as improved the quality of fruits after harvest. Chen et al. (2020) used a chitosan–carboxymethyl cellulose (CMC) layer-by-layer assembly system to evaluate the postharvest quality of lemon. The combined application of formulations improve the firmness, vitamin C content, morphology as well as significantly control the weight loss and flesh browning.

Rodrigues et al. (2018) evaluate the efficacy of edible coatings based on linseed mucilage, alginate, and fructooligosaccharide containing *Lactobacillus casei* on the shelf life of fresh-cut yacon cubes. This coating appears to be an efficient probiotic carrier and maintains the viable microbe cell count of about 8 log CFU/g, however, in gastrointestinal tract, reduction in viable count was observed to 2.96 log CFU/g, suggesting that this may be an effective carrier of potential probiotics. Droby et al. (2003) evaluated the efficiency of calcium propionate, sodium bicarbonate, and sodium ethylenediaminetetraacetic acid (EDTA) against postharvest pathogens of apple and peach, and also evaluated their efficacy against the biocontrol product aspire. Application of all three materials had differential effects on *Botrytis cinerea* and *Penicillium expansum* in vitro. The inhibitory effect increased with increases in the concentration of each of the materials. On apple, application of sodium bicarbonate and EDTA exhibited curative activity against infection caused by *B. cinerea* whereas application of Ca-propionate, exhibited both protective as well as curative effects against infections caused by *B. cinerea*. However, in case of *P. expansum* infection, sodium bicarbonate up to 0.4% and several EDTA concentrations did not show any protective or curative effects in apple. Sodium bicarbonate (2%) was the most effective agent against the fungal pathogen when used singly. Although mixed application of 2% sodium bicarbonate with Aspire, exhibited consistent biocontrol activity in terms of both protective as well as curative actions against *Botrytis* and *Penicillium* rot in apple and *Monilinia* and *Rhizopus* rot in peach.

18.4 Microbes and their probiotic activities

The term probiotics is used for any known concentration of living microorganisms that confers health benefits to the host (Pop et al., 2020). Currently numerous living microorganism have been used to improve the quality of food or fresh produce (Sanders, 2008; Holzapfel et al., 1998). Probiotics support the growth of indigenous microbes via controlling the growth of phytopathogens and also elicit immune response of the body after consumption (Gismondo et al., 1999).

In the food industry some of the bacterial genera have been used over the last few decades in probiotic mixes. Lactic acid bacteria (LAB) are the most prominent of these probiotic bacteria. These bacteria are commonly present as inhabitants in the oral cavities and intestinal tracts of human, where they regulate certain physiological processes and also inhibit the growth of some pathogenic microbiota (Bosch et al., 2012; Guimarães et al., 2018). Although *Lactobacillus acidophilus*, *Lactobacillus rhamnosus*. *Bifidobacteria* (Saarela et al., 2008), *Enterococcus* (Douglas and Sanders, 2008), *Escherichia*, *Bacillus*, *Propionibacterium* are the most commonly used bacterial genera in probiotics.

However; some members of genera *Enterococcus* and *Streptococcus* have also been reported as opportunistic pathogens (Quirós-Sauceda et al., 2014).

Similarly yeast genera have been also frequently used in probiotics. *Saccharomyces cerevisiae* has been reported as among the first probiotics consumed by human beings, and this might be due to diverse and frequent use in dairy sector (De Prisco and Mauriello, 2016). Various yeast species, such as *Debaryomyces hansenii, Torulaspora delbrueckii, Pichia membranaefaciens, Candida* sp., *Metschnikowia pulcherrima*, and *Saccharomyces boulardii* have been reported as probiotics (Kumura et al., 2004; Marín et al., 2016; Zhou et al., 2016; Parafati et al., 2016; Ceugniez et al., 2017).

However, selection of effective probiotics depends on viability of the microorganism and capacity to be maintained at an optimum cell count ($\sim 10^9$ viable cells) throughout the process, from the time of formulation to reaching the consumer (Guimarães et al., 2018; Pop et al., 2020). Probiotics also must withstand adverse environmental conditions, such low pH, bile; they must have good colonization of the gut, and must be tolerated by immune system and also reach in bioactive compounds that inhibit the growth of phytopathogens (Parvez et al., 2006; Saarela et al., 2000).

18.5 Probiotic incorporation in edible films/coatings

To improve the quality, functionality, shelf life of food products, incorporation of probiotics with edible films or coatings is one of the latest strategies in the food industry (Pavli et al., 2018). Several reports are available in which beneficial microbial strains incorporated in biological films, not only enhance the stability or shelf life of the fruit, but potentially serve as biocontrol agents against various phytopathogens during postharvest storage (Potjewijd et al., 1995; Fan et al., 2009; Yinzhe and Shaoying, 2013; Pavli et al., 2018; Guimarães et al., 2018) (Table 18.2).

In the food industry, in probiotics a concentration of 10^{8-9} viable cells per day dose has been recommended to confer health benefits (Pop et al., 2020). However, it is challenging task to maintain the viability of microbial cells in the coating from the time of formulation, through storage to consumption. The viability of probiotics is influenced by several factors such as food matrix, physiological status of microbes, media culture, temperature, pH, formulation as well as storage conditions (Pop et al., 2019, 2020). The chemical composition of packaging materials is also an important factor for probiotic growth because it might influence moisture content, gaseous transport between the food particles and coating materials (Rivera-Espinoza and Gallardo-Navarro, 2010).

Although plyometric material used for microencapsulation is also a limiting factor due to nature of the bile and lower pH of gastrointestinal tract, which act as barrier to the encapsulated probiotics in terms of their optimum concentration (De Prisco and Mauriello, 2016; Pinto et al., 2015). Although biomaterials also contain some bioactive agents that may provide additional health benefits with food products to the consumers (Lopez-Rubio et al., 2006). However, it has been mentioned that mechanical and barrier properties of films and coatings may sometimes be affected by incorporation of microorganisms (Kanmani and Lim, 2013; Sánchez-González et al., 2013, 2014).

TABLE 18.2 Microbial strains as probiotics carriers in edible coating and their impact after application.

Microbial strain	Fruits/crops	Edible coating/film	Disease control or pathogen	Advantage	References
Streptococcus salivarius subsp. thermophilus, *Lactobacillus delbrueckii* subsp. bulgaricus, *Lactobacillus acidophilus*	*Triticum aestivum* (wheat bread)	Sodium alginate, whey and glycerol	Mold spoilage by *Aspergillus* and *Penicillium*	Increase shell life and protect from spoilage	Gregirchak et al. (2020)
Meyerozyma caribbica	*Persea americana Mill. cv. Hass*	Sodium alginate	Anthracnose (*Colletotrichum gloeosporioides*)	Reduce weight loss and control of postharvest diseases	Iñiguez-Moreno et al. (2020)
Lactobacillus plantarum	*Vitis vinifera L., cultivar Red Globe*	Pregelatinized potato starch or sodium caseinate	Gray mold (*Botrytis cinerea*)	Reduced *Botrytis* disease incidence	Marín et al. (2019)
Pichia guilliermondii	*Citrus reticulata* Blanco cv. Shogun	Aloe vera gel	Fruit decay (*Penicillium digitatum*)	Enhanced fruit quality and inhibit fruit decay	Jiwanit et al. (2018)
Wickerhamomyces anomalus	*Citrus sinensis* L. Osbeck	Chitosan and locust bean gum (LBG)	Green mold (*P. digitatum*)	Reduced disease incidence	Kharchoufi et al. (2018)
Bacillus subtilis	*Fragaria × ananassa* Duch, cv. Albion	Guar gum, candelilla wax, glycerol, and gallic acid	Fruit decay (*Rhizopus stolonifera*)	Reduction of *R. stolonifera* severity index	Oregel-Zamudio et al. (2017)
Candida sake	*Vitis vinifera*	Sodium caseinate and starch	Bunch rot-causing pathogen/gray mold (*B. cinerea*)	Improve control of disease	Marín et al. (2016)
Pichia membranaefaciens	*Citrus sinensis* L. Osbeck cv. Jincheng 447	Chitosan	Anthracnose (*C. gloeosporioides*)	Biocontrol activity	Zhou et al. (2016)
W. anomalus, *Metschnikowia pulcherrima*, and *Aureobasidium pullulans*	*Citrus reticulata* Blanco	Polysaccharide matrix, LGB	Postharvest decay (*P. digitatum* and *Penicillium italicum*)	Control of postharvest decays	Parafati et al. (2016)
B. subtilis	*Citrus reticulate* Blanco cv. Shogun	Chitosan and crude extract	Green mold (*P. digitatum*)	Reduction of fruit decay	Waewthongrak et al. (2015)

(*Continued*)

TABLE 18.2 (Continued)

Microbial strain	Fruits/crops	Edible coating/film	Disease control or pathogen	Advantage	References
W. anomalus	*Citrus sinensis* L. Osbeck	Sodium alginate and LBG	Green mold (*P. digitatum*)	Reduced green mold and preservation	Aloui et al. (2015)
Candida utilis	*Solanum lycopersicum*	Chitosan	Fruit rot by (*Alternaria alternata* (Fr.) Keissler and *Geotrichum candidum*)	Controlling postharvest decay of tomato	Sharma et al. (2006)

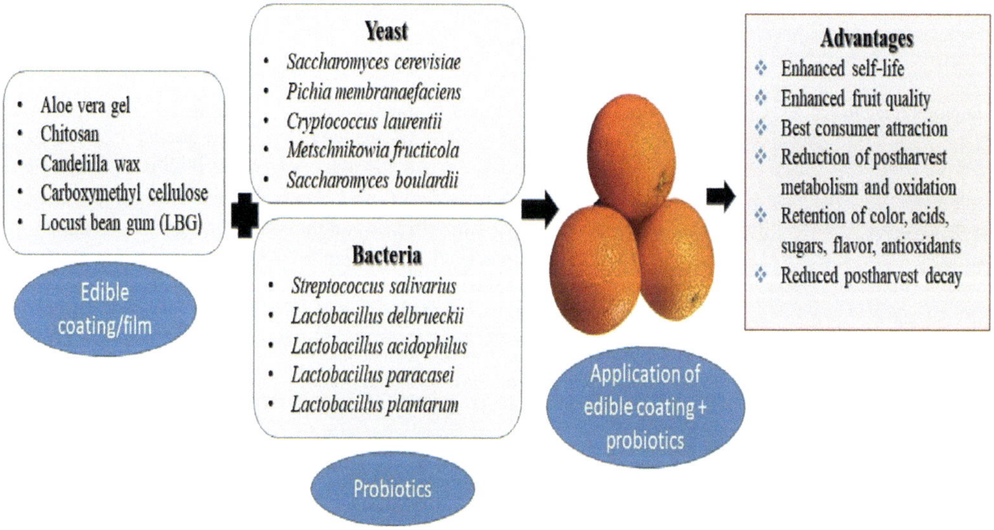

FIGURE 18.1 Overview of probiotics in edible coating/film and their potential application.

18.6 Probiotics in edible coatings and their potential applications

Over the last few years, various biomaterials including chitosan, gelatine, CMC, aloe vera, sodium alginate (NaAlg), and locust bean gum (LBG), have been frequently used as carriers of bacterial and yeast probiotics to improve the quality, functionality, shelf life of the fresh produce (Aloui et al., 2015; Fan et al., 2009; Sharma et al., 2006; Yinzhe and Shaoying, 2013) (Fig. 18.1). These microbial strains have been isolated very commonly from the epiphytic or endophytic regions of fruits and potentially applied as antagonistic agents against a range of phytopathogens to reduce postharvest decay and enhanced shelf life (Droby, 2005, Lutz, et al., 2013; Zhimo et al., 2020). The most common antagonistic mechanism of microbial strains including bacteria and yeasts against phytopathogens are secretion of bioactive compounds, antibiotic synthesis, induction of host resistance, competition for space and nutrients, phytohormone modulation, volatile production, extra polysaccharides synthesis, hydrogen cyanide, siderophore and biofilm formation, and these factors may be regulated by modifying environmental factors and conditions (Liu et al., 2013; Singh et al., 2017, 2020).

18.6.1 Yeast as probiotics in the edible coating/film

The incorporation of yeast microbial cells in edible coatings has been reported used to impart resistance against various phytopathogens during postharvest storage. Viability of yeast cells in edible films/coatings and their impact on management practices can be affected by nature of the yeast, coating materials, temperature, pH as well as the surrounding environment (Potjewijd et al., 1995). Initially *Candida* spp. was incorporated as antagonistic microorganisms in cellulose coatings to inhibit the growth of phytopathogens (Guimarães et al., 2018). However, recently various coating materials and films have been formulated, which can be feasible and efficient in delivery of yeast probiotics without affecting their viability, concentration from the time of formulation or application for consumption. In the different studies researchers have used various yeast strains along with different coating materials against a range of phytopathogens to evaluate their antagonistic potential.

Zhou et al. (2016) evaluated the efficacy of *P. membranaefaciens* alone and in combination with chitosan against *Colletotrichum gloeosporioides* causal pathogen of anthracnose in citrus fruits. Combined application of *P. membranaefaciens* (1×10^8 cells/mL) and chitosan (0.01%, w/v) showed more effective antagonistic activity than individual applications of yeast strains. During this study the author reported low concentrations of chitosan maintained the growth of yeast in vivo and in vitro conditions. Chitosan inhibited spore germination and mycelial growth of *C. gloeosporioides*. Further, Kharchoufi et al. (2018) evaluated the efficacy of chitosan, LBG, with chemically synthesized water pomegranate peel extract (WPPE) and methanol pomegranate peel extract (MPPE) with yeast *Wickerhamomyces anomalus* to control the growth of *Penicillium digitatum* in oranges. The application of chitosan along with WPPE and MPPE inhibited the growth of *P. digitatum* at all concentrations during in vitro trials, whereas LBG showed antagonistic activity only at highest MPPE or WPPE concentrations. The application of MPPE-based coatings slightly inhibited the growth of *W. anomalus*, whereas no significant effects of WPPE were observed on the growth of yeast. In vivo trials on oranges artificially inoculated with *P. digitatum* showed, addition of 0.361 g dry WPPE/mL along with chitosan and LBG coatings, significantly reduced the disease incidence by 49% and 28%, respectively, in comparison to controls. However, addition of *W. anomalus* cells with chitosan and LBG + WPPE, showed strong antagonistic activity against the pathogen and showed disease incidence up to 95% and 75%, respectively, with respect to controls.

Fan et al. (2009) applied a film of *Cryptococcus laurentii* along with combination of alginate (2%, w/v), glycerol (2%, w/v), palmitic acid (0.5%, w/v), glycerol monostearate (0.5%, w/v), and β-cyclodextrin (0.5%, w/v) on strawberries. The viability of *C. laurentii* in the biofilm was remained stable and no any significant changes were observed up to 20 days of storage; however, application of *C. laurentii* along with different concentrations of coating materials inhibited the growth of a mold pathogen and maintained the commercial quality of strawberries. Parafati et al. (2016) reported efficacy of *W. anomalus*, *M. pulcherrima*, and *Aureobasidium pullulans* along with a polysaccharide matrix, LBG against postharvest decays caused by *P. digitatum* and *Penicillium italicum* on mandarins. The application of yeasts, polysaccharides matrix, and with or without LBG, enhanced the viability of microbial cells over time and enhanced peroxidase activity and reduced the rate of decrease of superoxide dismutase as well as significantly controlled postharvest pathogens of mandarin fruits.

Aloui et al. (2015) evaluated efficiency and viability of *W. anomalus* cells along with NaAlg and LBG, based biofilm for the control of *P. digitatum*. Application of NaAlg and LBG films maintained the survivality of more than 85% of the *W. anomalus* population and completely inhibited the growth of *P. digitatum* in a synthetic medium. Although application of *W. anomalus* with NaAlg and LBG had been shown effective in reducing weight loss and maintaining firmness of "Valencia" oranges during storage, and also reduced green mold in the inoculated fruits by more than 73% after 13 days of storage. Marín et al. (2016) applied a formulation of sodium caseinate (NaC) and starch as coating material with surfactants and *Candida sake* as a biocontrol agent to evaluate their effectiveness against the pathogen *B. cinerea* in grapes. The application of coating materials enhanced the survivality of *C. sake* and effectively reduced disease incidence of *B. cinerea*. Addition of the surfactant did not have any significant effect in survivality or disease control, but played an essential role in enhancing dispersability of formulation on the grape surface.

González-Estrada et al. (2017) evaluated the viability, and biocontrol activity of *D. hansenii* mixed with arabinoxylan (AX) matrix on limes. The application of AX as bioactive coatings maintains the yeast viable by more than 97%, and mixed application with yeast cells inhibited the growth of *P. italicum*. Coatings maintained the viability of yeast cells up to 100% and showed effectivity in reducing weight loss, but had no any significant effect on lime color. Iñiguez-Moreno et al. (2020) evaluated the antagonistic potential of *Meyerozyma caribbica* entrapped with NaAlg to control anthracnose caused by *C. gloeosporioides* in avocado fruit. These authors reported NaAlg maintained viability of the yeast cells after coating on the fruit surface. However, minimum reduction in viable cell counts ranged between 0.39 and 1 log10 CFU/mL was recorded in storage conditions. The edible films with *M. caribbica* were capable of inhibiting *C. gloeosporioides* growth in vitro, while preventive treatments were more effective than curative treatments for anthracnose. In vitro control at 25°C decreased by up to 50%; In vivo disease severity reached to 100% at 6°C. Jiwanit et al. (2018) evaluated antagonistic potential of yeast *Pichia guilliermondii*, mixed with *Aloe vera* gel (AG) and applied as a coating on the surface of mandarin. The application of the coating significantly elicited the expression of defense-related genes, enzyme activities, as well as accumulation of bioactive compounds in flavedo tissues of mandarin. These authors further reported that coating of *P. guilliermondii*, with AG significantly inhibited fruit decay caused by *P. digitatum* and application of *P. guilliermondii* with AG significantly maintained fruit quality over the storage period.

Karabulut et al. (2003) evaluated the efficiency of *Metschnikowia fructicola* with ethanol, and sodium bicarbonate (alone or in combinations), applied onto table grapes 24 h before harvest to control postharvest diseases. The application of bioformulations in four experiment significantly reduced the total number of decayed berries caused by *B. cinerea*, *Alternaria* spp., or *Aspergillus niger* after storage for 30 days at 1°C followed by 2 days at 20°C.

18.6.2 Bacteria as probiotics in the edible coating/film

Bacteria have also been incorporated in films or coatings to either confer probiotic ability to the food product or to act as an antimicrobial agent. In general, it is reported that incorporation of probiotic bacteria in edible films and coatings increases their survival rate and enhances shelf life of food or fresh produce (Altamirano-Fortoul et al., 2012; Soukoulis et al., 2017; Concha-Meyer et al., 2011).

There are several reports published that showed bacteria cells as probiotic carried entrapped within edible coatings and applied to fresh produce to enhance the food quality, shelf life, and inhibit the pathogen during postharvest storage. Gregirchak et al. (2020) reported a comparative study of edible coatings on the wheat bread containing LAB. The study revealed coating with *Streptococcus salivarius* subsp. *thermophilus*, *L. delbrueckii* subsp. *bulgaricus*, *L. acidophilus* along with NaAlg, whey and glycerol, showed better and effective protective properties against microbial spoilage. The viability of LAB was found higher in the coating containing alginate and whey, however, after 120 h of storage slightly reduced the viability from initial concentration of 10^8-10^9 CFU/g. The application of edible coatings containing LAB to wheat bread diminished the number of mesophilic aerobic and facultative aerobic bacteria in the bread crust and protected them from contamination by *Aspergillus* and *Penicillium*.

Valerio et al. (2020) evaluated effectiveness of *Lactobacillus paracasei* with citrus pectin coating on dehydrated apple snack. The strain was incorporated into apple slices using vacuum impregnation (IP1) or soaking/stirring (IP2); apple slices were then coated with citrus pectin and dehydrated at 4°C for 14 days. The coating application significantly imbedded microbial populations on the apple surface, and the probiotic products showed increased antioxidant activity and total phenol content. Marín et al. (2019) used *Lactobacillus plantarum* as a probiotic and pregelatinized potato starch (PS) or NaC as coating matrices with oleic acid as surfactants to improve grapes quality during postharvest storage. Application of NaCs based coating enhanced the survivality of *L. plantarum* after 7 days of storage. However, PS-based coating without adding surfactant showed better inhibition potential against disease incidence caused by *Botrytis* in comparison to the NaC formulation and water.

Probiotics viability can also be improved by adding some additional nutrients. In a study Shigematsu et al. (2019) evaluated the effect coconut water on the viability of *L. acidophilus* mixed with NaAlg edible coating. The addition of coconut water improved the viability of *L. acidophilus* for 7 days and provided three log cycle more microbes to the coating than those without the coconut water. However, after 21 days of treatment, viability of bacterial cells was reported to be above log 6 CFU, a level considered adequate for health benefits. The mixed application also conserves the sensory characteristics of minimally processed carrots.

Oregel-Zamudio et al. (2017) reported the impact of *Candelilla* wax alone or in combination with *B. subtilis* to evaluate their effect on the shelf life of strawberry. The application of *Candelilla* wax with *B. subtilis* reduced the decay percentage up to 100% after 6 days of treatment in comparison to the control. Similar trends were observed after application against pathogenic strain *Rhizopus stolonifer* after 24 h. The single application of *B. subtilis* showed maximum effectivity against the pathogens, but did not show such an effect in controlling weight loss or enhancing the shelf life of strawberry.

Barragán-Menéndez et al. (2020) evaluated the effect of *Aloe vera* addition on the viability and antifungal activity of *L. paracasei* bacteria immobilized in chitosan films. They used different proportions of chitosan, *A. vera*, and carbon sources at several pH levels to evaluate optimum viability of microorganisms. After the study they reported optimum conditions of viability to be 100% *A. vera*, pH 4.5 and 0.1 M glucose. The formulation with 10 g/L glycerol and 100% *A. vera*, showed 60% inhibition of *C. gloeosporioides*.

18.7 Conclusion

Coatings containing probiotics are recent and efficient strategies to enhance fruit quality, shelf life, decrease the rate of deterioration of fresh produce. However, inclusion of probiotic microbes has an additional benefit in terms of eliciting immune system, and providing bioactive compounds to humans. Therefore optimum reach of probiotics after consumption can be achieved by an effective delivery system. Searching for bioactive compounds produced in biopolymers of edible coatings or biofilms will confer extra health benefits to the consumers.

The incorporation of probiotics into edible coatings can be influenced by nature of biomaterials, temperature pH, which after encapsulation can alter the potential effects of microorganism used. Therefore compatibility of edible coatings with probiotic microbes needs to be studied for efficient and effective bio formulation. The microencapsulation of probiotics presents several aspects to be considered, such as the type of probiotic microorganisms, the methods of encapsulation, and the coating materials. There is need to explore noble biometrics that may act as inducers of effective bioactive compounds which not only maintain the optimum probiotic concentrations but also provide health benefits after consumption. In addition, screening of effective biocontrol probiotics and their compatibility with the biometrics must be studied for the effective control of phytopathogens during storage and enhancing the shelf life of fresh produce.

References

Abu-Shama, H.S., Abou-Zaid, F.O.F., El-Sayed, E.Z., 2020. Effect of using edible coatings on fruit quality of Barhi date cultivar. Sci. Hortic. 265, 109262.

Adetunji, C.O., Ojediran, J.O., Adetunji, J.B., Owa, S.O., 2019. Influence of chitosan edible coating on postharvest qualities of *Capsicum annuum* L. during storage in evaporative cooling system. Croat. J. food Sci. Technol. 11 (1), 59–66.

Aloui, H., Licciardello, F., Khwaldia, K., Hamdi, M., Restuccia, C., 2015. Physical properties and antifungal activity of bioactive films containing *Wickerhamomyces anomalus* killer yeast and their application for preservation of oranges and control of postharvest green mold caused by *Penicillium digitatum*. Int. J. Food Microbiol. 200, 22–30.

Alsaggaf, M.S., Moussa, S.H., Tayel, A.A., 2017. Application of fungal chitosan incorporated with pomegranate peel extract as edible coating for microbiological, chemical and sensorial quality enhancement of Nile tilapia fillets. Int. J. Biol. Macromol. 99, 499–505.

Altamirano-Fortoul, R., Moreno-Terrazas, R., Quezada-Gallo, A., Rosell, C.M., 2012. Viability of some probiotic coatings in bread and its effect on the crust mechanical properties. Food Hydrocoll. 29 (1), 166–174.

Barragán-Menéndez, C., Gálvez-López, D., Rosas-Quijano, R., Salvador-Figueroa, M., Ovando-Medina, I., Vázquez-Ovando, A., 2020. Films of chitosan and aloe vera for maintaining the viability and antifungal activity of *Lactobacillus paracasei* TEP6. Coatings 10 (3), 259.

Bosch, M., Nart, J., Audivert, S., Bonachera, M.A., Alemany, A.S., Fuentes, M.C., et al., 2012. Isolation and characterization of probiotic strains for improving oral health. Arch. Oral. Biol. 57 (5), 539–549.

Bourtoom, T., 2008. Edible films and coatings: characteristics and properties. Int. Food Res. J. 15 (3), 237–248.

Cerqueira, M.A., Bourbon, A.I., Pinheiro, A.C., Martins, J.T., Souza, B.W.S., Teixeira, J.A., et al., 2011. Galactomannans use in the development of edible films/coatings for food applications. Trends Food Sci. Technol. 22 (12), 662–671.

Ceugniez, A., Coucheney, F., Jacques, P., Daube, G., Delcenserie, V., Drider, D., 2017. Anti-Salmonella activity and probiotic trends of kluyveromyces marxianus S-2-05 and kluyveromyces lactis S-3-05 isolated from a French cheese, Tomme d'Orchies. Res. Microbiol. 168 (6), 575–582.

Chen, F., Zhang, J., Chen, C., Kowaleguet, M.G., Ban, Z., Fei, L., et al., 2020. Chitosan-based layer-by-layer assembly: towards application on quality maintenance of lemon fruits. Adv. Polym. Technol. 2020, 1–10.

Concha-Meyer, A., Schöbitz, R., Brito, C., Fuentes, R., 2011. Lactic acid bacteria in an alginate film inhibit Listeria monocytogenes growth on smoked salmon. Food Control 22 (3–4), 485–489.

De Prisco, A., Mauriello, G., 2016. Probiotication of foods: a focus on microencapsulation tool. Trends Food Sci. Technol. 48, 27–39.

Debeaufort, F., Quezada-Gallo, J.A., Voilley, A., 1998. Edible films and coatings: tomorrow's packagings: a review. Crit. Rev. Food Sci. 38 (4), 299–313.

Douglas, L.C., Sanders, M.E., 2008. Probiotics and prebiotics in dietetics practice. J. Am. Diet. Assoc. 108 (3), 510–521.

Droby, S., 2005. Improving quality and safety of fresh fruits and vegetables after harvest by the use of biocontrol agents and natural materials. In: I International Symposium on Natural Preservatives in Food Systems, vol. 709. pp. 45–52.

Droby, S., Wisniewski, M., El Ghaouth, A., Wilson, C., 2003. Influence of food additives on the control of postharvest rots of apple and peach and efficacy of the yeast-based biocontrol product aspire. Postharvest Biol. Technol. 27 (2), 127–135.

Droby, S., Wisniewski, M., Macarisin, D., Wilson, C., 2009. Twenty years of postharvest biocontrol research: is it time for a new paradigm? Postharvest Biol. Technol. 52 (2), 137–145.

Fan, Y., Xu, Y., Wang, D., Zhang, L., Sun, J., Sun, L., et al., 2009. Effect of alginate coating combined with yeast antagonist on strawberry (fragaria × ananassa) preservation quality. Postharvest Biol. Technol. 53 (1–2), 84–90.

Firdous, N., Khan, M.R., Butt, M.S., Shahid, M., 2020. Application of aloevera gel based edible coating to maintain postharvest quality of tomatoes. Pak. J. Agric. Sci. 57 (1).

Frazão, G.G.S., Blank, A.F., de Aquino Santana, L.C.L., 2017. Optimisation of edible chitosan coatings formulations incorporating *Myrcia ovata* Cambessedes essential oil with antimicrobial potential against foodborne bacteria and natural microflora of mangaba fruits. LWT Food Sci. Technol. 79, 1–10.

Galie, S., Garcia-Gutierrez, C., Miguelez, E.M., Villar, C.J., Lombo, F., 2018. Biofilms in the food industry: health aspects and control methods. Front. Microbiol. 9, 898.

Galus, S., Kadzińska, J., 2015. Food applications of emulsion-based edible films and coatings. Trends Food Sci. Technol. 45 (2), 273–283.

Gao, Y., Kan, C., Chen, M., Chen, C., Chen, Y., Fu, Y., et al., 2018. Effects of chitosan-based coatings enriched with cinnamaldehyde on mandarin fruit cv. ponkan during room-temperature storage. Coatings 8 (10), 372.

Gismondo, M.R., Drago, L., Lombardi, A., 1999. Review of probiotics available to modify gastrointestinal flora. Int. J. Antimicrob. Agents 12 (4), 287–292.

González-Estrada, R.R., Carvajal-Millán, E., Ragazzo-Sánchez, J.A., Bautista-Rosales, P.U., Calderón-Santoyo, M., 2017. Control of blue mold decay on Persian lime: application of covalently cross-linked arabinoxylans bioactive coatings with antagonistic yeast entrapped. LWT Food Sci. Technol. 85, 187–196.

Gregirchak, N., Stabnikova, O., Stabnikov, V., 2020. Application of lactic acid bacteria for coating of wheat bread to protect it from microbial spoilage. Plant Foods Hum. Nutr. 1–7.

Grosso, A.L., Asensio, C.M., Grosso, N.R., Nepote, V., 2020. Increase of walnuts' shelf life using a walnut flour protein-based edible coating. LWT 118, 108712.

Guilbert, S., Gontard, N., Cuq, B., 1995. Technology and applications of edible protective films. Packag. Technol. Sci. 8, 339–346.

Guimarães, A., Abrunhosa, L., Pastrana, L.M., Cerqueira, M.A., 2018. Edible films and coatings as carriers of living microorganisms: A new strategy towards biopreservation and healthier foods. Compr. Rev. Food Sci. Food Saf. 17 (3), 594–614.

Gunaydin, S., Karaca, H., Palou, L., De La Fuente, B., Pérez-Gago, M.B., 2017. Effect of hydroxypropyl methylcellulose-beeswax composite edible coatings formulated with or without antifungal agents on physicochemical properties of plums during cold storage. J. Food Qual. 2017.

Holzapfel, W.H., Haberer, P., Snel, J., Schillinger, U., in't Veld, J.H.H., 1998. Overview of gut flora and probiotics. Int. J. Food Microbiol. 41 (2), 85–101.

Ibrahim, M.A., Sharoba, A.M., El Waseif, K.H., El Mansy, H.A., El Tanahy, H.H., 2017. Effect of edible coating by chitosan with lemongrass and thyme oils on strawberry quality and shelf life during Storage. J. Food Technol. Nutr. Sci. 3 (007).

Iñiguez-Moreno, M., Ragazzo-Sánchez, J.A., Barros-Castillo, J.C., Sandoval-Contreras, T., Calderón-Santoyo, M., 2020. Sodium alginate coatings added with *Meyerozyma caribbica*: Postharvest biocontrol of *Colletotrichum gloeosporioides* in avocado (Persea americana Mill. cv. Hass). Postharvest Biol. Technol. 163, 111123.

Jiwanit, P., Pitakpornpreecha, T., Pisuchpen, S., Leelasuphakul, W., 2018. The use of Aloe vera gel coating supplemented with *Pichia guilliermondii* BCC5389 for enhancement of defense-related gene expression and secondary metabolism in mandarins to prevent postharvest losses from green mold rot. Biol. Control 117, 43–51.

Kader, A.A., 1992. Postharvest biology and technology: an overview. Postharvest Technol. Hortic. Crop.

Kanmani, P., Lim, S.T., 2013. Development and characterization of novel probiotic-residing pullulan/starch edible films. Food Chem. 141 (2), 1041–1049.

Karabulut, O.A., Smilanick, J.L., Gabler, F.M., Mansour, M., Droby, S., 2003. Near-harvest applications of *Metschnikowia fructicola*, ethanol, and sodium bicarbonate to control postharvest diseases of grape in central California. Plant. Dis. 87 (11), 1384–1389.

Kharchoufi, S., Parafati, L., Licciardello, F., Muratore, G., Hamdi, M., Cirvilleri, G., et al., 2018. Edible coatings incorporating pomegranate peel extract and biocontrol yeast to reduce *Penicillium digitatum* postharvest decay of oranges. Food Microbiol. 74, 107–112.

Krochta, J.M., 2002. Proteins as raw materials for films and coatings: definitions, current status, and opportunities, Protein-based Films and Coatings., vol. 1. pp. 1–40.

Kumura, H., Tanoue, Y., Tsukahara, M., Tanaka, T., Shimazaki, K., 2004. Screening of dairy yeast strains for probiotic applications. J. Dairy Sci. 87 (12), 4050–4056.

Liu, J., Sui, Y., Wisniewski, M., Droby, S., Liu, Y., 2013. Utilization of antagonistic yeasts to manage postharvest fungal diseases of fruit. Int. J. Food Microbiol. 167 (2), 153–160.

Lopez-Rubio, A., Gavara, R., Lagaron, J.M., 2006. Bioactive packaging: turning foods into healthier foods through biomaterials. Trends Food Sci. Technol. 17 (10), 567–575.

Lutz, M.C., Lopes, C.A., Rodriguez, M.E., Sosa, M.C., Sangorrín, M.P., 2013. Efficacy and putative mode of action of native and commercial antagonistic yeasts against postharvest pathogens of pear. Int. J. Food Microbiol. 164 (2–3), 166–172.

Marín, A., Cháfer, M., Atarés, L., Chiralt, A., Torres, R., Usall, J., et al., 2016. Effect of different coating-forming agents on the efficacy of the biocontrol agent *Candida sake* CPA-1 for control of *Botrytis cinerea* on grapes. Biol. Control 96, 108–119.

Marín, A., Plotto, A., Atarés, L., Chiralt, A., 2019. Lactic acid bacteria incorporated into edible coatings to control fungal growth and maintain postharvest quality of grapes. HortScience 54 (2), 337–343.

Melo, N.F.C.B., de MendonçaSoares, B.L., Diniz, K.M., Leal, C.F., Canto, D., Flores, M.A., et al., 2018. Effects of fungal chitosan nanoparticles as eco-friendly edible coatings on the quality of postharvest table grapes. Postharvest Biology and Technology 139, 56–66.

Mohamed, C., Elise, N.G., Etienne, T.V., Loiseau, G., Montet, D., 2020. Antifungal activity of edible coating made from chitosan and lactoperoxidase system against phomopsis sp. RP257 and pestalotiopsis sp. isolated from mango. J. Food Saf. e12785.

Oregel-Zamudio, E., Angoa-Pérez, M.V., Oyoque-Salcedo, G., Aguilar-González, C.N., Mena-Violante, H.G., 2017. Effect of candelilla wax edible coatings combined with biocontrol bacteria on strawberry quality during the shelf-life. Sci. Hortic. 214, 273–279.

OuYang, Q., Okwong, R.O., Chen, Y., Tao, N., 2020. Synergistic activity of cinnamaldehyde and citronellal against green mold in citrus fruit. Postharvest Biol. Technol. 162, 111095.

Palou, L., Smilanick, J.L., Droby, S., 2008. Alternatives to conventional fungicides for the control of citrus postharvest green and blue moulds. Stewart Postharvest Rev. 2 (2), 1–16.

Parafati, L., Vitale, A., Restuccia, C., Cirvilleri, G., 2016. The effect of locust bean gum (LBG)-based edible coatings carrying biocontrol yeasts against *Penicillium digitatum* and *Penicillium italicum* causal agents of postharvest decay of mandarin fruit. Food Microbiol. 58, 87–94.

Parvez, S., Malik, K.A., Ah Kang, S., Kim, H.Y., 2006. Probiotics and their fermented food products are beneficial for health. J. Appl. Microbiol. 100 (6), 1171–1185.

Paul, S.K., 2019. Edible films and coatings for fruits and vegetables. Reference Module in Materials Science and Materials Engineering. Elsevier, Amsterdam, The Netherlands.

Pavli, F., Tassou, C., Nychas, G.J.E., Chorianopoulos, N., 2018. Probiotic incorporation in edible films and coatings: Bioactive solution for functional foods. Int. J. Mol. Sci. 19 (1), 150.

Pinto, S.S., Verruck, S., Vieira, C.R., Prudêncio, E.S., Amante, E.R., Amboni, R.D., 2015. Influence of microencapsulation with sweet whey and prebiotics on the survival of Bifidobacterium-BB-12 under simulated gastrointestinal conditions and heat treatments. LWT Food Sci. Technol. 64 (2), 1004–1009.

Pinzon, M.I., Sanchez, L.T., Garcia, O.R., Gutierrez, R., Luna, J.C., Villa, C.C., 2020. Increasing shelf life of strawberries (Fragaria ssp) by using a banana starch-chitosan-Aloe vera gel composite edible coating. Int. J. Food Sci. Technol. 55 (1), 92–98.

Pop, O.L., Pop, C.R., Dufrechou, M., Vodnar, D.C., Socaci, S.A., Dulf, F.V., et al., 2020. Edible films and coatings functionalization by probiotic incorporation: a review. Polymers 12 (1), 12.

Pop, O.L., Socaci, S.A., Suharoschi, R., Vodnar, D.C., 2019. Pro and prebiotics foods that modulate human health. The Role of Alternative and Innovative Food Ingredients and Products in Consumer Wellness. Elesvier, Amsterdam, The Netherlands, p. 283.

Potjewijd, R., Nisperos, M.O., Burns, J.K., Parish, M., Baldwin, E.A., 1995. Cellulose-based coatings as carriers for *Candida guillermondii* and *Debaryomyces* sp. in reducing decay of oranges. HortScience 30 (7), 1417–1421.

Quirós-Sauceda, A.E., Ayala-Zavala, J.F., Olivas, G.I., González-Aguilar, G.A., 2014. Edible coatings as encapsulating matrices for bioactive compounds: a review. J. Food Sci. Technol. 51 (9), 1674–1685.

Raghav, P.K., Agarwal, N., Saini, M., 2016. Edible coating of fruits and vegetables: a review. Int. J. Sci. Res. Mod. Educ. 1 (1), 188–204.

Ramos, Ó.L., Fernandes, J.C., Silva, S.I., Pintado, M.E., Malcata, F.X., 2012. Edible films and coatings from whey proteins: a review on formulation, and on mechanical and bioactive properties. Crit. Rev. Food Sci. Nutr. 52 (6), 533–552.

Rikhotso, M.M., Magwaza, L.S., Tesfay, S.Z., Mditshwa, A., 2019. Evaluating the efficacy of chitosan and CMC incorporated with moringa leaf extracts on reducing peteca spot incidence on 'Eureka'lemon. J. Food Sci. Technol. 56 (11), 5074–5086.

Rivera-Espinoza, Y., Gallardo-Navarro, Y., 2010. Non-dairy probiotic products. Food Microbiol. 27 (1), 1–11.

Rodrigues, F.J., Cedran, M.F., Garcia, S., 2018. Influence of linseed mucilage incorporated into an alginate-base edible coating containing probiotic bacteria on shelf-life of fresh-cut yacon (*Smallanthus sonchifolius*). Food Bioprocess. Technol. 11 (8), 1605–1614.

Rojas-Graü, M.A., Soliva-Fortuny, R., Martín-Belloso, O., 2009. Edible coatings to incorporate active ingredients to fresh-cut fruits: a review. Trends Food Sci. Technol. 20 (10), 438–447.

Saarela, M., Mayrhofer, S., Domig, K.J., Danielsen, M., Mayo, B., Mättö, J., 2008. Susceptibility of bifidobacteria originating from different sources to tetracycline, erythromycin, streptomycin and vancomycin. Int. J. Probiotics Prebiotics 3, 269–270.

Saarela, M., Mogensen, G., Fonden, R., Mättö, J., Mattila-Sandholm, T., 2000. Probiotic bacteria: safety, functional and technological properties. J. Biotechnol. 84 (3), 197–215.

Sánchez-González, L., Saavedra, J.I.Q., Chiralt, A., 2013. Physical properties and antilisterial activity of bioactive edible films containing *Lactobacillus plantarum*. Food Hydrocoll. 33 (1), 92–98.

Sánchez-González, L., Saavedra, J.I.Q., Chiralt, A., 2014. Antilisterial and physical properties of biopolymer films containing lactic acid bacteria. Food Control 35 (1), 200–206.

Sanders, M.E., 2008. Probiotics: definition, sources, selection, and uses. Clin. Infect. Dis. 46 (Supplement_2), S58–S61.

Santa Catharina, C.M., Bretschneider, F.G.B., Piva, B., Weber, C.I., Stadler, F., Romio, A.P., et al., 2020. Edible coating containing carvacrol for postharvest microbiological conservation of tomato/Revestimento comestível contendo carvacrol para conservação microbiológica pós-colheita de tomate. Braz. J. Dev. 6 (1), 4786–4794.

Sharma, N., Verma, U., Awasthi, P., 2006. A combination of the yeast Candida utilis and chitosan controls fruit rot in tomato caused by Alternaria alternata (Fr.) Keissler and Geotrichum candidum Link ex Pers. J. Hortic. Sci. Biotechnol. 81 (6), 1043–1051.

Shigematsu, E., Dorta, C., Santos, D.N., Ferreira, K.A., Góes-Favoni, S.P., Oshiiwa, M., et al., 2019. Edible coating with coconut water to preserve probiotic strains and sensory characteristics of minimally processed carrots. Int. Food Res. J. 26 (4).

Singh, V.K., Singh, A.K., Kumar, A., 2017. Disease management of tomato through PGPB: current trends and future perspective. 3 Biotech. 7 (4), 255.

Singh, M., Srivastava, M., Kumar, A., Singh, A.K., Pandey, K.D., 2020. Endophytic bacteria in plant disease management. Microbial Endophytes. Woodhead Publishing, pp. 61–89.

Soukoulis, C., Behboudi-Jobbehdar, S., Macnaughtan, W., Parmenter, C., Fisk, I.D., 2017. Stability of *Lactobacillus rhamnosus* GG incorporated in edible films: Impact of anionic biopolymers and whey protein concentrate. Food Hydrocoll. 70, 345–355.

Spadaro, D., Droby, S., 2016. Development of biocontrol products for postharvest diseases of fruit: the importance of elucidating the mechanisms of action of yeast antagonists. Trends Food Sci. Technol. 47, 39–49.

Šuput, D.Z., Lazić, V.L., Popović, S.Z., Hromiš, N.M., 2015. Edible films and coatings: Sources, properties and application. Food Feed Res. 42 (1), 11–22.

Valencia-Chamorro, S.A., Palou, L., del Rio, M.A., Pérez-Gago, M.B., 2011. Antimicrobial edible films and coatings for fresh and minimally processed fruits and vegetables: a review. Crit. Rev. Food Sci. Nutr. 51 (9), 872–900.

Valerio, F., Volpe, M.G., Santagata, G., Boscaino, F., Barbarisi, C., Di Biase, M., et al., 2020. The viability of probiotic *Lactobacillus paracasei* IMPC2. 1 coating on apple slices during dehydration and simulated gastro-intestinal digestion. Food Biosci. 34, 100533.

Vieira, M.G.A., da Silva, M.A., dos Santos, L.O., Beppu, M.M., 2011. Natural-based plasticizers and biopolymer films: a review. Eur. Polym. J. 47 (3), 254–263.

Waewthongrak, W., Pisuchpen, S., Leelasuphakul, W., 2015. Effect of *Bacillus subtilis* and chitosan applications on green mold (*Penicilium digitatum* Sacc.) decay in citrus fruit. Postharvest Biol. Technol. 99, 44–49.

Xu, D., Qin, H., Ren, D., 2018. Prolonged preservation of tangerine fruits using chitosan/montmorillonite composite coating. Postharvest Biol. Technol. 143, 50–57.

Xu, L., Zhang, B., Qin, Y., Li, F., Yang, S., Lu, P., et al., 2020. Preparation and characterization of antifungal coating films composed of sodium alginate and cyclolipopeptides produced by *Bacillus subtilis*. Int. J. Biol. Macromol. 143, 602–609.

Yinzhe, R., Shaoying, Z., 2013. Effect of carboxymethyl cellulose and alginate coating combined with brewer yeast on postharvest grape preservation. ISRN Agron. 2013.

Zhimo, V.Y., Biasi, A., Kumar, A., Feygenberg, O., Salim, S., Vero, S., et al., 2020. Yeasts and bacterial consortia from kefir grains are effective biocontrol agents of postharvest diseases of fruits. Microorganisms 8 (3), 428.

Zhou, Y., Zhang, L., Zeng, K., 2016. Efficacy of *Pichia membranaefaciens* combined with chitosan against *Colletotrichum gloeosporioides* in citrus fruits and possible modes of action. Biol. Control 96, 39–47.

CHAPTER

19

Microbial volatiles: Prospects for plant defense and disease management

Ajay Kumar[1], V. Yeka Zhimo[1], Antonio Biasi[1], Oleg Feygenberg[1], Shoshana Salim[1], James Francis White[2], Michael Wisniewski[3] and Samir Droby[1]

[1]Department of Postharvest Science, Agricultural Research Organization (ARO), Volcani Center, Rishon LeZion, Israel [2]Department of Plant Biology, Rutgers University, New Brunswick, NJ, United States [3]Appalachian Fruit Research Station, Agricultural Research Service, United States Department of Agriculture, Kearneysville, WV, United States

19.1 Introduction

Microbes are ubiquitous in nature and interact with large numbers of different microbes; interactions include bacteria–bacteria, bacteria–fungi, fungi–fungi, fungi–bacteria, and form complex multispecies networks (Hung et al., 2015). However, these microbial communities produce a range of volatiles or secondary metabolites (Schulz and Dickschat, 2007; Morath et al., 2012; Müller et al., 2013) in which some of the metabolites, whose diffusible nature act as signaling or infochemical molecules may affect population as well as functional genomics of the microorganisms and facilitate microbial interactions, plant–microbe interactions, and functioning as growth promotors as well as potential biocontrol agents (Ryan and Dow, 2008; Frey-Klett et al., 2011; Haq et al., 2014).

Volatile organic compounds (VOCs) are low-molecular-weight carbon–containing compounds (100–500 Da) having lower range of boiling point and high vapor pressure (Schmidt et al., 2016; Lemfack et al., 2014). The easy diffusion as well as high evaporation rate of the volatiles may cause easy access up to long-range distances, below the ground, in between the soil space, liquids, or in the atmosphere (Schmidt et al., 2015). In the recent past, >1000 volatiles from different microbes, including bacteria (Kai et al., 2009), yeasts

DOI: https://doi.org/10.1016/B978-0-12-821843-3.00021-0

(Ljunggren et al., 2019), fungi (Schnürer et al., 1999; Effmert et al., 2012) have been isolated and applied in the various fields of agriculture as growth regulators, potential biocontrol agents, or health care pharmaceuticals related with human beings (Garbeva et al., 2014a,b; Schmidt et al., 2015).

Microbial volatiles generally constitute a diverse range of chemical functional groups like alkanes, alcohols, aldehydes, ketones, esters, ethers, terpenoids, benzenoids, etc. (Piechulla and Degenhardt, 2014; Schmidt et al., 2015) and the composition of microbial volatiles depends upon various factors such growth stage of the microorganism, media constituents, source of energy, moisture contents, temperature, and pH of the growing media (Wheatley, 2002; Insam and Seewald, 2010; Romoli et al., 2014).

19.2 Factors affecting microbial volatile composition

Microbial volatiles are direct or indirect products of various biochemical pathways in which amino acid catabolism, fatty acids degradation, and terpenoids biosynthesis are the some most common pathways (Piechulla and Degenhardt, 2014; Peñuelas et al., 2014). Some of the microbial volatiles are very commonly detected in volatile spectrum of most of the microorganisms, however, some are strain specific and present in particular microorganisms and can vary quantitatively as well as qualitatively with changes in media composition, substrate, variation in pH, temperature, oxygen availability as well as physiological state of the microorganism (Garbeva et al., 2014a,b; Kai et al., 2009; Misztal et al., 2018).

In a study, Misztal et al. (2018) evaluated limiting factors, which influence diversity and abundance of microbial volatiles. They found diversity as well abundance largely dependent upon the environmental factors. Emission rate and composition of volatiles also depend upon physiological state, media constituents, and growth stage of the microbes. The growth stage of microbial cells was observed to be correlated with microbial metabolism and it varied with the different stages, that is, lag, log, and stationary of the microbes (Bunge et al., 2008).

Zareian et al. (2018) reported detailed volatile spectrum of *Pseudomonas* and enhancement of glucose content in the medium from 0.5% to 1 % was found to increase the cyclic compounds along with some alcohols, aldehydes, and ketones. However, addition of egg white protein was noticed to enhance sulfur derivatives like methanethiol, dimethyl disulfide, methanol, 1-propanol, 2-methylpropanol as well as benzene derivatives, whereas mixed supplement of egg white protein with glucose enhanced the short-chain carboxylic acid in the volatile spectrum. Kai et al. (2009), reviewed addition of glucose in the nutrient broth of *Stenotrophomonas rhizophila* responsible for the variation of gas chromatography (GC) profiles, both qualitatively and quantitatively. They also reported that betaphenylethanol and dimethyl pyrazine emitted irrespective of glucose in the media, however, trimethyl pyrazine, tetramethyl pyrazine, and betaphenylethyl acetate had only been present in the absence of glucose in the media. In another study, different forms of glucose (L or D form) were also reported to affect the volatile profile. Media supplementation with L-glucose led to reduced amounts of volatile emissions as compared to D-glucose is firmly established (Fiddaman and Rossall, 1993, 1994). However, the study conducted by Ryu et al. (2004) reported that addition of sucrose as the main carbohydrate led to enhanced production of 2,3-butanediol.

Hamilton-Kemp et al. (2005) reported that supplementation of fatty acids in media composition was effective to enhance aliphatic alcohols such as 1-octanol, 1-decanol, and 1-dodecanol in the volatile spectrum of Enterobacteriaceae members. Duponnois and Kisa (2006) reported that addition of trehalose was effective in influencing the volatiles released from *Pseudomonas monteilii*.

Garbeva et al. (2014a) observed volatile emissions by *Pseudomonas fluorescens* and its utilization by strains of *Collimonas pratensis* and *Serratia plymuthica*, as energy sources for the growth. Similarly Owen et al. (2007) reported carbon containing volatile production by rhizosphere microbes acting as energy resources for other bacterial strains. Kleinheinz et al. (1999) reported that degradation products of alpha-pinene released by plants were utilized as sole energy sources by the strain *P. fluorescens*. Meanwhile, microbial strains also helped in degradation of alpha-pinene.

Garbeva et al. (2014b) reported volatile release from several soil bacteria affecting growth, antibiotic production, and gene expression of other bacterial strains. Furthermore, application of volatiles released from four phylogenetically different bacterial strains *C. pratensis*, *S. plymuthica*, *Paenibacillus* sp., *Pedobacter* sp. in the growth media of *P. fluorescens*, stimulating growth of *P. fluorescens*, and all the four strains differentially affected gene expression of *P. fluorescens*. Moreover, volatile composition produced by mixtures of four bacterial strains was different from individual bacterial species grown under monoculture conditions.

19.3 Extraction of volatile compounds

With the latest advancement in technology, a range of volatile compounds has been detected, which are present even in very low quantities. The advanced separation capacity and sensitive detection power of GC coupled with mass spectroscopy (MS) appeared as a standard approach for the analysis of microbial volatiles (Farag et al., 2006; Farag and Wessjohann, 2012). Meanwhile, sensitive extraction or trapping materials enhance the detection probability of novel volatiles (Wenke et al., 2012). For extraction of preconcentrate volatiles from aqueous samples, a closed-loop stripping apparatus is preferred (Meruva et al., 2004) but, separation or purification of volatiles from solvents is difficult, however, solid-phase microextractions is a rapid sample preparation method and also allows to combination of all steps, including sampling, extraction, concentration, and injection into the analytical instrument in a single step and also free from the solvent; and it is considered as an advanced technique because there is no requirement for solvent volatile quantification as well as detection (Goupry et al., 2000; Marilley and Casey, 2004).

Identification of volatile compounds is generally carried out by comparing the mass spectra of volatile compounds present in a sample with spectra from different databases present in the libraries like the Wiley or NIST. However, the compounds can only be considered as identified if they demonstrate identical Kovat's indices on two columns of different polarity and display mass spectra coincident with the library (Schulz-Bohm et al., 2017; Audrain et al., 2015). Online databases of microbial volatiles are also available (http://bioinformatics.charite.de/mvoc), listing more than 1000 microbial volatile compounds along with the detailed information of producing or secreting microorganisms, biosynthetic pathways as well as their biological effects

(Lemfack et al., 2014). However, for novel volatiles not mentioned in the database, chemical structures can be elucidated through nuclear magnetic resonance (NMR) analysis.

19.4 Diversity of volatiles

All living organisms produce volatiles, which some play essential roles in microbial interaction, plant–microbe interaction, and growth promotion as well as defense (Garbeva et al., 2004). The spectrum of microbial volatiles are highly diverse in nature in terms of chemical constituents and are dominated by hydrocarbons, alcohols, ketones, benzenoids, terpenes, pyrazines, esters, acids, and each have specific functional attributes such as pathogenesis, signal transduction (Piechulla and Degenhardt, 2014). Some of the important constituents of volatiles are discussed in the following:

Hydrocarbons: Hydrocarbon-derived volatiles are produced from fatty acid biosynthetic pathways and are classified into long-chain and short-chain hydrocarbons. Long-chain hydrocarbons are generally synthesized by bacteria and cyanobacteria; short-chain hydrocarbons are usually synthesized by fungi (Tellez et al., 2001). In a study Audrain et al. (2015) reviewed short-chain alkanes such as from decane to tetradecane that are not frequently synthesized by microbes, however, long-chain hydrocarbons such as hexadecane were abundantly present in cyanobacteria.

Ketones/alcohols: The alcohol and ketone derivatives of volatiles are generally produced via fatty acid decarboxylation. Ryu et al. (2003) reported 3-hydroxy-2-butanone and its oxidized form 2,3-butanedione derived from the fermentation of pyruvate under anaerobic conditions. However, Hamilton-Kemp et al. (2005) observed that long-chain alcohol derivatives frequently found in Gram-negative bacteria which may have resulted from beta and alpha oxidation of fatty acids, while short-chain alcoholic volatiles were detected in the firmicutes and proteobacteria (Farag et al., 2013; Whiteson et al., 2014).

Acids: The content of acids in bacterial volatile spectrum is generally less in comparison to ketones. However, presence of several short-chain acids such as acetic acids, propionic acid, or butyric acids is also reported (Schulz and Dickschat, 2007). These aliphatic organic acids represent major by-products of anaerobic metabolism and formed during bacterial fermentation of carbohydrates.

Terpenes: Terpenes are derived from the mevalonate pathway or deoxyxylulose pathway and are generally monoterpenes (C10) and sesquiterpenes, and their derived volatiles are detected in bacterial volatiles (Audrain et al., 2015).

Sulfur-containing compounds: These compounds are produced during catabolic degradation of methionine. However, biogenesis of methionine derived volatiles such as dimethyl sulfide and 1-(methylthio)-3 pentanone are often mediated by bacteria (Audrain et al., 2015).

19.5 Microbial volatiles: Approaches to plant defense

Microbes and their volatiles perform several functions and have been described to contribute in the various aspects of plant growth promotion or biotic stress management. The release of microbial volatiles directly protect plants by inhibiting growth of pathogenic organisms, however, they indirectly elicit plant defense response by enhancing immune response against pathogen attack. However, in sustainable agriculture, microbial VOCs have been broadly used as plant growth promotors and potential biocontrol agents against a broad range of phytopathogens during developmental or postharvest storage (Tables 19.1 and 19.2).

To combat biotic stress and pathogen attack, plants defend themselves via two main resistance pathways, including salicylic acid-mediated systemic acquired resistance (SAR) or the ethylene/jasmonic acid-mediated systemic resistance (EA-ISR) (Garbeva and Weisskopf,

TABLE 19.1 Summary of volatile organic compounds (VOCs) and their effects on plant growth and development.

Microbial species	Plant	Volatile compounds	Effect	References
Arthrobacter agilis	*Medicago sativa*	*N,N*-dimethyl-hexadecanamine	Modulation growth and morphogenetic processes	Velázquez-Becerra et al. (2011)
Bacillus subtilis	*Solanum lycopersicum L.*	Albuterol and 1,3-propanediol	Growth promotion	Tahir et al. (2017a,b)
Pseudomonas fluorescens	*Nicotiana tabacum*	13-Tetradecadien-1-ol, 2-butanone, and 2-methyl-n-1-tridecane	Growth promotion	Park et al. (2015)
Bacillus megaterium	*Arabidopsis thaliana*	2-Pentylfuran	Growth promotion	Zou et al. (2010)
Podosphaera pannosa	*Rosa chinensis Jacq.*	Hexadecanol, octadecanol, tetradecanol, *n*-butyl hexadecanoate, *n*-butyl stearate	Used to detect mildew-infected rose plants	Yang et al. (2019)
Bacillus sp., *Serratia* sp., *Staphylococcus* sp.	*Lactuca sativa*	Acetoin	Growth promotion	Fincheira et al. (2016)
Bacillus amyloliquefaciens and *Bacillus atrophaeus*	*N. tabacum*	2-Pentylfuran, 13-tetradecadien-1-ol, 2-butanone, and 2-methyl-*n*-1-tridecene	Growth promotion and phytopathogens control	Tahir et al. (2017a,b)
B. amyloliquefaciens	*Lycopersicon esculentum, cv. Jiangshu*	2-Undecanone, undecanal, 2-tridecanone	tomato wilt pathogen management	Raza et al. (2016a,b)
Streptomyces sp.	*Centella asiatica L.*	2,5-Bis(hydroxymethyl) furan monoacetate	Antifungal properties, protect plants from infection	Phuakjaiphaeo et al. (2016)
Paenibacillus polymyxa	*Cucumis sativus*	1-Octen-3-ol, benzothiazole, citronellol	Mycelial growth inhibition and of pigmentation	Zhao et al. (2011)
Fusarium graminearum	*Triticum aestivum*	(Z)-3-Hexenyl acetate	Increased resistance	Ameye et al. (2015)

TABLE 19.2 Summary of volatile organic compounds (VOCs) employed for controlling pathogens causing plant diseases and their effects on plant.

Volatiles compounds	Microbes	Pathogens	Plant	Disease	Effect	References
3-methyl-1-butanol, 2-methyl-1-butanol, and phenylethyl alcohol	*Streptomyces lavendulae*	*Ceratocystis fimbriata*	*Ipomoea batatas*	Black-spot disease	Pathogen control	Li et al. (2020)
Phenylethyl alcohol, α-curcumene, and E (β) farnesene	*Muscodor brasiliensis sp. nov.*	*Penicillium digitatum*	*Schinus terebinthifolius*	Green mold symptoms	Prevention of green mold symptoms development, reduction in use of fungicides	Pena et al. (2019)
Ethanol 3-methyl-1-butanol, 2-methyl-1-butanol and phenylethyl alcohol, ethyl acetate and ethyl octanoate	*Saccharomyces cerevisiae*	*Sclerotinia sclerotiorum*	*Phaseolus vulgaris*	White mold	Reduction in disease, control the pathogen	Fialho et al. (2011)
1,3,5,7-cyclooctatetraene and 3-methyl-1-butanol	*Candida intermedia strain C410*	*Botrytis cinerea*	*Fragaria* × *ananassa* Duch.	Fruit rot	Control of postharvest fruit diseases	Huang et al. (2011)
Sesquiterpenes (α-bisabolene and bulnesene)	*Anthracophyllum discolor*. Sp4	*Mucor miehei*	*Solanum tuberosum*	White-rot	Controlled fungal plant diseases	Schalchli et al. (2015)
Trans-1,10-dimethyl-trans-9-decalol, 1,4-dimethyladamantane, 1H-benzocyclohepten-7-ol	*Streptomyces platensis* F-1	*Rhizoctonia solani, S. sclerotiorum, Botrytis cinerea*	*Oryza sativa, Brassica napus, Fragaria* × *ananassa*	Leaf blight/seedling blight, fruit rot	Used as biofumigant to control fungal plant diseases	Wan et al. (2008)
Disulfide, dimethyl trisulfide and acetophenone	*Streptomyces globisporus JK-1*	*Botrytis cinerea*	*Solanum lycopersicum*	Postharvest gray mold	Potential in control of postharvest grey mold	Li et al. (2004)
Decyl alcohol and 3,5,5-trimethylhexanol	*Bacillus strain D13*	*Xanthomonas oryzae* pv. *oryzae* (*Xoo*)	*O. sativa*	Bacterial leaf blight	Helpful in understanding biocontrol mechanisms of *Bacillus*	Xie et al. (2018)

2,4-di-*tert*-butylphenol	*Flavobacterium johnsoniae strain GSE09*	*Phytophthora capsici*	*Piper nigrum*	Crown blight	Disease suppression	Sang and Kim (2012)
2, 6-di-tert-butyl-4-methylphenol, and 2,4-di-tert-butylphenol	*Bacillus siamensis G-3*	*Botrytis cinerea* and *Rhizopus stolonifer*	*Rubus idaeus*	Raspberry postharvest mixed diseases	Disease control	Zhang et al. (2020)
2-ethylhexanol, 1-nonanol and 6-pentyl-2H-pyran- 2-one	*Trichoderma asperellum* T1	*Corynespora cassiicola* and *Curvularia aeria*	*Lactuca sativa*	Leaf spot fungal pathogens	Controlled plant diseases and enhanced crop yields	Wonglom et al. (2020)
2-phenethyl	*Aureobasidium pullulans* strains L1 and L8	*Monilinia fructicola, Monilinia fructigena*	*Prunus avium, Prunus persica, and Prunus armeniaca*	Control of brown rot	Reduced disease	Di Francesco et al. (2020)
Volatile organic compound	*Bacillus* sp. MB015 and *Pseudomonas* sp. MB108	*Fusarium oxysporum*	*Physalis peruviana L*	Vascular wilt caused	Reduced disease severity	Toloza-Moreno et al. (2020)
3-methyl-1-butanol, phenylethyl alcohol and 2-methyl-1-butanol	*Pseudomonas chlororaphis* subsp. *aureofaciens* SPS-41	*C. fimbriata*	*I. batatas*	Black rot disease	Reduced weight loss rate and disease severity	Zhang et al. (2019)
1-(2-Aminophenyl) ethanone and benzothiazole	*Bacillus amyloliquefaciens* PP19, *Exiguobacterium acetylicum* SI17, *B. pumilus* PI26	*Peronophythora litchii*	*Litchi chinensis*	Downy blight	Reduced the severity of litchi downy blight	Zheng et al. (2019)
Ethylene, hydrogen cyanide, aldehydes, and ketones	*Trichoderma harzianum*	*F. oxysporum* f. sp. *cicero*	*Cicer arietinum* L.	Wilting of chickpea	Strong inhibitory effect on the mycelial growth	Kumar et al. (2019)
Benzothiazole and CF-3	*Bacillus subtilis* CF-3	*M. fructicola*	*P. persica*	Peach fruit decay	Prevented decay	Zhou et al. (2019)
Tricholin, peptaibols, massoilactone, gliovirin, heptelidic acid	*Trichoderma isolates* and *B. subtilis*	*Phytophthora drechsleri*	*Cucumis melo*	Root rot	Percent disease incidence was decreased while percent plant survival, dry root, and shoot weight increased	Anjum et al. (2019)

(*Continued*)

TABLE 19.2 (Continued)

Volatiles compounds	Microbes	Pathogens	Plant	Disease	Effect	References
3-methyl-1-butanol and 2-methyl-1-butanol	*S. cerevisiae*	*Phyllosticta citricarpa*	*Citrus sinensis*	Citrus black spot	Reduced mycelial growth, germination, and appressorium formation from conidia	Toffano et al. (2017)
Pyrazine (2,5-dimethyl), benzothiazole, phenol (4-chloro-3-methyl), and phenol-2,4-bis (1,1-dimethylethyl)	*Bacillus velezensis* ZSY-1	*Alternaria solani*, *Botrytis cinerea*	*S. lycopersicum*	Early blight and grey mold	Inhibited the spore production rate	Gao et al. (2017)
Benzenes, ketones, aldehydes, alkanes, acids, one furan and naphthalene compound	*B. amyloliquefaciens T-5*	*Ralstonia solanacearum*	*Lycopersicon esculentum cv. Jiangshu*	Bacterial wilt pathogen	Controlled plant pathogens	Raza et al. (2016a,b)
1,3 pentadiene, acetoin (3-hydroxy-2-butanone) and thiophene	*B. amyloliquefaciens CPA-8*	*M. fructicola*	*Prunus avium*	Fruit pathogen decays	Suppressed mycelial growth	Gotor-Vila et al. (2017)
Cycloocta-2,4-dien-1-ol	*Trichoderma koningiopsis* YIM PH30002	*Phoma herbarum*, *Fusarium flocciferum*, *Scytalidium lignicola*, and *Epicoccum nigrum*	*Panax notoginseng*	Root-rot disease	Disease suppression	Chen et al. (2016)
Mesityl oxide, 2-methylpropyl ester, 4-methyldecane, 4-penten-2-one,4-methyl, o-Xylene, Toluene	*Staphylococcus sciuri MarR44*	*Colletotrichum nymphaeae*	*Fragaria* × *ananassa*	Strawberry anthracnose	Promising biofumigant for management of strawberry anthracnose	Alijani et al. (2019)

2020). SAR is generally induced after attack of pathogenic microbes, however, EA-ISR responds after colonization with beneficial microbes or their metabolites or "O"antigen of the lipopolysaccharides existing in outer membranes (Meziane et al., 2005). Several bacterial volatiles such as butanediol secreted from *Bacillus* (Ryu et al., 2004) and tridecane secreted from *Paenibacillus polymyxa* have been reported to actively elicit defense response of plants against pathogen attack. Lee et al. (2012) reported application of tridecane in pure form to upregulate defense-related genes and significant reduction in disease symptoms caused by *Pseudomonas syringae* on host *Arabidopsis thaliana*. But, here concentration used for the upregulation of defense genes were high (or in the millimolar range). However, in the volatile spectrum range, tridecane did not elicit defense gene upregulation. The author further suggested that presence of different volatiles might have synergistic effects on the regulation of defense genes (Lee et al., 2012; Garbeva and Weisskopf, 2020).

Recently, Garbeva and Weisskopf (2020) reviewed the role of microbial volatiles in plant protection, generally due to induction of plant defense mechanisms rather than by the pathogen inhibition (Sharifi and Ryu, 2016). However, some strains showed both properties of defense induction as well as pathogen inhibition (Tahir et al., 2017a,b). Furthermore, they stated that induction of plant resistance immune response as well as pathogen inhibition is strain, compound, and dose specific.

19.6 Application of microbial volatiles in plant disease management

19.6.1 Fungal volatiles in disease management

Fungi secrete large numbers of volatile compounds having stimulatory as well as inhibitory functions with potential opportunities in growth promotion and plant disease management (Schalchli et al., 2016; Raza et al., 2015). Application of volatiles against fungal pathogens may inhibit mycelial growth, conidia formation, as well as virulence gene suppression in pathogens (Minerdi et al., 2009). Although effectiveness of volatiles are strain specific, but also depends on various factors such as physiological state and nutrient source in the growth media (Schalchli et al., 2011; Strobel et al., 2001). The most common and preferred medium for fungal growth is potato dextrose agar and its composition significantly affects volatiles spectra of microbes grown on it (Ezra and Strobel, 2003; Schieber and Saldaña, 2009). Demain (1992) reported that white-rot fungal polysaccharides and lipids were the most efficient carbon sources for the biosynthesis of various secondary metabolites in comparison to dextrose.

Currently a range of volatiles have been isolated and applied against various pathogens. Schalchli et al. (2015) reported volatiles isolated from the white-rot fungi *Anthracophyllum discolor* and evaluated their efficacy against mycelial growth of *Botrytis cinerea, Fusarium oxysporum*, and *Mucor miehei*. The volatiles secreted by the strains of *A. discolor* showed inhibitory potential against pathogenic strains *M. miehei* (76%), *B. cinerea* (20%), and *F. oxysporum* (10%). The volatile spectrum of strains showed sesquiterpenes, especially α-bisabolene and bulnesene and the chlorinated aromatic compounds such as 1,5-dichloro-2,3-dimethoxybenzene, 3,5-dichloro-4-methoxybenzaldehyde, and 3-chloro-4-methoxybenzaldehyde as the most abundant volatiles revealed in the GC–MS volatile spectrum of a strain of *A. discolor*.

Mercier and Jiménez (2004) used *Muscodor albus* fumigation for the control of various pathogens, including *Botrytis, Colletotrichum, Geotrichum, Monilinia, Penicillium,* and *Rhizopus,* the main casual agents of postharvest diseases of fresh fruits. The application of *M. albus* completely inhibited the growth of blue mold (*Penicillium expansum*) and gray mold (*B. cinerea*) in wound-inoculated fruits. The volatile spectrum of *M. albus* showed 2-methyl-1-butanol and isobutyric acid as abundant volatiles and might be responsible for the inhibition of pathogenic strains growth. Similarly Huang et al. (2011) used a *Candida intermedia* strain against *B. cinerea* and application significantly suppressed the conidial germination and mycelial growth of *B. cinerea*. The volatile spectrum of *C. intermedia* showed 49 volatiles of various chemical constituents, in which 1,3,5,7-cyclooctatetraene and 3-methyl-1-butanol were the most abundant volatiles that might help in significant reduction of fruit rot in strawberry incidence caused by *B. cinerea*. Sesquiterpenes and monoterpenes were reported as major volatile compounds produced by endophytic fungal strain *Hypoxylon anthochroum* and the application of volatiles significantly inhibited the growth of *F. oxysporum* on cherry tomatoes with percentages over 50% (Macías-Rubalcava et al., 2018).

19.6.2 Bacterial volatiles in plant disease management

A range of bacterial volatiles has been reported and it has been assumed that volatiles are released from primary or secondary metabolism of bacterial physiology and the specificity of the bacterial volatiles are strain specific (Kai et al., 2009). Bacterial communities associated with different plant organs as epiphytes or endophytes play important roles in plant growth promotion, phytopathogen management, as well as mitigation of various biotic and abiotic stresses via various mechanisms including nitrogen fixation, nutrient acquisition, and phytohormone synthesis. In addition, various secreted volatiles play an important role in growth promotion, phytopathogen inhibition, plant–microbe interaction inter and intraorganismic communication between both below ground and above ground microorganisms (Gutiérrez-Luna et al., 2010; Maffei, 2010; Maffei et al., 2011; Lemfack et al., 2014; Kanchiswamy et al., 2015).

Over the last few decades a range of bacterial volatiles, including 2-pentanone, 3-hexanone, 4-heptanone, 1-hexanol, indole, pentadecane 2-heptanol, 2-undecanone, 2-tridecanone, 2-pentadecanone, acetophenone tridecanal, tetradecanal, 6,10,14-trimethyl 2-pentadecanone, and benzaldehyde, have been obtained for various bacteria (Schulz and Dickschat, 2007; Gutiérrez-Luna et al., 2010; Blom et al., 2011; Groenhagen et al., 2013; Weise et al., 2014). Some of these like 1-hexanol, indole, and pentadecane are well-known for the growth promotion in plants (Blom et al., 2011). However, volatile 3-hexanone, produced by *Burkholderia ambifaria*, significantly increased the biomass of *Arabidopsis* (Groenhagen et al., 2013). The differential emission of acetophenone, tridecanal, tetradecanal, 6,10,14-trimethyl 2-pentadecanone, and benzaldehyde has been reported in different rhizobacterial strains of lemon and effects observed in *Arabidopsis* roots was proportional to the type and amount of compounds produced by the bacteria (Gutiérrez-Luna et al., 2010).

The GC–MS spectrum of volatiles from *Streptomyces albulus* showed 4-methoxystyrene, 2-pentylfuran, and anisole as most abundant volatile compounds which exhibited strong antifungal activity against *Sclerotinia sclerotiorum* and *F. oxysporum*. Volatile application significantly inhibited mycelial growth of *S. sclerotiorum* (100%) and *F. oxysporum* (25%)

(Wu et al., 2015). However, the GC–MS spectrum of *Streptomyces alboflavus* showed 2-methylisoborneol as the most abundant volatiles among 27 reported volatiles and their application significantly reduced the growth of various phytopathogens including *Fusarium moniliforme*, *Aspergillus flavus*, *Aspergillus ochraceus*, *Aspergillus niger*, and *Penicillium citrinum* (Wang et al., 2013a). Danaei et al. (2013) reported 2-methyl-1-butanol, 2-pentane 3-methyl, acetaldehyde, *n*-hexane, and palmitic acid, in which 2-methyl-1-butanol was the most abundant volatile in the GC–MS spectrum of *Streptomyces coelicolor*, and their application significantly inhibited the growth of *B. cinerea* and *Penicillium chrysogenum*. In another study, Danaei et al. (2014) reported 20 volatiles from the strains of *Streptomyces griseus* which significantly inhibited spore germination and mycelial growth of both *P. chrysogenum* and *B. cinerea*. Phenol, 2-methyl-5-1methylethyl (carvacrol) was the most abundant volatile, however, isocyclocitral and benzene, 1,2-dimethoxy-4-1-methylethenyl were present in lower concentrations, and authors reported activity of phenol, 2-methyl-5-1methylethyl, dimethyl sulfoxide (DMSO), cyclohexanol, and naphthalene antifungal agents against the fungal pathogens. Li et al. (2020) reported volatiles from *Streptomyces lavendulae* and evaluated antagonistic activity against *Ceratocystis fimbriata*, the causal agent of "black-spot disease" in sweet potato during postharvest storage. The GC–MS spectrum of *S. lavendulae* showed heptadecane, tetradecane, and 3-methyl-1-butanol to be the most abundant volatiles. However, compounds 2-methyl-1-butanol, 3-methyl-1-butanol, pyridine, and phenylethyl alcohol had also showed strong antifungal potential against *C. fimbriata*. The volatiles inhibited growth of mycelia as well as sporulation of *C. fimbriata* under in vitro; while, in vivo exposure to the *S. lavendulae* strain significantly decreased lesion diameter and rate of water loss in sweet potato.

In another study, Wan et al. (2008) studied the volatiles spectrum of *Streptomyces platensis* and evaluated their efficacy against pathogenic strains of *Rhizoctonia solani*, *S. sclerotiorum*, and *B. cinerea*. Application of *S. platensis* significantly reduced the pathogenicity of all the three pathogenic strains. The volatile profile of *S. platensis* showed geosmin, as the most abundant and phenylethyl alcohol. However, (+)-epi-bicyclesesquiphellandrene (antifungal compounds) were also significantly present in the volatile spectrum of *S. platensis*.

Wang et al. (2013b) reported 19 volatiles [o-xylene; 2-heptanone; benzene; 2-propenyl; benzene; 1,4-dichloro; undecane; 1,2-methyl; acetophenone; 2-nonanone; nonanone; 1-(4-methylphenyl) ethanone; 2-decanone; naphthalene; 2-undecanone; tridecane; 2-dodecanone; tetradecane; 2-tridecanone; pentradecane; hexadecane; and tetradecane] in the GC–MS spectrum of *Bacillus amyloliquefaciens*. Strain showed effective antagonistic effects against *F. oxysporum*. Similarly, Chaurasia et al. (2005) reported volatiles of *Bacillus subtilis*, which also showed antagonistic activity as well as structural deformities against six pathogenic fungi during in vitro evaluation.

Xie et al. (2018) reported decyl alcohol and 3,5,5-trimethylhexanol among 12 volatiles isolated from a *Bacillus* strain having antagonistic activity against *Xanthomonas oryzae*. Application of volatiles significantly reduced the colony diameter as well as cell motility of *X. oryzae*. Application of decyl alcohol and 3,5,5-trimethylhexanol showed strong antagonistic activity and inhibited the growth of pathogens at minimum inhibitory amounts equivalent to 0.48 and 2.4 mg, respectively. Furthermore, they reported that application of these volatiles repressed transcriptional expression of virulence-associated genes. Similarly, Fiddaman and Rossall (1993) reported volatiles from strain of *B. subtilis*, having antifungal activity against *R. solani* and *Pythium ultimum*.

In a study, Yuan et al. (2012) reported 36 distant volatiles from strains of *B. amyloliquefaciens*, in which 11 volatiles completely inhibited growth and spore germination of *F. oxysporum*. The volatile spectrum showed 12 benzenes, 7 alkyls, 3 alcohols, 7 ketones, 2 aldehydes, 3 naphthyls, 1 ester, and 1 ether compound, in which all the benzene compounds exhibited antifungal activity, especially against *F. oxysporum*. benzothiazole (BTH), phenol, and 2,3,6-trimethylphenol were the most abundant volatiles. Furthermore, they also reported that some volatiles, including toluene, ethylbenzene, propylbenzene, isopropylbenzene pentadecane, and tetradecane exhibited less effectiveness, and dodecane and tridecane did not show any antifungal activity.

Massawe et al. (2018) reported volatiles of an endophytic *Bacillus* sp. strain and its potential application against *S. sclerotiorum*. Volatile application significantly inhibited mycelial growth of *S. sclerotiorum*. Moreover, volatiles 2-undecanone, BTH, 1,3-butadiene, and *N,N*-dimethyldodecylamine exhibited strong inhibition with percentage inhibition reaching up to 98.1%, 73.8%, 72.0%, and 58.9%, respectively, against *S. sclerotiorum* mycelia in in vitro assays. Pentadecane and IR-(+)-α-pinene exhibited moderate levels of inhibition (19.6%−34.6%), however, dodecane and tridecane exhibited only 6.4% and 11.2% inhibition. However, Zheng et al. (2019) reported volatiles 1-(2-aminophenyl) ethanone (EA) and BTH produced by several strains of *Bacillus* having inhibitory activity against pathogen *Peronophythora litchii* on plate assays as well as on litchi fruit, while one of the volatiles α-Farnesene was able to suppress pathogen growth under in vivo conditions but not under in vitro conditions. Ul Hassan et al. (2019) reported 3-methyl-1-butanol as the most abundant volatile in GC−MS spectra of *Bacillus licheniformis*, which showed antagonistic activity against mycotoxigenic strains of *Aspergillus* and *Penicillium*. In vitro application of volatiles showed strong inhibition potential against *Aspergillus westerdijkiae* followed by *Aspergillus carbonarius*, *Penicillium verrucosum*, *A. niger*, *A. flavus*, *Aspergillus parasiticus*, and minimum inhibition against *A. ochraceus*. Zhang et al. (2013) reported O-anisaldehyde as the most abundant volatiles produced by the strain of *Bacillus atrophaeus* and showed inhibitory activity against fungal pathogen *B. cinerea*.

Ting et al. (2011) reported volatile compounds from endophytic bacterial strains of *Herbaspirillum* sp. having inhibitory activity against *F. oxysporum*. Application of *Herbaspirillum* sp. showed highest effectiveness with 20.3% inhibition against the pathogen *F. oxysporum*. However, analysis of the volatile spectrum revealed 2-pentane 3-methyl, methanethiol and 3-undecene to exhibit the most significant antagonistic activity, and the extent of inhibition was attributed to the presence or synergistic action of components. The GC−MS spectrum of *Pseudomonas chlororaphis* showed, 3-methyl-1-butanol, followed by phenylethyl alcohol as the most abundant volatiles in which volatile 2-methyl-1-butanol showed strong inhibitory potential against *C. fimbriata*, the causal agent of black rot disease in sweet potato (Zhang et al., 2019).

19.7 Conclusion

Volatiles of microbial origin show interspecific and intraspecific interactions microbes and host, and the extent of interaction is much longer and influential than the nonvolatile compounds (Tirranen and Gitelson, 2006). However, there are few studies on the molecular mechanisms involved in these interactions. Molecular mechanistic studies of volatiles might provide deep insights of potential application in the various aspects of growth promotion and phytopathogen management (Chen, 2008). Diversity and composition of microbial volatiles depend considerably upon the substrate as well as medium constituents. There is an urgent need to study substrates that support maximum amounts of volatiles. Microbial volatiles are being broadly used in plant disease management. Research on formulation of volatiles to combat plant disease management is also needed. Although, methods for production of large amounts of volatile compounds in fermentation culture have not been achieved, there is a need to develop this technology to produce volatiles for application in agricultural practice.

References

Alijani, Z., Amini, J., Ashengroph, M., Bahramnejad, B., 2019. Antifungal activity of volatile compounds produced by *Staphylococcus sciuri* strain MarR44 and its potential for the biocontrol of Colletotrichum nymphaeae, causal agent strawberry anthracnose. Int. J. Food Microbiol. 307, 108276.

Ameye, M., Audenaert, K., De Zutter, N., Steppe, K., Van Meulebroek, L., Vanhaecke, L., et al., 2015. Priming of wheat with the green leaf volatile Z-3-hexenyl acetate enhances defense against *Fusarium graminearum* but boosts deoxynivalenol production. Plant Physiol. 167, 1671–1684.

Anjum, M.Z., Ghazanfar, M.U., Hussain, I., 2019. Bio-efficacy Of *Trichoderma isolates* and *Bacillus subtilis* against root rot Of muskmelon *Cucumis melo* L. caused by *Phytophthora drechsleri* under Controlled and field conditions. Pak. J. Bot. 51, 1877–1882.

Audrain, B., Farag, M.A., Ryu, C.M., Ghigo, J.M., 2015. Role of bacterial volatile compounds in bacterial biology. FEMS Microbiol. Rev. 39, 222–233.

Blom, D., Fabbri, C., Connor, E.C., Schiestl, F.P., Klauser, D.R., Boller, T., et al., 2011. Production of plant growth modulating volatiles is widespread among rhizosphere bacteria and strongly depends on culture conditions. Env. Microbiol. 13, 3047–3058.

Bunge, M., Araghipour, N., Mikoviny, T., Dunkl, J., Schnitzhofer, R., Hansel, A., et al., 2008. On-line monitoring of microbial volatile metabolites by proton transfer reaction-mass spectrometry. Appl. Env. Microbiol. 74, 2179–2186.

Chaurasia, B., Pandey, A., Palni, L.M.S., Trivedi, P., Kumar, B., Colvin, N., 2005. Diffusible and volatile compounds produced by an antagonistic *Bacillus subtilis* strain cause structural deformations in pathogenic fungi in vitro. Microbiol. Res. 160, 75–81.

Chen, H.W., 2008. Microbial volatile organic compounds: generation pathways and mass spectrometric detection. China Biotechnol. 28, 124–133.

Chen, J.L., Liu, K., Miao, C.P., Sun, S.Z., Chen, Y.W., Xu, L.H., et al., 2016. Salt tolerance of endophytic *Trichoderma koningiopsis* YIM PH30002 and its volatile organic compounds (VOCs) allelopathic activity against phytopathogens associated with *Panax notoginseng*. Ann. Microbiol. 66, 981–990.

Danaei, M., Baghizadeh, A., Porseiedi, S., Amini, J., Yaghoobi, M.M., 2013. Effect of volatile substances of *Streptomyces coelicolor* on control of *Botrytis cinerea* and *Penicillium chrysogenum*. Casp. J. Appl. Sci. Res. 2, 11.

Danaei, M., Baghizadeh, A., Pourseyedi, S., Amini, J., Yaghoobi, M.M., 2014. Biological control of plant fungal diseases using volatile substances of *Streptomyces griseus*. Eur. J. Exp. Biol. 4, 334–339.

Demain, A.L., 1992. Regulation of secondary metabolism. In: Finkelstein, D.B., Ball, Ch (Eds.), Biotechnology of filamentous fungi: technology and products. Newnes, Stoneham, MA, 536-500.

Di Francesco, A., Di Foggia, M., Baraldi, E., 2020. Aureobasidium pullulans volatile organic compounds as alternative postharvest method to control brown rot of stone fruits. Food Microbiol. 87, 103395.

Duponnois, R., Kisa, M., 2006. The possible role of trehalose in the mycorrhiza helper bacterium effect. Botany 84, 1005–1008.

Effmert, U., Kalderas, J., Warnke, R., Piechulla, B., 2012. Volatile mediated interactions between bacteria and fungi in the soil. J. Chem. Ecol. 38, 665–703.

Ezra, D., Strobel, G.A., 2003. Effect of substrate on the bioactivity of volatile antimicrobials produced by *Muscodor albus*. Plant Sci. 165, 1229–1238.

Farag, M.A., Ryu, C.M., Sumner, L.W., Paré, P.W., 2006. GC–MS SPME profiling of rhizobacterial volatiles reveals prospective inducers of growth promotion and induced systemic resistance in plants. Phytochemistry. 67, 2262–2268.

Farag, M.A., Wessjohann, L.A., 2012. Volatiles profiling in medicinal licorice roots using steam distillation and solid-phase microextraction (SPME) coupled to chemometrics. J. Food Sci. 77, C1179–C1184.

Farag, M.A., Zhang, H., Ryu, C.M., 2013. Dynamic chemical communication between plants and bacteria through airborne signals: induced resistance by bacterial volatiles. J. Chem. Ecol. 39, 1007–1018.

Fialho, M.B., Moraes, M.H.D.D., Tremocoldi, A.R., Pascholati, S.F., 2011. Potential of antimicrobial volatile organic compounds to control *Sclerotinia sclerotiorum* in bean seeds. Pesqui. Agropecu. Bras. 46, 137–142.

Fiddaman, P.J., Rossall, S., 1993. The production of antifungal volatiles by *Bacillus subtilis*. J. Appl. Bacteriol. 74, 119–126.

Fiddaman, P.J., Rossall, S., 1994. Effect of substrate on the production of antifungal volatiles from *Bacillus subtilis*. J. Appl. Bacteriol. 76, 395–405.

Fincheira, P., Venthur, H., Mutis, A., Parada, M., Quiroz, A., 2016. Growth promotion of *Lactuca sativa* in response to volatile organic compounds emitted from diverse bacterial species. Microbiol. Res. 93, 39–47.

Frey-Klett, P., Burlinson, P., Deveau, A., Barret, M., Tarkka, M., Sarniguet, A., 2011. Bacterial-fungal interactions: hyphens between agricultural, clinical, environmental, and food microbiologists. Microbiol. Mol. Biol. Rev. 75, 583–609.

Gao, Z., Zhang, B., Liu, H., Han, J., Zhang, Y., 2017. Identification of endophytic *Bacillus velezensis* ZSY-1 strain and antifungal activity of its volatile compounds against Alternaria solani and *Botrytis cinerea*. Biol. Control. 105, 27–39.

Garbeva, P., Hordijk, C., Gerards, S., De Boer, W., 2014a. Volatiles produced by the mycophagous soil bacterium *Collimonas*. FEMS Microbiol. Ecol. 87, 639–649.

Garbeva, P., Hordijk, C., Gerards, S., De Boer, W., 2014b. Volatile-mediated interactions between phylogenetically different soil bacteria. Front. Microbiol. 5, 289.

Garbeva, P., Voesenek, K., Elsas, J.D.V., 2004. Quantitative detection and diversity of the pyrrolnitrin biosynthetic locus in soil under different treatments. Soil. Biol. Biochem. 36, 1453–1463.

Garbeva, P., Weisskopf, L., 2020. Airborne medicine: bacterial volatiles and their influence on plant health. N. Phytologist 226 (1), 32–43.

Gotor-Vila, A., Teixidó, N., Di Francesco, A., Usall, J., Ugolini, L., Torres, R., et al., 2017. Antifungal effect of volatile organic compounds produced by *Bacillus amyloliquefaciens* CPA-8 against fruit pathogen decays of cherry. Food Microbiol. 64, 219–225.

Goupry, S., Rochut, N., Robins, R.J., Gentil, E., 2000. Evaluation of solid-phase microextraction for the isotopic analysis of volatile compounds produced during fermentation by lactic acid bacteria. J. Agric. Food Chem. 48, 2222–2227.

Groenhagen, U., Baumgartner, R., Bailly, A., Gardiner, A., Eberl, L., Schulz, S., et al., 2013. Production of bioactive volatiles by different *Burkholderia ambifaria* strains. J. Chem. Ecol. 39, 892–906.

Gutiérrez-Luna, F.M., López-Bucio, J., Altamirano-Hernández, J., Valencia-Cantero, E., De La Cruz, H.R., Macías-Rodríguez, L., 2010. Plant growth-promoting rhizobacteria modulate root-system architecture in *Arabidopsis thaliana* through volatile organic compound emission. Symbiosis. 51, 75–83.

Hamilton-Kemp, T., Newman, M., Collins, R., Elgaali, H., Yu, K., Archbold, D., 2005. Production of the long-chain alcohols octanol, decanol, and dodecanol by *Escherichia coli*. Curr. Microbiol. 51, 82–86.

Haq, I.U., Zhang, M., Yang, P., Van Elsas, J.D., 2014. The interactions of bacteria with fungi in soil: emerging concepts. Adv. Appl. Microbiol. 89, 185–215.

Huang, R., Li, G.Q., Zhang, J., Yang, L., Che, H.J., Jiang, D.H., et al., 2011. Control of postharvest *Botrytis* fruit rot of strawberry by volatile organic compounds of *Candida intermedia*. Phytopathology. 101, 859–869.

Hung, R., Lee, S., Bennett, J.W., 2015. Fungal volatile organic compounds and their role in ecosystems. Appl. Microbiol. Biotechnol. 99, 3395–3405.

Insam, H., Seewald, M.S.A., 2010. Volatile organic compounds (VOCs) in soils. Biol. Fertil. Soils 46, 199–213.

Kai, M., Haustein, M., Molina, F., Petri, A., Scholz, B., Piechulla, B., 2009. Bacterial volatiles and their action potential. Appl. Microbiol. Biotechnol. 81, 1001–1012.

Kanchiswamy, C.N., Malnoy, M., Maffei, M.E., 2015. Bioprospecting bacterial and fungal volatiles for sustainable agriculture. Trends Plant Sci. 20, 206–211.

Kleinheinz, G.T., Bagley, S.T., John, W.S., Rughani, J.R., McGinnis, G.D., 1999. Characterization of alpha-pinene-degrading microorganisms and application to a bench-scale biofiltration system for VOC degradation. Arch. Env. Con. Toxicol. 37, 151–157.

Kumar, M., Kumar, V., Rana, M., Srivastava, S., 2019. Effect of volatile and non volatile compounds of trichoderma Spp. against *Fusarium isolates* causing chickpea wilt in Punjab. Plant Arch. 19, 159–162.

Lee, B., Farag, M.A., Park, H.B., Kloepper, J.W., Lee, S.H., Ryu, C.M., 2012. Induced resistance by a long-chain bacterial volatile: elicitation of plant systemic defense by a C13 volatile produced by *Paenibacillus polymyxa*. PLoS One. 7, 11.

Lemfack, M.C., Nickel, J., Dunkel, M., Preissner, R., Piechulla, B., 2014. mVOC: a database of microbial volatiles. Nucl. Acids Res. 42, D744–D748.

Li, H.F., Imai, T., Ukita, M., Sekine, M., Higuchi, T., 2004. Compost stability assessment using a secondary metabolite: geosmin. Env. Technol. 25, 1305–1312.

Li, X., Li, B., Cai, S., Zhang, Y., Xu, M., Zhang, C., et al., 2020. Identification of rhizospheric actinomycete *Streptomyces lavendulae* SPS-33 and the inhibitory effect of its volatile organic compounds against *Ceratocystis Fimbriata* in postharvest sweet potato (*Ipomoea Batatas* (L.) Lam.). Microorganisms 8, 319.

Ljunggren, J., Borrero-Echeverry, F., Chakraborty, A., Lindblom, T.U., Hedenström, E., Karlsson, M., et al., 2019. Yeast volatomes differentially affect larval feeding in an insect herbivore. Appl. Env. Microbiol. 85, e01761–19.

Macías-Rubalcava, M.L., Sánchez-Fernández, R.E., Roque-Flores, G., Lappe-Oliveras, P., Medina-Romero, Y.M., 2018. Volatile organic compounds from *Hypoxylon anthochroum* endophytic strains as postharvest mycofumigation alternative for cherry tomatoes. Food Microbiol. 76, 363–373.

Maffei, M.E., 2010. Sites of synthesis, biochemistry and functional role of plant volatiles. S. Afr. J. Botany 76 (4), 612–631.

Maffei, M.E., Gertsch, J., Appendino, G., 2011. Plant volatiles: production, function and pharmacology. Nat. Prod. Rep. 28, 1359–1380.

Marilley, L., Casey, M.G., 2004. Flavours of cheese products: metabolic pathways, analytical tools and identification of producing strains. Int. J. Food Microbiol. 90, 139–159.

Massawe, V.C., Hanif, A., Farzand, A., Mburu, D.K., Ochola, S.O., Wu, L., et al., 2018. Volatile compounds of endophytic *Bacillus* spp. have biocontrol activity against *Sclerotinia sclerotiorum*. Phytopathology 108, 1373–1385.

Mercier, J., Jiménez, J.I., 2004. Control of fungal decay of apples and peaches by the biofumigant fungus *Muscodor albus*. Postharvest Biol. Technol. 31, 1–8.

Meruva, N.K., Penn, J.M., Farthing, D.E., 2004. Rapid identification of microbial VOCs from tobacco molds using closed-loop stripping and gas chromatography/time-of-flight mass spectrometry. J. Ind. Microbiol. Biotechnol. 31, 482–488.

Meziane, H., Van Der Sluis, I., Van Loon, L.C., Höfte, M., Bakker, P.A., 2005. Determinants of *Pseudomonas putida* WCS358 involved in inducing systemic resistance in plants. Mol. Plant Pathol. 6, 177–185.

Minerdi, D., Bossi, S., Gullino, M.L., Garibaldi, A., 2009. Volatile organic compounds: a potential direct long-distance mechanism for antagonistic action of *Fusarium oxysporum* strain MSA 35. Env. Microbiol. 11, 844–854.

Misztal, P.K., Lymperopoulou, D.S., Adams, R.I., Scott, R.A., Lindow, S.E., Bruns, T., et al., 2018. Emission factors of microbial volatile organic compounds from environmental bacteria and fungi. Env. Sci. Tech. 52, 8272–8282.

Morath, S.U., Hung, R., Bennett, J.W., 2012. Fungal volatile organic compounds: a review with emphasis on their biotechnological potential. Fungal Biol. Rev. 26, 73–83.

Müller, A., Faubert, P., Hagen, M., zu Castell, W., Polle, A., Schnitzler, J.P., et al., 2013. Volatile profiles of fungi–chemotyping of species and ecological functions. Fungal Genet. Biol. 54, 25–33.

Owen, S.M., Clark, S., Pompe, M., Semple, K.T., 2007. Biogenic volatile organic compounds as potential carbon sources for microbial communities in soil from the rhizosphere of Populus tremula. FEMS Microbiol. Lett. 268, 34–39.

Park, Y.S., Dutta, S., Ann, M., Raaijmakers, J.M., Park, K., 2015. Promotion of plant growth by *Pseudomonas fluorescens* strain SS101 via novel volatile organic compounds. Biochem. Biophys. Res. Commun. 461, 361–365.

Pena, L.C., Jungklaus, G.H., Savi, D.C., Ferreira-Maba, L., Servienski, A., Maia, B.H., et al., 2019. *Muscodor brasiliensis* sp. nov. produces volatile organic compounds with activity against *Penicillium digitatum*. Microbiol. Res. 221, 28–35.

Peñuelas, J., Asensio, D., Tholl, D., Wenke, K., Rosenkranz, M., Piechulla, B., et al., 2014. Biogenic volatile emissions from the soil. Plant Cell Env. 37, 1866–1891.

Phuakjaiphaeo, C., Chang, C.I., Ruangwong, O., Kunasakdakul, K., 2016. Isolation and identification of an antifungal compound from endophytic *Streptomyces* sp. CEN 26 active against *Alternaria brassicicola*. Lett. Appl. Microbiol. 63, 38–44.

Piechulla, B., Degenhardt, J., 2014. The emerging importance of microbial volatile organic compounds. Plant Cell Env. 37, 811–812.

Raza, W., Ling, N., Yang, L., Huang, Q., Shen, Q., 2016a. Response of tomato wilt pathogen *Ralstonia solanacearum* to the volatile organic compounds produced by a biocontrol strain *Bacillus amyloliquefaciens* SQR-9. Sci. Rep. 6, 1–13.

Raza, W., Wang, J., Wu, Y., Ling, N., Wei, Z., Huang, Q., et al., 2016b. Effects of volatile organic compounds produced by *Bacillus amyloliquefaciens* on the growth and virulence traits of tomato bacterial wilt pathogen *Ralstonia solanacearum*. Appl. Microbiol. Biotechnol. 100, 7639–7650.

Raza, W., Yuan, J., Ling, N., Huang, Q., Shen, Q., 2015. Production of volatile organic compounds by an antagonistic strain *Paenibacillus polymyxa* WR-2 in the presence of root exudates and organic fertilizer and their antifungal activity against *Fusarium oxysporum* f. sp. niveum. Biol. Control 80, 89–95.

Romoli, R., Papaleo, M.C., De Pascale, D., Tutino, M.L., Michaud, L., LoGiudice, A., et al., 2014. GC–MS volatolomic approach to study the antimicrobial activity of the antarctic bacterium *Pseudoalteromonas* sp. TB41. Metabolomics 10, 42–51.

Ryan, R.P., Dow, J.M., 2008. Diffusible signals and interspecies communication in bacteria. Microbiology 154, 1845–1858.

Ryu, C.M., Farag, M.A., Hu, C.H., Reddy, M.S., Kloepper, J.W., Paré, P.W., 2004. Bacterial volatiles induce systemic resistance in *Arabidopsis*. Plant Physiol. 134, 1017–1026.

Ryu, C.M., Farag, M.A., Hu, C.H., Reddy, M.S., Wei, H.X., Paré, P.W., et al., 2003. Bacterial volatiles promote growth in *Arabidopsis*. Proc. Natl. Acad. Sci. U.S.A. 100, 4927–4932.

Sang, M.K., Kim, K.D., 2012. The volatile-producing *Flavobacterium johnsoniae* strain GSE09 shows biocontrol activity against *Phytophthora capsici* in pepper. J. Appl. Microbiol. 113, 383–398.

Schalchli, H., Hormazabal, E., Becerra, J., Birkett, M., Alvear, M., Vidal, J., et al., 2011. Antifungal activity of volatile metabolites emitted by mycelial cultures of saprophytic fungi. Chem. Ecol. 27, 503–513.

Schalchli, H., Hormazábal, E., Becerra, J., Briceño, G., Hernández, V., Rubilar, O., et al., 2015. Volatiles from white-rot fungi for controlling plant pathogenic fungi. Chem. Ecol. 31, 754–763.

Schalchli, H., Tortella, G.R., Rubilar, O., Parra, L., Hormazabal, E., Quiroz, A., 2016. Fungal volatiles: an environmentally friendly tool to control pathogenic microorganisms in plants. Crit. Rev. Biotechnol. 36, 144–152.

Schieber, A., Saldaña, M., 2009. Potato peels: a source of nutricionally and farmacologically interesting compounds-a review. Food 3, 23–29.

Schmidt, R., Cordovez, V., De Boer, W., Raaijmakers, J., Garbeva, P., 2015. Volatile affairs in microbial interactions. ISME J. 2329–2335.

Schmidt, R., Etalo, D.W., de Jager, V., Gerards, S., Zweers, H., de Boer, W., et al., 2016. Microbial small talk: volatiles in fungal–bacterial interactions. Front. Microbiol. 6, 1495.

Schnürer, J., Olsson, J., Börjesson, T., 1999. Fungal volatiles as indicators of food and feeds spoilage. Fungal Genet. Biol. 27, 209–217.

Schulz, S., Dickschat, J.S., 2007. Bacterial volatiles: the smell of small organisms. Nat. Prod. Rep. 24, 814–842.

Schulz-Bohm, K., Martín-Sánchez, L., Garbeva, P., 2017. Microbial volatiles: small molecules with an important role in intra-and inter-kingdom interactions. Front. Microbiol. 8, 2484.

Sharifi, R., Ryu, C.M., 2016. Are bacterial volatile compounds poisonous odors to a fungal pathogen *Botrytis cinerea*, alarm signals to *Arabidopsis* seedlings for eliciting induced resistance, or both? Front. Microbiol. 7, 196.

Strobel, G.A., Dirkse, E., Sears, J., Markworth, C., 2001. Volatile antimicrobials from *Muscodor albus*, a novel endophytic fungus. Microbiology 147, 2943–2950.

Tahir, H.A.S., Gu, Q., Wu, H., Niu, Y., Huo, R., Gao, X., 2017a. Bacillus volatiles adversely affect the physiology and ultra-structure of *Ralstonia solanacearum* and induce systemic resistance in tobacco against bacterial wilt. Sci. Rep. 7, 1–15.

Tahir, H.A.S., Gu, Q., Wu, H., Raza, W., Hanif, A., Wu, L., et al., 2017b. Plant growth promotion by volatile organic compounds produced by *Bacillus subtilis* SYST2. Front. Microbiol. 8, 171.

Tellez, M.R., Schrader, K.K., Kobaisy, M., 2001. Volatile components of the cyanobacterium *Oscillatoria perornata* (Skuja). J. Agric. Food Chem. 49, 5989–5992.

Ting, A.S.Y., Mah, S.W., Tee, C.S., 2011. Detection of potential volatile inhibitory compounds produced by endobacteria with biocontrol properties towards *Fusarium oxysporum* f. sp. cubense race 4. World J. Microbiol. Biotechnol. 27, 229–235.

Tirranen, L.S., Gitelson, I.I., 2006. The role of volatile metabolites in microbial communities of the LSS higher plant link. Adv. Space Res. 38, 1227–1232.

Toffano, L., Fialho, M.B., Pascholati, S.F., 2017. Potential of fumigation of orange fruits with volatile organic compounds produced by Saccharomyces cerevisiae to control citrus black spot disease at postharvest. Biol. Control. 108, 77–82.

Toloza-Moreno, D.L., Lizarazo-Forero, L.M., Uribe-Vélez, D., 2020. Antagonist capacity of bacteria isolated from cape gooseberry cultures (Physalis peruviana L.) for biological control of *Fusarium oxysporum*. Trop. Plant Pathol. 1–12.

Ul Hassan, Z., Al Thani, R., Alnaimi, H., Migheli, Q., Jaoua, S., 2019. Investigation and application of *Bacillus licheniformis* volatile compounds for the biological control of toxigenic *Aspergillus* and *Penicillium* spp. ACS Omega 4, 17186–17193.

Velázquez-Becerra, C., Macías-Rodríguez, L.I., López-Bucio, J., Altamirano-Hernández, J., Flores-Cortez, I., Valencia-Cantero, E., 2011. A volatile organic compound analysis from *Arthrobacter agilis* identifies dimethylhexadecylamine, an amino-containing lipid modulating bacterial growth and *Medicago sativa* morphogenesis in vitro. Plant Soil. 339, 329–340.

Wan, M., Li, G., Zhang, J., Jiang, D., Huang, H.C., 2008. Effect of volatile substances of *Streptomyces platensis* F-1 on control of plant fungal diseases. Biol. Control 46, 552–559.

Wang, C., Wang, Z., Qiao, X., Li, Z., Li, F., Chen, M., et al., 2013a. Antifungal activity of volatile organic compounds from *Streptomyces alboflavus* TD-1. FEMS Microbiol. Lett. 341, 45–51.

Wang, B., Yuan, J., Zhang, J., Shen, Z., Zhang, M., Li, R., et al., 2013b. Effects of novel bioorganic fertilizer produced by *Bacillus amyloliquefaciens* W19 on antagonism of *Fusarium* wilt of banana. Biol. Fert. Soils 49, 435–446.

Weise, T., Thürmer, A., Brady, S., Kai, M., Daniel, R., Gottschalk, G., et al., 2014. VOC emission of various Serratia species and isolates and genome analysis of *Serratia plymuthica* 4Rx13. FEMS Microbiol. Lett. 352, 45–53.

Wenke, K., Weise, T., Warnke, R., Valverde, C., Wanke, D., Kai, M., et al., 2012. Bacterial volatiles mediating information between bacteria and plants. Biocommunication of Plants. Springer, Berlin, Heidelberg, pp. 327–347.

Wheatley, R.E., 2002. The consequences of volatile organic compound mediated bacterial and fungal interactions. Antonie Van Leeuwenhoek 81, 357–364.

Whiteson, K.L., Meinardi, S., Lim, Y.W., Schmieder, R., Maughan, H., Quinn, R., et al., 2014. Breath gas metabolites and bacterial metagenomes from cystic fibrosis airways indicate active pH neutral 2, 3-butanedione fermentation. ISME J. 8, 1247–1258.

Wonglom, P., Ito, S.I., Sunpapao, A., 2020. Volatile organic compounds emitted from endophytic fungus Trichoderma asperellum T1 mediate antifungal activity, defense response and promote plant growth in lettuce (*Lactuca sativa*). Fungal Ecol. 43, 100867.

Wu, Y., Yuan, J., Yaoyao, E., Raza, W., Shen, Q., Huang, Q., 2015. Effects of volatile organic compounds from *Streptomyces albulus* NJZJSA2 on growth of two fungal pathogens. J. Basic. Microbiol. 55, 1104–1117.

Xie, S., Zang, H., Wu, H., Uddin Rajer, F., Gao, X., 2018. Antibacterial effects of volatiles produced by *Bacillus* strain D13 against *Xanthomonas oryzae* pv. oryzae. Mol. Plant Pathol. 19, 49–58.

Yang, F., Dong, W., Zhang, X., Li, Y., Zhou, S., Zhu, G., et al., 2019. Volatile-organic compound changes in rose twigs consequent to infection with rose powdery mildew. Chil. J. Agric. Res. 79, 596–608.

Yuan, J., Raza, W., Shen, Q., Huang, Q., 2012. Antifungal activity of *Bacillus amyloliquefaciens* NJN-6 volatile compounds against *Fusarium oxysporum* f. sp. cubense. Appl. Env. Microbiol. 78, 5942–5944.

Zareian, M., Silcock, P., Bremer, P., 2018. Effect of medium compositions on microbially mediated volatile organic compounds release profile. J. Appl. Microbiol. 125, 813–827.

Zhang, X., Gao, Z., Zhang, X., Bai, W., Zhang, L., Pei, H., et al., 2020. Control effects of *Bacillus siamensis* G-3 volatile compounds on raspberry postharvest diseases caused by *Botrytis cinerea* and *Rhizopus stolonifer*. Biol. Control 141, 104135.

Zhang, Y., Li, T., Liu, Y., Li, X., Zhang, C., Feng, Z., et al., 2019. Volatile organic compounds produced by *Pseudomonas chlororaphis* subsp. aureofaciens SPS-41 as biological fumigants to control *Ceratocystis fimbriata* in postharvest sweet potatoes. J. Agric. Food Chem. 67, 3702–3710.

Zhang, X., Li, B., Wang, Y., Guo, Q., Lu, X., Li, S., et al., 2013. Lipopeptides, a novel protein, and volatile compounds contribute to the antifungal activity of the biocontrol agent *Bacillus atrophaeus* CAB-1. Appl. Microbiol. Biotechnol. 97, 9525–9534.

Zhao, L.J., Yang, X.N., Li, X.Y., Wei, M.U., Feng, L.I.U., 2011. Antifungal, insecticidal and herbicidal properties of volatile components from *Paenibacillus polymyxa* strain BMP-11. Agric. Sci. China 10, 728–736.

Zheng, L., Situ, J.J., Zhu, Q.F., Xi, P.G., Zheng, Y., Liu, H.X., et al., 2019. Identification of volatile organic compounds for the biocontrol of postharvest litchi fruit pathogen *Peronophythora litchii*. Postharvest Biol. Technol. 155, 37–46.

Zhou, M., Li, P., Wu, S., Zhao, P., Gao, H., 2019. *Bacillus subtilis* CF-3 Volatile Organic Compounds Inhibit *Monilinia fructicola* Growth in Peach Fruit. Front. Microbiol. 10, 1804.

Zou, C., Li, Z., Yu, D., 2010. *Bacillus megaterium* strain XTBG34 promotes plant growth by producing 2-pentylfuran. J. Microbiol. 48, 460–466.

Further reading

Bitas, V., Kim, H.S., Bennett, J.W., Kang, S., 2013. Sniffing on microbes: diverse roles of microbial volatile organic compounds in plant health. Mol. Plant Microbe Int. 26, 835–843.

Field, J.A., Verhagen, F.J., de Jong, E.D., 1995. Natural organohalogen production by basidiomycetes. Trends Biotechnol. 13, 451–456.

Gray, C.M., Monson, R.K., Fierer, N., 2010. Emissions of volatile organic compounds during the decomposition of plant litter. J. Geophys. Res. Biogeosci 115, G3.

Jünger, M., Vautz, W., Kuhns, M., Hofmann, L., Ulbricht, S., Baumbach, J.I., et al., 2012. Ion mobility spectrometry for microbial volatile organic compounds: a new identification tool for human pathogenic bacteria. Appl. Microbiol. Biotechnol. 93, 2603–2614.

Müller, T., Thißen, R., Braun, S., Dott, W., Fischer, G., 2004a. (M) VOC and composting facilities part 2:(M) VOC dispersal in the environment. Env. Sci. Pollut. Res. 11, 152.

Müller, T., Thißen, R., Braun, S., Dott, W., Fischer, G., 2004b. (M) VOC and composting facilities Part 1:(M) VOC emissions from municipal biowaste and plant refuse. Env. Sci. Pollut. Res. 11, 91–97.

Romoli, R., Papaleo, M.C., de Pascale, D., Tutino, M.L., Michaud, L., LoGiudice, A., et al., 2011. Characterization of the volatile profile of Antarctic bacteria by using solid-phase microextraction-gas chromatography-mass spectrometry. J. Mass. Spectrom. 46, 1051–1059.

CHAPTER

20

Molecular techniques used in plant disease diagnosis

Swapnil Sapre[1], *Iti Gontia-Mishra*[1], *Vishwa Vijay Thakur*[2], *Sumana Sikdar*[1] *and Sharad Tiwari*[3]

[1]Biotechnology Centre, Jawaharlal Nehru Agriculture University, Jabalpur, India [2]ICAR-Indian Institute of Natural Resins and Gums, Namkum-Ranchi, India [3]Department of Plant Breeding and Genetics, Jawaharlal Nehru Agriculture University, Jabalpur, India

20.1 Introduction

There is an urgent need to double the agricultural production by the year 2050 in order to fulfill the expected food and nutritional demands of ever increasing population (Narvaez et al., 2017). The agricultural crops are exposed to many biotic stresses in the environment, leading to plant diseases. Globally, the plant diseases are regarded as an imperative biotic constraint, which are accountable for crop yield and monetary losses in the agriculture. The Food and Agriculture Organization assesses that fungal, viral, and bacterial diseases, pests, and weeds leads to ~25% of crop losses (Martinelli et al., 2015). The disease in plants is mainly instigated by phytopathogenic microorganisms such as fungi, bacteria, viruses, viroids, phytoplasma, and nematodes (Khater et al., 2017). Specifically, the fungal diseases cause voluminous losses to crop production in agriculture. The detection of plant pathogen at an early stage lays foundation for proper disease control and management under greenhouses and field conditions. The detection strategies largely relay on visual assessment of symptoms and further laboratory examinations, which are usually labor-intensive, tedious, and time consuming. In the recent years, the development of molecular procedures for plant disease identification has resolved the inadequacies of conventional methods, by providing the means for fast and precise detection of plant diseases (Khaled et al., 2018).

This chapter discusses the impact of biotic stress, that is, phytopathogens on yield losses, inefficiency of conventional methods for plant disease detection, and the current molecular methods for proper and timely disease diagnosis. Moreover, the chapter also

Food Security and Plant Disease Management
DOI: https://doi.org/10.1016/B978-0-12-821843-3.00001-5

encompasses the up-to-date progress on the plant disease management practices such as introduction of transgenic plants, development of disease-resistant plant using molecular breeding such as marker-assisted selection (MAS) and quantitative trait locus (QTL). Further, it also addresses application of beneficial microbes for biocontrol of plant diseases, along with the -omics-based solution for disease diagnosis.

20.2 Impact of biotic stress on crop yield

Biotic stress is an adverse condition in which plant cannot sustain its normal growth due to the interaction with deleterious microorganisms (fungi, bacteria, viruses, viroids, phytoplasma, and nematodes). These microorganisms primarily grow on or inside plant tissues and cause various type of symptoms like chlorosis, stunting, rotting, or local lesions formation (Balodi et al., 2017). Infection by a combination of fungi, bacteria, and viruses together are common and cause severe disease symptoms, rather than infections by individual pathogens (Pandey et al., 2017). The interaction between plant and pathogen produces various morphological, biochemical, and molecular response and suggest that it induces various signaling pathways. The biotic stress causes a terrific yield loss as viral, fungal, and bacterial infects leaves resulting in decreased level of photosynthesis in all the major crops. According to Wang et al. (2013), the biotic stresses are responsible for approximately, 28.2%, 37.4%, 31.2%, 40.3%, 26.3%, and 28.8% yield losses in wheat, rice, maize, potatoes, soybeans, and cotton crops, respectively. Furthermore, losses due to plant diseases not only hinder the crop productivity but also pose a huge risk to food and nutritional security, worldwide (Miller et al., 2009).

20.3 Conventional methods for detection and diagnosis of plant diseases

It is the need of the hour to accurately diagnose the incidence, severity, and adverse effect of plant diseases on important field and horticultural crop plants. The conventional ways of detecting plant pathogens is mainly by visual inspection. The visual symptoms often develop when a possibly great destruction has already occurred to the crop. The most readily used methods for plant disease identification includes (1) visual inspection of plant diseases and (2) microscopic assessment of morphological characteristics to detect the pathogens.

20.3.1 Visual inspection of plant diseases

The recognition of affected plants is the essential prerequisite for plant disease diagnosis. The visual estimates of disease identification are often based on characteristic plant disease symptoms. These symptoms of different plant diseases can be grouped as follows.

20.3.1.1 Distortion of plant tissues or organs

Some symptoms listed such as reduced growth of shoots and roots, short internodes length, deformed structures of leaves, chlorosis, and reduced number of flowers and fruits are indicator of disease incidence in plants.

20.3.1.2 *Abnormal growth of tissues or organs*

In some disease some over growth symptoms like gall formation and profuse flowering indicates disease incidence.

20.3.1.3 *Necrosis and deviation from normal appearance of plant parts*

These are important and noticeable symptoms which help in determining the causative pathogen. Such symptoms include spots on leaves, stems and fruits, rotting of stems or roots, wilt, or diebacks. The alteration of normal appearance of plant parts includes development of mosaic patterns on leaves and change in the color of vegetative (leaves, stems, etc.) and reproductive (flower and fruit) parts of plants.

20.3.2 Microscopic evaluation of morphological features to identify pathogens

Visual inspection of symptoms in field condition is often challenging and necessitates prior information of taxonomy and pathology to recognize causing pathogen. To overcome these difficulties, microbiological techniques can be useful in which pathogen can be isolated and cultured on suitable medium in artificial conditions. The direct agar plating method usually utilizes acidified potato dextrose agar medium or malt extract medium for culturing fungi from symptomatic plant tissues (Afouda et al., 2009). Nonetheless, fungal reproductive structures are seldom formed on these medium; hence for fungal spore/condia formation, often medium with lower nutrient content is suggestive, such as water agar or corn-meal agar medium (Mancini et al., 2016). Furthermore, by obtaining an optimum growth, microscopic observations of fungal mycelium and spores make it easier to identify the invading pathogen. Similar, techniques can also be useful for bacterial disease diagnosis. Diagnosticians with experience in morphological characteristics of plant pathogens can identify the isolated organisms.

20.4 Modern approaches for plant disease diagnosis

The precise and timely detection of plant pathogens are the keystones for efficient disease control in agriculture leading to reduced yield losses (Doddaraju et al., 2019). The conventional detection system relies on morphological identification of causative organism of plant diseases, which is often challenging and time-taking phenomenon. It also needs the thorough knowledge and know-how of identifying detailed fungal and bacterial structural characteristics. An effective diagnostic tool must possess some characteristics like high sensitivity and specificity, reproducibility, and rapidity. If any technique is capable of multiplexing and cost effective, it is more suitable for diagnostic purpose. All diagnostic techniques rely on either specific sites of nucleic acid or specific proteins to detect the plant pathogen. Detection of these molecules is essential and several technologies have been developed time to time to achieve the goal.

20.4.1 DNA-based techniques

Polymerase chain reaction (PCR) is a standard tool for detection of diverse pathogens of plants. PCR technique is required to amplify particular region of DNA to make multiple copies within minimum time. Success of PCR relies on binding of specific oligonucleotide sequence known as primers with template DNA. For disease diagnostics, pathogen nucleic acid specific primers can amplify in the PCR hence, demonstrate presence of pathogen. Although, PCR is a cheap and rapid technique, it can also lead to false results due to cross contamination in sample and high dependency on the specificity of primers (Lau and Botella, 2017). There are many types of PCR such as nested PCR, cooperative PCR, multiplex PCR, and DNA fingerprinting which are readily applied for identification of plant disease causing agent (Martinelli et al., 2015). To overcome the limitations of PCR techniques, some other amplification methods have develop but these also has their strengths and limitations. These includes loop-mediated isothermal amplification (Notomi et al., 2000), helicase-dependent amplification (Vincent et al., 2004), rolling circle amplification (Fire and Xu, 1995), and recombinase polymerase amplification (Piepenburg et al., 2006).

In advance to PCR technology, DNA microarray is also a useful technique for multiple pathogen detection simultaneously infecting same plant and producing complex disease symptoms. In microarray, DNA fragment specific to pathogen labeled with fluorescent dyes are hybridized with probes to detect pathogen. The advent of DNA sequencing techniques and subsequent bioinformatic analysis validates the accurate identification of pathogen.

20.4.2 RNA-based techniques

Various viruses causing diseases in crops have RNA as genetic material and PCR techniques can be used to detect by converting RNA into DNA. Reverse transcriptase PCR (RT-PCR) is utilized to convert RNA in complementary DNA (cDNA). RT-PCR can also be helpful to detect living status of fungi (Aslam et al., 2017). In nature, mRNA gets easily degraded in dead cells and cannot be converted into cDNA (Capote et al., 2012). Real-time PCR (qRT-PCR) is presently regarded as the benchmark process for identification of plant pathogens. This method permits simultaneous screening of the reaction along with the amplification process by employing fluorescent signal that escalates proportionately to the amount of amplicons produced and number of pathogens present in the sample (Belete and Boyraz, 2019). qRT-PCR can be used for identification and quantification of pathogen load at a particular point of time. This technique has been utilized for precise detection of various phytopathogenic fungi (Black, 2009). Serial analysis of gene expression is a method which utilizes specific gene sequence for the quantitative gene expression profiling of host–pathogen interaction which provides the identification of multiple transcripts, instantaneously (Sharma et al., 2007). This method was initially used to identify fungal pathogen *Blumeria graminis* on the barley leaves (Dawei and Peng, 2014).

20.4.3 Protein-based techniques

Disease diagnostics at protein level mostly based on antigen and antibody reaction which is a specific, sensitive, and accurate process. But this reaction is invisible, so labeling using fluorescent dyes is an essential requirement. The enzyme-linked immunosorbent assay (ELISA) is such a tool in which antigen–antibody reaction takes place and enzyme linked with secondary antibody convert substrate into a colored component (Edwards and Cooper, 1985). For detection of any pathogen, a specific primary antibody against pathogen specific protein is prerequisite to perform ELISA. The sensitivity of ELISA depends on various factors such as the pathogen, duration at which was sample and pathogen load (Schaad et al., 2003). Thus far ELISA procedures can use both poly and monoclonal antibodies which are manufactured for many phytopathogenic bacteria and fungi, similarly, quick detection kits are commercially accessible (Martinelli et al., 2015). There are multiple kits available for various viral diseases of field crops and horticultural crops. Besides, ELISA several other techniques such as Western blotting, immune-strip assays, dot-blot immune-binding assays, and serologically specific electron microscopy are also known to detect the phytopathogen.

20.5 Omics approaches for plant disease diagnosis

The -omics is a comprehensive term implies to the study of various biomolecules such as DNA, RNA, proteins, and metabolites, etc. The -omics technologies comprising of genomics, transcriptomics, proteomics, metabolomics, next-generation sequencing–based genotyping, and mass spectrometry, were introduced in agricultural research particularly for the improvement of crops for better disease resistance (Hasin et al., 2017). In the last decade, studies implying next-generation sequencing and metabolomics have fetched the discovery of biomarkers and unique genes related to diseases. -Omics tools are mainly used to identify probe markers of disease resistance and detection (Guillemin et al., 2016). The application of -omics technologies for enhanced crop yield, protection, and plant disease control has currently gained ample consideration (Pathak et al., 2018). The scientists in India, are conducting research work on decoding the complete genome sequence of a fungal pathogen (*Tilletia indica*) causing Karnal bunt of wheat, for deciphering important genes/proteins related to pathogenesis (Kumar et al., 2017). These kinds of studies will encourage the cumulative scientific activities in molecular breeding, disease diagnosis, and management. Similarly, comparative gene expression analysis could be utilized to generate the regulatory evidence through transcriptomic analysis to get insight on the expression of gene taking part in stress modulations. Transcriptomics researches deal with high-throughput methods comprising either microarray technology or RNA-sequencing (RNA-seq)-based approaches, which provide the comparative transcriptomic analysis of diseased versus healthy plants (AbuQamar et al., 2016). Thus the information derived from these experiments can be subsequently used for disease diagnosis and further for development of disease resistance in crop plants. The main aim of proteomics investigation is to comprehend the functional and structural work forces of proteins and deducing their role and functions in contribution to crop productivity and disease resistance. The proteomic studies identify molecular abundance and activities of protein at the various functional stages (Pathak et al., 2018). Proteomic studies are largely employed to explore the deviations in protein levels under challenging extreme abiotic or biotic stress conditions. Moreover, protein microarrays, for instance, were excellently used to distinguish the target proteins networking with calmodulin and calmodulin-like proteins in diseased plants, making it a potential method for early disease diagnosis (AbuQamar et al., 2016). Similarly, metabolomics can be applied to find and characterize diversified amount of biomolecules produced by a healthy and or diseased plant to identify the active plant defense metabolites (Kumar et al., 2015). These metabolomic technologies have the proficiency to quantify a wider range of tiny functional molecules when plants are exposed to extreme biotic stresses (Zhao et al., 2016). Thus metabolomics can be prudently applied for early disease diagnosis. Taken together the -omic technologies are paving a way to modern plant disease diagnostic system.

20.6 Management of plant diseases

20.6.1 Transgenic plants for disease resistance

Transgenic plants are those which have their genome engineered using recombinant DNA technology by adding an alien gene or removal unfavorable gene. The major goal to develop a transgenic plant is to achieve a trait which is not present in target plant such as resistance to abiotic and biotic stress, improved nutritional status or change in the physiology of plant (Lai et al., 2012). Plant diseases and its resistance in crops are continuously changing phenomena as pathogens and crops are evolving with the time but pathogens are more rapidly evolving entity. Fungi, bacteria, viruses, or nematodes instigate many plant diseases which impacted a significant loss to the crop yield. There are several ways to manage crops diseases. The conventional breeding techniques have also been contributed to develop disease-resistant crops but emerging diversity of pathogens and resistance toward chemicals limits the scope of these lengthy and labor-intensive processes. Another option to prevent crop diseases requires following agronomical practices or cultural practices like crop rotation or manipulation in crops' growing period. There are several chemicals available which reduce the disease incidents in crops but also create serious problems to environment, human, and animal health. As an alternative, transgenic crops offer several benefits compared to traditional breeding techniques. It allows the integration, deletion, editing, or regulation of specific genes within minimum time and unnecessary changes in the genome of the target crop (Dong and Ronald, 2019). After development of first transgenic tobacco plant for antibiotic resistance (deFramond et al., 1983), many transgenic plants have been produced in last three decades for improvement of various agronomic traits including disease and insect resistance.

20.6.1.1 Strategies to develop disease resistance in crops through transgenic technology

With increasing knowledge of cellular machinery responding toward pathogens and emergence of new technologies of molecular biology, various strategies were developed to achieve the goal of disease resistance in plants. Some of the breakthrough technologies by which transgenic plants have been developed and exhibited appropriate results under laboratory or field trails are summarized below. The example of transgenic plants developed for resistance against various pathogens using different strategies is presented in Table 20.1.

a. *Identification of resistance genes and incorporation in target plants*

Gene for gene theory (Flor, 1956) has established the understanding of plant–microbe interaction. Using these R-genes, resistance was drawn in various crops from its compatible donors. Field trials in tomato expressing *Bs2* gene isolated from

TABLE 20.1 Various strategies to develop transgenic plants resistant against various pathogens.

Gene/s	Donor	Recipient	Effect	References
R-genes introduced in susceptible crop				
Rpi-vnt1.1	*Solanum venturii*	Potato	Exhibited protection against to potato late blight disease	Foster et al. (2009)
Pm3e	Wheat	Wheat	Expressed powdery mildew resistance and no fitness costs associated with high transgene expression	Koller et al. (2019)
Multiple R-genes	Wild potato relatives	Potato	R-genes transformation turned susceptible variety into a resistant variety	Haverkort et al. (2016)
Rxo1	Maize	Rice	*Rxo1* gene conditions a resistance reaction to a diverse collection of pathogen strains	Zhao et al. (2005)
Integration of stacked genes				
Rpi-sto1, *Rpi-vnt1.1*, and *Rpi-blb3*	*Solanum stoloniferum*, *S. venturii*, and *Solanum bulbocastanum*	Potato	The resistance spectrum of triple *Rpi* gene transformants against *Phytophthora infestans* demonstrated a collective resistance against all the *Rpi* genes	Zhu et al. (2012)
RB, *Rpi-blb2*, and *Rpi-vnt1.1*	*S. bulbocastanum* and *S. venturii*	Potato	Field trial showed total resistance toward local blight races	Ghislain et al. (2019)
Rpi-vnt1.1, *RB*, and *Rpi-blb2*	*S. bulbocastanum*	Potato	Exhibited complete protection against late blight disease was observed in detached-leaf assays and whole plant assays	Webi et al. (2019)
Multiple alleles of *Pm3* locus	Wheat	Wheat	Conferred protection toward wheat powdery mildew	Brunner et al. (2012)
Transgenic line expressing of chitinase and glucanase				
DrChit	*Drosera rotundifolia*	Tobacco	Antifungal activity was observed against *Trichoderma viride*	Durechova et al. (2019)
CmCH1	*Coniothyrium minitans*	Soybean	Reduced symptoms were observed on the leaves of transgenic plants	Yang et al. (2020)
PnGlu1	*Panax notoginseng*	Tobacco		Taif et al. (2020)

(*Continued*)

TABLE 20.1 (Continued)

Gene/s	Donor	Recipient	Effect	References
			Protection against *Fusarium solani* infection was observed along with high levels of jasmonic acid gene expression	
endo α-1, 3-Glucanase	*Trichoderma harzianum*	Rice	Demonstrated an exceptional protection against sheath blight disease	Kumar et al. (2019)
AGLU1 and *RCH10*	Alfalfa and rice	Oil palm	Transgenic lines showed resistance toward *Ganoderma boninense*	Hanin et al. (2020)
RNAi for disease resistance				
CP3 gene	Rice tungro spherical virus	Rice	Virus transmission was reduced	Malathi et al. (2019)
ODC	*Fusarium oxysporum* f. sp. *lycopersici*	Tomato	Confer resistance against fungal spores exposure	Singh et al. (2020)
Mi-msp2	–	Arabidopsis	*Mi-msp2* gene can be used as a potential target for root-knot nematodes management in crops of economic importance	Joshi et al. (2019)
SlJAZ2	–	Tomato	Tomato plants designed for bacterial speck resistance	Ortigosa et al. (2019)
CsLOB1		Citrus	Conferred resistance to citrus canker	Jia et al. (2017)
eIF4G		Rice	Transgenic lines displayed resistance to rice tungro spherical virus	Macovei et al. (2018)
OsERF922		Rice	Conferred enhanced rice blast resistance	Wang et al. (2016)

pepper showed considerable resistance toward bacterial spot disease caused by *Xanthomonas* sp. (Horvath et al., 2015; Kunwar et al., 2018). Similarly, rice demonstrated resistance against bacterial streak transformed with *Rxo1* gene from maize in vitro (Zhao et al., 2005).

b. *Stacking of R-genes to obtain broad spectrum resistance*

Resistance offered by single transgene integration often lacks durability and later on the crop render susceptible to pathogen. The main reason for this is rapid evolution of pathogen against single R-gene and mutates itself to avoid recognition by host plants. This problem can be better addressed by integration of multiple R-genes in single host as stacked genes. Transgenic expression of different alleles of *Pm3* locus in wheat significantly exhibited resistance toward powdery mildew in field condition (Brunner et al., 2012) whereas potato plants over-expressing RB, *Rpi-blb2*, or *Rpi-vnt1.1* confer resistance against late blight of potato (Zhu et al., 2012; Ghislain et al., 2019; Webi et al., 2019). The resistance obtained using stacked up R-genes is assumed to be a persistent phenomenon because simultaneous mutation on multiple R-genes is a rare event.

c. *Targeting fungal cell wall molecules*

Chitin and glucans are major entities of fungal cell wall and hydrolysis of these components using specific enzymes can hinder the development of fungal pathogen, thus prevent the disease occurrence (Neuhaus, 1999). There are several reports which demonstrated the production of chitinases and glucanases in transgenic plants and confer resistance toward fungal pathogens (Durechova et al., 2019; Kumar et al., 2019; Hanin et al., 2020; Taif et al., 2020; Yang et al., 2020).

d. *Small RNA-mediated resistance*

It has been earlier known that expression of viral genes in plants demonstrates immunity toward viral diseases (Lomonossoff, 1995). Recent researches demonstrated that such events are mediated by RNAi which exhibits a vital part in development of virus resistant transgenic plants (Dong and Ronald, 2019). The RNAi approach inhibits the synthesis of a key protein of virus which is necessary to complete their life cycle. Presently, this approach is also applicable to develop resistance toward other disease causing organisms like fungi, bacteria, and nematodes.

e. *CRISPR-Cas as a tool for disease resistance*

As a modern genome editing tool, clustered regularly interspaced short palindromic repeats (CRISPR)-CRISPR-associated (Cas) technology originated from bacteria has been recognized as an outstanding prospective to obtain disease resistance in plants. In bacterial cell, it works as machinery for viral resistance. Its specificity to cut at targeted site makes it important tool in molecular biotechnology. In the past decade, the CRISPR/Cas system has been used to retort various agricultural experiments, comprising the improvement in biotic stress resistance (Arora and Narula, 2017). The implementation of CRISPR/Cas tools has largely been focused to generate resistance toward viral diseases, but recently consolidated attempts are taken up to develop fungal and bacterial disease resistance (Borrelli et al., 2018).

20.6.2 Molecular breeding (marker-assisted selection/quantitative trait locus)

Intending to generate disease resistance in plants through conventional breeding approach, numerous disease-resistant varieties have been designed against several diseases are developed since 20th century (Pathania et al., 2017). Although, this approach involves the selection of superior progenies from the segregating generations based on the phenotype alone. This was laborious, time taking, and difficult to conduct primarily due to factors like genotype–environment interactions, linkage drags, etc. Also, plant breeders faced a major problem of the resistance breakdown due to loss of effectiveness of the resistance gene (R-gene). To overcome above-mentioned problem, breeders made a constant effort to innovate an advanced breeding approach that would make trait selection more effective and reliable. The improvement of breeding programs based on molecular markers approach and their application is a novel panorama for the development of disease-resistant crops. Molecular markers, which facilitate selection by allowing detection of differences between genotypes at the molecular level, can significantly minimize the time required for development of crop varieties (Collard et al., 2005; Hasan et al., 2015). The discovery of tightly linked molecular markers to R-genes facilitates the efficient selection because molecular markers are not affected due to epistatic interactions and the environmental effects, which also shorten the time for varietal development (Nzuki et al., 2017). The widespread implementation of molecular markers in different aspects of plant science like germplasm evaluation, gene tagging/mapping, characterization of traits of interest, have proved their utility in crop improvement (Xu, 2010; Jiang, 2013). Since last two decades, with the help of PCR-based random amplified polymorphic DNA, sequence-tagged sites, and microsatellite markers (Simple Sequence Repeats), several R-gene resistant to a number of plants disease have been mapped in various crops (Collard et al., 2005; Panigrahi and Thangadurai, 2011). With the advent of molecular markers in crops for numerous traits, especially biotic stresses have opened the avenues for MAS. MAS can be used in several breeding programs including the selection of quantitative trait loci from breeding lines/populations, marker-assisted recurrent selection, marker-assisted backcrossing, introgression of useful genes from wild relatives, and breeding lines for pyramiding of genes. The applications of MAS approach in crop improvement have been highlighted by many researchers (Farokhzadeh and AliFakeheri, 2014; Das et al., 2017; Pathania et al., 2017). In the present scenario, MAS is an extensively applied approach for introgression of disease-resistant genes (oligogenes and QTLs along with oligogenes) in targeted plants through backcross-breeding and gene pyramiding. Many disease-resistant varieties have been developed using MAS in and around the world. In the subsequent section, the basic steps of MAS for the improvement of disease-resistant variety will be discussed and also highlight some success stories of the development of a disease-resistant variety of different crops using this approach.

20.6.2.1 Discovery of R-genes

The first step of MAS for resistance breeding is identification of effective disease R-genes followed by its isolation and cloning. The molecular characterizations of first R-gene, that is, *Hm1*, confer resistance against *Cochliobolus carbonum* in maize was discovered by Johal and Briggs (1992). Afterward a number of R-genes were identified and cloned from various plant species. Once the R-gene is characterized at molecular level, the subsequent step is to decipher its nature and pattern of inheritance in upcoming generations.

20.6.2.2 Discovery of R-genes linked molecular markers (gene tagging)

The next step after deciphering properties and pattern of inheritance of targeted R-gene is to discover the tightly linked marker(s) associated with targeted R-gene. These markers can be applied for indirect selection in MAS. The identification of DNA markers, genes, and QTLs linked to specific R-gene is carried out through QTL mapping.

20.6.2.3 Selection of parents

The characteristic features of parents crossed to prepare mapping population are important as they should differ for one or more traits of interest. The parent with preferred traits, such as high yield and improved quality, but vulnerable to certain disease are called recipient or recurrent parent. On other hand, donor parents are tolerant to that particular disease but may not perform as well as an elite variety in other areas. The phenotypic characters of each individual plants are controlled either qualitatively (by monogenes/oligogenes) or quantitatively (by polygenes). The tagging of quantitatively inherited traits is not easy because they are mostly affected by the environment and genetic background recipient parent. Another alternative to overcome this problem is to cross donor parent with multiple recipient lines.

20.6.2.4 Mapping population development

Different types of mapping population like F_2 population, back cross population, recombinant inbred lines (RILs), double haploid (DH), etc. could be used for MAS. The preference of the mapping population to be developed mainly relied on the target trait. In case of qualitatively inherited traits, F2, F3, or BC1F1 population can be used. However, when trait of interest is quantitatively inherited near-isogenic lines, RILs, and DHs are better suitable options.

20.6.2.5 Phenotyping of population for selected trait

The phenotyping procedure depends on nature of inheritance of targeted trait, that is, quantitative or qualitative. For instance the biotic stress resistance, the mapping population infested with particular the pest or pathogen and scoring the plants for resistance or susceptibility.

20.6.2.6 Discovery of polymorphic markers

Thereafter the above two processes, the subsequent step is to detect the markers showing polymorphisms among parental lines and tightly linked with targeted traits. Later on, the bulked-segregant analysis may be used to detect markers, which inherit with target trait.

20.6.2.7 Development of linkage map

Moreover, after deferring the inheritance behavior of markers, the construction of linkage maps is carried out. The linkage map helps us to decipher the exact locus and relative genetic distances between the markers. A number of statistical software like Mapmanager, Cartographer, Mapmaker, Linkage, and JoinMap are available for this purpose. The foremost step of utmost importance is validation of identified markers and their linkage distances in alternate populations.

20.6.2.8 Success stories

Samba Mahsuri (BPT5204) is a promising rice variety which is prevalent among Indian farmers but susceptible to bacterial leaf blight (BLB) caused by *Xanthomonas oryzae* pv. *oryzae*. This disease is responsible for heavy losses in rice production in India. To develop BLB resistant variety five resistant genes from donor line SS1113 were selected and introgressed into Samba Mahsuri (Sundaram et al., 2008). The F1 line obtained after crossing between donor line and Samba Mahsuri was backcrossed with Samba Mahsuri. With each subsequent step, tightly linked markers were detected and foreground selection was performed. To know the genome recovery percentage of the recipient parent genome, background selection was performed using microsatellite markers up to BC4 generations. After self-cross of BC4F1 plants the obtained BC4F2 plants showed the different combinations of the target gene. The recipient parent which has high number of gene pyramids exhibited resistance against the BLB pathogen was selected. Eventually, commercially released line RPBio-226 (IET 19,046), had three resistance genes (Xa21, Xa13, and Xa5) against BLB was named "Improved Samba Mahsuri." Similarly, stripe rust is a severe disease of barley and a number of genes showing resistant were discovered by few researchers (Castro et al., 2002; Toojinda et al., 2000). Castro et al. (2003) pyramided three QTLs present on different chromosomes and observed resistance at the seedling stage.

In addition, Fusarium head blight (FHB) caused by multiple species of *Fusarium* and lead to substantial loss in wheat production. In past two decades, mapping of QTL that provided resistant against various disease are main concern of different research groups. Approximately 100 QTL showed resistant against FHB have been identified. The QTL Fhb1 was recognized for resistant against FHB. A QTL Fhb1 identified in Chinese wheat variety "Sumai 3" (Dweba et al., 2017) showed promising level of resistant against FHB was mapped by Cuthbert et al. (2006). Later on this QTL was validated as a major QTL against FHB disease in Chinese wheat line W14 (Chen et al., 2007).

20.6.3 Plant growth–promoting rhizobacteria-assisted disease management

In the line with the recent trends for improved agriculture technologies the use of chemical fungicides are mostly being substituted by biological control agents. This could be attributed to following possible reasons such as a threat to the development of fungicide-resistant fungal strains, impact of fungicides on environment due to their recalcitrant nature (Khaledi and Taheri, 2016) and potential negative effects on other beneficial microbes including plant growth–promoting rhizobacteria (PGPR) and mycorrhiza (Singh et al., 2010). In the present scenario, disease management of pathogenic fungi, through biological suppression/control plays a vital role. The recent researches suggest that the biological control agents can grow, propagate, colonize, and protect the plants against plant diseases. For these listed reasons, biocontrol of plant diseases has gained ample attention in disease control strategy. The PGPR are bacteria residing in and around the roots many plants species that produce favorable effects on plant. There are two ways in which PGPR can assist plant growth improvement. The indirect methods of plant growth promotion used by PGPR encompasses, the prevention from the harmful effects of pathogens on plants by producing growth inhibitory substances such as siderophores, antibiotics, lytic enzymes, or by enhancing the innate resistance of the host (van Loon, 2007; Gontia-Mishra et al., 2017). Besides PGPR, *Trichoderma* spp. hase been employed as biocontrol agents in order to combat a various of plant diseases (Mukherjee et al., 2012). The consortium of *Bacillus* spp. and *Serratia* sp. effectively controlled the prevalence of phytophthora blight on sweet pepper as well as improved the fruit quality (Zhang et al., 2019). Similarly, siderophore and cyanogen producing *Pseudomonas aeruginosa* was used as biocontrol agent for *Macrophomina phaseolina* in crops and considerably reduced the pathogen attack

(Singh et al., 2010). Microbes facilitate plant growth and control plant diseases in crops which are effectively utilized as an alternative eco-friendly approach. There are several benefits for the application of antagonistic microbes as biocontrol agents over chemical fungicides which possess specificity against target plant pathogens, low cost, low doses, and eco-friendly nature. Though, these biological agents are much safer than chemical agents but are potentially vulnerable to environmental factors.

20.7 Future prospects and concluding remarks

With the global climate change or other environmental shifts, newer plant pathogens are emerging and evolving, which causes severe plant diseases. These plant pathogens can be detected by employing either the conventional or relatively modern methods. The precise and early detection of plant pathogen is important for active surveillance and effective management of the diseases. The conventional detection system are largely symptoms based, which are tedious, exhaustive, and even may not detect disease on asymptomatic plants. Whereas, the modern diagnostic tools uses DNA/RNA/protein-based as indicator of pathogen are reliable, precise, accurate, and quick. These diagnostic tools not only reveal the presence of pathogens but could exactly quantify the biomass of pathogens, providing the evidences for disease severity. The recent techniques such as DNA barcoding, DNA/protein microaaray, and RNA-seq can aid in overcoming the challenges of convention detection methods. With the advent of -omic technologies, even the phylogenetic relationships of pathogen at different taxonomic levels can be generated by the comparative analysis of genomic sequences. The efficient plant disease management can be practiced by developing transgenic plants resistant to various diseases, or by generating disease-resistant crop varieties using molecular breeding techniques. Besides, the application of potential biological control agents can also effectively assist in plant disease management. Furthermore, the future goal of the pathologists and biotechnologists is to decrease the cost of molecular procedures, increase the dimensions to process the adequate number of sample and envisage easier protocols.

Acknowledgement

This work was supported by funds provided by Science and Engineering Research Board, New Delhi, India, grant number # PDF/2017/001001, received to author Iti Gontia-Mishra.

References

AbuQamar, S.F., Moustafa, K., Tran, L.S.P., 2016. 'Omics' and plant responses to *Botrytis cinerea*. Front. Plant Sci. 7, 1658.

Afouda, L., Wolf, G., Wydra, K., 2009. Development of a sensitive serological method for specific detection of latent infection of *Macrophomina phaseolina* in cowpea. J. Phytopathol. 157, 15–23.

Arora, L., Narula, A., 2017. Gene editing and crop improvement using CRISPR-Cas9 system. Front. Plant Sci. 8, 1932.

Aslam, S., Tahir, A., Aslam, M.F., Alam, M.W., Shedayi, A.A., Sadia, S., 2017. Recent advances in molecular techniques for the identification of phytopathogenic fungi–a mini review. J. Plant Interact. 12, 493–504.

Balodi, R., Bisht, S., Ghatak, A., Rao, K.H., 2017. Plant disease diagnosis: technological advancements and challenges. Indian Phytopathol. 70, 275–281.

Belete, T., Boyraz, N., 2019. Biotechnological tools for detection, identification and management of plant diseases. Afr. J. Biotechnol. 18, 797–807.

Black, J., 2009. Quantitative real-time polymerase chain reaction (qPCR) of filamentous fungi in carpet. RTI Press, Research Triangle Park, NC.

Borrelli, V.M.G., Brambilla, V., Rogowsky, P., Marocco, A., Lanubile, A., 2018. The enhancement of plant disease resistance using CRISPR/Cas9 technology. Front. Plant Sci. 9, 1245.

Brunner, S., Stirnweis, D., Quijano, C.D., Buesing, G., Herren, G., Parlange, F., et al., 2012. Transgenic *Pm3* multilines of wheat show increased powdery mildew resistance in the field. Plant Biotechnol. J. 10, 398–409.

Capote, N., Aguado, A., Pastrana, A.M., Sanchez-Torres, P., 2012. Molecular tools for detection of plant pathogenic fungi and fungicide resistance. INTECH Open Access Publisher, Valencia.

Castro, A.J., Capettini, F., Corey, A.E., Filichkina, T., Hayes, P.M., Kleinhofs, A., et al., 2003. Mapping and pyramiding of qualitative and quantitative resistance to stripe rust in barley. Theor. Appl. Genet. 107, 922–930.

Castro, A., Chen, X.M., Hayes, P.M., Knapp, S.J., Line, R.F., Toojinda, T., et al., 2002. Co-incident QTL that determine seedling durable resistance to wheat rusts and adult plant resistance to stripe rust in barley. Crop Sci. 42, 1701–1705.

Chen, X., Faris, J.D., Hu, J., Stack, R.W., Adhikari, T., Elias, E.M., et al., 2007. Saturation and comparative mapping of a major Fusarium head blight resistance QTL in tetraploid wheat. Mol. Breed. 19, 113–124.

Collard, B.C.Y., Jahufer, M.Z.Z., Brouwer, J.B., Pang, E.C.K., 2005. An introduction to markers, quantitative trait loci (QTL) mapping and marker-assisted selection for crop improvement: the basic concepts. Euphytica 142, 169–196.

Cuthbert, P.A., Somers, D.J., Thomas, J., Cloutier, S., Brule-Babel, A., 2006. Fine mapping Fhb1, a major gene controlling Fusarium head blight resistance in bread wheat (*Triticum aestivum* L.). Theor. Appl. Genet. 112, 1465–1472.

Das, G., Patra, J.K., Baek, K.H., 2017. Insight into MAS: a molecular tool for development of stress resistant and quality of rice through gene stacking. Front. Plant Sci. 8, 985.

Dawei, W., Peng, Y., 2014. Assessing the impact of dominant sequencing base gene expression profiling techniques (SGEPTs) on phytopathogenic fungi. Chiang Mai J. Sci. 41, 9422–9444.

deFramond, A.J., Bevan, M.W., Barton, K.A., Flavell, R., Chilton, M.D., 1983. Mini-Ti plasmid and a chimeric gene construct: new approaches to plant gene vector construction. In: Downey, K., Voellmy, R.W., Ahmad, F. (Eds.), Advances in Gene Technology: Molecular Genetics of Plants and Animals. Miami Winter Symposium, 20th ed. Elsevier, Florida, pp. 159–170.

Doddaraju, P., Kumar, P., Gunnaiah, R., Gowda, A.A., Lokesh, V., Pujer, P., et al., 2019. Reliable and early diagnosis of bacterial blight in pomegranate caused by *Xanthomonas axonopodis* pv. punicae using sensitive PCR techniques. Sci. Rep. 9, 10097.

Dong, O.X., Ronald, P.C., 2019. Genetic engineering for disease resistance in plants: recent progress and future perspectives. Plant Physiol. 180, 26–38.

Durechova, D., Jopcik, M., Rajninec, M., Moravcikova, J., Libantova, J., 2019. Expression of *Drosera rotundifolia* chitinase in transgenic tobacco plants enhanced their antifungal potential. Mol. Biotechnol. 61, 916–928.

Dweba, C.C., Figlan, S., Shimelis, H.A., Motaung, T.E., Sydenham, S., Mwadzingeni, L., et al., 2017. Fusarium head blight of wheat: Pathogenesis and control strategies. Crop Prot. 91, 114–122.

Edwards, M.L., Cooper, J.I., 1985. Plant virus detection using a new form of indirect ELISA. J. Virol. Methods 11, 309–319.

Farokhzadeh, S., AliFakeheri, B., 2014. Marker-assisted selection for disease resistance: applications in breeding. Int. J. Agric. Crop Sci. 7, 1392–1405.

Fire, A., Xu, S.Q., 1995. Rolling replication of short DNA circles. Proc. Natl. Acad. Sci. USA 92, 4641–4645.

Flor, H.H., 1956. The complementary genic systems in flax and flax rust. Adv. Genet. 8, 29–54.

Foster, S.J., Park, T., Pel, M., Brigneti, G., Sliwka, J., Jagger, L., et al., 2009. *Rpi-vnt1.1*, a *Tm-22* homolog from *Solanum venturii*, confers resistance to potato late blight. Mol. Plant Microbe Interact. 22, 589–600.

Ghislain, M., Byarugaba, A.A., Magembe, E., Njoroge, A., Rivera, C., Roman, M.L., et al., 2019. Stacking three late blight resistance genes from wild species directly into African highland potato varieties confers complete field resistance to local blight races. Plant Biotechnol. J. 17, 1119–1129.

Gontia-Mishra, I., Sapre, S., Kachare, S., Tiwari, S., 2017. Molecular diversity of 1-aminocyclopropane-1-carboxylate (ACC) deaminase producing PGPR from wheat (*Triticum aestivum* L.) rhizosphere. Plant Soil 414, 213–227.

Guillemin, N., Horvatić, A., Kules, J., Galan, A., Mrljak, V., Bhide, M., 2016. Omics approaches to probe markers of disease resistance in animal sciences. Mol. BioSyst. 12, 2036–2046.

Hanin, A.N., Parveez, G.K.A., Rasid, O.A., Masani, M.Y.A., 2020. Biolistic-mediated oil palm transformation with alfalfa glucanase (*AGLU1*) and rice chitinase (*RCH10*) genes for increasing oil palm resistance towards *Ganoderma boninense*. Ind. Crops Prod. 144, 112008.

Hasan, M.M., Rafii, M.Y., Ismail, M.R., Mahmood, M., Rahim, H.A., Alam, M.A., et al., 2015. Marker-assisted backcrossing: a useful method for rice improvement. Biotechnol. Biotechnol. Equip. 29, 237–254.

Hasin, Y., Seldin, M., Lusis, A., 2017. Multi-omics approaches to disease. Genome Biol. 18, 83.

Haverkort, A.J., Boonekamp, P.M., Hutten, R., Jacobsen, E., Lotz, L.A.P., Kessel, G.J.T., et al., 2016. Durable late blight resistance in potato through dynamic varieties obtained by cisgenesis: scientific and societal advances in the DuRPh project. Potato Res. 59, 35–66.

Horvath, D.M., Pauly, M.H., Hutton, S.F., Vallad, G.E., Scott, J.W., Jones, J.B., et al., 2015. The pepper *Bs2* gene confers effective field resistance to bacterial leaf spot and yield enhancement in Florida tomatoes. Acta Hortic. 1069, 47–51.

Jia, H., Zhang, Y., Orbović, V., Xu, J., White, F.F., Jones, J.B., et al., 2017. Genome editing of the disease susceptibility gene *CsLOB1* in citrus confers resistance to citrus canker. Plant Biotechnol. J. 15, 817–823.

Jiang, G.L., 2013. Molecular markers and marker-assisted breeding in plants. In: Anderson, S.B. (Ed.), Plant Breeding from Laboratories to Fields. InTech, Croatia, pp. 45–83.

Johal, G.S., Briggs, S.P., 1992. Reductase activity encoded by the HM1 disease resistance gene in maize. Science 258, 985–987.

Joshi, I., Kumar, A., Singh, A.K., Kohli, D., Raman, K.V., Sirohi, A., et al., 2019. Development of nematode resistance in Arabidopsis by HD-RNAi-mediated silencing of the effector gene *Mi-msp2*. Sci. Rep. 9, 17404.

Khaled, A.Y., Abd Aziz, S., Bejo, S.K., Nawi, N.M., Seman, I.A., Onwude, D.I., 2018. Early detection of diseases in plant tissue using spectroscopy–applications and limitations. Appl. Spectros. Rev. 53, 36–64.

Khaledi, N., Taheri, P., 2016. Biocontrol mechanisms of *Trichoderma harzianum* against soybean charcoal rot caused by *Macrophomina phaseolina*. J. Plant Prot. Res. 56, 21–31.

Khater, M., de la Escosura-Muñiz, A., Merkoçi, A., 2017. Biosensors for plant pathogen detection. Biosens. Bioelectron. 93, 72–86.

Koller, T., Brunner, S., Herren, G., Sanchez-Martin, J., Hurni, S., Keller, B., 2019. Field grown transgenic *Pm3e* wheat lines show powdery mildew resistance and no fitness costs associated with high transgene expression. Transgenic Res. 28, 9–20.

Kumar, R., Kumari, K., Hembram, K.C., Kandha, L., Bindhani, B.K., 2019. Expression of an endo α-1, 3-Glucanase gene from *Trichoderma harzianum* in rice induces resistance against sheath blight. J. Plant Biochem. Biotechnol. 28, 84–90.

Kumar, A., Pandey, V., Singh, M., Pandey, D., Saharan, M., Marla, S.S., 2017. Draft genome sequence of Karnal bunt pathogen (*Tilletia indica*) of wheat provides insights into the pathogenic mechanisms of quarantined fungus. PLoS One 12, e0171323.

Kumar, A., Pathak, R.K., Gupta, S.M., Gaur, V.S., Pandey, D., 2015. Systems biology for smart crops and agricultural innovation: filling the gaps between genotype and phenotype for complex traits linked with robust agricultural productivity and sustainability. OMICS 19, 581–601.

Kunwar, S., Iriarte, F., Fan, Q., daSilva, E.E., Ritchie, L., et al., 2018. Transgenic expression of EFR and *Bs2* genes for field management of bacterial wilt and bacterial spot of tomato. Phytopathology 108, 1402–1411.

Lai, K.S., Yusoff, K., Mahmood, M., 2012. Heterologous expression of hemagglutinin-neuraminidase protein from Newcastle disease virus strain AF2240 in *Centella asiatica*. Acta Biol. Crac. Ser. Bot. 54, 142–147.

Lau, H.Y., Botella, J.R., 2017. Advanced DNA-based point-of-care diagnostic methods for plant diseases detection. Front. Plant Sci. 8, 2016.

Lomonossoff, G.P., 1995. Pathogen-derived resistance to plant viruses. Annu. Rev. Phytopathol. 33, 323–343.

Macovei, A., Sevilla, N.R., Cantos, C., Jonson, C., Slamet-Loedin, G.B., Cermak, I., et al., 2018. Novel alleles of rice eIF4G generated by CRISPR/Cas9-targeted mutagenesis confer resistance to Rice tungro spherical virus. Plant Biotechnol. J. 16, 1918–1927.

Malathi, P., Muzammil, S.A., Krishnaveni, D., Balachandran, S.M., Mangrauthia, S.K., 2019. Coat protein 3 of rice tungro spherical virus is the key target gene for development of RNAi mediated tungro disease resistance in rice. Agri Gene 12, 100084.

Mancini, V., Murolo, S., Romanazzi, G., 2016. Diagnostic methods for detecting fungal pathogens on vegetable seeds. Plant Pathol. 65, 691–703.
Martinelli, F., Scalenghe, R., Davino, S., Panno, S., Scuderi, G., Ruisi, P., et al., 2015. Advanced methods of plant disease detection: a review. Agron. Sustain. Dev. 35, 1–25.
Miller, S.A., Beed, F.D., Harmon, C.L., 2009. Plant disease diagnostic capabilities and networks. Annu. Rev. Phytopathol. 47, 15–38.
Mukherjee, M., Mukherjee, P.K., Horwitz, B.A., Zachow, C., Berg, G., Zeilinger, S., 2012. *Trichoderma*–plant–pathogen interactions: advances in genetics of biological control. Indian J. Microbiol. 52, 522–529.
Narvaez, F.Y., Reina, G., Torres-Torriti, M., Kantor, G., Cheein, F.A., 2017. A survey of ranging and imaging techniques for precision agriculture phenotyping. IEEE-ASME T. Mech. 22, 2428–2439.
Neuhaus, J.M., 1999. Plant chitinases (PR-3, PR-4, PR-8, PR-11). In: Datta, S.K., Muthukrishnan, S. (Eds.), Pathogenesis-related Proteins in Plants. CRC Press, Boca Raton, FL, pp. 77–105.
Notomi, T., Okayama, H., Masubuchi, H., Yonekawa, T., Watanabe, K., Amino, N., et al., 2000. Loop-mediated isothermal amplification of DNA. Nucleic Acids Res. 28, E63.
Nzuki, I., Katari, M.S., Bredeson, J.V., Masumba, E., Kapinga, F., Salum, K., et al., 2017. QTL mapping for pest and disease resistance in cassava and coincidence of some QTL with introgression regions derived from *Manihot glaziovii*. Front. Plant Sci. 8, 1168.
Ortigosa, A., Gimenez-Ibanez, S., Leonhardt, N., Solano, R., 2019. Design of a bacterial speck resistant tomato by CRISPR/Cas9-mediated editing of SlJAZ2. Plant Biotechnol. J. 17, 665–673.
Pandey, P., Irulappan, V., Bagavathiannan, M.V., Senthil-Kumar, M., 2017. Impact of combined abiotic and biotic stresses on plant growth and avenues for crop improvement by exploiting physio-morphological traits. Front. Plant Sci. 8, 537.
Panigrahi, J., 2011. Molecular mapping and map based cloning of genes in plants. In: Thangadurai, D. (Ed.), Plant Biotechnology and Transgenics. Bentham Science Publishers, Potomac, USA.
Pathak, R.K., Baunthiyal, M., Pandey, D., Kumar, A., 2018. Augmentation of crop productivity through interventions of omics technologies in India: challenges and opportunities. 3 Biotech. 8, 454.
Pathania, A., Rialch, N., Sharma, P.N., 2017. Marker-assisted selection in disease resistance breeding: a boon to enhance agriculture production current developments in biotechnology and bioengineering: crop modification, nutrition, and food production. Available from: https://doi.org/10.1016/B978-0-444-63661-4.00009-8.
Piepenburg, O., Williams, C.H., Stemple, D.L., Armes, N.A., 2006. DNA detection using recombination proteins. PLOS Biol. 4, e204.
Schaad, N.W., Frederick, R.D., Shaw, J., Schneider, W.L., Hickson, R., Petrillo, M.D., et al., 2003. Advances in molecular-based diagnostics in meeting crop biosecurity and phytosanitary issues. Ann. Rev. Phytopathol. 4, 305–324.
Sharma, P.C., Matsumura, H., Terauchi, R., 2007. Use of serial analysis of gene expression (SAGE) for transcript profiling in plants. In: Varshney, R.K., Tuberosa, R. (Eds.), Genomics-Assisted Crop Improvement. Springer, Dordrecht, pp. 227–244.
Singh, N., Kumar, S., Bajpai, V.K., Dubey, R.C., Maheshwari, D.K., Kang, S.C., 2010. Biological control of *Macrophomina phaseolina* by chemotactic fluorescent *Pseudomonas aeruginosa* PN1 and its plant growth promotory activity in chir-pine. Crop Prot. 29, 1142–1147.
Singh, N., Mukherjee, S.K., Rajam, M.V., 2020. Silencing of the ornithine decarboxylase gene of *Fusarium oxysporum* f. sp. *lycopersici* by host-induced RNAi confers resistance to Fusarium wilt in tomato. Plant Mol. Biol. Rep. 38, 419–429. Available from: https://doi.org/10.1007/s11105-020-01205-2.
Sundaram, R.M., Vishnupriya, M.R., Biradar, S.K., Laha, G.S., Reddy, G.A., Shoba Rani, N., et al., 2008. Marker assisted introgression of bacterial blight resistance in Samba Mahsuri, an elite indica rice variety. Euphytica 160, 411–422.
Taif, S., Zhao, Q., Pu, L., Li, X., Liu, D., Cui, X.M., 2020. A β-1,3-glucanase gene from *Panax notoginseng* confers resistance in tobacco to *Fusarium solani*. Ind. Crops. Prod. 143, 111947.
Toojinda, T., Broers, L.H., Chen, X.M., Hayes, P.M., Kleinhofs, A., Korte, J., et al., 2000. Mapping quantitative and qualitative disease resistance genes in a doubled haploid population of barley (*Hordeum vulgare*). Theor. Appl. Genet. 101, 580–589.
van Loon, L.C., 2007. Plant responses to plant growth-promoting rhizobacteria. Eur. J. Plant Pathol. 119, 243–254.
Vincent, M., Xu, Y., Kong, H.M., 2004. Helicase-dependent isothermal DNA amplification. EMBO Rep. 5, 795–800.

Wang, F., Wang, C., Liu, P., Lei, C., Hao, W., Gao, Y., et al., 2016. Enhanced rice blast resistance by CRISPR/Cas9-targeted mutagenesis of the ERF transcription factor gene OsERF922. PLoS One 11, e0154027.

Wang, M., Zheng, Q., Shen, Q., Guo, S., 2013. The critical role of potassium in plant stress response. Int. J. Mol. Sci. 14, 7370–7390.

Webi, E.N., Daniel, K., Johnson, K., Anne, N., Marc, G., Eric, M., 2019. Extreme resistance to late blight disease by transferring 3 R genes from wild relatives into African farmer-preferred potato varieties. Afr. J. Biotechnol. 18, 845–856.

Xu, Y., 2010. Molecular Plant Breeding. CAB International.

Yang, X., Yang, J., Li, H., Niu, L., Xing, G., Zhang, Y., et al., 2020. Overexpression of the chitinase gene *CmCH1* from *Coniothyrium minitans* renders enhanced resistance to *Sclerotinia sclerotiorum* in soybean. Transgenic Res. 29, 187–198.

Zhang, L., Wang, D., Hu, Q., Dai, X., Xie, Y., Li, Q., et al., 2019. Consortium of plant growth-promoting rhizobacteria strains suppresses sweet pepper disease by altering the rhizosphere microbiota. Front. Microbiol. 10, 1668.

Zhao, L., Huang, Y., Hu, J., Zhou, H., Adeleye, A.S., Keller, A.A., 2016. 1H NMR and GC-MS based metabolomics reveal defense and detoxification mechanism of cucumber plant under nano-Cu stress. Environ. Sci. Technol. 50, 2000–2010.

Zhao, B., Lin, X., Poland, J., Trick, H., Leach, J., Hulbert, S., 2005. A maize resistance gene functions against bacterial streak disease in rice. PNAS 102, 15383–15388.

Zhu, S., Li, Y., Vossen, J.H., Visser, R.G.F., Jacobsen, E., 2012. Functional stacking of three resistance genes against *Phytophthora infestans* in potato. Transgenic Res. 21, 89–99.

CHAPTER

21

Biomanagement of *Fusarium* spp. associated with pulse crops

Sampat Nehra[1,*], *Raj Kumar Gothwal*[1,2,*], *Alok Kumar Varshney*[1,2], *Pooran Singh Solanki*[1,2], *Shivani Chandra*[3], *Poonam Meena*[4], *P.C. Trivedi*[4] *and P. Ghosh*[1]

[1]Birla Institute of Scientific Research, Statue Circle, Jaipur, India [2]Department of BioEngineering, Birla Institute of Technology, Mesra, Ranchi, Jaipur Campus, Jaipur, India [3]Amity Institute of Biotechnology, Amity University, Noida, India [4]Department of Botany, University of Rajasthan, Jaipur, India

21.1 Introduction

Pulses, belonging to the family Leguminosae, are edible dry seeds of plants. They are consumed in the form of flour, split grain, dehulled split grain, whole seed, and also as of green vegetables. They are grown virtually in every corner of the world map and have a strong history of nourishing people for centuries. Pulses are significant source of proteins, vitamins, and minerals and are rich in micro and macronutrients and therefore, they are considered to be best fit in sustainable cropping arrangement. Pulses are preferred for irradiating protein malnutrition and add to the nutritional security of the world. Common bean, pea, chickpea, lentil, cowpea, mung bean, black gram, and pigeon pea are majorly produced and consumed pulses globally. Furthermore, different parts of the globe cultivate and consume a large number of minor pulses. Although the pulses are primarily grown for human consumption, they are also used as animal feed substantially. Pulses are grown in all three different seasons in India (Table 21.1).

Along with the early cereal grains, pulses were among the very first crops cultivated in the fields as far back as 11,000 years ago. As summarized in Tables 21.2 and 21.3, in the past

* Content is equally contributed by the authors.

Food Security and Plant Disease Management
DOI: https://doi.org/10.1016/B978-0-12-821843-3.00010-6

TABLE 21.1 Growing seasons of pulses in Indian subcontinent (Sinha et al., 2018).

Growing season	Pulses
Kharif season (July–October)	*Cajanus cajan* (L.) Millsp., arhar (tur)
	Vigna mungo (L.) Hepper, urd (black gram)
	Vigna radiata (L.) Wilezek, moong (green gram)
	Vigna unguiculata (L.) Walp, lobia (cowpea)
	Macrotyloma uniflorum (Lam.) Verdc., kulthi (horse gram)
	Vigna aconitifolia (Jacq.) Marechal., moth
Rabi season (October–March)	*Cicer arietinum* (L.), gram
	Lens culinaris Medik, lentil
	Pisum sativum (L.), pea
	Phaseolus vulgaris (L.), rajmash
Summer season (crops are grown in between March and July)	*V. radiata* (L.) Wilezek, green gram
	V. mungo (L.) Hepper
	Black gram and *V. unguiculata* (L.) Walp, cowpea

TABLE 21.2 Global production of pulses in million tonnes after triennium endings of 1971, 1981, 1991, 2001, 2011, and 2014 (Rawal and Navarro, 2019).

Year	Dry bean	Chickpea	Dry pea	Cowpea	Lentil	Pigeon pea	Faba bean	Other pulses	Total pulses
1971	12	7	9	1	1	2	4	6	43
1981	14	6	8	1	1	2	4	4	41
1991	17	7	15	2	2	3	4	6	57
2001	18	8	11	4	3	3	4	6	56
2011	23	11	10	6	4	4	4	6	68
2014	24	13	11	7	5	4	4	7	77

50 years, world pulse production has grown rapidly. In the past one decade alone, the world has produced between 70 and 80 million tonnes of pulses every year. The major countries growing pulses are Argentina, Brazil, Australia, Canada, the United States, Ethiopia, Myanmar, Mexico, China, and India, whereas, Argentina, France, Ethiopia, and Turkey are important exporters. The gross margin in production of pulses worldwide has been depicted in Table 21.4. The annual global production of pulses, in the triennium ending 2014, was about 77 million tonnes. Among the pulses, the production of the dry bean pulses was about 24 million tonnes, around 11 million tonnes of dry pea, around 13 million tonnes of chickpea, and around 7 million tonnes of cowpea. The annual production of lentil in the same period for pigeon pea and faba beans was about 5 and 4 million tonnes, respectively. All other pulses

TABLE 21.3 Global yield of pulses in kilograms per hectare after triennium endings of 1971, 1981, 1991, 2001, 2011, and 2014 (Rawal and Navarro, 2019).

Year	Dry bean	Chickpea	Dry pea	Cowpea	Lentil	Pigeon pea	Faba bean	Other pulses	Total pulses
1971	532	652	1164	237	605	694	966	670	1789
1981	555	624	1099	377	611	676	1177	669	2195
1991	628	706	1696	402	752	722	1508	819	2690
2001	750	766	1773	414	827	722	1501	848	3101
2011	784	901	1600	520	1040	779	1755	889	3597
2014	824	956	1616	614	1152	718	1807	929	3784

TABLE 21.4 Gross margin in production of pulses (average per hectare) in major pulse producing countries worldwide expressed in USD per hectare.

Country	Year	Chickpea	Mung bean	Red lentil	Common bean	Urd bean	Reference
India	2013–14	204	109	265	–	68	CACP (2015)
Bangladesh	2011	–	–	141	–	–	Rahman et al. (2013)
Myanmar	2016	139	468	–	–	248	Zorya et al. (2016)
Australia	2015	294	542	369	–	–	Rawal and Navarro, 2019
Turkey	2009	379	–	–	–	–	Orum et al. (2009)
Saskatchewan, Canada	2016	–	–	919	–	–	Rawal and Navarro, 2019
The United States (Southeast, North Dakota)	2015	–	–	–	174	–	Rawal and Navarro, 2019
Brazil	2015	–	–	–	220	–	CONAB (2015)
Mexico	2015–16	–	–	–	309	–	AARFS (2013)
Kenya	2010–13	–	–	–	303	–	Fintrac Inc. (2013)
Uganda	2012	–	–	–	147	–	Kilimo Trust (2012)

that were not classified separately including lupins, bambara bean, vetches, and other minor pulses together produced around 7 million tonnes (Tables 21.1 and 21.2).

In 2011–13 (average), global pulse production in Mexico accounted 73.3 million metric tonnes (Joshi and Rao, 2017) whereas in India the production was around 19.3 million tonnes (ESI, The Economic Survey, 2014–2015; Singh et al., 2017).

India stands on the top in pulse production and has come a long way from a pulse-deficient country to self-sufficient one. According to FAOSTAT data, in India during the triennium 2012–14, total pulses were grown on 25,586 thousand hectares and account for over 60%–70% share in total world production (Table 21.5). Chickpea and pigeon pea

TABLE 21.5 Average production, yield, and area harvested of pulses in India during 2012–14 (Rawal and Navarro, 2019).

Crop	Production (million tonnes)	Yield (kilograms per hectare)	Area harvested (thousand hectares)	Share in world production (%)
Dry pea	0.61	831	732	5.44
Pigeon pea	3.0	652	4584	67.4
Chickpea	8.8	922	9553	67.3
Mung bean	1.48	469	3187	–
Urd bean	1.79	569	3183	–
Lentil	1.1	722	1520	21.8
Phaseolus and *Vigna* pulses	3.8	1353	2827	12.0

TABLE 21.6 Average net returns in kilograms per hectare for chickpea in selected north Indian states (Rawal and Navarro, 2019).

State	1981–82	1995–96	2004–05	2013–14
Uttar Pradesh	88	71	129	−2
Madhya Pradesh	81	44	104	26
Rajasthan	22	22	33	32
Haryana	24	43	36	14

were shared as 67.4% individually in global pulse production and were first in world map pulse production, whereas lentil was shared as 21.8%, dry pea as 5.44%, and both *Phaseolus* and *Vigna* pulses shared as 12% in total world pulse production (Table 21.4).

The production and profitability of pulse crops are more among other agricultural production in Indian climatic conditions. Statistically, Indians consume 30% of the world's total pulse and produces a quarter of the total world's pulses, accounting for one-third of the total average under pulses. However, India has not been able to keep up the production of the pulses with the population growth. The main pulses grown in India are chickpea (*Cicer arietinum* L.), pigeon pea (*Cajanus cajan* (L.) Millsp.), pea (*Pisum sativum*), green gram (*Vigna radiata*), black gram (*Vigna mungo*), and lentil (*Lens culinaris* Medikus). These pulses account for over 80% of the total pulse production in the country. The major pulse-growing areas in the country are Uttar Pradesh, Madhya Pradesh, Maharashtra, Rajasthan, Karnataka, Andhra Pradesh, Gujarat, Tamil Nadu, Jharkhand, and Bihar that account for more than 75% of total production.

The Directorate of Economics and Statistics report pointed out that growing of pulses is economically beneficial as compared to cereal crops, for example, ratio of average net returns from cultivation of chickpea was more as compared to average returns from cultivation of wheat from selected north Indian states as shown in Table 21.6.

Several phytopathological diseases infect the pulses and other annual leguminous crop, which affects the quality of the produce and reduces the crop yield, which in turn affects the economy of the region. As summarized in Table 21.7, there are many diseases and pests that affect the pulse production adversely. In pulse crop production, soilborne pathogens, that survive in the soil and spread the diseases through soil, cause significant economic losses.

Among all biotic stresses, fungal phytopathogens are important because of their ability to infect all plant parts at all growth stages. By an estimate, legumes are known to be affected by more than 100 fungal diseases all around the world. *Fusarium* spp. are

TABLE 21.7 Major diseases and pests associated with pulses.

Crop	Diseases and pests
Chickpea	• Wilt (*Fusarium oxysporum* f. sp. *ciceris* Matuo and Sato) • Blight (*Mycosphaerella pinodes* B. and Blox) • Ascochyta blight (*Ascochyta rabiei*) • Rust (*Uromyces ciceris-arietinii* (Grogn.) Jacz. and Boy.) • *Helicoverpa* pest • Botrytis gray mold (BGM) • Dry root rot (*Rhizoctonia bataticola* or *Macrophomina phaseolina*)
Pigeon pea	• White mold (Sclerotinia rot) (*Sclerotinia sclerotum*) • Blight (*Alternaria alternata*) • Anthracnose; *Colletotrichum* spp. • Cercospora leaf spot (*Cercospora cajani*) • Wilt (*F. oxysporum* f. sp. *udum* (Butler) Snyd. and Hans.) • Sterility mosaic
Pea	• Powdery mildew (*Erysiphe polygoni* DC) • Wilt (*F. oxysporum* f. sp. *pisi*) • Rust (*Uromyces viciae-fabae* (pers.) Schroet) • *Ascochyta* blight • Bacterial blight
Lentil	• Rust (*U. viciae-fabae* [Pers.] J. Schrot) • Collar rot (*Sclerotium rolfsii*: *Athelia rolfsii* [*teleomorph*] or *Corticium rolfsii*) • Wilt (*F. oxysporum* f. sp. *lentis*) • BGM • *Ascochyta* blight
Mung bean	• Root rot (*R. bataticola* (Taub.) Butler (*M. phaseolina* Tassi Goid)) • Mung bean yellow mosaic virus • Powdery mildew (*E. polygoni*) • Cercospora leaf spot • Helicoverpa pest, white fly, bean, aphid, leaf hopper, green and brown mirids, bean pod borer and thrips
Urd bean	• Root rot (*R. bataticola* (Taub.) Butler (*M. phaseolina* Tassi Goid)) • Leaf spot (Cercospora leaf spot) • Anthracnose • Powdery mildew (*E. polygoni*) • Rust (*Uromyces phaseoli typica* (Syn: *U. appendiculatus*)) • Stem canker and Macrophomina blight • Bacterial leaf blight • Yellow mosaic disease

pathogenic soilborne ascomycete fungi affecting many pulses in world by causing wilt that is a lethal vascular syndrome in the affected plants (Flood, 2006; Dita et al., 2018). Economic losses caused by *Fusarium* wilt of common stock pathogens are estimated at 50%–75% of the attainable yield for many crops (Navas Cortes et al., 2000; Mihajlovic et al., 2017). Up to 100% yield losses have been reported due to wilt disease (Choudhary et al., 2013).

21.1.1 The primary hindrance

Soilborne pathogens that cause wilts and root rots have been primary concern for many years as it hinders the sustainable production. The complex nature of these wilts further aggravates the problem. In this chapter, we will discuss the different wilts and root rots that cause considerable losses to pulses and their biological management strategies. Host–parasite relationship, and recent methods used for early diagnosis and timely detection will also be discussed.

21.2 Major diseases associated with pulses

Pulse production in India is self-sufficient, however, in supply and demand chain always observes an increasing gap. Protein requirement of vegetarian population worldwide is fulfilled by pulses that are rich in protein. A large number of fungal diseases infect pulses all around the world, limiting the crop yield to a great extent. Pigeon pea, chickpea, lentil, and pea record considerable damage of crops as they are attacked by several diseases (Table 21.7). Pigeon pea is very important legume crop in India, and more than 100 phytopathogens such as bacteria, fungi, nematodes, and viruses infect the crop, and among which the vascular wilt pathogenic fungus *Fusarium oxysporum* f. sp. *udum* (Butler) Snyd. and Hans. is economically damaging (Kumar et al., 2020). Pigeon pea is also infected by sterility mosaic. Chickpea is commonly affected by wilt (*F. oxysporum* f. sp. *ciceris* Matuo and Sato), blight (*Mycosphaerella pinodes* B. and Blox), and rust (*Uromyces ciceris-arietinii* (Grogn.) Jacz. and Boy.). Lentil is infected by rust (*Uromyces viciae-fabae* [Pers.] J. Schrot), collar rot (*Sclerotium rolfsii or Athelia rolfsii [teleomorph] or Corticium rolfsii*), and wilt (*F. oxysporum* f. sp. *lentis*). Pea is infected by powdery mildew (*Erysiphe polygoni* DC), wilt (*F. oxysporum* f. sp. *pisi*), and rust (*U. viciae-fabae* (pers.) Schroet). Among the fungal diseases, *Fusarium* spp. has proved to be the most destructive disease affecting a wide range of commercially domesticated pulse crops. *Fusarium* wilt caused by *Fusarium* spp. is a peril to the pulses growers. *Fusarium* vascular wilt is a seed and soilborne disease.

21.3 *Fusarium* wilts: An overview

Presently, about 80% of plant diseases in agriculture can be related to fungal pathogens. *Fusarium* wilts are very important and damaging among these fungal diseases. *Fusarium* vascular wilts are caused by pathogenic strains of several *Fusarium* species, including *F.*

avenaceum, F. eumartii, F. oxysporum, F. solani, F. sulphureum, and *F. tabacinum* (Okungbowa and Shittu, 2012), which are usually very host-specific in nature. *Fusarium* wilt is a very harmful fungal disease, which blocks the xylem cells that are helpful in conducting water, as a result the plant wilts and finally dies.

Fusarium spp. is abundant and has frequent association with plant roots. These soil-borne fungi are present either as parasites or saprophytes (Nelson et al., 1994). It is distributed worldwide covering almost all types of climatic conditions. For the better understanding, the life cycle of *Fusarium* spp. can be divided into three different stages: dormant, parasitic, and saprophytic stages. Most species are saprobes, which are harmless; some species are parasitic in nature, which produce mycotoxins on host plants. The pathogenic strains are very host-specific in their action, which has led to the development of the "formae specialis" concept.

Fusarium species are filamentous Ascomycetes fungi (Family Hypocreaceae) (Pitt et al., 1994). These species mainly produce microconidia, macroconidia, mycelia, and chlamydospores that serve as propagules in infecting host plants. Presently, there are over 100 different formae speciales described, causing disease in a wide range of dicot and monocot plant species. For example, *Fusarium* wilts of cotton (*F. oxysporum* f. sp. *vasinfectum*), *Fusarium* wilt of date palm (*F. oxysporum* f. sp. *albedinis*), banana wilt, panama wilt (*F. oxysporum* f. sp. *cubense*), *Fusarium* wilt of *Callistephus* (*F. oxysporum* f. sp. *callistephi*), *Fusarium* wilt of sweet potato (*F. oxysporum* f. sp. *batatas*), *Fusarium* wilt of tomato (*F. oxysporum* f. sp. *lycopersici*), and *Fusarium* yellows of common beans (*F. oxysporum* f. sp. *phaseoli*) (Okungbowa and Shittu, 2012).

Mainly the wilts are caused by *F. oxysporum* in pulses and other crops of economic importance. *Fusarium* species are vascular wilt phytopathogens, which are among the most important toxigenic fungi. They are considered to be very toxic because they produce mycotoxins such as trichothecenes, fumonisins, and zearalenones, which contaminate the seeds. Through seeds they enter the food chain, and hence affect both human and animal health and are thus hazardous to agricultural products, wildlife, livestock, and humans (Balali and Iranpoor, 2006; Arif et al., 2011; Wang et al., 2011). Table 21.8 shows various wilts and root rots of major pulse crops grown in India.

21.4 Occurrence of *Fusarium* wilts in pulses worldwide

Table 21.9 shows that pulses grown worldwide are infected with *Fusarium* wilt. Thus *Fusarium* wilt is an important fungal disease globally. Occurrence of pea wilt is also global as it is common in many countries worldwide such as the United States, Canada, France, South Australia, Tropical Africa, Taiwan, Philippines, India, Pakistan, Bangladesh, China, Brazil, Thailand, Russia, and many other countries. Wilt disease in pigeon pea has been reported from 15 countries (Kiprop et al., 2002). This disease has spread over in some areas of South Africa in recent survey, Southern Zambezian province (Biswas and Ghosh, 2016). Pigeon pea wilt is considered to be more significant in India and Eastern Africa. Chickpea wilt is widely found in almost all the countries it is grown, such as Iran, Mexico, Syria, Ethiopia, Spain, India, Pakistan, Nebraska (USA), and Sudan. Most of the countries that are growing lentil face the problem of lentil wilt. Some of the countries where the

TABLE 21.8 Wilts and root rots associated with major pulse crops in India (Kumari and Katoch, 2020).

Pulse crop	Disease	Pathogenic fungi	References
Chickpea (*Cicer arietinum* L.)	Wilt	*Fusarium oxysporum* f. sp. *ciceris*	Booth (1971a), Haque and Ghaffar (1992), Sharma (1995), Akhtar (1955), Srivastava et al. (2011), Khan et al. (2016), Tawfik and Allam (2004), Singh et al. (2016), Knights and Hobson (2016), Jendoubi et al. (2017), Pandey et al. (2017), Sankar et al. (2018)
	Black root rot	*Fusarium solani*	Ghosh et al. (2013)
	Dry root rot	*Rhizoctonia bataticola* (Taub.) Butler (Pycnidial stage: *Macrophomina phaseolina* Tassi Goid)	Kadam et al. (2018), Deepa et al. (2018)
	Collar rot	*Sclerotium rolfsii* (Teleomorph: *Athelia rolfsii* (Curzi) Tu and Kimbrough)	Ghosh et al. (2013), Ahsan et al. (2018)
Pigeon pea (*Cajanus cajan* (L.) Millsp.)	Root rot/wilt	*F. oxysporum* f. sp. *udum*	Mahesh et al. (2010), Devi and Chhetry (2012), Singh et al. (2016), Sharma et al. (2018), Birajdar et al. (2018)
	Dry root rot	*R. bataticola* (Taub.) Butler (*M. phaseolina* Tassi Goid)	Maruti et al. (2017)
Pea (*Pisum sativum*)	Root rot	*Rhizoctonia solani*	Nongmaithem et al. (2017)
	Wilt	*F. oxysporum*, *Fusarium orthoceras* var. *pisi*	Hagedorn (1991), Skovgaard et al. (2002), Leslie and Summerell (2006), Smith (2007), Sardana et al. (2007), Tariq et al. (2015)
	Wilt/root rot complex	*F. oxysporum* f. sp. *pisi*; *F. solani* f. sp. *pisi* (Jones) Synder and Hansen	Nongmaithem et al. (2017)
Lentil (*Lens culinaris* Medikus)	Wilt	*F. oxysporum* f. sp. *lentis*	Erskine et al. (1990), Chaudhary and Amarjit (2002), Sharfuddin and Mohanka (2012), Singh et al. (2017); Bedasa (2018), Arya and Kushwaha (2019)
	Root rot	*R. solani*	Tiwari et al. (2018)
	Collar rot or root rot	*S. rolfsii* (*A. rolfsii*)	Kushwaha (2016), Tiwari et al. (2018)
Green gram (*Vigna radiata*)	Root rot	*R. bataticola* (Taub.) Butler (*M. phaseolina* Tassi Goid)	Shahid and Khan (2016), Mallaiah and Rao (2016)
	Seedling rot and web/leaf blight	*R. solani* Kuhn (*Thanatephorus cucumeris*)	Singh et al. (2013)
Black gram (*Vigna mungo*)	Dry root rot	*R. bataticola* (Taub.) Butler (*M. phaseolina* Tassi Goid)	Tetali et al. (2015)
Cowpea/soybean (*Vigna unguiculata* L.)	*Fusarium* wilt	*F. oxysporum* f. sp. *tracheiphilum*, *F. semitectum*	Armstrong and Armstrong (1980), Rodrigues and Menezes (2006), Khare et al. (2016)
Beans (*Phaseolus vulgaris* L.)	*Fusarium* wilt	*F. oxysporum* f. sp. *phaseoli*	Abawi et al. (2006), Mukankusi et al. (2010), Kalyebara and Kassozi (2005), Alwathnani and Perveen (2012), Saul-Jose et al. (2018)

TABLE 21.9 Occurrence of *Fusarium* wilts in pulses in different countries worldwide.

Disease	Affected countries	References
Chickpea wilt	Spain	Jimenez-Diaz et al. (1990)
	Syria	Haware et al. (1990)
	India	Andrabi et al. (2011)
	Pakistan	Hameed et al. (2012)
	Mexico	Arvayo-Ortiz et al. (2012)
	Nebraska	Harveson (2011)
	The United States	Harveson (2011)
	Iran	Moradi et al. (2012)
	Sudan	Ali et al. (2002)
Pigeon pea wilt	15 countries including, India, East Africa	Nene et al. (1989), Kiprop et al. (2002)
	Malawi	Nene et al. (1989)
	Bangladesh, Grenada, Indonesia, Mauritius, Myanmar, Nepal, Nevis, Venezuela, Trinidad, Tobago	Kannaiyan et al. (1984), Reddy et al. (1993), Marley and Hillocks (1996)
	Southern Africa reaching areas in Mozambique (Southern Zambezian province)	Gwata et al. (2006), Biswas and Ghosh (2016)
Pea wilt	India, Bangladesh, Brazil, Philippines, South Australia, Taiwan, Thailand, Tropical Africa, France, the United States, Pakistan, China, Russia, Canada, and many other countries	Kuldhar et al. (2013)
Lentil wilt	America	Wilson and Brandsberg (1965)
	Bangladesh	Baker (1991)
	Argentina	Ravenna and Negri (1979)
	Brazil	Veiga et al. (1974)
	India	Vasudeva and Srinivasan (1952)
	Italy	Tosi and Cappelli (2001)
	France	Moreoau (1978)
	Czechoslovakia	Ujevic et al. (1965)
	Myanmar and Pakistan	Bahl et al. (1991)
	Nepal	Karki (1991)
	Syria	Bayaa et al. (1986), EI-Ahmed and Mouselli (1986)
	Uruguay	Carrera (1940)
	Russia	Kotova et al. (1965), Ansari (2003)

disease is found are India, Pakistan, Nepal, Bangladesh, Myanmar, Brazil, Argentina, Czechoslovakia, Italy, Russia, France, Uruguay, and Syria (Sinha et al., 2018).

21.5 Economic losses due to *Fusarium* infection in pulses

Most susceptible plants for *Fusarium* species are solanaceous crops such as banana, sweet potato, cucurbits, and most importantly, pulses (Miller et al., 2011). Nevertheless, this fungus is infectious for other herbaceous plants, cotton, ornamental plants, palms, China aster (*Callistephus chinensis*), carnations (*Dianthus*), and French/runner beans, peas as well and stands responsible for tremendous losses in the affected crops.

The wilt disease causes substantial losses in almost all the major pulse-growing countries globally. Table 21.10 depicts the losses in yield occurred in different pulses. Since *Fusarium* spp. is soilborne in nature, the pathogen infects the roots of the plant and spreads in the xylem vessels along with the water transport system and subsequently reaches the stem and foliage. *Fusarium* wilt is a major concern in pulse farming as it causes great economical loss in a varied diversity of pulse crops. The symptoms of *Fusarium* wilt range in pulses are stunted growth, reddish discoloration of the xylem vessels (visible in cross section under microscope as dots or lines inside the stem), wilting of the leaves, and leaf color changing to yellow. The stem and root decay and presence of fungus in the form of white, pink, or orange growth (particularly in wet conditions) are other symptoms of *Fusarium* wilt (Okungbowa and Shittu, 2012).

Wilt diseases caused by *Fusarium* spp. not only reduces the crop yield but also affects the quality of the pulses. Besides crop losses, there are other losses that result in low marketable yield. These may be the legumes that cannot be sold because of their small size, distorted shape, little nutritional value, decolorization of legumes, and broken or cracked legume. There are also some other hidden losses resulting from increased costs with the product associated with fumigation or other soil treatments, roguing or replanting dead plants, and labor and fuel charges.

Once an existing field becomes infested with the pathogenic fungus, it cannot be used. Costs associated with preparing new land for production are also counted as increased cost with the yield.

TABLE 21.10 Yield losses due to wilt in pulses.

Disease	Estimated yield losses	References
All pulses—commonly	50%	Sinha et al. (2018)
Chickpea	Up to 100% (under favorable condition)	Navas Cortes et al. (2000), Halila and Strange (1996), Patra and Biswas (2017)
Pea	25%–50%	Nongmaithem et al. (2017)
Pigeon pea	30%–100% (worldwide)	Biswas and Ghosh (2016)
Lentil	20%–24%	Saxena and Johansen (1990), Ali (2007), Jaruhar and Prasad (2011)

21.6 Control of *Fusarium* vascular wilt in pulses

Fusarium is difficult to control in pulses or other agricultural crops, as physical, chemical, and cultural methods of control are not only ineffective but also expensive. The reasons for the widespread distribution of *Fusarium* species are (1) their capability to grow and develop on a wide range of substrates and (2) their efficient mechanisms for spore dispersal (Nelson et al., 1994). The fungus is soilborne phytopathogen, thus it can survive in soil and plant debris for numerous years. The infected soil moves from one place to another through tools used in the fields and by animals. This can lead to outbreaks in new areas. The disease is capable of surviving in variety of environmental conditions. The amount and distribution of infected inoculum in the field and stage of plant growth when the disease occurs decide how much yield reduction will occur (Muimba-Kankolongo, 2018). In chickpea, dry and warm growing conditions are most favorable for the growth of *Fusarium* wilt. This poses a great problem in the Indian subcontinent regions and in areas where the crop is sown in spring season (Knights and Hobson, 2016).

The parasitic phase in the *Fusarium* spp. includes the penetration of host roots, colonization of fungus in the root cortex and endodermis, and movement of the fungus to the xylem tissues, colonization of the xylem vessels of stems and leaves, expression of the symptoms, and finally death of the host tissues and plant. The formation of resting structures in the dead host forms the saprophytic stage of *Fusarium* spp. (Schnathorst, 1981). After penetration into the host it is very difficult to manage the growing population of the fungus.

In the dormant stage of the *Fusarium* spp. the germination of mycelia, chlamydospores, microconidia, and macroconidia (fungal propagules) that are present in infected soil is inhibited because of microbiostasis or mycostasis (Huisman, 1982). Wilts caused by *Fusarium* are commonly assumed to be monocyclic in nature that means that the disease does not spread from one plant to another during the season (Egel and Martyn, 2007). The reason behind this is that the propagules that are capable of dissemination of disease to other plants to cause secondary infections are only formed very late in the crop season. So to manage *Fusarium*'s dispersal and growing population is a very difficult task under field levels.

21.7 Host–pathogen interaction studies for *Fusarium* species

Host–pathogen interaction with respect to understanding the disease incidence and its severity is of great concern. Walter et al. (2009) observed enhanced intercellular penetration process due to hydrolyzing enzymes secreted by *Fusarium*. *Fusarium* species being the soilborne plant pathogens, which infect the water-conducting vessels of host plants colonizes within the host root and the cortex by emerging mycelia (Bishop and Cooper, 1983), that reaches to the endodermal cells and finally invade the xylem vessels through the pits. The mycelia remain in the xylem vessels where they invade the plant in an upward direction (colonization) through the stem and top crown of the plant (Rodriguez-Galvez and Mendgen, 1995; Agrios, 2005). Inoue et al. (2002) have claimed experimentally that effective fungal colonization is genetically controlled.

Microconidia that are produced by mycelia, within the xylem vessels, move upward in the transpiration stream, until they are trapped in pit cavities or at vessel end walls. The microconidia spread the infection by germinating into new hyphae that penetrate the vessels adjacent to it and colonize. Thus the infection increases in the population inside the host (Schnathorst, 1981). At this stage, significant cytological alterations are caused as the *Fusarium* hyphae spread inside the cell apoplast and results in expression of the symptoms (Walter et al., 2009). A sequence of xylem vessel clogging by mycelia, spores (fungal) and gels, gums, tyloses and crushing of the vessels by proliferating adjacent parenchyma cells from the host plant in an attempt to defend itself plug vessels and become the reason for the breakdown of the water economy of infected host plant (Agrios, 2005; Moreau et al., 1978; VanderMolen et al., 1977; Walter et al., 2009; Okungbowa and Shittu, 2012).

As a result of this lower branches begin to wilt, followed by the entire plant, which eventually leads to death of the plant. The new spores can be disseminated to new plants or areas by wind or water or can go back to the soil due to the decomposition of the plant. In the whole cycle, conidia are also formed in sporodochia on dead leaves. Hyphae and chlamydospores are also formed extensively. The chlamydospores are mixed and remained in the soil, when the infected plant residues decay. They remain dormant for several years until their favorable time for germination comes. Again, they germinate by parasitic or saprophytic colonization of a new host (Fassihiani, 2000).

Mechanism of Pathogenicity of *Fusarium* spp.: There is remarkably very little data available about the molecular mechanism of pathogenicity or pathogenicity genes required by *Fusarium* species to cause disease and how hosts tolerate or combat the pathogen. The whole-genome sequencing and comparative genomic analyses of some *Fusarium* species have proved that pathogenic *Fusarium* species consists of a larger number of proteins. These proteins were basically related to pathogenicity-related protein families such as hydrolytic enzymes, transcription factors, and transmembrane transporters, which play significant roles in pathogenicity, as compared to nonpathogenic species (Okungbowa and Shittu, 2012).

21.8 Symptomatology and epidemiology of *Fusarium* wilts

Fusarium occurs predominantly as a saprophyte and parasite. It is one of the most widely recognized, major fungi of the cultivated soil, and is known to be cosmopolitan.

One of the economic importance of *Fusarium* species is their role as plant pathogens. *Fusarium* causes diseases such as crown rot, head blight, and scab on cereal grains; vascular wilts on a wide range of horticultural crops; root rots; cankers; and other diseases such as pokkah-boeng on sugarcane and bakanae disease of rice (Booth, 1971a) and root rot and wilt on pulses (Singh et al., 2016; Saul-Jose et al., 2018; Abhijeet et al., 2018). *Fusarium* wilt occurs at the early seedling stage of plants but is mainly visible at flowering and fruiting stages. Seedling and adult plants show the symptoms of *Fusarium* wilt with yellowing and drying of leaves followed by whole plant wilting, for example, in the pigeon pea field, wilted plants can be easily identified in the field with wilt patches (Kumar et al., 2020).

The common symptoms of *Fusarium* wilts are stunted growth, reddish discoloration of the xylem vessels, which are visible under the microscope inside the stem as dots (cross section) or lines (longitudinal section). Other symptoms are pink, white, or orange growth of fungus on the outside of infected stems, particularly in wet conditions, root or stem decay (Mueller and Beckman, 1988; Miller et al., 2011; Okungbowa and Shittu, 2012). The amount of symptomatology depends upon a number of factors such as inoculum density. If the inoculum density of fungal organism is more in the field, the severity of leaf damage will be more accordingly. In young plantlets or at the early seedling stage or cotyledons, all the leaves wilt and drop, thus the stems are bare or naked.

Early identification of *Fusarium* wilt may be difficult because early symptoms may resemble some types of seedling disease and symptoms developing much later may be very similar to some other diseases. For example, symptoms of crown or root rot, stem cankers, drought, pest injury, nutrient deficiency, bacterial, and *Verticillium* wilts are very similar to wilt (Plant Health Research and Diagnostics, 2007; Elliot, 2009; Okungbowa and Shittu, 2012).

Therefore sometimes it is difficult to identify wilt based on its symptomatology in the field at an early stage. However, the common symptoms of the wilt diseases in pulses are summarized in Table 21.11.

Spreading of disease: Macroconidia formed on the decaying plant parts can be carried to short distances through air and spread the disease. Spreading of *Fusarium* disease from

TABLE 21.11 Symptoms of the wilt diseases in pulses.

Disease	Symptoms	References
Fusarium wilt	• Wilting, chlorosis, necrosis, premature leaf drop, browning of the vascular system, stunted growth, and damping off	Reddy et al. (1990)
	• Symptoms do not appear in early seedling stage, symptoms are clearly visible at later stages	Jain and Reddy (1995)
	• The interveinal clearing as initial symptom and loss of turgidity in leaves	Biswas and Ghosh (2016)
Pea wilt	• Drooping and withering of young seedlings and cotyledons, yellowing of lower leaves, stunted plant growth, curling of leaflet margins downward and inward	Virgin and Walker (1940), Walker (1935), Doling (1963), Nyvall and Haglaund (1972)
	• Swollen and brittle stems near the soil, brown discoloration in xylem vessels, internal woody stem tissue turning lemon brown to orange brown due to discoloration	
	• External appearance of the root is healthy; secondary root rots develop in plants that are wilted for long periods	
	• Death of wilted plant	
Chickpea wilt	• The leaves droop down, turn yellow, and dry out; vascular system is discolored, scattered yellow color patches	Luthra et al. (1943), Kumari and Khanna (2018), Singh et al. (2007)
	• *Fusarium oxysporum* remains as mycelium and chlamydospores in seed and soil, more than 5–6 years	Nelson (1981), Kumari and Khanna (2018)
	• *F. oxysporum* penetrates a host root through wound or root apices	Nelson (1981)

(Continued)

TABLE 21.11 (Continued)

Disease	Symptoms	References
Pigeon pea wilt	• Xylem vessels turn brown and black with a purple band extending upward from the base of the main stem • Symptoms commonly appear in flowering or podding phase of crop and rarely seen at early seedling stage • Young plants (1–2 months old) die from wilt without any visible symptoms	Allen and Lenne (1998), Biswas and Ghosh (2016)
Lentil wilt	• Plant affected at three stages: preemergence, postemergence, and adult stage • Mortality is recorded at early and early postemergence stage infection	Vasudeva and Srinivasan (1952), Claudius and Mehrotra (1973)
	• Red lesions produced in some young plants; older plants turn yellow, droop, and wither • In advance stage—discoloration of xylem cell walls, containing brown fungal hyphae • In advanced stage—wilt fungus spreads into the adjacent parenchyma producing conidia and chlamydospores	Khare et al. (1979), Ansari (2003)

one field to another can occur when agricultural equipment and infected plants are moved from one place to another, which is very difficult to control.

21.9 Identification of *Fusarium* species associated with pulse crops in the fields

The genus *Fusarium* presently contains over 20 species (Wang et al., 2011). The most common species infectious for the plants include *F. solani, F. oxysporum, F. equiseti*, and *F. chlamydosporum* (Burgess et al., 1994; Chimbekujwo, 2000). The morphological and cultural characters of *Fusarium* are similar in between isolates that cause wilt of pulses, the fungus is host-specific so called as *formae speciales* (f. sp.). Each *formae speciales* causes different visible or invisible symptoms.

In *F. oxysporum* f. sp. *ciceris* that infects chickpea, the races can be differentiated based on their ability to incite specific reactions. These reactions result in two types of symptoms: (1) yellowing syndrome and (2) wilting syndrome. In chickpea, the latter is more destructive and economically more important than former (Knights and Hobson, 2016).

Conventionally, indirect methods have been used to study fungal colonization of plant roots. These methods such as microbial isolation lack direct observation of infection sites and also do not include the interactions between fungal pathogens and their antagonists (Olivain et al., 2006).

The key features for the differentiation of *Fusarium* species are microscopic and macroscopic such as, color of the colony; length and shape of the macroconidia; the number, shape, and arrangement of microconidia; and presence or absence of chlamydospores

(Larone, 1995; De Hoog et al., 2000). Morphological characterization of *Fusarium*, *F. oxysporum*, and some of its *formae species* are given in Table 21.12.

Molecular methods for identification: For the rapid identification of *Fusarium* strains up to subspecies levels, molecular methods, such as 28S rRNA gene sequencing are used (Hennequin et al., 1999). Some other popular methods for identification are polymerase chain reaction-based rDNA detection method (Lacmanova et al., 2009), detection of protein banding patterns by SDS-PAGE, and esterase isozyme electrophoresis (El-Kazzaz et al., 2008).

With the help of whole-genome sequence of some *Fusarium* species, the host-specific *Fusarium* species can be identified. For example, genome of maize pathogen, *Fusarium verticillioides*, with a genome size of 41.7 Mb, organized into 12 chromosomes (Xu and Leslie, 1996); genome of the cereal pathogen, *Fusarium graminearum*, which has a genome size of 36.2 Mb, organized into 4 chromosomes (Cuomo et al., 2007); and genome of tomato pathogen, *F. oxysporum* f. sp. *lycopersici*, with a genome size of 59.9 Mb, organized into 15 chromosomes (Ma et al., 2010).

Comparative genome analyses have proved the presence of 96 mobile pathogenicity chromosomes in *Fusarium* (Ma et al., 2010). The genomes of pathogenic *Fusarium* species contain a larger number of proteins in the pathogenicity-related protein families, compared to nonpathogenic fungi. These are very helpful for the identification of pathogenic and nonpathogenic strains nowadays.

21.10 History: *Fusarium* associated with pulses

The devastating fungus *Fusarium* was first reported by E.J. Butler in early 19th century in pigeon pea crop, in India (Booth, 1971a,b). *Fusarium udum* is host-specific to pigeon pea. The name *F. udum* was known as an imperfect state because of the presence of a well-distinguished prominent hook in the macroconidia (Padwick, 1940; Subramanian, 1963; Booth, 1971b; Patel and Patel, 2012). In 1918 Butler discovered another f. sp. of *F. oxysporum* that causes wilt disease in gram (*C. arietinum* L.) and named *F. oxysporum* f. sp. *ciceri s* (McRae, 1932). In the pea, *Fusarium* wilt was first recognized during 1918 by Bisby in Minnesota (Chupps and Sherf, 1960; Ali et al., 2014). Lentil wilt caused by *F. oxysporum* f. sp. *lentis* was first reported by Prissyajnyak from Russia in 1931 (Anonymous, 1934; Ansari, 2003). In 1934 the incidence of such a disease was reported in pea for the first time from West Bengal, India (Chupps and Sherf, 1960; Ali et al., 2014). Some brief history of the research on *Fusarium* infection in pulses has been summarized in Table 21.13.

21.11 Management of *Fusarium* wilt in pulses

21.11.1 Chemical control

One of the major concerns among the pulse growers is that the plant growth and development are affected due to the infection of plant pathogenic microbes. In the last few decades, the diseases were tried to control through chemicals.

TABLE 21.12 *Fusarium, Fusarium oxysporum* and its formae speciales—morphological characterization (Sinha et al., 2018).

Fungus	Whole colony morphology	Macroconidia morphology	Microconidia morphology	References
Fusarium—common features of *Fusarium* wilt	Woolly to cottony white, cream, tan, salmon cinnamon, red, yellow, violet, pink, or purple; on the reverse, it may be colorless, red, tan, dark purple, or brown	Long, slender, rather pointed at both ends, dorsoventrally curved, sickle-shaped, septate	Appearing as small, usually single-celled spores, oval shaped, in some species apiculate, tear-drop, or pear shaped and sometimes spherical	Refai et al. (2015)
Fusarium oxysporum	Floccose, sparse, color range may be white to pale violet	Short to medium length, straight, slightly curved, slender and thin walled, usually 3 septate, abundant in sporodochia	Oval, elliptical, or kidney shaped and usually aseptate, abundant in the aerial mycelia	Leslie and Summerell (2006)
F. oxysporum f. sp. *udum*	Hyaline, slender, branched, white to salmon pink (in mass)	Straight to falcate and thin walled, 1–5 septate, but predominantly 3 septate, curved to almost hooked, abundant in sporodochia	Fusiform to reniform or oval, white to salmon pink, 0–1 septa	Pandey et al. (2013)
F. oxysporum f. sp. *lentis*	Woolly, white, sometimes turning pink	Straight to fusiform, falcate, slender, thin walled, single celled with indistinct septa (rarely 1–5 septate)	Single celled, hyaline, ovoid, cylindrical, oblong, or slightly curved	Vasudeva and Srinivasan (1952)
F. oxysporum f. sp. *ciceris*	Floccose, sparse, white	Straight to slightly curved, slender and thin walled, 3 or 4 septa	Ellipsoidal, 0 to 1 septa	Lodhi et al. (2016)
F. oxysporum f. sp. *pisi*	Aerial mycelium white, usually becoming purple	Fusiform, slightly curved, pointed at the tip, 3–5 septate	Abundant, ellipsoidal to cylindrical, straight, or curved	ATCC (2013)

TABLE 21.13 History of research on *Fusarium* infection in pulses.

Pulse	Causal organism	Research	References
Pigeon pea (*Cajanus cajan* (L.) Millsp.)	*Fusarium*	First reported in early 19th century by E.J. Butler in India.	Booth (1971b)
	Fusarium udum	Name *F. udum* was accepted as an imperfect state.	Booth (1971b)
	F. udum	Presence of macroconidia, having well-distinguished prominent hook. *F. udum* is host-specific to pigeon pea.	Padwick (1940), Subramanian (1963), Booth (1971b), Patel and Patel (2012)
Chickpea (*Cicer arietinum* L.)	*Fusarium oxysporum* f. sp. *ciceris*	1918—Butler discovered another f. sp. *F. oxysporum*, which causes wilt disease in chickpea and named *F. oxysporum* f. sp. *ciceris*.	McRae (1932)
		F. oxysporum f. sp. *ciceris* reported as pathogenic to chickpea by Prasad and Padwick, later accepted all over the world as the causal agent of chickpea.	Prasad and Padwick (1939), Booth (1971c), Kaiser et al. (1994), Lodhi et al. (2016)
Pea (*Pisum sativum*)	*Fusarium* spp.	1918—Bisby in Minnesota, first time recognized Fusarium wilt of pea.	Chupps and Sherf (1960), Ali et al. (2014)
Lentil (*Lens culinaris* Medikus)	*F. oxysporum* f. sp. *lentis*	1931—Prissyajnyak first reported lentil wilt caused by *F. oxysporum* f. sp. *lentis* from Russia. 1934—The occurrence of lentil wilt was reported from West Bengal, India.	Anonymous (1934), Ansari (2003)

According to a study, so far the most successful way of fighting the disease is to be resistant to the pathogen, as being a soilborne pathogen it can remain in the soil for very long period of time. Fungicidal dressings of seed are only effective against seedborne infection (Knights and Hobson, 2016). Resistance can be induced by using chemicals such as prochloraz and carbendazim (Song et al., 2004), salicylic acid (Amel et al., 2010), and bavistin (Alam et al., 2010). However, these chemicals are hazardous to human health and environment.

These chemical fungicides were first used in 18th century. The use of synthetic chemicals as fungicides in agricultural production provided the required relief in control of phytofungal diseases. Haider et al. (1978) reported that vascular wilt disease of pigeon pea can significantly controlled by using Benomyl (50WP) and Carbendazim (50WP). Maheshwari et al. (2008) tested seven fungi toxicants against *F. oxysporum* f. sp. *lentis*. After observing the fungal growth on plants they concluded that carbendazim is the most efficient toxicant for checking the fungal growth. The other toxicants that gave good results after carbendazim were captan, hexaconazole, and diniconazole. Drenching the soil with carbendazim under in vivo soil treatment reduced wilt incidence and plant mortality to a great extent and hence it is considered to be effective. Carbendazim gave highest grain yield followed by captan. Hexaconazole and diniconazole were found least effective than other fungi toxicants. It was reported that application of boron (Bo), zinc (Zn) or manganese (Mn), and methyl bromide (CH3Br) to the soil substantially reduces the disease event of *Fusarium* wilt (Biswas and Ghosh, 2016). Raxil Ultra (Tebuconazole), Topsin-M (Thiophanate Methyl), Score (Difenoconazole), Derosil (Carbendazim),

Hombre (Imidacloprid Tebuconazole), and Dividend fungicide (Metalaxyl M + Difenoconazole) show fungicidal property against pea wilt (Khan et al., 2016).

The frequent use of fungicides for long period of time is harmful and toxic to other non-targeted organisms such as microbes, earthworms, and humans (genotoxicity), causing disbalance in the ecosystems (Patel et al., 2014; Nega, 2014; Shuping and Eloff, 2017). Furthermore, the enormous use of chemicals for long period as the solitary method of disease control may indeed result in nontarget effects on soil microbial populations (Kumari and Katoch, 2020). With the growing cognizance about health hazards caused by the intake of pesticide-rich food, nowadays we are shifting from only chemicals to integrated disease management approaches.

21.11.2 Biological control

Biological control can be defined as a way to suppress or manage the disease by the application of a biological agent to a host plant. Thus, by applying biological agent, the development of disease by a pathogen can be prevented. The bacterial or fungal strains can be used as biological agents with regard to plant diseases. The suppression of disease by using biological agents is as good as using chemicals.

There are many traditional strategies as well as chemical and biological approaches for managing the wilt disease in pulses. The best method of control is breeding for resistant cultivars. A better economic and environment-friendly alternative to chemicals is the introduction of soil microbes in the field. These soil microbes such as *Trichoderma harzianum*, *Trichoderma koningii*, *Trichoderma asperellum*, *Streptomyces griseoviridis*, and *Penicillium* spp. are present in the rhizosphere of the plants, which stimulate the growth of the plant by the degradation of the complex substrates. They also have the ability to restrict the pathogen attacks (Osuide et al., 2002; Syed et al., 2010; Borrero et al., 2012).

Sharfuddin and Mohanka (2012) observed that various *Fusarium* species cause massive losses in lentils, which are an economically important crop (Elad, 2000; Howell, 2002; Eziashi et al., 2007; Shalini and Kotasthane, 2007). They performed their experiments under in vitro conditions. They tested 19 strains of *Trichoderma* spp. among the different isolates. Of all the isolates, two *Trichoderma* isolates, 5 and 7 of *T. harzianum* (Th), 2 and 18 of *T. viride* (Tv), and isolate 9 of *T. koningii* (Tk) were found to be more efficient among all. All these isolates showed improved antagonism against the tested phytopathogen. The isolate Th-5 caused maximum inhibition (82.8%) followed by Th-7 (82.3%), Tv-2 (79.2%), Tv-18 (74.4%), and Tk-9 (71.0%). *Trichoderma* spp. has been recorded as an important antagonistic soil fungus. These species have the ability to reduce disease occurrence caused by soilborne phytopathogenic fungi (Misra, 2006; Srivastava et al., 2011; Kucuk and Kivanc, 2003; Chang et al., 2006; Narayanan et al., 2015; Shanmugapackiam et al., 2016).

Mahesh et al. (2010) reported that the combined effect of chemical and antagonist was very effective in suppressing the wilt disease in pigeon pea. They reported that application of chemical carbendazim along with soil application of *Pseudomonas fluorescens* and *Trichoderma viride* resulted in minimum mean wilt incidence. The next best treatment was carbendazim seed treatment with *T. viride* alone. Also, carbendazim seed treatment with $ZnSO_4$ and *T. viride* seed treatment were proven to be beneficial in controlling wilt incidence and also in increasing total yield compared to untreated control. This research work was also supported by the results of Ingole et al. (2005), Mandhare and Suryawanshi (2005), Narayanan et al. (2015), Shanmugapackiam et al. (2016). Upadhyay (1992) and Singh et al. (2002) revealed that *Fusarium* can also be eradicated from soil due to production of certain substances/antibiotics by the phenomenon soil fungistasis. Prasad et al. (2002) who have reported soil application of *T. viride* and *Trichoderma harzianum* is more effective in reducing wilt and wet root rot of chickpea. Nikam et al. (2007) reported management of *F. oxysporum* f. sp. *ciceris*. through soil application of *T. viride* (@ 25 kg/ha) under pot culture studies. They concluded that wilt diseases can be managed by integration of various practices such as chemical treatment of seeds, use of resistant varieties, and soil application of chemicals.

The effectiveness of *Trichoderma* strains against a pathogen is dependent upon several factors. These factors include environmental conditions such as temperature, pH, nutrition, and ion content of the soil; target fungus; and the crop plant (Parthasarathy et al., 2017). Otadoh et al. (2011) screened five isolates of *Trichoderma* viz., *T. koningii, T. reesei, T. atroviride, T. asperellum*, and *T. harzianum* and found that these isolates effectively inhibit the growth of *F. oxysporum* f. sp. *phaseoli* causing wilt in beans. The highest inhibitory effect on growth of *F. oxysporum* f. sp. *phaseoli* was achieved by *T. reesei* (32.2%) followed by *T. atroviride* and *T. koningii* both at 28.8% while isolate of *T. asperellum* showed the lowest inhibitory effect to *F. oxysporum* f. sp. *phaseoli* (20.3%). Similar findings were observed by Howell (2002); Abo-Elyousr, Kamal and Hashem (2009), Radwan et al. (2007), Vinale et al. (2004) and confirmed the biocontrol potential of these biological managers or agents.

Pandey and Upadhyay (1999) stated that *T. viride* was found to be most effective to control the *Fusarium* wilt of pigeon pea. Anand and Reddy (2009) isolated 42 strains of *Trichoderma* sp. from cultivated lands and analyzed for their antagonistic potential against *S. rolfsii* and *Fusarium ciceris*. The potential of biocontrol agents ultimately lies in their capacity to control pathogens in vivo. Kashem et al. (2011) also studied the efficacy of *Trichoderma* sp. against *F. oxysporum* causing foot and root rot of lentil (*L. culinaris* Medik). All these biocontrol agents suppress plant pathogens through a variety of mechanisms particularly competition, mycoparasitism, induced resistance, antibiosis, etc. (Sinha et al., 2018).

From past many decades it has been accepted by researchers that plant and plant extracts have significant potential to manage fungal pathogens. Chandra and Singh (2005) studied the efficacy of five plant extracts viz., *Azadirachta indica, Eucalyptus globules, Jatropha multifida, Allium sativum*, and *Calotropis procera* and found that all the extracts have potential for reducing wilt incidence in *C. arietinum*. Mukhtar (2007) observed that *A. indica* and *Datura metel* plant extracts showed higher fungitoxicity against *F. oxysporum* f. sp. *ciceris* whereas others equally inhibited the growth of the pathogen in vitro. Acetone extracts of *Datura stramonium* have been reported to have antifungal activity against several fungi including *F. oxysporum* (Gaire and Subedi, 2013). Leaf of *A. indica* at very high concentration completely inhibited germination of pathogen spores (Lodhi et al., 2016).

Devi and Chhetry (2012) reported the antifungal effect of aqueous extracts of locally available plant species against *F. udum*, the causal agent of pigeon pea wilt at different concentrations, that is, 5%, 10%, 15%, and 20% using poisoned food technique. Among the tested plants, that is, *Allium cepa, A. sativum, Carica papaya, Citrus limon, Eryngium foetidum, Gynura angulosa, Aloe vera, A. indica, Spilanthes acmella, Adhatoda vasica*, higher inhibition was recorded with *A. sativum, A. indica, S. acmella,* and *A. vasica.* Among them *A. sativum* at 20% alone recorded 100% inhibition of mycelial growth of *F. udum.* In botanicals *Solanum nigrum* L., *Tagetes erecta, Polyalthia longifolia, Parthenium hysterophorus* L., *Zingiber officinale* were also proved very effective against *F. udum* (Chaudhary et al., 2013). Shukla and Dwivedi (2012) from their study revealed that plant extracts were most promising results against wilt pathogens of pigeon pea and chickpea, that is, *F. udum* and *F. oxysporum* f. sp. *ciceris* at 5%, 10%, and 15% concentration. They reported that at 15% concentration the growth of *F. udum* was reduced by turmeric (89.2%) followed by garlic (88.26%) and black pepper (82.22%). In case of *F. oxysporum* f. sp. *ciceris,* at same concentration garlic, turmeric, and black pepper reduced the growth by 94.63%, 87.96%, and 77.74%, respectively.

The extracts of *A. sativum, Z. officinale,* leaf extracts of *A. indica, C. procera, Moringa oleifera,* and *P. hysterophorus* L. have shown good fungitoxic effects against *F.* f. sp. *pisi.* Antibiosis, competition, mycoparasitism, induced resistance, etc. could be the right justification for mechanism of action by these biocontrol agents. Okungbowa and Edema (2006) and Okungbowa (2011) have reported the possible use of plant extracts for the control of *F. oxysporum.* Strains of *F. oxysporum,* isolated from healthy roots, are termed as nonpathogenic. These strains are interesting, since some of them can protect plants against the pathogenic strains (Shishido et al., 2005). Work on nonpathogenic and protective strains should be more prioritized. Also, rhizosphere fungi such as *Penicillium* spp., *T. asperellum, T. harzianum, T. koningii,* and *S. griseoviridis* have been used to control the disease from last few decades and more work should be done on integrated management practices.

Lesser emphasis on the conventional strategies and increasing ecological, economic, and social problems arising with excessive application of chemical pesticides emphasize the need to understand molecular interactions between the host and the pathogen and application of biotechnological techniques along with biological approaches for managing *Fusarium* spp. in pulse crop fields.

There are many disadvantages associated with the use of the biocontrol measures. Biopesticides and biofungicides are commonly composed of living organisms and as a result, their efficacy may vary under a wide range of biotic, environmental, and cultural conditions. Conventional fungicides are often affordable and easier to apply and their efficacy is constant except in few cases with very severe conditions. Furthermore, many chemical fungicides are effective to control soilborne pathogens and possess competitive advantages, not yet available with their biofungicide counterparts. The best option is to cultivate resistant varieties. It is commonly practiced as well as maintaining good hygiene (such as prevention of contamination, during planting and seedling transplanting, irrigation, and clearing of debris from previous year's planting; Jones and Woltz, 1981). Due to many utility problems faced by the users, integrated pest management strategies offer bright opportunities of success, provided that the mode of action of the various components is clear and well understood. With the help of modification of the soil pH and fertilizer composition, disease multiplication and further development can be significantly reduced (AVRDC, 2005; Borrero et al., 2012). Crop rotation is commonly suggested and used by farmers but it is commonly ineffective due to the effective survival strategies of the pathogen. The development of new races that may overcome cultivar resistance is the possibility of future to control the *Fusarium* disease (AAFC, Canada, 2006).

21.12 Conclusions

Fusarium wilt poses a major problem as it causes massive economic losses in important crops such as pulses globally. Although there are measures in effect to prevent and control the disease, still a lot more needs to be done. The major obstacles in combating the fungus are ever-changing weather conditions, several agronomic limitations such as contamination of agricultural soils and the products, infestation of insect–pests and diseases, and inapt marketing of pulses. The use of resistant cultivars has been accepted as the best method to control the disease. However, development of resistant varieties is time consuming and tedious. Therefore alternative control approaches are to be investigated. The primarily and most focused alternative method is to develop biocontrol methods. Further research should be focused at determining the actual role of root exudates, plant extracts, genetic management of spore germination as well as whether it is vascular occlusion or presence of toxins that causes death of the plant, are of great necessity for a better understanding of the biology of the organism.

In last 15 years, plant breeders and agricultural scientists have played a major role in facilitating and improving the growth of pulse production. In their joint efforts with CGIAR institutions and national agricultural research systems, government has laid down specific and focused areas of improvement that are—crop production technology, postharvest technology, support to farmers. and appropriate marketing of the pulses. The forthcoming 10-Year Research Strategy document should prioritize the major areas of improvements and also facilitate investment for research on pulse crops and associated problems. National and international research institutions should join hands in the collection and preservation of genetic resources for pulses.

Pulses are the major source of nutrition worldwide. They are rich sources of protein, minerals, and micronutrients. The low levels of cholesterol and glycemic index make them virtuous for human health. The genetic diversity of the pulses can be explored in order to increase their production and profit. To achieve this goal efficient pulse-breeding programs are required. Breeding strategies should aim to develop new improved varieties taking into consideration various biotic–abiotic stresses. Plant breeders should develop various resistant varieties against harmful phytopathogen. Emphasis should be given on elaborating uses of pulses as food, feed, and fuel in all pulse-growing countries worldwide.

Training and education of farmers should be emphasized that directs them to adapt various methods that increase the yield of the crop such as crop rotation, intercropping, and introduction of newer pulse–cereal complementary cropping systems. They should also be trained for better soil preparation, timely sowing and harvesting, use of disease-resistance varieties, application of biocontrol agents, and other farm management practices.

The need of hour is to develop strong, multitasking global international pulse research community that is well informed about their own regions and countries so the community together can take decisions based on problems of individual region and countries that also apply globally. A pulse research community that assimilates work across all countries and regions, is capable of meeting regional to global needs. An integrated, well–linked, and large agricultural science community is important to the vision mentioned here. Collaboration of global and regional networks of scientists, government partners, and industry players is necessary to evaluate socioeconomic dimensions, nutrition and health, and spatial analysis of pulses for improved productivity and sustainability.

Finally, region-specific challenges and opportunities should be addressed. Integrated research programs should be prioritized to increase the yield and nutritional value of the pulses. Important issues of pulses are breeding, use of relevant pulse species, cultivars for specific growing conditions and uses; region-specific agronomic practices and production technologies; new type of sustainable, diversified cropping systems optimized for specific regions; socioeconomic dimensions of production and consumption; locally and regionally relevant policies and enabling environments; and national level capacity to undertake research, for example, through intra- and cross-regional partnerships.

References

AAFC, Canada, 2006. Crop profile for greenhouse tomato in Canada. Agriculture and Agri-Food Canada. <http://dsp-psd.pwgsc.gc.ca/collection_2009/agr/A118-10-24-2006E>.

AARFS, 2013. <http://www.aarfs.com.mx/imagenes/COSTO_PRODUCCION_FRIJOL_2015-16.pdf>; Fintrac Inc.

Abawi, G.S., Ludwig, J.W., Gugino, B.K., 2006. Bean root rot evaluation protocols currently used in New York. Annu. Rep. Bean Improvement Cooperative 49, 83–84.

Abo-Elyousr, Kamal, A.M., Hashem, M.M., 2009. Biological control of *Fusarium* wilt in tomato by plant growth-promoting yeasts and rhizobacteria. Plant Pathol. J. 25 (2), 199–204.

Agrios, G.N., 2005. Plant Pathology, fifth ed. Elsevier Academic Press, Burlington.

Ahsan, M.S., Kumar, M., Upadhyay, J.P., Hussain, M.A., Gupta, P.K., Singh, A., 2018. Effect of different doses of *Trichoderma harzianum* and fungicides for the management of collar rot of chickpea caused by *Sclerotium rolfsii*. Int. J. Pure Appl. Biosci. 6 (1), 1656–1660.

Akhtar, C.M., 1955. Further Studies on Wilt Disease of Gram (*Cicer arietinum*) in the Punjab (M.Sc. thesis). University of the Punjab, Lahore.

Alam, S.S., Sakamoto, K., Amemiya, Y., Inubushi, K., 2010. Biocontrol of soil-borne Fusarium wilts of tomato and cabbage with a root colonizing fungus, *Penicillium* sp. EU0013. In: Proceedings of the 19th World Congress of Soil Science: Soil Solutions for a Changing World. 1–6 August, Brisbane, Australia, pp. 20–22.

Ali, M., 2007. Augmentation of pulses production. In: Mission. 2007: IIPR Kanpur, 24, pp. 16–21.

Ali, M.E.K., Inanaga, S., Sugimoto, Y., 2002. Sources of resistance to *Fusarium* wilt of chickpea in Sudan. Phytopathol. Mediterr. 41 (3), 163–169.

Ali, M., Jain, S.K., Lal, M., Zuhaib, M., Kumar, S., Srivastva, A.S., 2014. Survey, media requirement and management of *Fusarium* wilt of pea. Bioscan 9 (3), 1213–1216.

Allen, D.J., Lenne, J.M. (Eds.), 1998. The Pathology of Food and Pasture Legumes. CAB International, Wallingford, Oxon.

Alwathnani, H.A., Perveen, K., 2012. Biological control of *Fusarium* wilt of tomato by antagonist fungi and cyanobacteria. Afr. J. Biotechnol. 11 (5), 1100–1105.

Amel, A.H., Soad, M.A., Ahmed, A.I., 2010. Activation of tomato plant defense response against *Fusarium* wilt disease using *Trichoderma harzianum* and salicylic acid under greenhouse conditions. Res. J. Agric. Biol. Sci. 6 (3), 328–338.

Anand, S., Reddy, J., 2009. Biocontrol potential of *Trichoderma* sp. against plant pathogens. Int. J. Agric. Sci. 2, 30–39.

Andrabi, M., Vaid, A., Razdan, V.K., 2011. Evaluation of different measures to control wilt causing pathogens in chickpea. J. Plant Prot. Res. 51 (1), 55–59.

Anonymous, 1934. Annual Report. Department of Agriculture Bengal Part II, 51.

Ansari, S., 2003. Ecofriendly Management of Vascular Wilt of Lentil *Lens culinaris* Medik (Ph.D. thesis). G.B. Pant University of Agriculture and Technology, Pantnagar.

Arif, M., Pani, D.R., Zaidi, N.W., Singh, U.S., 2011. PCR-based identification and characterization of *Fusarium* sp. associated with mango malformation. Biotechnol. Res. Int. Available from: https://doi.org/10.4061/2011/141649.

Armstrong, G.M., Armstrong, J.K., 1980. Cowpea wilt *Fusarium oxysporum* f. sp. *tracheiphilum* race 1 from Nigeria. Plant Dis. 64, 954–955.

Arvayo-Ortiz, R.M., Esqueda, M., Acedo-Felix, E., Gonzalez-Rios, H., Vargas-Rosales, G., 2012. New lines of chickpea against *Fusarium oxysporum* f. sp. *ciceris* wilt. Am. J. Appl. Sci. 9 (5), 686–693.

Arya, A., Kushwaha, K.P.S., 2019. Management of lentil wilt through host resistance. Int. J. Curr. Microbiol. Appl. Sci. 8 (3), 438–444.

ATCC. American Type Culture Collection, 2013. <www.atcc.org>.

AVRDC, 2005. The World Vegetable Centre. *Fusarium Wilt*. Fact Sheet, 5, 627. <http://www.avrdc.org/pdf/tomato/Fusarium>.

Bahl, P.N., Lal, S., Sharma, B.M., 1991. An overview of the production and problems of lentil in South Asia. In: Paper Presented in ICAR-ICARDA Seminar "Lentil in South Asia". New Delhi, 11–15 March, 10.

Baker, M.A., 1991. Plant protection of lentil in Bangladesh. In: Paper Presented in ICAR ICARDA Seminar "Lentil in South Asia". New Delhi 11–15 March, 20.

Balali, G.R., Iranpoor, M., 2006. Identification and genetic variation of *Fusarium* species in Isfahan, Iran, using pectic zymogram technique. Iran. J. Sci. Technol. Trans. A 30 (1), 91–102.

Bayaa, B., Erskine, W., Khoury, L., 1986. Survey of Wilt Damage on Lentils in Northern Syria. Arab. J. Plant Prot. 4, 118–119.

Bedasa, T., 2018. Distribution and Management of *Fusarium* Wilt (*Fusarium oxysporum* f.sp. *lentis*) of Lentil (*Lens culinaris* Medikus) in Central Highlands of Ethiopia (M.Sc. thesis in Agriculture (Plant Pathology)). Haramaya University, Haramaya. <https://hdl.handle.net/10568/9834>.

Birajdar, A.S., Deshmukh, A.S., Chavhan, R.L., Kadam U.S., 2018. Analysis of genetic diversity of Fusarium udum causing wilt of pigeon pea. Int. J. Curr. Microbiol. Appl. Sci. 6, 843–852.

Bishop, C.D., Cooper, R.M., 1983. An ultrastructural study of vascular colonization in three vascular wilt diseases I. Colonization of susceptible cultivars. Physiol. Plant Pathol. 23 (3), 323–343.

Biswas, K., Ghosh, P., 2016. Recent advancements and biological management of *Fusarium udum*: a causative agent of pigeon pea wilt. Int. J. Appl. Nat. Sci. 5 (3), 57–72.

Booth, C., 1971a. The Genus *Fusarium*. Commonwealth Mycological Institute, Kew, p. 237.

Booth, C., 1971b. The Genus *Fusarium*. Commonwealth Mycological Institute, Kew, p. 114.

Booth, C., 1971c. The Genus *Fusarium*. Commonwealth Mycological Institute, Kew, p. 137.

Borrero, C., Trillas, M.I., Delgado, A., Aviles, M., 2012. Effect of ammonium/nitrate ratio in nutrient solution on control of *Fusarium* wilt of tomato by *Trichoderma asperellum* T34. Plant Pathol. 61, 132–T139.

Burgess, L.W., Summerell, B.A., Bullock, S., Gott, K.P., Backhouse, D., 1994. Laboratory Manual for *Fusarium* Research, third ed. University of Sydney, Australia.

Carrera, C.J.M., 1940. The genus *Fusarium* in the Argentina republic study and systematic classification second contribution. Rev. Argent. Agron. VII. 4, 272–296.

Chandra, H., Singh, S., 2005. Control of chickpea wilt *Fusarium oxysporum* f. sp. *ciceri* using bioagents and plant extracts. Indian J. Agric. Sci. 75 (2), 115–116.

Chang, K.F., Hwang, S.F., Wang, H., Turnbull, G., Howard, R., 2006. Etiology and biological control of *Selecrotinia* blight of cone flower using *Trichoderma* species. Plant Pathol. J. 5, 15–19.

Chaudhary, R.G., Amarjit, K., 2002. Wilt disease as a cause of shift from lentil cultivation in Sangod Tehsil of Kota, Rajasthan. Indian J. Pulse Res. 15, 193–194.

Chaudhary, K., Singh, S.R., Singh, A.K., Mala, S., 2013. Evaluation of fungicides, botanicals, neem products and bio-agents against wilt of pigeon pea caused by *Fusarium udum* Butler. TECHNOFAME - J. Multidiscip. Adv. Res. 2 (2), 1–6.

Chimbekujwo, I.B., 2000. Frequency and pathogenicity of fusarium wilts (*Fusarium solani* and *Fusarium equiseti*) of cotton (*Gossypium hirsutum*) in Adamawa in Nigeria. Rev. Biol. Trop. 48 (1), 1–5.

Choudhary, A.K., Kumar, S., Patil, B.S., Bhat, J.S., Sharma, M., Kamal, S., et al., 2013. Narrowing yield gaps through genetic improvement for *Fusarium* wilt resistance in three pulse crops of the semi-arid tropics. SABRAO J. Breed. Genet. 45 (3), 341–370.

Chupps, C., Sherf, A.F., 1960. Vegetable Disease and Their Control. The Ronald Press Company, New York, p. 693.

Claudius, G.R., Mehrotra, R.S., 1973. Wilt and root rot disease of lentil *Lens esculenta* at Sagar. Indian Phytopathol. 26, 268–273.

Commission for Agricultural Costs and Prices (CACP), 2015. Price policy for rabi crops: the marketing season 2016-17. Commission for Agricultural Costs and Prices, Ministry of Agriculture. Government of India, New Delhi.

CONAB, 2015. National Food Supply Company (CONAB) of Brazil.

Cuomo, C.A., Güldener, U., Xu, J.-R., Trail, F., Turgeon, B.G., Di Pietro, A., 2007. The *Fusarium graminearum* genome reveals a link between localized polymorphism and pathogen specialization. Science 317, 1400–1402.

Deepa, Sunkad, G., Sharma, M., Mallesh, S.B., Mannur, D.M., Sreenivas, A.G., 2018. Distribution and severity of dry root rot of chickpea caused by *Rhizoctonia bataticola* in parts of North Karnataka, India. Int. J. Curr. Microbiol. Appl. Sci. 7 (4), 194–200.

De Hoog, G.S., Guarro, J., Gene, J., Figueras, M.J., 2000. second ed. Atlas of Clinical Fungi, vol. 1. Centraalbureau voor Schimmelcultures, Utrecht, The Netherlands.

Devi, T.R., Chhetry, G.K.N., 2012. Rhizosphere and non-rhizosphere microbial population dynamics and their effect on wilt causing pathogen of pigeon pea. Int. J. Sci. Res. Publ. 2 (5), 1–4.

Dita, M., Barquero, M., Heck, D., Mizubuti, E.S.G., Staver, C.P., 2018. *Fusarium* wilt of banana: current knowledge on epidemiology and research needs toward sustainable disease management. Front. Plant Sci. 9, 1–21.

Doling, D.A., 1963. Effect of root damage on the development of *Fusarium* wilt in peas. Trans. Br. Mycol. Soc. 46, 577–584.

Egel, D.S., Martyn, R.D., 2007. *Fusarium* wilt of watermelon and other cucurbits. Plant Health Instructor. Available from: https://doi.org/10.1094/PHI-I-2007-0122-01.

Elad, Y., 2000. Biological control of foliar pathogens by means of *Trichoderma harzianum* and potential mode of action. Crop Prot. 19, 709–714.

El-Ahmed, M., Mouselli, N., 1986. Lentil wilt in Southern Syria. Arab. J. Plant Prot. 4 (1), 30.

El-Kazzaz, M.K., El-Fadly, G.B., Hassan, M.A.A., El-Kot, G.A.N., 2008. Identification of some *Fusarium* spp. using molecular biology techniques. Egypt. J. Phytopathol. 36 (1–2), 57–69.

Elliot, M.L., 2009. *Fusarium* wilt of Canary Island date palm, Fact Sheet, 215. University of Florida Plant Pathology, pp. 1–7.

Erskine, W., Rihawe, S., Capper, B.S., 1990. Variation in lentil straw quality. Ann. Food Sci. Technol. 28, 61–69.

ESI, 2015. The Economic Survey 2014-2015. The Economic Survey of India, New Delhi.

Eziashi, E., Omamor, I.B., Odigie, E.E., 2007. Antagonism of *Trichoderma viride* and effect of extracted water soluble compounds from *Trichoderma* species and benlate solution on *Ceratocystis paradoxa*. Afr. J. Biotechnol. 6, 388–392.

Fassihiani, A., 2000. Symptomless carriers of the causal agent of tomato wilt pathogen. J. Agric. Sci. Technol. 2, 27–32.

Fintrac Inc., 2013. Kenya Horticulture Competitive Project: Pulse Value Chain Analysis, 2010–2013. United States Agency for International Development Kenya Horticulture Competitive Project, Nairobi.

Flood, J., 2006. A review of *Fusarium* wilt of oil palm caused by *Fusarium oxysporum* f sp. *elaeidis*. Phytopathology 96 (6), 660–662.

Gaire, B.P., Subedi, L., 2013. A review on the pharmacological and toxicological aspects of *Datura stramonium* L. J. Integr. Med. 11 (2), 73–79.

Ghosh, R., Sharma, M., Telangre, R., Pande, S., 2013. Occurrence and distribution of chickpea diseases in central and southern parts of India. Am. J. Plant Sci. 4, 940–944.

Gwata, E.T., Silim, S.N., Mgonja, M., 2006. Impact of a new source of resistance to *Fusarium* wilt in pigeon pea. J. Phytopathol. 154 (1), 62–64.

Hagedorn, D.J., 1991. Handbook of Pea Diseases A1167. University of Wisconsin-Extension, Madison.

Haider, M.G., Singh, R.K., Prasad, H., Nath, R.P., Sharma, R.N., 1978. Effect of some common fungicides on the incidence of pigeon pea wilt. Phytopathol. Notes 31, 511–512.

Halila, M.H., Strange, R.N., 1996. Identification of the causal agent of wilt of chickpea in Tunisia as *Fusarium oxysporum* f. sp. *ciceri* race 0. Phytopathol. Mediterr. 35 (2), 67–74.

Hameed, A., Saddiqa, A., Nadeem, S., Iqbal, N., Atta, B.M., Shah, T.M., 2012. Genotypic variability and mutant identification in *Cicer arietinum* by seed storage protein profiling. Pak. J. Bot. 44 (4), 1303–1310.

Haque, S.E., Ghaffar, A., 1992. Efficacy of *Trichoderma* species and *Rhizobium meliloti* in the control of root-rot of fenugreek. Pak. J. Bot. 24 (2), 217–221.

Harveson, R.M., 2011. Soil borne root and stem diseases of dry beans in Nebraska. <http://extension.unl.edu/publication>.

Haware, M.P., 1990. *Fusarium* wilt and other important diseases of chickpea in the Mediterranean area. Options Mediterra neennes – Ser. Seminaires 9, 61–64.

Hennequin, C., Abachin, E., Symoens, F., Lavarde, V., Reboux, C., Nolard, N., et al., 1999. Identification of *Fusarium* species involved in human infections by 28S rRNA gene sequencing. J. Clin. Microbiol. 37, 3586–3589.

Howell, C.R., 2002. Cotton seedling pre emergence damping-off incited by *Rhizopus oryzae* and *Pythium* spp. and its biological control with *Trichoderma* spp. Phytopathology 92, 177–180.

Huisman, O.C., 1982. Interactions of root growth dynamics to epidemiology of root invading fungi. Annu. Rev. Phytopathol. 20, 303–327.

Ingole, M.N., Ghawade, R.S., Raut, B.T., Shinde, V.B., 2005. Management of pigeon pea wilt caused by *Fusarium udum* Butler. Crop Prot. Product. 1 (2), 67–69.

Inoue, I., Namiki, F., Tsuge, T., 2002. Plant colonization by the vascular wilt fungus *Fusarium oxysporum* requires FOW1, a gene encoding a mitochondrial protein. Plant Cell 14, 1869–1883.

Jain, K.C., Reddy, M.V., 1995. Inheritance of resistance to *Fusarium* wilt in pigeon pea *Cajanus cajan* L. millsp. Indian J. Genet. Plant Breed. 55 (4), 434–437.

Jaruhar, H.B., Prasad, A., 2011. Effect of different pH levels on the growth and sporulation of *Fusarium oxysporum* schlecht. f. sp. *lentis* (Vasudeva and Srinivasan) the causal organism of wilt disease of lentil. Bioscan 6 (1), 289–291.

Jendoubi, W., Bouhadida, M., Boukteb, A., Mohamed Beji, M., Kharrat, M., 2017. *Fusarium* wilt affecting chickpea crop. Agriculture 7, 23.

Jimenez-Diaz, R.M., Singh, K.B., Trapero-Casas, A., Trapero-Casas, J.L., 1990. Resistance in kabuli chickpea to *Fusarium* wilt. Plant Dis. 75, 914–918.

Jones, J.P., Woltz, S.S., 1981. *Fusarium* incited diseases of tomato and potato and their control. In: Nelson, P.E., Toussoun, T.A., Cook, R.J. (Eds.), *Fusarium: Diseases, Biology and Taxonomy*. Pennsylvania State University Press, University Park, pp. 157–168.

Joshi, P.K., Rao, P.P., 2017. Global pulse scenario: status and outlook. Ann. N. Y. Acad. Sci. 1392, 6–17.

Kadam, A.M., Chavan, S.S., Dhutraj, D.N., Rewale, K.A., 2018. Survey of dry root rot of chickpea incidence in Marathwada region. J. Pharmacogn. Phytochem. 7 (1S), 3004–3008.

Kaiser, W.J., Alcala-Jimenez, A.R., Hervas-Vargas, A., Trapero Casas, J.L., Jimenez- Diaz, R.M., 1994. Screening of wild cicer species for resistance to races 0 and 5 of *F. oxysporum* f. sp. *ciceris*. Plant Dis. 78, 962–967.

Kalyebara, R., Kassozi, M.S., 2005. The impact of improved bean varieties in Uganda. CIAT Annual Report. CIAT Bean Programme, Cali.

Kannaiyan, J., Nene, Y.L., Reddy, M.V., Ryan, J.G., Raju, T.N., 1984. Prevalence of pigeon pea diseases and associated crop losses in Asia, Africa and the Americas. Tropical Pest. Manag. 30 (1), 62–71.

Karki, P.B., 1991. Present status and future prospects of plant protection research in Nepal. In: Paper Presented in ICAR-I CARDA Seminar 'Lentil in South Asia', New Delhi. 13p.

Kashem, M.A., Hossain, I., Hasna, M.K., 2011. Use of *Trichoderma* in biological control of foot and root rot of lentil (*Lens culinaris* Medik). Int. J. Sustain. Crop Prod. 6 (1), 29–35.

Khan, S.A., Awais, A., Javed, N., Javaid, K., Moosa, A., Haq, I.U., et al., 2016. Screening of Pea germplasm against *Fusarium oxysporum* f. sp. *pisi* and in vitro management through chemicals. Pak. J. Phytopathol. 28 (2), 127–131.

Khare, M.N., Agarwal, S.C., Jain, A.C., 1979. Lentil Diseases and Their Control. Technical Bulletin, Jawaharlal Nehru Krishi Vishwa Vidyalaya (JNKVV), Jabalpur, pp. 1–29.

Khare, K.B., Loeto, D., Wale, K., Salani, M., 2016. Seed-borne fungi of cowpea [*Vigna unguiculata* (L.) Walp] and their possible control in vitro using locally available fungicides in Botswana. Int. J. Bioassays 5 (11), 5021–5024.

Kilimo Trust, 2012. Development of Inclusive Markets in Agriculture and Trade (DIMAT): The Nature and Markets of Bean Value Chains in Uganda. United National Development Programme, Kampala.

Kiprop, E.K., Baudoin, J.P., Mawangombe, A.W., Kimani, P.M., Mergeal, G., Maquet, A., 2002. Cultural characteristics, pathogenicity and vegetative compatibility of *Fusarium udum* isolation from pigeon pea [*Cajanus cajan* L. Millsp.] in Kenya. Eur. J. Plant Pathol. 1082, 147–154.

Knights, E.J., Hobson, K.B., 2016. *Fusarium* wilt. Chickpea Overview, Reference Module in Food Science. Elsevier Publication. Available from: https://doi.org/10.1016/B978-0-08-100596-5, 00035-4p.

Kotova, V.V., Khaleeva, J.N., Shikuhova, E.G., 1965. Fungus diseases of legume. In: Distribution of Pests and Diseases of Crops in the USSR in 1964. Trudy Vses. Inst. Zasch. Rast. 25, 159–164.

Kucuk, C., Kivanc, M., 2003. Isolation of *Trichoderma* spp. and determination of their antifungal, biochemical and physiological features. Turk. J. Biol. 27, 247–253.

Kuldhar, D.P., Badgujar, S.L., Dey, U., 2013. Screening of pea varieties, germplasm lines and genotypes against pea wilt *Fusarium oxysporum* f. sp. *pisi*. Int. J. Plant Prot. 6 (1), 209–210.

Kumar, G., Bajpai, R., Teli, B., Meher, J., Rashid, M.M., Sarma, B.K., 2020. Management of *Fusarium udum* causing wilt of pigeon Pea. In: Singh, B.P., Singh, G., Kumar, K., Nayak, S.C., Srinivasa, N. (Eds.), Management of Fungal Pathogens in Pulses: Current Status and Future Challenges. Springer International Publishing, pp. 191–204.

Kumari, S., Khanna, V., 2018. Biological management of vascular wilt of chickpea (*Cicer arietinum* L.) incited by *Fusarium oxysporum* f. sp. *ciceris* by antagonistic rhizobacteria co-inoculated with native *Mesorhizobium*. Int. J. Curr. Microbiol. Appl. Sci. 7 (1), 920–941.

Kumari, N., Katoch, S., 2020. Wilt and root rot complex of important pulse crops: their detection and integrated management. In: Singh, B.P., Singh, G., Kumar, K., Nayak, S.C., Srinivasa, N. (Eds.), Management of Fungal Pathogens in Pulses: Current Status and Future Challenges. Springer International Publishing, pp. 93–119.

Kushwaha, S.K., 2016. Studies on Collar Rot of Lentil Caused by *Sclerotium rolfsii* Sacc. (Ph.D. thesis). JNK Vishwa Vidyalaya, Jabalpur.

Lacmanova, I., Pazlarova, J., Kostelanska, M., Hajslova, J., 2009. PCR-based identification of toxigenic *Fusarium* species. Czech J. Food Sci. 27 (2), 90–94.

Larone, D.H., 1995. Medically Important Fungi - A Guide to Identification, third ed. ASM Press, Washington, DC.

Leslie, J.F., Summerell, B.A., 2006. The *Fusarium* Laboratory Manual. Blackwell Publishing Ltd, Oxford.

Lodhi, N.A.K., Abbas, A., Waris, W., Asad, M., Aslam, M.M., 2016. Chickpea wilt and its management strategies - a review paper. Imp. J. Interdiscip. Res. (IJIR) 2 (11), 1281–1290.

Luthra, J.C., Sattar, A., Bedi, K.S., 1943. Further studies on control of gram blight. Indian FMG 4, 413–416.

Ma, L.-J., Van der Does, H.C., et al., 2010. Comparative genomics reveals mobile pathogenicity chromosomes in *Fusarium*. Nature 464, 367–373.

Mahesh, M., Saifulla, M., Sreenivasa, S., Shashidhar, K.R., 2010. Integrated management of pigeon pea wilt caused by *Fusarium udum* Butler. J. Biol. Sci. 2, 1–7.

Maheshwari, S.K., Bhat, N.A., Masoodi, S.D., Beig, M.A., 2008. Chemical control of lentil wilt caused by *Fusarium oxysporum* f.sp. *lentis*. Ann. Plant Prot. Sci. 16 (2), 419–421.

Mallaiah, B., Rao, K.V., 2016. Survey and virulence studies on dry root rot of green gram (*Vigna radiata* (L.) Wilczek) incited by *Macrophomina phaseolina* (Tassi.) Goid. Int. J. Trop. Agric. 34 (3), 521–527.

Mandhare, V.K., Suryawanshi, A.V., 2005. Application of *Trichoderma* species against pigeon pea wilt. JNKVV Res. J. 32 (2), 99–100.

Marley, P.S., Hillocks, R.J., 1996. Effect of root knot nematodes (*Meloidogyne* spp.) on *Fusarium* wilt in pigeon pea *Cajanus cajan*. Field Crop Res. 46 (1), 15–20.

Maruti, S., Gururaj, A.S., Amaresh, Y.S., 2017. In vitro efficacy of fungicides and bioagents against dry root rot of pigeon pea caused by *Rhizoctonia bataticola* (Taub.) Butler. Int. J. Pure Appl. Biosci. 5 (6), 1341–1347.

McRae, W., 1932. Report on Imperial Mycologists. Science Agriculture Research Institute Pusa, pp. 31–78.

Mihajlovic, M.V., Rekanovic, E., Hrustic, J., Grahovac, M., Tanovic, B., 2017. Methods for management of soilborne plant pathogens. Pest. Phytomed. 32 (1), 9–24.

Miller, S.A., Rowe, R.C., Riedel, R.M., 2011. *Fusarium* and *Verticillium* wilts of tomato, potato, pepper, and eggplant. Fact Sheet. Ohio State University, Columbus.

Misra, A.K., 2006. Wilt of guava a disease of national importance. Indian Phytopathol. 59, 269–280.

Moradi, H., Bahramnejad, B., Amini, J., Siosemardeh, A., Haji-Allahverdipoor, K., 2012. Suppression of chickpea (*Cicer arietinum* L.) *Fusarium* wilt by *Bacillus subtillis* and *Trichoderma harzianum*. Plant Omics J. 5 (2), 68–74.

Moreoau, B., 1978. Diseases and Pests of Lentil. Institute National de Vulgarisation Pourles Fruits, Legumes et Champignons, pp. 95–100.

Moreau, M., Catesson, A., Peresse, M., Czaninski, Y., 1978. Dynamique comparee des reactions cytologiques du xyleme de l'oeillet en presence de parasites vascularies. J. Phytopathol. 91, 289–306.

Mueller, W.C., Beckman, C.H., 1988. Correlated light and electron microscope studies of callose deposits in vascular parenchyma cells of tomato plants inoculated with *Fusarium oxysporum* f. sp *lycopersici*. Physiol. Mol. Plant Pathol. 33, 201–208.

Muimba-Kankolongo, A., 2018. Root and tuber crops, Food Crop Production by Smallholder Farmers in Southern Africa: Challenges and Opportunities for Improvement, first ed. Academic Press, pp. 123–172.

Mukankusi, C.M., Melis, R., Derera, J., Laing, M., Buruchara, R.A., 2010. Identification of sources of resistance to *Fusarium* root rot among selected common bean lines in Uganda. J. Anim. Plant Sci. 7 (3), 876–891.

Mukhtar, I., 2007. Comparison of phytochemical and chemical control of *Fusarium oxysporum* f.sp. *ciceri*. Mycopathology 5 (2), 107–110.

Narayanan, P., Vanitha, S., Rajalakshmi, J., Parthasarathy, S., Arunkumar, K., Nagendran, K., et al., 2015. Efficacy of biocontrol agents and fungicides in management of mulberry wilt caused by *Fusarium solani*. J. Biol. Control. 29 (2), 107–114.

Navas Cortes, J.A.N., Hau, B., Jimenez-Diaz, R.M., 2000. Yield loss in chickpeas in relation to development of *Fusarium* wilt epidemics. Phytopathology 90 (11), 1269–1278.

Nega, A., 2014. Review on concepts in biological control of plant pathogens. J. Biol. Agric. Healthc. 27 (4), 33–35

Nelson, P.E., 1981. Life cycle and epidemiology of *Fusarium oxysporum*. In: Mace, M.E., Bell, A.A., Beckman, C.H. (Eds.), Fungal Wilt Disease of Plants. Academic Press, New York, pp. 51–80.

Nelson, P.E., Dignani, M.C., Anaisse, E.J., 1994. Taxonomy, biology, and clinical aspects of *Fusarium* species. Clin. Microbiol. Rev. 7 (4), 479–504.

Nene, Y.L., Sheila, V.K., Sharma, S.B., 1989. A world list of chick pea Cicer arietinum L. and pigeon pea *Cajanus cajan* L. Millsp. pathogens. Legumes Pathology Progress Report-7. ICRISAT, Patancheru, 23p.

Nikam, P.S., Jagtap, G.P., Sontakke, P.L., 2007. Management of chick pea wilt caused by *Fusarium oxysporium* f. sp. *ciceri*. Afr. J. Agric. Res. 2, 692–697.

Nongmaithem, N., Basudha, C., Sharma, S.K., 2017. Incidence of rust, powdery mildew and wilt in pea and broad bean plant of Manipur, India. Int. J. Curr. Microbiol. Appl. Sci. 6 (8), 2611–2616.

Nyvall, R.F., Haglaund, W.A., 1972. Site of infection of *F. oxysporum* f.sp. *pisi* race 5 on peas. Phytopathology 62, 1419–1424.

Okungbowa, F.I., 2011. Screening of four food spices of plant origin for antifungal activity against a *Fusarium oxysporum* soil isolate. J. Teknologi Industri Pangan 22 (1), 92–95.

Okungbowa, F.I., Edema, N.E., 2006. Antifungal activity of six asteraceous plants extracts against *Fusarium oxysporum*. Nigerian J. Bot. 20 (1), 45–49.

Okungbowa, F.I., Shittu, H.O., 2012. *Fusarium* wilts: an overview. Environ. Res. J. 6 (2), 83–102.

Olivain, C., Humbert, C., Nahalkova, J., Fatehi, J., Floriane, H., Alabouvette, C., 2006. Colonization of tomato root by pathogenic and nonpathogenic *Fusarium oxysporum* strains inoculated together and separately into the soil. Appl. Environ. Microbiol. 72 (2), 1523–1531.

Orum, J.E., Pedersen, S.M., Yazar, A., Sezen, M.S., Jacob, S.E., 2009. Report on Local Crop Production in Different Regions with Gross Margin Analysis on Field Sites, Sustainable Water Use Securing Food Production in Dry Areas of the Mediterranean Region, University of Copenhagen, Copenhagen. <http://swup-med.dk/deliverables/d5.2.pdf>.

Osuide, M.I., Aluya, E.I., Emoghene, A.O., 2002. Control of *Fusarium* wilt of tomato (*Lycopersicon esculentum* Mill) by *Trichoderma* species. Acta Phytopathol. Entomol. Hungarica 37 (1), 47–55.

Otadoh, J.A., Okoth, S.A., Ochanda, J., Kahindi, J.P., 2011. Assessment of *Trichoderma* isolates for virulence efficacy on *Fusarium oxysporum* f. sp. *Phaseoli*. Trop. Subtrop. Agroecosyst. 13, 99–107.

Padwick, G.W., 1940. The genus *Fusarium*, V *Fusarium udum* Butler, *F. vasinfectum* Atk and *F. lateritium* Nees var. *uncinatum* Wr. Indian J. Agric. Sci. 10, 863–878.

Pandey, K.K., Upadhyay, J.P., 1999. Comparative study of chemical, biological and integrated approach for management of *Fusarium* wilt of pigeon pea. J. Mycol. Plant Pathol. 29, 214–216.

Pandey, S., Sharma, M., Guvvala, G., 2013. An updated review of biology, pathogenicity, epidemiology and management of wilt disease of pigeon pea *Cajanus cajan* L. Millsp. J. Food Legumes 26 (1–2), 1–14.

Pandey, P., Irulappan, V., Bagavathiannan, M.V., Senthil-Kumar, M., 2017. Impact of combined abiotic and biotic stresses on plant growth and avenues for crop improvement by exploiting physio-morphological traits. Front. Plant Sci. 8, 537.

Parthasarathy, S., Ramalingam, K., Prabakar, K., 2017. Soil borne diseases of major pulses and their biological management. Innov. Farming 2 (1), 1–11.

Patel, S.I., Patel, B.M., 2012. Pigeon pea wilt and its management: a review. AGRES. - Int. e-J. 1 (4), 400–413.

Patel, N., Desai, P., Patel, N., Jha, A., Gautam, H.K., 2014. Agronanotechnology for plant fungal disease management: a review. Int. J. Curr. Microbiol. Appl. Sci. 3 (10), 71–84.

Patra, S., Biswas, M.K., 2017. Eco-friendly management of *Fusarium oxysporum* f. sp. *ciceri* the causal agent of chick pea wilt disease under in-vitro condition. Int. J. Curr. Microbiol. Appl. Sci. 6 (3), 1852–1858.

Pitt, J.I., Hocking, A.D., Bhudhasamai, K., Miscamble, B.F., Wheeler, K.A., Tanboon-Ek, P., 1994. The normal mycoflora of commodities from Thailand. 2. Beans, rice, small grains and other commodities. Int. J. Food Microbiol. 23, 35–43.

Prasad, N., Padwick, G.W., 1939. The genus *Fusarium* 11: a species of *Fusarium* as a cause of wilt of gram (*C. arietinum* L). Indian J. Agric. Sci. 9, 731–735.

Prasad, R.D., Rangeshwaran, R., Anuroop, C.P., Rashni, H.J., 2002. Biological control of wilt and root rot of chickpea under field conditions. Ann. Plant Prot. Sci. 10 (1), 72–75.

Radwan, M.B., Al-Mahareeq, F., Ali-Shtayeh, M.S., AL-Masri, M.I., 2007. Biological control of *Rhizoctonia solani* by indigenous *Trichoderma* spp. isolates from Palestine. Hebron Univ. Res. J. 3 (1), 1–15.

Rahman, M.S., Hossain, M.A., Sarker, M.J.U., Bakr, M.A., 2013. Adoption and profitability of Bari lentil varieties in some selected areas of Bangladesh. Bangladesh J. Agric. Res. 37 (4), 593–606.

Ravenna, N.A., Negri, R.C., 1979. Lentil wilt caused by the *Fusarium* sp. in Rosario. San Miguel de Tucuman. 2, 759–765.

Rawal, V., Navarro, D.K., 2019. The Global Economy of Pulses. Food and Agriculture Organization of the United Nations, Rome.

Reddy, M.V., Sharma, S.B., Nene, Y.L., Hall, S.D., Sheila, V.K., 1990. Pigeon pea: disease management. The Pigeon Pea, Nene, Y.L., Hall, S.D., Shiela, V.K., (Eds), C.A.B. International Wallingford Oxon OX10 8 DEUK, 303–347.

Reddy, M.V., Raju, T.N., Sharma, S.B., Nene, Y.L., McDonald, D., 1993. Handbook of Pigeon Pea Diseases. Information Bulletin No. 14. International Crops Research Institute for the Semi-Arid Tropics, Patancheru, p. 64.

Refai, M., Hassan, A.A., Hamed, M., 2015. Monograph on the Genus *Fusarium*. https://doi.org/10.13140/RG.2.1.3104.2728.

Rodrigues, A.A.C., Menezes, M., 2006. Identification and pathogenic characterization of endophytic *Fusarium* species from cowpea seeds. Mycopathologia 3, 203–215.

Rodriguez-Galvez, E., Mendgen, K., 1995. The infection process of *Fusarium oxysporum* in cotton root tips. Protoplasma 189, 61–72.

Sankar, P.M., Vanitha, S., Kamalakannan, A., Anantha Raju, P., Jeyakumar, P., 2018. Biological management of *Fusarium* wilt in chick pea (*Cicer arietinum* L.) caused by *Fusarium oxysporum* f. sp. *ciceris*. Legume Res. – Int. J. 42 (6), 838–843.

Sardana, S., Mahajan, R.K., Gautam, N.K., Ram, B., 2007. Genetic variability in pea (*Pisum sativum* L.) germplasm for utilization. SABRAO J. Breed. Genet. 39 (1), 31–41.

Saul-Jose, R.-P., Ochoa-Marquez, R., Rosales-Serna, R., Ibarra-Perez, F.J., Mendez-Aguilar, R., Hernandez-Delgado, S., et al., 2018. Analysis of *Fusarium*-common beans pathosystem in Aguascalientes, Mexico. In: Askun, T. (Ed.), Fusarium – Plant Diseases, Pathogen Diversity, Genetic Diversity, Resistance and Molecular Markers. IntechOpen Publisher, pp. 101–117.

Saxena, N.P., Johansen, C., 1990. Advances in biotic and abiotic stress in pulse crops. Indian Inst. Pulse Res., Kanpur 24, 211–219.

Schnathorst, W.C., 1981. Life cycle and epidemiology of *Verticillium*. In: Mace, M.E., Bell, A.A., Beckman, C.H. (Eds.), Fungal Wilt Diseases of Plants. Academic Press, New York, pp. 81–111.

Shahid, S., Khan, M.R., 2016. Biological control of root-rot on mung bean plants incited by *Macrophomina phaseolina* through microbial antagonists. Plant Pathol. J. 15 (2), 27–39.

Shalini, S., Kotasthane, A.S., 2007. Parasitism of *Rhizoctonia solani* by strains of *Trichoderma* spp. Electron. J. Environ. Agric. Food Chem. 6 (8), 2272–2281.

Shanmugapackiam, S.S., Parthasarathy, M., Jebaraj, D., Christopher, D.J., 2016. Exploitation of compost amended biological agents for the management of vascular wilt *Fusarium oxysporum* f. sp. *lycopersici* in tomato. Adv. Life Sci. 5 (7), 2933–2941.

Sharfuddin, C., Mohanka, R., 2012. *In vitro* antagonism of indigenous *Trichoderma* isolates against phytopathogen causing wilt of lentil. Int. J. Life Sci. Pharma Res. 2 (3), L195–L202.

Sharma, K., 1995. M.Sc. thesis, Department of Plant Pathology, S. K. N. College of Agriculture, Jobner, R.A.U. Bikaner, 48p.

Sharma, R.L., Mishra, T., Bhagat, R., Swarnkar, V., Kumar, K., 2018. Integrated disease management for pigeon pea wilt caused by *Fusarium udum*. Int. J. Chem. Stud. 6 (5), 2748–2751.

Shishido, M., Miwa, C., Usami, T., Amemiya, Y., Johnson, K.B., 2005. Biological control efficiency of *Fusarium* wilt of tomato by nonpathogenic *Fusarium oxysporum* Fo-B2 in different environments. Phytopathology 95 (9), 1072–1080.

Shukla, A., Dwivedi, S.K., 2012. Bioefficacy of plant extracts against *Fusarium* species causing wilt in pulses. IOSR J. Eng. 2 (1), 136–144.

Shuping, D.S.S., Eloff, J.N., 2017. The use of plants to protect plants and food against fungal pathogens: a review. Afr. J. Tradit. Complement. Altern. Med. 14 (4), 120–127.

Singh, R., Singh, B.K., Upadhyay, R.S., Rai, B., Lee, Y.S., 2002. Biological control of *Fusarium* wilt disease of pigeon pea. Plant Pathol. J. 18, 279–283.

Singh, G., Chen, W., Rubiales, D., Moore, K., Sharma, Y.R., Gan, Y., 2007. Diseases and their management. In: Yadav, S., Redden, R., Chen, W., Sharma, B. (Eds.), Chickpea Breeding and Management. CABI Publishing, pp. 497–519.

Singh, J., Mishra, K.K., Singh, A.K., 2013. Current status of web blight of mung bean. Asian J. Soil Sci. 8, 495–504.

Singh, D., Sinha, B., Rai, V.P., Singh, M.N., Singh, D.K., Kumar, R., et al., 2016. Genetics of *Fusarium* wilt resistance in pigeon pea (*Cajanus cajan*) and efficacy of associated SSR markers. Plant Pathol. J. 32 (2), 95–101.

Singh, P., Shahi, B., Singh, K.M., 2017. Trends of pulses production: a study on current scenario and strategies in India with special reference to Bihar. Econ. Aff. 62 (3), 389–398.

Sinha, P., Rizvi, G., Parashar, R., 2018. Management of wilt disease of pulses: a review. Int. J. Pure Appl. Biosci. 6 (4), 696–708.

Skovgaard, K., Bodker, L., Rosendahl, S., 2002. Population structure and pathogenicity of members of the *Fusarium oxysporum* complex isolated from soil and root necrosis of pea (*Pisum sativum* L.). FEMS Microbiol. Ecol. 42 (3), 367–374.

Smith, S.N., 2007. An overview of ecological and habitat aspects in the genus *Fusarium* with special emphasis on the soil-borne pathogenic forms. Plant Pathol. Bull. 16, 97–120.

Song, W., Zhou, L., Yang, C., Cao, X., Zhang, L., Liu, X., 2004. Tomato *Fusarium* wilt and its chemical control strategies in a hydroponic system. Crop Prot. 23 (3), 243–247.

Srivastava, S., Singh, V.P., Kumar, R., Srivastava, M., Sinha, A., Simon, S., 2011. In vitro evaluation of carbendazim 50%, WP, antagonists and botanicals against *Fusarium oxysporum* f.sp. *psidii* associated with rhizosphere soil of guava. Asian J. Plant Pathol. 5 (1), 46–53.

Subramanian, S., 1963. Fusarium wilt of pigeon pea I. Symptomatology and infection studies. Proc. Indian Acad. Sci., Sect. B 57, 134–138.

Syed, S.A., Kazunori, S., Yoshimiki, A., et al., 2010. Biocontrol of soil-borne *Fusarium* wilts of tomato and cabbage with a root colonizing fungus, *Penicillium* sp. Eu0013. In: Proceedings of the 19th World Congress of Soil Science, Soil Solutions for a Changing World, 1–6 August, Brisbane, Australia.

Tariq, R.M.S., Sahil, S.T., Ahmad, T., Hannan, A., 2015. Changes in mineral profile and morphological traits of pea (*Pisum sativum* L.) as influenced by *Fusarium oxysporum* f. sp. *pisi* under natural conditions. J. Plant Pathol. Microbiol. 5 (5), 1–6.

Tawfik, A.A., Allam, A.D.A., 2004. Improving cumin production under soil infestation with Fusarium wilt pathogen: II - field trial of different landraces and seed treatments. Assiut Univ. Bull. Environ. Res. 7 (2), 47–63.

Tetali, S., Karpagavalli, S., Pavani, S.L., 2015. Management of dry root rot of blackgram caused by *Macrophomina phaseolina* (Tassi) Gold using bioagent. Plant Arch. 15 (2), 647–650.

Tiwari, N., Seid, A., Kumar, S., Sarker, A., 2018. *Fusarium* wilt: a killer disease of lentil. In: Askun, T. (Ed.), Fusarium – Plant Diseases, Pathogen Diversity, Genetic Diversity, Resistance and Molecular Markers. IntechOpen Publisher, pp. 119–138.

Tosi, L., Cappelli, C., 2001. First report of *Fusarium oxysporum* f. sp. *lentis* of lentil in Italy. Plant Dis. 85 (5), 562.

Ujevic, I., Kovacikova, E., Urosevic, B., 1965. *Botrytis parasitica* and *Fusarium oxysporum* as dangerous parasites of lentil. Polnopospodarstvo 118, 584–594.

Upadhyay, R.S., 1992. Ecology and biological control of *Fusarium udum* in relation to soil fungistatis of antagonistic microorganism. J. Plant Prot. Tropics 9, 1–9.

VanderMolen, G.E., Beckman, C.H., Rodehorst, E., 1977. Vascular gelation: a general response phenomenon following infection. Physiol. Plant Pathol. 11, 95–100.

Vasudeva, R.S., Srinivasan, K.V., 1952. Studies on the wilt disease of lentil *Lens esculenta* Moench. Indian Phytopathol. 5, 23–32.

Veiga, P., Manara, W., Manara, N.T.F., Tarrago, M.F.S., 1974. Occurrence of *Ascochyta lentis* Bond. and VassiI on lentil, *In* Rio Grande do SuI. Revista do Centro de CienciasRurais. 4, 239–242.

Vinale, F., D'Ambrosio, G., Abadi, K., Scala, F., Marra, R., Turra, D., et al., 2004. Application of *Trichoderma harzianum* (T22) and *Trichoderma atroviride* (P1) as plant growth promoters, and their compatibility with copper oxychloride. J. Zhejiang Univ. (Agric. Life Sci.) 30 (4), 2–8.

Virgin, W.J., Walker, J.C., 1940. Relation of the near wilt fungus to the pea plant. J. Agric. Res. 60, 241–248.

Walker, J.C., 1935. A study of resistance to *Fusarium* wilt in Alaska peas. Am. J. Bot. 22, 849–857.

Walter, S., Nicholson, P., Dooha, F.M., 2009. Action and reaction of host and pathogen during *Fusarium* head blight disease. N. Phytol. 185, 54–66.

Wang, H., Xiao, M., Kong, F., Chen, S., Dou, H., Sorrell, T., et al., 2011. Accurate and practical identification of 20 *Fusarium* species by seven-locus sequence analysis and reverse line blot hybridization, and an in vitro antifungal susceptibility study. J. Clin. Microbiol. 49 (5), 1890–1898.

Wilson, V.E., Brandsberg, J., 1965. Fungi isolated from diseased lentil seedlings in 1963-64. Plant Dis. Report 49, 660–662.

Xu, J.R., Leslie, J.F., 1996. A genetic map of *Gibberella fujikuroi* mating population A (*Fusarium moniliforme*). Genetics 143, 175–189.

Zorya, S., Eliste, P., Dawe, D., Aye, D.K.S.S., 2016. Myanmar: Analysis of farm production economics, Economic and Sector Work Report No. 100066-MM, The World Bank, Washington, DC.

Index

Note: Page numbers followed by "*f*" and "*t*" refer to figures and tables, respectively.

C

D

W

X

Y

Z

9780128218433